HIGH PRESSURE ENGINEERING

HIGH PRESSURE ENGINEERING

The Second International Conference
sponsored by The Institution of Mechanical Engineers,
The European High Pressure Research Group,
The High Pressure Technology Association, and
The Association Internationale pour l'Avancement de la
Recherche et de la Technologie de Haute Pressions

University of Sussex, Brighton, 8–10 July 1975

Edited by

H. Ll. D. Pugh

Published by
Mechanical Engineering Publications Limited for
The Institution of Mechanical Engineers
LONDON

First published 1977

This volume is complete in itself

There is no supplementary discussion volume

ISBN 0 85298 354 9

Made and printed in Great Britain
by Acolortone Limited, Ipswich, Suffolk

Bound by
Fowler-Wilson Bookbinders,
Sandy, Bedfordshire

CONTENTS

Introduction, by H. Ll. D. Pugh 7

Formal Opening, by Sir Hugh Ford 9

MATERIALS

Fracture characteristics of 3½% NiCrMoV steel for use in pressure vessels, by E. F. Walker, E. H. Watson, D. B. James, P. McIntyre, and M. J. May 13

Cemented carbide in high pressure equipment, by B. Zetterlund 35

Review of methods and equipment for material testing under pressure, by H. Ll. D. Pugh and E. F. Chandler 41

The deformation and fracture under pressure of a number of metals used in high pressure equipments, by H. Ll. D. Pugh and E. F. Chandler 53

Materials and processes considerations in the design of pressure vessels, by T. E. Davidson, B. B. Brown, and D. P. Kendall 63

The creep of thick walled 2½% NiCrMo cylinders under internal pressure at 350°C: development of a testing machine and preliminary results, by W. J. Skelton, A. Salin, and R. G. Patton 73

DESIGN

The development of kilobar pressure equipment from the laboratory to the industrial scale. (Some reminiscences of the work at the ICI Northwich establishment, 1928-39), by W. R. D. Manning 101

A review of the present state of the art in high pressure container design, by H. Ll. D. Pugh and B. Crossland 115

The development of a new type of compound tubing designed for high fatigue strength, by J. Rogan 131

On the elastic-plastic expansion of tapered pressure vessel assemblies, by R. L. Hewitt, M. C. de Malherbe, and J. M. Alexander 137

The elastohydrodynamic lubrication of high pressure plain gland clearance seals: the effect of seal geometry and ram velocity on seal leakage and ram stability, by M. J. Gray and W. J. Skelton 147

Some investigations on the behaviour of high pressure packings used in secondary compressors for low density polyethylene production, by A. Traversari and E. Giacomelli 161

The development and performance of a 340 MPa (50,000 lbf/in^2) pumping system, by C. G. Ludlow, R. Pakseresht, and B. Crossland 175

Some perspective elaborations in the field of hydraulic extrusion, by Yu. S. Konyaev 191

Design and performance of water jet systems for slotting of concrete, by A. G. Norsworthy, U. H. Mohaupt, and D. J. Burns 195

Experience with an internally-heated gas-medium deformation apparatus to 500 MPa (5 kbar), by M. S. Patterson 209

MEASUREMENT

Transducers for very high pressures, by V. E. Bean 217

Transducers and fluids for high pressure (1.4 GPa) systems, by A. W. Birks 221

Some considerations of piston gages, by D. H. Newhall and L. H. Abbot 237

Stress analysis of high pressure vessels by means of photo-elastic investigations, by A. Kuske, W. Steinchen, and J. Zech 247

Fracture toughness measurement and ultrasonic crack measurement in thick-wall cylinder geometries, by D. P. Kendall, J. H. Underwood, and D. C. Winters 255

FATIGUE AND FRACTURE

Fatigue and fracture of thick-walled pressure vessels, by J. S. C. Parry and R. W. E. Shannon 267

The fatigue and burst properties of high pressure tubing made from stainless steels, by J. Rogan 281

Fatigue strength and mode of fracture of high pressure tubing made from low-alloy high-strength steels, by J. Rogan 287

Fatigue design of containers for very high pressure, by M. Nishihara, Y. Yamaguchi, and S. Hattori 297

Fatigue behaviour of container components submitted to internal pulsating pressure, by P. Hestermans 307

A novel method for determining the fatigue life of thick walled cylinders. Design of the machine and initial results, by A. J. W. Buckland and Bela Lengyel 315

Effect of internal surface flow on the fatigue strength of a thick-walled cylinder under cyclic internal pressure, by K. Nishioka and K. Hirawaka 325

Failure problems in high pressure equipment, by B. Tomkins 331

The application of linear elastic fracture mechanics to fatigue and fracture of thick wall cylinders, by T. G. Crymble, B. D. Goldthorpe, and B. A. Austin 341

A comment related to the Crymble, Goldthorpe and Austin ring fracture tests, by J. H. Underwood 349

CLOSURES

Fatigue of threaded end closures of high pressure vessels, by M. G. Gray and R. J. Pick 353

Threaded closures for large high pressure vessels, by D. M. Fryer and C. W. Smith 367

SAFETY

Inspection and safety aspects of design, by D. E. Witkin 377

Additional safety devices for pressure vessels, by H. D. B. Raes 383

Review of design for adequate gas relief from vessels at high pressure, by R. Stephenson 391

Some safety aspects on hydrostatic extrusion in production plants, by K. Zander 401

The HPTA high pressure safety code, by G. Saville and B. G. Cox 409

ntroduction

'he wide interest in high pressure arising out of the pioneering work of Prof. P. W. Bridgman has led to the organisation of many onferences in this field over the past decade or so, most of which have been concerned with the effects of high pressure on the behaviour f matter and on a variety of phenomena of interest to a wide range of disciplines as geophysics, chemistry, solid state physics, engineering nd metallurgy. In general, however, little attention has been devoted in these conferences to studies of methods of containing high ressure, to the engineering design of equipment and to techniques associated with high pressure including pressure transmitting media, eals, transducers, etc. Since the common interest of all high pressure workers is in high pressure engineering, there is a need for con- erences in this field to be held from time to time and the first conference on this subject was held in London in 1967. Since that time, rogress has been rapid not only in the development of equipment for containing ever higher pressures but also in the increasing use f high pressure as an industrial tool. Thus at the time of the first conference the maximum static pressure which had been attained vas stated to be 500 kb, whereas today some 3 Mb are quoted in the literature. On the industrial side, in addition to the developments 1 chemical engineering, the production of polythene, the synthesis of diamonds and of carbonado diamonds, and the isostatic ompaction of powders, hydrostatic extrusion has now been taken up by a number of firms as a new industrial process. These industrial pplications enable us to reap the benefits of research and at the same time are powerful incentives to augment further research in the igh pressure field.

In view of these developments in the last eight years and others concerned with the increasing importance of safety and progress in acture mechanics, it became apparent that there was a need for the Second International Conference on High Pressure Engineering. 'his conference was held under the chairmanship of Dr H. Ll. D. Pugh, National Engineering Laboratory, Scotland at the University f Sussex on 8th to 10th July 1976 and was sponsored by the Applied Mechanics Group of the Institution of Mechanical Engineers vith co-sponsorship by AIRAPT (International Association for High Pressure Science and Technology), EHPRG (European High ressure Research Group) and HPTA (High Pressure Technology Association). The conference was opened by Professor Sir Hugh 'ord, FRS Imperial College, London and the guest speaker at the conference dinner was Professor J. L. M. Morrison, formerly of Jniversity of Bristol, both of whom have made substantial contributions to high pressure engineering.

The assembling of a diversity of experts from academic and industrial establishments, from various countries and representing a range f interests to present and discuss their contributions in the field of high pressure engineering did much for the success of the conference. f a total of 38 papers, 17 were by authors from industry, which was a substantial increase on the first conference, and some two-thirds f the 200 delegates were from industry. It was a pleasure to see so many, some 45%, of the delegates from abroad representing 19 ountries. The strong industrial participation is indicative of the rapidly growing interest in the industrial use of high pressure over a ide range of applications from chemical engineering to mechanical working. The industrial interest is also reflected by the fact that at eir request, an exhibition of commercial high pressure engineering products was arranged and despite very short notice, some nine rms participated.

In order to allow authors to include their most recent results, there was no requirement for authors to produce their full papers rior to the conference. Authors were asked to submit their full papers after the conference if they wished publication. It is gratifying at the present volume incorporates all the papers presented at the conference. A few authoritative review papers which summarize e state of the art in some aspects of high pressure engineering have been included but most of the papers are original contributions hich extend our knowledge in existing and into new fields. The volume also includes a most interesting account by Mr W. R. D. Ianning, formerly of Imperial Chemical Industries, describing for the first time the work for which he was primarily responsible on e design and construction of the high pressure equipment used in the studies leading to the first polymerization of ethylene and that sed in the industrial production of polythene during the first few years.

The book is divided into six sections, dealing with, Materials, Design, Measurement, Fatigue and Fracture, Closures, and Safety, ıd having the ultimate objective of facilitating the design of safe and reliable equipment both for use in the laboratory and especially r work under industrial conditions. A number of papers are devoted to the application of fracture mechanics to container design and the question of the safety of high pressure equipment. This latter subject is now of prime importance both for industrial and labo- atory high pressure equipment with the passing of the Health and Safety Act in the United Kingdom.

The Planning Panel wish to express their gratitude to the many people who contributed to the success of the conference. Particular anks are due to Professor Sir Hugh Ford and Professor J. L. M. Morrison, to Professor M. W. Thompson, Pro-Vice Chancellor of e University of Sussex for making available the accommodation and facilities of that University and for his kind words of welcome, the authors for their excellent papers, to the contributors to the discussion and to the Conference Department of the Institution of lechanical Engineers who were responsible for the detailed organisation of the conference.

Finally, I wish to record my personal thanks to my colleagues on the Planning Panel for their untiring efforts and enthusiasm in anning and organising this conference and for acting as chairmen of the technical sessions.

ast Kilbride, Scotland
ıne 1976

H. Ll. D. Pugh,
Chairman, Planning Panel,
c/o Institution of Mechanical Engineers, London

Formal Opening By Professor Sir Hugh Ford FRS

There is a saying that if you make the best mousetraps the world will beat a pathway to your door. I don't know whether it's true, but it seems that if you run Conferences on High Pressure Engineering, the world is sufficiently interested and concerned to attend it.

We have delegates here from just about every country in which high pressure is used, in one process or another. My first duty is the very pleasant one of welcoming all of you. The organisers are, I am sure, very gratified at the response. Over 180 people have registered, over half from abroad. There are important groups from Japan, USA, Russia, Germany and France.

Although High Pressure Engineering is a specialised field, it has spread into a wide range of sophisticated industrial applications, and this is reflected in the fact that two thirds of those attending are from Industry.

I am sure you will forgive me if I mention one delegate in particular. I am delighted to be able to welcome my former boss and the man who introduced me to the pleasures and hazards of High Pressure. Mr W. R. D. Manning is a name known all over the world for his critical contributions to high pressure design of polyethylene plants, and it is an honour—of which I am justly proud—to have been a member of that very small team of engineers who were responsible for building the first commercial plant for polyethylene manufacture.

It is a far cry from the dark days of 1939. And it is also, perhaps, too long since our first conference in 1967. On that occasion, I had the pleasure of welcoming you to my own Department of Mechanical Engineering in Imperial College. If you have managed to find your way here to the University of Sussex in what used to be the beautiful Sussex countryside, and if you have not been drenched in the thunderstorm, I am sure you would wish to join me in thanking the authorities of the University for allowing us to hold the Conference in these excellent surroundings.

There is one preoccupation of everybody who works in high pressure. It is that of safety: health and safety at work has become a major concern of all industrial activity lately, but it has always been a first consideration in high pressure work. I think the record of safe performance of high pressure equipment is something for which we can take some satisfaction, and it is very proper that the first session today should be on the subject of safety in high pressure plant. The record of safety in such plant both compared with that in industry generally and certainly as regards death and accidents on the roads and in the home, is extremely good, but nobody can afford to be complacent. Often the incidents are not related to the high pressure equipment itself but to the auxiliaries. Quite apart, therefore, from the design and manufacture of the high pressure components, we have to look upon our objective as the design of a system.

In the afternoon session, we are to discuss a draft code of practice for high pressure engineers. Clearly such a code could be a helpful guide, but you may conclude that the introduction of a code could detract from, rather than the reverse, the engineer's recognition of his constant responsibility for the safety and effectiveness of everything he designs.

Later sessions are concerned with materials of design and the behaviour of materials in high pressure components; fatigue, creep, fracture, embrittlement and the like; also with specific applications in chemical plant, hydrostatic extrusion, isostatic compaction and compressor systems; and finally with measurement and stress analysis, a continuing activity of all those concerned in the improvement and extension of high pressure applications.

We have a concentrated programme during the next three days and in welcoming all of you to this Conference, it is my confident wish that you will have found it well worth while to have attended it.

Chairman, may I declare this conference formally opened.

MATERIALS

FRACTURE CHARACTERISTICS OF $3\frac{1}{2}\%$ Ni Cr Mo V STEEL FOR USE IN PRESSURE VESSELS

E. F. WALKER*, CEng., MRAeS,. E. H. WATSON†, AMet., AIM., D. B. JAMES*, BSc.,
P. McINTYRE*, BSc., PhD and M. J. MAY*, BSc., PhD, FIM
*Physical Metallurgy Centre, British Steel Corporation
+River Don Works, British Steel Corporation

SYNOPSIS Mechanical property data are presented for a $3\frac{1}{2}\%$ NiCrMoV (NCMV) steel heat treated to produce martensitic and bainitic structures representing the extreme conditions found in thick sections. The martensitic condition had both superior strength and toughness. It has been demonstrated that although isothermal and temper embrittlement can result in appreciable loss in impact resistance, the fracture toughness is only marginally affected. A study of the environmental fracture resistance of the martensitic steel has indicated that this alloy has good resistance to hydrogen induced stress corrosion cracking.

INTRODUCTION

1. The development of improved steels for use in the manufacture of pressure vessels is motivated by the pace of industrial developments and the increased awareness that design and the economics of the end product are closely related. This momentum is usually manifest in the first instance in demands for steels of greater strength followed by expressions of concern regarding the possibility of an increased risk of fracture in service. The development of pressure vessel steels is no exception to this.

2. A recent survey[1] of the properties of pressure vessels manufactured of $3\frac{1}{2}\%$ NCMV is being supplemented by a continuing research effort directed towards a detailed understanding of the effects of both mechanical and environmental variables on the properties of $3\frac{1}{2}\%$ NiCrMoV steels. Particular emphasis has been given to the study of the effects of heat treatment on fracture toughness and the consequences of exposure of the steel to hydrogen containing environments. In addition the influence of isothermal and temper embrittlement on fracture toughness, impact resistance and environmental fracture properties is discussed.

FRACTURE TOUGHNESS OF $3\frac{1}{2}\%$ NCMV

Material

3. Material was obtained from two 165 mm ($6\frac{1}{2}$ in) thick end slices of a forged cylinder of vacuum poured $3\frac{1}{2}\%$ NCMV steel made by the basic electric arc process. The chemical composition (wt. %) was as follows:-
C 0.33, Si 0.16, Mn 0.25, Ni 3.55, Cr 1.12, Mo 0.54, V 0.09, S 0.007, P 0.006, Sn 0.011, Sb 0.0064 and As 0.031.

4. The forged cylinder had an internal diameter of 518 mm (20.4 in) and a wall thickness of 305 mm (12 in). Transverse tensile, Charpy V-notch, and fracture toughness (compact tension type) test pieces were machined from the tangential direction so that fracture occurred in the radial direction. The thickness of the compact tension specimens varied from 38 mm ($1\frac{1}{2}$ in) for the least tough, to 100 mm (4 in) for the toughest materials tested. Test pieces were machined from material which had been subjected to the various heat treatments cited in the next section.

Heat treatment

5. In thick sections, the usual heat treatment for $3\frac{1}{2}\%$ NCMV steel can result in microstructural gradients ranging from tempered martensite in the surface region to tempered bainite. Investigation has shown, that oil quenching and air cooling 102 mm (4 in) thick sections can reasonably simulate the cooling rates near the surface and mid-wall positions respectively of a 508 mm (20 in) internal diameter cylinder of 318 mm ($12\frac{1}{2}$ in) wall thickness water quenched from 850°C. Test material was, therefore, heat treated in the form of 102 mm thick blanks or as smaller samples strapped together to make the required thickness and tests conducted to examine the influence of these microstructural features on properties.

The heat treatments employed to produce the basic martensitic and bainitic conditions were as follows:-

Soak 6 h 850°C oil quenched (martensitic)
Soak 6 h 850°C air cooled (bainitic)

7. The effect of tempering temperature was investigated by subsequent treatments of 12 h duration at 540, 560, 580 and 600°C, followed by air cooling. Although the usual cooling rates from the tempering

temperature do not normally present a tempering embrittlement problem with pressure vessels it was proposed to slow cool material after tempering in order to determine the effect of extreme embrittling conditions on fracture toughness. Selected test blanks were, therefore, allowed to cool in the furnace (at 5°C/h) after tempering at 600°C.

8. The influence of prolonged ageing at a service temperature of 300°C was also of interest, and material tempered at 600°C was subsequently aged for 1,000 h at 300°C prior to testing.

Results of fracture studies

9. Effect of tempering temperature

The influence of tempering temperature on tensile properties is shown in Table 1 and Fig. 1. As would be expected, the martensitic material has the superior tensile strength and 0.2% proof stress with a slightly inferior ductility compared to the bainitic material. Raising the tempering temperature lowers both the tensile strength and proof stress, the latter falling from 1195 N/mm^2 (77.4 $tonf/in^2$) to 1030 N/mm^2 (66.7 $tonf/in^2$) for the martensitic material and from 1020 N/mm^2 (66.1 $tonf/in^2$) to 930 N/mm^2 (60.2 $tonf/in^2$) for the bainitic material, as the tempering temperature is raised from 540 to 600°C. The elongation of the martensitic material is hardly affected by the tempering temperature whilst that of the bainitic steel increases slightly with increasing tempering temperature.

10. The Charpy V-notch impact transition curves are illustrated in Figs. 2 and 3 and the 50% FATT, 54 J (40 ft lb) and 34 J (25 ft lb) temperatures are listed in Table 2. It can be seen, that the impact properties are improved by raising the tempering temperature. As is well known, the impact strength of the tempered martensite structure is superior to that for the tempered bainite structure.

11. The fracture toughness in both the bainitic and martensitic conditions increases markedly as the tempering temperature is raised from 540 to 580°C, but shows little further increase after tempering at 600°C. The martensitic material, although possessing the higher proof and tensile strengths, also has a much superior fracture toughness. The relevant details are summarised in Table 1 and Fig. 4.

12. The results suggest that for the best combination of fracture toughness and strength the tempering temperature should not be less than 580°C. This gave a plane strain fracture toughness (K_{IC}) greater than 125 $MNm^{-3/2}$ for the steel in the present programme simulating the extreme situation of a fully bainitic condition at the mid-wall position, and substantially greater than 155 $MNm^{-3/2}$ for the near surface position.

13. Effect of test temperature

Test material in the 600°C tempered condition was tested at temperatures from below -100°C to 300°C. The results are shown in Figs. 5-7. The tensile strength and proof stress increase with decreasing temperature the most rapid increase taking place below 0°C. Elongation is not markedly affected down to -100°C, but for the higher strength martensitic material a rapid reduction follows between -100 and -150°C. The bainitic material was not tested at -150°C.

14. The plane strain fracture toughness (K_{IC}) of the bainitic material increases rapidly from about 52 $MNm^{-3/2}$ to 127 $MNm^{-3/2}$ as the temperature is increased from -80°C to 20°C (Fig. 7). An upper shelf level in toughness appears to exist above 50°C. Although the toughness results obtained at these higher temperatures were approximate K_Q values an upper shelf K_{IC} value of 150 $MNm^{-3/2}$ would probably be a conservative estimate. In this context, upper shelf K_{IC} values (valid according to the ASTM criteria) greater than 200 $MNm^{-3/2}$ have been reported between 93°C (200°F) and 260°C (500°F) for a 3½% NCMV steel of lower carbon content (0.28%) but similar yield strength to the present case(2).

15. The martensitic material possesses a superior toughness to the bainitic material and does not exhibit a rapid transition in toughness until the temperature falls below -20°C. Then the toughness is lowered from a K_Q value of 180 $MNm^{-3/2}$ to a K_{IC} of 78 $MNm^{-3/2}$ at -80°C.

16. It must be remembered that although the strength and fracture toughness is lower in the mid-wall position, service stresses are less severe in this region than near the surfaces of a pressure vessel. Also buried defects have a less serious effect than similar sized surface defects.

17. Effect of embrittling treatments

The effect of slow cooling from the tempering temperature (furnace cooling at 5°C/h) and ageing for 1,000 h at 300°C, on the properties of material tempered at 600°C, is illustrated in Figs. 8-12. The tensile strength and proof stress of the furnace cooled material were below those of the air cooled steel (Fig. 8). A marked deterioration in the impact properties of the martensitic material is evident, although the lower strength bainitic material is not affected (Figs. 9-11). Thus an increase of over 100°C occurs in the FATT of the martensitic steel, which is still superior, however, to the bainitic condition (Fig. 11). This confirms that, under extreme slow cooling conditions, the steel is subject to temper embrittlement as indicated by the impact test. However, the fracture toughness for both the bainitic and martensitic materials is not greatly affected by the slow cooling treatment (Fig. 11). This suggests that the effect of temper embrittlement on toughness may be strain rate sensitive.

18. Although the tensile strength and proof stress of the material aged for 1,000 h at 300°C are lower than for the un-aged material, (Fig. 8), the impact properties show signs of deterioration (Figs. 9, 10, 12). The FATT of the martensitic steel is raised from -100 to -60°C and that of the bainitic steel from 40 to 48°C. The fracture toughness remains excellent in both cases, which suggests that any isothermal embrittlement is also more sensitive to high strain rate conditions.

19. It must be appreciated, however, that a 1,000 h exposure time is short compared with a normal service life. Little information exists for 3½% NCMV steel regarding the kinetics of embrittlement for long times at 300°C, although extrapolation of existing impact data on other materials(3) suggests that increased embrittlement may be expected at exposure times in excess of 10,000 h (Fig. 13). On the other hand, tests to date indicate that a very serious degree of embrittlement, as manifested by impact tests, needs to be present before a marked deterioration in fracture toughness occurs.

20. Observations on service performance

A number of studies have been undertaken which indicate that 3½% NCMV and other pressure vessel steels can suffer some limited reduction in apparent fracture resistance during extended service at temperatures as low as 290°C. For example, it has been noted that the room temperature impact energy of a steel possessing the composition shown in Table 3 reduced by about 50% from the range 31J to 64J to the range 13J to 16J during a period of approximately 15 years in service.

21. Another pressure vessel, manufactured from 3½% NCMV steel with the composition shown in Table 4, was also found to have suffered a reduction in impact energy from the range 27J to 33J to the range 13J to 27J following a period of 4 years in service with wall temperatures of between 165°C at the outer surface and 210-350°C at the inner surface. Reference to Fig. 13 shows that an extrapolation of available data relating to the kinetics of isothermal embrittlement would confirm the feasibility of such an effect. However, the present work has shown that although predominantly intergranular fracture may be produced as the result of an impact test, the fracture resistance of such steel measured by a fracture mechanics test indicates only a marginal reduction in toughness from 157 $MNm^{-3/2}$ to 130 $MNm^{-3/2}$. These high fracture toughness values suggest that vessels would normally continue to operate satisfactorily. Experience, therefore, supports the suggestion that the impact test results in an over emphasis of the effects of embrittlement.

ENVIRONMENTAL FRACTURE OF 3½% NCMV

Influence of isothermal embrittlement on environmental fracture

22. The susceptibility of the 3½% NCMV pressure vessel steel given in Table 4 to hydrogen-assisted cracking has been evaluated by sustained load tests in a dry hydrogen sulphide gas test environment using fatigue pre-cracked CKS test pieces of 25.4 mm and 38.0 mm thickness. Hydrogen sulphide was used because this gas is known to accelerate the rate of entry of hydrogen into the steel, and hence produces much more rapid crack growth rates than are observed in hydrogen gas(4). The threshold stress intensity at which the embrittled steel became susceptible to crack growth was found to be 40 $MNm^{-3/2}$ at 20°C in gas at 133 mbar pressure. A similar, un-embrittled test piece gave a threshold stress intensity value of 50 $MNm^{-3/2}$. Embrittlement is thus shown to exert a much smaller influence on susceptibility to environmental fracture than on impact energy. One important effect of the embrittlement was, however, on the fracture mode observed during testing in hydrogen sulphide gas. While the un-embrittled steel displayed the usual quasi-cleavage fracture mode observed with 3½% NCMV steels in hydrogen-containing environments as shown in Fig. 14, after embrittlement this became predominantly intergranular as shown in Fig. 15.

Hydrogen induced cracking

23. Strength level is recognised to be one of the most important parameters controlling the susceptibility of steels towards hydrogen-induced cracking. This relationship has been studied for 3½% NCMV steel of the composition given in Table 5, heat treated to two strength levels. CKS test pieces of 13, 25 and 38 mm thickness were utilised.

24. For the higher strength steel the dependence of time to failure of fatigue pre-cracked test pieces on initial stress intensity level for tests under constant load in pure hydrogen gas at atmospheric pressure and ambient temperature is shown by the Brown curve in Fig. 16. This indicates a threshold stress intensity value of about 57 $MNm^{-3/2}$. A slightly lower threshold value of 50 $MNm^{-3/2}$ was determined for the same steel by tests conducted in hydrogen sulphide gas. For reasons stated below in the section on 'Effects of stress state' this more conservative value is believed to be the valid threshold for hydrogen induced cracking. The fracture toughness of this steel at 25°C was 122 $MNm^{-3/2}$, which exceeds twice the threshold stress intensity at which environmental cracking can occur and illustrates the need for designers to consider these factors rather than fracture toughness values where pressure vessel steels encounter aggressive environments during service.

A significantly higher threshold stress intensity value of 70 $MNm^{-3/2}$ was determined by tests in hydrogen sulphide gas for the lower strength steel. This inverse dependence of resistance to hydrogen-assisted failure on proof stress confirms the usually observed behaviour for 3½% NCMV steel and emphasises the need to restrict the strength of steels which are to be exposed to hydrogen at near ambient temperature.

25. Because many pressure vessels operate at temperatures above ambient, a study has been made of the influence exerted by temperature over the range 25-130°C on the susceptibility to cracking and crack growth rate behaviour of the higher strength 3½% NCMV steel. As temperature was increased from 25°C to 71°C and 130°C, the threshold stress intensity determined for crack growth to initiate in pure hydrogen rose from 57 to 75 and 96 $MNm^{-3/2}$, respectively. This behaviour was confirmed by tests in hydrogen sulphide gas which gave a threshold value of 98 $MNm^{-3/2}$ at 125°C. Since the fracture toughness (K_Q) value of the steel at this temperature is 96 $MNm^{-3/2}$, this indicates an effective resistance to cracking at 125°C. Despite the reduced susceptibility to cracking in hydrogen at elevated temperatures, the results of crack growth rate studies presented in Fig. 17 show an increase in equilibrium crack growth rate achieved once initiation has occurred. The observation of steady state crack growth rates at intermediate stress intensity levels at each temperature suggested the presence of a thermally activated rate controlling step in the cracking process. From the equilibrium values of crack growth rate, an activation energy for this process of 16,750 ± 4,180 Joules/gram-atom was calculated. This relatively low value may indicate that the diffusion of hydrogen atoms through the iron lattice limits the rate of cracking at a given temperature.

26. Very high crack growth rates of about 10 mm/s were measured in hydrogen sulphide gas at ambient temperature. These are attributed to the higher availability of atomic hydrogen at the crack tip region in hydrogen sulphide gas which acts as a poison to the recombination of atoms liberated by its reaction with the stressed steel surface. The threshold stress intensity values for cracking to occur in hydrogen sulphide gas, were, nevertheless, generally similar to those determined in hydrogen because in both cases it is necessary for sufficient crack tip deformation to occur to expose a clean metal surface with which the gas can react before crack growth is possible. The influence of contamination of the pure hydrogen test environment by additions of oxygen, nitrogen and water vapour has been investigated. Nitrogen was found to exert no influence on the cracking process while a 0.5% addition of water vapour to hydrogen at 1000 mbar pressure resulted in a 50% reduction in crack velocity, but no change in threshold stress intensity. Oxygen additions produced no effect on cracking in hydrogen until the concentration exceeded 1%. A 2% addition of oxygen was sufficient to completely arrest crack growth, presumably by the formation of a protective surface layer at the crack tip (Fig. 18). None of these contaminants was found to influence crack growth in hydrogen sulphide gas to a significant extent.

27. Fatigue crack growth studies conducted with higher strength 3½% NCMV steel given in Table 5 have shown the influence of hydrogen on rates of cracking shown in Fig. 19. Only when the R value (ratio of minimum to maximum fatigue stress intensities, K min/K max) reached a value of 0.9 was hydrogen found to accelerate the rate of progagation at a cyclic frequency of 113 Hz. Examination of the fracture surfaces confirmed that hydrogen exerts no effect until R is high enough for K Min to exceed the static load threshold stress intensity (KISCC). When this occurs the mod of separation changes from wholly transgranular as observed in air, to partially quasi cleavage and transgranular.

Cracking in high purity steam

28. High purity condensing steam at 1000 mbar pressure and a temperature of approximately 100°C, generated from de-ionised water with the typical analysis shown in Table 6, has been used as a test environment for the 3½% NCMV steel of 1000 N/mm^2 proof stress.

29. Analysis of the steam condensate for dissolved oxygen showed this to be present at below 1 ppm.

30. The test pieces used for this work were of the 25.4 mm thick fatigue precracked CKS type and were exposed under constant load at initial stress intensity levels of 80 and 120 $MNm^{-3/2}$ for periods of approximately 2,500 h. Following the exposure, the cracks were extended by fatigue to define any sub-critical flaw growth which had occurred. For purposes of comparison, tests were also conducted with similar test pieces of the steel in an environment of hydrogen sulphide gas at 133 mbar pressure to ascertain the threshold stress intensity values for hydrogen-induced cracking at temperatures of 25 and 100°C. A summary of the results obtained is given in Table 7.

31. Examination by scanning electron microscopy of the test piece tested at ambient temperature in hydrogen sulphide gas showed a predominantly intergranular failure mode with some accompanying transgranular quasi-cleavage separation.

32. In contrast, the test piece which failed in hydrogen sulphide at 100°C, showed only limited sub-critical flaw growth over a distance of approximately 100 μm before the onset of unstable propagation by microvoid coalescence. The threshold

stress intensity for hydrogen-induced cracking is thus very close to the fracture toughness value for this material at 100°C.

33. No evidence of cracking was found on the test piece exposed to condensing steam with an initial stress intensity of 80 $MNm^{-3/2}$. However, the test piece tested under similar conditions with an initial stress intensity of 120 $MNm^{-3/2}$ showed irregular crack growth in the central (plane strain predominant) region of the test piece up to a maximum distance of 0.9 mm. Growth was found to have occurred by a mixed mode of failure with small areas of environmentally-assisted brittle intergranular and quasi-cleavage separation in addition to larger intermediate regions of ductile microvoid coalescence which was predominantly transgranular in nature and associated with fine carbide particles. Fig. 20 illustrates the complex fractography observed. Optical metallography confirmed the largely ductile transgranular mode of cracking, as shown in Fig. 21.

34. The partly intergranular nature of the fracture produced in hydrogen sulphide may indicate some degree of temper embrittlement in this steel since 3½% NCMV usually fails by a transgranular mode.

35. Since no growth occurred in condensing steam at an initial stress intensity level of 80 $MNm^{-3/2}$, it is concluded that this is below the threshold value for this environment. The growth observed at an initial stress intensity level of 120 $MNm^{-3/2}$ would not have been expected on the basis of a hydrogen induced mechanism since this is below the threshold value of 140 $MNm^{-3/2}$ determined in hydrogen sulphide. The fractographic observations support the conclusion that this is not a straightforward example of environmentally-assisted failure. The large amount of fine scale microvoid coalescence is indicative of a thermally activated plastic deformation mode of cracking. The closeness of the applied stress intensity to that required for unstable propagation may have enabled ductile deformation to occur during the prolonged period of the test, thus allowing exposure of a fresh metal surface to the steam which may then have dissociated to give hydrogen which would have been responsible for the localised brittle fracture modes.

36. These findings contrast with an earlier study by Parker and Saddler(5) on 3Cr½Mo acid open hearth steel of 695 N/mm^2 proof stress which had been used for a turbine disc on the Hinkley Point 'A' turbine generator and in which cracks were found to form intergranularly during service. This steel, which was temper embrittled, was tested in condensing high purity steam at 120°C and 1990 mbar pressure. Cracks formed more quickly in pre-cracked IT WOL test pieces at an initial stress intensity level of 66 $MNm^{-3/2}$ than with an initial stress intensity of 110 $MNm^{-3/2}$ and were found to be predominantly intergranular rather than ductile in nature. The maximum crack growth rate of 4×10^{-7} mm/s was rather higher than found in this investigation. This apparently lower resistance to crack growth in steam may be associated with the more severe temper embrittlement in that material, and differences in the steam composition, since this was bled off an operating power plant. Both of these factors could have contributed to a higher susceptibility towards an active path mechanism of stress corrosion failure. The steel used in the current study, having a much higher strength level, would be more prone to a hydrogen-induced failure mechanism.

Effect of stress state

37. Despite the major influence which stress state is recognised to exert over the behaviour of steels during fracture toughness testing, little effort has been made to assess how environmental fracture can be influenced by similar considerations. A study has, therefore, been conducted to determine the dependence of environmental fracture behaviour on test piece thickness (and hence stress state since the predominance of plane strain to plane stress at the crack tip would fall as test piece thickness is reduced). An 817M40 steel with the composition and properties shown in Table 8 has been used for this work. Although this is not generally used as a pressure vessel steel today, there is no reason to believe that the observations made would not apply to other steels.

38. Test pieces of thicknesses ranging from 0.8 mm to 25.4 mm but with otherwise standard 13 mm CKS dimensions, were tested in both hydrogen sulphide gas and 3.5% sodium chloride solution. This range of thicknesses allowed threshold stress intensity measurements to be made in each environment as the stress state varied from plane stress predominance to plane strain predominance. Fig. 22 shows the influence of thickness on threshold stress intensity values in these environments. While the behaviour in hydrogen sulphide gas is little affected by test piece thickness, it is evident that in the less aggressive aqueous environment, the threshold value is strongly dependent on stress state. This influence of environment on the effect of changes of stress state was explained when the fracture appearance of the test pieces was examined. Those tested in the less aggressive sodium chloride environment were found to possess shear lips at the edges of the test pieces in the sub-critical flaw growth regions. These occupied a significant proportion of the overall thickness of the test pieces as the thickness was reduced and this is clearly shown for a 3.17 mm thick test piece in Fig. 23. In hydrogen sulphide gas, which is a much more potent source of atomic hydrogen than sodium chloride solution, no shear lips were observed on any of the test pieces as shown for a 3.17mm thick sample in Fig. 24. Evidently sufficient hydrogen can be occluded from an aggressive test environment like hydrogen sulphide gas to completely overcome the effects of stress relaxation at the test piece surfaces and

thus yield a valid threshold stress intensity, however low the test piece thickness, since cracking can proceed whatever the stress state. In sodium chloride solution the availability of hydrogen is not sufficient to promote cracking until the stress intensity is sufficiently high to both create a clean metal surface at the crack tip and allow the formation of shear lips where relaxation of constraint occurs at the edges of the test piece. A much thicker test piece is therefore needed to determine a threshold stress intensity value which is valid for large sections since, as in the case of fracture toughness testing, the proportion of the fracture surface occupied by shear lips must remain small. Current tests show that the behaviour in hydrogen gas is intermediate between that in hydrogen sulphide gas and sodium chloride solution. This may explain the slightly higher threshold stress intensity value determined for 3½% NCMV steel at ambient temperature in hydrogen than in hydrogen sulphide.

39. Another consequence of this work is that it highlights the possibility of hydrogen induced cracking occurring under predominantly plane stress conditions providing that the availability of hydrogen and susceptibility towards cracking of the steel are sufficiently high. This raises the likelihood of hydrogen-induced cracking occurring at stressed surfaces other than sharp pre-existing defects. Indeed, initial work has shown that under certain circumstances cracks can initiate more readily at the surface of stressed notches than at a fatigue pre-crack.

SUMMARY

40. For optimum fracture toughness 3½% NCMV steel should be tempered at the highest temperature which is consistent with the required strength level. For maximum strength and toughness martensitic materials are superior to bainitic materials.

41. Martensitic 3½% NCMV steel is readily temper embrittled by furnace cooling although this is not reflected in the fracture toughness properties, the embrittlement appearing to be strain rate sensitive and can produce a substantial increase in the Charpy V-notch impact fracture appearance transition temperature.

42. Although martensitic 3½% NCMV steel is susceptible to hydrogen-induced sub-critical flaw growth under static load at ambient temperature, the degree of susceptibility is reduced as the proof strength is lowered. Furthermore, the threshold condition for hydrogen induced failure approaches the critical fracture toughness level at higher temperatures.

43. Martensitic 3½% NCMV steel is only susceptible to accelerated fatigue crack growth in hydrogen at frequencies of about 100 Hz when the R ratio value is sufficiently high for the minimum cyclic stress intensity to exceed the threshold stress intensity value for hydrogen-induced crack growth under static load.

44. Where the availability of hydrogen is sufficient, hydrogen-induced cracking can occur under predominantly plane stress conditions.

45. Tests in high purity condensing steam have shown that cracks can propagate when the applied stress intensity is close to the fracture toughness of the steel. Cracking proceeds by a mixed ductile mode associated with plastic deformation and by brittle inter- and trans-granular modes associated with the environment.

ACKNOWLEDGEMENTS

The authors wish to thank Mr. W.H. Bailey, Chief Metallurgist, River Don and Associated Works, and Dr. K.J. Irvine, Head of Research BSC Special Steels Division, for permission to publish this paper.

REFERENCES

1. Watson, E.H. Autoclave reactors for polyethylene production . Amer. Inst. Chem. Eng., Chem. Eng. Progress Technical Manual on Safety in Polyethylene Plants, 1974, 2, 35-47.

2. Begley, J.A. and Toolin, P.R. Fracture toughness and fatigue crack growth rate properties of a NiCrMoV steel sensitive to temper embrittlement, Int. J. Fract. Mech., 1973, 9, 243.

3. McMahon, C.J. Temper embrittlement in steel, ASTM STP-407, Aug. 1968, 131.

4. McIntyre, P. The relationship between stress corrosion cracking and sub-critical flaw growth in hydrogen and hydrogen-sulphide gases , paper presented at the Int. Conf. on Stress Corrosion Cracking and Hydrogen Embrittlement of Iron-Base Alloys, 42 Unieux-Firminy, France, 10-16th June, 1973.

5. Parker, J.G. and Saddler, M.A. Stress corrosion cracking of a low alloy steel in high purity steam, Corr. Sci., 1975, 15, 57.

LIST OF SYMBOLS

J	=	Joules
F.A.T.T.	=	Fracture appearance transition temperature
C.K.S.	=	Compact 'K' Specimen
K_I	=	Plane strain stress intensity. The elastic stress field intensification at the tip of a crack subjected to opening mode displacement is uniquely defined by K_I. It is a function of applied force and test piece geometry having dimensions of stress x length$^{\frac{1}{2}}$.
K_{Ic}	=	A critical value of K_I. The plane strain fracture toughness
K_Q	=	Stress intensity for 5% deviation from elastic behaviour (fracture toughness under non-valid conditions)
K_ISCC	=	Threshold stress intensity above which the material is susceptable to stress corrosion cracking
K_IH_2S	=	Threshold stress intensity for susceptability to cracking in hydrogen sulphide gas.
IT WOL	=	1" thick T type wedge opening loaded test piece.
R	=	Fatigue cycle mode. Algebraic ratio of minimum to maximum stress intensities in fatigue cycle.
K MIN	=	Minimum stress intensity in the fatigue cycle.
K MAX	=	Maximum Stress intensity in the fatigue cycle.
Δ K	=	Range of stress intensity in the fatigue cycle.

TABLE 1

Influence of Tempering Temperature on Strength and Fracture Toughness of $3\frac{1}{2}$%NiCrMoV Steel

(a) Martensitic Condition

Temperature °C	0.2% Proof Stress N/mm^2	Tensile Strength N/mm^2	Elongation % on $5.65\sqrt{A}$	Reduction of Area %	K_{IC} MNm$^{-3/2}$	
540	1195	1290	15.0	51	144.5	148.9
560	1170	1260	13.5	40	156.2	159.4
580	1115	1225	14.0	44	189.4*	188.5*
600	1030	1140	14.4	54	186.6*	

(b) Bainitic Condition

Temperature °C	0.2% Proof Stress N/mm^2	Tensile Strength N/mm^2	Elongation % on $5.65\sqrt{A}$	Reduction of Area %	K_{IC} MNm$^{-3/2}$	
540	1020	1200	14.0	41	68.3	68.8
560	1015	1190	15.5	53	70.1	110.6
580	980	1165	17.0	54	133.5	126.6
600	930	1125	16.0	53	127.7	

Non-valid K_{ic} (i.e. K_a)

TABLE 2

Effect of Tempering Temperature on Charpy V-Notch Transition Temperature

Tempering Temperature °C	Microstructure	Transition Temperature °C		
		50% FATT	54J	34J
540	Martensitic Bainitic	-35 +78	>+25 +98	-36 +76
560	Martensitic Bainitic	-82 +80	-55 +98	-84 +60
580	Martensitic Bainitic	-88 +55	-65 +83	-90 +44
600	Martensitic Bainitic	-100 +40	-70 +62	-95 + 2

TABLE 3

Chemical Composition of an NCM Pressure Vessel Steel

Element, wt%											
C	Si	Mn	S	P	Ni	Cr	Mo	V	Sn	Sb	As
0.30	0.23	0.67	0.032	0.029	2.68	0.64	0.60	0.003	0.010	0.014	0.033

TABLE 4

Chemical Composition of a $3\frac{1}{2}$% NCMV Pressure Vessel Steel

Element, wt%											
C	Si	Mn	S	P	Ni	Cr	Mo	V	Sn	Sb	As
0.34	0.20	0.44	0.012	0.011	3.43	1.56	0.62	0.14	0.015	0.003	0.030

TABLE 5

Chemical Compositions of Two 3½% NCMV Steels Possessing 0.2% Proof Stresses of 1000 and 1250 N/mm^2

0.2% Proof Stress N/mm^2	Tensile Strength N/mm^2	Element, wt.%.								
		C	Si	Mn	S	P	Ni	Cr	Mo	V
1000	1200	0.31	0.16	0.41	0.007	0.006	3.55	1.10	0.55	0.12
1250	1320	0.39	0.30	0.60	0.007	0.004	3.20	0.82	0.62	0.19

TABLE 6

De-Ionised Water Analysis, ppm

CO_2	Not Detected
Na	0.02
NH_3	0.02
SO_4	0.02
Cl	0.01
TDS*	1.0

* Total Dissolved Solids

TABLE 7

Test Data Determined in Condensing Steam and in Hydrogen Sulphide Gas

Test Environment	Temperature °C	Pressure mbar	Test Result
H_2S Gas	25	133	$K_I\ H_2S^* = 70.4\ MNm^{-3/2}$
H_2S Gas	100	133	$K_I\ H_2S = 140\ MNm^{-3/2}$
Condensing Steam	≏ 100	1000	No growth with $K_I = 80\ MNm^{-3/2}$
Condensing Steam	≏ 100	1000	0.9 mm max. growth with $K_I = 120\ MNm^{-3/2}$ in 2520 h. (Max. growth rate = 1×10^{-7} mm/s)

* Threshold stress intensity for cracking in hydrogen sulphide gas

TABLE 8

The Chemical Composition and Mechanical Properties of 817M40 Steel

0.2% Proof Stress N/mm^2	Tensile Strength N/mm^2	Element, wt%.								
		C	Si	Mn	S	P	Ni	Cr	Mo	V
1080	1190	0.34	0.23	0.55	0.045	0.030	1.60	1.13	0.30	0.01

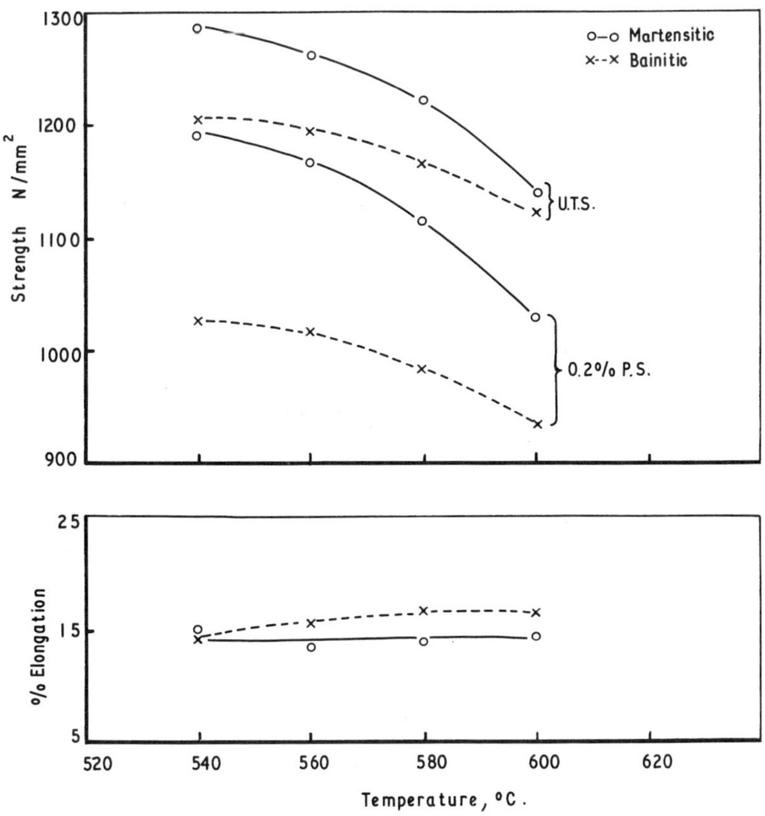

Fig. 1: Effect of tempering temperature on tensile properties of 3½ NCMV

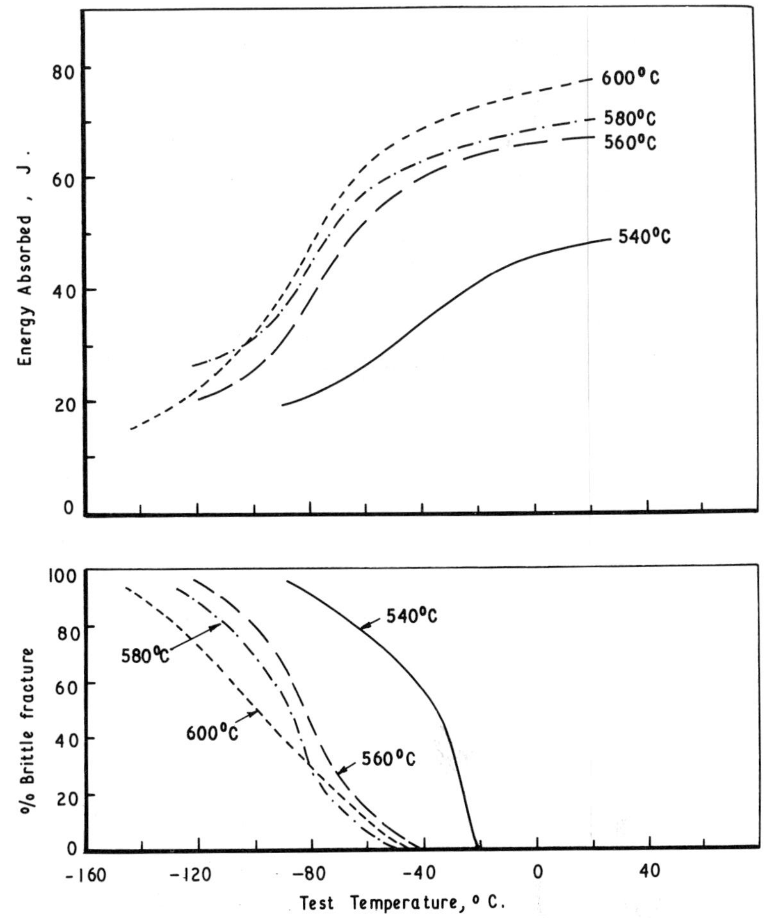

Fig. 2: Effect of tempering temperature on impact properties of 3½ NCMV in martensitic condition

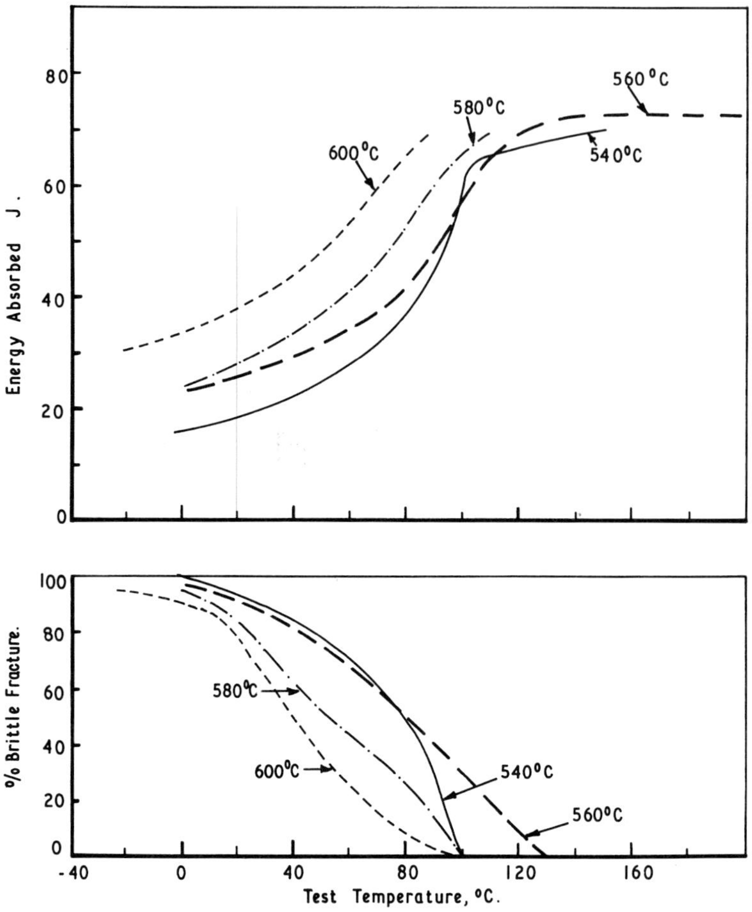

Fig. 3: Effect of tempering temperature on impact properties of 3½ NCMV in bainitic condition

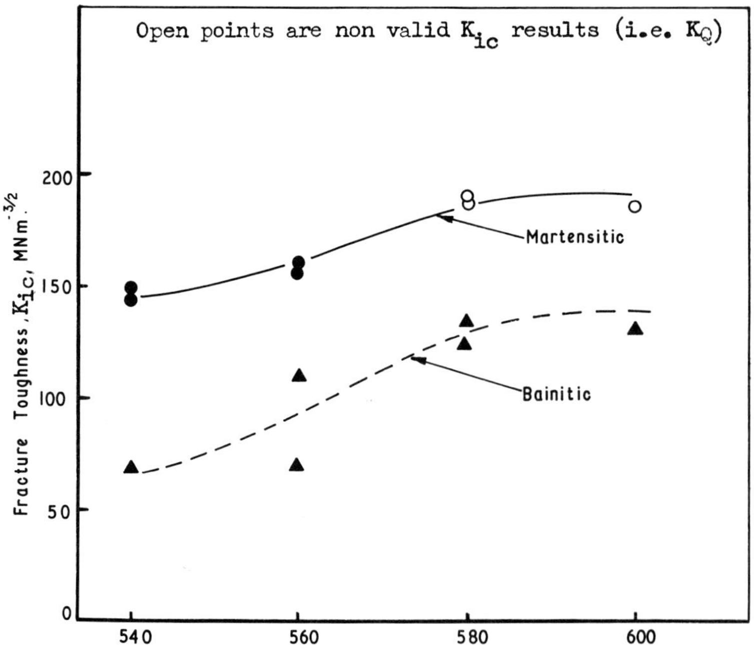

Fig. 4: Effect of tempering temperature on fracture toughness at 20°C

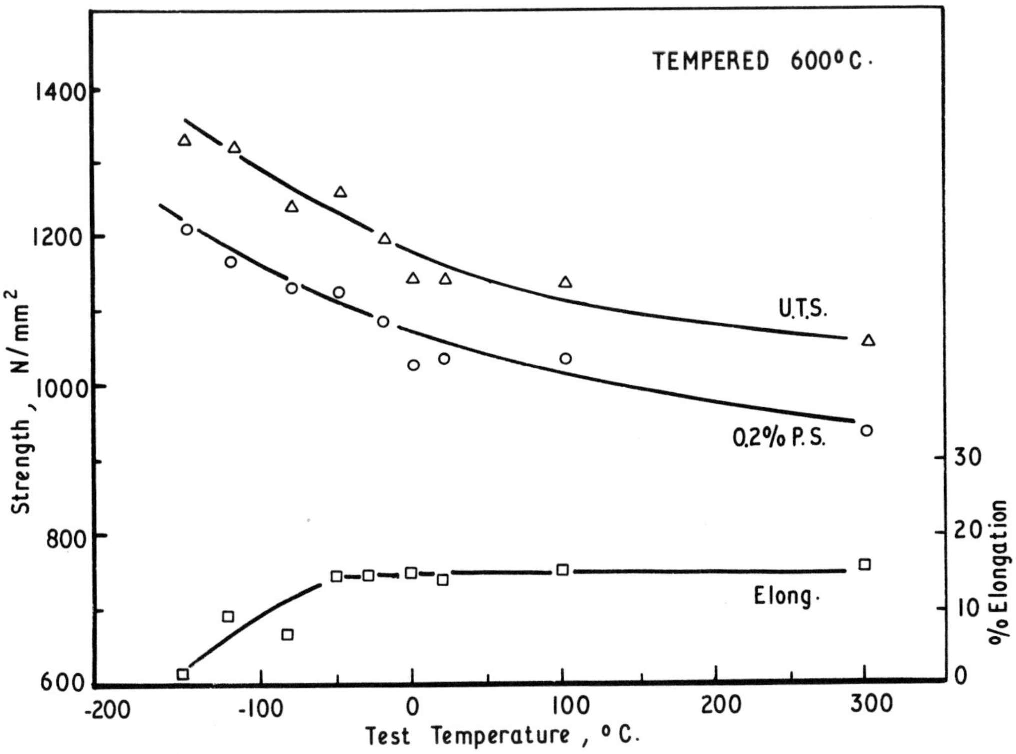

Fig. 5: Effect of temperature on tensile properties of $3\frac{1}{2}$ NCMV in martensitic condition

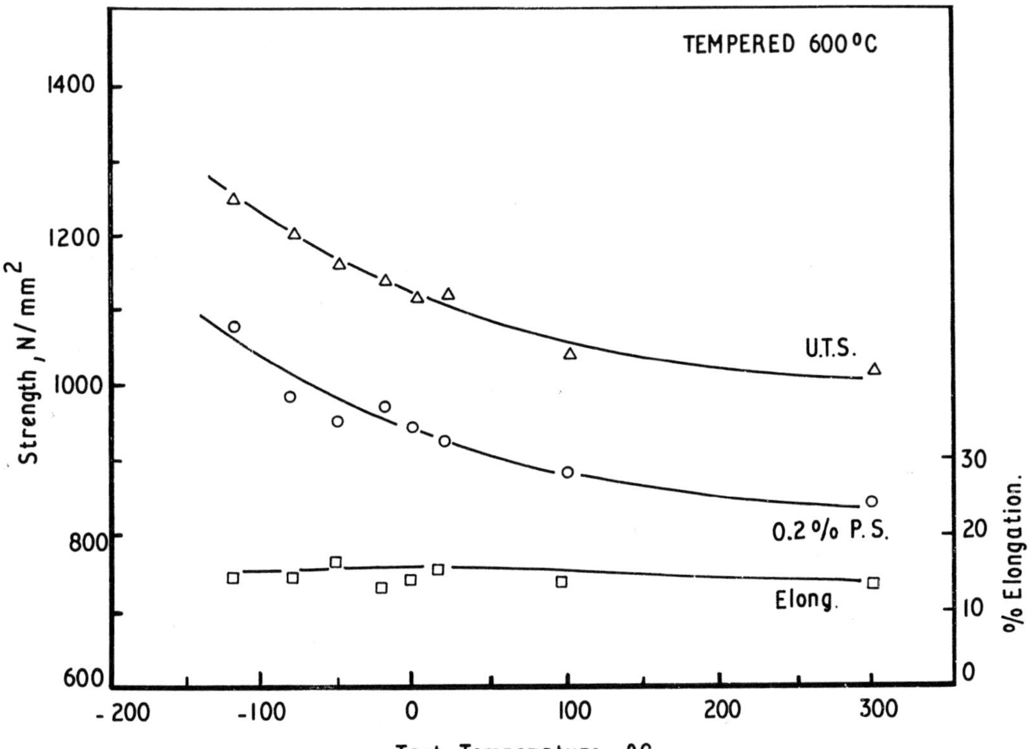

Fig. 6: Effect of temperature on tensile properties of $3\frac{1}{2}$ NCMV in bainitic condition

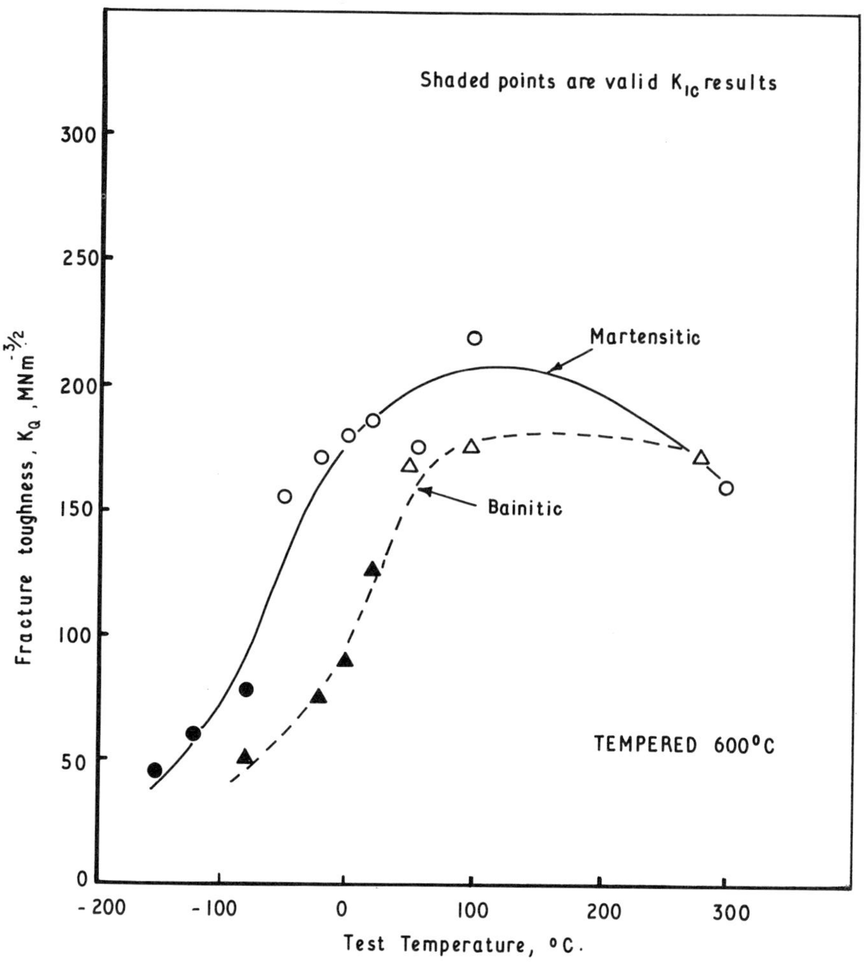

Fig. 7: Effect of temperature on fracture toughness of $3\frac{1}{2}$ NCMV

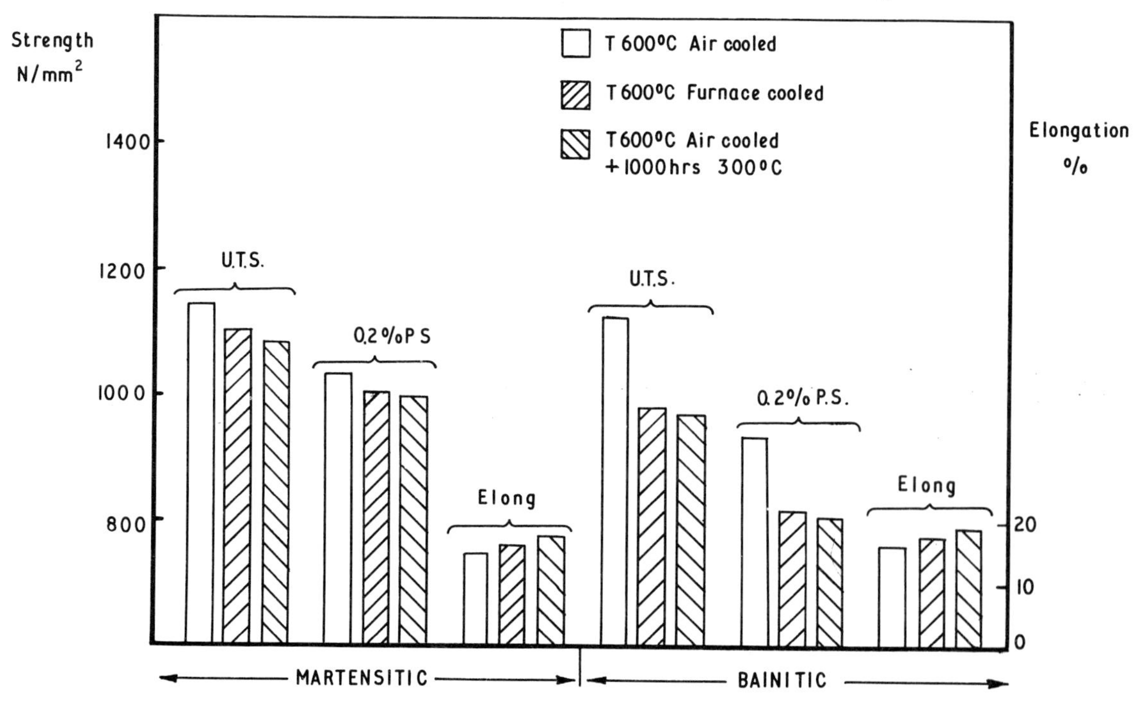

Fig. 8: Effect of embrittling treatment on tensile properties of $3\frac{1}{2}$ NCMV

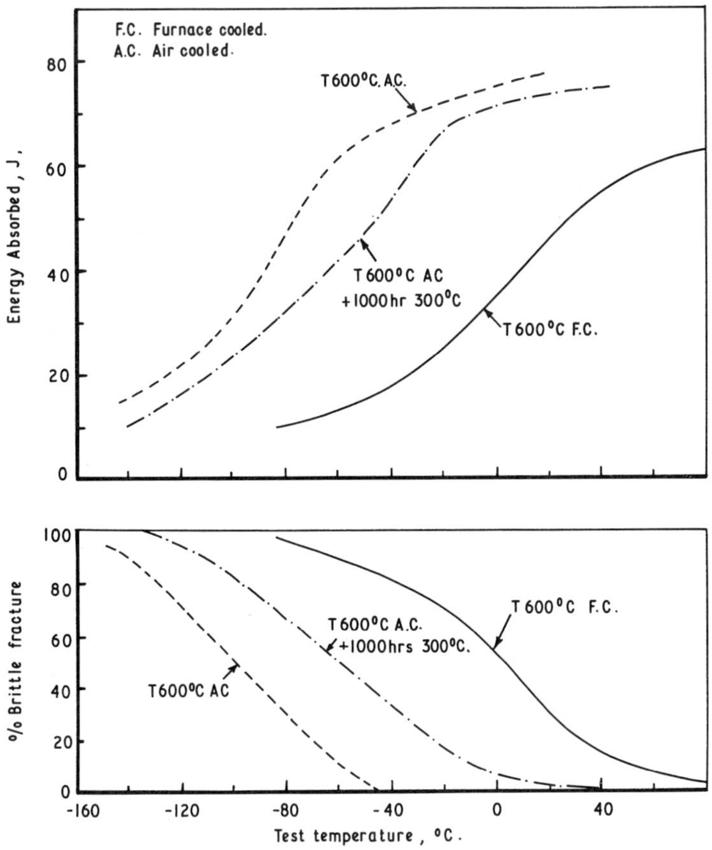

Fig. 9: Effect of embrittling treatments on impact properties of 3½ NCMV in martensitic condition

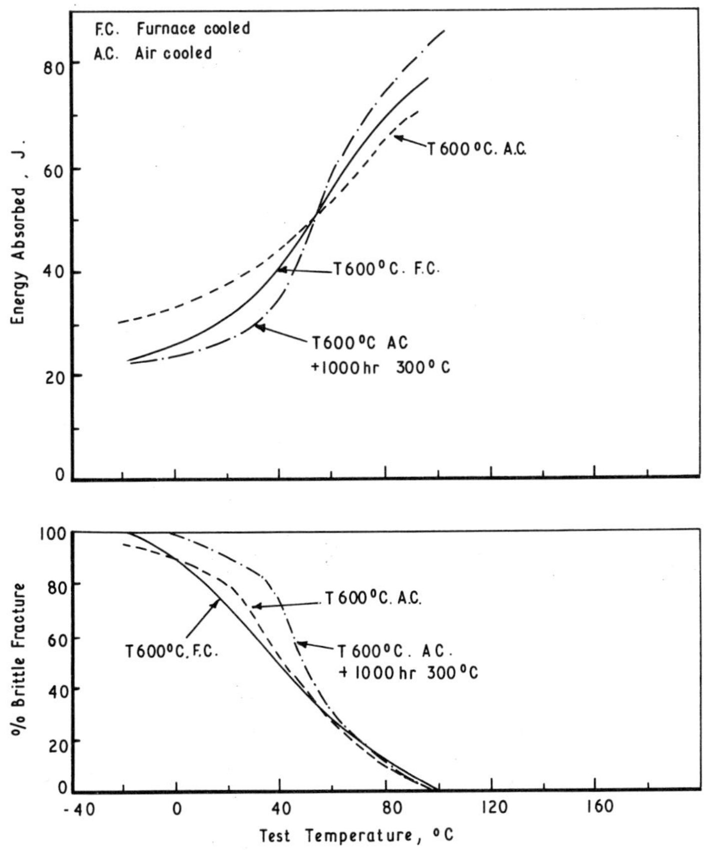

Fig. 10: Effect of embrittling treatments on impact properties of 3½ NCMV in bainitic condition

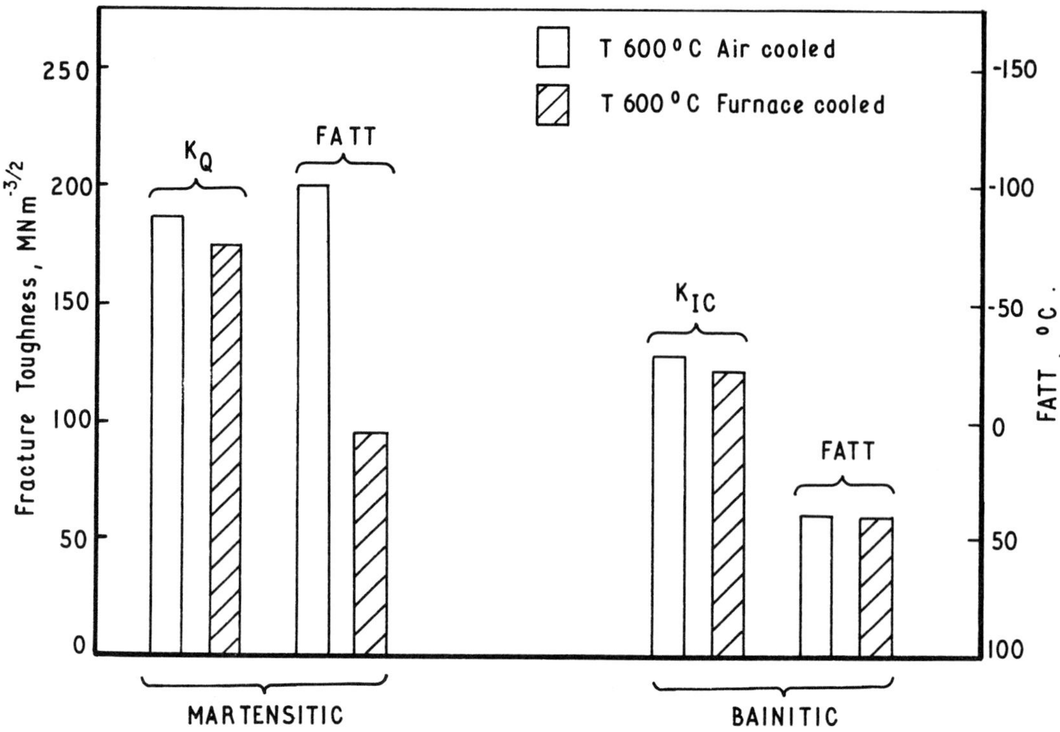

Fig. 11: Effect of slow cooling from tempering temperature on toughness

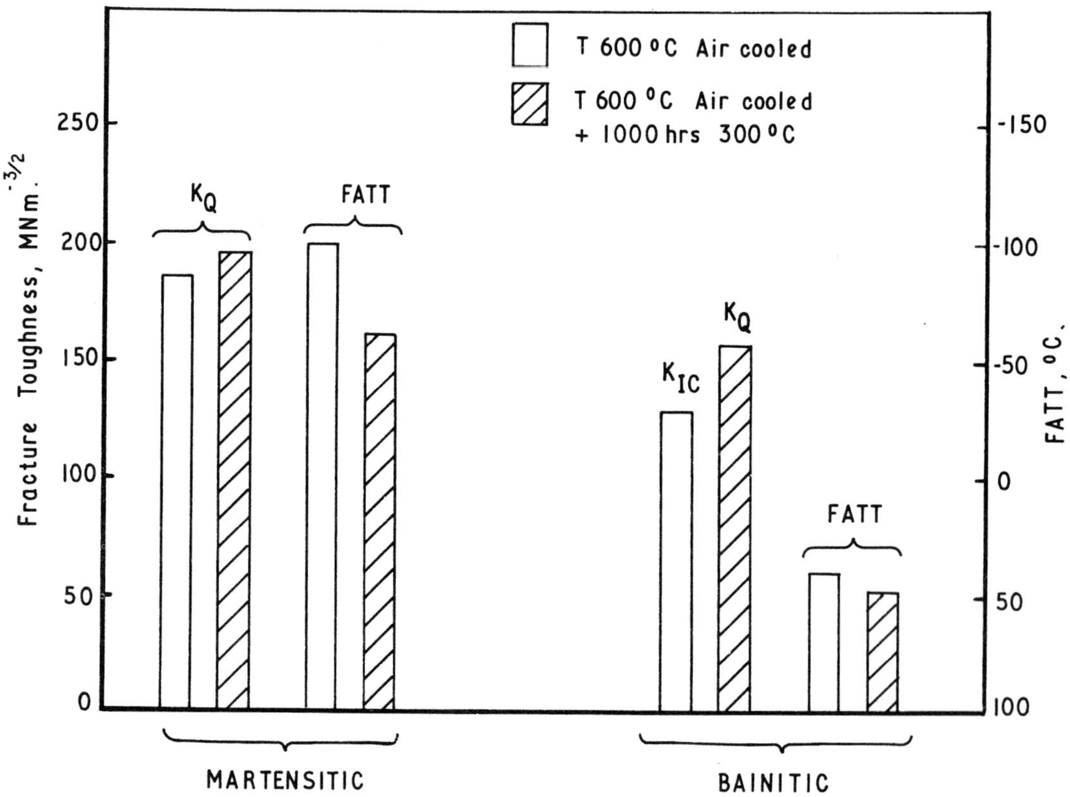

Fig. 12: Effect of ageing at 300 °C on toughness

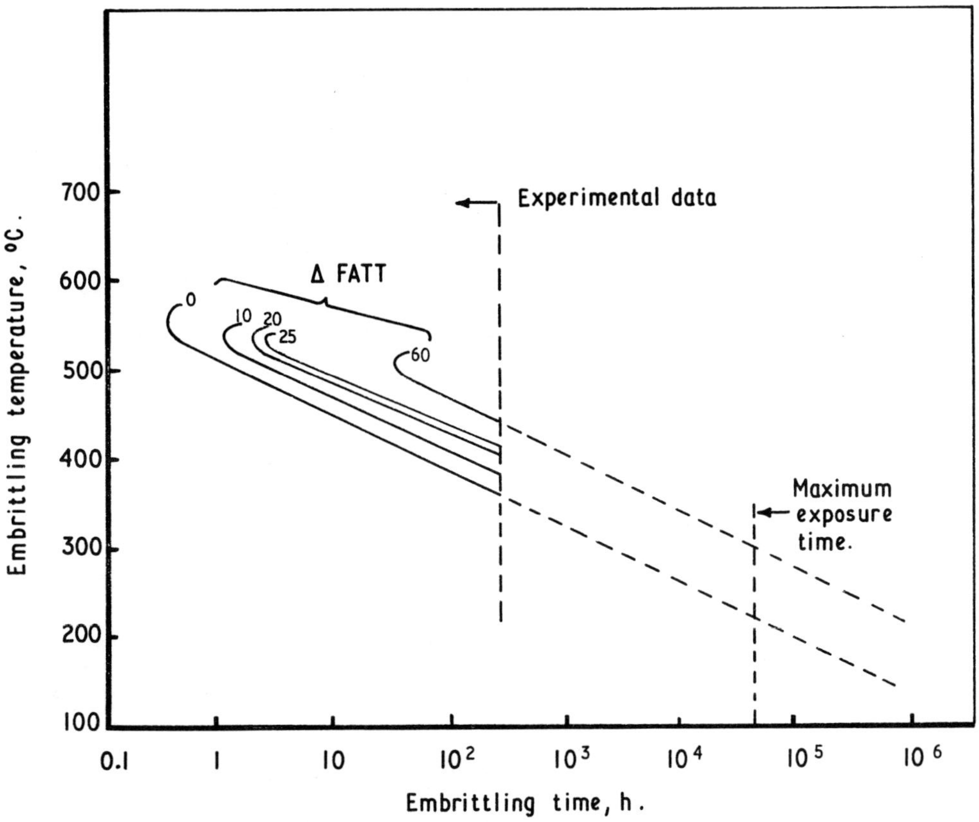

Fig. 13: Influence of temperature on time required for embrittlement to occur (Ref. 3)

Fig. 14: Quasi-cleavage cracking in an unembrittled $3\frac{1}{2}$ NCMV steel tested in hydrogen sulphide gas

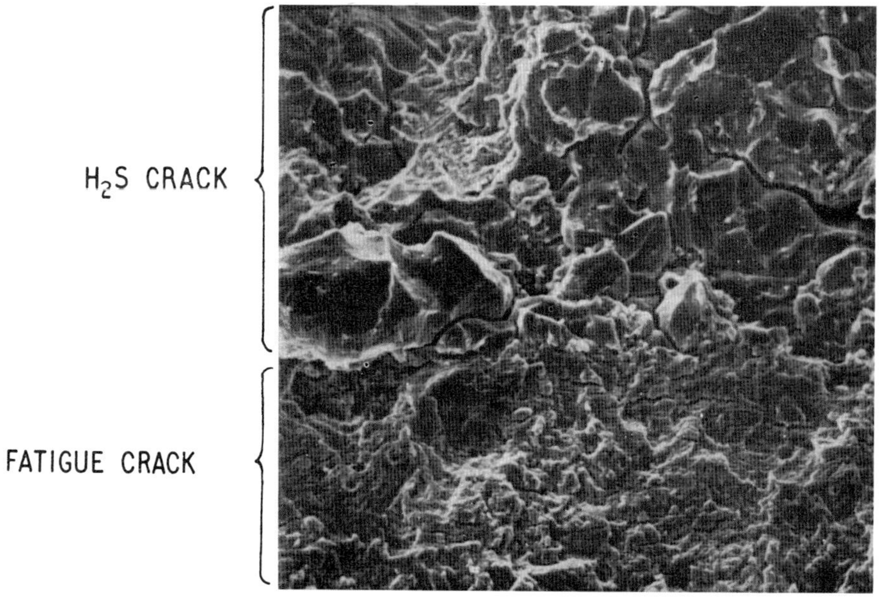

Fig. 15: Intergranular cracking in an embrittled $3\frac{1}{2}$ NCMV steel tested in hydrogen sulphide gas

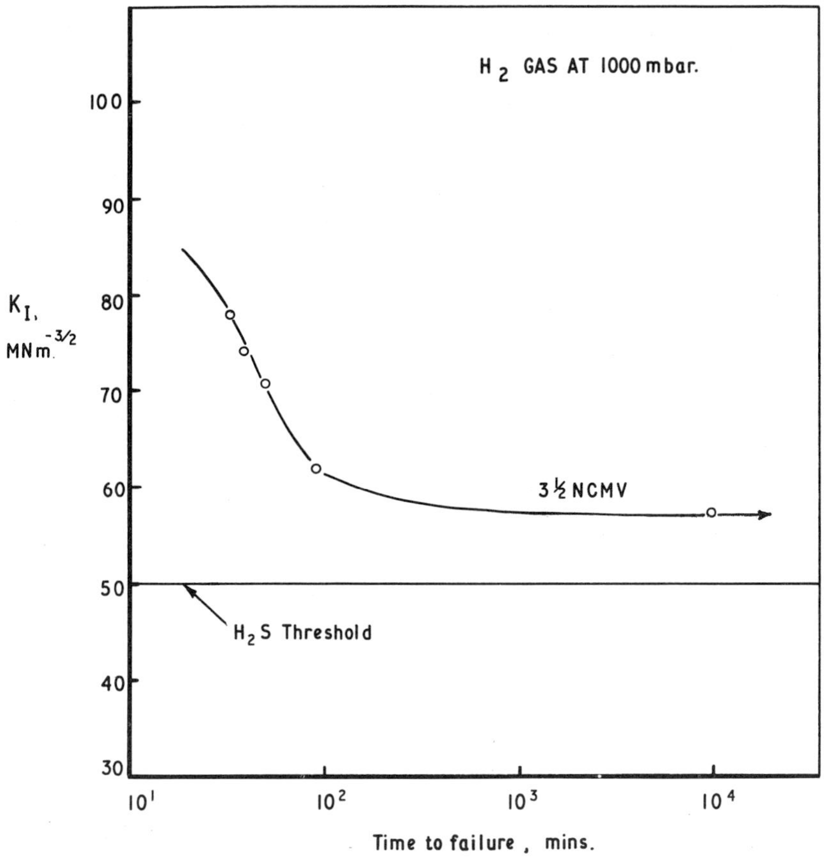

Fig. 16: Brown curve for 3½ NCMV steel of 1250 N/mm² proof stress

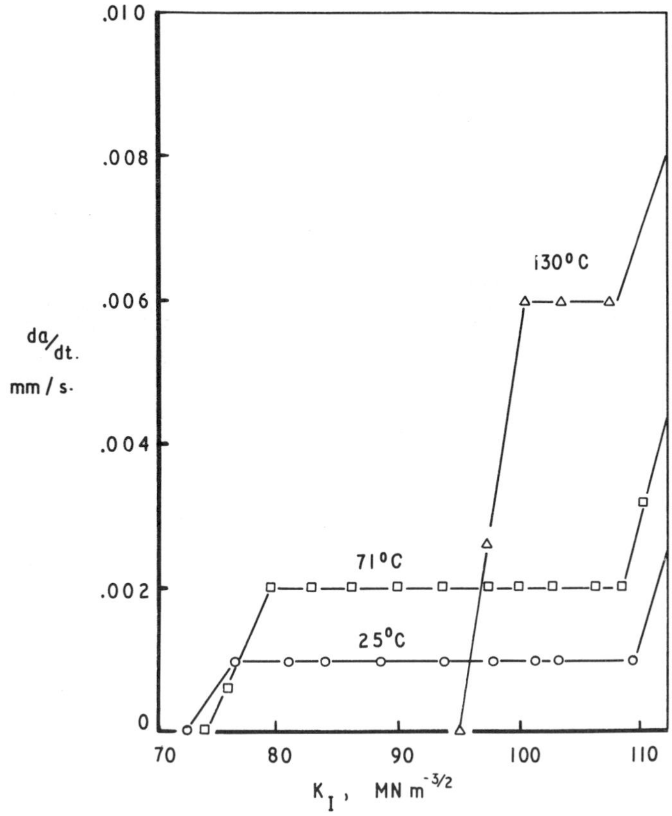

Fig. 17: Effect of temperature on crack growth rate of 3½ NCMV steel of 1250 N/mm² proof stress in hydrogen

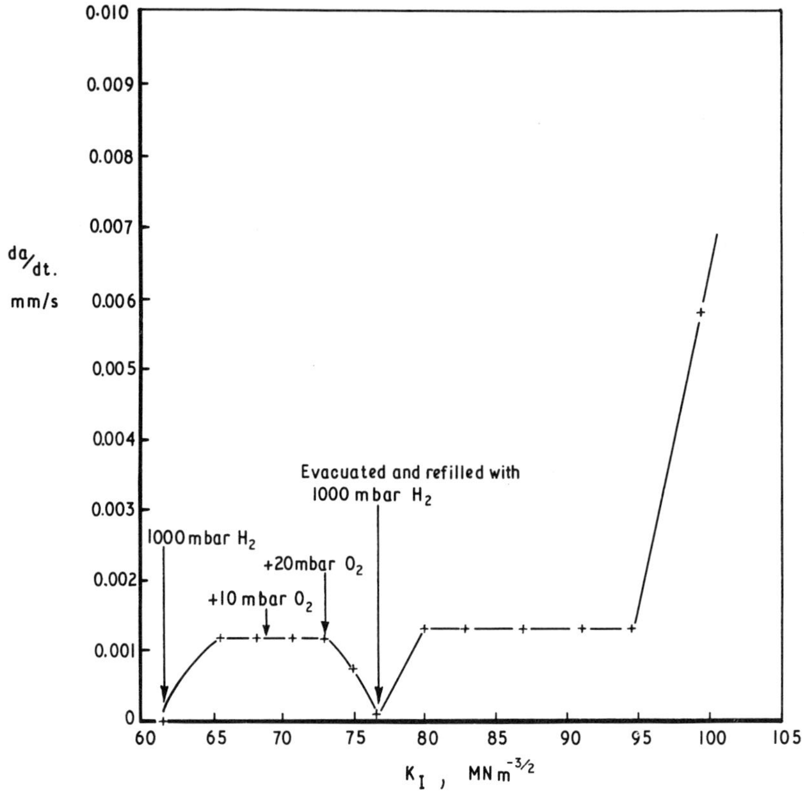

Fig. 18: Influence of oxygen additions on crack growth of 3½ NCMV steel in hydrogen at 25°C

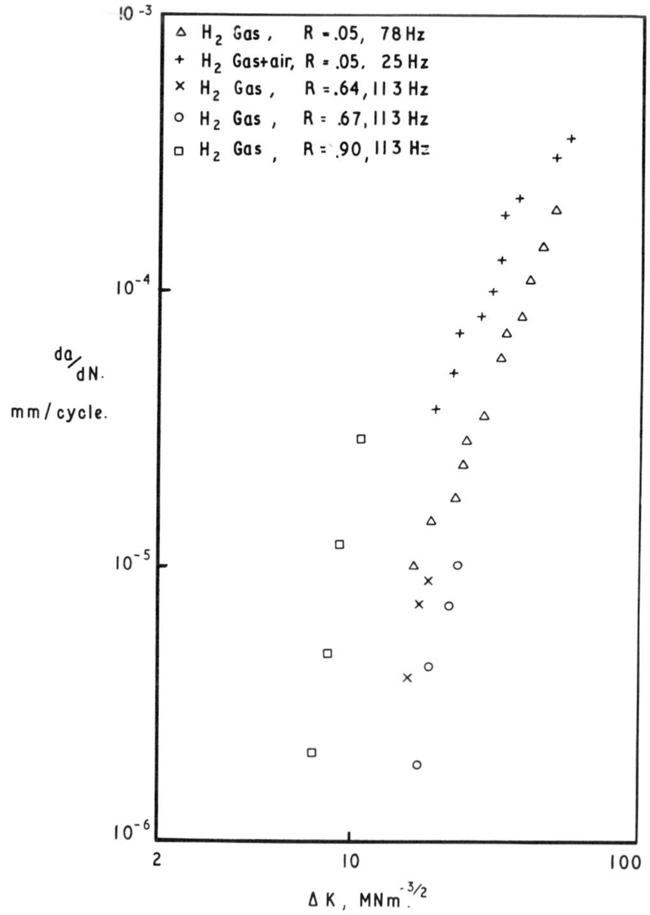

Fig. 19: Influence of hydrogen on fatigue crack growth rate in 3½ NCMV

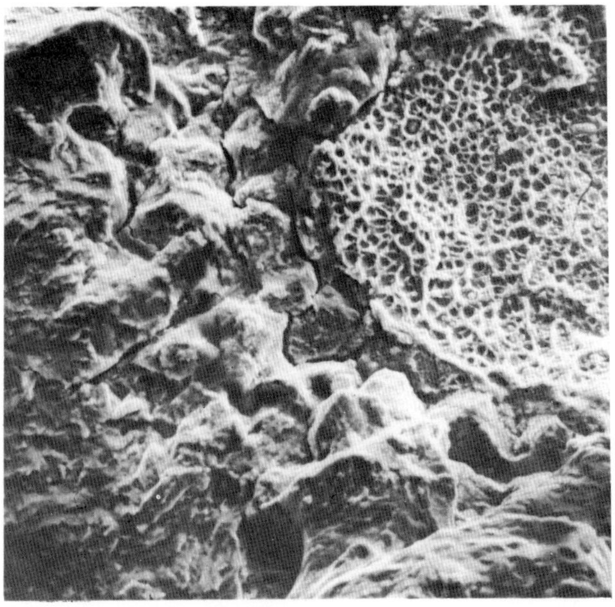

x 490

Fig. 20: Mixed cleavage, intergranular and ductile fracture modes on a crack grown in condensing steam

x 450

Fig. 21: Ductile transgranular cracking produced in condensing steam

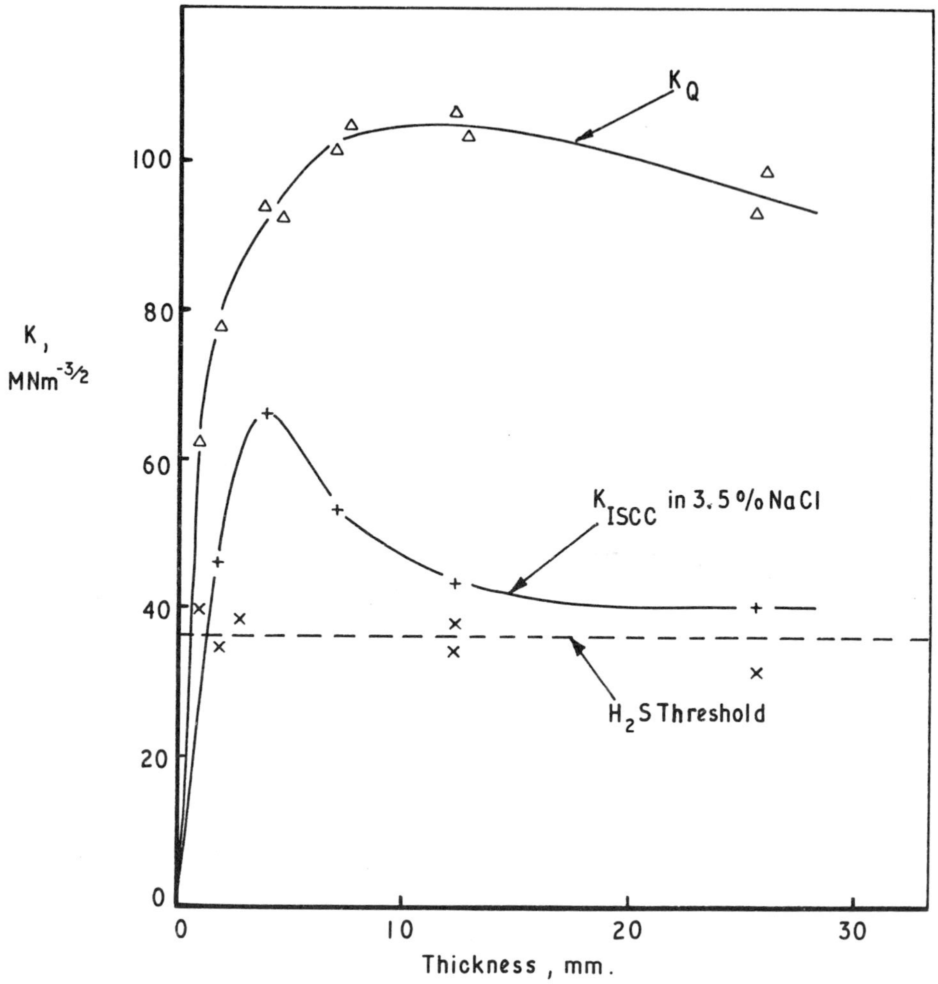

Fig. 22: Influence of test piece thickness on properties of 817M40 steel

x 20

Fig. 23: Shear lips on a 3.17 mm thick test piece tested in 3.5% sodium chloride solution

x 20

Fig. 24: Absence of shear lips on a 3.17 mm thick test piece tested in hydrogen sulphide gas

CEMENTED CARBIDE IN HIGH PRESSURE EQUIPMENT

B. ZETTERLUND, MSc Metallurgy
SANDVIK AB, Coromant Wear Parts, Sweden

INTRODUCTION

Cemented carbide is a material associated with a very high hardness and a relatively low toughness. The hardness is coupled to a very high compressive strength and this property along with the exceptionally high Young's modulus are the main features making cemented carbide interesting in high pressure engineering.

One aim of this paper is to submit some basic information about the pertinent properties of cemented carbide. A second is to give examples of the use in high pressure equipment.

It does not seem quite clear where the lower limit of 'high pressure' should be set. Considering pressures of the order 10 000 MPa (100 kbar) the number of applications for cemented carbide is not large. The only regular large-scale use is in tooling for the manufacture of synthetic diamonds. However, if pressures down to the order 100 MPa (1 kbar) are regarded as 'high pressures' there are a large number of applications.

Accepting this very wide range one application from the very high pressure range which will be discussed below is the anvils for the manufacture of diamonds. As to the low end of the range tools for plastic forming and plungers and liners for compressors and pumps mainly used in chemical industry will be mentioned.

PERTINENT PROPERTIES OF CEMENTED CARBIDE

Conventional cemented carbide contains tungsten carbide, WC, as the hard constituent and cobalt, Co, as the soft binder phase. The mechanical properties are primarily determined by features related to the structure, such as carbide grain size and cobalt content, but also by manufacturing variables and raw materials. The strength of a cemented carbide component depends on these parameters as well as on design and prestressing.

The important properties when discussing high pressure components are:

—hardness,
—compressive strength,
—fatigue strength,
—Young's modulus,
—toughness and ductility.

For grain sizes of WC from about 1 μm to about 4 μm and Co-contents varying from 6 to 40 vol %, the range covered may be 800 to 1800 HV in hardness and 3300 to 6900 N/mm^2 in compressive strength (Fig. 1 and 2). The exceptionally high modulus of elasticity varies between 450 000 and 650 000 N/mm^2 and decreases with increasing Co-content as does the hardness and compressive strength. The hardest grades of cemented carbide thus surpass high-strength steel by a factor of 2 to 3 in respect of these three properties.

On the other hand and as a consequence, the toughness and ductility is usually inferior to steel. Discussions of toughness of cemented carbide are often based on the required energy to fracture in transverse rupture tests (bend test). The toughness measured this way increases with increasing Co-content as illustrated in Fig. 3, and with increasing WC-grain size, at least in the lower and medium Co-range.

The concept of fracture toughness has recently been applied to cemented carbide as a toughness criterion. According to K_{IC}-determinations at SANDVIK the fracture toughness of the common WC–Co-grades fall in the interval 10–20 MNm$^{-3/2}$. This level may on the one hand be compared to high-speed steels which roughly fall in the 15–25 MNm$^{-3/2}$ range and on the other hand to ceramics, the fracture toughness of which scarcely exceed 3 MNm$^{-3/2}$.

The fatigue strength, finally, is extremely high as long as the stress cycle is purely compressive (Fig. 4). As a consequence of the limited ability of cemented carbide to withstand tensile stresses the fatigue strength in case of pulsating tensile or alternating loads is much lower.

HIGH PRESSURE APPLICATIONS

Pressure range 10 000 *MPa* (100 *kbar*)

Cemented carbide has for a long time been accepted as the only economic material for dies and anvils in the manufacture of man-made diamonds. The demands on the tool material in this process are extreme. For the anvils it is necessary to use a grade with a very high compressive strength and consequently a low Co-content. One viable grade for this purpose holds approximately 10 vol % of Co and has a compressive strength of about 6700 N/mm^2. The anvils are normally subjected to a stress state of a compressive character, whereas the dies have to withstand high tensile loads. Hence the need for ductility and low notch sensitivity and higher tensile strength is pronounced for the dies. The choice of grades is therefore more towards a higher Co-content, i.e. around 15 vol %.

It is a widespread experience that the life of the tools operating in the high-pressure range show a wide scatter in life. For instance, it is not uncommon that the maximum life in a batch of anvils is 500 to 1000 times the minimum life. From a physical or a mechanical point of view this is not surprising but it is not characteristic of cemented carbide. It is probably, at least partly, a consequence of the extremely high stress level which makes very small variations in load distribution and material properties strongly influence the life.

Assuming the fractures are a consequence of fatigue the scatter in life may easily be understood. Reverting to Fig. 4, the Wöhler diagram showing compressive fatigue strength versus number of cycles to failure, it is evident that the anvils are used at a stress level where the fatigue curve is only declining a little and a small change in stress or strength will drastically influence the life.

In fact, considering the grade used for anvils and assuming a nominal life of 100 cycles, an increase in stress or a decrease in fatigue strength by 2% will lead to a mean life of only 5 cycles instead of 100. Similarly a 2% decrease in stress or increase in fatigue strength results in a mean life of 425 cycles.

Thus, very small variations in load as well as strength may have tremendous effects on the life of the cemented carbide tools.

A great deal of work has therefore been done to create the most consistent and well controlled manufacture of high pressure tools. It is equally important for the user to establish a well controlled process. However, the effects of these efforts are only small and a scatter of the order of a factor much less than 100 does not seem attainable.

The mode of fracture of anvils for diamond synthesis has been subjected to many studies. These have shown that the most common fracture mode seems to be the one illustrated in Fig. 5A, although crack patterns like the ones in Fig. 5B and 5D also occur frequently along with small breaks at the top edge, Fig. 5C.

Attempts have been made to interpret these crack patterns partly with the help of optical and electron microscopy and partly theoretically by calculating stresses. Fractography indicates that the initiation of the first crack may occur close to the centre of the plane of fracture in Fig. 5A. Secondary cracks then propagate radially and perpendicular to this plane in the conical part of the anvil. This type of crack initiation is supported by calculations of the stress pattern with the help of the finite element method starting from approximations of the distribution of load (Fig. 6). This is of course a simplified model not taking account of temperature gradients, residual stresses and other important factors.

According to these results it seems likely that central cracks in these high pressure anvils may be initiated through plastic deformation in compression resulting in tensile stresses on unloading. The stress cycle may accordingly not be entirely on the compressive side which would be preferable. The influence of a tensile stress component on the fatigue strength of a cemented carbide grade commonly used for high pressure anvils is illustrated by the Haigh diagram in Fig. 7. With the number of cycles to failure as parameter half the stress amplitude, σ_a, is shown as a function of the mean stress, σ_m. The precise form of the curves for $N \neq \frac{1}{4}$ is not known; the curves in Fig. 7 have been drawn by analogy with known data for other brittle materials. The diagram may however serve to illustrate the decrease in fatigue strength in terms of the maximum allowable stress amplitude ($2\sigma_a$) for a given life when the mean stress (σ_m) approaches the tensile side. This may serve to emphasize the importance of retaining a hydrostatic compressive stress state.

Pressure range 100–1000 *MPa* (1–10 *kbar*)

Examples of cemented carbide as a tool material in this range are found in the field of plastic forming of metals, i.e. cold forging and extrusion. In fact rock drilling tools could also be included since the carbide bits in that application work under specific loads of the order 2000 MPa (20 kbar).

Tools for cold forging and extrusion may vary considerably in size, design and other characteristics. Wear is usually the principal life limiting factor and hence the hardness and wear resistance of cemented carbide are the important properties in their competition with other materials. However, the high compressive strength, which makes high loads without plastic deformation tolerable, as well as the high Young's modulus, which contributes to the dimensional accuracy of the forged material, are also properties that favour the use of cemented carbide. Due to the comparatively low ductility it is also in these applications necessary to design and prestress the tools in order to minimize or avoid tensile stresses.

As an interesting application of cemented carbide in the lower end of this pressure interval it is worth mentioning sealing components such as plungers and liners used in compressors and pumps in the chemical industry for extrusion, transport and feeding of different chemicals into reactors etc. The pressures are of the order 500 MPa (5 kbar) and less. The main requirement on the material in this application is to provide a wear resistance and surface finish good enough to retain an accurate fit and a seal over a long period of time. Improvements in life of 15–30 times have been achieved in comparison to steel.

In many cases the sizes of these components are considerable. The length of plungers may exceed 2000 mm and the OD may be around 400 mm. This coupled with the extreme requirements on low porosity and low inclusions necessitates the most advanced manufacturing and inspection techniques.

Figures 8 and 9 serve as an illustration to this application. Plungers may either be of solid carbide or of the compound type with a carbide sleeve mechanically fitted on to a steel core. As in other cases where cemented carbide is used due consideration must be taken to minimize tensile stresses, which in this application may be due to bending and/or differences between steel and cemented carbide in thermal expansion.

IMPROVED PERFORMANCE AND EXTENDED USE

For applications in the very high pressure range, i.e. the order 10 000 MPa (100 kbar), increased life of the cemented carbide tools might be attainable if the properties in different parts of one and the same tool could be optimized more independently. For instance one common way of initiating cracks in anvils is probably by exceeding the elastic limit in compression in the centre. On unloading the plastic deformation that has occurred will give rise to tensile stresses which may initiate a crack. If a crack is not immediately initiated it is very likely to occur through fatigue after a number of cycles. A material with a higher compressive strength would lessen the risk of cracks in the centre, but would on the other hand increase the risk of initiation of surface cracks. Thus one way of improving the performance would be to create a combination using two grades of materials, i.e. a compound anvil. The manufacture of components with, for example, a continuous decrease in Co-content from the surface to the centre does, however, create very difficult and so far unsolved problems.

The use of cemented carbide in high-pressure equipment may possibly be extended to larger sizes of tools and other parts with the help of advanced methods of prestressing, such as strip winding and hot isostatic pressing as a technique for decreasing the risk of critical defects in the material.

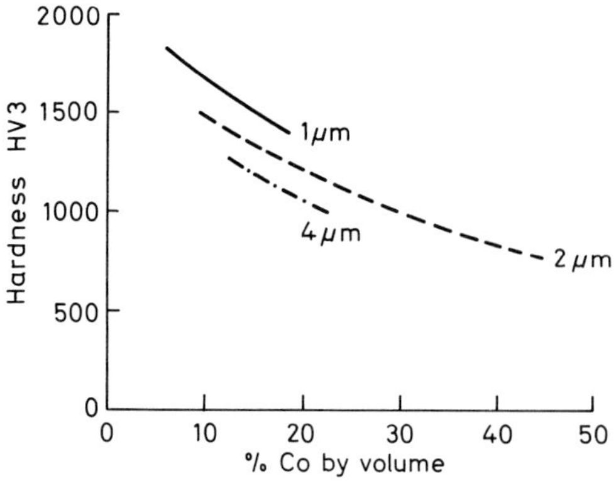

Fig. 1: Vickers hardness as a function of the cobalt content for different WC- grain sizes in WC-Co-carbides

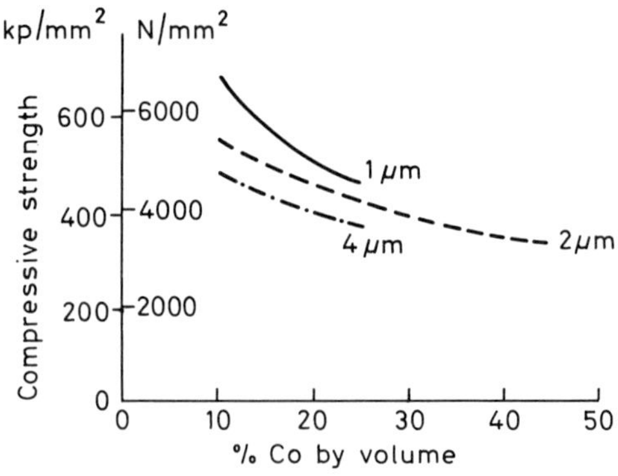

Fig. 2: Compressive strength as a function of the cobalt content for different WC grain size in WC-Co carbides

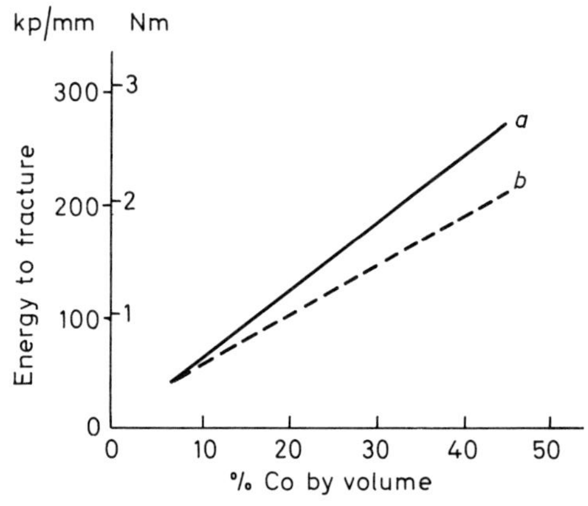

Fig. 3: Toughness, expressed as absorbed energy to fracture, as a function of the cobalt content in WC-Co carbides
(a) Toughness determined by bending a 4-point loaded test specimen
(b) Impact toughness valid for unnotched specimen

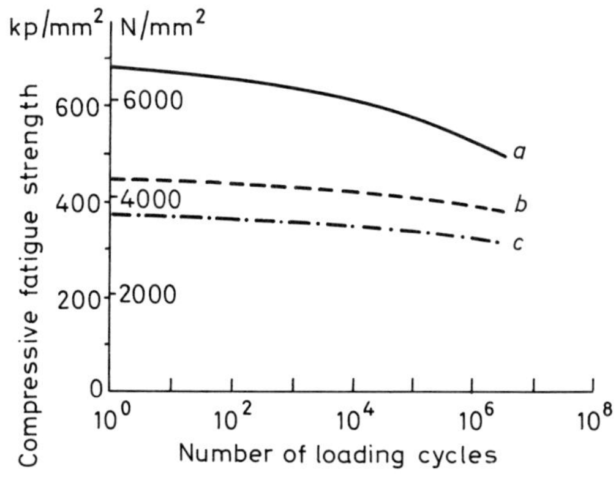

Fig. 4: Examples of Wohler curves from fatigue testing of WC-Co-carbides. Lower load limit 25 kp/mm² (36 000 psi)
(a) 1 μ m (40 μ in) WC, 10% Co by volume
(b) 4 μ m (160 μ in) WC, 13% Co by volume
(c) 4 μ m (160 μ in) WC, 24% Co by volume

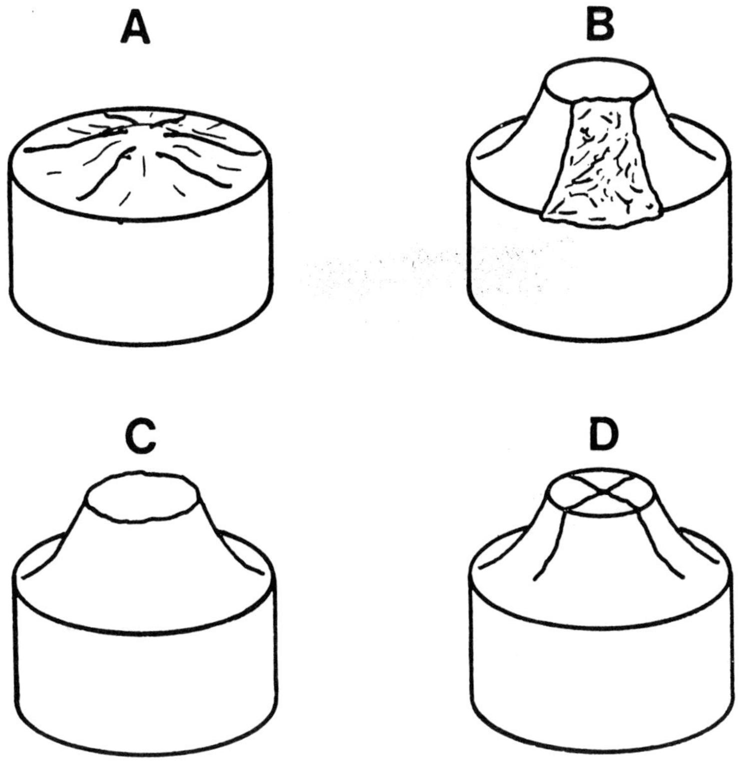

Fig. 5.

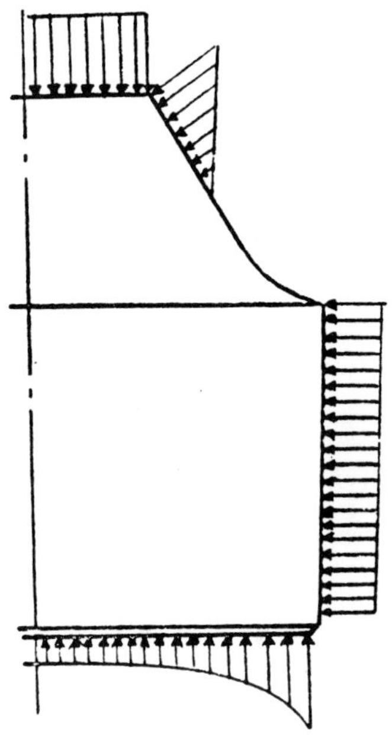

Fig. 6.

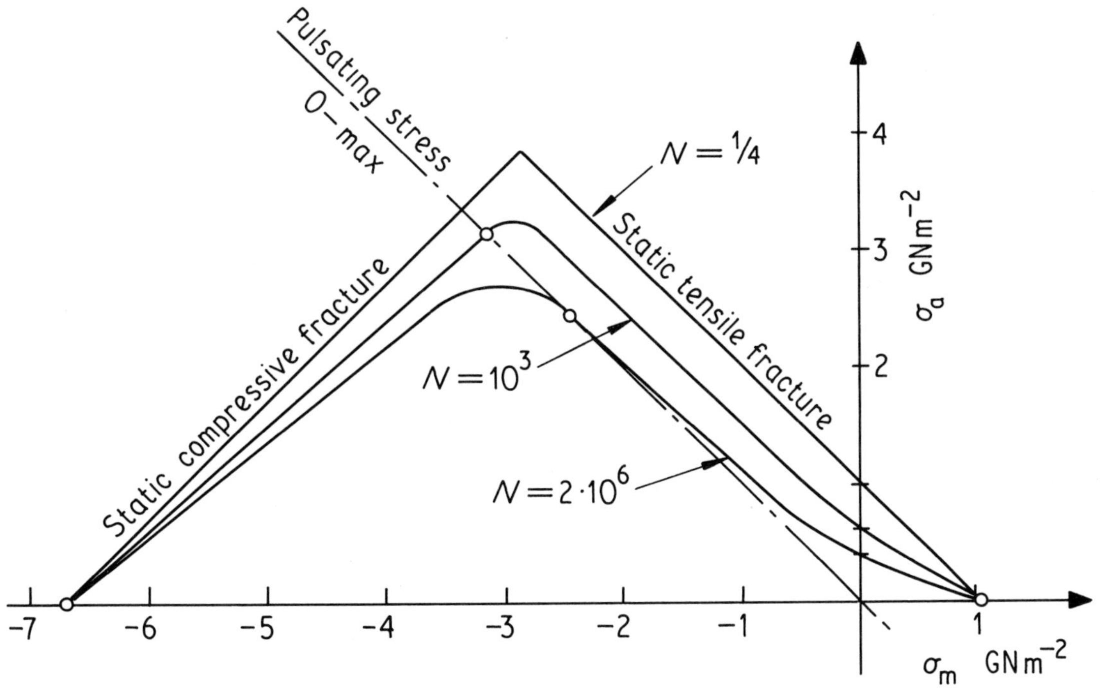

Fig. 7: Haigh Diagram for a typical anvil grade

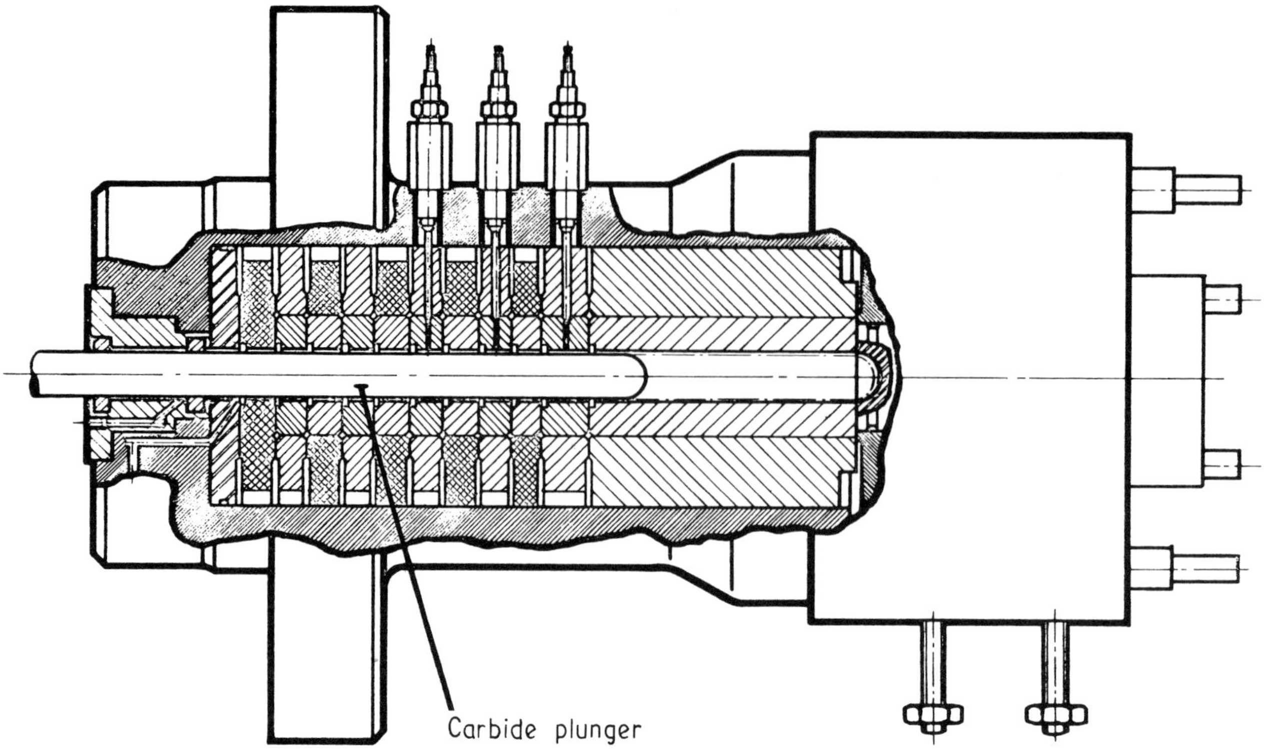

Fig. 8.

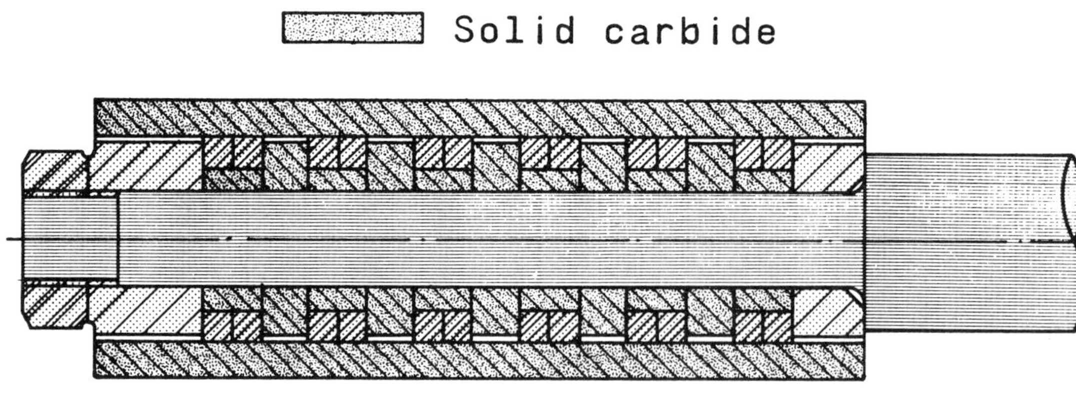

Fig. 9: Spacer ring assembly

REVIEW OF METHODS AND EQUIPMENTS FOR MATERIAL TESTING UNDER PRESSURE

H. Ll. D. PUGH, DSc, BSc, CEng, FInstP, FIMechE, FIM and E. F. CHANDLER, BA, CEng, MIMechE
National Engineering Laboratory, East Kilbride, Glasgow

SYNOPSIS This paper reviews the equipments and techniques designed and developed for providing information on the mechanical properties of materials under a superimposed pressure. Different designs of tension, torsion, compression, and shear apparatus are described and their capabilities assessed. Methods of determining stress-strain relations for materials under pressure are examined. Details are given of apparatus for carrying out tests at high pressure and high temperature.

INTRODUCTION

1. Knowledge of the mechanical properties of materials is fundamental for any form of design of structures and equipments. Although in the past useful information has been obtained from uniaxial tension or compression stressing or straining tests, in practice, such simple stress systems rarely occur. Stresses in structures and in materials undergoing forming processes are usually complex in character and the mechanical properties of the materials involved can be influenced by the prevailing stress system. It is expedient, therefore, to have information on the mechanical behaviour of materials under such conditions. This can only be obtained by investigating the properties of the materials under known conditions of combined stress. One approach is to carry out standard mechanical tests, such as tensile tests, in the presence of hydrostatic pressure. These investigations require the use of the specialized equipments and techniques which are reviewed in this paper.

BRIDGMAN STYLE TENSILE YOKE APPARATUS

2 The most commonly used tensile loading method is based on the tensile 'yoke' apparatus originated by Bridgman (ref. 1) in the early 1940s. Bridgman's high pressure container was capable of operating up to test pressures of 30 kbar and his complete apparatus is illustrated diagrammatically in Fig. 1. It consists basically of the high pressure container sealed at the lower end by a plug and at the upper end by a pressurizing plunger. The dumb-bell tensile specimen is held in an opposed yoke arrangement, the lower end of which rests on a load cell located on the sealed end. The top part of the yoke is contacted by the pressurizing plunger, which forces it downwards thus applying a tensile force to the specimen. The specimen extension is obtained by measuring the movement of the pressurizing/loading plunger and making suitable allowances for the deflection of the plunger, yoke and load cell. The pressurizing fluid is normally a liquid, with isopentane being used for the highest test pressures, although nitrogen has been used successfully up to 14 kbar by Carpentier D. et al (ref. 2). The hydrostatic pressure is usually determined by measuring the change in electrical resistance of a manganin wire gauge (ref. 3) or by a standard pressure gauge for tests in the lower pressure range.

3. Bridgman's original tensile loading system has been modified by a number of workers (refs 4,5). The modification, in which the yoke consists of opposed forks, allows larger diameter specimens to be tested and reduces the degree of non-axial loading. The shape of the specimen ends, however, are rather more complex. During the loading of the specimen there is also the possibility of large frictional effects developing between the moving fork and both the static fork and the container bore.

4. A major shortcoming of Bridgman's original apparatus is that pressure increases throughout the tensile test. Thus every part of the stress/strain curve corresponds to a different hydrostatic pressure. It is however possible to maintain a constant test pressure by adopting one of the following methods.

a Piping a constant pressure into the high pressure test chamber from an adjacent pumping system (ref. 6).

b Using top and bottom plungers of identical diameters, which are coupled together mechanically (ref. 7).

c Sealing the lower end of the container by a billet which is hydrostatically extruded through a die at an approximately constant pressure (refs 8,9).

d Using a counterbored container with a retractable lower piston instead of the 'straight through' container with a static lower plug (refs 10-12).

DIRECTLY COUPLED TENSILE TESTING APPARATUS

5. In the directly coupled tensile testing apparatus one end of the tensile test specimen is held securely with respect to the container whilst the other end is connected to the bottom plunger. As pressure is built up inside the container the

outward movement of the plunger is resisted by a ram or jack. When the test pressure is reached the ram and hence the bottom plunger is retracted at the required speed and a tensile load is applied to the specimen. Constant test pressure can be maintained by either an external pumping system or by controlling the relative movement of the top and bottom plungers. With appropriate grips and connectors most forms of tensile specimen can be tested in this type of equipment. There are various high hydrostatic pressure testing equipments (refs 11,13,14) which use the above method of loading and one of the more sophisticated types (ref. 11) is shown in Fig. 2.

STRESS AND STRAIN MEASUREMENT

6. The load being applied to the specimen can be measured by either a load cell outside the container or a load cell inside the container where it is subjected to the hydrostatic pressure. The external load cell although easier to use and less subject to failure has a number of shortcomings. It records the total of the load being applied to the specimen, the friction at the plunger seal, and the load on the plunger due to the pressure in the container. At high pressures the applied tensile load is only a small portion of the total load recorded by the external load cell and this factor coupled with variations of pressure during the actual test makes accurate determination of the tensile load extremely difficult. To reduce friction and thus improve the accuracy of the load measurement some equipments use the Johnson and Newhall (ref. 15) controlled clearance seal or the Morrison (ref. 16) seal for the loading plunger.

7. The internal load cell utilizes either capacitance (ref. 17) or more usually electrical resistance strain gauge (refs 18-20) techniques for determining load. The latter type of cell, consisting normally of a tube or round bar upon which are mounted a number of linear strain gauges, is connected in series with the specimen. To compensate for pressure and temperature effects a second tube carrying a similar arrangement of strain gauges but not subjected to the load on the specimen is also placed inside the container.

8. The problem with the internal load cell is the validity of applying the atmospheric calibration when the load cell is used under pressure. Pugh (ref. 18), using the apparatus shown in Fig. 3 was able to calibrate strain-gauged load cells under pressure. A pressure differential across the tension plunger applied a tensile load to the load cell while it was subjected to a known and constant test pressure. This pressure differential measured by calibrated manganin gauges was recorded simultaneously with the load cell output. The absence of hysteresis between the loading and unloading calibration curves confirmed that the friction between the packingless plunger and the cylinder was negligible and had no influence on the calibrations. The diameter of the plunger being known, the calibrations under pressure could be correlated with the atmospheric calibrations. In the pressure range atmosphere to 10 kbar, it was found that the load calibrations agreed within $\pm 1\frac{1}{2}$ per cent. Brandes (ref. 19) calibrated his load cell under pressure by using a split specimen method of Bridgman (ref. 1) and found little difference from the atmospheric calibration. Other workers reply on the fact the pressure has little effect (3 per cent at 10 kbar) on Young's Modulus and derive the load from strain measurements.

9. Strain, in the form of percentage elongation, is usually obtained from measurements of the movement of the plunger applying the tensile load to the specimen. For the evaluation of true strain ($\ln A_0/A$) and true stress, measurements of the reducing diameter of the specimen are required. One method by which measurements can be effected is the 'interrupted' test method (ref. 1). A load is applied to the specimen in the container at the appropriate test pressure, the load and pressure are removed, the specimen is removed from the container, and the new diameter of the specimen is measured. The procedure is repeated for a number of load increments and obviously is both slow and cumbersome. In an alternative method suggested by MacGregor (ref. 21), a tapered specimen with a series of circumferential marks along its length is used. After a known load has been applied to the specimen the new diameter at each mark is measured and true stress and strain at each position evaluated. The best method is to obtain a continuous photographic record of the changing diameter of the specimen (refs 11,22,23). For this a cross-bored high pressure container fitted with windows (Fig. 2) can be used. An optical system fitted to the container enables a silhouette of the deforming specimen to be projected on a screen. A galvanometer displaying load is placed adjacent to the screen and a continuous record of both load and specimen profile can be photographed (Fig. 4). The method also enables the radius of curvature at the neck to be obtained and hence corrections for non-uniformity of stress at the neck. Toughened plate glass windows are suitable for pressures up to 13 kbar whilst sapphire windows can be used for even higher pressures.

COMPRESSION APPARATUS

10. The simplest form of compression apparatus is that used by Bridgman (ref. 1). A static bottom plug or piston in the high pressure container supports the compression specimen. A top plunger is used to pressurize the fluid as well as to deform the test piece. The pressure will of course continue to increase throughout the test. An improvement on this equipment is the use of a stepped bore container with a moveable bottom plunger (ref. 24) and a platen/load cell located on the bore shoulder supporting the specimen.

11. An alternative arrangement is that originated by Griggs (ref. 25) and shown schematically in Fig. 5. A stepped bore container is again used, but the top and bottom plungers have the same diameter. The high pressure container is pressurized by the upward movement of the lower plunger, the upper plunger being fixed. Once the test pressure has been reached the plungers can be assumed to be mechanically coupled so that the distance between them will not alter during the test. The container is then raised so that the specimen contacts the top piston and is deformed with the fluid pressure remaining constant. Variations on this method include having the plungers yoked together, piping pressurized fluid into the container, and loading the specimen by the movement of either the yoke (ref. 26) or the container (ref. 27).

2. A compression apparatus, designed by Crossland id Ludlow (ref. 28), which uses a moving plunger ithin the high pressure chamber is shown in Fig. 6. ie required test pressure is generated by a high essure piston pushed into the lower cylinder by hydraulic ram. Pressure is transmitted through ı axial hole in the low pressure piston into the pper test chamber. The compression specimen is ounted between an anvil resting on the low pressure iston and a load cell connected to the upper half f the test chamber. Compressive force is applied o the specimen by the low pressure piston, which s actuated by a second hydraulic pump. With the wo stems of this piston being of identical dia- eters its movement should not affect the pressure ithin the high pressure container.

3. Methods of measuring the applied compressive oad in the above equipments are similar to those escribed for tensile testing. If measurements are ade externally, a standard compression load cell s used and if internally, a load cell utilizing oil strain gauges is designed to fit into the high ressure chamber. To reduce friction at the seals f the loading plunger and to improve the accuracy f measurement made with an external load cell a umber of experimentalists (refs 29,30) have used he controlled clearance seal. Compressive strain, hich is a function of reduction of height of the pecimen, is obtained by recording externally the ovement of the loading plunger or from a dis- lacement transducer located within the chamber ref. 31).

4. In most compression equipments the major imiting factor is the very high stress applied to he loading plunger (test pressure + applied ompressive stress) if large deformations of the pecimen take place.

5. Although the above equipments have been esigned essentially for compression testing they an be used for tensile tests by installing a ridgman tensile yoke arrangement.

ORSION APPARATUS

6. Apparatus for carrying out torsion tests under superimposed hydrostatic pressure can be divided nto two basic designs - the power unit applying he torque to the test piece is located either utside or inside the high pressure container.

7. There are a number of different versions of the first type of apparatus (refs 32-35), but n all of them one end of the torsion specimen is eld secure with respect to the high pressure ontainer whilst the other end is connected to a haft or piston which passes through a high pressure eal. The moveable end of the shaft is then either (a) connected to a motor gearbox arrangement such as a conventional torsion testing machine and the orque applied to the test specimen directly through the shaft (refs 32,33) or (b) secured to a fixed platen and the container rotated by some driving arrangement (refs 34,35). Fig. 7 shows one such example of type 'a' apparatus.

18. The torque applied to the specimen is measured by a torque tube placed in series with the specimen and located either inside or outside the pressure container. Most torque tubes consist of chevron-shaped foil strain gauges mounted on a thin tube or rod and a continuous record of torque can be made with them. In an alternative method by Crossland (ref. 32) the torque tube is positioned so that its outside surface is subjected to a hydrostatic pressure whilst the bore is open to the atmosphere. This requires a compromise between sensitivity of the torque tube and its strength to withstand external pressure. Torque in the tube and thus the specimen is determined from optical measurements of the relative twist of the calibrated tube (Fig. 7). Torsional strain is derived from recordings of the rotation of the driving head or container.

19. Pressurized fluid is normally piped into the container from a secondary intensifier or pump. The maximum test pressure is essentially governed by the strength of the container and difficulties with friction at the moving seal. To eliminate or reduce friction a Morrison or controlled clearance seal is often used.

20. With the second design (refs 36,37), the internal power unit, the complete torsion apparatus is accommodated inside the container and subjected to the test pressure. This type of rig, shown in Fig. 8, consists of a small synchronous motor, a gearbox to reduce the speed of rotation, a specimen housing, and a strain gauged torque tube. To reduce the overall length of the torsion unit it is customary to fit the specimen inside the torque tube. The angle of twist is determined from the number of 'makes and breaks' between a contact and the tip of the teeth on one of the gear wheels.

21. Control and operation of this type of apparatus is easier than that of the system with the external power unit. It has however some limit- ations. Practical limits on the size of the pressure container determine the size of the motor and thus the maximum torque capacity. Because of viscous drag on the motor and gearing due to the pressure transmitting fluid, fluids with low viscosity at high pressures, such as isopentane or kerosene, have to be used. Fluids with low viscosities have poor lubrication properties and introduce sealing problems. The limitations restrict the practical maximum pressure for operation of torsion apparatus to about 10 kbar.

HIGH PRESSURE, HIGH TEMPERATURE MATERIAL TESTING EQUIPMENT

22. The number of equipments specifically designed and used for testing materials in tension, or com- pression, or torsion under both high temperature and high hydrostatic pressure is rather limited. Most of them have been used by geologists.

23. For high pressure research at relatively low temperatures - up to about 200°C - the complete pressure apparatus can be placed inside a thermal bath. The temperature of the bath can be accurately controlled, but the main problem is the length of time needed to heat the specimen and stabilize it at the required temperature, particularly if the mass of the high pressure container is large. A change in the test temperature, even a few degrees, tends to be time consuming. The thermal bath is more usually used for working at temperatures below ambient temperature.

24. In the more usual form of external heating

a furnace is placed around the high pressure container or an electric heating tape is wrapped around it. Using this first method, Beresnev (ref. 9) was able to carry out tensile tests up to 400°C and at pressures up to 20 kbar. Few pressure liquids can be used at temperatures greater than 400°C and it is necessary therefore to resort to a gas such as argon, nitrogen or CO_2 with the greater problems of sealing and safety.

25. Fig. 9 shows an apparatus for testing materials at high pressures, which has an external heating source. It was developed by Griggs, Turner and Heard (ref. 29) and can operate at temperatures of 500°C and 5 kbar pressure. The design is based on Griggs' (ref. 25) compression apparatus (Fig. 5). The pressure medium, CO_2, is piped in from a 5 kbar pump. Although the heating element runs longitudinally with hairpin turns top and bottom around the test cylinder, it is shown wound circumferentially for clarity in Fig. 9. Thick thermal insulation surrounds the furnace and is also placed at the top and bottom surfaces of the container to reduce heat loss. With external heating the major difficulties include a reduction in the strength of the container and accelerated creep. Steels with low tempering temperature such as En 30 cannot be used for the container or plugs.

26. The temperature inside the container is measured by a thermocouple and corrected for pressure effects. Internal load and pressure measurements, however, provide obvious difficulties. The load on the specimen is therefore obtained from a standard external load cell, which is not subjected to the elevated temperature (Fig. 9). A controlled clearance seal is used to reduce friction at the loading plunger and so improve the accuracy of the load recording. The pressure gauge is incorporated in the pumping or intensifier circuit, which operates at essentially room temperature.

27. For operating temperatures above 500°C an internal furnace has to be used because of the reduced strength and performance of most high-strength steels in this temperature range. Fig. 10 shows the material testing rig of Griggs incorporating an internal furnace (ref. 29). Designed initially for operation at 800°C and 5 kbar, this particular apparatus has been uprated to 10 kbar and 1000°C (ref. 38) by the use of a 10 kbar pumping system and argon instead of CO_2 as the pressure medium. The internal furnaces consist of standard heating elements mounted in a pyrophyllite cylinder to insulate thermally the bore of the container from the elements. The outside wall of the container is water cooled. The internal furnace produces a number of problems. The first is that its use either means reducing the size of the test specimen or increasing the size of the container bore which in turn necessitates lower test pressures or larger containers. Convection inside the container is usually restricted by the use of baffles or by packing all available spare space with a material of low permeability. Temperature variation along the specimen is limited by winding the furnace with a number of independent heating zones.

28. Although the bore of the container can be kept relatively cool, sections of the loading pistons may be subjected to the furnace temperature, and therefore high temperature, high strength materials have to be used.

29. If an internal furnace is used, the time taken to heat the specimen to the required test temperature is greatly reduced. Load and pressure gauges are located, remotely from the test chamber and operate at room temperature.

SHEAR APPARATUS

30. Shear apparatus are normally used for examining the torsional shear properties of materials under pressures greater than 30 kbar. The basic principle of this type of apparatus originated by Bridgman (ref. 3) is shown in Fig. 11. It consists essentially of two opposed pistons separated by an anvil. Two thin disc-shaped specimens are placed between the faces of the pistons and the central anvil and subjected to a normal pressure between the platens of a hydraulic press. The surfaces in contact with the specimens are truncated cones embedded in steel support rings. They are usually made from tungsten carbide. As the end load is applied, the specimen will tend to extrude. The central anvil, driven through either a motor/gear train or hydraulic system, shears the specimen as it rotates about the axis of the stationary pistons.

31. An alternative design uses only one pair of opposed anvils and thus one specimen. One anvil is fixed to a rigid platen of the press whilst the other, the rotating anvil, is connected directly to the press's loading ram, which is essentially floating on oil and, ignoring seal friction, free to rotate (ref. 29). In another version, (ref. 39) the rotating anvil is connected through a conical thrust bearing to the ram (Fig. 12).

32. For both of these designs the specimen thickness is small, up to 0.3 mm thick by 7 mm diameter and, under pressure, contact between the anvils can occur and not always be detected. With test specimens of high electrical resistance, however, any contact can be monitored by a simple electrical circuit connected to the opposed anvils.

33. Abey and Stromberg (ref. 40) were able to increase the size of the specimen to 3 mm thick and 12 mm diameter by modifying Bridgman's arrangement as shown in Fig. 13. The surfaces of the rotating platen have tapered blank-ended recesses to receive the specimens and the conical pistons. To reduce torsional losses between the tapered surfaces of the pistons and cylinders and to minimize specimen extrusion a combination gasket is inserted. This gasket consists of three layers lava, mild steel, and indium. Friction between the lava and the platen is high and restricts extrusion of the sample. The indium layer has a low shear strength and allows relative rotation between the platen and pistons with only a small frictional effect. The mild steel layer acts as a separator between the lava and indium layers.

34. The average pressure applied to the specimen in all these rigs is determined from a standard compression load cell connected in series with the piston. In practice the pressure varies across the specimen and reaches a maximum at the centre (ref. 41). The torque applied to the disc can be derived from a torque tube fitted to the piston as shown in Fig. 12, from compression load cells or compression springs incorporated in the rotational drive system, or by oil pressure readings in hydraulically driven systems.

:KNOWLEDGEMENT

ıis paper is published by permission of the irector, National Engineering Laboratory, epartment of Industry. It is Crown copyright.

EFERENCES

. BRIDGMAN P.W. Studies in large plastic flow nd fracture. New York: McGraw-Hill, 1952.

. CARPENTIER D., CONTRE M., DAUMAS R., RECH J.P. nd FRANCOIS D. The deformation of beryllium nder hydrostatic pressure of up to 15 kbars (in rench). Conf. int. sur la metallurgie du eryllium, pp. 445-454. Paris: Presses niversitaires de France, 1966.

. BRIDGMAN P.W. The physics of high pressure. ondon: Bell, 1949.

. BRANDES M. An analysis of the methods of tretching testpieces under high hydrostatic ressure. Pr. Inst. mech. Precyzyjnej, 1962, 10(4), p. 19-22. English Translation: NEL Translation 294a. East Kilbride, Glasgow: National Engineering aboratory.

. BRANDES M. and SZLACHCIC H. Simple device for tretching testpieces under high hydrostatic ressures. Rev. scient. Instrum., 1965, 36(7), 991.

. RATNER S.I. Change of mechanical properties of etals under hydrostatic pressure (in Russian). Zh. Tekhn. Fiz., 1949, 19(3), 408.

7. NISHIHARA M., TANAKA K. and MORAMATSU T. Effect of hydrostatic pressure on mechanical behaviour of metals. In Soc. Mat. Sci. Proc. 7th Japan congress on testing materials, p. 154. Kyoto: Soc. Mat. Sci., 1964.

8. PELCZYNSKI T. and PAWLAK I. Tensile testing of some metals under high hydrostatic pressure. Pr. Zakl. Plast. Polytech. Warsz., 1959.

9. BERESNEV B.I., BULYCHEV D.K. and MARTYNOV E.D. Zawodsk Lab., 1964, 8, 1017.

10. PUGH H.Ll.D. and GREEN D. Progress report on the behaviour of materials under high hydrostatic pressure. MERL Plasticity Report No 103. East Kilbride, Glasgow: National Engineering Laboratory, 1954.

11. PUGH H.Ll.D. The mechanical properties and deformation characteristics of metals and alloys under pressure. In ASTM Special Tech. Pub. No 374, Phila. Pa: American Society for Testing and Materials, 1964.

12. PUGH H.Ll.D. and GREEN D. The effect of hydrostatic pressure on the plastic flow and fracture of metals. Proc. Instn mech. Engrs, 1964-65, 179(Pt 1,3) 415-429.

13. GLADKOVSKI W.A. and OLEINIK M.I. Apparatus for investigating mechanical qualities of metals under high hydrostatic pressure (in Russian). Fiz. Metal. i Metalloved., 1957, 4, 531. (English Translation: Physics Metals Metallogr., N.Y., 1957, 4, 118-121).

14. YOSHIDA S. and OGUCHI A. An apparatus for tensile testing under high hydrostatic pressure. Trans Nat. Res. Inst. Metals, 1969, 11, 347.

15. JOHNSON D.P. and NEWHALL D.H. The piston gauge as a precise measuring instrument. Trans. Amer. Soc. mech. Engrs, 1953, 73, 301-310.

16. CROSSLAND B. The effect of fluid pressure on the shear properties of metals. Proc. Instn mech. Engrs, 1954, 168(40), 935-946.

17. CARPENTIER D. and CONTRE M. Description of an apparatus allowing mechanical tests under hydrostatic pressure up to 15 kilobars. Rev. scient. Instrum., 1970, 41(2), 189-192.

18. PUGH H.Ll.D. and GUNN D.A. A strain gauge load cell for use under high hydrostatic pressure. Proc. Instn mech. Engrs, 1967-68, 182(Pt 3C), 260.

19. BRANDES M. A method of recording the full load-extension curve while studying mechanical properties of materials under hydrostatic pressures up to 30 000 kg/cm^2. Proc. Instn mech. Engrs, 1967-68, 182(Pt 3C), 277-282.

20. INOUE N., DAMIANO V., HANAFEE J. and CONRAD H. Effects of hydrostatic pressure on mechanical behaviour of polycrystalline beryllium. Trans. Amer. Inst. min. (metall.) Engrs, 1968, 242, 2081-2089.

21. MACGREGOR C.W. Two-load method of determining average true stress/strain curve in tension. J. appl. Mech., 1939, 6(4), A156-A158.

22. PUGH H.Ll.D., LEES J., ASHCROFT K. and GUNN D. High pressure research at the National Engineering Laboratory, Engineer, 1961, 212(5508), 258-262.

23. PUGH H.Ll.D., HODGSON G. and GUNN D.A. Tensile strain measurement under high hydrostatic pressure using an optical method. J. scient. Instrum, 1963, 40, 221-224.

24. CHANDLER E.F. and MAIR W.M. Behaviour of metals under triaxial compressive stress. Proc. Instn mech. Engrs, 1967-68, 182(Pt 3C), 122-131.

25. GRIGGS D.T. Deformation of rocks under high confining pressures. J. Geol., 1936, 44, 541.

26. PATERSON M.S. Triaxial testing of materials at pressures up to 10 000 kg/sq cm (150 000 lbf/sq in). J. Instn Engrs. Aust., 1964, 36, 23-29.

27. GRIGGS D.T. and MILLER W.B. Deformation of Yule marble: Part I. Compression and extension experiments on dry Yule marble at 10 000 atm confining pressure, room temperature. Bull. geol. Soc. Amer., 1951, 62, 853-906.

28. CROSSLAND B. and LUDLOW C. Design and development of a machine for compression testing of rock samples under a constant pressure of up to 0.7 GPa (7 kbar). High Temp.-High Press., 1973, 5, 509-514.

29. GRIGGS D.T., TURNER F.J. and HEARD H.C. Deformation of rocks at 500° to 800°C. In GRIGGS D. and HANDIN J. (Eds). Rock formation, Memoir 79, pp. 39-104. Rochester, N.Y.: Geological Society of America, 1960.

30. HANDIN J. An application of high pressure in geophysics: experimental rock deformation. Trans. Amer. Soc. mech. Engrs., 1953, 75, 315-324.

31. CHANDLER E.F. and LITTLESON M.P. An internal stroke gauge for operation under hydrostatic pressure. NEL Report No 356. East Kilbride, Glasgow: National Engineering Laboratory, 1968.

32. CROSSLAND B. and MITRA A.K. Torsion machine for use under superimposed pressures of 1 GPa, and preliminary results. High Temp.-High Press., 1974, 6, 165-172.

33. OHMORI M., YOSHINAGA Y., KAWAHATA T. and SANEMASU Y. Plastic deformation of metals under high hydrostatic pressure. Proc. Jap. Cong. Mat. Res., 1970, pp. 139-142.

34. ERBEL S. Behaviour of materials twisted under hydrostatic pressure. Presented at conference organized by Department of Metallurgy, Polish Academy of Sciences, April 1966.

35. HANDIN J., HIGGS D.V. and O'BRIEN J.K. Torsion of Yule marble under confining pressure. In GRIGGS D. and HANDIN J. (Eds). Rock formation Memoir 79, pp. 245-74. Rochester, N.Y.: Geol. Soc. Amer., 1960.

36. PUGH H.Ll.D. and GUNN D. An apparatus for torsion tests under high hydrostatic pressures. NEL Report No 159. East Kilbride, Glasgow: National Engineering Laboratory, 1964.

37. ABEY A.E. Effect of hydrostatic pressure on the stress strain curves of OFHC copper. J. appl. Phys 1971, 42(10), 4085-4087.

38. HEARD H.C. Trans. Amer. Geophys. Uh., 1965, 46, 698.

39. TOWLE L.C. Shear-strength and friction measurement in thin layers under high pressure. J. appl. Phys., 1971, 42(6), 2368-2376.

40. ABEY A.E. and STROMBERG H.D. 70 kilobar shear apparatus. Rev. scient. Instrum., 1969, 40(4), 557-559.

41. LEES J. The design and performance of UHP equipment. In BRADLEY R.S. (Ed). Advances in high pressure research, Vol.1, pp. 1-83. London: Academic Press, 1966.

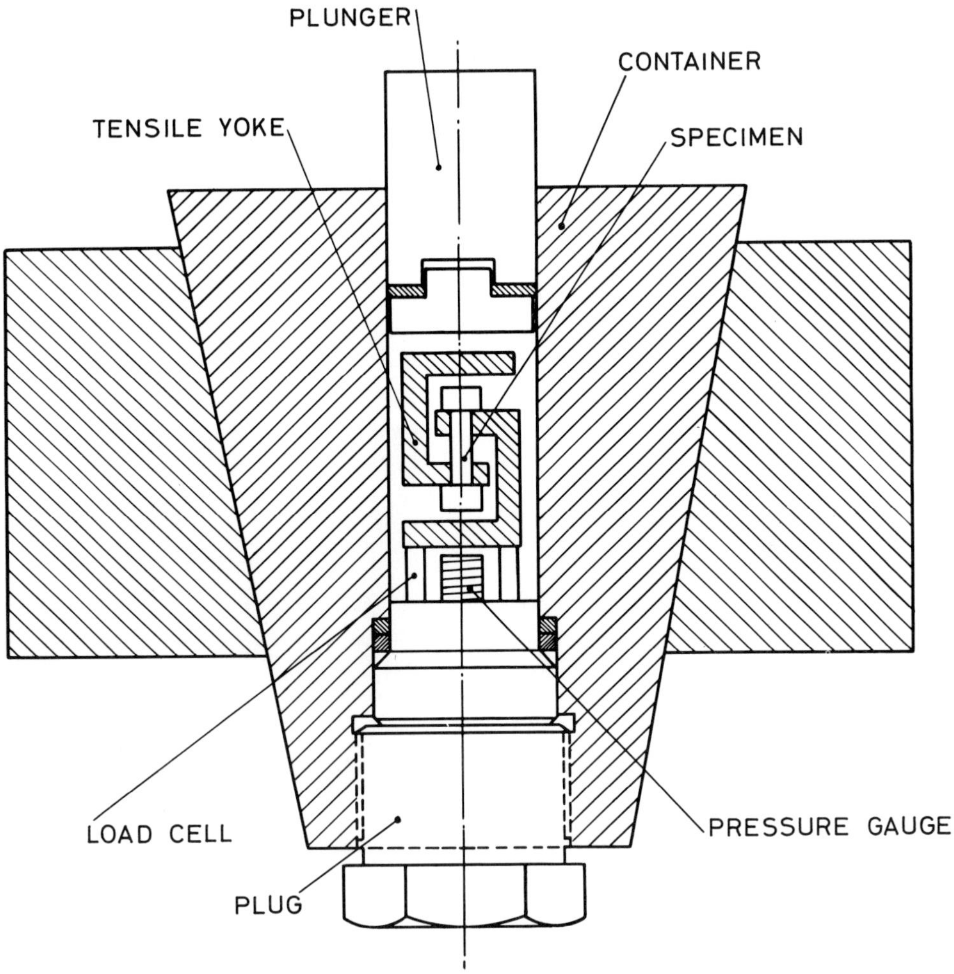

Fig. 1: Bridgman yoke apparatus for tensile tests under hydrostatic pressure (Ref. 1)

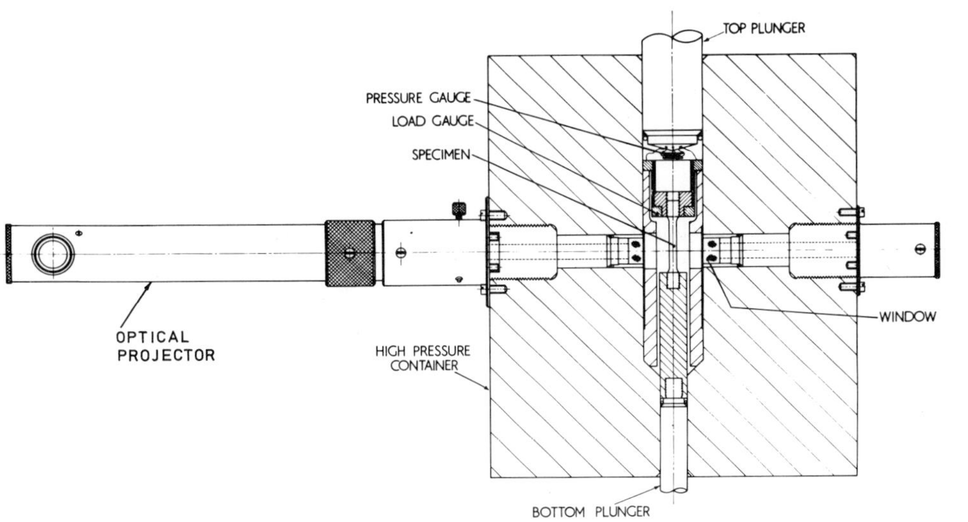

Fig. 2: Pugh's tensile apparatus for obtaining true stress/true strain curves under hydrostatic pressure (Ref. 11)

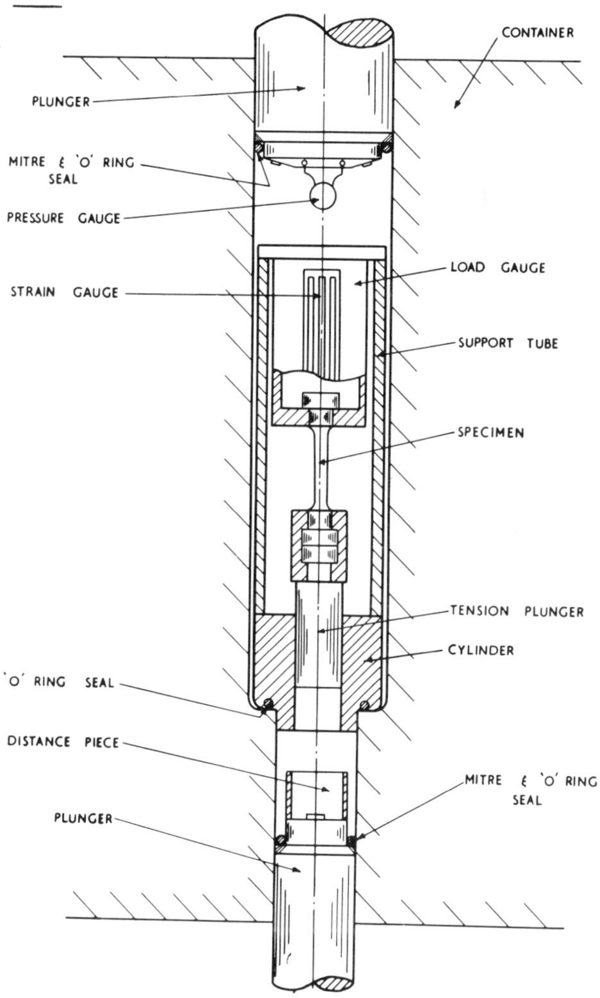

Fig. 3: Pressure differential apparatus for tensile tests under hydrostatic pressure (Ref. 18)

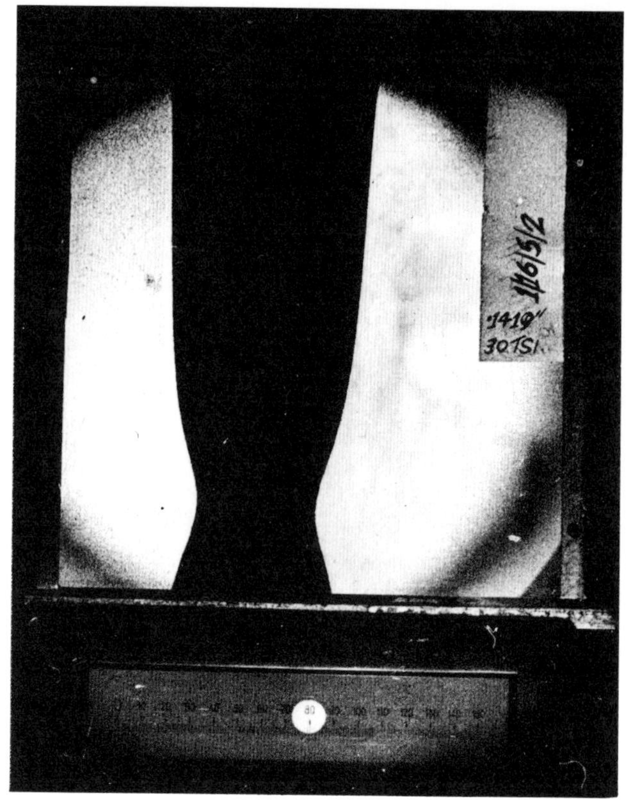

Fig. 4: Tensile test under pressure showing necked specimen and applied load

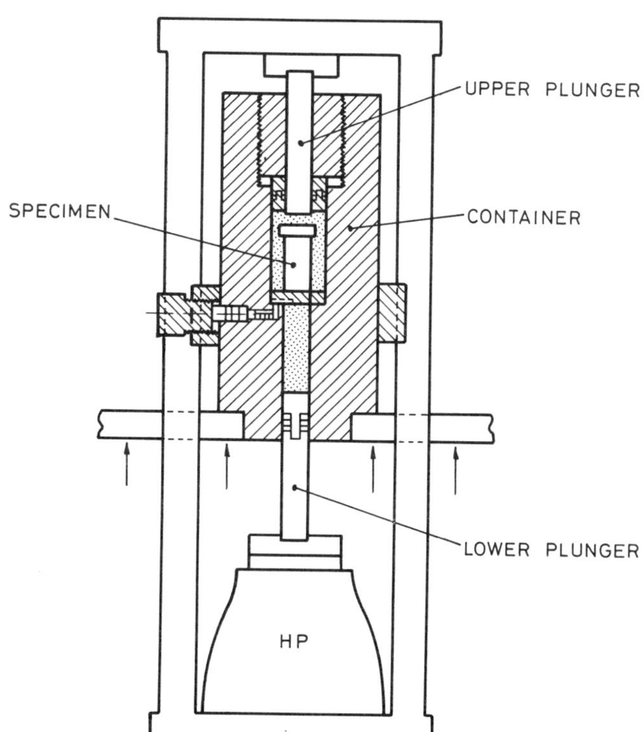

Fig. 5: Grigg's original high pressure compression apparatus (Ref. 25)

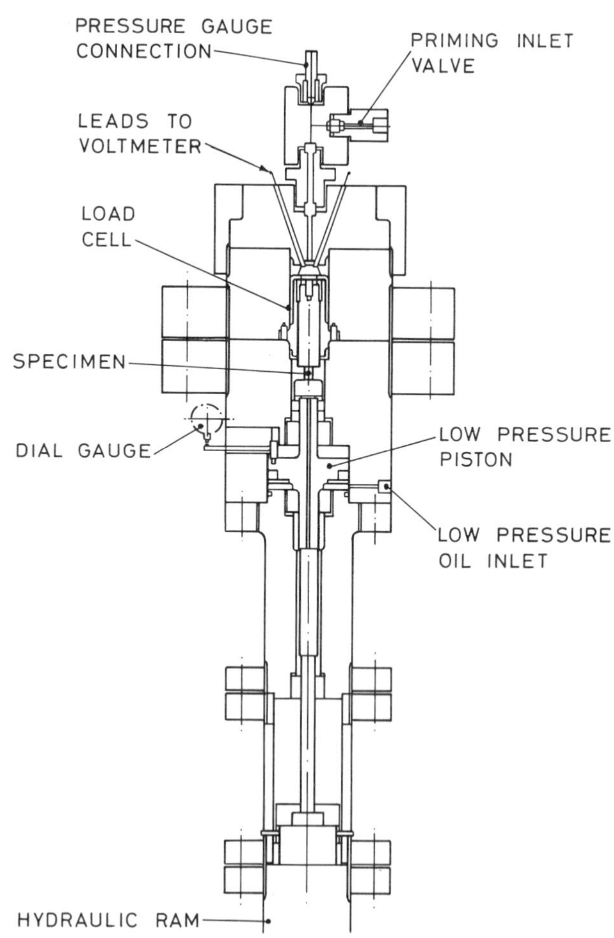

Fig. 6: Compression testing apparatus for pressures up to 7 kbar (Ref. 28)

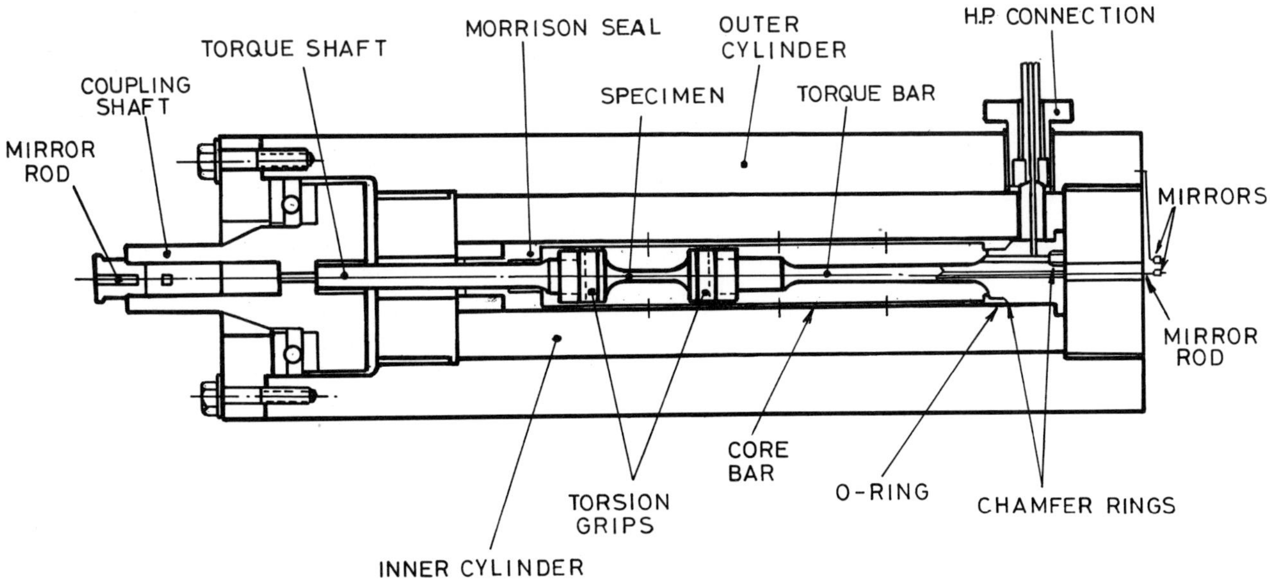

Fig. 7: Crossland's torsion testing machine for pressures up to 1.1 kbar (Ref. 32)

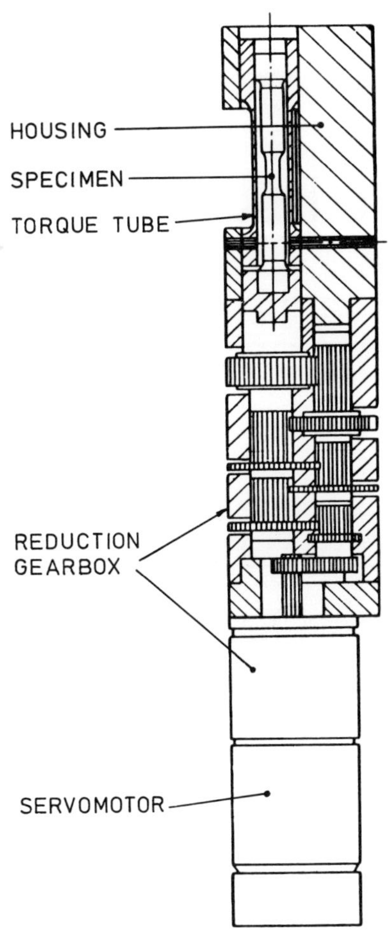

Fig. 8: Pugh's motorized apparatus for torsion tests under pressure (Ref. 36)

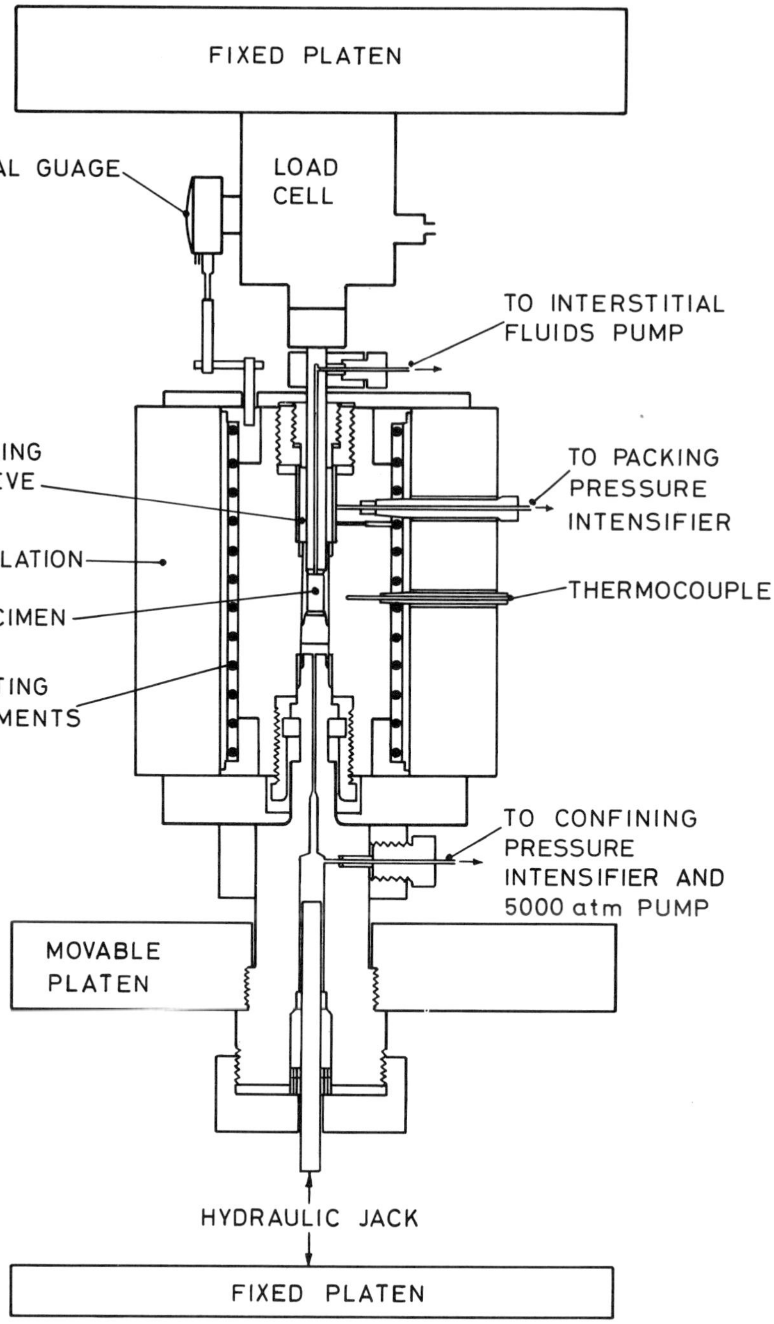

Fig. 9: 500°C, 5 kbar apparatus with external furnace (Ref. 29)

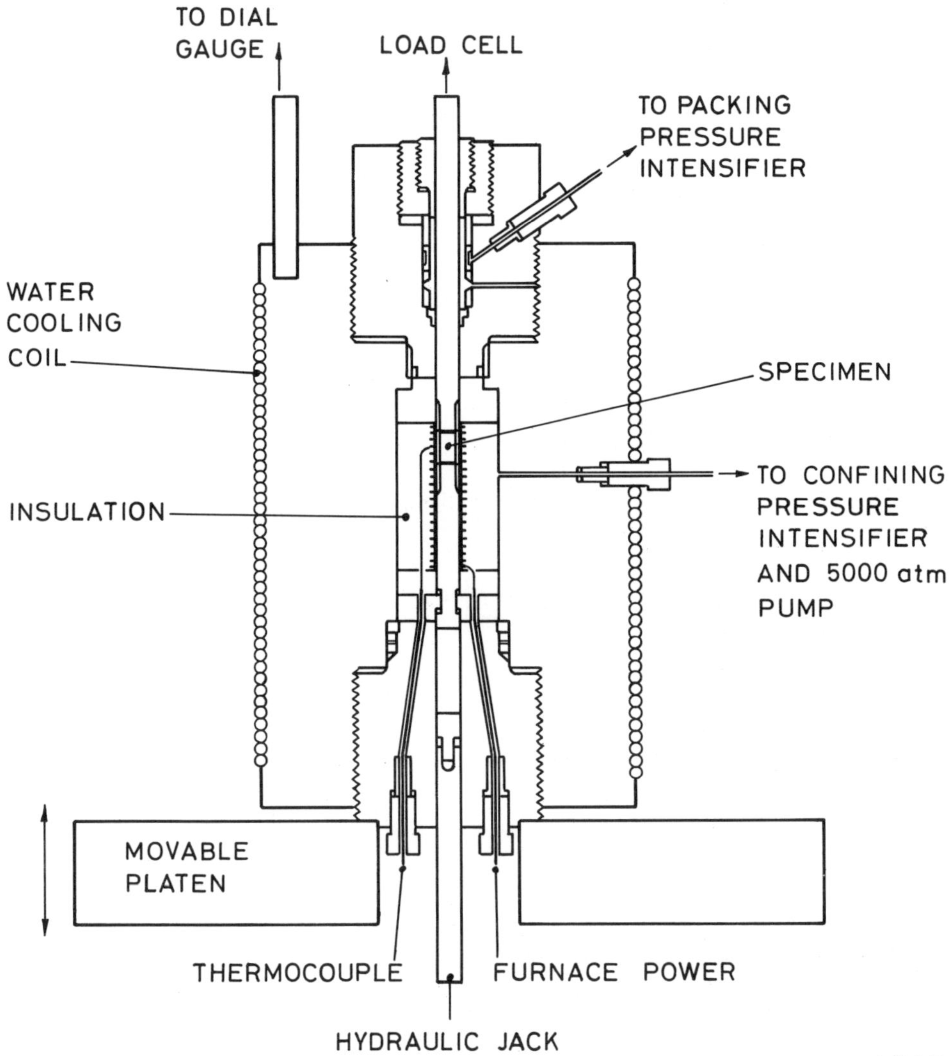

Fig. 10: 500°C, 5 kbar apparatus with internal furnace (Ref. 29)

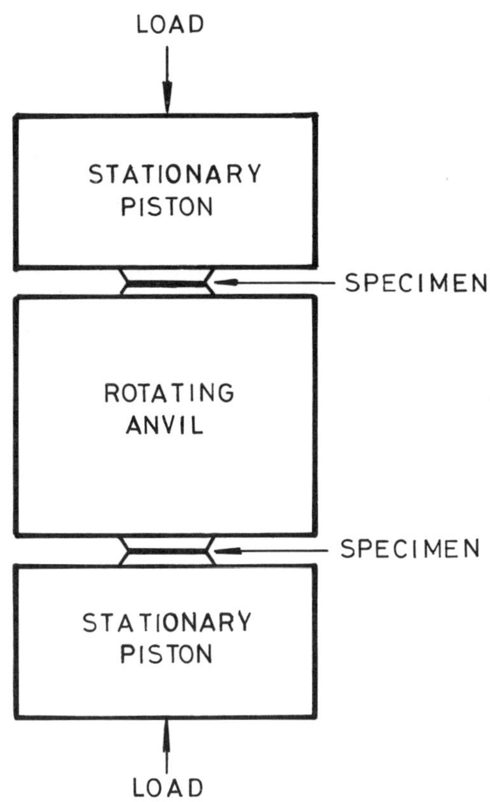

Fig. 11: Bridgman type shear apparatus (Ref. 3)

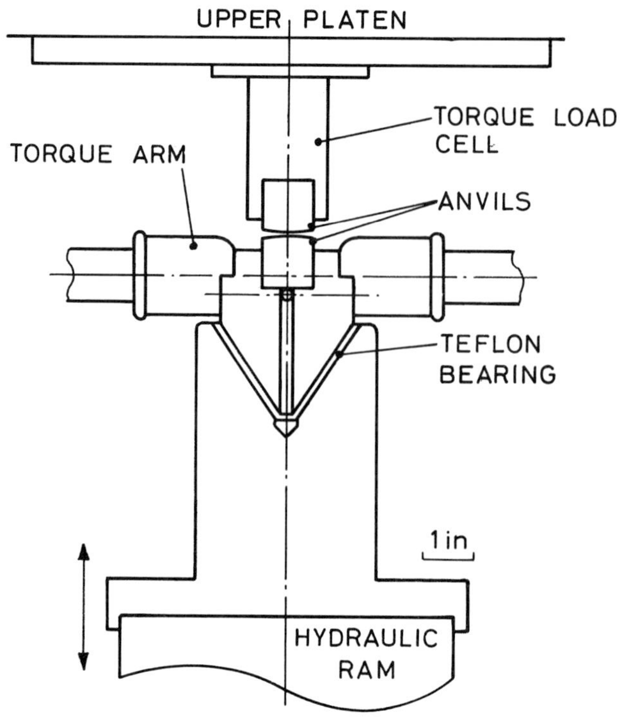

Fig. 12: Shear apparatus with a single pair of anvils (Ref. 39)

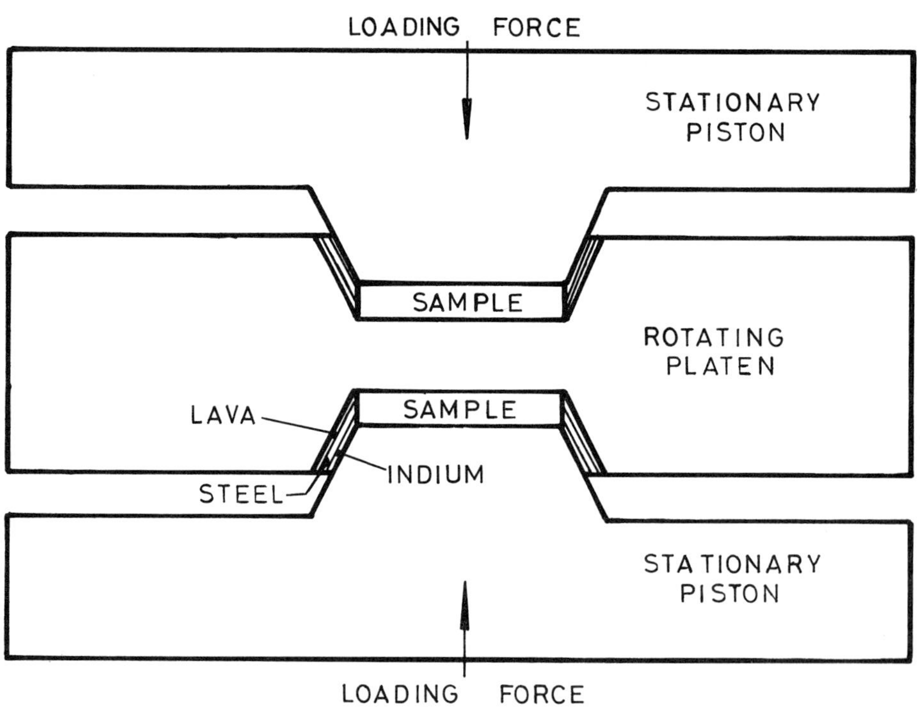

Fig. 13: Shear apparatus with a webbed girdle die (Ref. 40)

THE DEFORMATION AND FRACTURE UNDER PRESSURE OF A NUMBER OF METALS USED IN HIGH PRESSURE EQUIPMENTS

H. Ll. D. PUGH, DSc, BSc, CEng, FInstP, FIMechE, FIM and E. CHANDLER, BA, CEng, MIMechE
National Engineering Laboratory, East Kilbride

SYNOPSIS Investigations carried out on steels have shown that their ductility increased linearly with pressure but the general shape of their stress/strain curves were unaffected. There was evidence of two different criteria of fracture. Results are given of the effect of pressure on the tensile properties of M2 tool steel, subjected to different heat treatments, beryllium and molybdenum. It is shown how surface preparation has a critical effect on the properties of some metals. Preliminary results on the behaviour of En 26 notched bend specimens indicate that a superimposed hydrostatic pressure inhibits crack propagation.

INTRODUCTION

1. Many materials used in the construction of high pressure equipment exhibit the required high strength but only low ductility when tested in the conventional uniaxial mechanical testing machines. In order to minimize this problem of low ductility and the consequent liability to fracture high pressure containers and tooling are often designed so that the constructional materials are subjected to stress systems which have large hydrostatic components of compressive stress. Such stress arrangements enable larger strains in the high strength component to be sustained before fracture. For designs of this type information is required on the material's stress/strain characteristics such as flow stress, work hardening rate and strain to fracture, as a function of the hydrostatic component of stress. The information can be obtained from normal mechanical tests performed under a superimposed hydrostatic pressure.

2. Over the years the effect of pressure on the ductility of steels has been studied by a number of investigators (Refs 1-21) and some of these experimental results are shown in Fig. 1. All the tensile test investigations have shown that fracture strain increases with increasing pressure. For metals that have negligible ductility when tensile tested at atmospheric pressure the fracture strain/hydrostatic pressure graph consists of two sections. Initial increases in pressure have little effect on ductility until a particular pressure, which is dependent on the material, is reached. The fracture strain/pressure curve then rises relatively rapidly and linearly. For the more ductile steels, that is with a fracture strain greater than 0.2 at atmospheric pressure, the strain usually increases linearly with pressure. In general the greater the ductility of these steels at atmospheric pressure the greater the rate of increase in ductility with increasing pressure.

3. It has been found that a superimposed hydrostatic pressure has no significant effect on the stress/strain relationship for wrought metals such as steels when tested in tension (Refs 11, 17, 19-23) or torsion (Refs 2,5) other than to increase the strain to fracture. Since the yield represents one part of a stress/strain curve this confirms the universally accepted criterion of yielding of wrought metals, namely that yielding is independent of the hydrostatic component of stress and is dependent only on the deviatoric components of stress, that is the shear stresses. Pugh (Ref. 11) also found experimentally that the tensile strength and instability strain were independent of the value of the superimposed pressure. As a consequence he re-examined the widely accepted theoretical solution that a transverse pressure superimposed on a specimen in tension delayed the onset of necking and resulted in an increased uniform elongation. He found the original analysis to be erroneous. His corrected version showed that hydrostatic pressure had no effect on the instability point.

4. Bridgman (Ref. 3), who carried out tensile tests on a range of steels and tried to find a fracture stress criterion based on various combinations of stresses, concluded that, 'there is no universally valid criterion of fracture in terms of stresses only'. The two criteria which gave the closest agreement were that fracture occurred when (a) the maximum tensile stress at the centre of the neck reached a critical value and (b) the mean total hydrostatic component of stress reached a critical value.

5. Pugh (Ref. 24) however showed that a maximum deviatoric stress criterion gave a good representation of the data obtained from tension and torsion tests under pressure on cast iron (Refs 7,12) (Fig. 2) and from tension and compression tests under pressure on another cast iron (Ref. 24) (Fig. 3). In tensile tests under pressure this maximum deviatoric stress criterion is identical with the maximum tensile stress criterion. This latter criterion has also been found (Refs 11,19) to give a good representation of the results obtained from tensile tests under pressure on ductile steels and other ductile wrought materials, that is when the net tensile stress achieves a critical value fracture takes place. On the other hand a few exploratory tensile tests on two tool steels (hardness H_{V30} 583 and 950), which were still brittle (with a fracture strain less than 0.5) at pressures of 11 kbar (1.1 GN/m^2),

indicated that the net axial stress at fracture decreased with increasing pressure. Similar results have been obtained for AZ91 and two intermetallic compounds Titanium-Nickel and Nickel-Aluminium (Ref. 15).

6. It is possible that there may be two different criteria of fracture which apply to the results of tensile tests under increasing hydrostatic pressure. The first criterion is for brittle materials failing with negligible deformation under pressure. The second criterion, for ductile failure, applies when the material in question behaves in a ductile manner under the superimposed hydrostatic pressure. If this suggestion is correct the average net axial stress at fracture for a material which is brittle at atmospheric pressure and whose strain to fracture increases under pressure could generally be expected to decrease initially with increasing pressure and, after the hydrostatic pressure had reached a critical value, would not change with increasing pressure.

7. Tensile properties, under a superimposed compressive stress system, of three more metals used in the manufacture of high pressure equipment have been studied at the National Engineering Laboratory. The results of the investigations on these materials, M2 tool steel, beryllium, and molybdenum, are given below.

Tensile properties of M2 tool steel under pressure

8. M2 steel is a widely used general-purpose molybdenum high speed steel. A typical analysis for the steel is 0.83 per cent carbon, 4.2 per cent chromium, 5 per cent molybdenum, 6.25 per cent tungsten, and 2 per cent vanadium. Because of its good strength and toughness properties combined with good wear resistance performance it is normally used for both cutting tools, such as drills, and cold forging punches. At the National Engineering Laboratory it is also used for high pressure plungers and thin-walled hydrostatic extrusion dies. Following the failure of a few dies made from the material it was decided to obtain more detailed information on the effect of different heat treatments on the mechanical properties of M2 steel by carrying out tensile tests under hydrostatic pressure. The standard atmospheric tensile test on M2 steel provided only a limited amount of information because of the low ductility of the material.

9. Different steel manufacturers and reference books recommend differing heat treatments for M2 steel. It was decided to examine the effects of the three described below.

Treatment 1 - Recommended by the steel manufacturer. Preheat at 840°C, harden from 1210°C, oil quench, and double temper at 610°C. Hardness H_{v30} 763.

Treatment 2 - Recommended by GKN for hardness combined with toughness. Preheat at 840°C, harden from 1180°C, oil quench, and double temper at 590°C. Hardness H_{v30} 792.

Treatment 3 - To give good strength and toughness properties but at a much lower tempering temperature. Preheat at 840°C, harden from 1220°C, oil quench, and double temper at 260°C. Hardness H_{v30} 715.

The specimens were conventional screw-ended tensile testpieces with an average gauge length of 14.4 mm and a gauge diameter of 2.9 mm. The gauge length sections were ground and then polished.

10. The specimens were tested in the high pressure tensile rig described by Pugh (Ref. 11) over a range of pressures up to a maximum of 15 kbar (1.5 GN/m^2). The nominal stress/extension curves at atmospheric pressure and at the highest test pressure for the different heat treatments are shown in Fig. 4. For all three heat treatments the effect of increasing pressure was to extend the stress/strain curve of the steel. There were negligible differences between the shapes of the curves for Treatments 1 and 2; the tensile strengths of the specimens, which were unaffected by increasing pressure, were 26.5 kbar (2.65 GN/m^2). The curve for Treatment 3 differed significantly from those for 1 and 2. The tensile strength was lower, 24.1 kbar (2.41 GN/m^2), and necking was delayed and was far less pronounced.

11. The effect of pressure on ductility ($\ln A_o/A_f$) is shown in Fig. 5. Pressures below 5 kbar (500 MN/m^2) had negligible effect for Treatment 1 and 2, but beyond this pressure ductility started to increase and reached a reduction of area of 37 per cent at 15 kbar (1.5 GN/m^2). The GKN Treatment (2) was also found to give a slightly better ductility at the lower pressures. The ductility of the specimens subjected to Treatment 3 was hardly affected by increasing pressure but the results exhibited rather more scatter.

12. For all three heat treatments of the M2 tool steel the net axial stress at fracture decreased linearly with increasing pressure (Fig. 6). This decrease in fracture stress, which was approximately 0.7 x the increase in applied test pressure, agrees with previous findings for nominally brittle materials. For Treatments 1 and 2 there was some slight indication of changes in the slopes of the curves at the highest pressures.

13. Examination of the specimens after they had been tensile tested revealed longitudinal cracking originating from the fracture surfaces. The severity of the cracking increased with the increasing degree of deformation of the specimens obtained with increasing pressure.

Tensile properties of beryllium under pressure

14. In X-ray diffraction investigations the high pressure container and pistons are normally made from beryllium (Ref. 25). The beryllium can be in either cast and extruded or hot pressed powder form. It is known that the mechanical properties and behaviour of brittle materials such as beryllium are extremely sensitive to the type of surface preparation. In exploratory tests on beryllium (Ref. 26) it was noticed that if direct contact between the pressurizing medium and specimen surface was prevented, for example by sheathing, greater ductility was attained. Aladag, Pugh and Radcliffe (Ref. 27), however, suggested that if the surface of the beryllium was properly prepared, say by a very careful electro-polishing to remove surface damage, sheathing would be unnecessary. It was therefore decided to ascertain the effect of surface preparation and sheathing on the tensile properties of beryllium under hydrostatic pressure.

15. Two types of beryllium were tested - extruded ingot with grain size of 250 μm and hot pressed (QMV) powder material, grain size 50 μm. Three different surface finishes were examined.

a As machined, with a final lathe cut of 25 μm.

b As for (a) followed by chemically etching in 5 per cent sulphuric acid.

c As for (a) followed by electro-polishing.

Specimens of each surface finish were tested in the sheathed and unsheathed condition. Sheathing was by dipping the specimen in a mixture of latex rubber solution and ethyl methyl ketone to give a protective layer of 125 μm thick. The specimens were all tensile tested under a hydrostatic pressure of 7.7 kbar (770 MN/m^2).

16. The experiments showed (Fig. 7) that the machined specimens, as expected, tended to give lower ductilities than the other two types of surface finishes, which gave comparable results. For each surface preparation, however, ductility was increased by sheathing. The reduction of area of the chemically etched hot pressed powder material and of the cast and extruded beryllium rose respectively from 12.3 to 53.7 per cent and from 33.8 to 54.5 per cent.

17. Sheathing however had no effect on the stress/strain relations of the two types of beryllium in all three surface finishes other than increasing the strain to fracture (Ref. 28). This increase in ductility naturally had a significant effect on the fracture stress of the materials (Fig. 8). In the case of the chemically-etched hot-pressed material, sheathing the specimen converted the net axial stress at fracture from an apparent compressive stress of 1.76 kbar (176 MN/m^2) to a tensile stress of 3.18 kbar (318 MN/m^2).

18. A more comprehensive investigation was then made over a range of test pressures, up to a maximum of 12.4 kbar (1.24 GN/m^2) on the two types of beryllium. For these tests the gauge lengths of the specimens were chemically etched and then sheathed. From Fig. 9 it can be seen that the fracture strain ($\ln A_o/A_f$)/pressure relationships of the two differently fabricated berylliums were very similar. Both curves can be divided into two distinct sections. Initially the ductility was an increasing function of pressure, but above 5 kbar (500 MN/m^2) it became a linear function of pressure.

19. At lower pressures (up to 7 kbar (700 MN/m^2)) the net axial fracture stress for the extruded ingot beryllium was a decreasing linear function of pressure (Fig. 10). Beyond that pressure it is not possible to draw any firm conclusion because of the scatter in the experimental results. Difficulty in the accurate measurement of the fracture areas of the extruded ingot beryllium at higher pressures because of uneven deformation in the necked region is partly responsible for the scatter in the results. Another difficulty is that at the higher strains the specimens 'tear' apart thus making the accurate determination of the fracture load difficult. The overall results are similar to earlier tensile tests on cast iron (Ref. 15). Results for the hot pressed beryllium also showed an apparent small initial decrease in stress with increasing pressure. At a pressure of approximately 5 kbar (500 MN/m^2) there appeared to be an abrupt increase in fracture stress and beyond this pressure the net axial stress at fracture became an increasing linear function of pressure.

Tensile properties of molybdenum under pressure

20. In high pressure-high temperature research it is much simpler and less expensive to use an external heater to heat the high pressure container and its contents than to use an internal heater. For temperatures up to 1200°C containers manufactured from refractory metals such as molybdenum and tungsten and their alloys (Refs 29,30) can be used. The effect of pressure on the mechanical behaviour of one of these materials, molybdenum, was investigated.

21. The molybdenum examined had been fabricated by two different metal working processes. After sintering as an 18 mm square rod it was swaged down to 12 mm diameter rod and heat treated at 1500°C for 1 hour. Part of the rod was then hydrostatically extruded down to 8 mm diameter at a pressure of 12 kbar and a temperature of 250°C. Metallurgical examination of the materials revealed porosity in the structures of both forms of molybdenum but, as may be expected, the extruded material had the smaller amount of porosity. Tensile specimens were machined from the 12 mm diameter bar (batch 1) and the 8 mm bar (batch 2). The hardnesses of the two batches were respectively H_{V10} 183 and H_{V20} 274.

22. As in the earlier work on beryllium, sheathed and unsheathed specimens from batch 1 were tensile tested at pressures of 1.5 kbar (150 MN/m^2) and 4.6 kbar (460 MN/m^2). The results (Figs 11 and 12) showed that the ductilities of the unsheathed specimens were considerably less than the sheathed specimens. All subsequent specimens were therefore sheathed before being tested under pressure.

23. Both batches of molybdenum showed little ductility at atmospheric pressure. Batch 1 exhibited a gradual brittle/ductile transition with increasing pressure whereas with the extruded material a sharp brittle/ductile transition took place. Beyond this transitional region the fracture strain ($\ln A_o/A_f$)/pressure curves of the two forms of molybdenum appeared to be linear (Fig. 11) and the percentage elongation for the unextruded molybdenum was far greater than that for the extruded material for corresponding cross-sectional areas (Figs 11 and 12).

24. Under a superimposed pressure of less than 0.7 kbar (70 MN/m^2) the tensile strengths (that is, nominal stress at onset of necking) of both batches of molybdenum were attained. The tensile strengths for the unextruded (batch 1) and the extruded (batch 2) specimens were respectively 5.5 kbar (550 MN/m^2) and 8.2 kbar (820 MN/m^2) and were unaffected by increasing pressure.

25. The curves of net axial stress at fracture against hydrostatic pressure curve of the two molybdenums differed significantly as shown in Fig. 13. The fracture stress of the unextruded molybdenum increased linearly with pressure and at a rate of 1.6 x the applied pressure. The graph for the extruded molybdenum, batch 2, was in two distinct sections. At pressures below the brittle/ductile transition the axial fracture stress was constant, but beyond the transition pressure it increased abruptly by some 3.5 kbar (350 MN/m^2) and

then continued to increase at a rate of about 1.1 x the applied pressure. The difference in the slopes of these curves is due to some extent to the further compaction and work hardening of the material by the additional hydrostatic extrusion process.

Effect of pressure on the toughness of En 26 steel

26. In both research and industry there is a demand for equipments capable of operating at very high pressures in the search for new processes, new products, and greater efficiencies. To attain such pressures, materials of higher strengths but usually with reduced toughnesses have to be used. As mentioned earlier very few structures function under a simple uniaxial stress system but their designs are based on information obtained from uniaxial mechanical tests. One example is the use of tungsten carbide for the inner ring of a multi-ring container.

27. The only work known to the authors on the effect of a hydrostatic compressive stress on the fracture toughness of a material is that by Auger and Francois (Ref. 31). In their studies they carried out three-point bending tests on the aluminium alloy AZ6GU under pressures of up to 6 kbar and found an increase in fracture toughness with pressure.

28. Some preliminary results on the effect of pressure on the ductility of En 26 steel are given in this section. En 26 is a $2\frac{1}{2}$ per cent nickel-chromium-molybdenum high carbon steel. It has a typical composition of 0.4 per cent carbon, 0.6 per cent manganese, 2.6 per cent nickel, 0.65 per cent chromium, and 0.6 per cent molybdenum. Its heat treatment is reasonably straightforward and gives tensile strengths in the range of 10 kbar (1 GN/m^2) to 15 kbar (1.5 GN/m^2) with reasonable ductility. For a number of years the National Engineering Laboratory has used En 26 steel for the manufacture of monobloc pressure containers and as the outer ring of multi-ring containers.

29. The specimens used were single notch three-point bend testpieces with an overall length of 35.5 mm by 6.4 mm wide by 4.8 mm thick and a 60° notch 3.2 mm deep with a notch root radius of less than 0.1 mm. The loading span was 25.4 mm. The surfaces of the specimens were ground and the notch was spark machined. The specimens were divided into two batches and given the following heat treatments.

Batch 1 - The specimens were hardened at 820^oC, oil quenched, and tempered at 550^oC. Hardness was H_{v30} 389. This is the heat treatment normally used for En 26 to give a good combination of strength, ductility, and impact properties.

Batch 2 - The specimens were hardened at 820^oC, oil quenched, and tempered at 350^oC. Hardness was H_{v30} 470. This treatment gives higher strength properties but poor impact properties.

30. The tests were carried out using the apparatus (Ref. 11) normally used at the National Engineering Laboratory for tensile tests under pressure but with the tensile rig replaced by a three-point bending system. The load applied to the specimen was measured by an internal load cell and the deflection of the specimen by a linear displacement transducer connected to the bottom ram.

31. It is not possible to obtain valid fracture toughness, K_{IC}, values unless both crack depth and thickness of the test specimen are not less than $2.5(K_{IC}/y)^2$ where y is the 0.2 per cent proof stress of the material being tested (Ref. 32). As the En 26 specimens used in this work did not conform to this requirement, the results have been given in terms of the applied central load. The variation in dimensions between the specimens, which did not exceed 0.1 mm, has been ignored in the presentation of the experimental results. Figs 14 and 15 show the effect of various pressures on the relation between the applied load and the deflection of the two differently heat treated batches of En 26 steel.

32. Specimens from batch 1 (H_{v30} 389) exhibited no rapid crack propagation even at atmospheric pressure (Fig. 14). As the pressure was increased crack propagation was inhibited, that is for a particular deflection the applied load beyond the elastic range increased with pressure. Specimens from batch 2 (H_{v30} 470) failed, as expected, in a more brittle manner (Fig. 15). At the lower pressures the effect of increasing pressure was to extend the load/deflection curve, by delaying crack propagation. When the applied load reached a maximum at a pressure of about 4.5 kbar (450 MN/m^2), the effect of increasing pressure was similar to that for batch 1, that is for a particular deflection, load increased with pressure.

33. In Fig. 16 the values for maximum applied load for both series of tests are plotted against the superimposed hydrostatic pressure. For the range of pressures covered this load was higher for the En 26 steel tempered at 350^oC (batch 2). The results suggest that for the En 26 steel with hardness H_{v30} 389 (batch 1) the maximum value for the applied load showed a slight increase with pressure. The H_{v30} 470 steel showed an initial increase in applied load with pressure, but above a pressure of 4.5 kbar (450 MN/m^2) pressure again appeared to have little effect on the maximum load. These results are similar to those obtained from investigations on the effect of a superimposed pressure on the tensile strength of wrought metals (Ref. 1).

DISCUSSION

34. It had been found in previous investigations into the effects of a superimposed hydrostatic pressure on the mechanical properties of steels that the effect of increasing pressure was to increase their strain to fracture without altering the general shape of their stress/strain curve, that is their flow stress, tensile strength, and instability strain were independent of pressure. For nominally ductile steels, namely those that necked when tested under atmospheric pressure, fracture strain was an increasing linear function of pressure and fracture occurred when the net axial tensile stress reached a critical value. The few reported tests on brittle steels had suggested that increasing pressure initially had little effect on ductility but, once a particular pressure range was reached, ductility increased relatively rapidly and the net axial tensile stress at fracture apparently decreased with increasing pressure.

35. Results from the tensile tests on M2 tool steel confirmed the above findings for 'brittle' metals with the ductility 'transition' occurring at about 5 kbar and the net axial fracture stress decreasing as a linear function of pressure. Although the stress/strain and ductility/pressure curves for specimens of heat treatment 3 (H_{V30} 715) differed considerably from those for treatments 1 and 2 the rate of decrease in fracture stress with pressure was similar. At the highest test pressures, however, there was an indication that for treatments 1 and 2 the gradients of the fracture stress/pressure curves were reducing. These tests indicated also that the specimens heat treated to the GKN recommendation had better hardness, ductility and fracture stress properties than those subjected to the steel manufacturers recommended heat treatment.

36. It is evident from the investigations on the two forms of beryllium that, although the type of surface finish affected the ductilities when tested under pressure, sheathing had a dominant effect. This was most noticeable for the hot pressed powder material. These findings are consistent with the concept of the penetration of the pressure medium into surface cracks, which would tend to accentuate the stress concentrations at the tip of the crack and promote premature fracture. When the surface finish was improved, surface damage such as cracks was reduced, which enabled greater ductility and greater fracture stress to be obtained under pressure.

37. The ductility/pressure curves for the hot pressed powder and extruded ingot berylliums were similar in shape. Initially the ductilities of the two forms were increasing functions of pressure but, at about 5 kbar, became linear functions of pressure. Their fracture stress characteristics under pressure, however, differed. The net axial stress at fracture for the extruded ingot decreased with increasing pressure. The hot pressed powder material after showing an apparent small decrease in stress with pressure exhibited an abrupt increase followed by a linear increase in fracture stress with pressure. This transition pressure corresponded to that for the change in the ductility curve. These differing fracture stress/pressure curves indicate that the fracture stress behaviour of beryllium is governed to some extent by the structure or fabrication of the material.

38. This fact is again demonstrated by the results obtained from the tests on molybdenum, which are in many ways similar to those for beryllium. With the original batch of material, sintered and swaged, the net axial stress at fracture increased linearly with pressure and at a rate of about 1.6 x the applied pressure. The effect of additional working of the molybdenum, by hydrostatic extrusion was to reduce the rate of increase in fracture stress to about 1.1 of the applied hydrostatic pressure. The additional working also made the brittle ductile transition far more pronounced with the transition occurring over a small pressure range.

39. The effect of pressure on the mechanical behaviour of the En 26 steel notched bend specimens was to delay fracture by inhibiting crack propagation. Up to the maximum load increasing pressure had negligible effect on the slope of the load/deflection curve of the steel. Beyond the maximum load the effect of an increasing pressure was to raise the load/deflection curve, that is, for a particular deflection the applied load increased with pressure. This is an indication that the reduction in the cross-sectional area of the notched specimens was retarded by increasing pressure. As mentioned earlier the material did not conform to the requirements needed for standard fracture toughness applications. These preliminary tests have demonstrated, however, that a superimposed pressure should increase the fracture toughness of some brittle materials. The tensile investigations on beryllium as well as cast iron and molybdenum suggest that an increasing hydrostatic pressure may not have the same beneficial effect on the fracture toughness of these types of materials, that is, cast, hot pressed, or sintered. Indeed their toughness may well be reduced.

40. To date very little research has been carried out on the fracture toughness of materials under these conditions and it is obviously an area in which a great deal of important research needs to be carried out.

ACKNOWLEDGEMENTS

We wish to thank Dr Shishmintsev of the Institute of Physics of Metals, Sverdlovsk, USSR, for providing the molybdenum specimens. Part of the work on beryllium was carried out under a contract from AWRE, Aldermaston, to whom we are grateful for permission to publish. This present paper is published by permission of the Director of the National Engineering Laboratory, Department of Industry. It is Crown copyright.

REFERENCES

1. PUGH H.Ll.D. and CHANDLER E.F. Mechanical properties of materials under pressure. Professor Galerkin's Anniversary Volume 1974 and also NEL Report No 577. East Kilbride, Glasgow: National Engineering Laboratory, 1974.

2. COOK G. The effect of fluid pressure on the permanent deformation of metals by shear. Selected Engineering Papers, No 170. London: Institution of Civil Engineers, 1934.

3. BRIDGMAN P.W. Studies in large plastic flow and fracture. New York: McGraw-Hill, 1952.

4. PUGH H.Ll.D. and GREEN D. Behaviour of Materials under High Hydrostatic Pressure. MERL Plasticity Report No 103. National Engineering Laboratory, East Kilbride, Glasgow, 1954.

5. CROSSLAND B. The effect of fluid pressure on the shear properties of metals. Proc.Instn mech.Engrs, 1954, 168(40), 935-946.

6. GLADKOVSKII V.A. and OLEINIK M. Apparatus for investigating the mechanical qualities of metals under high hydrostatic pressure. Phys. Metals and Metallography, 1957, 4(3), 118-121.

7. CROSSLAND B. and DEARDEN W.H. Plastic flow and fracture of a 'brittle' material (grey cast iron) with particular reference to the effect of fluid pressure. Proc.Inst mech.Engrs, 1958, 172(26), 805-820.

8. RYABININ Yu.N., VERESCHAGIN L.F., BALASHOV D.B. and LIVSHITZ L.D. Instr. Exp. Tech., 1959, 2, 265.

9. BERESNEV B.I., VERESCHAGIN L.F., RYABININ Yu.N. and LIVSHITZ L.D. Some problems of large plastic deformation in metals under high pressure. Akad. Nauk., SSSR, Moscow, 1960. Translated by V M Newton, Pergamon Press, 1963.

10. PELCZYNSKI. T. The influence of hydrostatic pressure on the plastic deformation properties of metals. Arch. Hutnictiw., 1962, 7(1), 3-13.

11. PUGH H.Ll.D. The mechanical properties and deformation characteristics of metals and alloys under pressure. In Am. Soc. Test. Mater. Special Tech Pub. No 374. Phila, Pa: American Society for Testing and Materials, 1964.

12. PUGH H.Ll.D. and GREEN D. The effect of hydrostatic pressure on the plastic flow and fracture of metals. Proc. Instn mech.Engrs, 1964-65, 179(1), 415-437.

13. DAVIDSON T.E., UY J.C. and LEE A.P. The tensile fracture characteristics of metals under hydrostatic pressures to 23 kilobars. Acta metall., 1966, 14, 937-948.

14. ERBEL S. Behaviour of materials twisted under hydrostatic pressure. Presented at conference organised by Department of Metallurgy, Polish Academy of Sciences, April 1966.

15. CHANDLER E.F. Tensile properties of a number of materials under hydrostatic pressure. NEL Report No 306. East Kilbride, Glasgow: National Engineering Laboratory 1967.

16. NOLAN C.J. and DAVIDSON T.E. The effects of cold reduction by hydrostatic fluid extrusion on the mechanical properties of 18 per cent nickel maraging steels. Trans. Am. Soc. Metals, 1969, 62, 271-277.

17. BRANDES M. Mechanical properties. In PUGH, H Ll D (Ed). The mechanical behaviour of materials under pressure, pp 236-298. London: Elsevier, 1970.

18. YAJIMA M., ISHII M. and KOBAYASHI M. The effects of hydrostatic pressure on the ductility of metals and alloys. Int. J. Fracture Mech., 1970, 6(2), 139-150.

19. OHMORI M., YOSHIMAGA Y., SUKEGAWA and MARUYAMA Y. Deformation characteristics of mild steel under hydrostatic pressure. Proc. 16th Jap. Congr. Mater. Res., 1973, pp 120-125.

20. NOBUKI M. and OGUCHI A. Effect of hydrostatic pressure on the ductility of a quenched and tempered carbon steel. J. Jap. Inst. Met., 1974, 38, 5, 401-7.

21. FRENCH I.E. and WEINRICH P.F. The influence of hydrostatic pressure on the tensile deformation of a spheroidised 0.5 per cent C steel. Scr. metall., 1974, 8, 87-90.

22. CARPENTIER D. and FRANCOIS D. Mechanical properties of zinc and cobalt under pressure. In PUGH, H Ll D (Ed). Engineering solids under pressure, pp 61-68. London: Institution of Mechanical Engineers, 1971.

23. FUNG P.K., BURNS D.J. and LIND N.C. Yield under high hydrostatic pressure. To be published.

24. PUGH H.Ll.D. Fracture criterion for cast iron. Nature, 1968, 218(5145), 985.

25. WHATLEY L.S. and VAN VALKENBURG A. High pressure optics. In BRADLEY, R S (Ed). Advances in high pressure research. Vol 1, pp 327-371. London: Academic Press, 1966.

26. CHANDLER E.F. Effect of sheathing on the mechanical properties of a material under hydrostatic pressure. NEL Report No 255. East Kilbride, Glasgow: National Engineering Laboratory, 1966.

27. ALADAG E., PUGH H.Ll.D. and RADCLIFFE S.V. Mechanical behaviour of beryllium at high pressures. Acta metall., 1969, 17, 1467-1481.

28. CHANDLER E., LINDSAY H.M., PUGH H.Ll.D. and WHITE J.S. The effect of surface finish on the pressure-induced ductility of beryllium. J. Mat. Sci., 1973, 8, 1788-1794.

29. WILLIAMS D.W. Externally heated cold seal pressure vessels for use to 1200°C at 1000 bars. Min. Mag., 1966, 35, 1003.

30. WILLIAMS D.W. and HARRIS P.G. Performance and failure of refractory metal pressure vessels between 900 and 1300°C. Proc. Instn mech. Engrs, 1967-68, 182 (pt 3C).

31. AUGER J.P. and FRANCOIS D. Mesures de tenacite sur an alliage d'aluminium a haute resistance sous pression hydrostatique. Rev. Phys. App., 1974, 9, 637-649.

32. BRITISH STANDARDS INSTITUTION. Methods for plane strain fracture toughness (K_{1C}) testing. DD3 1971.

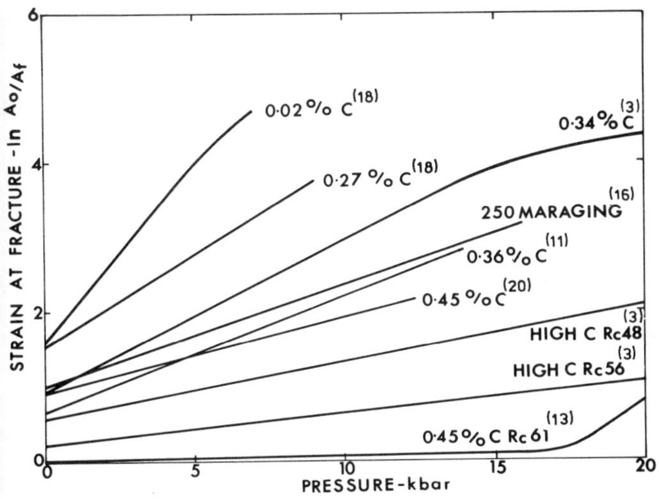

Fig. 1: Ductility of steels under pressure

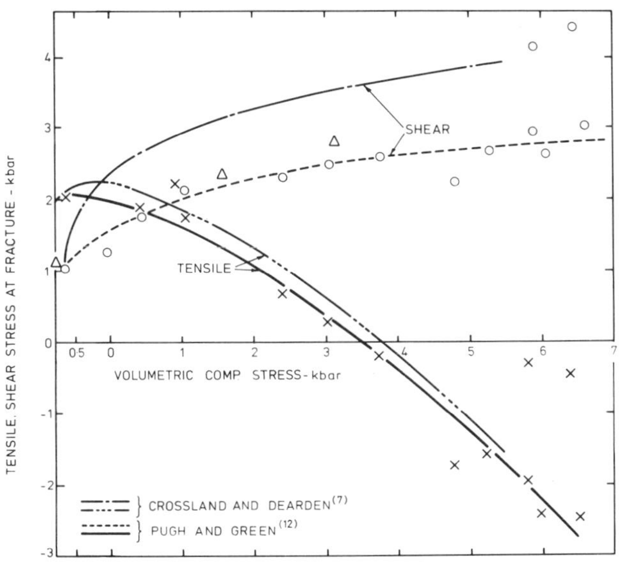

Fig. 2: Maximum shear and tensile stresses at fracture for cast iron tested under hydrostatic pressure

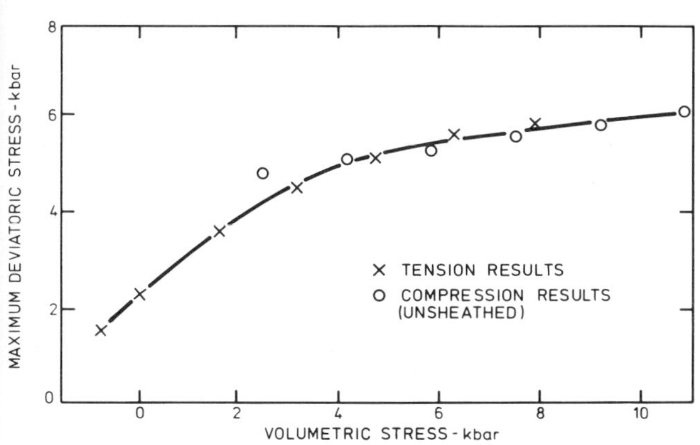

Fig. 3: Deviatoric stresses at fracture for cast iron tested under hydrostatic pressure (after Pugh Ref. 24)

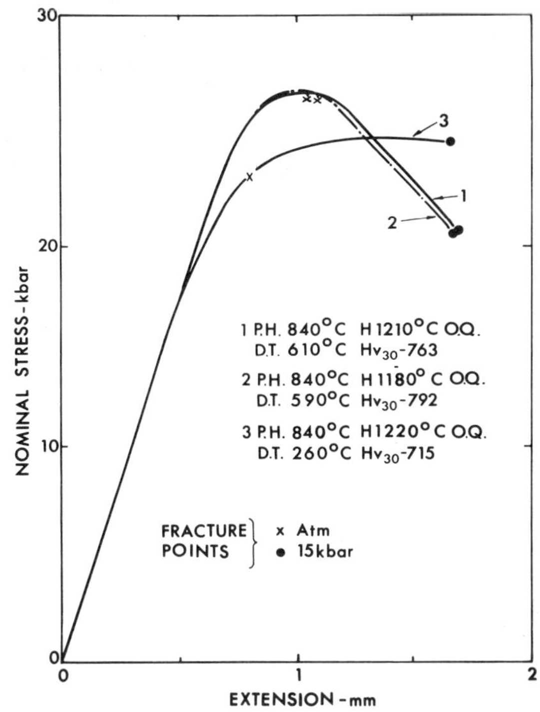

Fig. 4: Nominal stress/extension curves for M2 tool steel tested under pressures of 1 atmosphere and 15 kbar

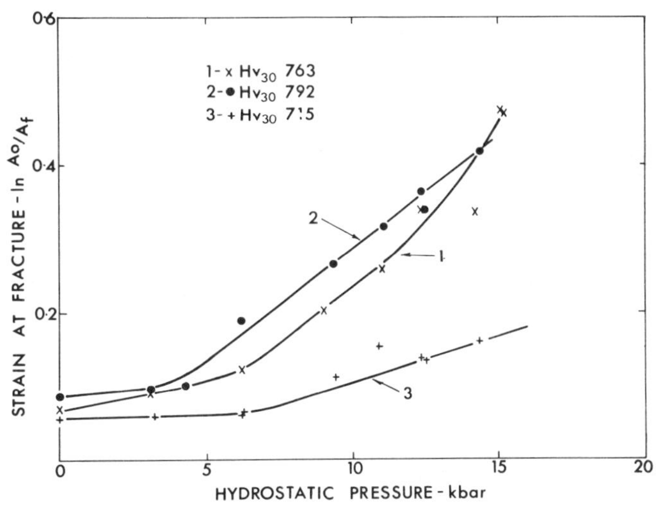

Fig. 5: Effect of pressure on the ductility of M2 tool steel

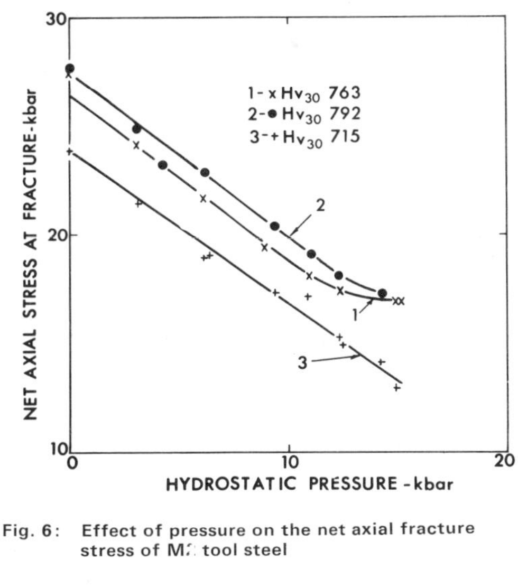

Fig. 6: Effect of pressure on the net axial fracture stress of M2 tool steel

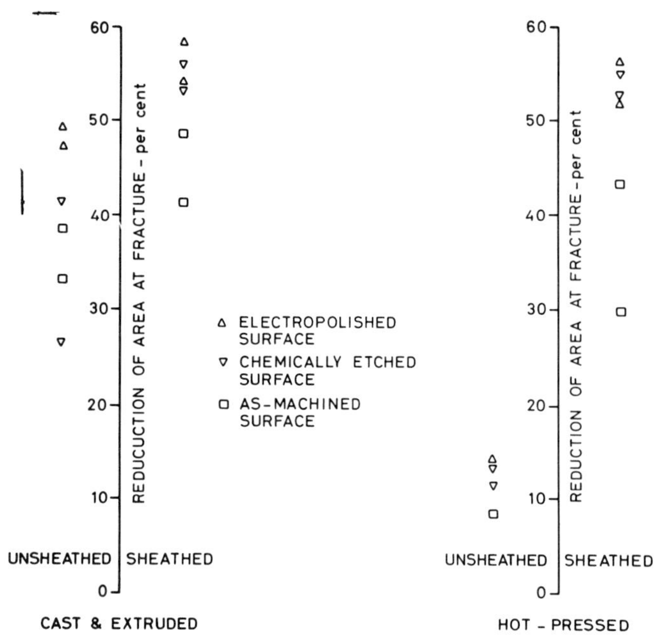

Fig. 7: Effect of surface preparation on the ductility of beryllium tested at 7.7 kbar

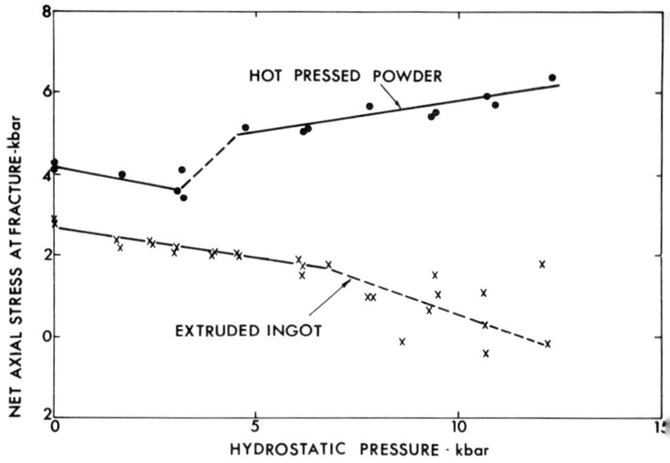

Fig. 10: Effect of pressure on the net axial fracture stress of beryllium

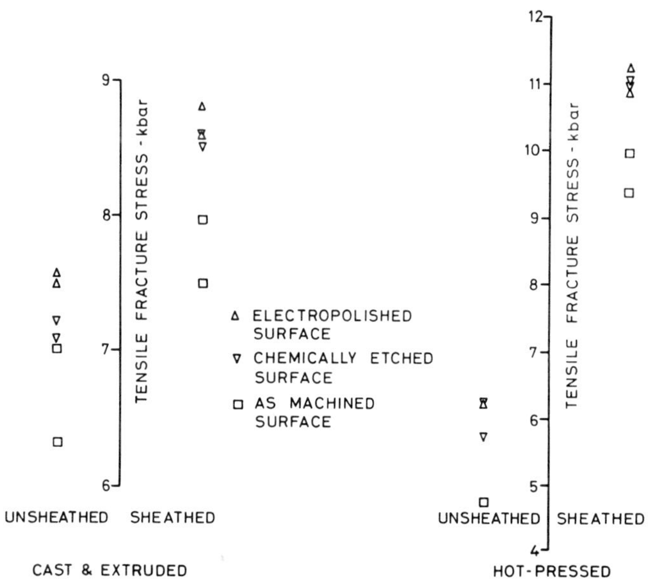

Fig. 8: Effect of surface preparation on fracture stress of beryllium tested at 7.7 kbar

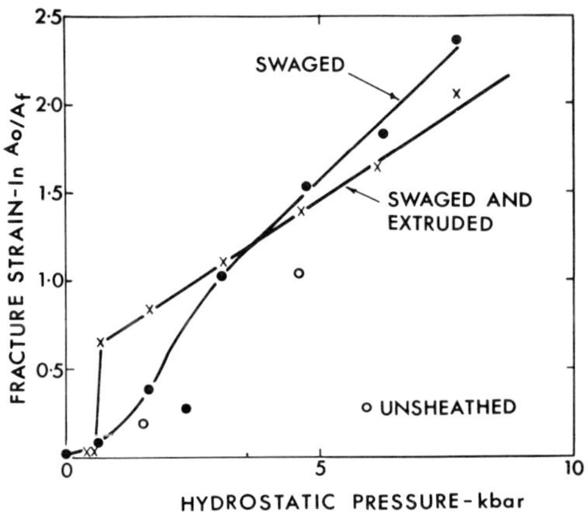

Fig. 11: Effect of pressure on the fracture strain of molybdenum

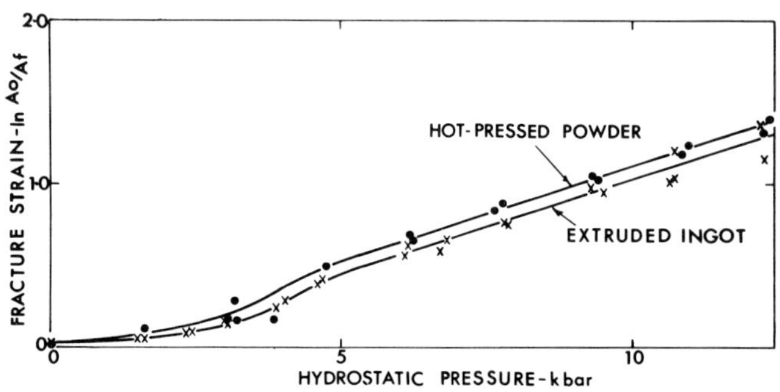

Fig. 9: Effect of pressure on the ductility of beryllium

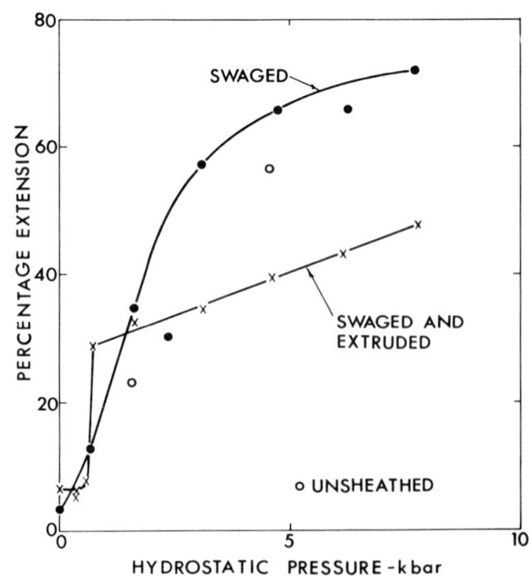

Fig. 12: Effect of pressure on the percentage extension of molybdenum

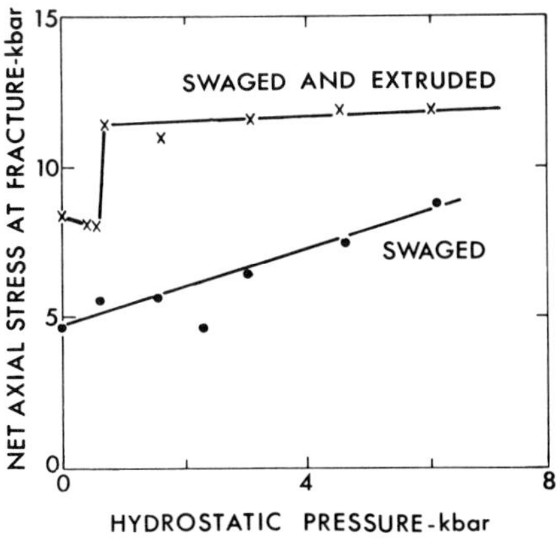

Fig. 13: Effect of pressure on the net axial fracture stress of molybdenum

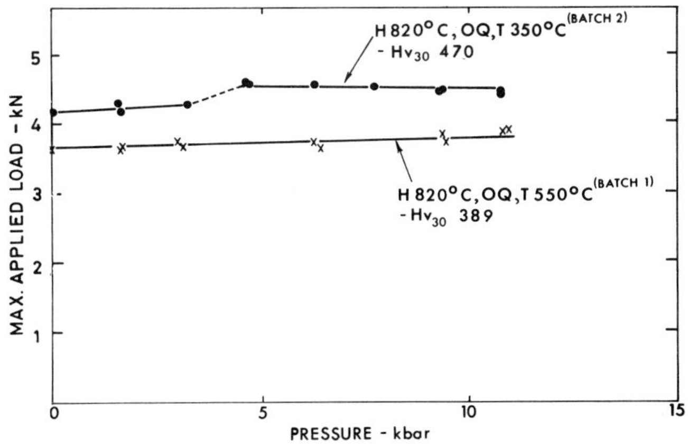

Fig. 16: Effect of pressure on the maximum applied load of En26 steel notched bend specimens

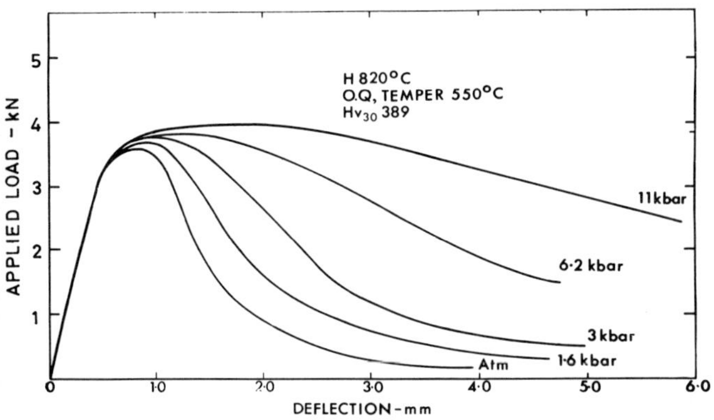

Fig. 14: Applied load/deflection curves for En 26 steel notched specimens, Hv 30 389, at various hydrostatic pressures

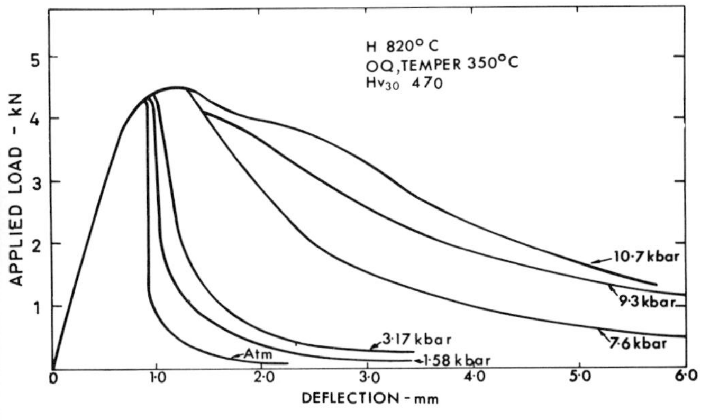

Fig. 15: Applied load/deflection curves for En 26 steel notched specimens, Hv 30 470, at various hydrostatic pressures

MATERIALS AND PROCESSES CONSIDERATIONS IN THE DESIGN OF PRESSURE VESSELS

T. E. DAVIDSON, PhD
Benet Weapons Laboratory, Watervliet Arsenal

B. B. BROWN, BSc
Benet Weapons Laboratory, Watervliet Arsenal

D. P. KENDALL, MSc
Benet Weapons Laboratory, Watervliet Arsenal

SYNOPSIS The incorporation and/or recognition of many factors during the design of thick-walled pressure vessels can have considerable impact on their fatigue and fracture resistance. Using some of the concepts of fracture mechanics, the importance of material selection, particularly in terms of strength versus toughness, will be described and examples presented and discussed showing the importance of toughness on fatigue and fracture behavior.

Based on the concepts of fracture mechanics, residual stresses can significantly effect crack growth rate and thus fatigue life. Through example, the use of autofrettage residual stresses will be described along with the dependence of the magnitude of such stresses on the autofrettage process.

Two types of stress concentrations often found in pressure vessels are crossholes and configurational discontinuities on the outside surface. The effects of crossholes can be minimized by orientation which will be described. Outside surface discontinuities cannot be neglected when residual stresses are present and examples will be shown where they can be quite harmful.

INTRODUCTION

The choices of geometry, materials and their processing in the design of pressure vessels are greatly dependent upon intended use. The design approach should differ between a vessel intended for static or infrequent loading as compared to a vessel to be subjected to frequent or cyclic loading. This paper addresses itself to recent concepts and results of designs for repeated loading and offers design approaches to provide resistance to fatigue cracking and to avoid catastrophic failures in vessels under these conditions. Specifically three major points will be considered: the use of imposed residual stresses to extend fatigue life; the significance of fracture toughness as a material property and a design factor; and finally the varying effects caused by stress concentrators in pressure vessels and methods of minimizing these effects.

It is not the purpose of this paper to intently dwell on the differences or commonality of design for various types of pressure vessels nor to examine design philosophy from the standpoint of elastic strength, yield criteria, endurance limits, etc. These aspects of design have been well investigated and are readily available in the literature.

In recent years there has been increased usage of cyclically loaded, highly stressed pressure vessels. This usage has introduced new critical engineering factors to be considered in design. The cannon barrel has been one area of early awareness of the need for new engineering approaches as designers have been pressed to accommodate increasing pressures in lighter weight weapons and is the field in which the authors have first approached these problems.

Historically, and in some instances even today, many high pressure vessels have been designed without consideration of material toughness and residual stresses. An oft traveled path has been to design based on material yield strength alone, the premise being: the stronger the material the better it will be and the longer it will last. This path has led to some rather unsatisfactory pressure vessels exhibiting short fatigue lives and catastrophic failures. We shall discuss some examples of this later on.

FRACTURE MECHANICS

Probably the best way to introduce the importance of material toughness and residual stresses is by considering the concept of fracture mechanics in a highly simplified form. The basis of this concept is the existence of a stress intensity factor K, which defines the stress state in the region of a crack tip. In generalized form, the stress intensity factor for a crack in a body is

$$K = Y\sigma\sqrt{\pi b} \qquad (1)$$

where Y is a parameter that is a function of crack shape and depth, σ is the nominal tensile stress normal to the crack plane, and b is the crack depth.

Since almost all high pressure vessels are cylinders we will look more closely at the application of fracture mechanics to a pressurized cylinder without residual stresses. The nominal stress in this case is the tangential bore stress, σ_t, defined as:

$$\sigma_t = P\,\frac{(W^2+1)}{W^2-1} \qquad (2)$$

where W is the ratio of outside to inside diameter.

For the case of a crack emanating from the inside surface of a thick-walled cylinder, Throop (1)(2) gives values for Y of 0.5, 1.5 and 1.0 for semi-circular, straight fronted and semi-elliptical crack shapes respectively. The latter crack shape is the one most encountered. Although Y varies as a function of crack depth, Throop has cited that assuming a constant value up to a crack depth approaching 1/3 of the wall thickness represents a reasonable approximation. It should be noted that under cyclic loading, growing a crack to 1/3 of the wall thickness would consume a major portion, if not all, of the usable fatigue life. The value of Y for deeper cracks is the subject of current studies.

Based on the above one can now look at both the fracture and the crack growth process in a pressurized cylinder without residual stresses as follows:

FRACTURE

$$K = K_{Ic} = Y\sigma_t\sqrt{\pi b_c} \quad (3)$$

$$b_c = \frac{1}{\pi Y^2}\left[\frac{K_{Ic}}{\sigma_t}\right]^2 \quad (4)$$

CRACK GROWTH

$$\frac{db}{dN} = C(\Delta K)^m \quad (5)$$

$$\Delta K = K_{max} = Y\sigma_t\sqrt{\pi b} \quad (6)$$

$$\frac{db}{dN} = C(Y\sigma_t\sqrt{\pi b})^m \quad (7)$$

where K_{Ic} is the material toughness, b_c is the critical crack depth, C is a material constant and m is a material based exponent as proposed by Paris (3) and Throop and Miller (4).

From the above we see that both crack growth and critical crack depth are dependent on the stress intensity, K, at the crack tip. Fracture will occur when K becomes equal to the fracture toughness of the material (K_{Ic}) which is a measurable material property. Thus, for a given fracture toughness and stress, equation 4 defines the critical crack or defect size (b_c). The rate of crack growth (db/dN) is predictable based on K, C, and m terms which are available.

So far we have dealt with cylinders without residual stresses and by extending this fracture mechanics approach we can introduce the importance of residual stress effects. The total stress in the stress intensity relationship is the sum of the tangential pressure stress (σ_{tp}) and any tangential residual stress (σ_{tr}). Equation (1) can be restated as

$$K = Y(\sigma_{tp} + \sigma_{tr})\sqrt{\pi b} \quad (8)$$

Referring to equation (7) then, depending on the sign and magnitude of the residual stresses, the crack growth rate, and therefore the fatigue life of a pressure vessel can be greatly increased or decreased.

FATIGUE AND FRACTURE

Consider now the practical implications of the above fracture mechanics concepts in the design and manufacture of pressure vessels. Let's first examine the concept of designing for strength alone and using the highest strength material available without regard for toughness. Fatigue failure is a three stage process consist-of initiation, propagation, and finally fracture. Fracture mechanics applies to the last two stages, i.e. propagation and fracture. Initiation is controlled by the shear strain range and thus is sensitive to yield strength. If the bore surface is smooth and undamaged in service, a high yield strength can possibly result in a usable fatigue life. However, if high strength is achieved by disregarding toughness, and it is low, the following can happen which will far outweigh any benefits of retarded initiation. The low fracture toughness will result in a small critical defect size which in the extreme can approach the intrinsic defect size of the material and fast fracture will occur perhaps even on the first cycle. The result will be brittle fracture or at best, a shortened fatigue life. Any type of surface damage or stress discontinuity will further aggravate the problem.

Basically then, the best combination of fatigue and fracture performance can be achieved by optimizing yield strength and toughness. Once the required yield strength is established, a material should be chosen with the highest practical fracture toughness. It may be necessary to actually lower the yield strength in order to obtain sufficient toughness. The goal is to have sufficient toughness so that at the operating stress level, the critical crack depth approaches or is equal to the wall thickness. This will not only generally result in an acceptable fatigue life but even more important a ductile rather than a brittle and catastrophic failure mode.

The following represents an example of how these concepts work in actual practice. Large caliber cannon barrels, which are nothing more than a special type of high-stored-energy pressure vessels, will be used as a vehicle of discussion.

A recent occurrence demonstrates what can happen when high strength, to the detriment of toughness is pursued in design and construction and the improvement that can be made by utilizing an appreciation of fracture mechanics and residual stresses. A barrel was designed which had a bore of 175 mm with a wall thickness of 89 mm. The designers relied on using steel at a high yield strength (1250 MPa or 180 ksi) to achieve the necessary elastic strength. At this yield strength level the steel in some barrels had a fracture toughness as low as 77 $MN/m^{3/2}$ or 70 $ksi\sqrt{in}$. This high strength, low toughness, barrel experienced short life brittle failures during initial testing.

Failure during firing and during subsequent hydraulic cycling in the laoratory occurred with critical crack depths in the range of 0.3 to 1.0

inch. The left side of Figure 1 shows the failure of a hydraulically cycled specimen of the original barrels. The upper left view is the specimen as failed with the crack running full length as typical of brittle fracture. The lower left view shows the fracture surfaces with the bore surfaces in the center. The fatigue crack as depicted by the light smooth area only extends in about one inch at which point it reached critical size and fast fracture occurred on the next cycle. The right side of Figure 1 shows the fracture experienced under the same test conditions by a barrel of the same geometry and material but with somewhat lower strength and higher fracture toughness (K_{Ic}). In this case failure was by a short ductile perforation as the crack passed through the outside wall while still in the fatigue growth mode. This leak-before-fracture type failure is the best failure mode that can be hoped for and occurs when the critical crack depth approaches or exceeds the wall thickness. The lower right view shows the fatigue crack progresses from the bore surface, in this case the top and bottom of the view, through the entire wall thickness until the outside wall, center of view, is perforated. Thus a critical crack depth equal to or greater than the wall thickness and a greatly preferred failure mode, was achieved by decreasing the material yield strength by about 17% which increased the fracture toughness (K_{Ic}) by 15%.

Figure 2 shows how crack propagation and critical crack depth are affected by this change in toughness. This is a plot of crack depth versus cycles for 175mm cylinders at a cyclic pressure of 345 MPa. Curve A is for a high strength-low toughness material and B for a slightly lower strength-higher toughness material. Comparing the two curves, one can see that the higher strength material reaches the rapid crack growth stage more rapidly and grows to a smaller crack depth at fracture. The lower strength material shows more uniform growth and grows to a greater depth i.e. the wall thickness resulting in a failure mode identical to that shown on the right of Figure 1 and exhibits a longer life.

RESIDUAL STRESSES

Let's now take a closer look at the question of beneficial residual stresses in pressure vessels. Residual stresses obtained by a shrunk multilayer or autofrettaged construction have long been used to enhance the elastic strength of vessels. Of equal, if not greater, importance is their effect upon fatigue.

The most effective means for inducing residual stresses is autofrettage, and an example of what benefit may be derived is shown by curve C which is for a cylinder overstrained 50% of its wall thickness. Even though the strength is lower than that of curve B, the life is doubled. Even further benefit would occur with higher amounts of overstrain.

It should be noted that the effects of residual stresses is mostly in the crack growth stage of fatigue fracture. Since crack initiation is a function of shear strain range, it will be little affected by residual stresses.

Why autofrettage residual stresses affect fatigue life through retarding crack growth rate is shown in Figure 3 which is a schematic plot of stress intensity versus relative crack depth for the cylinders shown in Figure 1 and 2. As previously stated, the stress in the stress intensity solution is the sum of the tangential stress due to pressure and any tangential residual stresses. So if the residual stresses are compressive the stress intensity will be lowered and crack growth rate decreased. This is shown in Figure 3. Up to the crack depths of 0.4 of the wall thickness, which comprises more than 90% of the total fatigue life, the stress intensity with residual stresses is lower than that without. Beyond 0.4, the residual stresses become tensile which increases the stress intensity over that in the non-residual stress case. This however has little effect on fatigue life when it is controled by a crack growing from the bore outward.

Some mention should be made concerning the magnitude of the autofrettage residual stresses since they are so important in extending fatigue life. It is well known that the magnitude of the stresses is dependent upon the amount of overstrain. Recent results have shown however that the magnitude is also a function of the type of autofrettage process used. This effect is shown in Figure 4 which is a plot of stress on the ordinate and radius on the abcissa. The solid lines show the tangential residual stresses introduced by hydraulic pressure as compared to those introduced by mechanical swaging (5). This process is accomplished by forcing an oversized conical mandrel through the bore of the tube to produce the required radial deformation. The curves were established using the Sach's boring-out, x-ray and a slitting technique. As can be seen, the compressive residual stresses near the bore are 60% greater in the mechanical swaged versus the hydraulically autofrettaged cylinder. This then substantially reduces the effective stresses as shown by the dotted curves which is the sum of pressure and residual stress. The residual stress results shown for the hydraulic process were obtained from a tube that was autofrettaged to produce about 65 percent overstrain, whereas the swaged tube was 100 percent overstrained. This, however, cannot account for the large difference in residual bore stress since the theoretical residual bore stress for 65 percent overstrain is only 10 percent less than that for 100 percent overstrain. The reason for this large difference is not clear at this time and is currently under investigation. Since autofrettage is accomplished in the swaging process by pressurizing only a short length of the tube at a time, the residual stresses are produced by a complex 3-dimensional deformation process. This has been studied by Chen (6) using a 3-dimensional, elastic-plastic finite element analysis for the case of a non-moving band of pressure loading. The results given in this reference do not give the residual stresses due to swaging since these are produced by a moving band of pressure. This produces a much different deformation history than that produced by the non-moving pressure used in this work. However, further development of this technique will consider the complete deformation history associated with swaging and should provide a basis for understanding the difference between the residual stresses produced by the two methods.

It has been shown analytically by Milligan (7) and experimentally by Kendall (8) that the bore residual stresses produced by hydraulic autofrettage can be of significantly lower magnitude than would be predicted by theory due to reverse yielding caused by the Bauschinger effect. It is possible that the triaxial deformation associated with the swage autofrettage process tends to result in less reverse yielding at the bore and thus higher residual stress magnitudes.

Before leaving this figure, one aspect that will become important in subsequent discussion of external stress concentrations should be pointed out. To gain compressive residual stresses near the bore (I.D.), one pays the penality of tensile residual stresses near the outside surface (O.D.). This can result in rather large effective tensile stresses at the O.D.

STRESS CONCENTRATIONS

The final point to be discussed is stress concentrations and their effect upon fatigue and fracture behavior. Various types of stress concentration are often found in pressure vessels and sometimes result in very unsatisfactory fatigue and fracture behavior. Two types will be examined --outside surface discontinuities and crossholes. Both are often seen in pressure vessels but their consequences are not often considered by the designer.

Normally, discontinuities at the O.D. do not receive much design attention since the operating stresses are much lower at the outside than near the I.D. However, as demonstrated in Figure 4, if one uses autofrettage or jacketing to gain compressive residual stresses near the bore, tensile residual stresses result near the O.D. Thus, although the tangential pressure stress at the O.D. may be low, the mean stress can be quite high. In view of this, O.D. stress concentration can be troublesome and their presence in vessels containing residual stresses should not be neglected.

Two examples of such O.D. initiated failures are shown in Figure 5. Both of these cylinders were 100% swage autofrettaged with fatigue fracture of the upper picture starting at a blind hole and in the lower at a thread interruption. These types of discontinuity not only shorten fatigue life but, due to the high tensile stress present, result in a small critical crack depth and much fast fracture as is seen particularly on the fracture surface of the interrupted thread fracture.

Severe O.D. stress discontinuities should be avoided in vessels with high residual stress levels. If this cannot be done, a possible approach is to reduce the amount of overstrain or interference in jacketing which will reduce the overall residual stress level and at least result in equal probability of inside and outside surface initiated failures.

The second type of stress discontinuity of concern involves holes intersecting the bore usually referred to as crossholes. This subject was first studied by Morrison (9) a number of years ago. Figure 6 shows some recent results of the authors on the effects of various types of crossholes and ways to minimizing their effect. This is a plot of pressure versus cycles to failure for smooth bore cylinders for three different hole configurations. The radial hole is the most common and its detrimental effect can be readily seen. Even worse is a hole at an angle to the longitudinal axis. A means for reducing their effect is shown by the curve marked tangential.

The reason there is a difference in effect upon fatigue of the various hole configurations is due to two factors. Cracks initiate where the hole intersects the bore surface and so the intersection profile becomes important. The highest stress concentration is where the major axis of the intersection profile is normal to the tangential stress direction. Thus one sees progressive increase in effect as the major axis orientation changes. The second point is that pressure acts in the hole itself making the hole act as a pressure vessel. One has these pressure stresses then superimposed on the gross tangential stresses of the cylinder. Fatigue cracks will grow radially, i.e. perpendicular to the principal tangential stress direction in the cylinder. For the case of the holes in the radial plane, then, cracks will always grow in a concentrated stress field. When the hole is out of the radial plane, as in the case of the tangential hole, the crack must move out of the plane of the hole to grow, thus away from the higher stress region.

Figure 7 shows the fracture surfaces of four specimens laid out so that the outside surface of each pair are in contact. On the lower left one sees the fatigue crack growth along the radial holes. In contrast to this, the upper right shows that in the case of tangential holes the fatigue crack which initiated at the hole-bore intersection grew in a radial plane away from the crosshole.

Based on the above, crossholes, if at all possible should be eliminated. However, if they are necessary, some things can be done to reduce their detrimental effect. First, use a tangential hole. Second, seal at or close as possible to the bore surface in order to keep pressure out of the hole and, if this is not possible, one might consider independent autofrettage of the hole itself.

SUMMARY

Many factors in the design and fabrication of a pressure vessel affect its success or failure. The authors have considered three such factors that have considerable bearing on the fatigue and fracture resistance of vessels but have a tendancy on occasion to be overlooked by designers.

Through the concepts of fracture mechanics and a practical example, the importance of fracture toughness in affecting fatigue life and controling fracture mode has been demonstrated.

It has been shown that even relatively small increases in toughness will extend fatigue life and through increasing the critical crack depth, will result in a ductile "leak-before-fracture" failure rather than brittle fracture. Achieving a reasonable balance of strength and toughness is quite important and a benefit can be gained by lowering strength to achieve such a balance.

Through reduction in the stress intensity, compressive residual stresses will markedly reduce crack growth rate and thus extend fatigue life. In the case of autofrettage residual stresses, their magnitude is a function of how they are introduced; higher levels of compressive stress resulting from mechanical as compared to hydraulic autofrettage.

Two types of common stress concentrators have been discussed. In the presence of residual stresses due to autofrettage or jacketing, outside surface stress concentrators can become important due to high effective stresses in this region. Such stress concentration can result both in low fatigue lives and small critical crack depths.

Crossholes are commonly found in vessels and their effect on fatigue life as a function of orientation demonstrated. Considerable advantage can be gained by orientating the hole in a tangential direction and by sealing near the bore surface so as not to permit pressure in the hole.

LIST OF SYMBOLS

b	=	crack depth
B	=	wall thickness
K	=	elastic stress intensity factor
K_{Ic}	=	plane strain fracture toughness
ΔK	=	range of K during a fatigue cycle
K_{max}	=	maximum value of K during a fatigue cycle (The minimum pressure during a fatigue cycle is essentially zero. The minimum value of K is assumed to be zero and therefore $\Delta K = K_{max}$)
m	=	a material constant
N	=	number of fatigue cycles
P	=	internal pressure
W	=	diameter ratio
Y	=	crack geometry parameter
σ	=	stress
		Subscripts
c	=	critical value
p	=	due to pressure
r	=	residual
t	=	tangential
y	=	at yield

APPENDIX

References

(1) Throop, J. F., "Fatigue Crack Tolerance in Thick Walled Cylinders", Technical Report No. WVT-7035, Benet Laboratories, Watervliet Arsenal, Watervliet, NY 12189, Nov. 1970.

(2) Throop, J. F., "Fatigue Crack Growth in Thick Walled Cylinders", Proceedings of the 28th National Conference on Fluid Power, Sep. 1972.

(3) Paris, P. C. The Fracture Mechanics Approach to Fatigue, in "Fatigue, and Interdisciplinary Approach", Burke, J. J. et al Syracuse University Press, 1964, pp. 107-132.

(4) Throop, J. F. and Miller, G. A., "Optimum Fatigue Crack Resistance", Achievement of High Fatigue Resistance in Metals and Alloys, ASTM STP-467, American Society for Testing and Materials, 1970, pp. 154-168.

(5) Davidson, T. E., Barton, C. S., Reiner, A. N. and Kendall, D. P., "A New Approach to the Autofrettage of High Strength Steel Cylinders", Experimental Mechanics, Vol. 2, no. 2, Feb 1962.

(6) Chen, P. C. T., "Elastic-Plastic Solution of a Two-Dimensional Tube Problem by the Finite Element Method", Watervliet Arsenal Technical Report WVT-TR-74038, Sept. 1974.

(7) Milligan, R. V., "The Influence of the Bauschinger Effect on Reverse Yielding of Thick-Walled Cylinders", Watervliet Arsenal Technical Report WVT-7036, Oct 1970.

(8) Kendall, D. P., "The Effect of Material Removal on the Strength of Autofrettaged Cylinders", Watervliet Arsenal Technical Report WVT-7003, Jan. 1970.

(9) Morrison, J. L. M., Crossland, B., and Parry, J. S. C., Jl. Mech. Eng. Science, 1, 207, 1959.

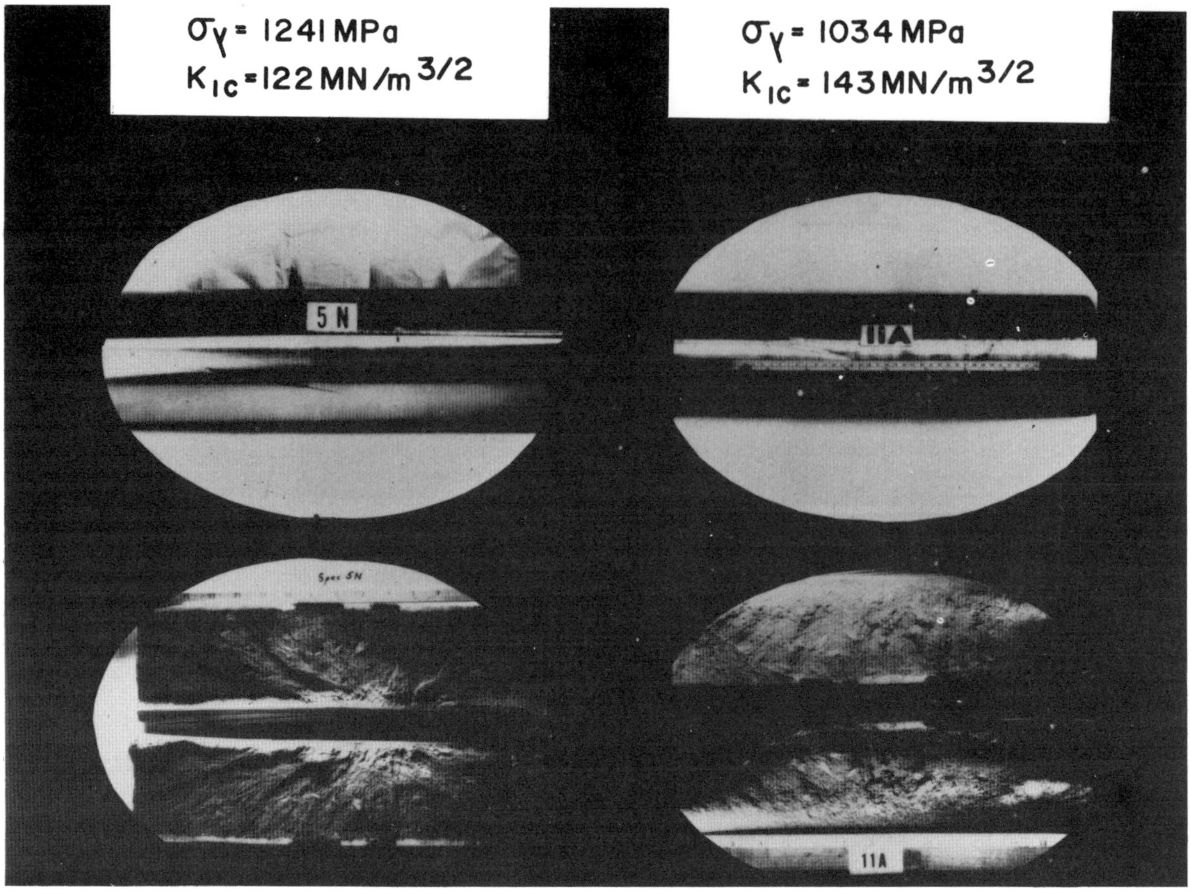

Fig. 1: Macroscopic fracture appearance P = 345MPa

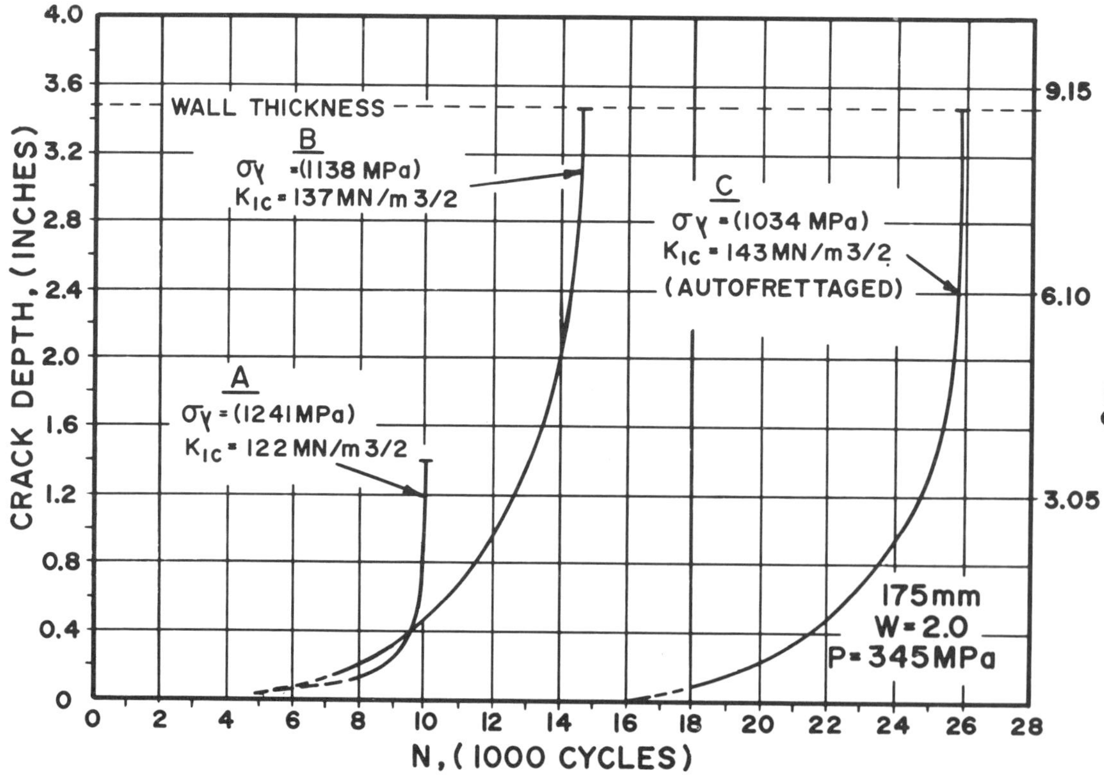

Fig. 2: Crack depth versus cycles

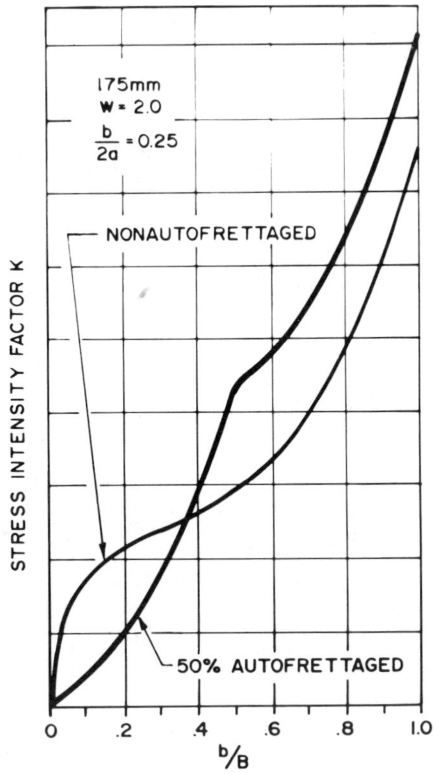

Fig. 3: Schematic of stress intensity factor versus relative crack depth showing effect of 50% autofrettage

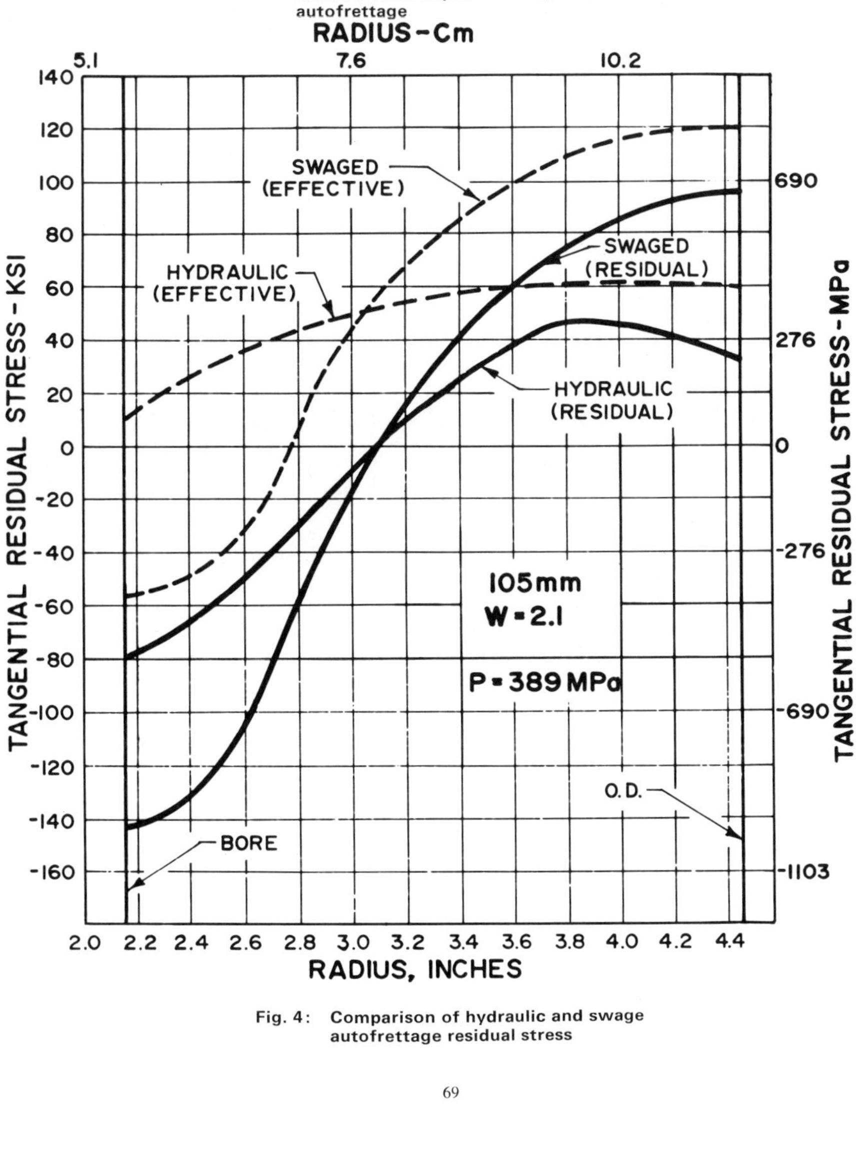

Fig. 4: Comparison of hydraulic and swage autofrettage residual stress

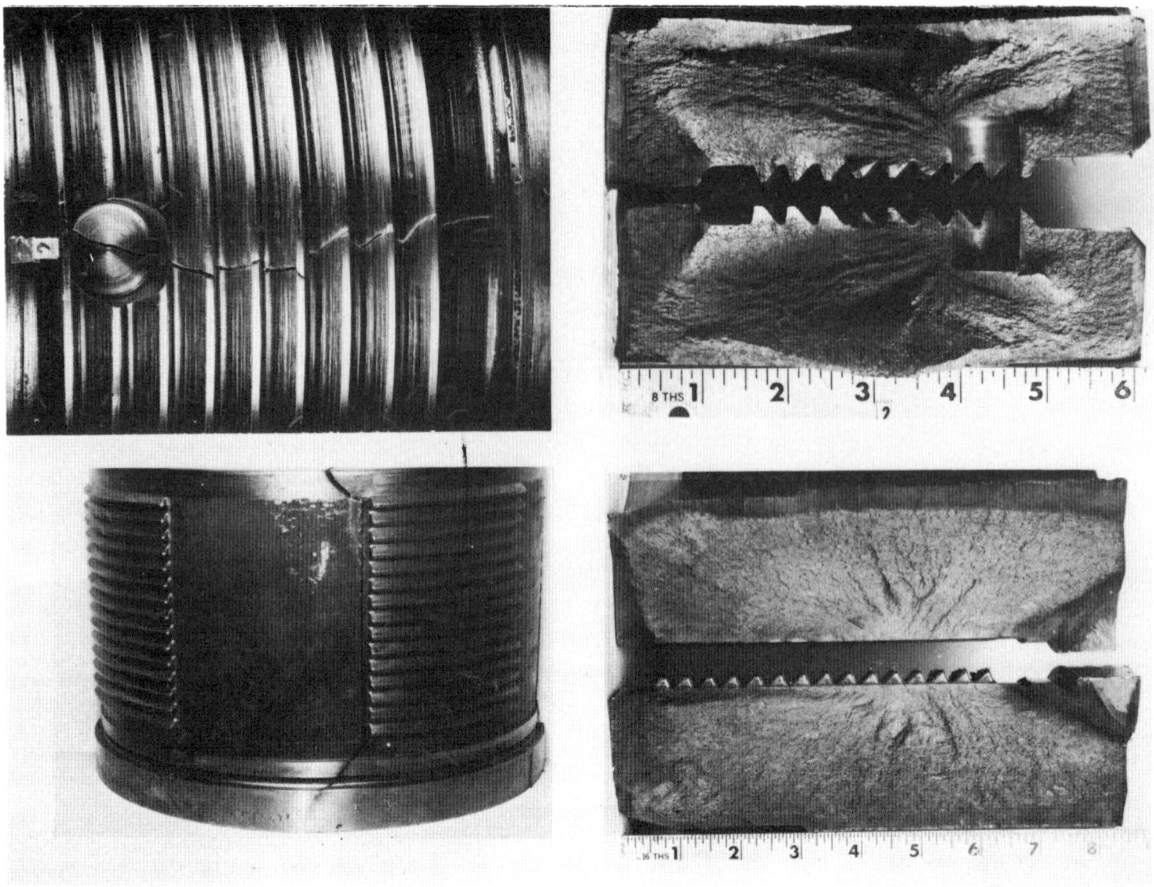

Fig. 5: External fatigue failure in autofrettaged cylinders

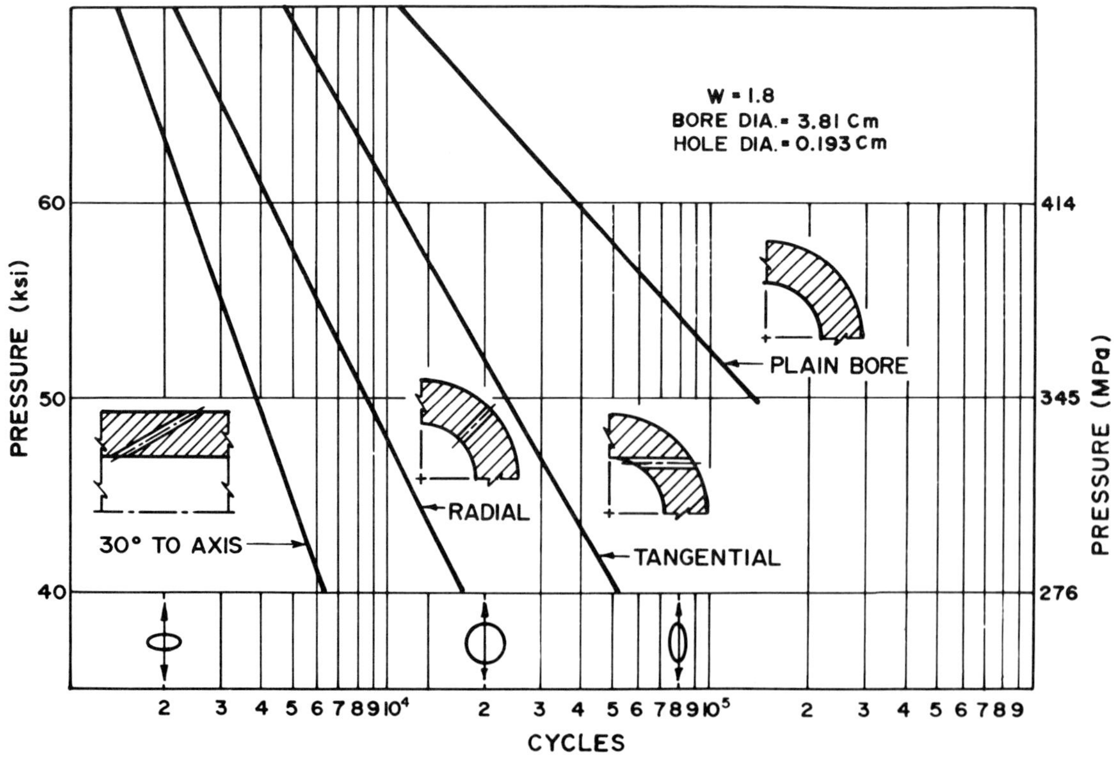

Fig. 6: Effect of holes intersecting the bore

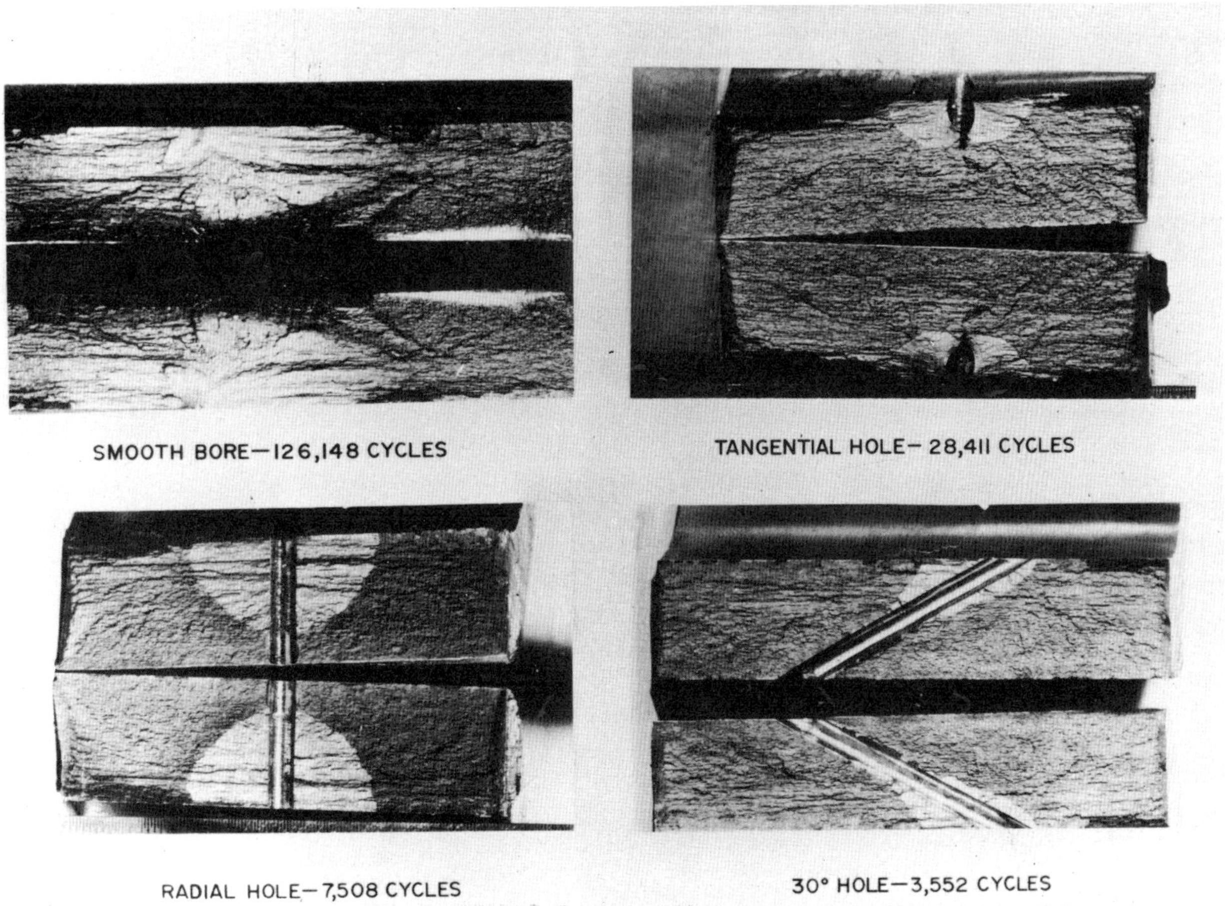

Fig. 7: Effect of cross holes on fatigue life P = 345MPa

THE CREEP OF THICK WALLED $2\frac{1}{2}\%$ Ni Cr Mo CYLINDERS UNDER INTERNAL PRESSURE AT 350° C: DEVELOPMENT OF A TESTING MACHINE AND PRELIMINARY RESULTS

W. J. SKELTON, BSc, PhD, CEng, MIMechE, Wh.Fell
Lecturer, Department of Mechanical Engineering, The Queen's University of Belfast
A. SALIM, BSc, MSc, PhD
Stress Analyst International Combustion Division, Clarke Chapman Ltd.
R. G. PATTON, BSc, PhD
Chief Fan Design Engineer, Davidson & Co. Ltd.

SYNOPSIS The paper describes a high pressure creep machine capable of operation at pressures up to 0.5 GN/m^2 and temperatures up to 370°C. The rig incorporates optical extensometers for measuring the circumferential and axial creep strains on the outside diameter of the specimen. Four tests on EN25 ($2\frac{1}{2}\%$ Ni. Cr. Mo.) steel cylinders (K = 1.67) at 350°C are reported. In all four tests the axial creep strain was negligible. A considerable amount of related tension and torsion creep data is presented and preliminary work on the correlation of this data with the thick-walled cylinder results is also reported.

1. INTRODUCTION

Over the past fifty years a considerable amount of attention has been devoted to the problem of creep and creep rupture in thick-walled cylinders at high temperature and numerous theoretical and experimental papers on this topic have been published. Skelton and Crossland (1)* gave a fairly comprehensive review of these papers up to 1967.

Most of the theoretical treatments predict that the axial creep strain or plastic flow is zero or at least negligible. This point has been substantially verified experimentally for lead cylinders at room temperature by Bailey (2) and for relatively thin-walled steel cylinders at elevated temperatures by Norton and Soderberg (3).

In previous studies conducted in the Department of Mechanical Engineering, Queen's University of Belfast (4) (5) (6), axial creep strains were not measured largely because of the additional experimental difficulties involved. In analysing the results of this work (particularly at large strains) there was always considerable debate as to the validity of the zero axial creep assumption and it was therefore finally decided to conduct a limited experimental study aimed at providing additional information in this area. In selecting a material for this investigation account was taken of the experimental work of Crossland and Kerr (7) relating to the elevated temperature static strength of thick-walled cylinders. One of the materials extensively investigated by these authors was EN25 (B.S.970) at various temperatures up to 370°C. This material is widely used and there is a considerable amount of basic mechanical properties data available. It was therefore felt that some data on its behaviour under creep conditions at 350°C might prove useful to designers. Thus the two basic aims of the investigation were firstly to provide some additional information on axial creep strain behaviour in a thick-walled closed end cylinder under internal pressure, and secondly to provide some tension, torsion and thick-walled cylinder creep data for EN25 steel at 350°C and to attempt correlations which might be of use to designers.

* References are given in Appendix 1.

2. NOTATION

A, m, n	Material Constants
P	Internal Pressure
$K = {}^{O.D}/_{I.D}$	Cylinder diameter ratio
t	time
σ_t	Tension creep stress
ε_t	Creep strain in simple tension
$\bar{\varepsilon}_t$	Total strain in simple tension i.e. (elastic + plastic + creep)
τ	Shear creep stress
γ	Shear creep strain in simple torsion
$\bar{\gamma}$	Total shear strain in simple torsion i.e. (elastic + plastic + creep)
ε_θ	Circumferential creep strain at outside diameter of cylinder
$\bar{\varepsilon}_\theta$	Total circumferential strain at outside diameter of cylinder i.e. (elastic + plastic + creep)
τ_m	Shear stress at mean diameter of cylinder
γ_m	Shear creep strain at mean diameter of cylinder
$\bar{\gamma}_m$	Total shear strain at mean diameter of cylinder i.e.(elastic+plastic+creep)
$\varepsilon_{\theta m}$	Circumferential creep strain at mean diameter of cylinder
$\bar{\varepsilon}_{\theta m}$	Total circumferential strain at mean diameter of cylinder i.e. (elastic + plastic + creep)

3. TENSION AND TORSION CREEP MACHINES

The tension creep machines used in this investigation were of the modified NPL design fitted with optical extensometry and capable of operating at specimen loads up to 2 tonf.(19.94kN). A detailed description of the machines is given by Skelton and Crossland (8). Fig. 1 shows a detailed drawing of the tension creep specimen used.

The torsion creep machines used were developed

in the Department of Mechanical Engineering, The Queen's University of Belfast. The design is based on the machines used by the late Dr. A.E. Johnson at NEL. The machines incorporate optical extensometry and fig. 2 shows the type of specimen used. A detailed description of this equipment is given by Patton et al (9).

4. DESIGN OF THICK-WALLED CYLINDER CREEP MACHINE

Fig. 3 shows a diagrammatic layout of the rig. The rig is capable of operating at pressures up to 30 tonf/in^2 (463.5 MN/m^2) and temperatures up to 370°C.

4.1 Pressure System

The pressurizing fluid used in the rig is Lockhead Heavy Duty Brake fluid. Initial fluid pressure in the specimen is generated by means of a small intensifier somewhat similar to that described by Skelton and Crossland (4). The low-pressure oil supply to the intensifier is delivered by a small electrically driven CAV fuel injection pump with a maximum operating pressure of 10,000 lbf/in^2 (69 MN/m^2).

During the intial pressurizing operation, stop valves 1, 2, 3 and 4 are all open and the desired test pressure is generated by the intensifier and controlled by means of the dead-weight pressure gauge. A microswitch, fitted to the dead-weight gauge, controls the electrical power input to the drive pump of the intensifier. Thus, when the test pressure is reached the pump stops automically.

Upon completion of the initial pressurizing operation, stop valves 1 and 2 are closed. Thus, the dead-weight pressure gauge and the intensifier are eliminated from the system and the test pressure is now registered by pressure gauge A which is fitted with two electrical contacts C and D. To allow for the gradual increase in internal volume of the system, due to creep of the specimen, a pressure compensator of the type shown in Fig. 4 is incorporated. The compensator is essentially an oil reservoir which is fitted internally with a 250 Watt, 110 Volt AC heating element. The electrical contact C on pressure gauge A is wired into the circuit of an electronic relay which, when closed, supplies power to the heating element. If the test pressure should fall below the correct value, power is thus automatically switched to the internal heater, slightly heating and expanding the fluid in the compensator, and thereby restoring the system to the correct test pressure. The pressure gauge contact C then opens and the heater is switched off. The continual making and breaking of the heater circuit results in a very slight cycling of pressure. However, with proper adjustment of the contacts and by using an extremely sensitive electronic relay, it has been possible to keep this cycling to $\pm$ 0.1% of the test pressure.

In the event of the compensator overheating, stop valve 2 is opened and the test pressure is controlled by the intensifier which is operated by means of the electrical contact E on pressure gauge B. When the compensator is sufficiently cool, stop valve 2 is closed and the compensator in conjunction with pressure gauge A is again used to control pressure. Should a fault occur in the compensator (such as a leak at the end plug or a breakdown of the heating element) then it can be isolated from the system by closing stop valve 4. Stop valve 2 is then opened and the test pressure controlled by the intensifier used in conjunction with pressure gauge B. The compensator can then be disconnected from the rig in order to facilitate any repair work. To reintroduce the compensator into the circuit, stop valve 3 is briefly closed. The compensator can then be installed and, with stop valve 4 open, it can be repressurized by means of the intensifier. With this achieved it is only necessary to close stop valve 2 and then open stop valve 3, the test pressure again being controlled by the compensator used in conjunction with pressure gauge A.

4.2 Oven: Temperature Control and Measurement

The oven is a heavy duty forced convection type (No. OV 300) supplied by A. Gallenkamp & Co. Ltd. The maximum operating temperature is about 370°C and a 'Satchwell' thermostat is used for temperature control.

Specimen temperature is measured by means of Platinum/Platinum + 10% Rodium thermocouples tied along the specimen gauge length. In the initial test runs, four thermocouples were used and it was found that even over prolonged periods the specimens temperature could be held at 350°C $\pm$ 1°C. Moreover, since the temperature distribution along the gauge length was found to be extremely good, it is now normal practice to use only two thermocouples located at each end of the gauge length.

4.3 Specimen Assembly

Fig. 5 shows a detail drawing of the test specimen (K = 1.667) and Fig. 6 shows the specimen assembly. Sealing at the ends of the specimen is effected by means of wave ring joints. The wave rings are made from a high-strength aluminium bronze (hydurax). One end of the specimen is blanked off by means of an end cap while the other end is fitted with a pressure supply pipe. The specimen assembly is mounted horizontally in the oven and the fluid supply pipe passes through a hole in the oven wall. The close fitting core bar greatly reduces the internal fluid volume of the specimen.

4.3 Extensometry

Since the door of the Gallenkamp oven is fitted with a window, it was decided to use optical extensometers for the measurement of creep strains. Four extensometers are fitted: two to measure axial strains and two for circumferential strain measurement.

The axial extensometers are of the 'Martens' type and utilize the optical lever principle. They are mounted on the specimen as shown in Fig. 7 and are held in position by means of Nimonic 90 spring clips. To prevent the possibility of slipping, the extensometer knife edges are located in sharp, but extremely shallow, notches in the surface of the specimen. All the mirrors are made from En58B stainless steel and are quite suitable for use over prolonged periods at 350°C.

Fig. 8 shows the diametral extensometers which are mounted at the centre section of the specimen on diameters which are at 90° to each other. The extensometer limbs are held on the specimen by means of Nimonic 90 springs, the ball seats of the limbs being located in four small indentations on the surface of the specimen. These spherical indentations are positioned on the specimen using a dividing head.

Both the axial and diametral extensometers can detect strains of 1 x 10^{-5}.

4.4 Safety Precautions

In the event of a specimen failure or even a serious leak occurring at one of the wave ring seals, high pressure oil at 350°C would be sprayed into the oven. This presents a distinct fire hazard and it was felt that the risk of damage should be minimized. Obviously a serious leak or specimen failure will cause a sharp drop in pressure. Should this occur the pointer on pressure gauge A will cross completely over contact C and will come to rest against contact D. The making of this contact actuates two relays which perform the following functions:-

(a) Cut power supply to pump motor and pressure compensator
(b) Cut power supply to oven
(c) Release the CO_2 extinguisher into the oven (see Fig. 3).

5. MATERIAL

The investigation was conducted on forged EN25 steel heat treated to give an ultimate tensile strength at room temperature of approximately 50 tonf/in^2 (770 MN/m^2). Table 1 summarizes the results of two tension tests conducted at 350°C. At this temperature the ultimate tensile strength is reduced to about 45.5 tonf/in^2 (700 MN/m^2).

0.2% Proof Stress	TONF/IN^2	32.9	32.9
	MN/m^2	508	508
U.T.S.	TONF/IN^2	45.32	45.60
	MN/m^2	700	704
YOUNG'S MODULUS	TONF/IN^2	12,220	12,240
	GN/m^2	189	189
% ELONGATION		12.3	15.8
% REDUCTION OF AREA		41.34	66.11

TABLE 1 TENSILE TEST DATA FOR FORGED EN25 STEEL AT 350°C

6. CREEP DATA AT 350°C

6.1 Tension Creep Data

Table 2 summarizes the details of the six tension creep tests completed to date. The test specimens were machined in the transverse direction relative to the longitudinal axis of the pressure creep specimens. The hot loading curves for these tests are shown in Fig. 9 and the creep strain/time curves are shown in Fig. 10. The nominal stress levels cover the range 30-38 tonf/in^2 (463.5-587.1 MN/m^2). Specimens B1/3 and B1/4 were both tested at 30 tonf/in^2 (463.5 MN/m^2) as a check on repeatability and, as can be seen from Fig. 10, the creep strain/time curves are in quite good agreement. The 50 tonf/in^2 (772.5 MN/m^2) curve shown in Fig. 10 was a preliminary test conducted on EN25 bar material.

All of these tests were conducted in constant load machines and as a consequence there is a continuous increase in stress associated with the gradual reduction in cross-sectional area as each test proceeds. Table 2 includes values for the average stress in each test this value being calculated from a knowledge of the initial and final strains. At the higher stress levels the difference between the nominal and average stress is very significant.

6.2 Torsion Creep Data

Table 3 summarizes the details of the six torsion creep tests completed to date. The hot loading curves for these tests are shown in Fig. 11 and the creep strain/time curves are shown in Fig. 12. The tests cover the shear stress range 18.18-22.11 tonf/in^2 (280.1-341.6 MN/m^2). The test specimens were machined in the longitudinal direction.

6.3 Pressure Creep Data

Table 4 summarizes the details of the four pressure creep tests conducted to date. The hot loading curves for these tests are shown in Fig. 13 and the plots of diametral creep strain and axial creep strain versus time are shown in Figs. 14 and 15 respectively.

PRESSURE SPECIMEN NO.		C1	C2	C3	C4
TEST PRESSURE	TONF/IN^2	21	20	22	21
	M Pa	324.4	309.0	339.9	324.4
% INITIAL PLASTIC DIAMETRICAL STRAIN ON O.D.		.800	.636	1.25	.815

TABLE 4 DETAILS OF PRESSURE CREEP TESTS

The test on specimen C2 at 20 tonf/in^2 (309 M Pa) was stopped after 1700 hours because of a leak in the pressure compensator. During the test on specimen C1 at 21 tonf/in^2 (324.4 M Pa) an electrical fault caused the contacts on the pressure gauge to weld together. This caused continuous heating of the pressure compensator and as a consequence the test pressure rose sharply leading to premature failure of the specimen after 835 hours. A photograph of the fracture is shown in Fig. 16a. A further test was conducted at 21 tonf/in^2 (324.4 M Pa) on specimen C4. This test ran very smoothly and was terminated after 2519 hours. As can be seen from Fig. 14, the two test results at 21 tonf/in^2 324.4 M Pa) pressure are in reasonable agreement although the secondary creep rates are somewhat different.

Specimen C3 was tested at an internal pressure of 22 tonf/in^2 (339.9 M Pa). After 320 hours the extensometry was removed from the specimen to avoid any damage due to specimen fracture. The specimen fractured after 540 hours. Figure 16b shows a photograph of the fracture.

In all four tests the axial creep strains. (Fig. 15) were positive but very small compared to the diametral creep strains (Fig. 14) i.e. (about 2% of the diametral strains). The axial strain readings gave no indication of a typically shaped creep curve i.e. (primary, secondary regions, etc.) despite the highly sensitive extensometers used. So far as could be determined the long term trend was fairly linear as shown in Fig. 15, with very small deviations often occurring from day to day. These results certainly support the validity of the assumption of zero axial creep strain.

7. CORRELATION OF TENSION, TORSION AND PRESSURE CREEP DATA

7.1 Bailey Theory

Bailey (10) (11) has published a relatively simple theory to predict the creep strain behaviour of thick-walled cylinders under internal pressure. The theory neglects elastic strains and it is assumed that the axial creep strain is zero. The effect of wall thinning at large strain values was ignored. Using the Bailey solution the following equation may be derived for the circumferential strain at the outside diameter:-

$$\varepsilon_\theta = \frac{A}{m}\left[\frac{P}{(n+1)(K^{\frac{2}{n+1}}-1)}\right]^{n+1} t^m \qquad \ldots (1)$$

Applying the Bailey theory to constant stress tension creep tests the creep strain is given by:-

$$\varepsilon_t = \frac{2A}{m.3^{\frac{n+2}{2}}}\sigma_t^{n+1} t^m \qquad \ldots (2)$$

In the case of torsion creep data the Bailey theory gives the shear creep strain as:-

$$\gamma = \left[\frac{2A}{m}\right]\tau^{\overline{n+1}} t^m \qquad \ldots (3)$$

Taking logs throughout equation (2) gives:-

$$\log \varepsilon_t = \log\left\{\frac{2A}{m.3^{\frac{n+2}{2}}}\right\} + \overline{n+1}\log\sigma_t + m.\log.t \qquad \ldots (4)$$

Similarly, taking logs throughout equation (3) gives:-

$$\log\gamma = \log\left\{\frac{2A}{m}\right\} + \overline{n+1}\log\tau + m.\log.t \qquad \ldots (5)$$

Thus, using either tension or torsion data, if log stress is plotted against log strain (with time constant) the slope of the linear portion of the graph yields a value of $\overline{n+1}$. Also, by plotting log time against log strain (with stress constant) a value of 'm' may be obtained. Knowing values for 'm' and 'n', the third constant 'A' can be found.

Bailey applied these equations to all three stages of creep by varying the value of the constant 'm'. More recently Larke and Parker (12) have shown how the Bailey equations can be used very successfully to cover all three creep regions by evaluating the constants 'A', 'm', and 'n' for each region. To date no attempt has been made to apply this latter method and instead only average values for the material constants covering the primary and secondary creep regions have been evaluated.

Figures 17 and 18 show log plots of the tension creep data. Both plots exhibit significant non-linearity which may be partly explained by the gradual increase in stress which is occurring as these tests progress. The plots yielded the following average values for the material constants 'A', 'm' and 'n':-

$$A = 3.34 \times 10^{-17}$$
$$m = 0.25$$
$$n = 9.0$$

and so the equation for tension creep strain can be written as:-

$$\varepsilon_t = 6.35 \times 10^{-19}\sigma_t^{10} t^{0.25} \qquad \ldots (6)$$

where σ_t is in tonf/in^2 and t is in hours. The theoretical curves predicted by equation (6) for the stress range 30-36 tonf/in^2 (463.5 - 556.2MN/m^2) are compared with the corresponding experimental curves in Fig. 10. In order to make some allowance for the gradual increase in stress which occurs in these tests, the theoretical curves are based on the average stress level throughout each test as given in table 2. The correlation is considered to be fairly good up to 36 tonf/in^2 (556.2 MN/m^2) but for higher stresses the discrepancy between the theoretical and experimental curves is considerable.

Figures 19 and 20 show log plots of the torsion creep data. These plots gave the following average values for the material constants 'A', 'm' and 'n':-

$$A = 2.9995 \times 10^{-13}$$
$$m = 0.219$$
$$n = 6.11$$

Thus the equation for the torsion creep strain can be written as:-

$$\gamma = 2.1 \times 10^{-12}\tau^{7.11} t^{0.219} \qquad \ldots (7)$$

where τ is in tonf/in^2 and t is in hours. The theoretical curves predicted by equation (7) are compared with the corresponding experimental curves in Fig. 12. With the exception of the test at 22.11 tonf/in^2 (341.6 MN/m^2) the correlation is considered good.

Clearly the tension and torsion creep data should yield the same values for the material constants 'A', 'm' and 'n'. The derived values for the constant 'm' are in reasonable agreement (0.25 for tension and 0.219 for torsion) but the values for the constant 'n' are in violent disagreement with a value of 9.0 for the tension data and 6.11 for the torsion data. The tension value is considered to be remarkably high but as can be seen from Fig. 10 this leads to a reasonable correlation over the stress range investigated. Largely as a result of this extremely high value for 'n' the constant 'A' derived from the tension data differs from the torsion derived value by a factor of approximately 6885, (i.e. 3.34 x 10^{-17} for tension and 2.2995 x 10^{-13} for torsion).

Using the torsion derived material constants

the following equation can be obtained for predicting the tension creep behaviour:-

$$\varepsilon_t = 2.44 \times 10^{-14} \sigma^{7.11} t^{0.219} \qquad \ldots (8)$$

Figure 21 compares the theoretical tension creep curves predicted by equation (8) with those predicted by equation (6) which is based on the tension-derived constants. Equation (6) predicts lower values for the creep strains at stress levels up to 34 tonf/in^2 (525.3 MN/m^2) and a higher creep strain at a stress of 36 tonf/in^2(556.2 MN/m^2). On face value it seems incredible that equations (6) and (8) can result in predictions for creep strains which are in such close agreement, however, this is clearly brought about by the interplay of the constants 'A' and 'n' over the fairly limited stress range considered.

Figure 22 compares the experimental pressure creep curves with the theoretical plots based on the Bailey analysis. For each test there are three theoretical plots. Two of these plots are the original Bailey analysis based on the material constants derived from the tension and torsion creep data. The third plot is a modified Bailey analysis described previously by Patton et al. (5). This analysis takes account of wall thinning and in this instance the torsion constants have been used. For test pressures of 21 and 22 tonf/in^2 (324.4 and 339.9 MPa.) the modified Bailey solution gives a much better correlation and as is to be expected the standard Bailey solutions (based on both tension and torsion constants) predict smaller creep strains. For the test at 20 tonf/in^2(309 MPa) pressure the modified Bailey curve lies significantly above the experimental plot. This is a somewhat surprising result which so far is without explanation. In all cases the basic shape of the theoretical curves is somewhat different from the shape of the experimental plots. This trend suggests a discrepancy in the value of 'm'.

7.2 Correlations Using Isochronous Shear Stress-Strain Curves

Manning (13) proposed a method for predicting the pressure-expansion curve of a metallic thick-walled cylinder at room temperature which takes account of wall thinning at large plastic strains. Patton et al. (12) have shown how this method may be extended for use under time dependent conditions. From a family of torsion creep curves a series of isochronous shear stress-shear strain curves can be constructed. Figure 23 shows a set of isochronous curves which have been derived from the torsion creep data given in Fig. 12 together with hot loading curve given in Fig. 11. If it is assumed that, no matter how the shear stress in the cylinder wall may vary with time, at a given time there is a unique relationship between stress and strain, then it is possible, using the Manning analysis, to predict a series of isochronous pressure-expansion curves for a thick-walled cylinder of any desired diameter ratio. Figure 24 shows the isochronous pressure-expansion curves derived from the isochronous shear stress -shear strain curves (Fig. 23). These curves can be used to construct the total strain-time curve for a particular value of internal pressure. The horizontal line AB in Fig. 24 gives the total strain values at various times for an internal pressure of 20 tonf/in^2 (309 MPa).

Parker (14) has developed a somewhat similar analysis for deriving isochronous pressure-expansion curves. This analysis assumes that the shear strain distribution across the cylinder wall is elastic and no allowance is made for wall thinning. Thus, for any particular value of circumferential strain at the outside diameter occurring at a particular time, the appropriate isochronous shear stress - shear strain curve (Fig. 23) can be used to determine the shear stress distribution across the wall and from this the required internal pressure can be calculated. Figure 25 shows the isochronous pressure-expansion curves derived using the Parker analysis.

Figure 26 compares the theoretical predictions based on the modified Manning analysis (P = 20 tonf/in^2) and the Parker analysis (P = 20 and 21 tonf/in^2) with the corresponding experimental curves. At a pressure of 20 tonf/in^2 the Parker analysis is in reasonable agreement with the experimental curve but predicts somewhat lower strain values. At a pressure of 21 tonf/in^2 the Parker curve again lies below the experimental curve and the discrepancy between the two plots increases with increasing time. It should be noted however, that the Parker analysis specifically relates to cases where the total circumferential strain is not sufficiently large to cause any significant wall thinning and with this in mind the trend displayed in Fig. 26 is readily explained.

The modified Manning curve at 20 tonf/in^2 pressure lies somewhat below the Parker curve and is also in reasonable agreement with the experimental plot. Unfortunately the isochronous shear stress - shear strain curves (Fig. 23) do not cover a sufficiently high stress range in order to calculate the modified Manning curve at 21 tonf/in^2 pressure. Clearly in this case the modified Manning curve would lie above the Parker curve but most likely would still be below the experimental plot.

7.3 Mean-Diameter Theory

A very simple method for estimating the creep strain occurring in a thick-walled cylinder is to treat it as being thin-walled and to calculate the equivalent shear stress at the mean diameter. This is very similar to an approach developed by Soderberg (15) for the secondary creep of cylinders based on tension creep data.

The shear stress at the mean diameter of a thin-walled cylinder can be written as:-

$$\tau_m = \frac{P}{4}\left(\frac{K+1}{K-1}\right) \qquad \ldots (9)$$

The corresponding shear strain γ_m is twice the circumferential strain, and thus, if a single torsion creep test is run at a shear stress given by equation (9) the circumferential creep strain at the mean diameter of a thick-walled cylinder will be given approximately by:-

$$\varepsilon_{\theta m} = \frac{\gamma m}{2} \qquad \ldots (10)$$

Assuming constancy of volume and zero axial creep the circumferential creep strain at the outside diameter can be written as:-

$$\varepsilon_\theta = \frac{\gamma m}{8}\left(\frac{K+1}{K}\right)^2 \qquad \ldots (11)$$

This method of analysis is clearly a gross oversimplification of the problem. However, Coffin et al (16) have shown that there is a radius in the cylinder wall where the effective stress or the second stress invariant J_2 remains more or less constant, and this has been proposed independently by Marriott and Leckie (17). This being the case, then a single creep test conducted at this effective stress should be adequate to predict the creep of a thick-walled cylinder under internal pressure.

In applying the mean-diameter analysis it was decided to work on a total strain basis using equation (11) to express the total circumferential strain at the outside diameter in terms of the total shear strain at the mean diameter:-

$$\varepsilon_\theta = \frac{\gamma_m}{8} \left(\frac{K+1}{K}\right)^2 \qquad ...(12)$$

Thus for any particular value of internal pressure and diameter ratio, equation (9) yields a value of the shear stress at the mean diameter. The isochronous shear stress - shear strain curves (Fig. 23) can then be used to give values of the total shear strain at the mean diameter for various times and equation (12) can then be used to predict corresponding values for the total circumferential strain at the outside diameter. The effect of wall thinning at large strains must also be taken into account in applying equations (9) and (12). To date this has been done very crudely by simply using an average K value for each pressure creep test which has been calculated from the experimental circumferential strain data. Fig. 26 shows the strain-time curves predicted by this approach for pressures of 20 and 21 tonf/in^2. The curve for 20 tonf/in^2 pressure is based on an average K ratio of 1.636 for the first 1600 hours of the test (specimen C2). The curve for 21 tonf/in^2 pressure is based on an average K ratio of 1.621 for the first 900 hours of the test (specimen C4). Both these curves are in excellent agreement with the corresponding experimental plots.

It is worth noting that these preliminary calculations depended upon a knowledge of the amount of wall thinning which occurred. It would seem that this problem can be readily overcome by using equations (9) and (12) in conjunction with an incremental strain solution which takes account of any significant wall thinning by updating the value of K ratio for each time interval. The technique developed by Patton et al (5) for use with the Bailey analysis could be adapted for this purpose. This approach is currently being evaluated.

8. CONCLUSIONS

(i) A pressure creep rig for operation at pressures up to 450 MPa and temperatures up to 350°C has been developed.

(ii) In all four pressure creep tests the axial creep strains were negligible and it is therefore concluded that the simplifying assumption of zero axial creep strain, made in most of the creep theories, is reasonable.

(iii) The preliminary correlation work using simple time hardening equations has not been entirely satisfactory and it is felt that a more detailed analysis of the data using the methods developed by Larke and Parker (12) is required.

(iv) The effect of wall thinning at large strains must be taken into account in developing a satisfactory correlation. Consequently the modified Bailey analysis and the modified Manning analysis give more accurate strain predictions at large strains than the standard Bailey solution and the Parker analysis. At lower strains the Parker analys is slightly better than the modified Manning approach.

(v) The mean diameter analysis with allowance made for wall thinning is most encouraging. Work is now in progress to incorporate this approach into an incremental strain solution.

9. ACKNOWLEDGEMENTS

The authors would like to acknowledge the advice and guidance given by Professor B. Crossland throughout the work.

The work was carried out in the Mechanical Engineering Laboratories of the Queen's University of Belfast and thanks are due to the staff of the Departmental Workshop and in particular Mr. R. H. Agnew and Mr. N. Alexander who assisted in the experimental work.

Appendix 1

References

(1) SKELTON, W.J. and CROSSLAND, B. 'Creep of thick walled cylinders subjected to internal pressure: review of literature' Conference on High Pressure Eng. 1967, Proc. Institution of Mechanical Engineers 1967-1968, Vol. 182 pt. 3C pp. 25-31

(2) BAILEY, R. W. 'Thick walled tubes under high pressures and temperatures' Engineering, Lond. 1930 129, 772; 785; and 818.

(3) NORTON, F.H. and SODERBERG, C.R. 'Report on tubular creep tests', Trans. Am. Soc. Mech. Engrs. 1942, 64, 769.

(4) SKELTON, W.J. and CROSSLAND, B. 'Creep of thick walled cylinders: development of a testing machine and preliminary results' Conference on High Pressure Eng. 1967, Proc. Institution of Mechanical Engineers 1967-1968, Vol. 182 pt. 3C pp. 139-150.

(5) PATTON, R.G., SKELTON, W.J. and CROSSLAND, B. 'Creep of thick walled cylinders based on torsion creep data for 0.18% carbon steel at 400°C' Pressure Vessel Design Conference Delft, Sept. 1969, Joint A.S.M.E. and Dutch Engineering Society Conference.

(6) SALIM, A. 'Creep of metals under constant and variable stress in relation to the creep of cylinders under internal pressure. Ph.D. thesis (The Queen's University of Belfast) November 1972.

(7) CROSSLAND, B. and KERR, W.F.K. 'The strength of thick walled steel cylinders at elevated temperatures' High Temperatures - High Pressures 1969, Vol. 1. 133-152.

(8) SKELTON, W.J. and CROSSLAND, B. 'Results of High Sensitivity Tensile Creep Tests on a 0.19% Carbon Steel Used for Thick Walled Cylinder Creep Tests' Conference on High Pressure Eng. 1967. Proc. Institution of Mechanical Engineers 1967-1968, Vol. 182 pt. 3C pp. 17-24.

(9) CROSSLAND, B., PATTON, R.G. and SKELTON, W.J., 'The Comparison of Torsion and Tension Creep Data for a 0.18% Carbon Steel' Advances in Creep Design. The A.E. Johnson Memorial Volume, Chapter 8. (Applied Science Publishers Ltd.)

(10) BAILEY, R.W., 'Steam piping for high pressures and high temperatures' Proc. Institution of Mechanical Engineers 1951 Vol. 164, 324-350.

(11) BAILEY, R.W., 'Creep relationships and their application to pipes tubes and cylindrical parts under internal pressure', Proc. I.Mech. Eng., Vol. 164, 425-431.

(12) LARKE, E.C. and PARKER, R.J., 'Mathematical Relationships for Primary, Secondary and Tertiary Creep and Their Use in Extrapolation of Tensile Creep Design. The A.E. Johnson Memorial Volume, Chapter 7. (Applied Science Publishers Ltd.)

(13) MANNING, W.R.D., 'The Over-strain of Tubes by Internal Pressure', Engineering 1945, Vol. 159, pp. 101, 183.

(14) PARKER, R.J., 'Expansion and Deformation in Thick Cylinders', Chem. Proc. Eng., November 1967.

(15) SODERBERG, C.R., 'The Interpretation of Creep Tests on Tubes'. Trans. A.S.M.E., 1941, Vol. 63, p. 737.

(16) COFFIN, L.F., SHEPLER, P.R. and CHERNIAK, G.S., 'Primary Creep in the Design of Internal Pressure Vessels'. J.Appl.Mech. 1949, Vol. 16, p. 229.

(17) MARRIOTT, D.L. and LECKIE, F.A., 'Some Observations on the Deflection of Structures during Creep'. Conference on Thermal Loading and Creep in Structures, Proc. Inst. Mech. Eng. 1963, V. 178, part 3L, p. 115.

TABLE 2.

Details of Tension Creep Tests

TENSION SPECIMEN NO.		B1/3	B1/4	B3/3	B3/4	B1/2	B1/1
NOMINAL TEST STRESS	$TONF/IN^2$	30	30	32	34	36	38
	MN/M^2	463.5	463.5	494.4	525.3	556.2	587.1
AVERAGE TEST STRESS	$TONF/IN^2$	30.063	30.053	32.150	34.347	36.740	39.653
	MN/M^2	464.50	464.3	496.7	530.7	567.6	612.6
YOUNG'S MODULUS (ROOM TEMP.)	$TONF/IN^2$	12,830	13,130	12,630	12,810	13,250	13,220
	GN/M^2	198	203	195	198	205	204
YOUNG'S MODULUS (350°C)	$TONF/IN^2$	11,300	11,660	11,100	11,410	11,720	11,550
	GN/M^2	175	180	171	176	181	178
% INITIAL PLASTIC STRAIN		0.0525	0.0486	0.195	0.500	0.7444	2.78

TORSION SPECIMEN NO.		B4/2	B4/1	D4/1	B4/3	D4/2	B4/4
SHEAR STRESS	$TONF/IN^2$	18.18	19.94	19.99	21.11	21.50	22.11
	MN/M^2	280.1	308.1	308.9	326.1	332.2	341.6
SHEAR MODULUS (ROOM TEMP.)	$TONF/IN^2$	5,210	5,230	5,320	5,360	5,200	5,200
	GN/M^2	80.5	80.8	82.2	82.8	80.3	80.3
SHEAR MODULUS (350°C)	$TONF/IN^2$	4,600	4,920	4,690	4,710	4,600	4,510
	GN/M^2	71.1	76.0	72.5	72.8	71.1	69.7
% INITIAL PLASTIC STRAIN		0.270	0.734	0.857	1.460	1.75	2.270

TABLE 3 DETAILS OF TORSION CREEP TESTS

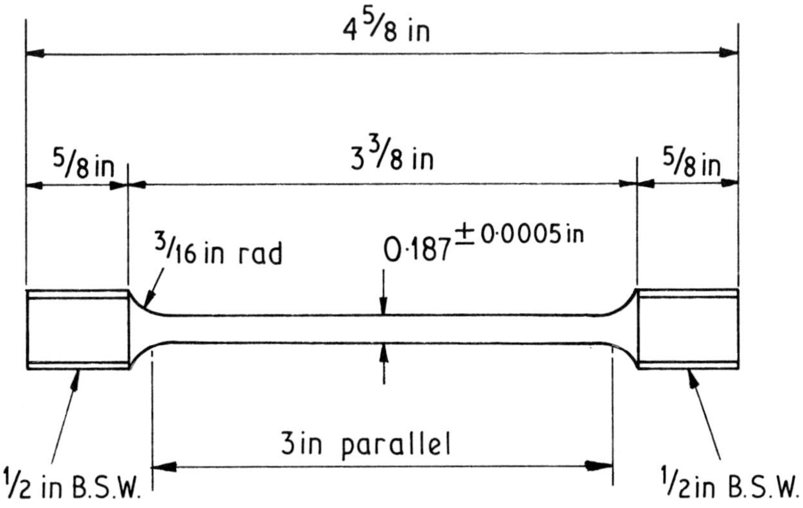

Fig. 1: Tension creep specimen for EN25

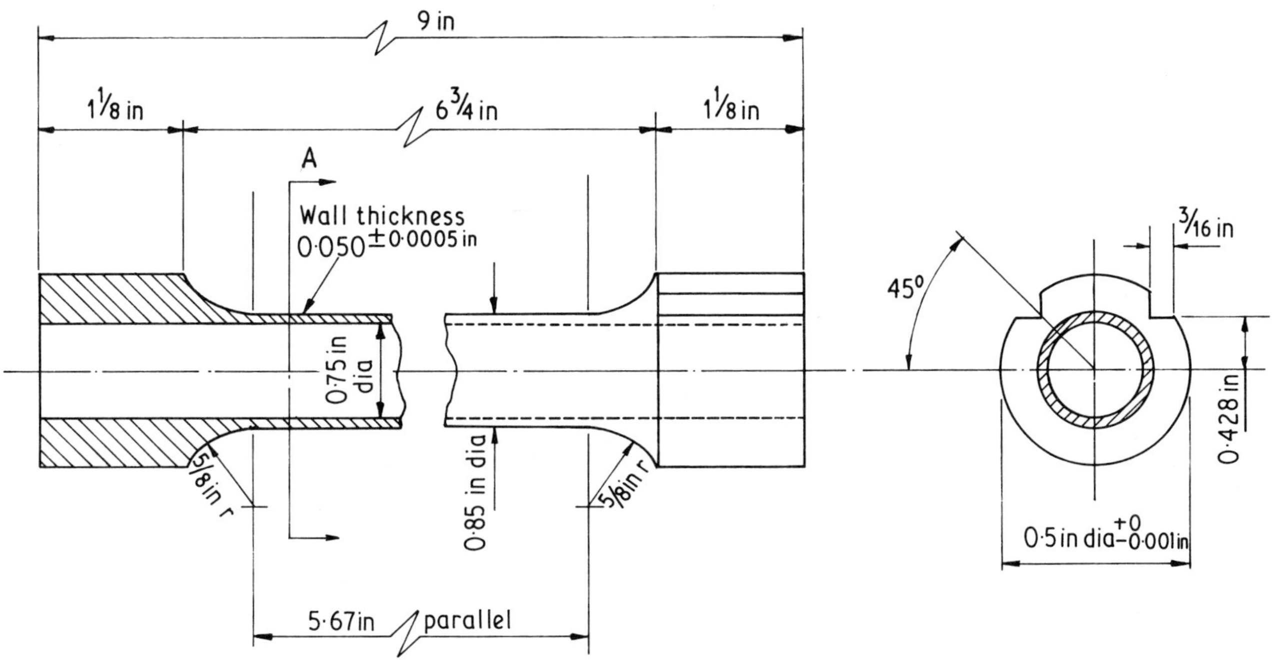

Fig. 2: Torsion creep specimen

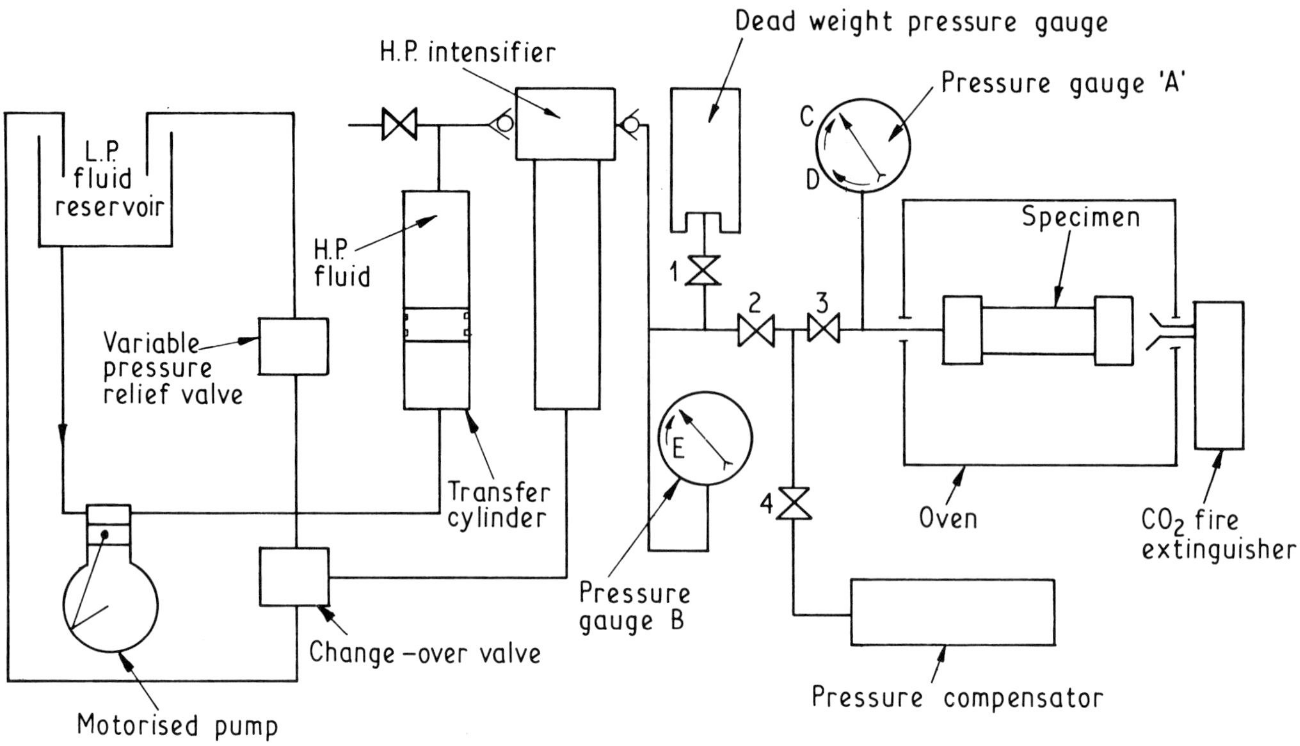

Fig. 3: Diagrammatic layout of high pressure creep rig

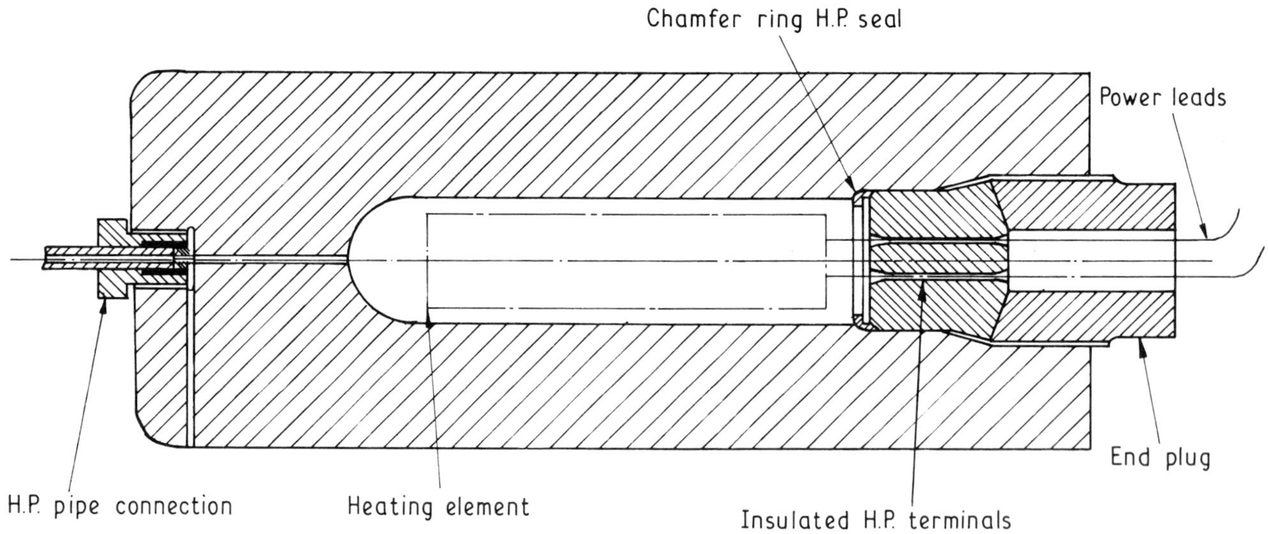

Fig. 4: Pressure compensator

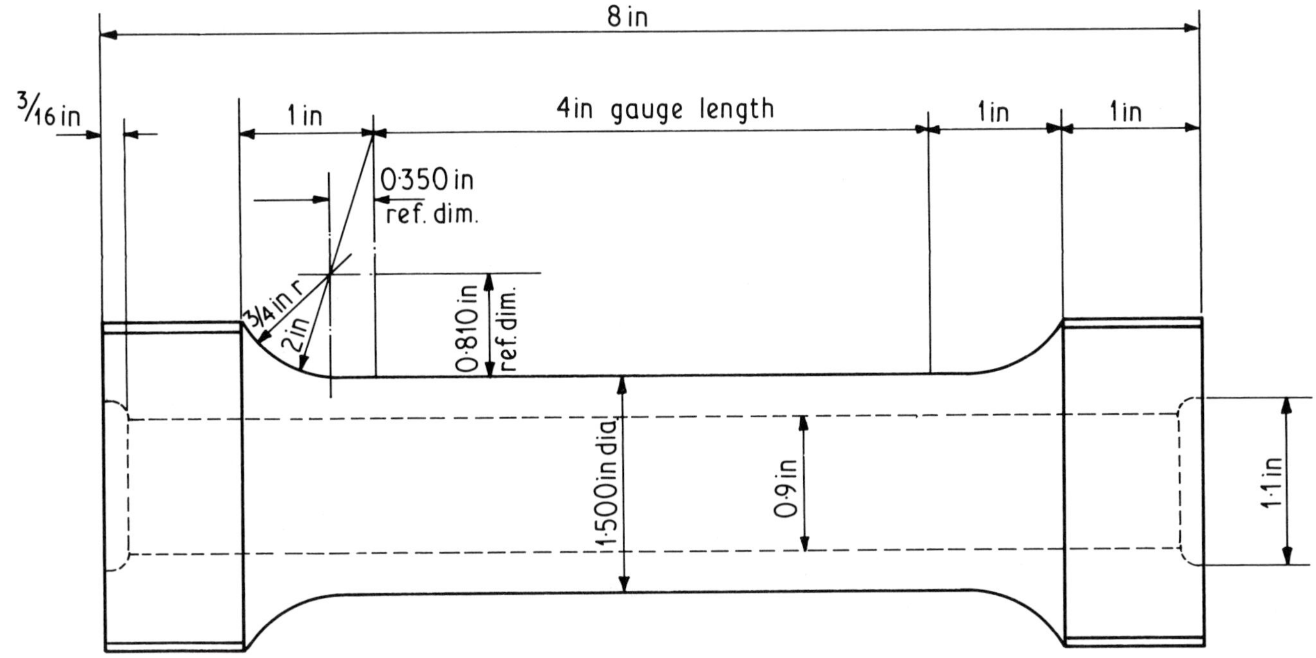

Fig. 5: Tubular creep specimen material EN25 K = 1.67
O.D. = 1.5

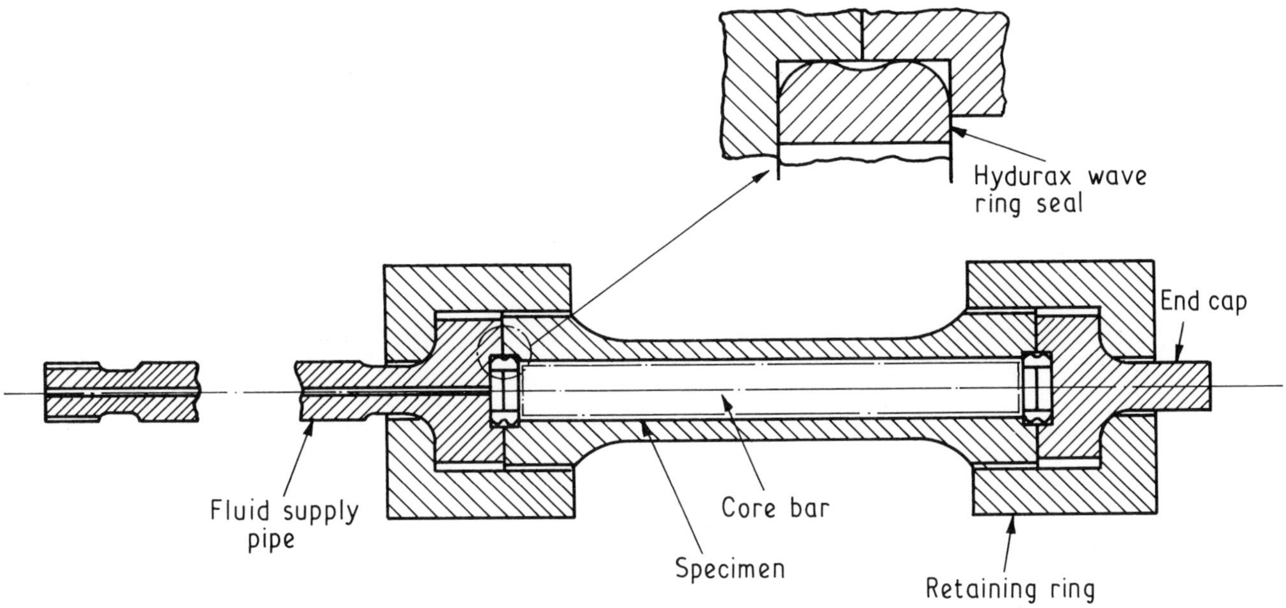

Fig. 6: Specimen Assembly

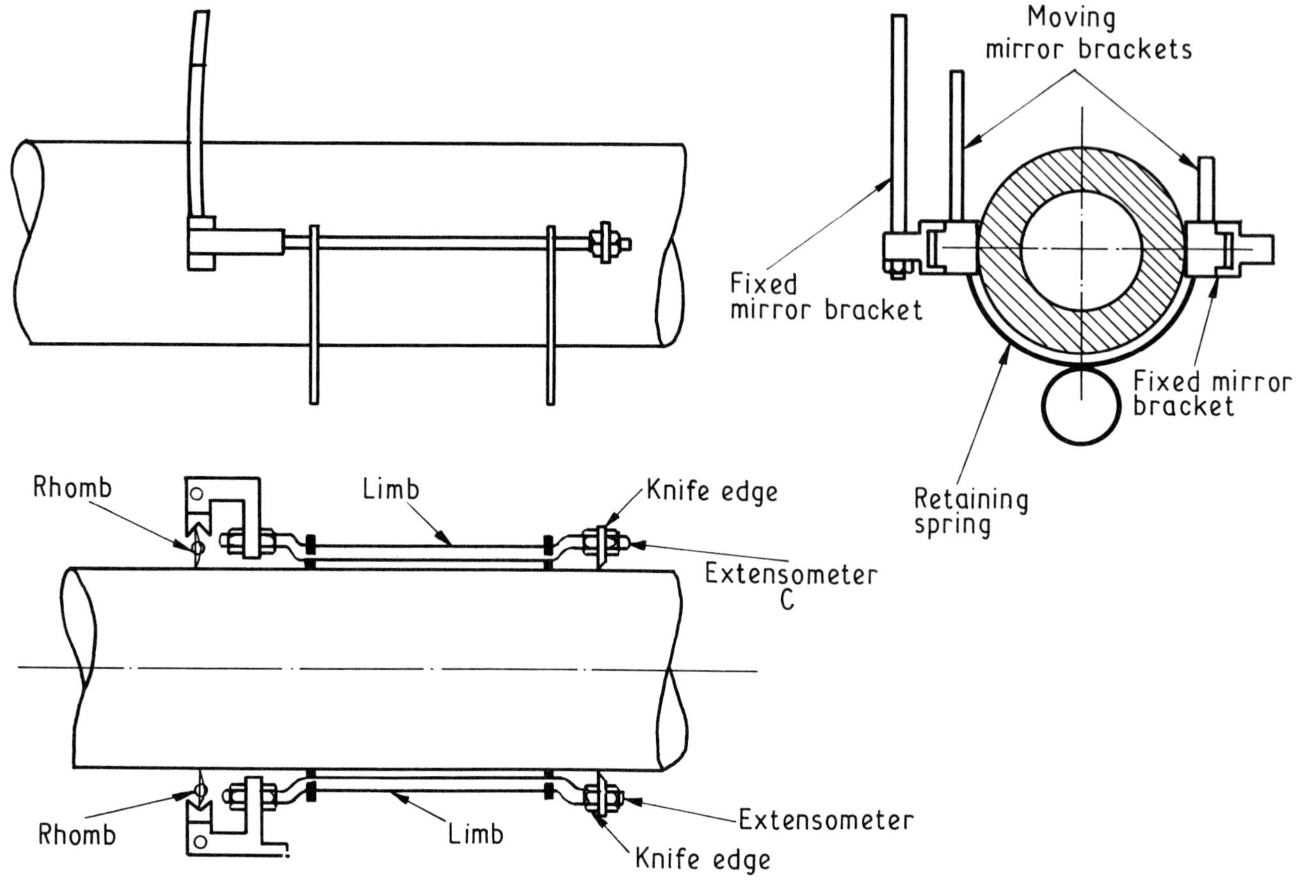

Fig. 7: **Axial creep extensometers**

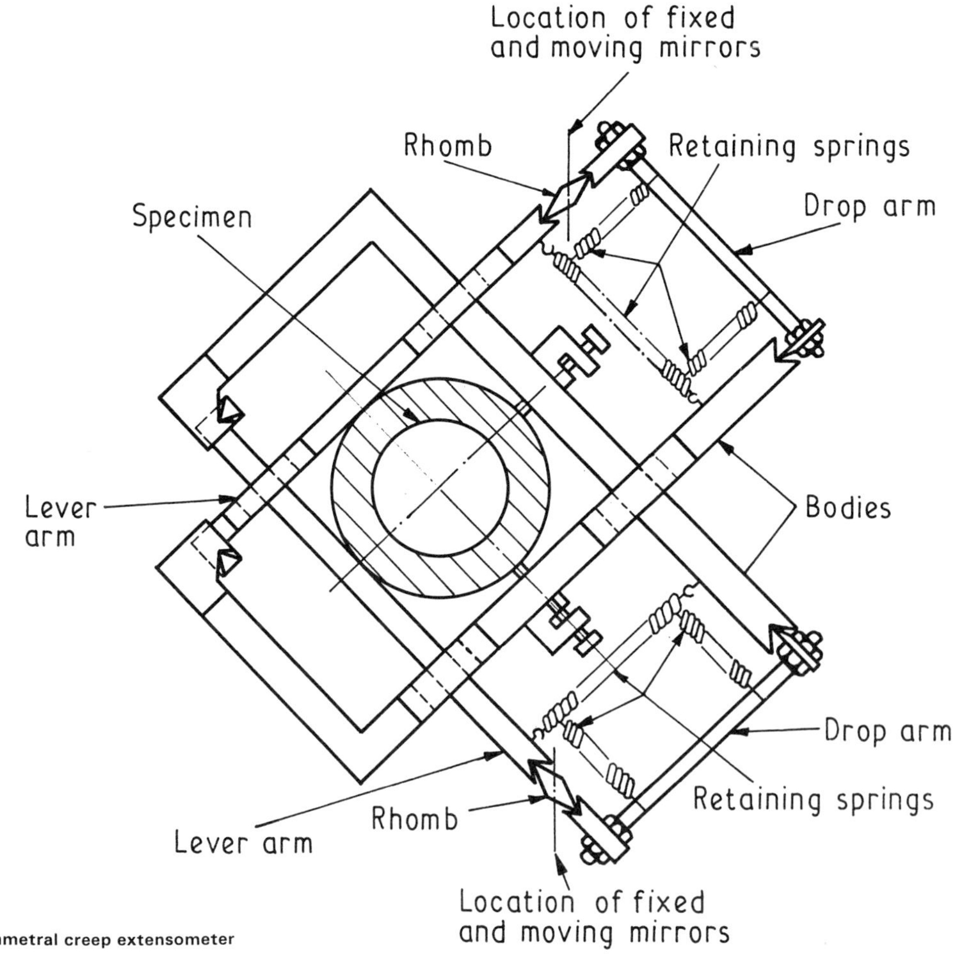

Fig. 8: **Diametral creep extensometer**

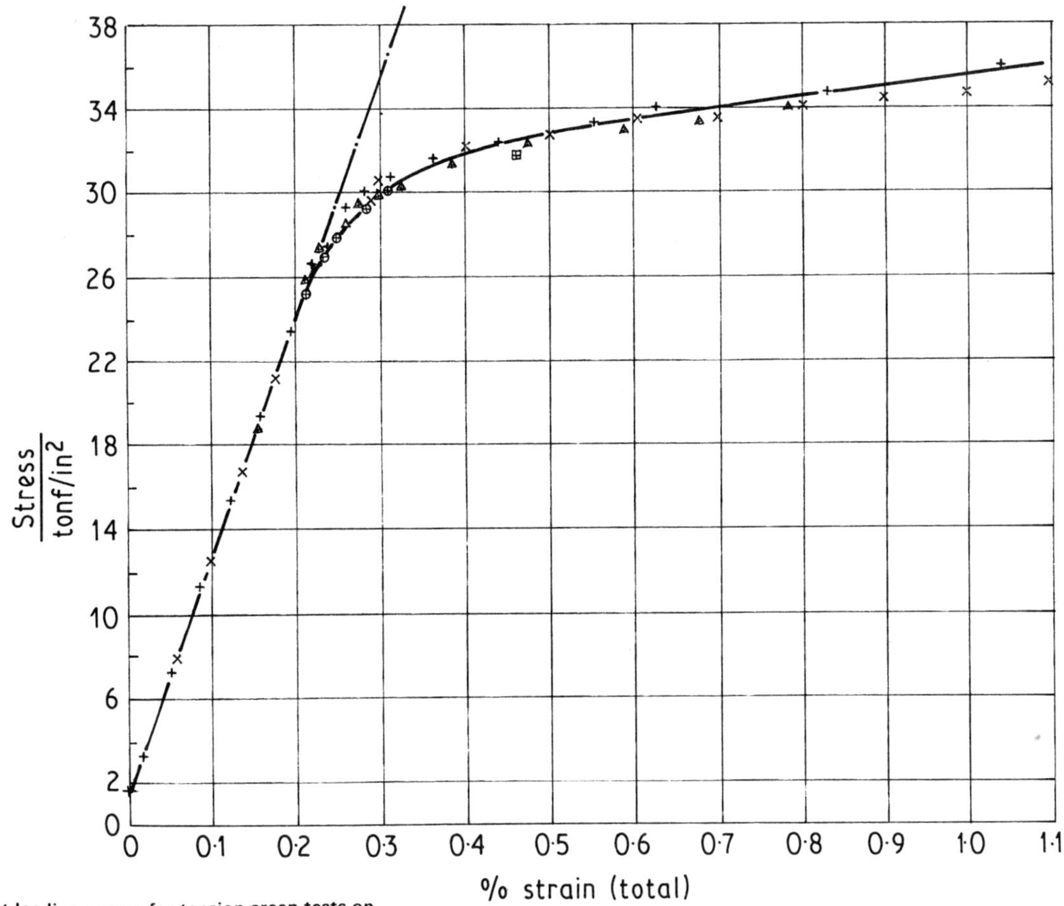

Fig. 9: Hot loading curves for tension creep tests on EN25 at 350°C

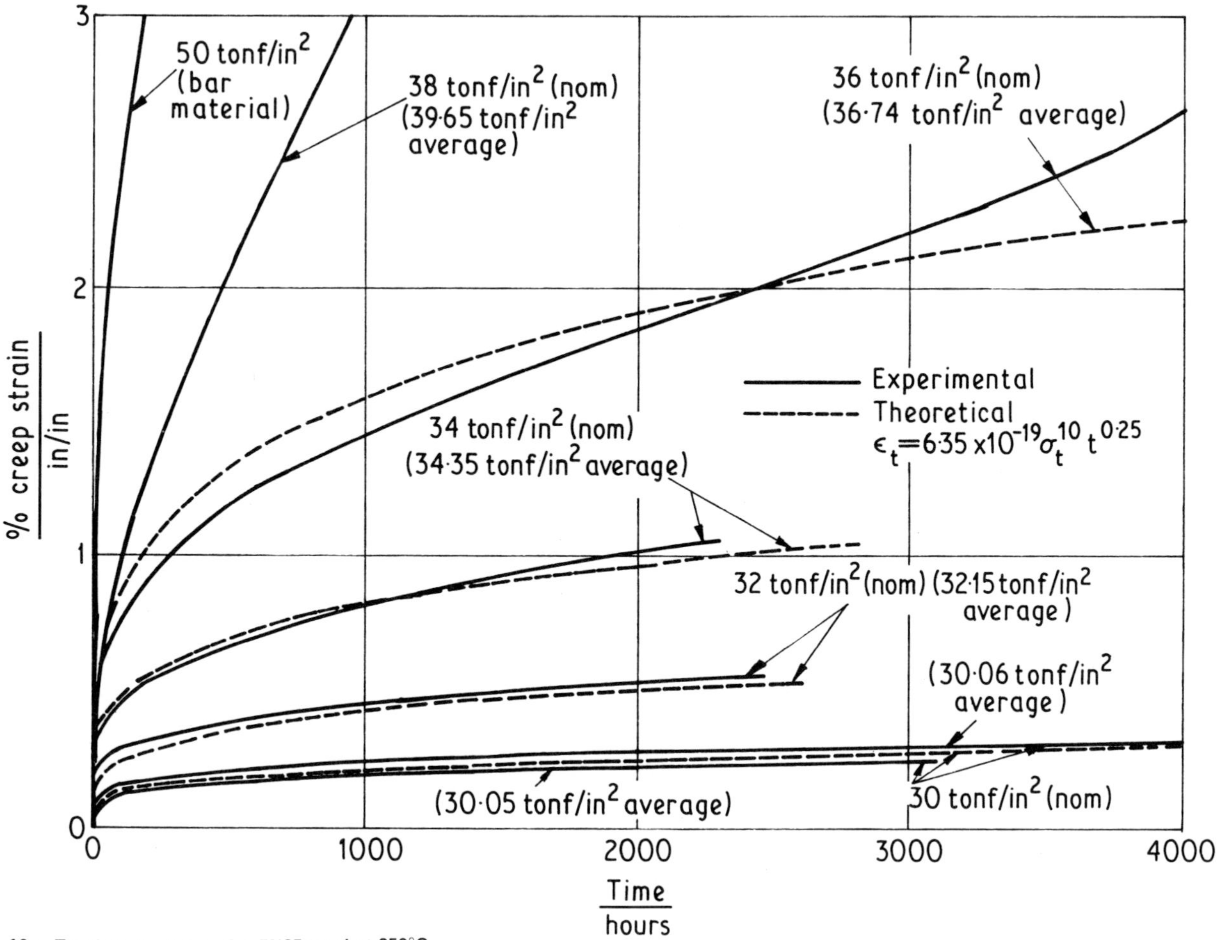

Fig. 10: Tension creep data for EN25 steel at 350°C

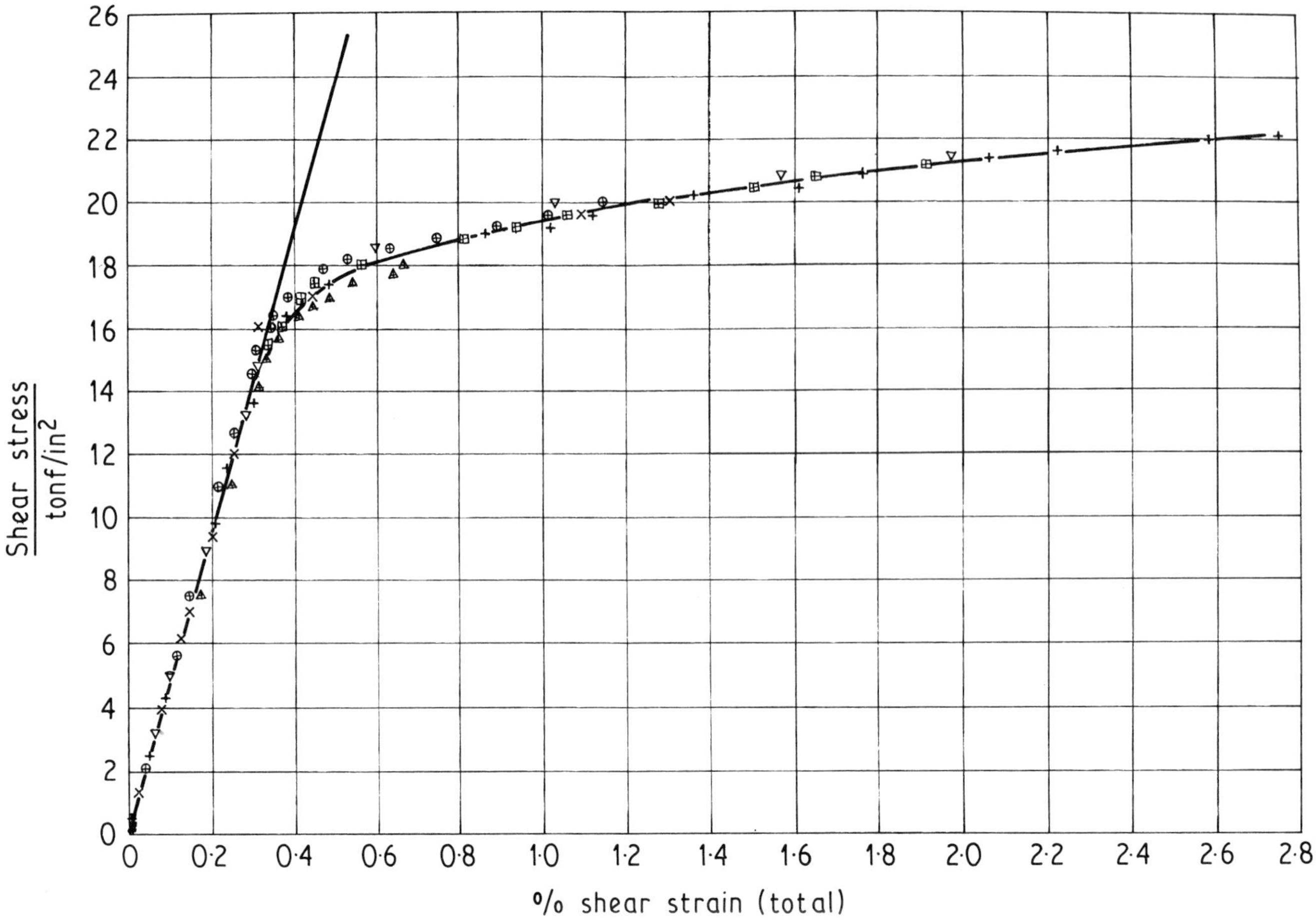

Fig. 11: Hot loading curves for torsion creep tests on EN25 at 350°C

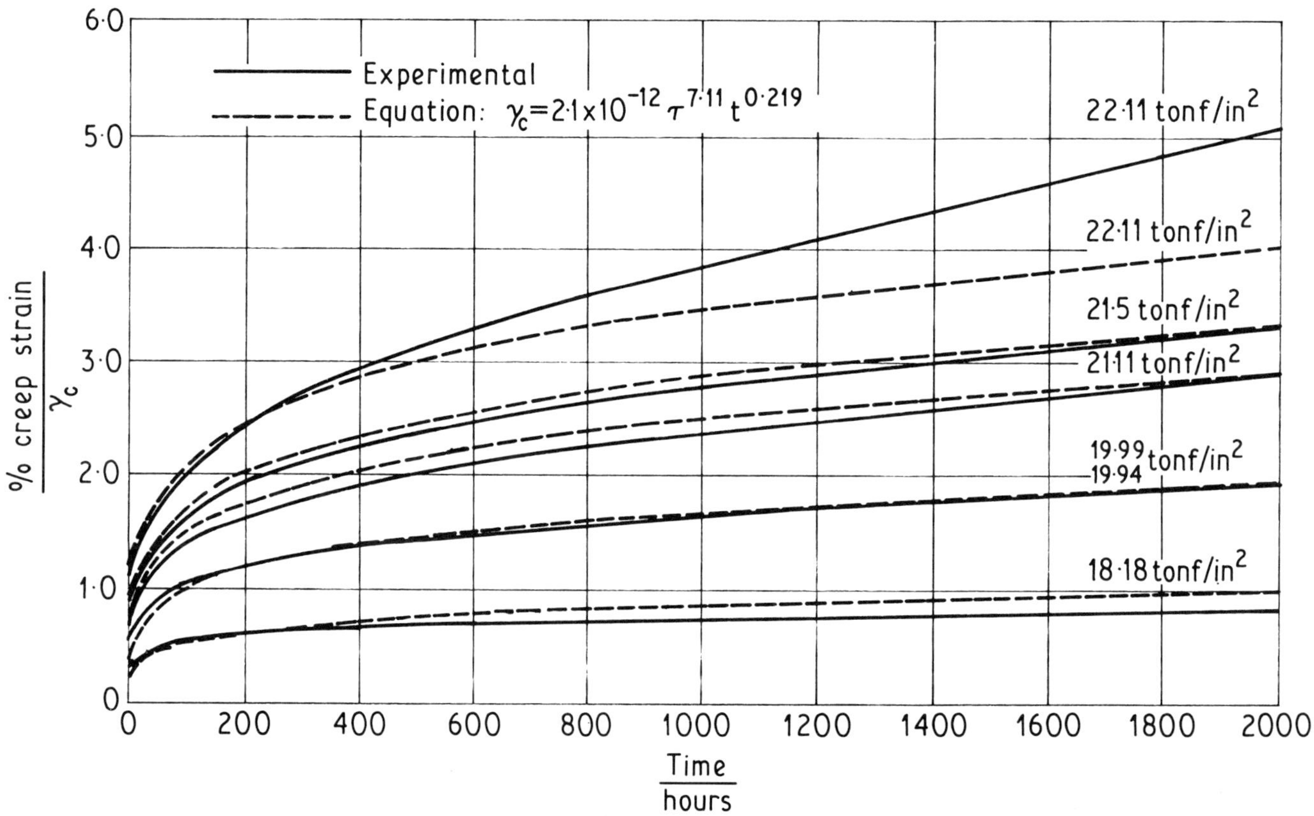

Fig. 12: Torsion creep data for EN25 steel at 350°C

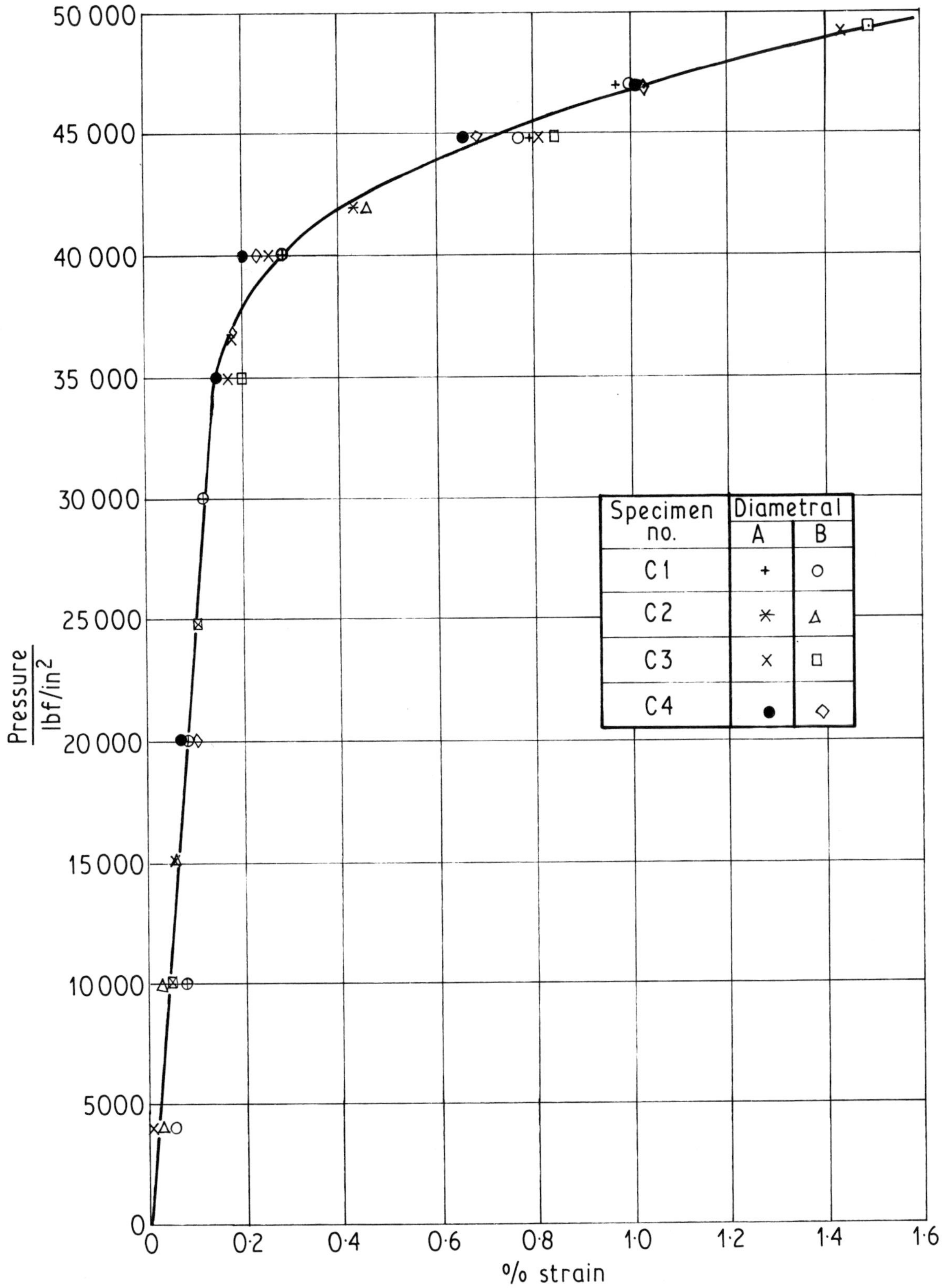

Fig. 13: Hot loading curves for pressure creep specimens (K = 1.67; EN25 steel at 350°C)

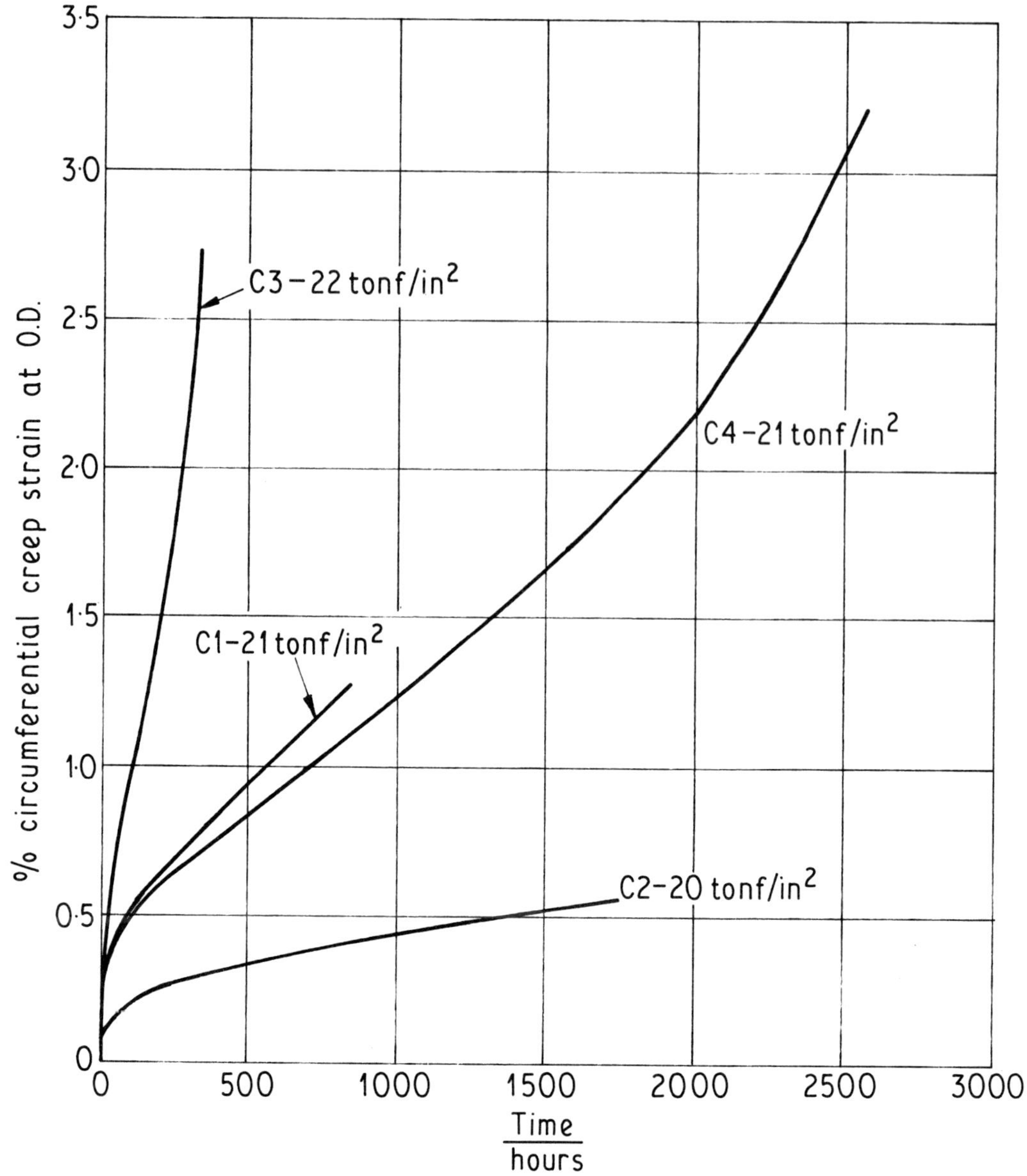

Fig. 14: Pressure creep tests on EN25 cylinders
K = 1.67, test temp = 350°C (diametral creep strains)

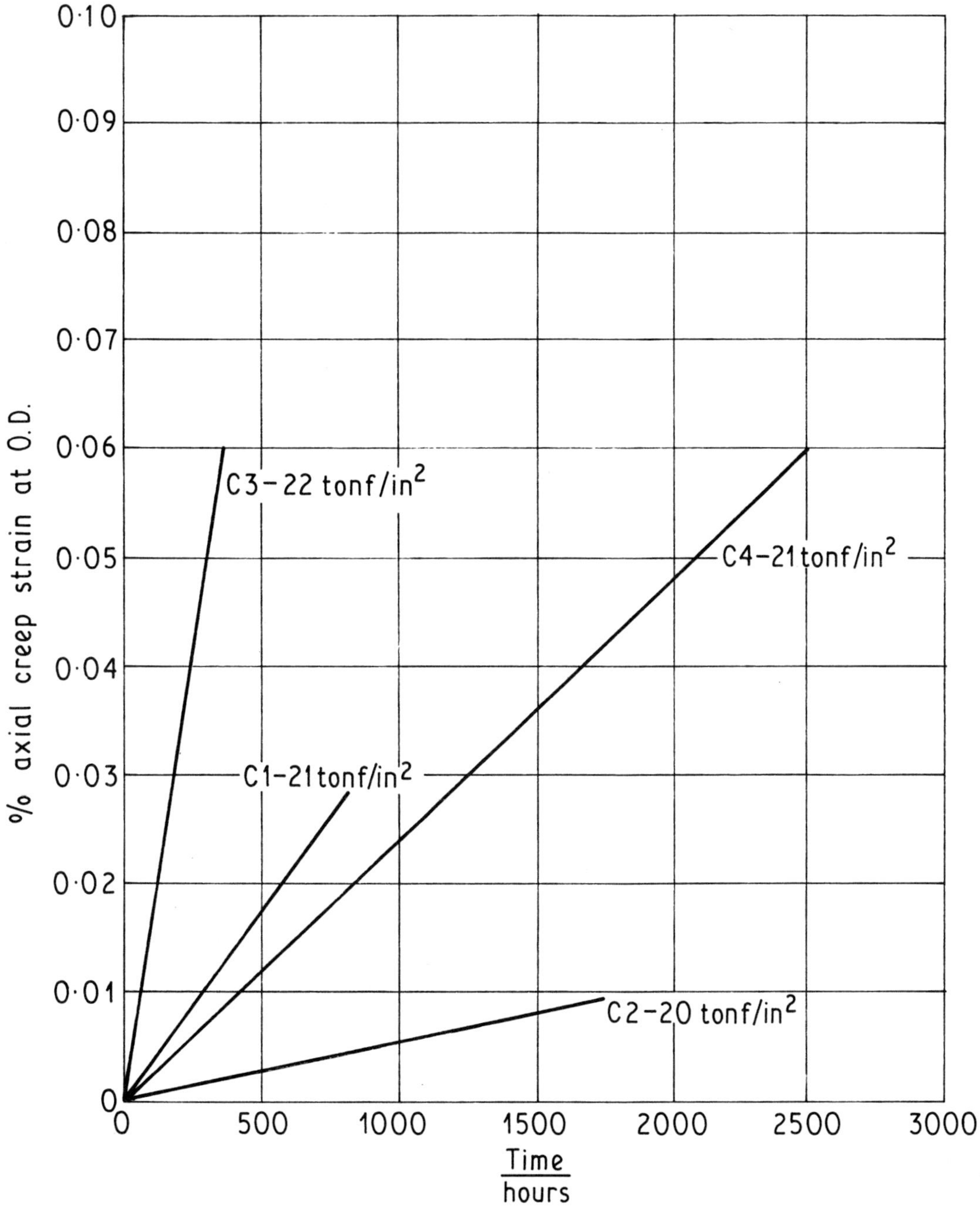

Fig. 15: **Pressure creep tests on EN25 cylinders K = 1.67, test temp = 350°C axial creep strains**

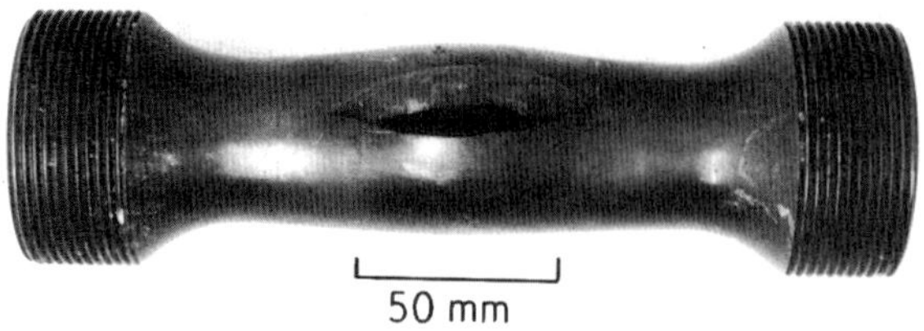

a Specimen C1 (P=21 tonf/in^2. T=350°C) premature failure after 835 hours (electrical fault)

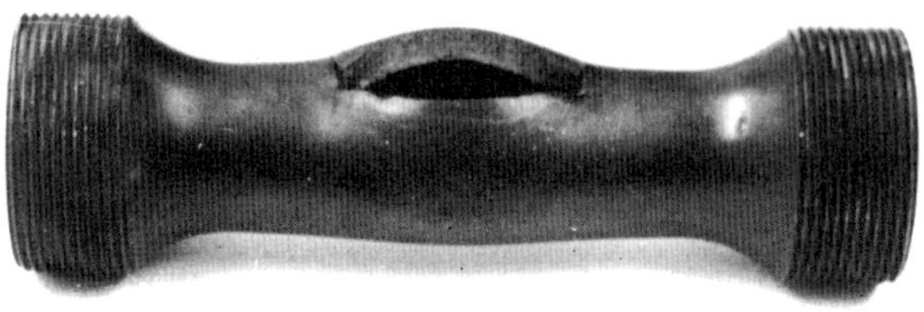

b Specimen C3 (P=22 tonf/in^2. T=350°C failure after 540 hours

Fig. 16: Fractured pressure creep specimens

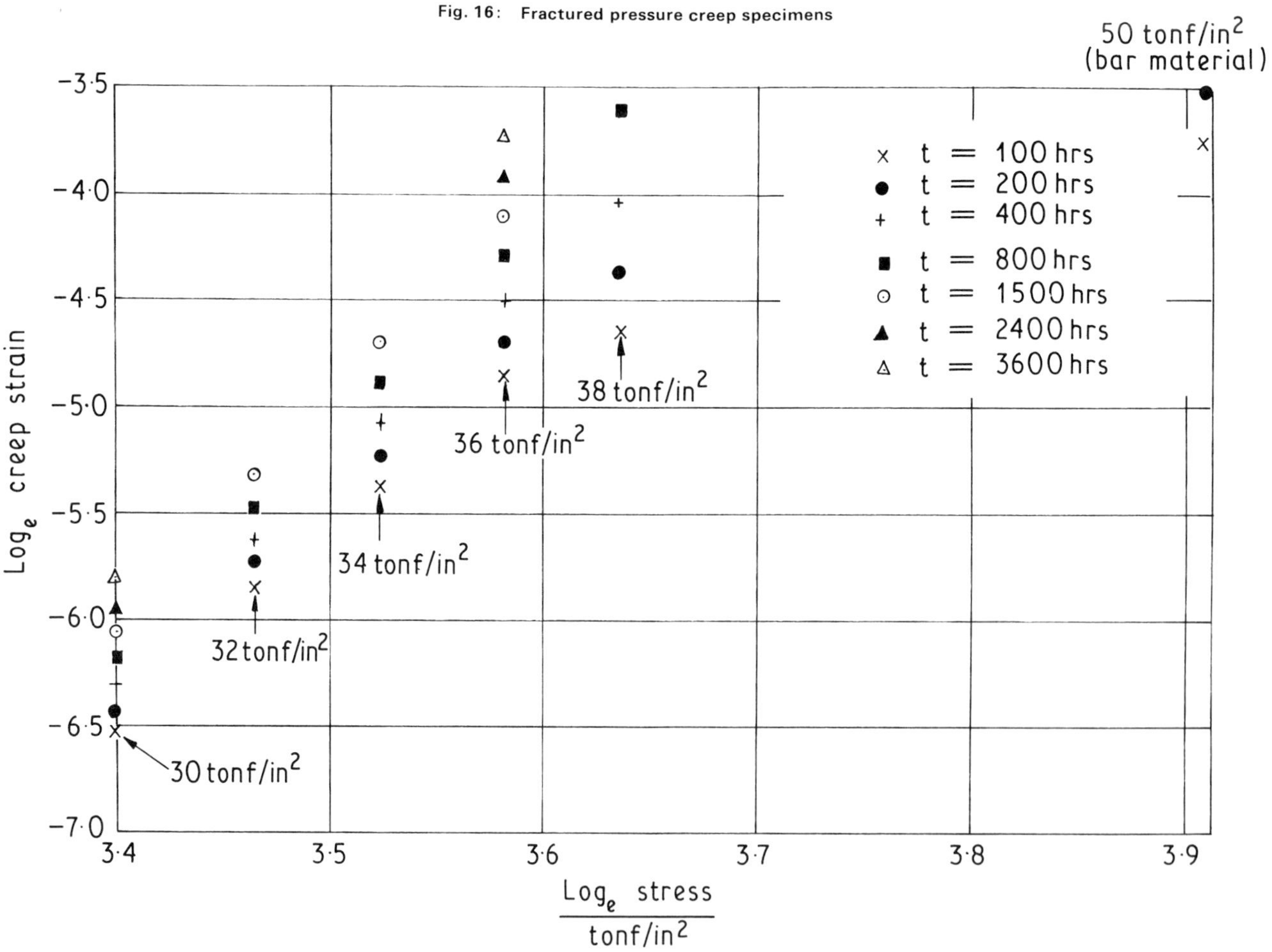

Fig. 17: Tension creep data for EN25 steel log_e creep strain vs log_e stress

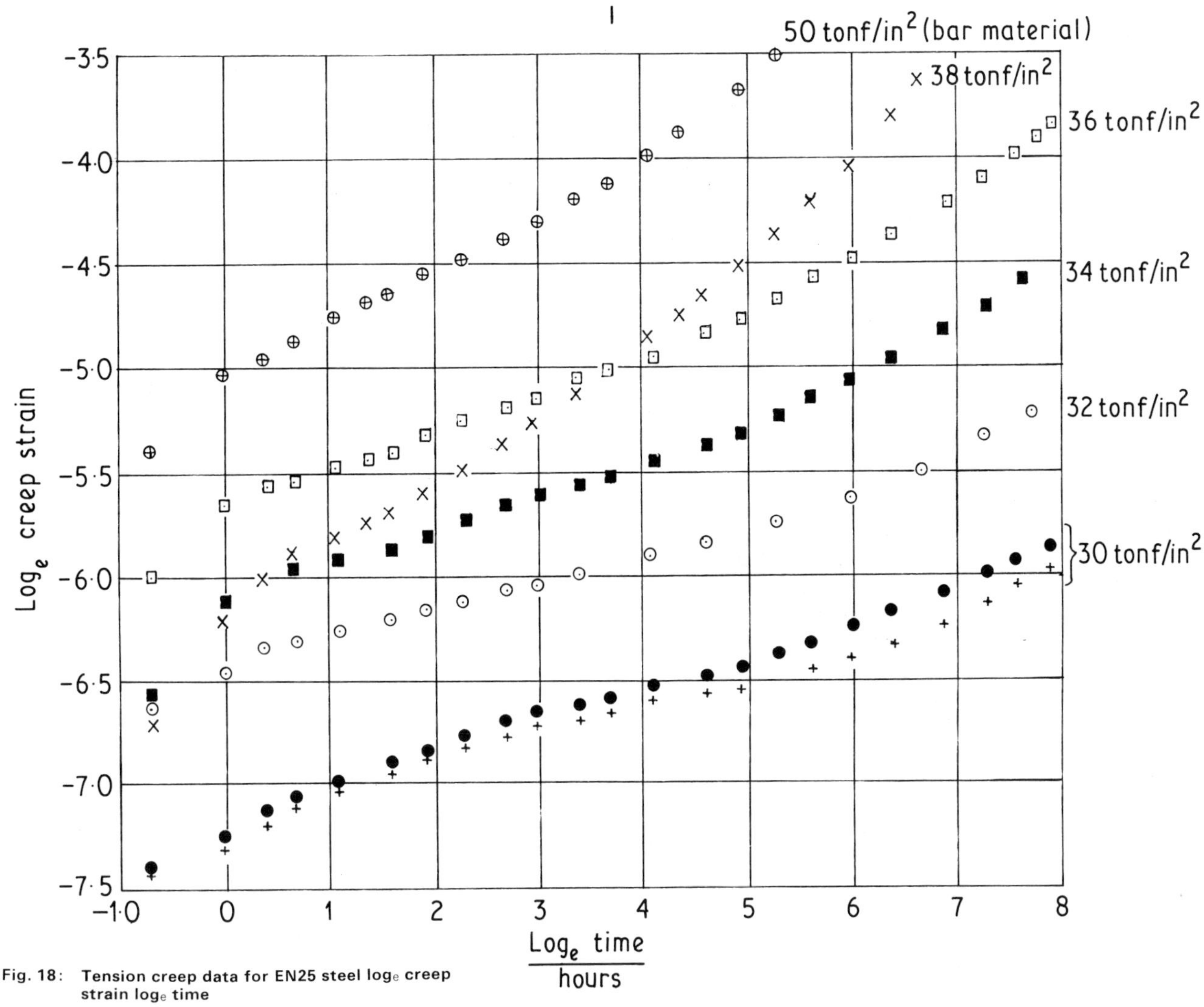

Fig. 18: Tension creep data for EN25 steel $\log_e$ creep strain $\log_e$ time

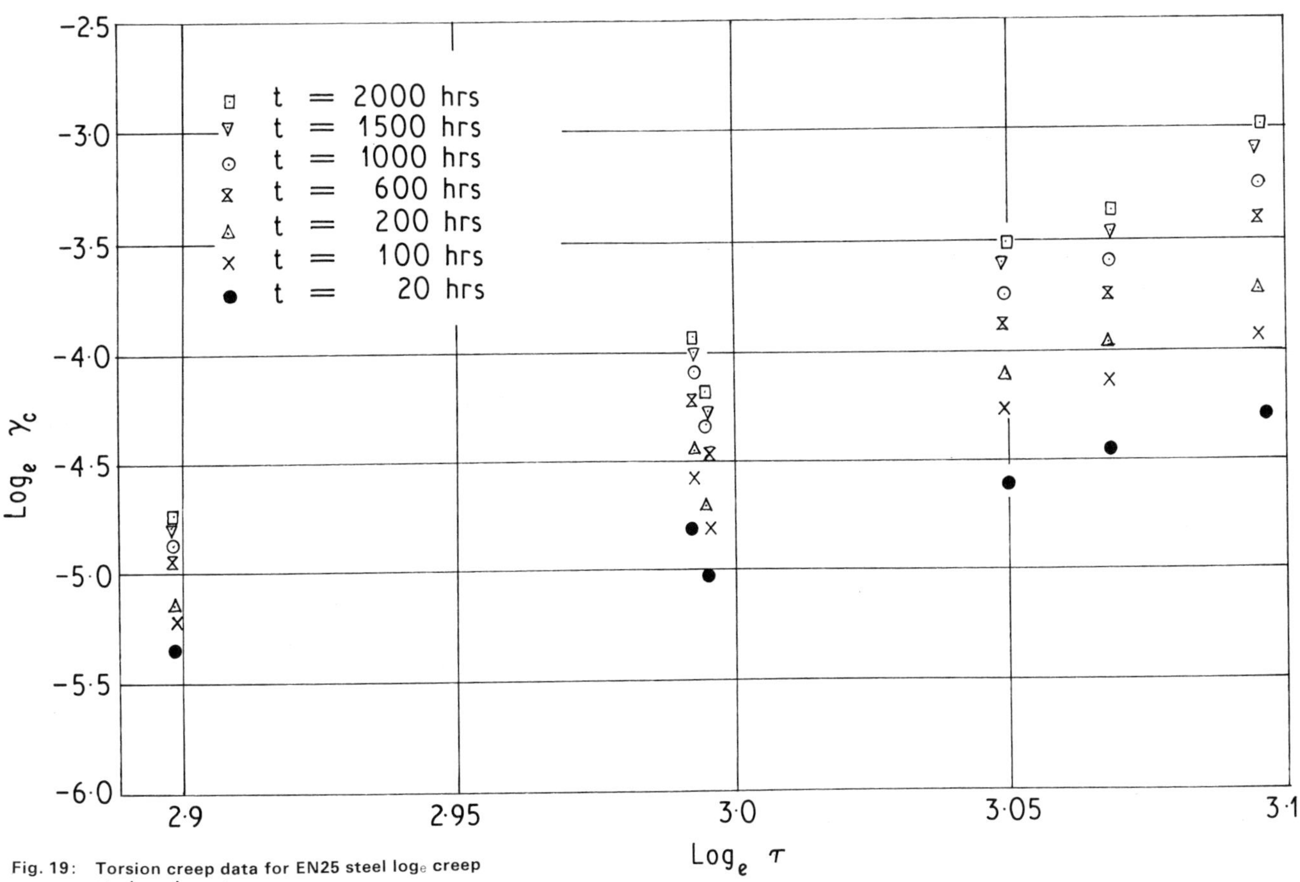

Fig. 19: Torsion creep data for EN25 steel $\log_e$ creep strain vs $\log_e$ stress

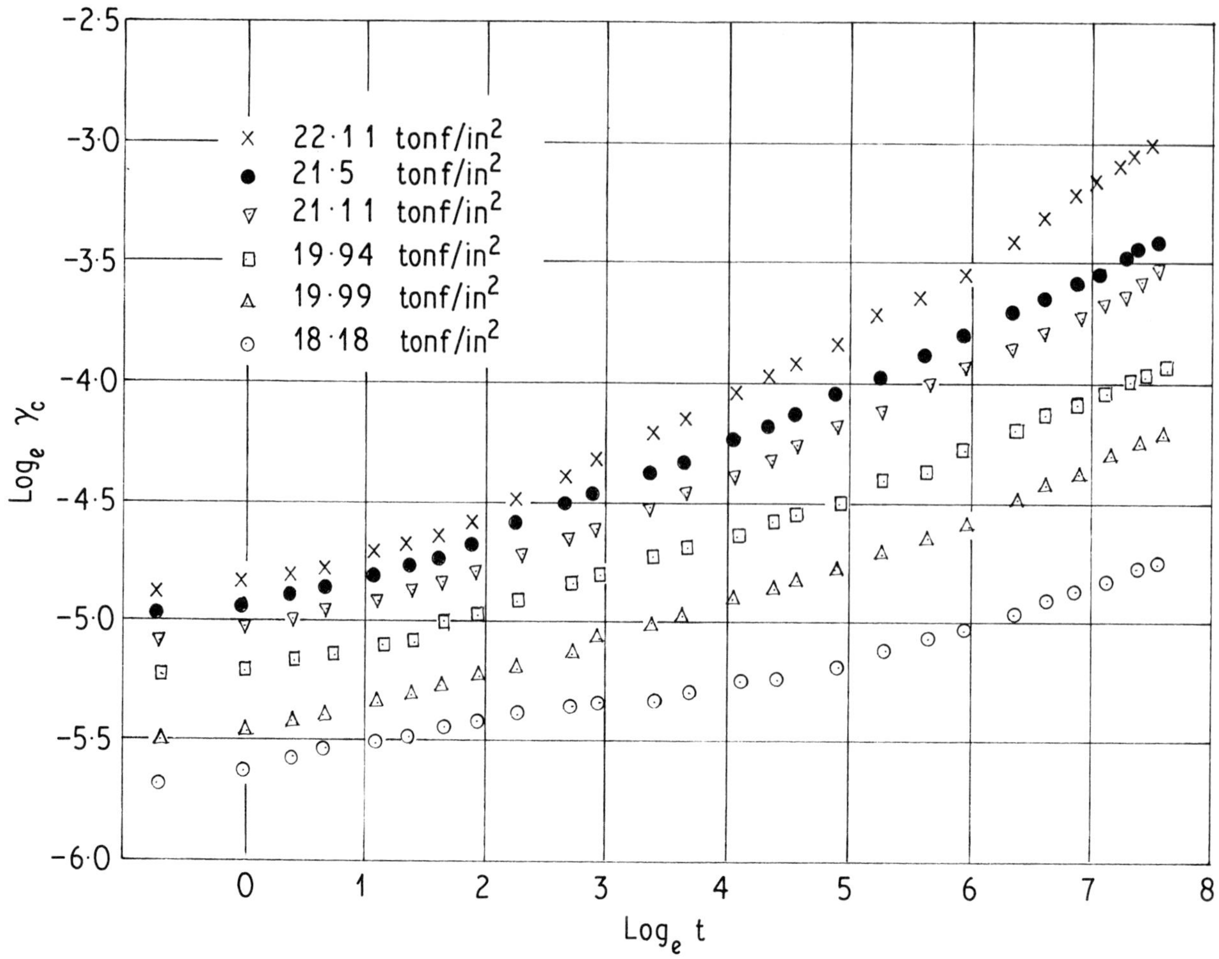

Fig. 20: Torsion creep data for EN25 steel log$_e$ creep strain vs log$_e$ time

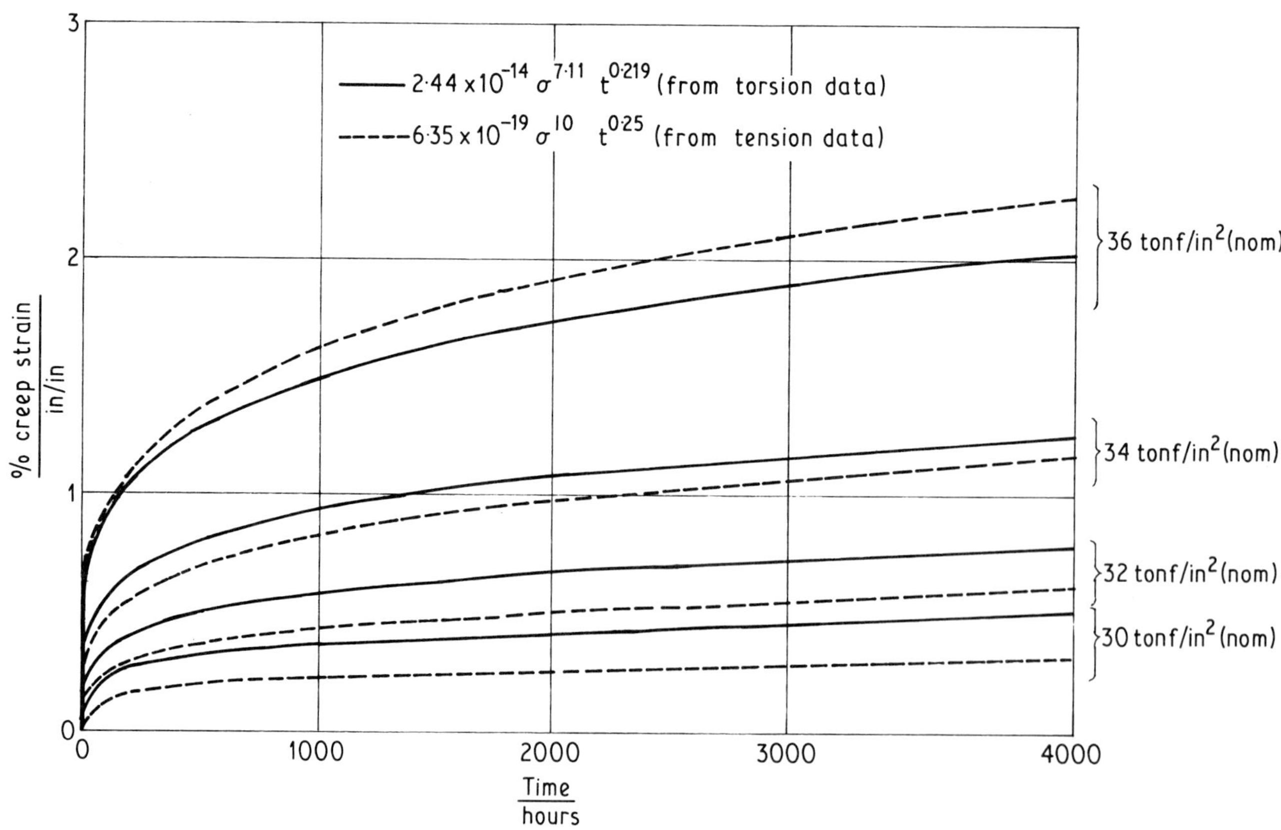

Fig. 21: Comparison of theoretical tension creep curves based on time hardening equations derived from tension and torsion creep data

Fig. 22: Comparison of experimental pressure creep curves with Bailey theory
(a) Pressure creep test on EN25 cylinder C2
K = 1.67, P = 20 tonf/in^2, test temp = 350°C
(b) Pressure creep test on EN25 cylinder C1
K = 1.67, P = 21 tonf/in^2, test temp = 350°C
(c) Pressure creep test on EN25 cylinder C4
K = 1.67, P = 21 tonf/in^2, test temp = 350°C
(d) Pressure creep test on EN25 cylinder C3
K = 1.67, P = 22 tonf/in^2, test temp = 350°C

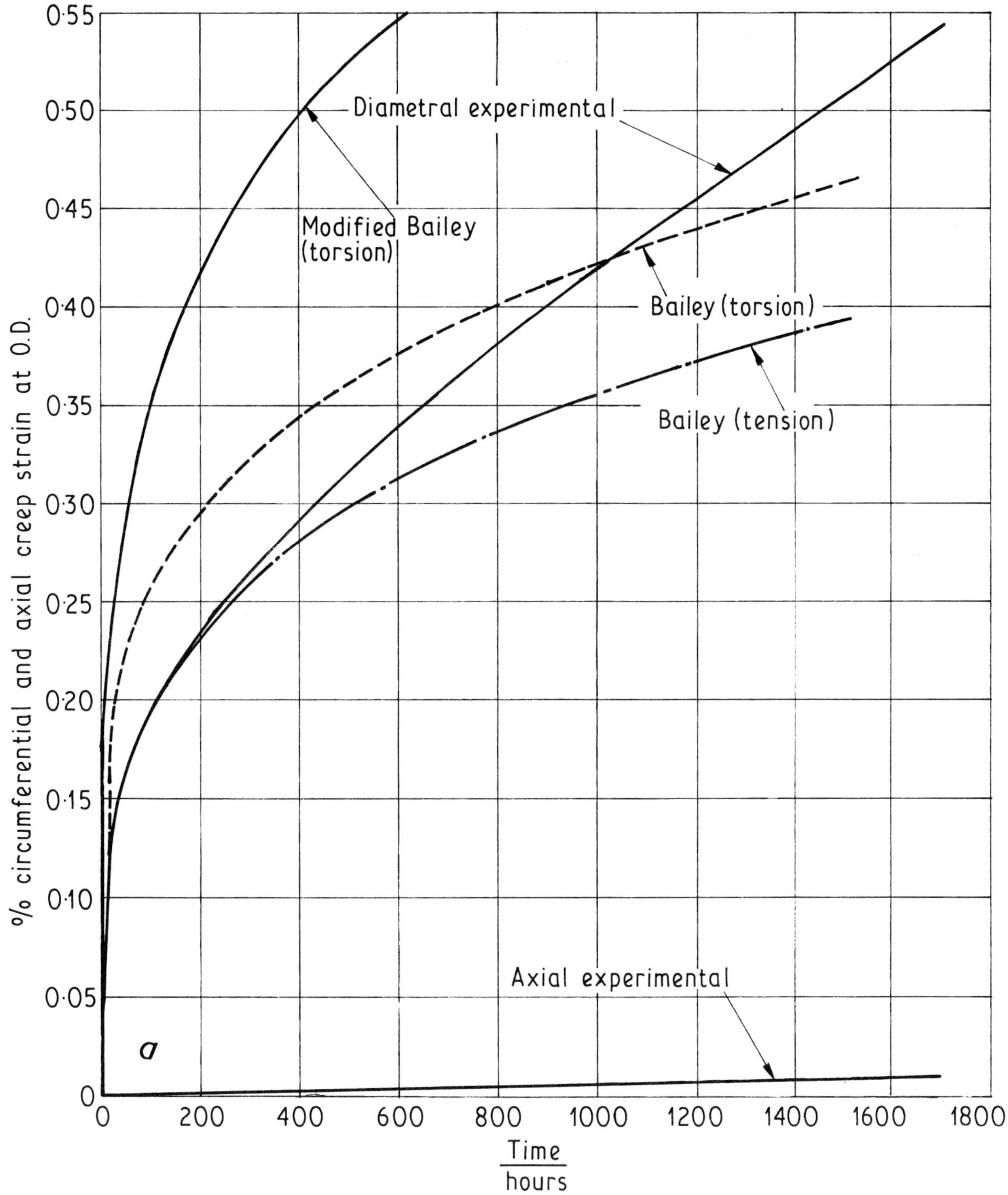

Fig. 22: (a)

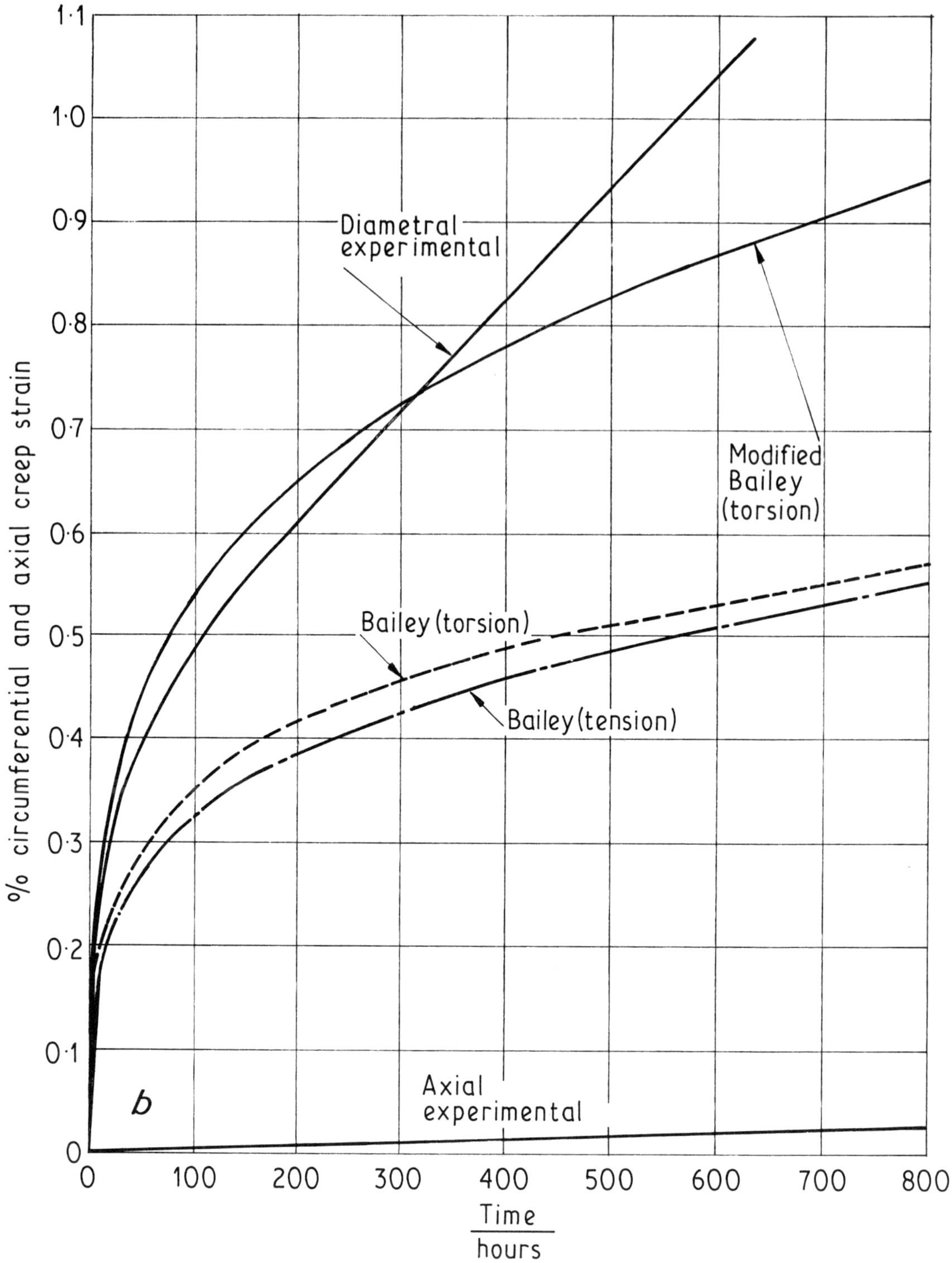

1·1
1·0
0·9
0·8
0·7
0·6
0·5
0·4
0·3
0·2
0·1
0
% circumferential and axial creep strain
Diametral experimental
Modified Bailey (torsion)
Bailey (torsion)
Bailey (tension)
Axial experimental
b
0
100
200
300
400
500
600
700
800
Time / hours

Fig. 22: (b)

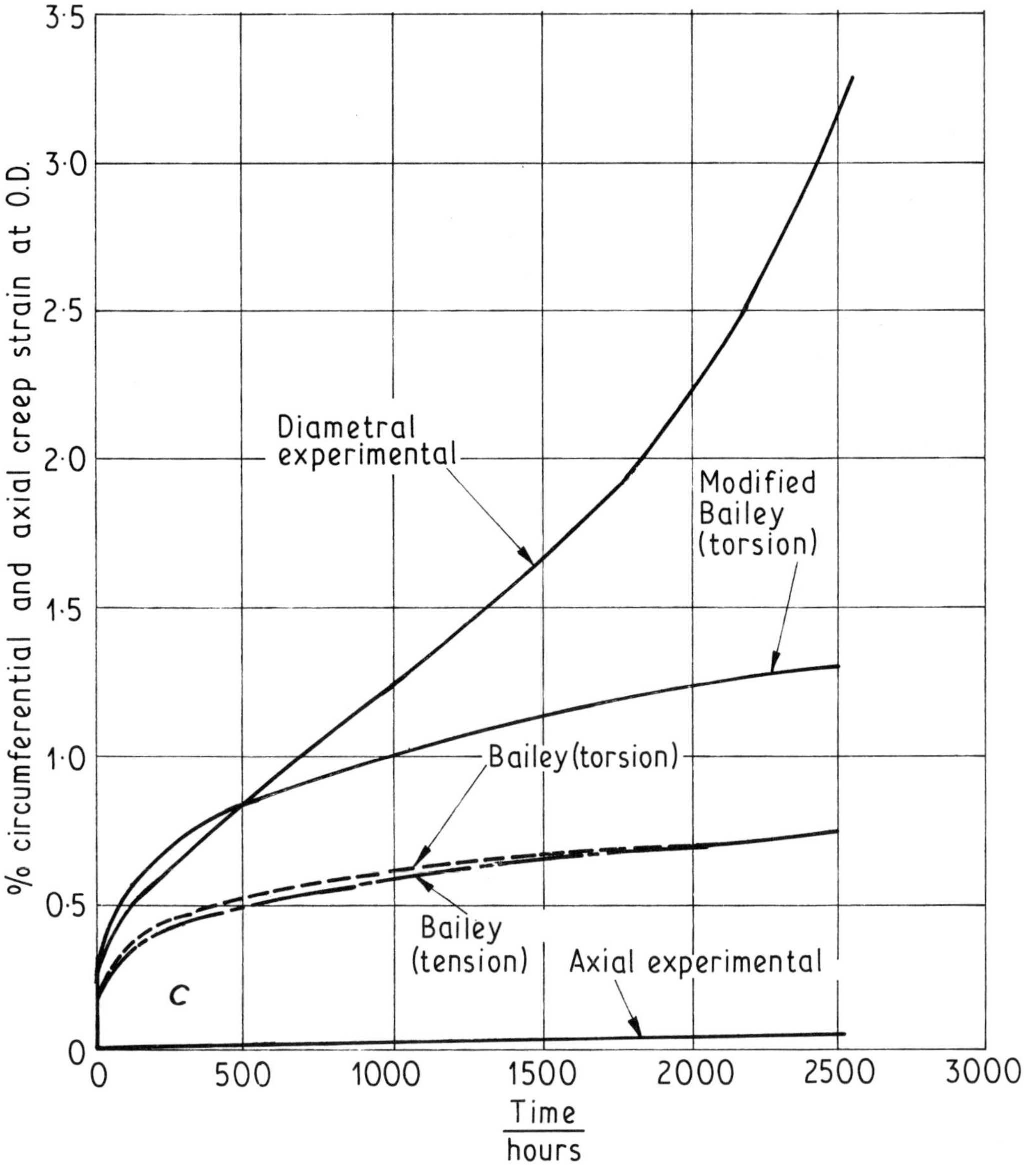

3·5
3·0
2·5
2·0
1·5
1·0
0·5
0
% circumferential and axial creep strain at O.D.
Diametral experimental
Modified Bailey (torsion)
Bailey (torsion)
Bailey (tension)
Axial experimental
c
0
500
1000
1500
2000
2500
3000
Time / hours

Fig. 22: (c)

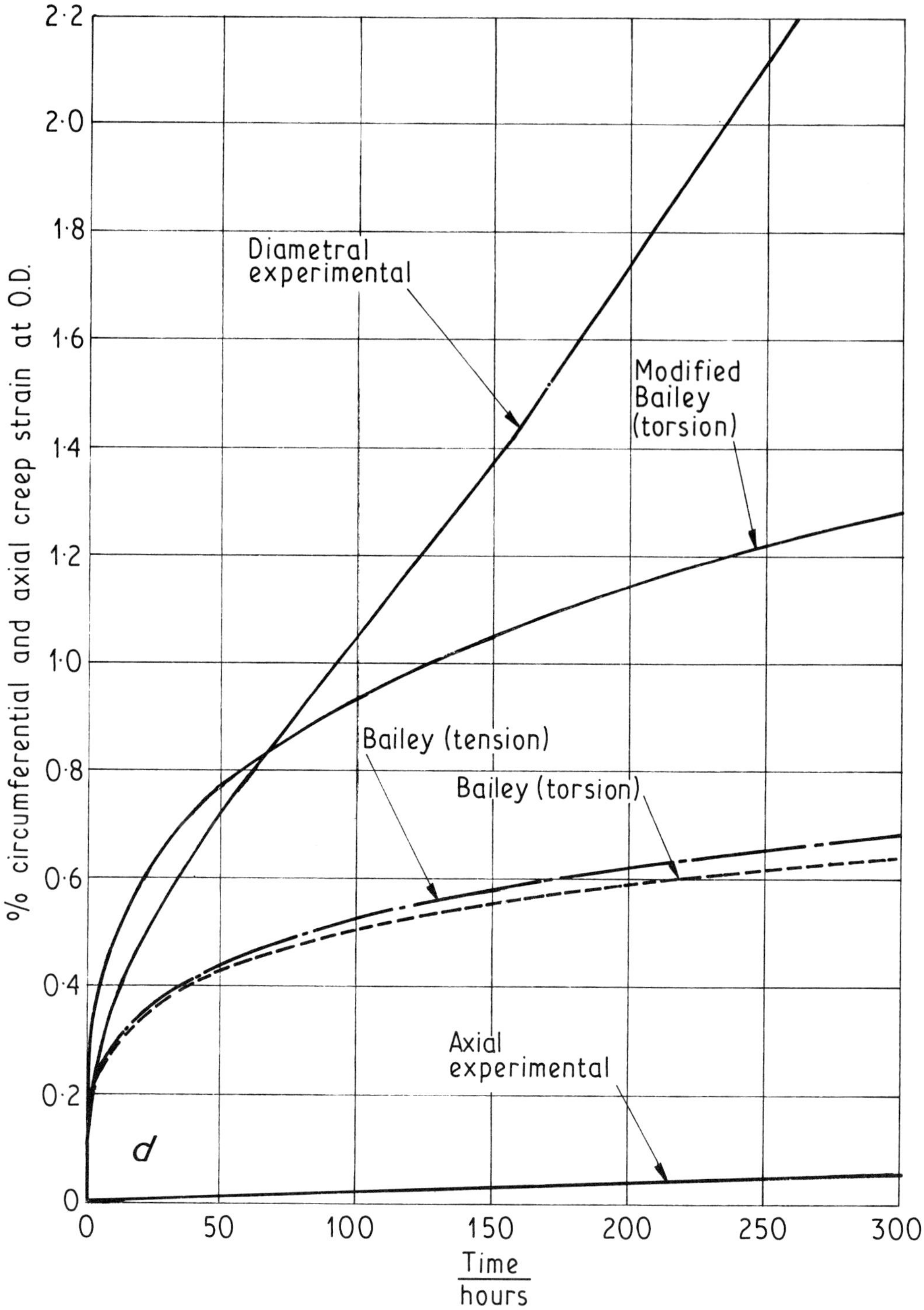

Diametral experimental
Modified Bailey (torsion)
Bailey (tension)
Bailey (torsion)
Axial experimental
d
% circumferential and axial creep strain at O.D.
Time / hours
0
0·2
0·4
0·6
0·8
1·0
1·2
1·4
1·6
1·8
2·0
2·2
0
50
100
150
200
250
300

Fig. 22: (d)

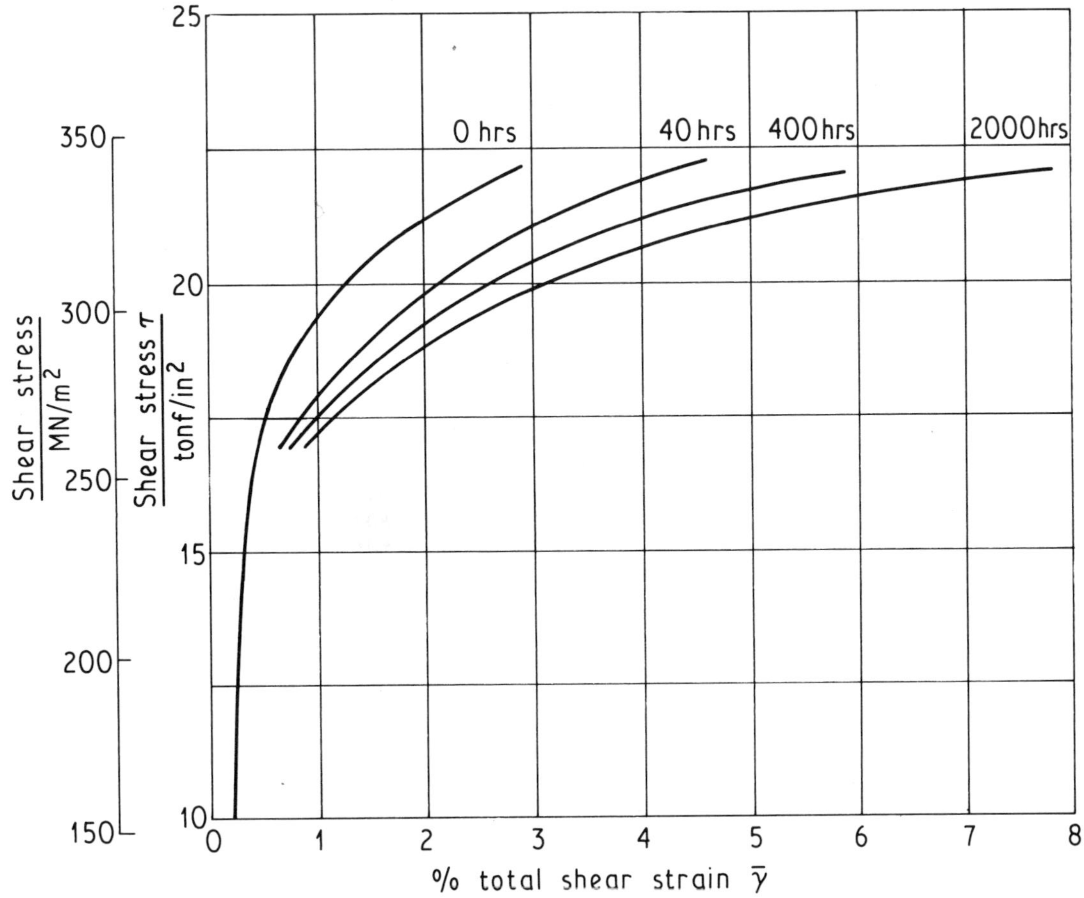

Fig. 23: Isochronous shear stress/shear strain curves

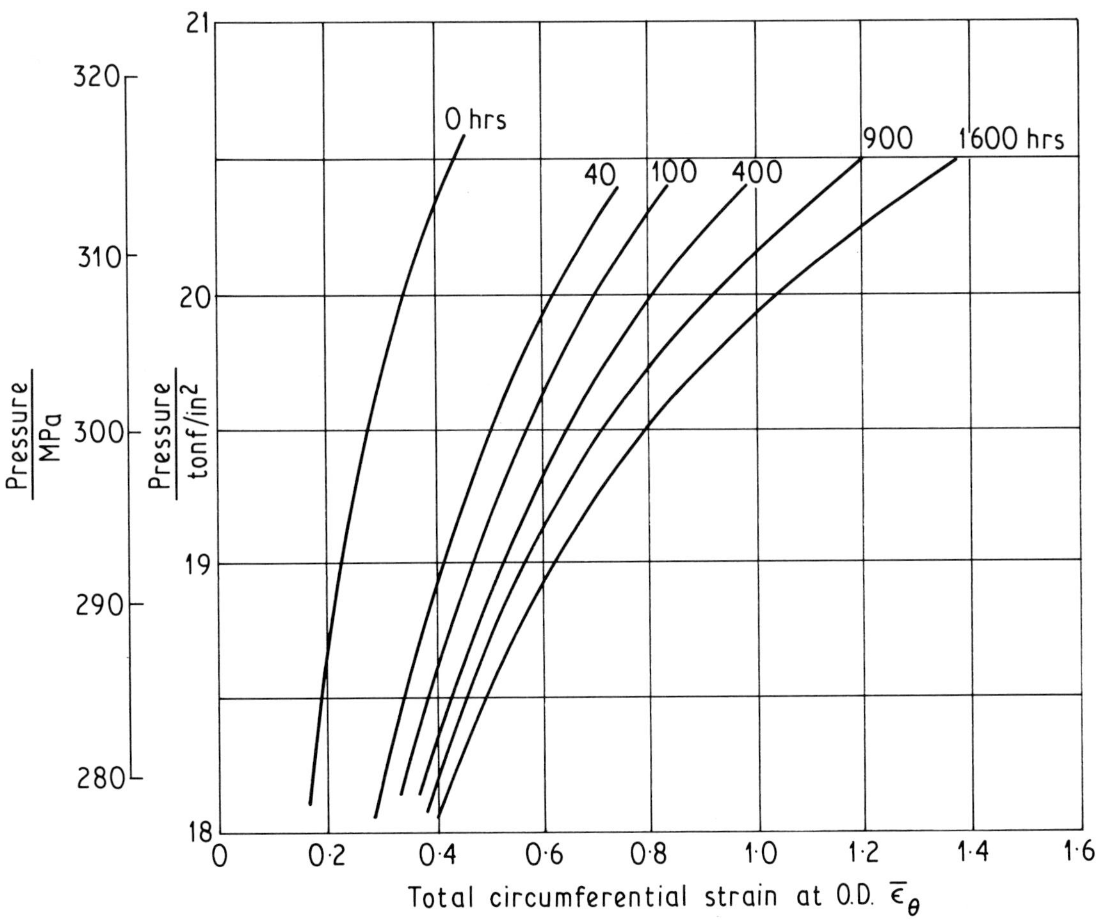

Fig. 24: Isochronous pressure/expansion curves (modified Manning analysis)

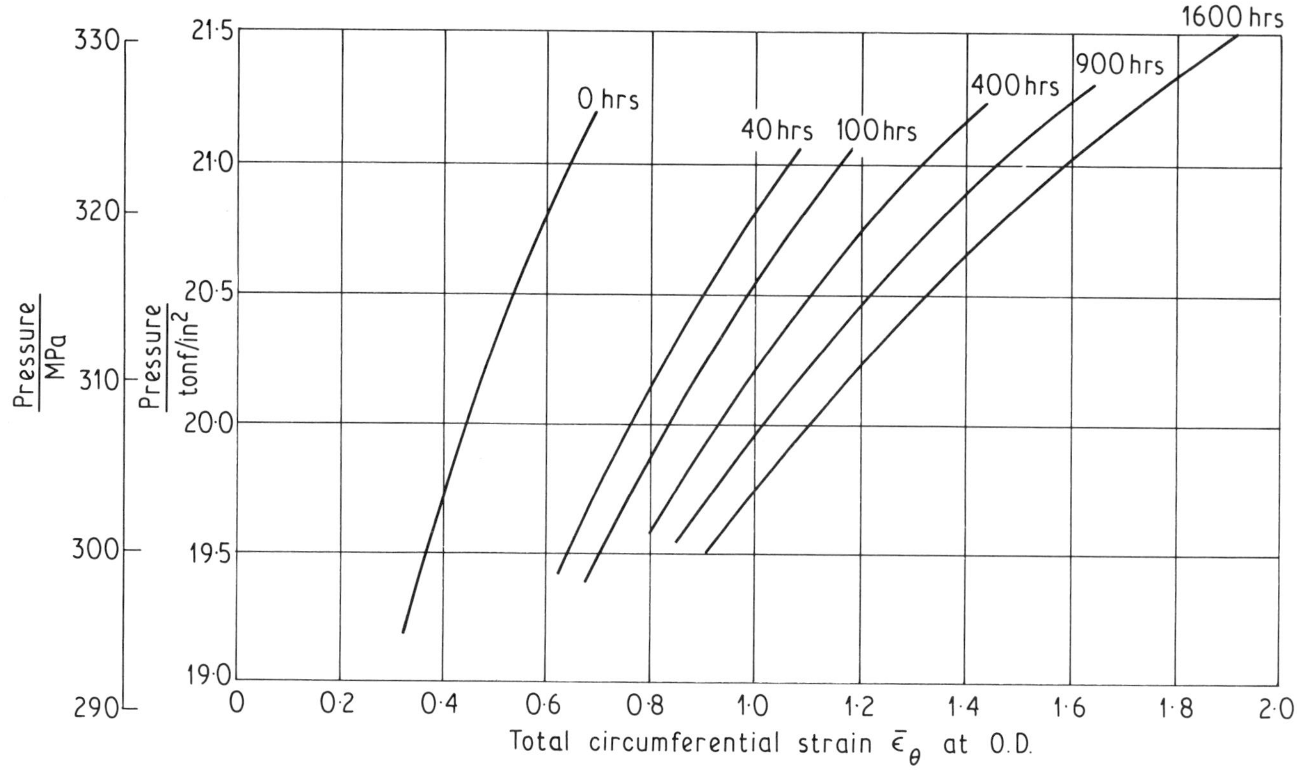

Fig. 25: Isochronous pressure/expansion curves (Parker analysis)

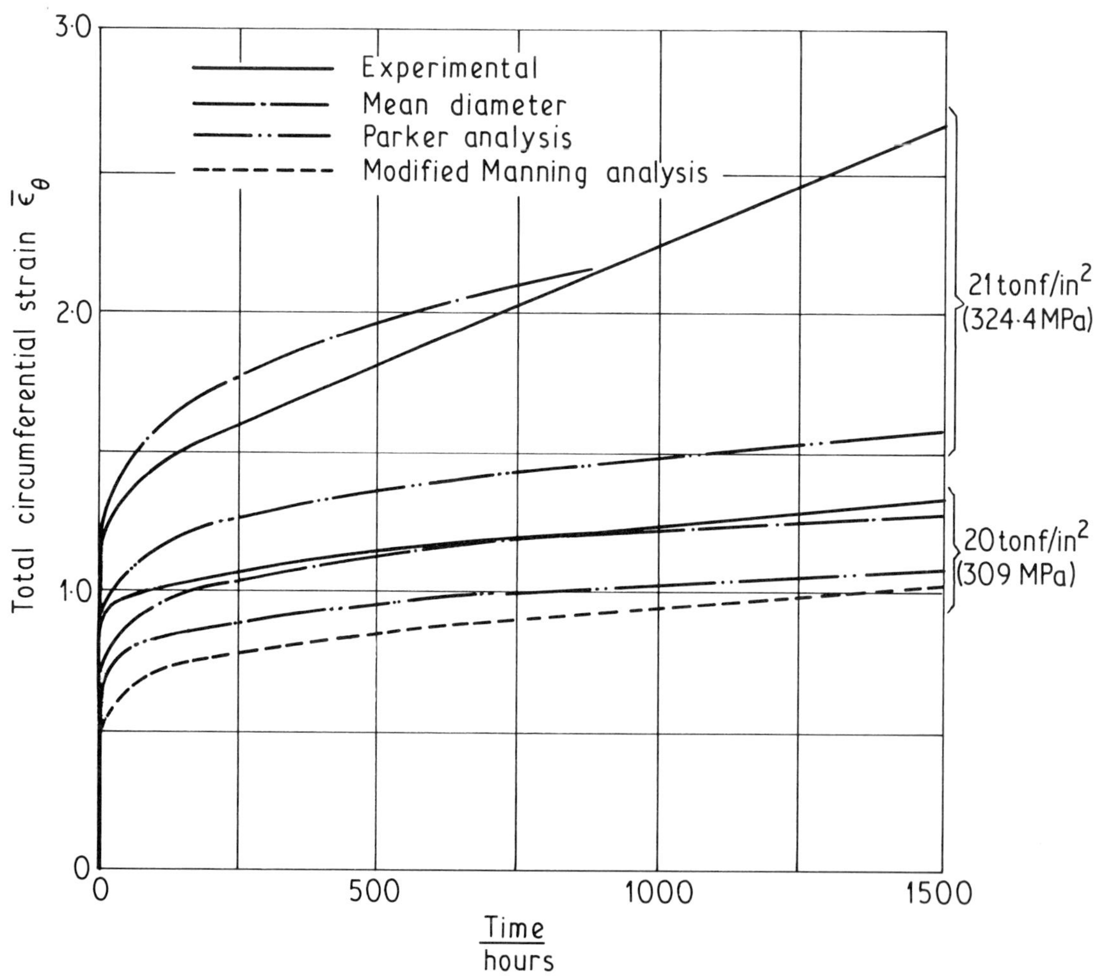

Fig. 26: Comparison of experimental pressure creep curves with various total strain theories

DESIGN

THE DEVELOPMENT OF KILOBAR PRESSURE EQUIPMENT FROM THE LABORATORY TO THE INDUSTRIAL SCALE (SOME REMINISCENCES OF THE WORK AT THE I.C.I. NORTHWICH ESTABLISHMENT, 1928-39)

W. R. D. MANNING, MA

1. The decision to undertake a programme of chemical studies at high pressure was taken in 1928. A pressure limit of $2\frac{1}{2}$ kbar was chosen, so it was said, because this was 10 times the working pressure of the Company's synthetic ammonia plant at Billingham, but only one tenth of that reported by Bridgman in his small scale work at Harvard; also it was about the pressure generated in the chambers of big guns when they fired.

2. The first action was taken in October of that year when a visit was paid to the late Dr. A. M. J. F. Michels, of the Van Der Waals University in Amsterdam, as the result of which he was engaged as a consultant, and a laboratory installation was ordered, to be made in Holland to his designs. It was early 1931 before this reached Northwich where it was erected in an open laboratory as seen in Fig. 1; as this shows, it was wholly hand-operated and had no safety protection whatsoever. In fact, the dangers were increased shortly after its assembly by the provision of oil thermostat baths for the reaction vessels, working up to 200°C in some cases, as was dramatically shown in 1932 when a joint failure deep in a bath smothered everything in the room in boiling oil: by great good fortune, it was empty at the time, though only minutes after a party of visiting V.I.P.'s had spent some time examining the installation.

3. This demonstrated the absolute necessity of providing proper safety protection for the people working on this job, and first priority was given to the design, construction and installation on a new site for high pressure work where the equipment could be operated with a minimum of risk, and also where the pressure - now raised to 3 kbar - could be power generated and operated by remote control. This project was entrusted to Mr. B. E. A. Vigers, who completed it in 4 months, an achievement which - in the present writer's opinion - never received the acknowledgement it deserved. A laboratory already equipped with safety cubicles had been vacated shortly before this and Vigers was able to adapt this very neatly to his scheme. This comprised an intermediate system at a pressure of 375 bars, fed by a motor-driven pump and controlled by an accumulator, which in turn fed the high pressure system through a 10 to 1 intensifier, from which the full 3 kbar pressure was piped to each cubicle. Fig. 2 is a view inside this room as it was in March 1933 when R. O. Gibson and E. W. Fawcett first discovered polyethylene while trying to react ethylene with benzaldehyde at 2 kbar.

4. The late Dr. J. C. Swallow, who was directing this work, had always insisted that we must be ready to develop to an industrial scale at short notice any commercially valuable reaction the chemists might come across. Strangely enough these first few milligrammes of polyethylene, together with the products of one or two subsequent experiments, were not recognised at the time for the importance they were soon to assume, but this was chiefly because there was so little of the substance and because the reaction was so difficult to control. Most of the subsequent attempts to prepare it ended in violent decompositions which sometimes damaged the equipment though fortunately without hurting anybody. And several other reactions were at the time looking more attractive, so that the high pressure research was not lacking in encouragement in 1933.

5. Two problems were already seen to face any attempt to work at pressures of this order industrially. The first of these was to elucidate the considerable mystery which then surrounded the problems of cylinder design: the second was to find some means of continuously raising a gas to these pressures. In fact, we were already coming up against the first in the small vessels we were then designing and making. Fig. 3 shows three of these, and Table I summarises their principle dimensions and features. It will be seen that the size has been progressively increased, while the diameter ratio has been correspondingly decreased, which might be interpreted either as over-confidence or foolhardiness. Fortunately, all these and several more like them gave good service without any sign of failure, and for this we doubtless owed more than we knew to the quality of the EN 25 type steel which had been used in all cases.

6. The tensile yield stress of this material was about 7.5 kbar and therefore all three vessels should have been fully elastic at the working pressure according to the maximum tensile stress criterion, which we had been using at that time. By 1934 however it was becoming fairly well established that the onset of overstrain in ductile materials was controlled by shear stresses, and this strongly suggested that such margin of safety as the later vessels possessed must lie mainly in the plastic region. It was important therefore to know how far this range extended; in other words to be able to predict the ultimate bursting strength of these cylinders.

7. Both Bridgman (1) and Michels (2) had reported a few bursting tests, although in neither case were details of the composition and mechanical properties of their materials included; nevertheless it seems that they had each based their designs on these results. The literature contained very little other data of this kind, although Cook and Robertson (3) described a series of bursting tests on mild steel specimens as long ago as 1911. We therefore undertook a few ourselves, using soft materials so that we could burst relatively thick cylinders, i.e. with diameter ratios of 2 or more, with the existing equipment. With these results, and with the help of Professor G. I. Taylor of Cambridge, a method was devised for calculating bursting pressures from torsion tests; this was applied to Cook and Robertson's mild steel tests and gave encouragingly good agreement as is shown in Fig. 4, which also contains two I.C.I. tests as well.

8. The next step was to produce a similar curve for the EN 25 steel we had been using, and for this a torsion specimen and a cylinder for bursting were machined from the same bar. Again the agreement was good, and a design curve was therefore constructed as shown in Fig. 5. Here the ordinates are the ultimate bursting pressures divided by 2.5, so that for a vessel to hold static pressure at normal temperatures with a safety factor of 2.5 the appropriate diameter ratio can be read off for the required pressure. The broken line in this diagram shows the pressure at which the innermost layers will first overstrain. It should be noted that this procedure assumes that there is no scale factor, i.e. that the bursting pressure depends on the ratio of the diameters only, (and not on their absolute size) as do the stresses and strains in an elastic cylinder.

9. This was regarded as a satisfactory solution to the first of these problems, and we continued to use it for the design of all subsequent vessels required for this work.

10. During the latter part of 1933 two important improvements to the small scale chemical studies were introduced. The first of these enabled the compression of gaseous reactants to be carried out in separate vessels so that the whole of the actual reaction vessel was available; in this way the reaction space was increased from about 10 ml. (as used in the original discovery of polyethylene) to 80 ml. The second was to heat the reaction vessels by means of copper jackets, wound for electric resistance heating, and fitting snugly round the vessel. In this way the temperature could be taken up to 300°C - or higher if the material of the vessel would stand it - and the dangers of spashing hot oil were eliminated. Fig. 6 shows one of the new arrangements; the large vessel at the back is used as a compressor to bring the pressure from about 700 bars up to the full 3 kbar if required and deliver it to the vessel seen next to it, partly hidden by the copper furnace which is mounted on a small hydraulic lift by which it can be raised or lowered by remote control from outside the cubicle. It was with this apparatus that M.W. (now Sir Michael) Perrin returned to the study of polyethylene in December 1935, and by a fortunate combination of the possible variables he obtained in less than half an hour about one hundred times the aggregate of all the previous yields. At once it became possible to characterise the material, and the electrical properties in particular amply fulfilled Perrin's hopes which he had held after studying the likely molecular structure of such a polymer.

11. From this point onwards polyethylene became increasingly important as various potential customers, mostly Government Departments, asked for samples. At first Perrin's success proved difficult to repeat, but gradually the chemistry was mastered and, with larger vessels, bigger yields were produced. Fig. 7 shows one of about ½ lb. being removed from a vessel of about ¾ litre capacity. This was a batch experiment as had been all those so far attempted, but the possibility of a continuous process which would avoid having to take the lid off a high pressure vessel to remove each batch was obviously attractive; and the first requirement before any serious testing of such a possibility could be undertaken was a continuous compressor. We were then working at pressures up to 2 kbar, although it had begun to appear that a product possessing most of the desirable properties could be made at 1½ kbar. Even so, this was far above any known commercial process then operating, and the few compressor manufacturers with whom we discussed the problem showed little inclination towards tackling it.

12. The chief difficulties encountered in raising the pressure of a reciprocating machine are usually (i) sealing the plungers, and (ii) preventing fatigue, and these tend to be much aggravated by increasing delivery pressures. In the case of ethylene moreover, there is the further problem that under pressure this gas can dissolve most lubricating oils into the gas phase, thus removing all lubricant from the rubbing surfaces and at the same time contaminating the product.

13. We had asked Dr. Michels some months previously to consider a somewhat similar problem, and his tentative solution was a mechanised version of Amagat's method in which the pressure is generated in a liquid such as a light lubricating oil and transferred to the gas by a mercury seal. We were, in fact, using this principle in our hand-controlled equipment, as in the vessel seen in the background of Fig. 6. The difficulty is that any leak past the oil plunger must be replenished after each stroke, since otherwise the seal will be broken and the oil and gas will meet one another. Michels however offered a very ingenious solution by making use of the high density of the mercury to float a steel valve on it and so raise it into the mouth of the connection leading to the oil plunger, thus preventing the mercury following the oil towards its plunger. Then, if any leak had taken place, there would be a rapid drop of pressure in the remaining oil and a make-up supply would automatically flow in.

14. No one at Northwich, least of all the present writer, had much faith in this idea, but it seemed the only hope at the time, and Michels was therefore asked to detail a prototype; in due course this was ordered, to be made in Holland from forgings obtained in Sheffield under the supervision of the I.C.I. inspection service. The machine was completed early in 1937 and shipped to Northwich after testing with nitrogen in Amsterdam. Fig. 8 shows it installed in one of the laboratory cubicles.

15. It comprised two separate units fitted onto the same base, each comprising an oil intensifier of about 6 to 1 ratio and a U-tube for the oil **seal**. The low pressure oil supply was obtained from a modified diesel engine fuel pump, and it was distributed to each intensifier in turn by means of a piston valve; this valve was moved by a solenoid energised by relays and contacts which in turn were actuated by the intensifier plungers. Both these plungers and also the piston valve were continuously rotated as they moved backward and forward, a development which Michels had patented as a means of reducing both the friction and the leakage. It was reasonably successful in both these respects, but whether the benefits justified the extra complication and resulting cost is somewhat doubtful.

16. This machine was at once used to test with longer continuous running the possibility of continuous extraction of the product by allowing it to escape with the unreacted gas from the bottom of a reaction vessel. This demonstrated the basis of a continuous process, and we were at once called upon to scheme out a pilot plant. This was to be located near the laboratory where the compressor could be made a part of the plant without having to move it from its position in one of the cubicles.

17. For a reaction vessel we had already ordered a vessel of about 10 litres capacity. Its design was simply a scaling up of the vessel shown in Fig. 3 (c) with the same quick opening lid, although this was no longer required now that continuous product extraction had been successfully demonstrated. Fig. 9 is a photograph of this vessel taken in the Research Department workshops at Northwich. As is seen it is mounted on a **bogie** with flanged wheels. This was to suit an existing **track** connecting various parts of the building, which was also given a branch into a concrete cubicle built alongside the laboratory wall, so that the vessel remained on its trolley while actually on stream. It was closely wrapped with square section copper tubing for cooling, while an electric muffle furnace, split on a diameter, was so mounted that the parts could be opened out or closed around the vessel as required by remote operation from a small adjacent cubicle which served as a control room.

18. What had not been allowed for however was a means of vigorous stirring, and this was now essential. After various trials it was found that the only safe way of achieving this was to have an induction motor in an adjacent space, also at the working pressure, with a shaft extending into the reaction space. Fortunately, our friends in Sheffield found a lump of steel, not of such high strength as the EN 25 steel of the vessel, but sufficient for 2 kbar working, and big enough to enclose an existing motor; and this was quickly bored out to take the $4\frac{3}{4}$" diameter of the stator and fitted with a top cover, and a stem at the bottom by which it was connected to the lid of the main vessel. This arrangement is also seen standing on the right hand side of the trolley in Fig. 9.

19. The gas entered through the top of the motor casing, passing through the 'air' gap of the motor, and down the stem. After reacting it was blown out through a valve at the bottom which was operated by hand from the control room through a reduction gear. From this valve the mixture of product and unreacted gas was led into a large steel box with a safety glass window. The gas separated from the liquid polymer and was led back to a gasometer from which it was recycled, while the polymer fell onto a tray where it solidified and could be removed as required.

20. The instrumentation was simple and the dials were located conveniently in the control room. The reaction pressure was measured by a bourdon gauge **whose** tube was located in the cubicle with a light spindle passing through the concrete wall to a pointer on the other side. Temperature measurement was by a thermocouple in the top of the reaction space, passing out through glass-insulated electrodes to a suitably calibrated milli-voltmeter. The current taken by the stirrer motor was also metered with a dial in the control room. The main process control was by means of a large hand wheel which operated the bottom valve; the object of the 'driver' was to keep the pressure as steady as possible, and when the reaction was fairly steady this could be done by judicious operation of the wheel. This was however quite a tricky business, and several times a decomposition reaction could not be prevented. Fortunately, however, we had in the meanwhile devised a form of safety valve which prevented the pressure from rising to dangerous levels. This is shown diagrammatically in Fig. 10; essentially it was a small intensifier with the small diameter exposed to the pressure in the vessel with a blank wave ring seal, while the larger was supported by oil pressure from the laboratory supply at 375 bars. Although it was very successful with this relatively small vessel, it proved too sluggish for larger units.

21. This plant began operations in July 1937 and was run for several periods of 48 hours or more, usually at $1\frac{1}{2}$ kbar. Mechanically it worked reasonably well, but the quality of the product was difficult to control. This was chiefly a chemical problem and is outside the scope of this paper, but it may be worth noting that the reaction was initiated in this plant by oxygen in the feed gas. This was extremely sensitive and had to be kept down to the order of 1 part in 10,000, or even less; moreover the exact concentration required depended both on the pressure and the temperature, which - incidentally - shows how extremely lucky Perrin had been in December 1935 to use a cylinder of ethylene which contained exactly the right amount of oxygen to react quickly at the conditions he had selected, but not so quickly that it got too hot and decomposed.

22. By the Autumn of 1937 the international situation was looking very ominous and Britain had been forced to pursue a vigorous rearmament programme. 'Polythene', to use the name we had then given this material, had a number of uses in this programme, which meant that the Government was calling for regular supplies of it. Our slender resources had therefore to be concentrated on the design of a commercial plant, which had to be on a separate site where ethylene had also to be produced. The plant for this latter, as well as all buildings and services, etc., were in the hands of the main Northwich Engineering Department, but the high pressure system was left to the small team in the Research Department which had been developing it.

23. Originally the quantity called for was 70 tons per annum, but before our schemes had got very far this had been increased to 100 tons and there was talk of this being doubled. Our experience with the 10 litre vessel in the pilot plant suggested that about 100 litres of reaction space would be needed to begin with, and as we did not want to take too big a jump in size with this still somewhat novel basis of design it was decided to begin with two 50 litre vessels, each with internal stirrer motors.

24. The compression problem was even more exacting, since there was still no reliable machine on the British market for pressures over a thousand bars, and, with the international situation as it then was, we obviously could not risk being dependent on an overseas supplier. An experimental 'booster' had been designed and built by Brotherhood's of Peterborough and installed at Northwich, which incorporated the ideas of both companies, but test running soon showed that there would be many problems to solve before a machine of this type could be recommended for plant service, as there would no doubt also be with a mercury lute compressor. After much deliberation it was finally decided that the latter looked the less speculative solution, and the job of designing and building an industrial version was given to a young man, Mr. E. W. Greensmith, who had just joined the Company, and - if the writer may be allowed a personal comment - seldom can anyone have been given a tougher entrance examination. Fig. 11 is a general view of his first machine. As can be seen, the intensifiers have been turned into the vertical plane and mounted behind the frame, while the four cylinders which together make up the two U-tubes are mounted in front. The piston valve system is placed horizontally at the back, and both this and the two intensifier plungers are continuously rotated. These machines were made entirely in the Northwich workshops of the company.

25. The gas supply at 250 bars was provided by an orthodox 4-stage Brotherhood compressor with a battery of large bottles to provide a buffer. The mercury compressors were controlled by the supply of oil at about 300 bars from Towler variable stroke pumps, one for each compressor. Their maximum rate was about 20 strokes a minute with the two intensifiers; this represented about 150 lb. per hour of ethylene.

26. Each compressor was directly connected to a reaction vessel, so that the supply to the latter was controlled by the output of the former. Apart from some fatigue cracking in the original U-tubes (which was overcome by slightly increasing the wall thickness and giving special care to the finish of the internal surfaces) these machines gave very reliable and consistent service. Virtually all the polyethylene made by I.C.I. during the war, some 4,000 tons, was made using compressors of this type, in the later stages with considerably larger versions, and some of these continued in service until 1950.

27. The reaction vessels were specified to have a length to diameter ratio of about 10 to 1, and to have a working pressure of 10 tons/in^2 ($1\frac{1}{2}$ kbar). Provision was also required for stirring with an induction motor which was to be placed in an integral space. We were advised that for the size we were likely to require the steel would have to be tempered to a lower ultimate strength if we still insisted on an impact value of at least 25 ft.lb. as measured by the Izod test. Our metallurgical adviser, Mr. E. W. Colbeck, the Chief Metallurgist at Northwich, considered that we should not reduce the impact value, and that we ought to design the cylindrical parts on the assumption that the steel would have a strength of between 50 and 55 tons/in^2 (7.7 to 8.5 kbar). We had no torsion data with which to construct a design diagram similar to that shown in Fig. 5, but as time was pressing we approximated by plotting a similar chart with the ordinated reduced by about 15%*. On this basis the diameter ratio for a 2.5 safety factor on the ultimate bursting pressure would be about 1.6. A further check was made however in view of the guesswork that had been involved in arriving at this figure. It was decided that a pressure test to 50% above working pressure should be called for, and that the $2\frac{1}{4}$ kbar which this would involve should just about raise the stress at the inner surface to the elastic limit. Under elastic conditions this would cause a maximum shear stress in the cylinder of about 3.7 kbar, which would correspond - according to the Maxwell (Von Mises) hypothesis - to a uniaxial tensile stress of 6.4 kbar, or 41.5 tons/in^2. The yield stress of the actual material was 42 tons/in^2.

28. We therefore accepted this and made the bore diameter $7\frac{1}{2}$" and the outside diameter 12", see Fig. 12. The reaction space was then made 6' 9" in parallel length, while the motor casing was about 2' long and of the same section. It was joined to the reaction space by a short link piece in which two hydraulically loaded relief valves were mounted, and which also held motor and stirrer shaft bearings in its central bore. There were wave ring seals at the two ends, and also at the top and bottom of the link piece, four in all; the middle joints were connected by split couplings with parallel buttress grooves as in some of the earlier vessels, e.g. that shown in Fig. 3 (d), except that these were split into two pieces along a diameter. For the rest of the design the technique was well tried and generally similar to that already described.

29. This design was submitted to three well known British steel companies with the request to tender for the supply of two such vessels. Two of them asked to be excused owing to pressure of other high priority work in their forges, but the third replied that they considered the proposed vessels dangerously weak for the very high specified working pressure: instead they suggested that the cylindrical parts should be made by shrinking three tubes over one another and offered to submit a suitable design, although they regretted that pressure of work on their design office would inevitably prevent this being done for some time. This would certainly have increased the weight considerably, and also the cost and the delivery time, and so the local I.C.I. Management decided, in view of the special circumstances of the international

* Subsequently Crossland and Bones (4) published an account of high precision bursting tests on an EN25 steel with a tensile strength of 8.43 kbar - very similar to the actual values obtained on the material of these vessels - and their bursting strengths were very close to those we guessed from the results of the stronger material.

situation then prevailing, to order these vessels on their own responsibility. And so they were duly ordered from the English Steel Corporation of Vickers Works, Sheffield, where they were finished and tested without trouble early in 1939. Altogether three more were ordered and they continued in service until 1946 when the plant, having been succeeded by a much larger plant on the other side of the Northwich site, was turned over to research and development work. In the course of this one of them was badly overstrained by a decomposition which could not be relieved because the relief valve had been incorrectly assembled. The result was a permanent swelling of $^1/_8$" on the outside diameter, but after a low temperature heat treatment (at 300°C) it was put back in service. One of the others suffered an even worse fate when it burst in the early days of 1947. In this case however the reaction in the gas (not pure ethylene) was very violent and it appeared that the pressure got up to something like 7 kbar before the vessel failed by splitting at each end of a diameter; fortunately it remained in one piece although it had been grossly deformed in the neighbourhood of the ruptures. The effect of this was, in fact, to confirm I.C.I.'s faith in the soundness of this method of design which they continue to employ.

30. The first two vessels were placed in concrete cubicles with steel doors with a smaller cubicle between them into which the product and unreacted gas from both vessels were piped and separated in a hopper, the latter being returned to a gas holder at atmospheric pressure, while the product in the form of a viscous liquid was withdrawn by means of a screw pump and allowed to solidify in rectangular moulds. The resulting blocks were then chopped up into small pieces of between $^1/_8$" and $^1/_4$" size, and sieved to remove any larger pieces. In this form it was then packed and sold.

31. Fig. 13 is a view looking into one of the cubicles through its open door, showing a vessel in position and connected up for operation. The reaction space is inside the part covered with heater windings, and the two relief valves can just be seen in the upper part of the photograph, between the two couplings of the link piece. The motor casing is almost invisible owing to bad lighting. Lower down and to the left is a sloping pipe which is carrying away the product and unreacted gas to the next door cubicle. It will also be seen that the vessel assembly is carried in a steel frame; this has pivots at the bottom, and when a vessel had to be removed to the workshop for servicing it was lowered in this frame to a horizontal position onto a bogie. Fig. 14 shows one being serviced, still on its bogie, with the motor casing nearest the camera. The relief valve casing is in the middle of the photograph, the hole at right angles to the axis of the vessel being that through which the valve plunger is inserted, while the large hole is for the vent pipe.

32. The erection of the plant was finished early in June 1939, and the various units were tested as they became available. Some polymer was actually made during that month and surprisingly few modifications were required. The control system which was similar to that in the pilot plant proved somewhat easier, probably due to the larger masses of reacting gas, but more than one decomposition occurred in the first few months; it was a considerable relief to the designers that these were blown off without damage to the equipment, although the loud report and the cloud of black smoke which followed it caused some anxiety in the neighbourhood when these were first heard and seen.

33. The commissioning trials proceeded throughout July and August while various minor troubles were sorted out. In spite of these, more than a ton of acceptable material was produced and shift crews were trained; it must be remembered also that we had to prepare to operate under 'black-out' conditions, which certainly did not make this work any easier. However, as the month of August went on, the progress was considered sufficiently satisfactory, and - no doubt stimulated by the imminent threat of war - the company's Production Department agreed to take over the responsibility for running the plant as from September 1st, 1939. Mr. D. W. Ginns was appointed plant manager, and 'Polythene' went onto the "Selling Range" of the company's products from that date.

34. The story of these early days of 'Polythene' has been told many times before and a number of people have been justly praised for their part in it. Perhaps however the engineering side has hardly received quite its fair share of these eulogies. For instance, in the years 1932 to 1936 when the High Pressure Section of the Research Department was a small team whose work excited little interest outside its own membership, the considerable technical progress that was made - and which the writer has endeavoured to describe in these pages - owed so much to the superb skill and experience of the craftsmen in the Research Department Workshop at Northwich. The writer has always felt that this debt has never been properly acknowledged, and he welcomes this opportunity to pay a long overdue tribute to these men and to their art which made it all possible.

35. At the present time when engineering craftsmanship seldom receives its true recognition, it is as well to remember that development work in this field always involves these skills, and woe betide the industrial community that forgets this fact and fails to ensure a succession of trained men (and women) with the ability to perform these tasks.

<u>POSTSCRIPT</u>: Design curves of the type shown in Fig. 5 were used extensively in the years following 1934, and it was not until the middle fifties that it was realised that temperatures well below those at which creep was detectable could nevertheless reduce the Ultimate Bursting Pressure appreciably. This problem was studied by Crossland and Kerr (5), and their experimental results indicate that, for instance, at 300°C with EN 25 steel, this pressure occurred some 5% lower than it did at room temperature. Thus, the factor of safety was probably nearer 2.3 than the assumed figure of 2.5. On the other hand, it should be added that some of the vessels operated by I.C.I. have endured working lives up to 100,000 hours at temperatures of this order and pressures well above the kilobar without showing any measurable dimensional changes.

REFERENCES:

1. Bridgman, P. W.: Phil. Mag., Vol. 24, p. 63 (1912)

2. Michels, A. M. J. F.: Vers. Akad. Wet. Amsterdam, Vol. 37, p. 498 (1928)

3. Cook, G. and Robertson, A.: 'Engineering', Vol. 92, p. 786 (1911)

4. Crossland, B. and Bones, J. A.: Proc. Inst. Mech. Engs., Vol. 172, p. 777 (1958)

5. Crossland, B. and Kerr, W. F. K.: 'High Temperatures - High Pressures', Vol. 1, p. 133, (1969).

Table 1 Details of 3 kbar Vessels

Year	Bore ins.	Dia. Ratio	Capacity ml.	Max. Stress kbar		Weight lb.	Seal Ring	Joint Connection
				Hoop	Stress			
1931	1.0	3.0	60	3.75	3.375	18.5	Lens	Screwed Plug
1932	1.5	2.25	250	4.48	3.74	37	Wave	Union Nut
1934	3.0	2.0	1,500	5.00	4.00	220	Wave	Split Collar

Fig. 1.

Fig. 2.

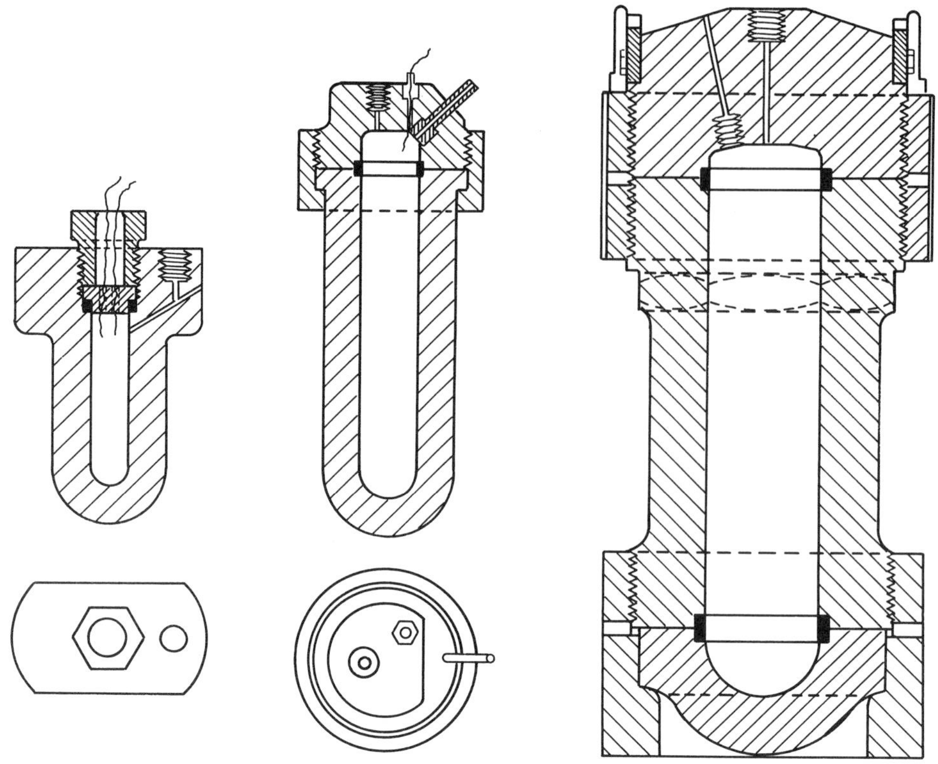

(a)

(b)

(c)

Fig. 3.

Fig. 3: (d) Split collar coupling being fitted on vessel (c) showing the buttress profile grooves about to engage. The coupling has to carry a load of 200 tons at full pressure (3 kbar), but only weighs 35 lb (16 kg)

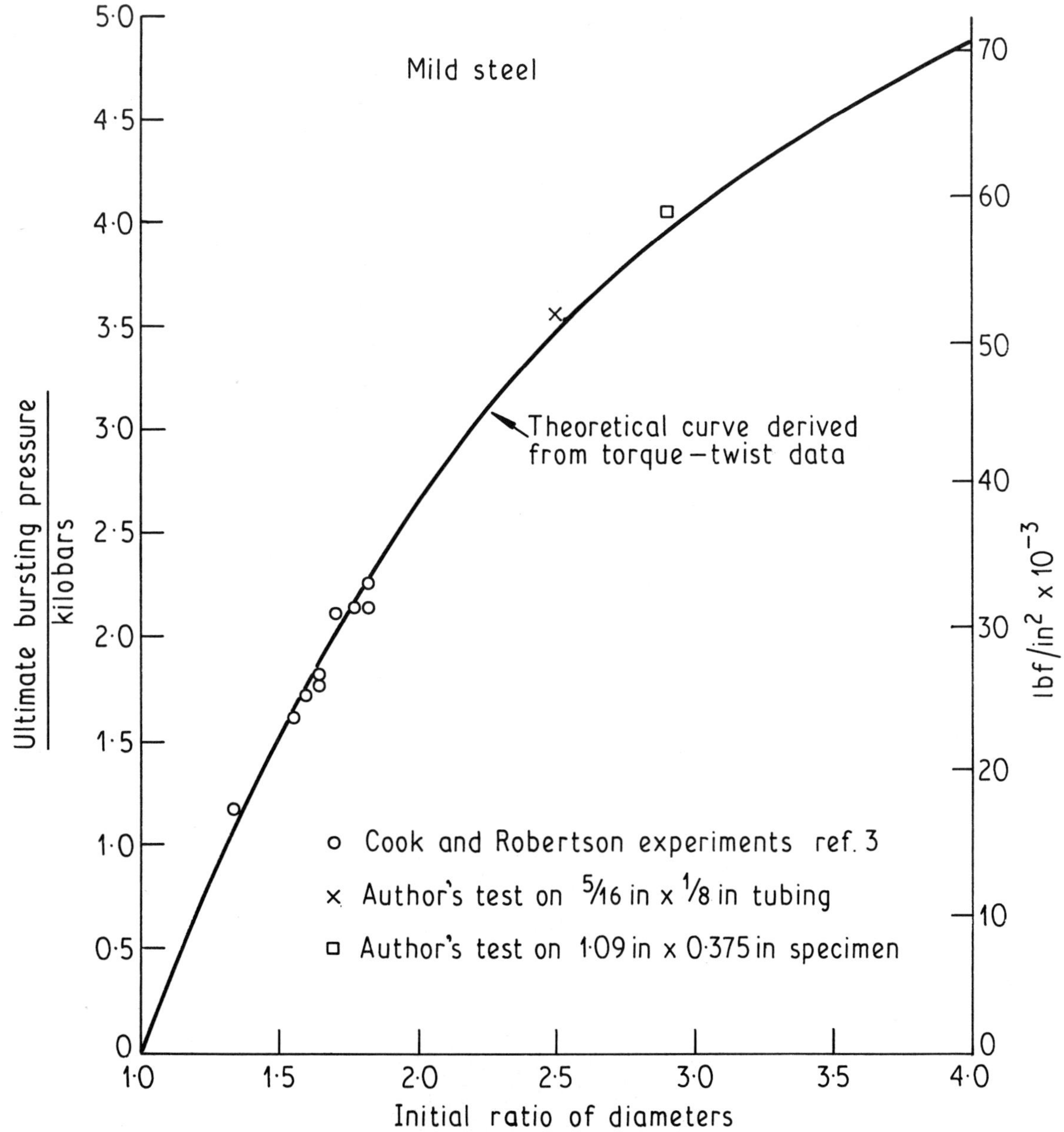

Mild steel
Ultimate bursting pressure / kilobars
lbf/in^2 x 10^{-3}
Theoretical curve derived from torque–twist data
o Cook and Robertson experiments ref. 3
× Author's test on 5/16 in x 1/8 in tubing
□ Author's test on 1·09 in x 0·375 in specimen
Initial ratio of diameters
0
0·5
1·0
1·5
2·0
2·5
3·0
3·5
4·0
4·5
5·0
10
20
30
40
50
60
70
1·0
1·5
2·0
2·5
3·0
3·5
4·0

Fig. 4.

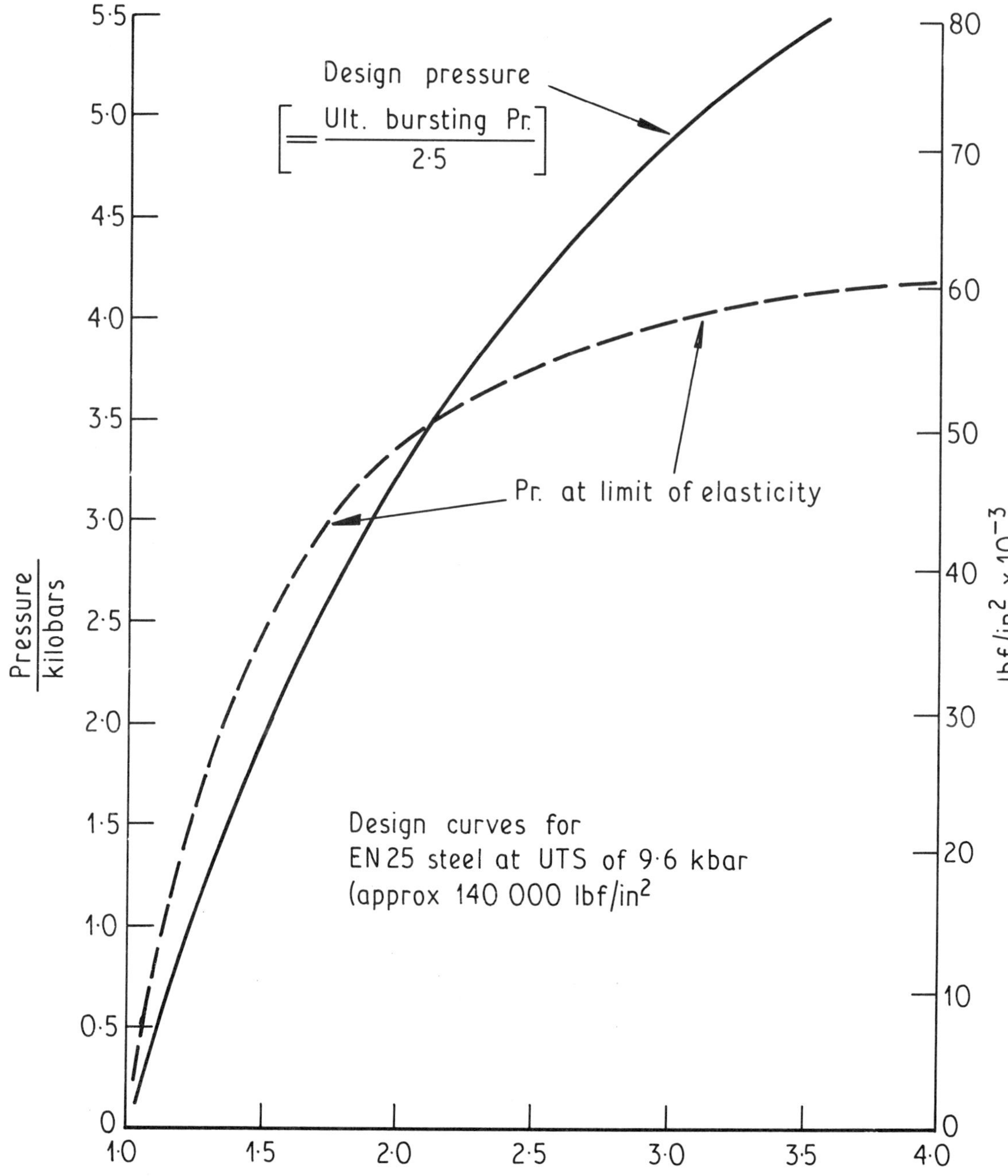

Design pressure
[= Ult. bursting Pr. / 2·5]
Pr. at limit of elasticity
Design curves for
EN 25 steel at UTS of 9·6 kbar
(approx 140 000 lbf/in²
Pressure / kilobars
lbf/in² x 10⁻³
Initial ratio of diameters
0
0·5
1·0
1·5
2·0
2·5
3·0
3·5
4·0
4·5
5·0
5·5
10
20
30
40
50
60
70
80

Fig. 5.

Fig. 6.

Fig. 7.

Fig. 8.

Fig. 9.

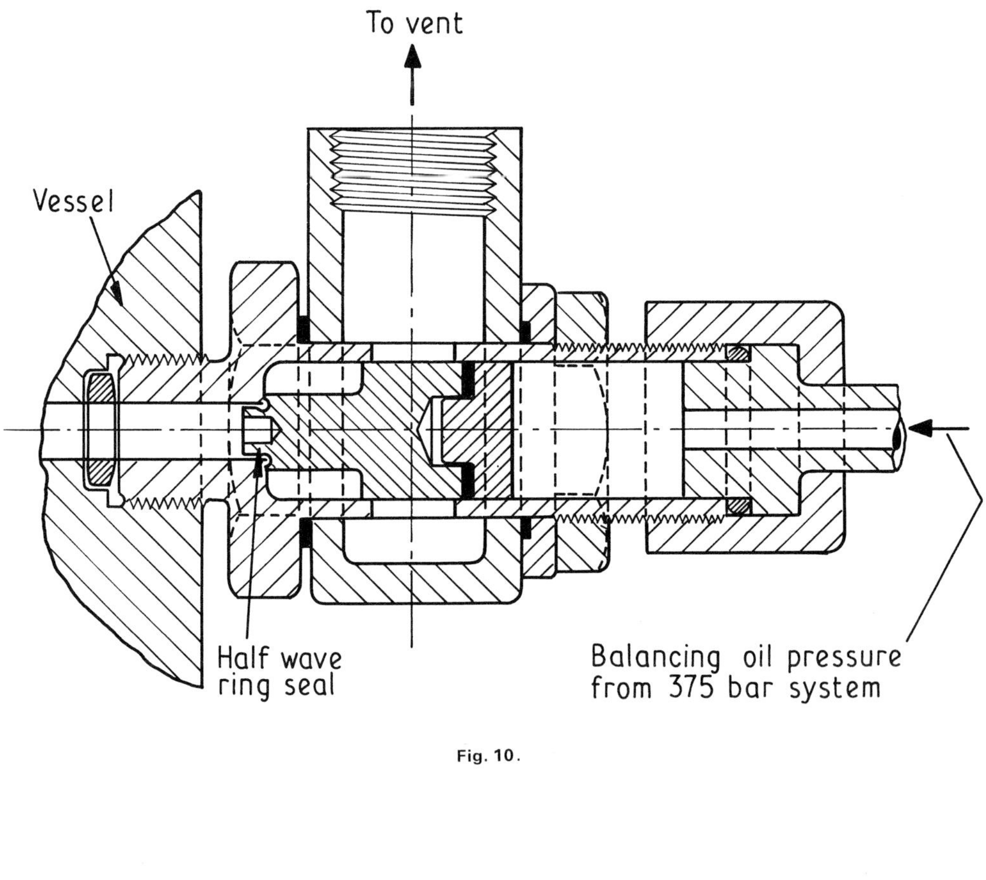

To vent
Vessel
Half wave
ring seal
Balancing oil pressure
from 375 bar system

Fig. 10.

OPEN
SHUT

Fig. 11.

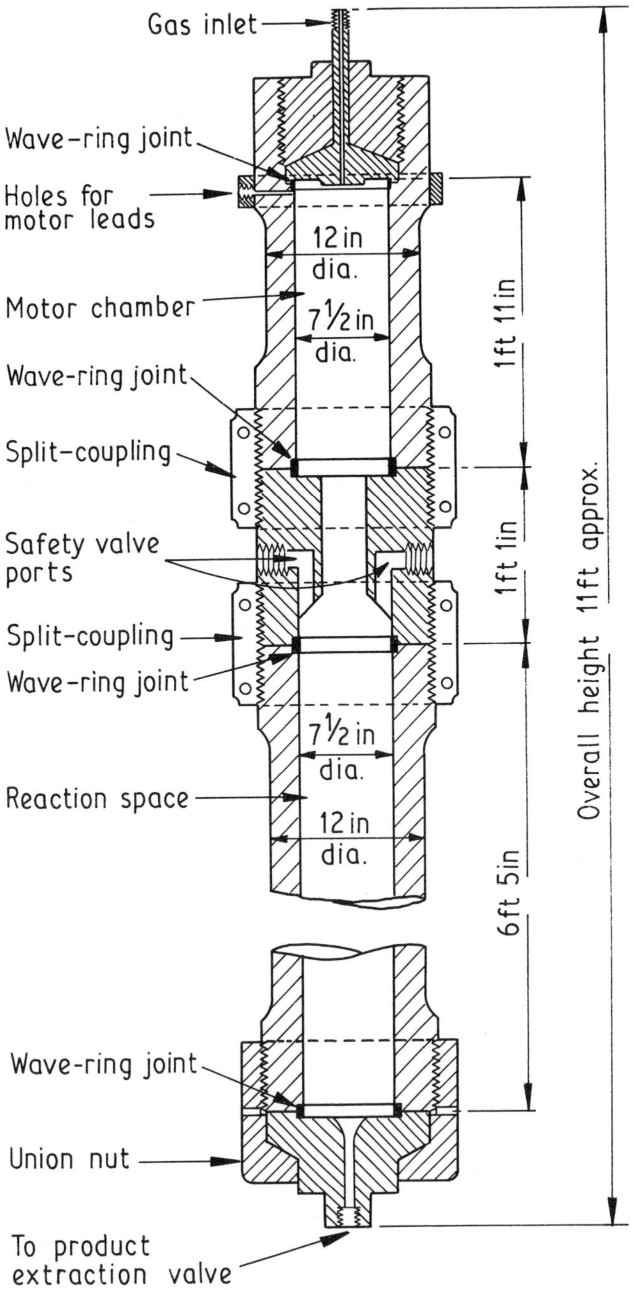
Gas inlet
Wave-ring joint
Holes for motor leads
12 in dia.
Motor chamber
7½ in dia.
Wave-ring joint
Split-coupling
Safety valve ports
Split-coupling
Wave-ring joint
7½ in dia.
Reaction space
12 in dia.
Wave-ring joint
Union nut
To product extraction valve
1 ft 11 in
1 ft 1 in
6 ft 5 in
Overall height 11 ft approx.

Fig. 12.

Fig. 13.

Fig. 14.

A REVIEW OF THE PRESENT STATE OF THE ART IN HIGH PRESSURE CONTAINER DESIGN

H. Ll. D. PUGH, Head of Hydrostatic Extrusion and High Pressure Division
National Engineering Laboratory, East Kilbride, Scotland
B. CROSSLAND, Professor and Head of Department of Mechanical and Industrial Engineering
The Queen's University of Belfast

SYNOPSIS This review paper is concerned with the design of high pressure containers. A brief review is given of the history of the subject and of the factors such as yield, fatigue, fracture toughness, etc. which influence the choice of materials and design stress levels. Obviously, it is very difficult with so many factors which may have to be allowed for to come to a completely satisfactory design procedure. In this paper the design objective is to keep all the stresses within the elastic limit under operating conditions. The designer can then assess his design for other modes of failure such as fatigue and fast fracture.

The constructions which have been considered in some detail are simple and autofrettaged monobloc cylinders, multi-ring containers made from materials of various yield strengths and moduli, sectored vessels, and wire wound containers. The essential equations for the design of such vessels and the limitation of the respective designs are considered in some detail. Some numerical examples of the design of containers are considered to draw attention to certain aspects of the design.

SYMBOLS

σ_θ principal hoop stress

σ_r principal radial stress

σ_z principal axial stress

ε_θ hoop strain

ε_r radial strain

ε_z axial strain

r any radius

r_b bore radius

r_e external radius

ν Poisson's Ratio

E Young's modulus

P_b internal pressure

P_e external pressure

Y tensile yield stress

S shear yield stress

P_Y yield pressure

P_c collapse pressure

P_m the internal pressure to cause yielding to a radius mr_b

$K = {}^{r_e}/_{r_b}$ = overall diameter ratio

$r_m = m\,r_b$ = radius of elastic/plastic interface

K_1, K_2, etc. diameter ratio of individual components in a multi-ring cylinder

Δ interference

$\ln x \equiv \log_e x$

1.0 INTRODUCTION

The behaviour of thick-walled cylinders under internal pressure has been of considerable interest since the invention and development of guns. However, it was not until 1833 that cylinder design was put on a more fundamental basis by Lamé and Clapeyron (1) who established the elastic stress distribution. Barlow (2) shortly after produced a more empirically based equation for the design of hydraulic cylinders, which has been used extensively. Turner (3) was the first to describe the yield and subsequent plastic yielding of a thick-walled cylinder for a typical ductile steel which exhibits an upper initial yield followed by a lower or plastic yield. This was a major step forward as it allows an estimate to be made of the residual stresses in an initially overstrained or autofrettaged cylinders. Since Turner many papers have been published concerned with the stress distribution in partially plastic thick-walled cylinders, for example, Hill, Lee and Tupper (4) (5), which have contributed to a detailed understanding of such matters as the axial stress distribution.

For a strain hardening material the maximum or ultimate pressure which a vessel can contain occurs at strains much larger than those associated with initial yield and the spread of yield through the wall thickness, and any meaningful theoretical treatment must take into account the dimensional changes and particularly the resulting reduction of the ratio of outer to inner diameter. If the cylinder material is sufficiently ductile then as it is expanded beyond complete yield the pressure will rise further until a maximum or 'ultimate' pressure is reached. This occurs when the rate of strengthening due to strain hardening is balanced by the rate of weakening due to the reduction of the diameter ratio. If the cylinder is expanded further it will exhibit a local instability in the form of 'ballooning' while the pressure begins to decrease until ultimately failure occurs. The pressure - expansion curve up to the ultimate pressure can be calculated from shear stress-strain data using a method proposed by Manning (6) which has been modified and extended to use true stress-strain data obtained from tension tests by Crossland and Bones (7) and Crossland, Jorgensen and Bones (8), though they emphasize the advantages of using torsion data.

The first significant experimental work on thick-walled cylinders is that of Cook and Robertson* (9) who showed that the initial yield pressure occurred at a critical value of the shear stress, and who also deduced an empirical equation for the ultimate pressure. Cook (10) (11) reported on further extensive work. He concluded that in a partially plastic closed-ended cylinder the axial stress was the mean of the hoop and radial stresses, and he also observed the non-symmetrical radial expansion in mild steel cylinders due to the large drop of stress at yield in such materials. Morrison (12) was the first to show that the differences in the shear yield stresses deduced from thick-walled cylinder, torsion and tension tests on mild steel could be explained by the effect of stress gradient on yield. Faupel and Furbeck (13) and Faupel (14) reported on an extensive programme of tests on alloy steel cylinders from which they concluded that yield was predicted by the Mises-Hencky yield criterion and they proposed an empirical equation for the ultimate pressure. Crossland and Bones (7) reported extensive data on mild and alloy steel cylinders including tests on cylinders with diameter ratios of up to 8. For a 0.15% carbon steel they showed that the initial yield was accurately predicted by the shear stress or Mises-Hencky yield criteria, if the effect of stress gradient was taken into account. The collapse pressure correlates well with the theory based on the lower yield stress, while at large strains there was remarkable agreement with theoretical curves given by an analysis based on that proposed by Manning (6). For EN25, an alloy steel, the pressure expansion curve up to complete through yielding of the wall agreed well with the theory proposed by Turner (3) which was expanded to predict strains, and the pressure - expansion curve at large strains also agreed with theory. Crossland and Bones measured axial strains and clearly showed that the assumption that the axial stress is the mean of the radial and hoop stresses is reasonable even at strains associated with those at the ultimate pressure. Crossland and Kerr (15) have reported further pressure testing of cylinders of two alloy steels and a mild steel at temperatures up to 370°C, and they again found reasonable agreement between experiment and theory. However, with mild steel at an elevated temperature the cylinders failed on a rising pressure/expansion curve before reaching the ultimate pressure, and at the higher temperatures it was observed that creep could not be ignored for any long term pressurization.

The review up to this point has been limited to reasonably ductile materials, but as the pressure is increased resort has to be made to increasingly brittle high strength materials such as tool steels or tungsten carbide, which have high compressive strength but low tensile strength. The failure of such materials under complex stress is still inadequately understood, and if they are used in high pressure vessel construction the aim of the design must be to force the material as far as possible into compression so that the material under working load remains essentially in compression. For example, in the belt equipment shown in fig. 1, which is used in diamond synthesis, the rams are loaded on their flanks by the sealing pressure in the gasket material, so that the ram material is essentially under compression. The inner most cylinder of the multiple shrink ring assembly is also in a state of high compressive stress. When the system is pressurised the hoop compressive stress is reduced and there is an additional axial component of stress due to the pressure developed in the sealing gaskets, and as a result the principal stresses remain compressive, and the ram is also subjected to compressive stresses on its end and flank surfaces.

So far the review has been limited to the design of cylinders to withstand steady pressure, but in all practical designs the container will have to withstand cycles of pressurization, so they must be able to withstand either long or short life fatigue. The subject of fatigue is considered in several papers in these proceedings, for instance, Parry and Shannon (16), but it is essential to briefly mention the subject here as it is one of the essential considerations in the design of high pressure containers. Early fatigue tests were carried out by Morrison and Cook (17) and Newhall and Kosting (18), but perhaps the first extensive research was that of Morrison, Crossland and Parry (19) who considered the long life fatigue of closed-ended cylinders of EN25, an alloy steel. This and subsequent work has been reviewed by Crossland (20). The important, though not conclusive, conclusions from this work is that the fatigue limit is dictated by a critical value of the shear stress at the bore of the cylinder, which was however appreciably lower than might have been expected from other fatigue data. In section 2 it will be shown that the state of stress in a close-ended thick-walled cylinder can be reduced to a simple shear stress with a superimposed triaxial or volumetric tensile stress, and in a thin-walled cylinder this triaxial tensile stress will be large while in a thick-walled cylinder it will be small. Consequently, it is surprising that both thick and thin-walled cylinders behave identically. It might have been expected that the thin-walled cylinder would have been weaker because of the higher repeated volumetric tensile stress. This has suggested that there may be an additional weakening factor related to the magnitude of the fluid pressure acting on the bore surface, and this is supported by the considerable increase in fatigue strength observed when there is a very thin layer of rubber at the bore of the cylinder or a very thin layer of heavily worked material at the bore.

Parry (21) has noted the considerable and beneficial effects on the fatigue behaviour of autofrettaging cylinders, especially those having cross-bores. As the amplitude of the

* It is of interest to note that at the time of writing Professor Andrew Robertson FRS is at the age of 92, the eldest surviving President of the Institution

shear stress and the associated volumetric tensile stress are unchanged it must be concluded that the residual compressive stresses at the bore are beneficial. This is supported by experimental data on torsional fatigue under a constant high pressure reported by Crossland (22) and Burns and Parry (23) which suggests that fatigue only occurs if there is a tensile principal stress acting during part of the cycle. If this suggestion is correct then it is extremely important as it indicates the possibility of achieving an adequate fatigue life at high pressure, by trying to get high induced compressive stresses.

In many applications the number of cycles of pressure applied is very limited. Low cycle fatigue of cylinders has been examined by Davidson, Eisenstadt and Reiner (24), Austin and Crossland (25), Crossland and Skelton (26), Austin, Reiner and Davidson (27), Terao and Yamashita (28) and Tomkins (29). In low cycle fatigue of ductile cylinders the behaviour is essentially stable at repeated pressures well above the initial yield pressure in unautofrettaged cylinders, but at pressures approaching the collapse pressure the cylinder behaves in an essentially unstable manner and the cylinder finally bursts in a manner akin to a static burst if the material has sufficient ductility. However, Crossland and Skelton (26) noted that as the hardness of the cylinder material was raised, and correspondingly its fracture toughness was reduced, the cylinder showed much less distortion before failure, and ultimately the failure became typical of a very brittle material. The fracture surface becomes completely transverse with only a slight indication of a shear lip at the outer and inner surface, and on breaking open the fracture surface there was either no indication of a fatigue crack or only a very small crack could be observed. Under such conditions the maximum cyclic pressure which could be contained, in fact, decreased as the hardness was raised, even though the pressure at the fatigue limit was raised. This demonstrates the importance of fracture toughness as a design parameter in high pressure vessel design.

The problem of fast fracture of the high strength steels which have had to be used for high pressure vessels is well known. There have been many examples of fast fracture of high pressure vessels particularly in the region of the end closures. Cases are known where cylinders designed for very high pressure have failed during the first pressurization at a pressure well below their design pressure, and even cases where a shrunk on design failed after shrinkage with no pressure, and these suggest that there were in these cases initial small defects which were larger than the critical crack size. Shannon and Austin (30) and Shannon (31) have reported experimental work concerned with longitudinal straight fronted radial cracks in which they have determined the stress intensity factor and compared these values with values computed from a finite element analysis. Perhaps the surprising discovery was that though extreme care was taken to grow sharp cracks, by reducing the pressure pulsations to a very low level towards the end of the crack growing phase, still the experimental values of stress intensity factor were some 15% less than the computed values, even though the material used was a high strength relatively low fracture toughness material. It suggests that linear elastic analysis even for such materials is excessively pessimistic, especially when it is noted that if the cylinder had been subjected to realistic pressure pulsations the discrepancy would have been even greater. Further work is reported in these proceedings by Parry and Shannon (16) and Crymble, Goldthorpe and Austin (32).

Fracture toughness is greatly influenced by the presence of hydrogen which causes a very great reduction in fracture toughness, and this can occur in practice, for instance, due to electroplating. Fracture toughness can also be considerably reduced by the pressurizing fluid used in the system. Though the effect of fluids on fracture toughness has been extensively studied, it has not been studied in high pressure systems where the fluid may be forced into the crack zone. Also in vessels running at modest pressures where some creep occurs there may be some deterioration in fracture toughness with creep strain, so that a vessel initially produced from an acceptable high fracture toughness material might conceivably become increasingly dangerous with time. This indicates yet again that as extensive as our knowledge is, there is still many regions of uncertainty.

In design it is desirable, if possible, to select a material and design so that if a fatigue crack grows it will grow stably through the wall thickness without fast fracture intervening, so that it will leak before fracture. A leak may still be an acceptable hazard since it is possible to provide a safe design, but if a fast fracture were to intervene then havoc could be wrought. As the working pressure is raised it becomes increasingly difficult and ultimately impossible to achieve leak before fast fracture. In such circumstances the vessel must be sheathed with a lower strength but higher fracture strength material, as for example in the vessel described by Crossland and Logan (33), and by Pugh, Wilkinson and Hodge (34), in which the high strength inner component is housed in an outer steel sheath which also takes the closure end load. The inner member which is a simple cylindrical component may fail but if so it can be replaced.

From the brief review it will be apparent that though much is understood about the design of thick-walled cylinders subject to internal pressure, nevertheless there are many aspects which are not so clearly understood or which are more difficult to quantify or to generalise in order to allow an optimum design to be achieved. Though the elastic and plastic analysis is well established the stress level which can be tolerated is ill defined in a fatigue situation or where leak before fast fracture is required or where the cylinder is required to operate at above ambient temperatures or where an unfavourable environment is involved. Like so many other design problems the designer must do the best he can with the analytical procedures available and then assess his design in the light of available data. He may, for instance, aim to keep all his stresses within the yield stress of the material he is using in his construction. At very high pressures he may have to resort to very high yield strength materials, together with a more complex design involving multi-component cylinders and/or overstrain to introduce favourable residual stresses. In more extreme circumstances he may be forced to use brittle materials which have very high compressive strength but relatively low tensile strength, so that it is necessary to induce high residual compressive stresses and to operate at pressures under which the stresses remain compressive or only become slightly tensile. Having designed the cylinder the designer will then be well advised to assess his design for other failure modes such as fatigue and fracture.

Medium or high strength materials are mainly used for the construction of high pressure containers, and in general they show a low rate of strain hardening and consequently the collapse pressure is not so very much less than the ultimate pressure. With such materials it is not unreasonable to consider them as elastic ideally plastic materials, which considerably simplifies the analysis and which allows generalised curves to be produced showing the performance of various cylinder designs which will be considered. Materials which might be considered in this category are alloy steels such as EN24, EN25, EN26 NCMV and Maraging or their equivalents. Many of these materials can be obtained in an electro-slag refined or vacuum remelted state which enhances their fracture toughness to strength ratio.

The object of the present paper is to present a review of the present state of the art of high pressure container design. It is obvious that the radial stress, σ_r, at the bore of a monobloc cylinder, must be compressive (i.e. negative) and equal to the fluid pressure, P, and it will be shown that for a thick-walled cylinder the tangential stress, σ_θ, is tensile and it approaches the same magnitude as the pressure. Consequently, even for a very large diameter ratio, K, the yielding of the cylinder according to the Tresca or maximum shear stress criterion occurs at the bore of the container when

$$\frac{\sigma_\theta - \sigma_r}{2} = \frac{Y}{2} \text{ or } S$$

where Y is the yield stress in uniaxial tension and S is the shear yield stress. However, for a large diameter ratio cylinder

$$\sigma_r = -P \text{ and } \sigma_\theta \simeq +P$$

$$\therefore \quad (\sigma_\theta - \sigma_r) \simeq 2P$$

so that at yield

$$P_Y = \frac{\sigma_\theta - \sigma_r}{2} = \frac{Y}{2} \text{ or } S$$

This means that a monobloc cylinder cannot contain a pressure greater than half the tensile yield stress or the torsion yield stress if it is to remain elastic, which even for a high strength Maraging steel would imply a pressure limitation of about 1 GPa*.

The necessity to contain much higher pressures for hydrostatic extrusion, isostatic compaction and research equipment has led to various design methods aimed at increasing the pressure containing capability. Since nothing can be done about the fact that the radial stress is always equal to the pressure all these methods are directed in one way or another to reducing or eliminating the circumferential tensile stresses. From fatigue considerations it is also desirable even for lower operating pressures to keep the circumferential stress compressive, as it appears likely that fatigue is inhibited if all the principal stresses remain compressive. The methods of reducing or eliminating the circumferential stress are:-

(1) Autofrettage or overstrain which produces favourable residual stresses.

(2) External support rings or cylinders which are shrunk on to the inner core cylinder. Alternatively, the two component cylinders are machined with matching tapers and they are then force fitted together with a suitable lubricant at the interface A convenient method which facilitates the assembly and dis-assembly of 2 ring containers is to pump oil under pressure usually through a small radial hole in the outer ring to the interface as proposed by Bratt (35).

(3) Using a sectored construction for the inner component; the sectors being forced into an initial state of compression by using an external shrink ring. In such a construction the sectors act as anvils under compressive load provided by the fluid pressure on their inner surface which is balanced by a compressive supporting load acting on the external surface. It is obviously an ideal situation for the use of materials with a high compressive strength, such as tungsten carbide, for the sectors.

(4) Wire winding of an inner core tube or a sectored vessel can provide very considerable advantages such as: high strength levels even with low alloy steel, as cold drawn steel wire has a very high yield strength; the wire will have no significant imperfections as with a forging and in any case it can be automatically inspected; fracture toughness though probably low is unimportant as the cross-section of the wire is small compared with the wall thickness of a monobloc cylinder; even if the wire fails slip will be prevented by the frictional forces and it will have no significant effect; the wire can be wound on with varying tension so nearly any initial stress distribution can be obtained; etc.

2.0 DESIGN OF MONOBLOC CONTAINERS

If it is assumed that a cylinder is sufficiently long so that end effects can be ignored then it follows by symmetry that any originally plane transverse sectionsremain plane and hence that the longitudinal strain ε_z is the same for all elements. The Lamé equations for radial equilibrium of an element of the cylinder with a bore pressure P_b and an external pressure P_e are

---

* $1\ {}^{N}/_{m^2} = 1$ Pa or 1 GPa $= 10^9\ {}^{N}/_{m^2} = 10$ kbar $=$ 145038 lbf/in^2 or psi

$$\sigma_r = \frac{P_b}{(K^2-1)}\left\{1-K^2\frac{r_b^2}{r^2}\right\} - \frac{P_eK^2}{(K^2-1)}\left\{1-\frac{r_b^2}{r^2}\right\}$$

$$\sigma_\theta = \frac{P_b}{(K^2-1)}\left\{1+K^2\frac{r_b^2}{r^2}\right\} - \frac{P_eK^2}{(K^2-1)}\left\{1+\frac{r_b^2}{r^2}\right\} \qquad (1)$$

Therefore

$$\sigma_r + \sigma_\theta = \frac{2}{(K^2-1)}(P_b - K^2P_e) = \text{a constant} \qquad (2)$$

Now Hooke's Law gives the following expressions for the strains

$$E\varepsilon_r = \sigma_r - \nu(\sigma_\theta + \sigma_z)$$
$$E\varepsilon_\theta = \sigma_\theta - \nu(\sigma_z + \sigma_r) \qquad (3)$$
$$E\varepsilon_z = \sigma_z - \nu(\sigma_r + \sigma_\theta)$$

Since as mentioned above ε_z is a constant and $(\sigma_r + \sigma_\theta)$ is a constant from equation 2, it follows that σ_z is also a constant so that for this monobloc there is simultaneous plane stress and plane strain. The above equations can be solved for all the unknowns if either the value ε_z or σ_z is known but this requires another condition to completely define the problem. This point is discussed later. The radial expansion resulting from the applied pressures is given by

$$u = r\varepsilon_\theta \qquad (4)$$

For the case of a closed ended monobloc cylinder subjected to internal pressure the axial stress must support the end load carried by the end closures. Hence

$$\sigma_z = \frac{P_b}{(K^2-1)} \qquad (5)$$

and the value of σ_r and σ_θ reduce to the first term on the **right**-hand side of equation 1. These equations can be interpreted in several ways, one of which is to realise that they reduce to a mean hydrostatic tensile component of stress

$$\sigma_m = \frac{1}{3}(\sigma_r + \sigma_\theta + \sigma_z) = \frac{P_b}{(K^2-1)} \qquad (6)$$

with a shear stress in the transverse plane

$$\tau = \frac{P_b}{(K^2-1)}K^2\frac{r_b^2}{r^2} \qquad (7)$$

which will have a maximum value at the bore of

$$\tau_{r=r_b} = \frac{K^2}{(K^2-1)}P_b \qquad (8)$$

It has, however, been shown by Crossland (36) for simple torsion under pressure and by Pugh and Green (37) for simple tension under pressure that pressure has no influence on the initial yield and plastic flow of wrought ductile materials except to extend the stress strain curve to larger strains before failure. A review of this work has been given by Brandes (38). Consequently, it is not unreasonable to expect for modest values of the mean hydrostatic tensile stress or what can be considered as a negative pressure that there will be no effect on the yield and plastic flow except some reduction in the strain to failure. If the shear yield determined from a torsion test is S then the internal pressure to cause yield will be:

$$P_y = \frac{(K^2-1)}{K^2}S = \frac{(K^2-1)}{K^2}\frac{Y}{2} \qquad (9)$$

If tensile data only is available then S will have to be replaced by $Y/2$ or $Y/\sqrt{3}$ depending on whether the Tresca or the Mises yield criterion is adopted.

For a closed- ended cylinder the diametral strain at the outer diameter is given by

$$(\varepsilon_\theta)_{r=r_e} = \frac{P}{E(K^2-1)}(2-\nu) \qquad (10)$$

while the axial strain which is uniform is given by

$$\varepsilon_z = \frac{P}{E(K^2-1)}(1-2\nu) \qquad (11)$$

For an open ended cylinder i.e. zero axial stress the pressure to cause initial yielding using the Tresca criterion is given by equation 9 and this is frequently the assumption made in design. However, several materials including high strength steels more nearly obey the Mises yield criterion which gives a different equation for the yield pressure viz

$$P_y = \frac{K^2-1}{\sqrt{(K^4+\frac{1}{3})}}S = \frac{(K^2-1)Y}{\sqrt{(3K^4+1)}} \qquad (12)$$

It will be seen from Fig. 2 that using the Mises criterion there is little difference between the pressures which an open ended and a closed ended cylinder can contain. In an open ended container, the end load on the closures must of course be supported externally and this is normally achieved by either a yoke or some form of press.

The open and closed ended conditions are those normally met in practice except in situations such as the Belt equipment used in diamond synthesis where an essentially hydrostatic state of stress is achieved by a very heavy shrink fit and a high compressive end load.

3.0 DESIGN OF AUTOFRETTAGED CYLINDERS

If the pressure in a cylinder is taken beyond its yield pressure then yield will spread through the wall of the cylinder as the pressure is increased until yield reaches the outer diameter, when the pressure can be raised no higher if the material is elastic-perfectly plastic which is not too far from reality with high strength steels. If the pressure is removed, then the bore material of the cylinder will be forced into a state of residual compressive stress, and when the cylinder is subsequently pressurised it will behave in a nearly perfectly elastic fashion up to the original pressure, as long as the range of stress variation is not excessive. The elastic behaviour of an over-strained cylinder is improved by subjecting the cylinder after its initial pressurization to a low temperature heat treatment at 200°-300°C, which has the effect of ironing out high local stress concentrations in the crystal structure by allowing a redistribution of the dislocations which pile up at grain boundaries.

It is doubtful if the state of stress in a partially plastic cylinder is yet fully described, and to establish the residual stresses requires a complete understanding of the Bauschinger effect. Franklin and Morrison (39) have examined the residual stresses in autofrettaged cylinders, and they explained these allowing for the Bauschinger effect by carrying out torsion tests on thin tubes beyond their yield point and then unloading the tubes and applying a reversed torque.

An adequate design procedure for **autofrettaged** cylinders can be established if it is assumed that the Tresca or maximum shear stress criteria describes the yield and subsequent plastic flow, or if the cylinder is closed-ended that the axial stress remains the mean of the radial and hoop stresses. Additionally, it is assumed that the elastic-plastic boundary is cylindrical and concentric with the bore of the cylinder. When the pressure is removed it is assumed that there is no Bauschinger effect so that the material behaves completely elastically during unloading, until the reversed stresses reach the same yield stress as that in virgin unstrained material. Consequently, the equation of radial equilibrium can be re-written as

$$\sigma_\theta - \sigma_r = 2S = r\frac{d\sigma_r}{dr} \qquad (13)$$

or $$d\sigma_r = 2S\frac{dr}{r}$$

Integrating this between the bore radius, r_b, and the radius of the elastic-plastic interface, mr_b, gives the difference in radial stress across the plastic core. Then consideration of the elastic outer shell between radius mr_b and radius Kr_b, which is on the point of yield at its bore, leads to the following equation for the pressure acting in the cylinder

$$P_m = S\left\{\ln m^2 + \frac{K^2 - m^2}{K^2}\right\} \qquad (14)$$

When the yield reaches the outer surface, i.e. m = K the pressure, P_c, is known as the collapse pressure or the pressure at which the cylinder will continue to expand without a further increase in pressure if there is no strain hardening.

$$P_c = S \ln K^2 \qquad (15)$$

where S can be stated in terms of the tensile yield as before.

If the strains in the elastic outer shell of a closed-ended cylinder are considered then it can be shown, Crossland (40), that

$$(\varepsilon_\theta)_{r=r_e} = \text{const} \times m^2 \qquad (16)$$

where the constant can be found from considering the cylinder on the point of yielding, i.e. m = 1, when from equations **10 and 9**

$$(\varepsilon_\theta)_{r=r_e} = \frac{P_Y}{E(K^2-1)}(2-\nu)$$

$$= \frac{S}{E}\frac{(2-\nu)}{K^2} \qquad (17)$$

If the cylinder behaves completely elastically during unloading then the residual stress at the bore will be

$$S - P_m\frac{K^2}{K^2-1}$$

which must not exceed the shear yield stress in reverse loading. For instance, if the cylinder is completely overstrained then the maximum diameter ratio which will not give reversed yielding on the removal of pressure will be given by

$$S - P_c\frac{K^2}{K^2-1} = -S$$

or $$S - \frac{K^2}{K^2-1} S\ln K^2 = -S$$

or $$\ln K = \frac{K^2-1}{K^2}$$

which gives a value of K = 2.218. Above a value of K of 2.218 a cylinder cannot be overstrained to the outer diameter without causing reversed yielding at the bore on unloading. Below a K of 2.218 a cylinder can be completely overstrained without reversed yielding occuring on unloading. So it can be concluded that for K < 2.218 the maximum pressure which a cylinder can contain elastically after autofrettage is given by the collapse pressure, P_c. For K > 2.218 the maximum pressure which can be contained after autofrettage is that which when removed gives rise to a residual shear stress at the bore of - S, which implies that as the pressure is removed the shear stress goes from + S to - S so the pressure will be

$$P = 2S\frac{(K^2-1)}{K^2} \qquad (18)$$

Fig. 3 shows the limiting value of pressure divided by the shear yield stress for an optimum autofrettaged cylinder. This is, of course, based on the assumption of no Bauschinger effect during unloading and patently this is not the case. Nevertheless, experience shows that cylinders autofrettaged to this optimum pressure and subsequently subjected to a low temperature heat treatment appear to behave elastically up to the autofrettage pressure.

If the material of construction is elastic ideally plastic then the maximum or ultimate pressure the cylinder can contain will be identical to the collapse pressure, and from Fig. 3 it will be seen that for K > 2.218 this pressure exceeds the optimum autofrettage pressure which provides a factor of safety against failure. Below K = 2.218 the optimum autofrettage pressure is equal to the ultimate pressure. In practice the materials used in the construction of even very high pressure vessels show a little strain hardening so that the ultimate pressure is, in fact, greater than the collapse pressure, which gives an additional factor of safety.

One of the problems of inducing residual stresses by autofrettage is the question of the stability of these stresses under cyclic loading. Morrison and Cook (17) found no deterioration of residual stresses after 2000 cycles of the autofrettage pressure, but Nishihara and Yamaguchi (41) have noted a considerable reduction in the residual stresses induced in maraging steel cylinders autofrettaged to 0.98 GPa and then subjected to 2 x 10^4 cycles of a pressure of 0.50 GPa. Obviously, this question of the stability of residual stresses is of great importance.

4.0 THE DESIGN OF MULTI-RING CONTAINERS

A method which is commonly used for containing high hydrostatic pressure is to support the container with external supporting rings in a multi-ring design. If the material of each ring were to be of the same material there would be no advantage of such a construction over an autofrettaged cylinder except that none of the material would be subject to yielding with the consequential uncertainties imposed by the Bauschinger effect. One of the main reasons for the use of compound cylinders stems from the inherent low fracture toughness of high strength materials. The use of such steels in an autofrettaged monobloc design would be dangerous, so that it is desirable to use it as the inner member of a compound vessel and to make the outer member or members of a lower strength but higher fracture toughness material. Another reason for the use of compound cylinders is the effect of size on the hardening of many steels, so that the smaller components can be hardened further than the larger ones.

The plastic-elastic analysis of thick walled multi-ring cylindrical containers has been given by Horne (42) whilst Parsons and Cole (43) have considered the optimum design of short composite duplex cylinders made of brittle materials using the Mohr criterion of failure. This paper is concerned with elastic design. Manning (44) developed a theory for the elastic design of a multi-ring container using the same material for all rings whilst Strub (45), Becker and Mollick (46) and Pugh (47) have analysed the case for an n-ring container with each component having a different yield stress. These analyses have been concerned with the maximum pressure that can be obtained in each design of container without yielding taking place at any point. Becker and Mollick also considered the question of the required interference fits between each ring. All the above analyses assumed that all the rings of the assembly had the same elastic constants, whereas in the analysis given below the optimum design is obtained for an n-ring container each component of which has different yield stresses and different elastic constants. It is a generalisation of Pugh's (47) analysis for a two ring container, each component having different properties.

Two criteria of yielding are commonly used namely the Von Mises or maximum energy of distortion criterion which involves all three principal stresses and the Tresca or maximum shear stress criterion which involves the maximum difference in principal stresses. The difference between the prediction of these criteria under the most general case is not more than ± 7½%. However, since the Tresca criterion does not require a knowledge of the value of the intermediate stress this criterion is used since it simplifies the mathematics. In a shrunk-on construction Manning and Labrow (48) have pointed out that the axial stresses are unknown so there are considerable doubts in using the Von Mises criterion, and this has been confirmed experimentally by Crossland and Burns (49). In general the Tresca criterion is only concerned with stresses in the transverse plane since the axial stress, σ_z, is in almost all cases the intermediate stress, and consequently the biggest stress difference is $(\sigma_\theta - \sigma_r)$. However, exceptions can be visualised so careful consideration should be given to each individual case.

Applying the Tresca criterion to a monobloc subject to a bore pressure P_b and an external pressure P_e it can be seen from equations (1) that

$$(\sigma_\theta-\sigma_r) = 2\,(P_b-P_e)\,\frac{K^2}{K^2-1}\,\frac{r_b^2}{r^2} \qquad (19)$$

which has a maximum value at the bore of

$$(\sigma_\theta-\sigma_r)_{r=r_b} = 2\,(P_b-P_e)\,\frac{K^2}{K^2-1} \qquad (20)$$

so that at yield

$$(\sigma_\theta-\sigma_r)_{r=r_b} = Y = 2\,S$$

$$\text{or } P_b-P_e = \frac{Y}{2}\left(1-\frac{1}{K^2}\right) \text{ or } S\left(1-\frac{1}{K^2}\right) \qquad (21)$$

The case of a container made up of an assembly of n-rings each having different yield stresses and different Young's Modulii and Poisson's ratios will now be considered. The container has a bore and interface radii of r_1, r_2, ... r_i .... r_n and an outside radius of r_{n+1}. The corresponding bore and interface pressures are P_1, P_2, ... P_i .... P_n respectively and the external pressure is taken as zero.

4.1 Maximum Pressure

Now the optimum design will be one in which all the rings reach their respective yield stresses simultaneously and applying the condition given in equation (21) above to the yielding of the bore of each ring in turn it follows that

$$\left.\begin{array}{ll}
\text{First ring} & P_1 - P_2 = \frac{Y_1}{2}\left(1-\frac{1}{K_1^2}\right) \\
\text{Second ring} & P_2 - P_3 = \frac{Y_2}{2}\left(1-\frac{1}{K_2^2}\right) \\
\cdots\cdots & \cdots\cdots \\
i\text{th ring} & P_i - P_{i+1} = \frac{Y_i}{2}\left(1-\frac{1}{K_i^2}\right) \\
\cdots\cdots & \cdots\cdots \\
n\text{th ring} & P_n - 0 = \frac{Y_n}{2}\left(1-\frac{1}{K_n^2}\right)
\end{array}\right\} \qquad (22)$$

Since these relations are simultaneously true, then by addition

$$P_1 = \frac{Y_1}{2}\left\{1-\frac{1}{K_1^2}\right\} + \frac{Y_2}{2}\left(1-\frac{1}{K_2^2}\right) + \cdots + \frac{Y_n}{2}\left(1-\frac{1}{K_n^2}\right). \qquad (23)$$

For a maximum value of P_1, $dP_1 = 0$ and by differentiation,

$$\frac{Y_1}{K_1^2}\frac{dK_1}{K_1} + \frac{Y_2}{K_2^2}\frac{dK_2}{K_2} + \dots \frac{Y_i}{K_i^2}\frac{dK_i}{K_i} \dots + \frac{Y_n}{K_n^2}\frac{dK_n}{K_n} = 0 \qquad (24)$$

For a fixed overall radius ratio K, we have by definition

$$K_1 \cdot K_2 \dots K_i \dots K_n = K \qquad (25)$$

which on differentiation becomes

$$\frac{dK_1}{K_1} + \frac{dK_2}{K_2} + \dots \frac{dK_i}{K_i} + \dots \frac{dK_n}{K_n} = 0 \qquad (26)$$

From comparison of equations (24) and (26) it can be seen that

$$\frac{Y_1}{K_1^2} = \frac{Y_2}{K_2^2} = \dots = \frac{Y_i}{K_i^2} = \dots \frac{Y_n}{K_n^2}$$

which for convenience can be expressed as

$$\frac{Y_1}{K_1^2} = \frac{Y_2}{K_2^2} = \dots = \frac{Y_i}{K_i^2} = \dots \frac{Y_n}{K_n^2} = \alpha^2 \qquad (27)$$

where

$$\alpha^2 = \left\{\frac{Y_1, Y_2 \dots Y_i \dots Y_n}{K^2}\right\}^{1/n} \qquad (28)$$

The maximum bore pressure P_{max} is from equation (23)

$$\left.\begin{aligned} P_{max} &= \frac{1}{2}\sum_{i=1}^{n} Y_i - \frac{n}{2}\left(\frac{Y_1Y_2\dots Y_i\dots Y_n}{K^2}\right)^{\frac{1}{n}} \\ P_{max} &= \frac{1}{2}\sum_{i=1}^{n} Y_i - \frac{n}{2}\alpha^2 \end{aligned}\right\} \qquad (29)$$

These relationships are similar to those derived by Becker and Mollick (46), Pugh (47) and Manning and Labrow (48) in their analysis of the multi-ring container and they are not dependent on the elastic constants of the rings.

The pressure at any interface when the container is up to pressure can be obtained from equation (22) viz

$$P_i = \frac{Y_i}{2}\left(1 - \frac{1}{K_1^2}\right) + \frac{Y_{i+1}}{2}\left(1 - \frac{1}{K_{i+1}^2}\right) + \dots \frac{Y_n}{2}\left(1 - \frac{1}{K_n^2}\right)$$

$$= \frac{1}{2}\sum_{i=1}^{n} Y_i - \frac{(n - i + 1)}{2}\alpha^2 \qquad (30)$$

A few examples using the above formulae have been worked out for a number of different multi-ring containers. For the purpose of these examples it is assumed that the axial stress is the intermediate stress and the overall diameter ratio, K, is taken to be 8. The residual stresses in these designs have been checked according to the analysis discussed in section 4.3 and it has been established that in none of the designs considered do these stresses exceed yield. The details of these designs, including their pressure carrying capacity, are given in table 1.

4.2 Radial Displacements

It may be seen from Section 4.1 that provided the Tresca criterion of yielding is used and that the axial stress is intermediate between the radial and circumferential stresses which is generally the case then the maximum pressure which a container can sustain is independent of the axial conditions. However, in considering radial displacements and interference fits it is necessary to define the axial condition or end loading.

Hill (5), for instance, has considered various end conditions for a monobloc cylinder subjected to internal pressure. If it is assumed that the cylinder is long and plane sections remain plane then it has been shown in section 2 that a state of plane stress also exists. In general use a high pressure cylinder is either closed at both ends by plugs attached to the cylinder itself or by floating plungers which are held in position by external forces applied usually by a press or some framework. Hill showed that a third case, that of zero axial strain, is intermediate between the first two, although nearer to the closed end condition to which it is strictly equivalent when the material is incompressible. The radial and circumferential stresses are identical in all three cases but the axial stresses are different and intermediate between the other two principal stresses.

4.2.1 Open Ended Cylinder (i.e. $\sigma_z = 0$)

From equation (3)

$$\varepsilon_\theta = \sigma_\theta - \nu\sigma_r \qquad (31)$$

and from equations (4) and (1) the displacement in the ith ring of a n ring container is

$$u = \frac{r}{E_i(K_i^2-1)}\left[P_i\left\{(1-\nu_i) + (1+\nu_i)K_i^2\frac{r_i^2}{r^2}\right\} - P_{i+1}K_i^2\left\{(1-\nu_i) + (1+\nu_i)\frac{r_i^2}{r^2}\right\}\right] \qquad (32)$$

4.2.2 Closed Ended Container

The axial load on the end closure is $\pi r_b^2 P_b$ but the question is how is this distributed in the case of a multi-ring container. If no shear stress exist at the interfaces between the rings then the end load would be taken by the ring to which the end closure is attached, probably the inner one. However, this is rather unrealistic and it is assumed that there is sufficient friction at the interfaces to allow the stress to be transferred between the rings, and at sections **remote** from the ends the stress may be regarded as uniform or

$$\sigma_z = \frac{P_b}{K^2-1} \qquad (33)$$

It can then be readily shown that the radial displacement in the ith ring is

$$u = \frac{r}{E_i(K_i^2-1)}\left[P_i\left\{(1-\nu_i) + (1+\nu_i)K_i^2\frac{r_i^2}{r^2}\right\} - P_{i+1}K_i^2\left\{(1-\nu_i) + (1+\nu_i)\frac{r_i^2}{r^2}\right\} - \nu P_b\right] \qquad (34)$$

4.2.3 Zero Axial Strain (i.e. $\varepsilon_z = 0$)

This end condition is sometimes considered owing to its relative mathematical simplicity. It corresponds to the case of a container held between two rigid end plates. Substituting this value in equation (3) and using equation (1) it follows that for the ith ring

$$u = \frac{r(1+\nu_i)}{E_i(K_i^2-1)}\left\{P_i\left(1-2\nu_i+K_i^2\frac{r_i^2}{r^2}\right) - P_{i+1}K_i^2\left(1-2\nu_i+\frac{r_i^2}{r^2}\right)\right\} \qquad (35)$$

4.2.4 Discussion

Becker and Mollick (46) have pointed out that in a long compound cylinder just as in a long monobloc a state of generalised plane strain may be assumed remote from the ends of each component. However, if ε_z is the same on both sides of an interface, then the longitudinal stresses on each side will differ. This would result in axial shear stresses at the interface which thereby violate the strict condition of generalised plane strain in all sections and the radial direction is no longer a principal direction. Babb (51) has also briefly discussed the question of axial shear stresses as a result of shrinking rings together. If on the other hand σ_z is the same on each side of an interface, then ε_z whilst constant within each ring would have a different value in each component.

4.3 Change of Interfacial Pressures due to Change in Internal Pressure

Consider the effect of change of the internal pressure (either an increase or decrease) on the interfacial pressures of an n-ring container assuming that the interfacial surfaces remain in contact at all times.

Consider a typical assembly of two adjacent rings in the middle of the n-ring assembly say the ith and the (i+1)th rings (Fig. 4). Since the interfacial surfaces remain in contact then the displacement of the outer surface (r_{i+1}) of the inner (i^{th}) ring must be the same as the radial displacement of the inner surface (r_{i+1}) of the outer (i+1)th ring. Using the above equations for the radial displacement the following relations between the changes in the interfacial pressure as a result of a change in the internal pressure by a value of P_1 may be obtained

4.3.1 Open Ended Cylinder

$$P_iN_i = P_{i+1}(L_i+M_{i+1}) - P_{i+2}K^2_{i+1}N_{i+1} \qquad (36)$$

where $$L = \frac{1}{E}\left[\frac{(K^2+1)}{(K^2-1)} - \nu\right]$$

$$M = \frac{1}{E}\left[\frac{(K^2+1)}{(K^2-1)} + \nu\right]$$

$$N = \frac{2}{E(K^2-1)}$$

The corresponding equations for all the rings in the assembly are as follows

$$\left.\begin{array}{l} P_1N_1 = P_2(L_1+M_2) - P_3K_2^2N_2 \\ P_2N_2 = P_3(L_2+M_3) - P_4K_3^2N_3 \\ \text{-------------------------------} \\ P_iN_i = P_{i+1}(L_i+M_{i+1}) - P_{i+2}K^2_{i+1}N_{i+1} \\ \text{-------------------------------} \\ P_{n-1}N_{n-1} = P_n(L_{n-1}+M_n) - 0 \end{array}\right\} \qquad (37)$$

From the last equation P_{n-1} may be obtained in terms of P_n. From the last but one equation P_{n-2} is then obtained in terms of P_n and so on until finally P_n may be obtained from the known value of P_1. Consequently, all the other values of change in interface pressure may be easily determined. If P_1 is made equal to P_b then these equations give the changes in interfacial pressures at each interface as a result of changing the bore pressure from P_b to zero.

4.3.2 Closed Ended Cylinder

Except for the last term the expression (equation 34) is identical with that (equation 32) for the open ended container. Corresponding equations therefore in this case will be similar to those for the open ended cylinder (equation 37) with an extra term at the end of each equation. Therefore, in this case the equation for the ith ring would be as follows

$$P_iN_i = P_{i+1}(L_i+M_{i+1}) - P_{i+2}K^2_{i+1}N_{i+1} + \frac{P_1}{(K^2-1)}\left(\frac{\nu_i}{E_i} - \frac{\nu_{i+1}}{E_{i+1}}\right) \qquad (38)$$

This set of equations can be solved in the same way as for equation 37. The only difference is an extra known term in P_1 in each equation.

4.3.3 Zero Axial Strain

Following the same procedure as for plane stress the pressures acting on each interface might be written as follows

$$
\left.
\begin{aligned}
P_1 . A_1 &= P_2 (B_2 + C_1) - P_3 . K_2^2 . A_2 \\
P_2 . A_2 &= P_3 (B_3 + C_2) - P_4 . K_3^2 . A_3 \\
&\cdots\cdots\cdots\cdots \\
P_i . A_1 &= P_{i+1} (B_{i+1} + C_i) - P_{i+2} K_{i+1}^2 \; A_{i+1} \\
&\cdots\cdots\cdots\cdots \\
P_{n-1} \; A_{n-1} &= P_n \; (B_n + C_{n-1}) - 0
\end{aligned}
\right\} \quad (39)
$$

where

$$A = \frac{2}{E} \; \frac{(1 - \nu)^2}{(K^2 - 1)} \qquad B = \frac{1 + \nu}{E} \left[1 + \frac{2(1 - \nu)}{K^2 - 1} \right]$$

$$C = \frac{1 + \nu}{E} \left[(1 - 2\nu) + \frac{2(1 - \nu)}{K^2 - 1} \right]$$

This is also a set of (n-1) equations to solve for (n-1) unknowns, which can be solved in the same way as for equation (37). The changes in interface pressures consequent on a change of P_1 in the bore pressure can thus be readily determined.

4.3.4 Discussion

Consider now the case of the optimum design discussed above in which the bore pressure is at its maximum value of P_{max}. If this bore pressure is now removed then all the interface pressures given by equation (30) will be reduced by values given by equations (37), (38) and (39) for the cases of open and closed ended containers and for zero strain respectively. The residual interface pressures P_i' will be those resulting from the interference fits during the assembly of the n-ring container.

4.4 Interference Fits

To determine the interference fits, imagine the disassembly of the n-ring container into its separate components. When disassembled, the two adjacent rings, the ith and the (i+1)th are free of stresses and the interface displacements at their common interface resulting from disassembly will now be different, the difference being the interference required at this interface to produce the residual pressure P'_{i+1} Consequently, the relative interference at the interface between the ith and the (i+1)th rings is as follows

$$\Delta_{r_{i+1}} = \frac{(u_i)_{r_{i+1}}}{r_{i+1}} - \frac{(u_{i+1})_{r_{i+1}}}{r_{i+1}} \qquad (40)$$

i.e. the difference in the relative radial displacements of the outer surface of the ith ring and of the inner surface of the (i+1)th ring when the residual stresses are removed.

Now the values to be used in equation (40) will depend on the conditions under which the assembly or dis-assembly of the multi-ring container are carried out. In general, however, the condition of the rings before assembly and after assembly is that they are not subjected to any axial stress. Consequently, although the residual interface stresses P_i' are different for the cases of open ended container, closed ended container and plane strain the formulae to be used to determine the interference fits in all cases will be those appropriate to the open ended container namely

$$\Delta_{r_{i+1}} = P'_i \; N_i - P'_{i+1} \quad (L_i + M_{i+1}) +$$

$$P'_{i+2} \quad K_{i+1}^2 N_{i+1} \qquad (41)$$

In conclusion, the maximum pressure which an assembly of n-components each with different elastic constants and yield stresses can withstand without yielding is given by equation (29). Equation (30) gives the values of the interface pressures when the maximum pressure is applied to the bore. The reduction in these interface pressures when the maximum pressure at the bore is removed is obtained by putting $P_1 = P_b$ in equations (37), (38) and (39), for open and closed ended containers and for plane strain respectively. The difference gives the residual interface pressures P'_i. The interference fits that are necessary are then obtained from equation (41).

5.0 LIMITATIONS AND IMPROVEMENTS IN MULTI-RING DESIGN

One of the limitations of the multi-ring container design described above is the limited extent to which a fixed support pressure can be applied. The residual stresses in the inner ring rises rapidly with an increase in the number of rings and eventually reaches a point when yielding due to circumferential compressive stresses may occur at the bore when there is zero internal pressure, or in one of the other rings depending on the elastic properties of the various rings.

This limitation is illustrated in Fig. 5 which has been given by Wilson and Skelton (52) using the analysis of multi-ring containers of the same material given by Manning (44). Since in this case, all the yield stresses are the same, it follows from equation (27) that all the K ratios are the same and from equation (29) that

$$P = n \; S \left[1 - \frac{1}{K^{2/n}} \right] \qquad (42)$$

The condition given by Manning for the reverse yielding of the inner ring under the action of the interference stresses is

$$P = 2 \; S \left[1 - \frac{1}{K^2} \right] \qquad (43)$$

If the Tresca criterion is used S could be replaced by Y/2 in both cases.

In this case the danger area in which reverse yielding would occur is shown shaded in Fig. 5. It can be seen that the optimum design of even a three-ring container of overall K ratio greater than about 4.6 would result in reverse yielding of the inner ring due to the residual stresses. It should be noted that if different materials are used for the individual rings as in the case of the first example in **Table** 1 then a larger K ratio can be used without yielding due to residual stress.

This difficulty can be overcome if the support pressure is made to vary in phase with the internal pressure such that at any time during loading or unloading any ring does not reach its yield stress. A simple method due to Bridgman (53) is illustrated diagrammatically in Fig. 6 in which the interface between the mating cylinders is tapered. The support pressure can be increased in phase with the bore pressure by either forcing the inner cylinder into the outer ring with an independent press force as in the Bridgman-Birch apparatus or alternatively automatically if the bore of the inner cylinder is stepped (Fig. 6). The bore pressure now exerts a force on the shoulder in the bore forcing the inner cylinder against the taper. Hence, the support pressure changes automatically in phase with the applied pressure its resulting magnitude depending on the geometry. Lubrication with a thin foil of soft metal facilitates the relative movement of the two cylinders. This arrangement has been used for a number of applications by Pugh et al (54, 55).

Another method is to separate each interface by a fluid layer which can be pressurized independently. The idea of fluid jacketing or cascading was used by Bridgman and a more recent application in the very high pressure field is the two stage apparatus for differential pressure extrusion described by Pugh (56) and shown in Fig. 7. Obviously, using this cascading arrangement it is possible to achieve a very high pressure using conventional materials since the yielding of each individual component is governed by equation (21). However, yielding of a component could readily arise because of insufficient axial stress, so that each component must be subjected not only to lateral but also to an axial support, which can be achieved by ensuring that the fluid at the interfaces also acts on the end faces. Apart from the practical difficulties of access to the inner container, it is possible with this design to achieve extremely high pressures such as those envisaged in the USSR for pressures up to 3 megabars. The cascading design has been investigated theoretically by Berman (57).

In fatigue situations such as in compressor liners there is a considerable advantage in jacketing the liner in order to keep the mean stress compressive. Manning (58) proposed a fluid supported cylinder for a high pressure pump which is shown in Fig. 8. In this arrangement the jacket can be pressurized to a pressure in excess of the delivery pressure to ensure that the linear is always in compression, or alternatively the pump delivery can be connected to the jacket. Lengyel and Alexander (59) have proposed dynamic fluid support systems in which the jacket pressure varies in step with the pressure in the pump cylinder, with the objective to keep the pressure difference across each cylinder within the fatigue limit.

6.0 WIRE WOUND CONTAINER DESIGN

The technique of pre-stressing pressure vessels by wire winding has long been used in practice for the wire winding of gun barrels. It appears that the technique was first advocated by Longridge (60, 61). Newitt (62) has given an analysis for winding under constant tension while Comstock (63) gave the solution of varying tension during winding in order to achieve constant hoop stress in the wire when the vessel is at working pressure. Further treatments have been given by Seely and Smith (64) and by Comings (65). Wilson and Skelton (52) pointed out that equations (42) and (43) which define the optimum design of a multi component cylinder also represent the optimum conditions for the strip wound design.

Wire wound containers and presses are manufactured by the Quintus Division of ASEA. The containers are used for isostatic compaction units and for hydrostatic extrusion equipments. A brief discussion of the container design has been given by Larker and Nisson (66). The basie principle of the design is not to allow tensile stresses in the core cylinder at any time. Consequently, the life of the wire wound container is determined by the fatigue life of the wire. They give the following stress analysis of the wire wound container.

A layer with thickness dr is wound on a cylinder with the radius r (Fig. 9). If the stress in the wire at winding is σ_w (r) of, the contact pressure between wire and cylinder is

$$P(r) = \frac{\sigma_w(r)dr}{r}$$

The increase of stresses at a point r_x in the material inside radius r due to P (r) are

radial stress

$$d\sigma_r(r_x) = \left\{\left(\frac{r_b}{r_x}\right)^2 - 1\right\}P(r)\frac{r^2}{r^2-r_b^2} = \left\{\left(\frac{r_b}{r_x}\right)^2 - 1\right\}\frac{\sigma_w(r)rdr}{r^2 - r_b^2} \quad (45)$$

tangential stress

$$d\sigma_\theta(r_x) = -\left\{\left(\frac{r_b}{r_x}\right)^2 + 1\right\}P(r)\frac{r^2}{r^2-r_b^2} = -\left\{\left(\frac{r_b}{r_x}\right)^2 + 1\right\}\frac{\sigma_w(r)rdr}{r^2 - r_b^2} \quad (46)$$

From these equations it is possible to calculate the pre-stress of a cylinder wound with a number of layers

$$\sigma_\theta(r_x) = -\left\{\left(\frac{r_b}{r_x}\right)^2 + 1\right\}\frac{\sigma_w(r)r}{r^2 - r_b^2}dr \int_{r_d}^{r_L} \quad (47)$$

$$\sigma_r(r_x) = -\frac{\left(\frac{r_b}{r_x}\right)^2 - 1}{\left(\frac{r_b}{r_x}\right)^2 + 1}\sigma_\theta(r_x) \quad (48)$$

For simplicity we assume the wire is wound with a stepwise constant stress σ_{w_i}

$$\sigma_\theta(r_x) = -\left\{\left(\frac{r_b}{r_x}\right)^2 + 1\right\}\left[\sigma_{w1}\int_{r_d}^{r_1}\frac{r}{r^2-r_b^2}dr + \sigma_{w2}\int_{r_1}^{r_2}\frac{r}{r^2-r_b^2}dr \ldots + \sigma_{w_n}\int_{r_{n-1}}^{r_L}\frac{r}{r^2-r_b^2}dr\right] \quad (49)$$

$$\sigma_\theta(r_x) = -\left\{\left(\frac{r_b}{r_x}\right)^2 + 1\right\}\left[\frac{\sigma_{w1}}{2}\ln\frac{r_1^2 - r_b^2}{r_d^2 - r_b^2} + \frac{\sigma_{w2}}{2}\ln\frac{r_2^2 - r_b^2}{r_1^2 - r_b^2} \dots + \frac{\sigma_{w_n}}{2}\ln\frac{r_L^2 - r_b^2}{r_{n-1}^2 - r_b^2}\right] \quad (50)$$

In the same way it is possible to calculate the stresses in the wire-mantle by integration from r to r_L taking into consideration the stress in the wire at winding.

If all the layers of wire are wound with the same stress the pre-stress at the bore ($r = r_b$) is

$$\sigma_e(r_b) = -\sigma_w \ln\frac{r_L^2 - r_b^2}{r_d^2 - r_b^2} \quad (51)$$

7.0 DESIGN OF A SECTORED VESSEL

In the normal methods of construction of multi-ring containers, it is difficult, if not impossible, to use the full compressive strength of hard brittle materials like tool steels and carbides because these materials are weak in tension and tend to fail prematurely at the bore due to the tensile hoop stresses which are developed there. A method of overcoming this difficulty is to construct the inner ring from sectors of such materials so that tensile hoop stresses cannot be generated.

The simplest form of design of such a container is shown in Fig. 10. It consists of an inner ring made up of sectors of a high strength material which are compressed by the action of an outer ring while a thin inner liner prevents the liquid in the bore from penetrating along the gaps between the sectors. The idea was first described by Poulter (67) while an improvement has been described by Ballhausen (68) in which the outer binding ring is replaced by a liquid pressure, thereby giving the apparatus more flexibility. A similar concept has also been described by von Platen (69) in an apparatus for generating very high pressure.

Consider the sectored vessel design (Fig. 10) in which the sectors are supported by a monobloc. The liner which prevents the fluid in the bore of the container from gaining access to the interfaces between the sectors is neglected as far as strength is concerned. In the initial assembly it is clear that as in the case of normal multi-ring containers there will be an interference fit between the sectors and the monobloc. Initially the sectored container acts as a conventional two ring container with compressive hoop stresses which tend to zero as the internal pressure increases. Thereafter the sectors move outward opening radial gaps between themselves. There is a limit to this movement determined by the tendency of the liner to extrude into the gaps. Thus, apart from the advantage of the use of higher strength materials in the sectors than in a conventional design, the range of pressure over which it is superior may be limited. A greater flexibility can be achieved by using a variable liquid pressure for support as suggested by Ballhausen (68).

The analysis of the sectored design has been given by Pugh (54). From equation (9) the maximum interface pressure which the monobloc can withstand without yielding is

$$P_2 = \frac{Y_2}{2}\left(1 - \frac{1}{K_2^2}\right) \quad (52)$$

Although initially the sectored ring behaves as an unbroken ring, the analysis is not the same as that for a monobloc subjected to internal and external pressure since both the radial and circumferential stresses are now compressive and the axial stress may not be intermediate between the radial and circumferential stresses. In this case the maximum stress difference used in the criterion of yielding may be between the axial stress and the compressive radial stress.

When the circumferential stresses in the sectored ring become zero, the internal pressure P_1 is given by

$$P_1 = K_1P_2 = K_1\frac{Y_2}{2}\left(1 - \frac{1}{K_2^2}\right) = \frac{K}{K_2}\frac{Y_2}{2}\left(1 - \frac{1}{K_2^2}\right) \quad (53)$$

By differentiating, the maximum pressure which can be obtained such that the bore of the monobloc is on the point of yielding is given by the condition

$$K_2 = \sqrt{3} \quad \text{and} \quad \frac{K_1}{K} = \frac{1}{\sqrt{3}} \quad (54)$$

and the maximum pressure P_{max} is given by

$$\frac{P_{max}}{Y_2} = \frac{K}{3\sqrt{3}} \quad (55)$$

For the failure of the segments which are considered to be loaded on their internal and external surfaces but completely unloaded on the tapered side faces Pugh (54) used a slip-line field solution due to Hill for a truncated wedge loaded only on its two parallel faces, viz

$$P_{max} = Y_1(1 + \alpha) \quad (56)$$

where α is the semi-angle of the sectors. The condition under which both the sectors and the monobloc reached their yield stress simultaneously is obtained by equating the value of P_{max} in equations (55) and (56) which gives

$$K = \frac{Y_1}{Y_2}(1 + \alpha)\, 3\sqrt{3} \quad (57)$$

$$\left.\begin{array}{l} \text{If} \quad K < \frac{Y_1}{Y_2}(1 + \alpha)\, 3\sqrt{3} \\ \text{then} \quad P_{max} = \frac{KY_2}{3\sqrt{3}} < Y_1(1 + \alpha) \end{array}\right\} \quad (58)$$

i.e. the sectors will not be stressed to their yield point when the bore of the outer ring is on the point of yielding and consequently advantage has not been taken of the full strength of the sectors.

$$\left.\begin{aligned} &\text{If}\quad K > \frac{Y_1}{Y_2}(1+\alpha)\,3\sqrt{3} \\ &\text{then}\quad P_{max} = Y_1(1+\alpha) < \frac{KY_2}{3\sqrt{3}} \end{aligned}\right\} \qquad (59)$$

The maximum pressure is now limited by the strength of the sectors and the outer ring is not stressed to its yield point.

The maximum pressure for a sectored container is greater than that for a conventional two ring container (Fig. 11), provided K is greater than some minimum value which in the case of equal yield stresses corresponds to a K ratio of about 3½. Even using the same materials, the sectored container of K = 8 would yield at a pressure 60 per cent higher than that for a conventional design. If brittle materials such as tungsten carbide of much superior compressive yield strength are used for the inner ring the sectored vessel will sustain without yielding larger pressures than for the conventional two ring container.

This construction combined with wire or strip winding would result in an efficient use of practical materials. Tool steels of compressive strengths of 3.1 GN/m^2 (450 KSI) or possibly carbides for small equipments, could be used for the sectors. When overwound by cold-drawn wires with tensile strengths up to 2.41 GN/m^2 (350 Ksi) containers could be constructed which would be able to contain very high pressures for a given overall diameter ratio.

8.0 SUMMARY

It is difficult to compare all the various designs considered in this paper. It is, for instance, difficult to generalise on the effect of varying the elastic properties of individual rings on the overall strength of a multi-ring container, or to compare the strength of a wire wound container in which the declared objective is to keep the stresses in the liner compressive with other designs in which the requirement is to keep the stresses elastic. However, some of the designs can be compared on a reasonable basis and such a comparison is given in Fig. 12 in which the strength of the materials used for the various components is assumed identical.

BIBLIOGRAPHY

1. Lamé, G. and Clapeyron, B. P. E. 'Memoire sur L'equilibre Interieur des Corps Solides Homogenes', Mem. Acad. Sci., Paris 1833, 4, 465.

2. Barlow, P. 'Force Exerted by Hydraulic Pressure in a Bramah Press', Trans. Instn. Civil Engrs., 1836, 1, 133.

3. Turner, L. B. 'The Stresses in Thick Hollow Cylinders subjected to Internal Pressure', Trans. Camb. Phil. Soc. 1909, 21, 377.

4. Hill, R., Lee, E. H. and Tupper, S. J. 'The Theory of Combined Plastic Deformation with Reference to a Thick Tube under Internal Pressure', Proc. Roy. Soc. A, 1947, 191, 278.

5. Hill R., Lee, E. H. and Tupper, S. J. 'Plastic Flow in a Closed-Ended Tube with Internal Pressure', Proc. First U.S. Nat. Congr. Appl. Mech., 1951, 561.

6. Manning, W. R. D. 'The Overstrain of Tubes by Internal Pressure', Engineering, 1945, 159, 101.

7. Crossland, B. and Bones, J. A. 'Behaviour of Thick-walled Cylinders Subjected to Internal Pressure', Proc. Inst. Mech. Engrs., 1958, 172, 777.

8. Crossland, B., Jorgensen, S. M. and Bones, J. A. 'The Strength of Thick-walled Cylinders', Trans. Amer. Soc. Mech. Engrs., 1958, 81, 95.

9. Cook, G. and Robertson, A. 'The Strength of Thick Hollow Cylinders under Internal Pressure', Engineering, 1911, 92, 786.

10. Cook, G. 'The Stresses in Thick-walled Cylinders of Mild Steel Overstrained by Internal Pressure', Proc. Inst. Mech. Engrs., 1934, 126, 407.

11. Cook, G. 'Some Factors Affecting the Yield Point in Mild Steel', Inst. Engrs. and Ship-Builders in Scotland, 1938, 81, 371.

12. Morrison, J. L. M. 'Yield of Mild Steel with Particular Reference to the Effect of Size of Specimen', Proc. Inst. Mech. Engrs., 1940, 142, 193.

13. Faupel, J. H. and Furbeck, A. R. 'The Influence of Residual Stresses on Behaviour of Thick-wall Closed-end Cylinders', Trans. Amer. Soc. Mech. Engrs., 1953, 73, p.345.

14. Faupel, J. H. 'Yield and Bursting Characteristics of Heavy-Wall Cylinders', Trans. Amer. Soc. Mech. Engrs., 1956, 78, 1031.

15. Crossland, B. and Kerr, W. F. K. 'The Strength of Thick-Walled Steel Cylinders at Elevated Temperatures', High Temperatures - High Pressures, 1969, 1, 133.

16. Parry, J. S. C. and Shannon, R. W. E. 'Fatigue and Fracture of Pressure Vessels', Proc. 2nd Int. Conf. on High Pressure Engineering, 1974 - to be published.

17. Morrison, J. L. M. and Cook, G. 'Report on Effect on an Autofrettaged Gun Tube of 2000 Repetitions of the Autofrettage Pressure', Declassified Paper of the Gun Design Committee U.K., 1945.

18. Newhall, D. H. and Kosting, P. R. 'Progressive Stress Damage and Strength of Centrifugally Cast CW Gun Tubes', Watertown Arsenal Report, 1949, No. 731/281.

19. Morrison, J. L. M., Crossland, B. and Parry, J. S. C. 'Fatigue under Triaxial Stress: Development of a Testing Machine and Preliminary Results', Proc. Inst. Mech. Engrs., 1956, 170, 697.

20. Crossland, B. 'Effect of Pressure on Fatigue', The Mech. Behaviour of Materials under Pressure, edited H. Ll. D. Pugh, 1970, Chapter 7, 299 (Elsevier Pub. Co. Ltd., London).

21. Parry, J. S. C. 'Fatigue of Thick Cylinders: Further Practical Information', Proc. Inst. Mech. Engrs., 1965, 180 (Pt. 1), 387.

22. Crossland, B. 'The Effect of Large Hydrostatic Pressures on the Torsional Fatigue Strength of an Alloy Steel', Proc. of the Int. Conf. on the Fatigue of Metals, published by Inst. Mech. Engrs., 1956, 138.

23. Burns, D. J. and Parry, J. S. C. 'Effect of Large Hydrostatic Pressures on the Torsional Fatigue Strength of Two Steels', J. Mech. Engng. Sci., 1964, 6, 293.

24. Davidson, T. E., Eisenstadt, E. and Reiner, A. N. 'Fatigue Characteristics of Open-end Thick-walled Cylinders Under Cyclic Internal Pressure', J. Basic Eng. Trans. A.S.M.E. Series D, no. 2, 1963, 85, 555.

25. Austin, B. A. and Crossland, B. 'Low Endurance Fatigue Strength of Thick-walled Cylinders: Development of a Testing Machine and Preliminary Results', Proc. Inst. Mech. Engrs. 1966, 180, 43.

26. Crossland, B. and Skelton, W. J. 'Effect of Varying Hardness on the Fatigue Strength of Thick-walled Cross-bore Cylinders of EN25, Proc. Inst. Mech. Engrs. 1967, 182 (pt. 3C), 106.

27. Austin, B. A., Reiner, A. N. and Davidson, T. E. 'Low Cycle Fatigue Strength of Thick-walled Pressure Vessels', Proc. Inst. Mech. Engrs., 1967, 182 (Pt. 3C), 91.

28. Terao, M. and Yamashita, Y. 'Low Cycle Fatigue Strength of Thick-walled Cylinders', Proc. 2nd Int. Conf. Pressure Vessel Technology, 1973, Pt. 2, 753.

29. Tomkins, B. 'Design of Very High Pressure Vessels for Repeated Pressure Applications', Proc. 2nd Int. Conf. Pressure Vessel Technology, 1973, Pt. 2, 835.

30. Shannon, R. W. E. and Austin, B. A. 'The Application of Fracture Mechanics to Failure of Thick-walled Cylinders', Proc. Conf. on Practical Application of Fracture Mechanics to Pressure Vessel Technology published by Inst. Mech. Engrs., 1971, paper 31, 109.

31. Shannon, R. W. E. 'Stress Intensity Factors for Thick-walled Cylinders', Int. Journal of Pressure Vessels and Piping, 1974, 2, 19.

32. Crymble, T. G., Goldthorpe, B. D. and Crossland, B., 'The Application of Linear Elastic Fracture Mechanics to Fatigue and Fracture of Thick-walled Cylinders', Proc. 2nd Int. Conf. of High Pressure Engineering, 1975 - to be published.

33. Crossland, B. and Logan, J. G. 'Review of the Design of Very High Pressure Containers', NEL/AIRAPT Int. Conf. on Hydrostatic Extrusion, 1973, Vol. 1, 237.

34. Pugh, H. Ll. D., Wilkinson, P. T. and Hodge, M. H. 'Machine for Hydrostatic Extrusion - Drawing of Wire', Proc. Inst. Mech. Engrs., 1972, 186, 499.

35. Bratt, E. 'Pressure-coupling Assembly by the Oil Injection Method', Mechanical World and Engineering Record, 1946, Vol. 119, p.399.

36. Crossland, B. 'The Effect of Fluid Pressure on the Shear Properties of Metals', Proc. Inst. Mech. Engrs., 1954, 168, 935.

37. Pugh, H. Ll. D. and Green, D. 'The Effect of Hydrostatic Pressure on the Plastic Flow and Fracture of Metals', Proc. Inst. Mech. Engrs., 1964-65, Vol. 179, p.3.

38. Brandes, M. 'Mechanical Properties of Materials under Hydrostatic Pressure', The Mech. Behaviour of Materials Under Pressure, Edited by H. Ll. D. Pugh, pub. Elsevier Pub. Co. Ltd., London, 1970, Chapter 6, 236.

39. Franklin, G. J. and Morrison, J. L. M. 'Autofrettage of Cylinders: Prediction of Pressure/External Expansion Curves and Calculations of Residual Stresses', Proc. Inst. Mech. Engrs., 1960, 174, 947.

40. Crossland, B. 'The Design of Thick-Walled Closed-Ended Cylinders Based on Torsion Data', Welding Research Council Bulletin, 1964, No. 94.

41. Nishihara, M. and Yamaguchi, Y. 'The Strength of Thick-Walled Cylinders under very High Pressure', Presented verbally at the 11th Annual Meeting of the European High Pressure Research Group, 1973.

42. Horne, M. R. 'The Elastic-Plastic Theory of Containers and Liners for Extrusion Presses', Proc. Inst. Mech. Engrs., 1955, 169, 107.

43. Parsons, B. and Cole, B. N. 'A Generalized Approach to the Optimum Design of Short Composite Cylinders', Proc. Inst. Mech. Engrs., 1967-68, 182, pt. 3C, 30.

44. Manning, W. R. D. 'The Design of Compound Cylinders for High Pressure Service', Engineering, 1947, 163, 349.

45. Strub, R. A. 'Distribution of Mechanical and Thermal Stresses in Multi-layer cylinders', Trans. Am. Soc. Mech. Engrs., 1953, 75, 73.

46. Becker, S. J. and Mollick, L. 'A Theory of the Ideal Design of a Compound Vessel', J. Eng. Ind., Trans. Am. Soc. Mech. Engrs., 1960, 82 series B, 136.

47. Pugh, H. Ll. D., Bullied Memorial Lectures, of the Untiversity of Nottingham 1965, Vol. IIIA.

48. Manning, W. R. D. and Labrow, S. 'High Pressure Engineering', Leonard Hill, (London), 1971.

49. Crossland, B. and Burns, D. J. 'Behaviour of Compound Steel Cylinders subjected to Internal Pressure', Proc. Inst. Mech. Engrs., 1961, 175, 1083.

50. Hill, R. 'The Mathematical Theory of Plasticity', published Oxford University Press, 1950.

51. Babb, S. E. 'Techniques in High Pressure Experimentation in Inorganic Chemistry', Interscience Publishers, New York, London and Sydney, 1967.

52. Wilson, W. R. D. and Skelton, W. J. 'Design of High Pressure Cylinders', Proc. Inst. Mech. Engrs., 1967, 182 (part 3C), 1.

53. Bridgman, P. W. 'Phase Diagram of Water to 45,000 kg/cm^2', Journal Chem. Phys., 1937, 5, 964.

54. Pugh, H. Ll. D. 'Symposium on Irreversible Effects of High Pressure and Temperature of Materials', A.S.T.M., 1964, Special Tech. Publication, 374.

55. Pugh, H. Ll. D., Lees, J. and Gunn, D. 'High Pressure Research at the National Engineering Laboratory', The Engineer, 1961, 212, 258.

56. Pugh, H. Ll. D., 'Cold Extrusion by a High Pressure Liquid', Proc. Int. Conf. on Prod. Engng. Research, Pittsburgh, A.S.M.E., 1963, p.394.

57. Berman, I. 'Design and Analysis of Commercial Pressure Vessels 500,000 psi', Trans. Am. Soc. Mech. Engrs., 1966, 88, 500.

58. Manning, W. R. D. 'High Pressure Engineering', Bullied Memorial Lectures of the University of Nottingham, 1963, Vol. 2.

59. Lengyel, B. and Alexander, J. M. 'Design of a Production Machine for Semi-continuous Hydrostatic Extrusion', Proc. Inst. Mech. Engrs., 1967, 182 (pt. 3C), 207.

60. Longridge, J. A. 'On the construction of artillery and other Vessels to Resist Great Internal Pressure', Proc. Inst. Civil Engrs., 1860, 19, 283.

61. Longridge, J. A. 'A Treatise on the Application of Wire to the Construction of Ordnance', E. and F. N. Spon, London, 1884.

62. Newitt, D. M. 'The Design of High Pressure Plant and the Properties of Fluids at High Pressure', Clarendon Press, Oxford, 1940.

63. Comstock, C. W. 'Some Considerations in the Design of Thick-wall Pressure Cylinders', Trans. Am. Inst. Chem. Engrs., 1943, 39, 299.

64. Seely, F. B. and Smith, J. O. 'Advanced Mechanics of Materials', John Wiley and Sons Inc., New York, 1952, 2nd edition.

65. Comings, E. W. 'High Pressure Technology', McGraw Hill Book Co., New York, 1956.

66. Larker, H. and Nilsson, S. Private Communication.

67. Poulter, T. C. U.S. Patent No. 2,554,499, 1951

68. Ballhausen, C. German Patent No. 1,142,341, 1963.

69. Von Platen, B. 'A Multiple Piston, High Pressure, High Temperature, Apparatus', Modern Very High Pressure Techniques, Edited by R. M. Wentorf, Butterworths, London, 1962, Chapter 6, 118.

TABLE 1

Example No.	Yield Strength (Y)		K Ratio for each Ring	Maximum Yield Pressure	
	(GN/m^2)	(Ksi)		(GN/m^2)	(Ksi)
1 3-ring container					
First ring	1.85	269	2.355)		
Second ring	1.39	202	2.040)	1.58	230
Third ring	0.925	134	1.667)		
2 4-ring container					
First ring	1.85	269	1.974)		
Second ring	1.54	224	1.802)	1.83	265
Third ring	1.23	179	1.612)		
Fourth ring	0.925	134	1.396)		
3 2-ring container					
First ring	1.85	269	3.364)	1.23	178
Second ring	0.925	134	2.378)		
4 2-ring container (Outer Ring auto-frettaged)					
First ring	1.85	269	2.828)	1.62	235
Second ring	1.85*	269*	2.828)		

* Outer Ring autofrettaged from 0.925 GN/m^2 (134 Ksi) up to 1.85 GN/m^2 (269 Ksi)

THE DEVELOPMENT OF A NEW TYPE OF COMPOUND TUBING DESIGNED FOR HIGH FATIGUE STRENGTH

J. ROGAN, BTech, DIC, PhD
Department of Mechanical Engineering, Imperial College of Science and Technology, London

SYNOPSIS The trend in the design of new plants for the production of low density polyethylene has been to increase the size of each reactor. This applies to both Autoclave and Tubular processes.

This has involved using high pressure tubing of larger and larger bore size which can pose problems with conventional monobloc tubing firstly in the limited tube length of the larger bore tubes and secondly in the danger of having poor fracture properties in the thicker walls of such tubing.

One way of overcoming both limitations with some possible additional advantages over monobloc tubing was to consider the production of multi-wall pressure tubing.

This paper describes the stages of the development of a two-wall (Duplex) tube which can be produced in long lengths and because of the residual stress distribution produced during manufacture has a very high fatigue strength and a safe mode of fatigue failure.

INTRODUCTION

1. A trial length of three-wall (Triplex) tubing produced by T.I. Chesterfield Ltd. was tested to establish its fatigue strength, mode of failure and pressure containment properties. The results showed that the Triplex tubing had very high fatigue and burst strengths. The mode of failure of the fatigue specimens was not satisfactory in that the outermost shell split from end-to-end following fatigue failure and, in the unlikely event of fatigue occurring in such tubing during plant operation, this could lead to gross leakage of the tube contents.

2. Measurements of the contact pressures at the interfaces of the Triplex tubing and an analysis of the results showed that a two-wall tube could be designed which would have high fatigue strength coupled with a safe leak-before-burst mode of fatigue failure.

3. A series of high pressure fatigue tests on shrunk compounded cylinders established the contact pressure at the interface of the two-component tube required to give optimum fatigue strength and fracture mode.

4. An extensive development project by T.I.C. into the process factors affecting the production of Duplex tubing with the required interface pressure, coupled with a research programme at Imperial College to evaluate the properties of the tubing has been in progress.

5. So far this has led to the manufacture of production quantities of Duplex tubing with bore sizes of 38 mm. and 50 mm. in lengths of more than 12 metres. The inner member is cold worked austenitic stainless steel and the outer shell is a conventional nickel chromium low alloy steel.

6. The results of a large number of fatigue tests on the Duplex tubing are presented and the fatigue strength of the Duplex tubing is much higher than that of non-autofrettaged monobloc tubing. The mode of fatigue failure is of the safe type and the pressure containment properties are high.

SHRUNK COMPOUND CYLINDERS

7. The shrunk compound cylinder tests were based on the 1.25 in. ID x 1.88 in. OD cold reduced type 304 material which had formed the inner shell of the original Triplex tubing.

8. These were cooled in liquid nitrogen and shrunk into outer shells machined from SAE 4340 tubing. The diameter ratio of the inner and outer shells was K = 1.5, giving a tube with 1.25 in. ID, 1.88 in. interface, and 2.81 in. OD with an overall diameter ratio K = 2.25.

9. By choosing a range of values for the initial diametral inteference and measuring the subsequent fatigue strength and mode of failure it was possible to establish the optimum value of interface pressure to give high fatigue strength combined with leak-before-burst type of failure.

10. A further series of tests on shrunk compound cylinders with a thinner wall Type 304 inner member, K_1 = 1.38, K_2 = 1.5 K overall = 2.06 established the optimum interface pressure for this alternative tube design.

11. The results of these shrunk compound tests will be discussed in detail in a future paper because there is very little published work upon the fatigue strength of compound cylinders. However some comments upon the results may be briefly made here.

12. Six strain gauges were bonded to the outside diameter of every outer shell. The hoop and axial strains induced as the cold inner shells were

shrunk into the outer shells were measured but the gauges WERE LEFT IN POSITION.

13. Any further changes in strain produced while tightening the flange joints during assembly of the shrunk cylinders for testing, and the changes in strain during pressure cycling were carefully measured. The fatigue tests were stopped periodically, strains measured, flange joints unbolted, strains measured, joints re-made, strains measured, and so on for all the specimens.

14. The results show that the actual stresses in a compound cylinder as set up on a fatigue testing machine with flange-cone ring joints differ considerably from the stresses computed from strain measurements due to shrinkage alone. The difference can be significant enough to affect the interpretation of the results of fatigue tests on compound cylinders such as those reported by Burns and Parry (Ref.1).

15. Still more important changes occur during the first 5,000 cycles and 10,000 cycles of internal pressure. The stress conditions tend to settle out and for example in a compound cylinder subjected to an internal pressure cycling from 0 to 46,000 Lbf/in², settled conditions were achieved after 50,000 cycles and the strain measurements remained constant until just before the specimen failed at 1.4 million cycles. These compound cylinder results are quite interesting and will be reported in full elsewhere.

DUPLEX TUBING

16. The optimum interface pressures for the Duplex tubing had been obtained from the shrunk compound cylinder results. The production techniques to produce such high interface pressures had been established after a development project by T.I.C.

17. Actual interface pressures were measured by strain gauging the outside diameter of a sample from each length of Duplex tubing and monitoring the change of strain as hydraulic pressure was applied to the interface. As the applied pressure reached the interface pressure, the strain readings changed quite markedly, thus the actual interface pressure could be measured to an accuracy of $\pm$ 500 Lbf/in² for a typical pressure of 13,000 Lbf/in².

18. The first trial production Duplex tubing was 0.945 in. ID x 2.125 in. OD, K = 2.25. It was not possible to achieve the full required interface pressure with this small-bore tubing due to the lack of rigidity of the inner tube. Yet the fatigue strength was higher than the best non-autofrettaged monobloc tubing ever tested by the writer, and the mode of fatigue failure was of the safe, leak-before-burst type as shown in Fig.1.

19. The pressure containment capacity of this Duplex tube was also very high. The tube burst at 119,800 Lbf/in² and failure was preceded by gross plastic deformation followed by a shear fracture spiralling from the bore. The small axial tear visible on the burst specimen in Fig.1 is no worse than the failure mode of monobloc high strength steel tubing with a K ratio of around 2.

20. After these encouraging first results, the next production runs were on 1.5 in. ID x 3.375 in. OD and 2.0 in. ID x 4.12 in. OD Duplex tubes in lengths of 11 to 12 metres.

21. The medium-carbon, nickel chromium molybdenum steel outer shells were unfortunately heat treated incorrectly to a lower strength level than the target strength.

22. This meant that the choice of interface pressure was limited to a magnitude which would not cause the material at the bore of the outer to be subjected to plastic strains during subsequent internal pressure fatigue testing. Thus the actual interface pressures and the resulting fatigue strengths of the Duplex tubes were not as high as would be the case with properly heat treated outers.

Fatigue Results

23. The 1.5 in. ID, $K_1 = K_2 = 1.5$, K overall = 2.25, Duplex tubes were called tube B, while the 2.0 in. ID, $K_1 = 1.38$, $K_2 = 1.5$, Koverall = 2.06 Duplex tubes were designated tube D.

24. The results of fatigue tests with the internal pressure cycling about a mean pressure of 30,000 Lbf/in² are shown in Fig.2 for Duplex tubes B and D. Also included on Fig.2 are some typical results for non-autofrettaged monobloc tubing made from high strength low-alloy steels.

25. A length of the 2 in. ID (50 mm) x 4.12 in. OD (104 mm) Duplex tube D was cold bent to a radius of 1 metre. This is a very sharp radius for tubing of such dimensions and most designers would use a radius of 1.5 or 2 metres.

26. However, fatigue tests on specimens cut from the 1 metre bend, called DB, showed that its fatigue strength at the fatigue limit was a range of pressure of 25,000 Lbf/in² about a mean pressure of 30,000 Lbf/in², which is higher than the strength of some straight monobloc tubing. The results for the bend DB are also shown in Fig.2.

27. To compare the fatigue results for repeated pressure and for high mean pressure tests on the whole range of tubing, the results have been redrawn on Fig.3. This expresses the effect of high mean shear stress upon the range of bore shear stress AT THE FATIGUE LIMIT. Shear stresses are calculated from $\tau = P. \frac{K^2}{K^2 - 1}$.

28. Fig.3 shows clearly the high fatigue strength of the Duplex tubing relative to non-autofrettaged monobloc tubing. If the outers had been heat treated correctly the fatigue strength of the Duplex tubing would have been as high as that of the shrunk compound tube and would be comparable with the high fatigue strength of properly autofrettaged high strength steel monobloc tubing.

Mode of Failure

29. The mode of fatigue failures of the 1.5 in. ID and 2.0 in. ID Duplex tubes B and D are shown in Fig.1.

30. The fatigue crack in the 1.5 in. Duplex grows radially through the inner and outer shells, pipping through to the outside surface in a safe, stable, leak-before-burst type of crack which is similar to the mode of fatigue failure in good quality monobloc tubing.

31. The interface pressures for the 2.0 in. Duplex had been chosen to give conditions in which a fatigue crack growing through the inner shell would not propagate into the outer shell. This is easily

arranged by choosing K inner, K outer and initial interface pressure so that the contact pressure during pressure cycling is sufficiently high to inhibit crack propagation from the inner to the outer shell.

32. The mode of failure is thus by leakage from the bore into the interface which is observed as leakage at the cone ring or lens ring joints at the ends of the tube.

33. Both modes of failure are perfectly suitable for operating processes in which gross leakage following fatigue failure is not permissible.

34. A further possible source of weakness in all types of compound or Duplex tube is the load transfer at the joints at the tube ends. Flange joints with cone or lens ring seals require very high joint loads for satisfactory operation. Indeed one of the first things which was studied on the original Duplex tube was the ability to make high pressure flange joints.

35. The compressive forces in the joint act on the inner shell while the tensile forces act on the outer shell. Possible transverse fatigue failures at the threads in the outer shell at the ends of the tubing had to be fully investigated.

36. In all of these high pressure fatigue tests upon Duplex tubing, transverse fatigue failures at the tube ends have never occurred, even though the ranges of internal pressure used in the fatigue machines are much greater than the cyclic pressures to which the tubes would be subjected in service conditions.

CONCLUSIONS

37. A comprehensive series of fatigue tests on shrunk compound cylinders provided the basis for design of a new type of high pressure tube.

38. Experiments with tube manufacturing techniques enabled long lengths of Duplex tubing to be produced with a specified and controlled high interface pressure between the inner and outer components of the tubing.

39. The inner shell is made from austenitic stainless steel and the outer shell from a low-alloy steel and the tubing can be made in very long lengths with bore sizes 2 in. diameter (50 mm) or larger.

40. The fatigue strength of the Duplex tubing is very high indeed and its mode of fatigue failure is completely satisfactory.

41. The combination of good fatigue properties and availability in long lengths should make this new type of tubing attractive for polyethylene plant designs since the cost saving in reducing the number of high pressure joints could be considerable.

ACKNOWLEDGEMENTS

Part of this work was financed by a research grant from Ethylene-Plastique, CdF Chimie, under the general supervision of Professor Sir Hugh Ford, F.R.S., to whom the author is indebted for encouragement and permission to use the facilities of the Mechanical Engineering Department at Imperial College.

A generous research grant from T.I. Chesterfield Ltd., Stainless Tube Division was gratefully received and thanks are due to Mr. G. Trotman, Mr. T. Thursfield and the management of that company.

REFERENCE

1. BURNS, D.J. and PARRY, J.S.C. Effect of mean shear stress on the fatigue behavior of thick-walled cylinders. 1st High Pressure Engineering Conference, I.Mech.E., London, 1967.

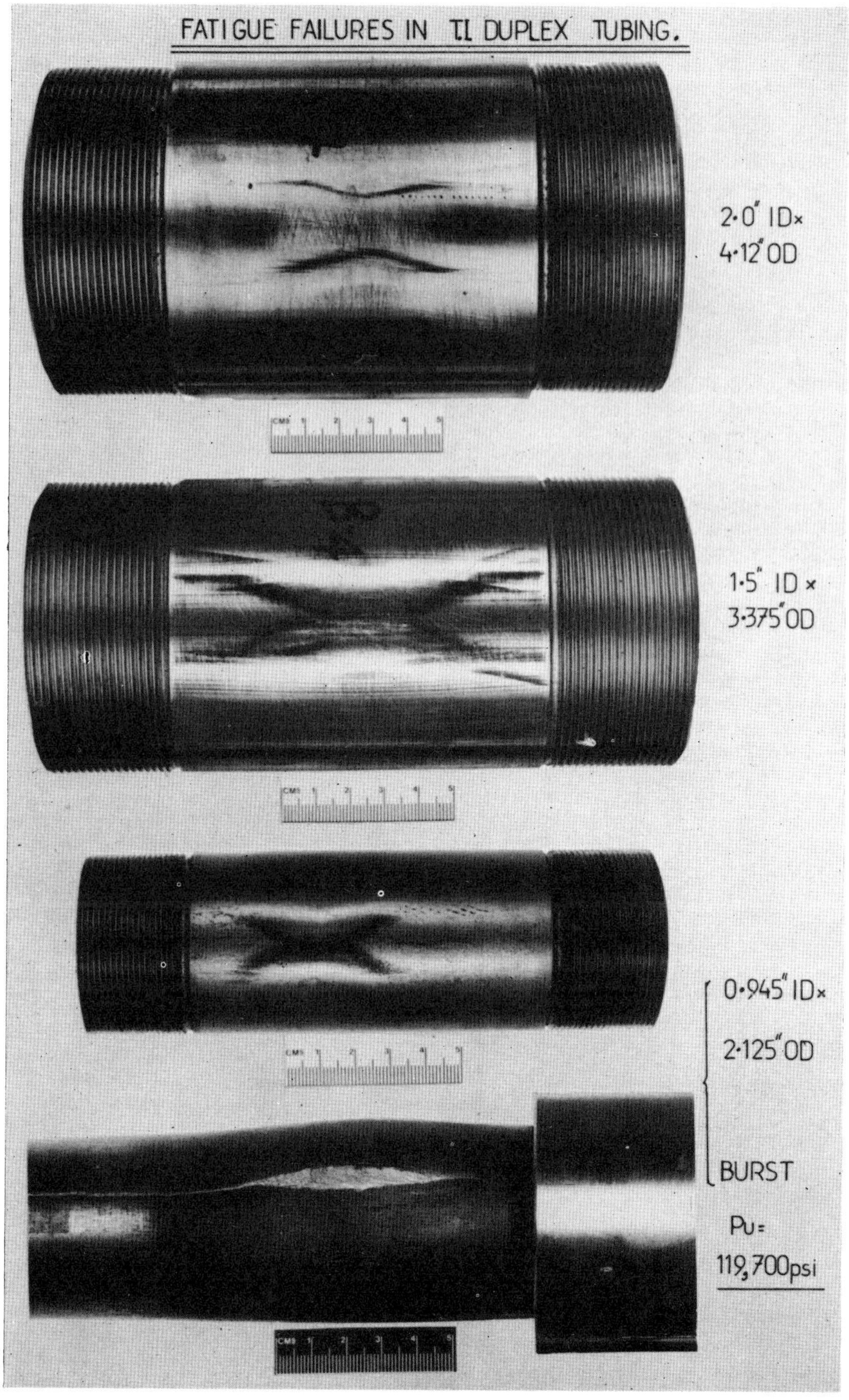

Fig. 1: Fatigue failures and burst failure in Duplex tubing

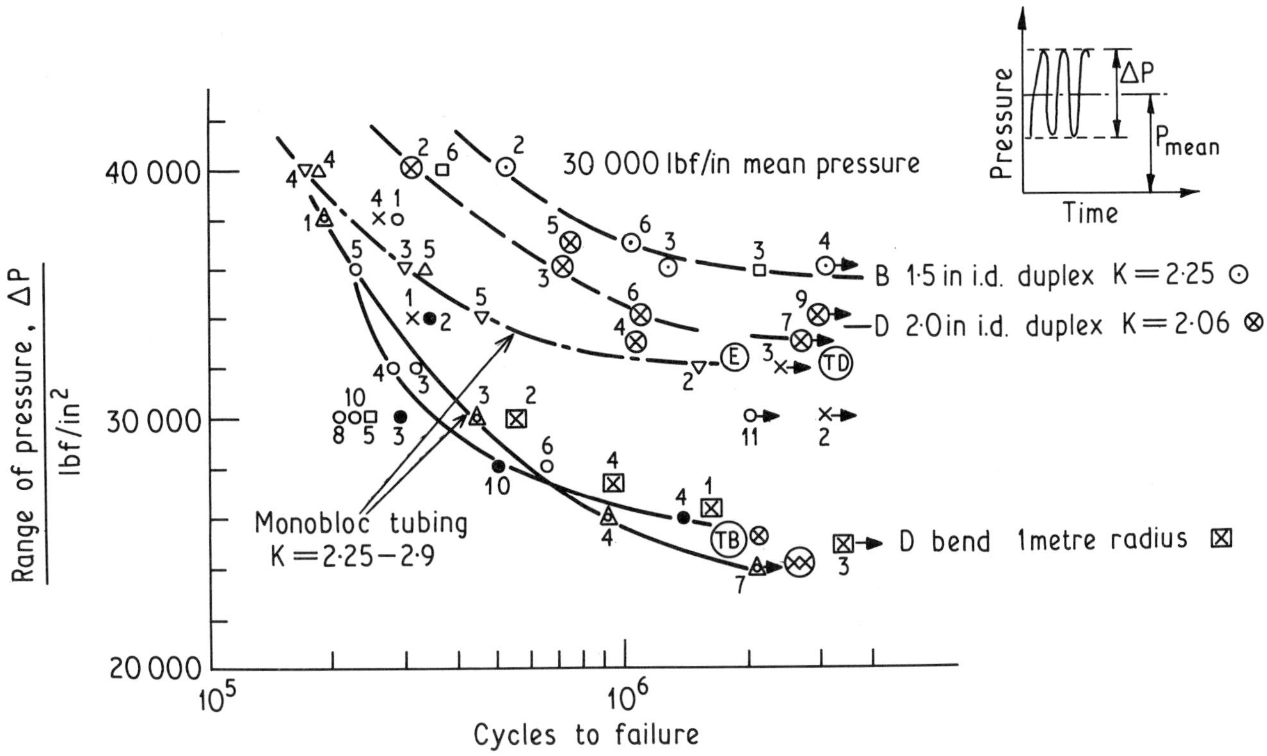

Fig. 2: Fatigue strength of Duplex tubing tested at 30,000 Lbf/in² mean pressure. Comparison with a selection of monobloc tubing with K ratios between 2.3 and 2.6

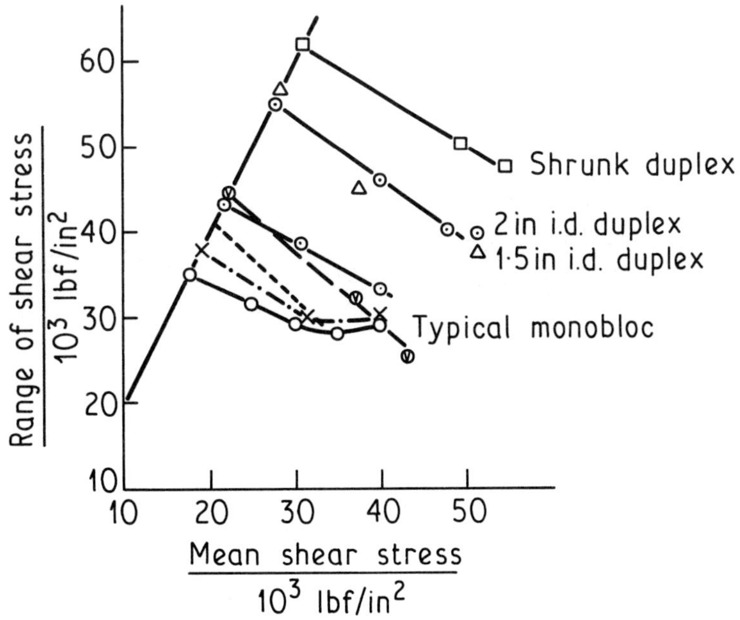

Fig. 3: Effect of mean shear stress upon the range of shear stress at the fatigue limit. Comparison of shrunk compounded, Duplex and monobloc tubing

ON THE ELASTIC-PLASTIC EXPANSION OF TAPERED PRESSURE VESSEL ASSEMBLIES

R. L. HEWITT
De Beers Diamond Research Laboratory, Johannesburg, South Africa
M. C. de MALHERBE
Dept. of Mechanical Engineering, McMaster University, Hamilton, Canada
J. M. ALEXANDER
Dept. of Mechanical Engineering, Imperial College, London, England

SYNOPSIS A numerical technique is presented for calculating the internal pressure required to produce a given bore strain in a tapered pressure vessel assembly. The method is applicable to open- or closed-ended vessels of elastic-plastic work-hardening material and the Mises criterion of yielding is assumed. Results are presented for several assemblies of different geometries, using stress-strain data for a given high strength low alloy steel.

NOTATION

A, B	Lamé constants
C, D	Constants defined by equation (32)
D'	Constant defined by equation (38)
E	Young modulus
E_p	Plastic tangent modulus = $d\bar{\sigma}/d\bar{\varepsilon}_p$
K	Ratio external radius/internal radius
k	Expression defined by equation (9)
L	Effective length of liner
m	Index in empirical stress-strain relationship
N	Normal stress on outside of liner due to taper
p	Internal pressure
p_{E1}	External pressure on liner due to taper alone
p_{E2}	External pressure on liner due to container without taper
p_{EX}	Total pressure on liner (= p_{E1} + p_{E2})
r_o	Initial radius to any point
r	Current radius to any point
r_{EL}	External radius of liner
r_{EC}	External radius of container
r_I	Internal radius of liner
u	Radial displacement
α_1, α_2	Constants defined by equations (25) and (26)
β	Constant in empirical stress-strain relationship
δ	Operator referring to a change in a quantity associated with a particular element in the material
Δ	Operator referring to a change in a quantity between two neighbouring points
ε_r, ε_θ, ε_z	Natural strains in radial, circumferential and axial directions
$\bar{\varepsilon}_p$	Equivalent plastic strain
θ	Semi-cone angle of taper
μ	Coefficient of friction
ν	Poisson ratio
σ_r, σ_θ, σ_z	Normal stresses in radial, circumferential and axial directions
σ_r', σ_θ', σ_z'	Deviatoric stresses in radial, circumferential and axial directions
σ_{r_n}	Radial stress at r = r_n
σ_{r_p}	Previous radial stress at r = r_{n-1} (i.e. when the interface was one step further in
$\bar{\sigma}$	Equivalent stress
σ_o	Equivalent stress at first yield

INTRODUCTION

1. It is advantageous in the design of high pressure cylinders to provide a variable external pressure which increases with the internal pressure. This can be achieved by placing the cylinder within another vessel or vessels and subjecting it directly to a controllable external hydrostatic pressure. A simpler method, proposed by Bridgman (Ref.1) and subsequently used by Beresnev (Ref.2) and Pugh (Ref. 3) among others, consists of making the cylinder with a tapered or conical outside surface, and fitting it into another cylinder having a matching taper on the bore. This container is supported as shown in Fig.1 and the action of the plunger pressurizes the fluid within the bore and forces the liner into the container, thus providing an external pressure on the liner.

2. There appear to be few reports of any stress analyses of such cylinders, however, and it is therefore difficult to achieve the optimum relationship between the taper and the K ratio of the liner and container. The present paper presents a method of analysis for such a tapered pressure vessel set which, although not optimizing the parameters, does allow the designer at least to check on his proposed design. The method is applicable to open- or closed-ended liners of elastic-plastic work-hardening materials in an elastic container and assumes the Mises criterion of yielding. The pressures required to produce an equivalent bore strain of ½% and 1% are calculated for pressure vessel sets of various lengths and semi-cone angles and with various assumed coefficients of friction between the liner and container.

Simplifications

3. In practice it is usually unnecessary to calculate the burst pressure of a vessel; it is normally sufficient to calculate the pressure required to produce yielding to a specified depth through the wall or to produce a specified bore strain.

4. This means that in the present problem it is sufficient to find a method which can account for expansion of the liner where the bore is plastic but the outside is still elastic.

5. Also, since the container is somewhat larger than the liner and is consequently more difficult to heat treat, it is made of a different material from the liner. This introduces an extra variable, i.e. the relationship between the flow characteristics of the liner and container materials. To avoid this added complication, it is assumed that the container material remains elastic. For each solution obtained the equivalent stress at the bore of the container is calculated to check the validity of this assumption.

6. Considering the pressure vessel shown in Fig.1, it is reasonable to assume that the liner will tend to expand more between X and X' as the pressure is increased. Thus most of the vertical force on the liner will be taken between X and X'. For simplicity it is assumed that the force is taken uniformly between these two points. Further, although this distance will decrease as the pressure is applied because of the fluid compressibility, a constant value is assumed in the calculations and is denoted by the effective length L of the liner. The liner can now be considered as a thick cylinder subjected to an internal and external pressure. The top of the liner is basically open-ended while the bottom is closed-ended, but only the latter is considered in the results since with an external pressure the closed-ended case generally allows the lower usable pressure.

External pressure on liner

7. The external pressure on the liner can be considered as being composed of two separate parts, one due to the taper and one to the restraining effect of the container. This latter pressure can be calculated by assuming the liner and container to be one vessel and finding the radial stress at the radius corresponding to the container bore. If the stresses at the elastic-plastic interface are known, this can easily be accomplished using the Lamé equations.

8. The external pressure due to the taper alone can be calculated by noting that the total downward load due to the fluid pressure on the closed end of the liner is supported by the taper of the container bore. Thus, simple equilibrium of the forces is sufficient to determine this pressure.

Method of solution

9. The problem has been reduced to that of the elastic-plastic expansion of an open- or closed-ended thick tube subjected to known internal and external pressures. The method used has been applied previously to both open- and closed-ended tubes with constant external pressure and has been chosen as it is particularly amenable to variations in boundary conditions of the type experienced in this particular problem.

10. A thick tube under pressure is considered, such that yielding has occurred. The elastic-plastic interface is supposed to have reached some point distant r from the centre and at this stage the stress and strain distributions throughout the tube are known. The interface is then allowed to move out a distance Δr.

11. The stresses and strains at the interface may be calculated using Lamé's equations and the yield criterion. Using this interface as a datum the stress and strain increment of a surface Δr inside the interface may be found using the Prandtl-Reuss equations, the equilibrium equation, the stress-strain relationship, yield criterion and the equation of strain compatibility.

12. Having obtained the stresses and strains for this surface, they may then be found for another surface Δr inside this one, and so on, until the bore is reached. Thus, the stress and strain distribution throughout the tube is known with the elastic-plastic interface at $(r + \Delta r)$. By starting with the elastic-plastic interface at the bore, it is possible to progress in the above manner until the tube becomes fully plastic. A matrix formulation of the equations is used based on the method put forward by Alexander and Gunasekera (Ref. 4).

THEORY

Prandtl-Reuss equations

13. The Prandtl-Reuss equations may be written in matrix form as

$$\begin{bmatrix} 1/E & -\nu/E & -\nu/E & 0 & 3\sigma'_r/2\bar{\sigma} \\ -\nu/E & 1/E & -\nu/E & -1 & 3\sigma'_\theta/2\bar{\sigma} \\ -\nu/E & -\nu/E & 1/E & 0 & 3\sigma'_z/2\bar{\sigma} \end{bmatrix} \begin{bmatrix} \delta\sigma_r \\ \delta\sigma_\theta \\ \delta\sigma_z \\ \delta\varepsilon_\theta \\ \delta\bar{\varepsilon}_p \end{bmatrix} = \begin{bmatrix} \delta\varepsilon_r \\ 0 \\ \delta\varepsilon_z \end{bmatrix} \quad (1)$$

where σ'_r, σ'_θ and σ'_z are the deviatoric stresses, $\bar{\sigma}$ is the equivalent stress, $\delta\bar{\varepsilon}_p$ is the equivalent plastic strain increment and the other symbols have their usual meanings.

Equilibrium equation

14. Equilibrium of the stresses requires that

$$d\sigma_r = \left(\frac{\sigma_\theta - \sigma_r}{r}\right) dr \quad (2)$$

This equation may be written in incremental form as

$$(\sigma_r)^r - (\sigma_r)^{r-\Delta r} = \left[\left(\frac{\sigma_\theta - \sigma_r}{r}\right)^r + \left(\frac{\sigma_\theta - \sigma_r}{r}\right)^{r-\Delta r}\right]\frac{\Delta r}{2} \quad (3)$$

which, when expanded and simplified, can be written as

$$\Delta\sigma_r(2r_n - 3\Delta r) + \Delta r\Delta\sigma_\theta = (\sigma_{\theta_n} - \sigma_{r_n})(2r_n - \Delta r)\frac{\Delta r}{r_n} \quad (4)$$

where σ_n is the stress at $r = r_n$ and $\Delta\sigma_r$ the change in stress associated with the change in position from $r = r_{n-1}$ to $r = r_n$. Thus it is possible to write

$$\sigma_{r_{n-1}} = \sigma_{r_n} - \Delta\sigma_r \quad (5)$$

and $$\sigma_{r_{n-1}} = \sigma_{r_p} + \delta\sigma_r \quad (6)$$

where σ_{r_p} is the previous radial stress at $r=r_{n-1}$

and $\delta\sigma_r$ is the change in radial stress at $r=r_{n-1}$ associated with a movement of the elastic-plastic interface. Eliminating $\sigma_{r_{n-1}}$ between equations (5) and (6) gives an expression for $\Delta\sigma_r$, i.e.

$$\Delta\sigma_r = (\sigma_{r_n} - \sigma_{r_p}) - \delta\sigma_r \qquad (7)$$

and similarly

$$\Delta\sigma_\theta = (\sigma_{\theta_n} - \sigma_{\theta_p}) - \delta\sigma_\theta \qquad (8)$$

Substituting equations (7) and (8) into equation (4) gives an equilibrium equation

$$-(2r_n-3\Delta r)\delta\sigma_r-\Delta r\delta\sigma_\theta = (\sigma_{\theta_n}-\sigma_{r_n})(2r_n-\Delta r)\frac{\Delta r}{r_n}$$

$$- (\sigma_{r_n}-\sigma_{r_p})(2r_n-3\Delta r)$$

$$- (\sigma_{\theta_n}-\sigma_{\theta_p})\Delta r \qquad (9)$$

On denoting the right-hand side by k,

$$-(2r_n - 3\Delta r)\delta\sigma_r - \Delta r\delta\sigma_\theta = k \qquad (10)$$

Stress-strain equation

15. An empirical stress-strain equation which closely fits many engineering materials can be written

$$\bar{\sigma} = \bar{\sigma}_o (1 + \beta\bar{\varepsilon}_p)^m \qquad (11)$$

where m and β are constants and $\bar{\sigma}_o$ is the equivalent stress at first yield. Differentiating this equation,

$$o = \delta\bar{\sigma} - E_p\,\delta\bar{\varepsilon}_p \qquad (12)$$

where

$$E_p = m\bar{\sigma}_o\beta(1 + \beta\bar{\varepsilon}_p)^{m-1} \qquad (13)$$

Yield criterion

16. The Mises yield criterion may be written

$$2\bar{\sigma}^2 = (\sigma_r-\sigma_\theta)^2 + (\sigma_\theta-\sigma_z)^2 + (\sigma_z-\sigma_r)^2 \qquad (14)$$

Differentiating this equation,

$$0 = 3\sigma_r'\delta\sigma_r+3\sigma_\theta'\delta\sigma_\theta+3\sigma_z'\delta\sigma_z-2\bar{\sigma}\delta\bar{\sigma} \qquad (15)$$

Strain compatibility

17. Compatibility of strains requires that

$$\delta\varepsilon_r = \frac{\partial}{\partial r}(\delta u) \qquad (16)$$

where u is the radial displacement. In incremental form this equation becomes

$$\delta\varepsilon_r = \frac{1}{\Delta r}[(\delta u)^{r+\Delta r} - (\delta u)^r] \qquad (17)$$

$$\text{But}\quad \delta\varepsilon_\theta = \frac{\delta u}{r} \qquad (18)$$

Thus, equation (17) can be written in terms of r_n and r_{n-1}, as

$$\delta\varepsilon_{r_{n-1}} = \frac{1}{\Delta r}[r_n\delta\varepsilon_{\theta_n} - r_{n-1}\delta\varepsilon_{\theta_{n-1}}] \qquad (19)$$

Solution matrix

18. Eliminating $\delta\bar{\sigma}$ between equations (12) and (15) and $\delta\varepsilon_r$ between equations (1) and (19) and combining the results with equation (10) gives

$$\begin{bmatrix} 1/E & -\nu/E & -\nu/E & r_{n-1}/\Delta r & 3\sigma_r'/2\bar{\sigma} \\ -\nu/E & 1/E & -\nu/E & -1 & 3\sigma_\theta'/2\bar{\sigma} \\ -\nu/E & -\nu/E & 1/E & 0 & 3\sigma_z'/2\bar{\sigma} \\ (3\Delta-r_n) & -\Delta r & 0 & 0 & 0 \\ 3\sigma_r' & 3\sigma_\theta' & 3\sigma_z' & 0 & -2E_p\bar{\sigma} \end{bmatrix} \begin{bmatrix} \delta\sigma_r \\ \delta\sigma_\theta \\ \delta\sigma_z \\ \delta\varepsilon_\theta \\ \delta\bar{\varepsilon}_p \end{bmatrix} = \begin{bmatrix} r_n\delta\varepsilon_{\theta_n}/\Delta r \\ 0 \\ \delta\varepsilon_z \\ k \\ 0 \end{bmatrix} \qquad (20)$$

External pressure on liner

19. Considering Fig.1 the external pressure on the liner due to the taper can be calculated as follows. For vertical equilibrium

$$P\pi r_I^2 = 2\pi r_{E1} L(N\sin\theta + \mu N\cos\theta)/\cos\theta \qquad (21)$$

where r_{E1} is the external radius of the liner, L is the effective length, θ the semi-angle of the taper, N the normal stress and μ the coefficient of friction between liner and container. Thus

$$N = Pr_I^2/2r_{E1} L(\tan\theta + \mu) \qquad (22)$$

Resolving horizontally the radial pressure on the outside of the liner due to the taper is therefore given by

$$P_{E1} = \frac{Pr_I^2(\cos\theta - \mu\sin\theta)}{2r_{E1} L(\tan\theta + \mu)} \qquad (23)$$

which can be written

$$P_{E1} = \alpha_1 P \qquad (24)$$

where α_1 is a function of the geometry of the system and the coefficient of friction.

20. The external pressure due to the restraining effect of the container can be found by considering the total pressure vessel set but without the taper. The pressure at the outside of the liner with the elastic-plastic interface at $r = r_{int}$ is given by the usual expression

$$P_{E2} = -\frac{\sigma r_{int} r_{int}^2(1 - r_{EC}^2/r_{EL}^2)}{(r_{int}^2 - r_{EC}^2)} \qquad (25)$$

where r_{EC} is the outside of the container.

Equation (25) may be rewritten as

$$P_{E2} = -\alpha_2 \sigma_{r_{int}} \quad (26)$$

where α_2 is a function of the geometry of the system and the position of the elastic-plastic interface. Denoting the total external pressure on the liner by P_{EX},

$$P_{EX} = \alpha_1 P - \alpha_2 \sigma_{r_{int}} \quad (27)$$

Boundary conditions

21. At the elastic-plastic interface, using the Mises yield criterion,

$$2\bar{\sigma}_o^2 = (\sigma_r - \sigma_\theta)^2 + (\sigma_\theta - \sigma_z)^2 + (\sigma_z - \sigma_r)^2 \quad (28)$$

and from the elastic stress-strain relationships

$$\sigma_z = \nu(\sigma_\theta + \sigma_r) + E\varepsilon_z \quad (29)$$

Combining equations (28) and (29) leads to the result

$$2\bar{\sigma}_o^2 = (\sigma_r - \sigma_\theta)^2 + [\sigma_\theta(1-\nu) - \nu\sigma_r - E\varepsilon_z]^2 + [\sigma_r(1-\nu) - \nu\sigma_\theta - E\varepsilon_z]^2 \quad (30)$$

For the elastic part of the tube, the Lamé equations are:

$$\sigma_r = A + B/r^2 \quad (31)$$

and
$$\sigma_\theta = A - B/r^2 \quad (32)$$

Substituting these two equations into equation (30) produces

$$\bar{\sigma}_o^2 = \frac{3B^2}{r_{int}^4} + (1-2\nu)^2 A^2 + (E\varepsilon_z)^2 - 2(1-2\nu)AE\varepsilon_z \quad (33)$$

At $r = r_{EL}$, σ_r is equal to the external pressure on the liner, P_{EX}. Thus, substituting equation (27) into equation (31) gives

$$A = \frac{B}{1-\alpha_2}\left(\frac{\alpha_2}{r_{int}^2} - \frac{1}{r_{EL}^2}\right) - \frac{\alpha_1 P}{(1-\alpha_2)} \quad (34)$$

which can be written

$$A = DB - CP \quad (35)$$

where C and D are constants which may be found from (34). Substituting this into equation (33) yields the following quadratic equation in B:

$$0 = B^2\left[\frac{3}{r_{int}^4} + (1-2\nu)^2 D^2\right] + B[-2DCP(1-2\nu)^2 - 2(1-2\nu)E\varepsilon_z D] + [C^2P^2(1-2\nu)^2 + (\varepsilon_z E)^2 + 2(1-2\nu)CPE\varepsilon_z - \bar{\sigma}_o^2] \quad (36)$$

Initially with the interface at the liner bore, $\sigma_{r_{int}} = -P$ and $r_{int} = r_I$, which substituted into equation (31) gives

$$P = -A - B/r_I^2 \quad (37)$$

Substituting this into equation (34) leads to

$$A = \frac{B}{(1-\alpha_1-\alpha_2)}\left(\frac{\alpha_1+\alpha_2}{r_I^2} - \frac{1}{r_{E1}^2}\right) \quad (38)$$

which may be written

$$A = BD' \quad (39)$$

where D' may be found from equation (38). Substituting equation (39) into (33) yields the following quadratic in B

$$0 = B^2\left[\frac{3}{r_I^4} + (1-2\nu)^2 D'^2\right] + B[-2(1-2\nu)E\varepsilon_z D'] + [(E\varepsilon_z)^2 - \bar{\sigma}_o^2] \quad (40)$$

NUMERICAL METHOD

22. The liner is initially split into a number of shells with undeformed inner radii designated by r_{on} where n is the index number of the shell. Initially, it is supposed that the elastic-plastic interface is at r_{n-1} where r is the current radius of the point r_o, and that the stress and strain distributions are known throughout the tube. The elastic-plastic interface is then allowed to move out a distance Δr to r_n. The stresses and subsequently the strains are then calculated at the interface using equations (36), (35), (32) and (31). Equation (36) contains P, the current internal pressure, which will not be known at this stage and so an iterative technique is required. Similarly the axial strain ε_z in equation (36) is unknown. Thus, values are assumed for P and ε_z (generally their previous values, i.e. when the interface was at r_{n-1}) and the stresses and strains calculated at the next inner surface, i.e. at r_{n-1}, using equation (20).

23. In this equation $E_p\bar{\sigma}$ is known from the previous equivalent strain at that surface, and the stresses σ_r etc. also refer to previous values at that surface. Thus the only unknown, apart from the stress and strain increments, is the current radius of the surface, i.e. r_{n-1}. This is assumed to have its previous value and a solution obtained. From the new value obtained for ε_θ, a better value of r_{n-1} is calculated and iteration confirmed until the desired degree of convergence is obtained. The stresses and strains are then calculated for the next inner surface in the same way, and so on until the bore is reached. The end load on the liner can then be calculated from the axial stresses and this is checked against zero for the open-ended case or against the product of the new internal pressure and the current bore, for the closed-ended case. The axial strain is then adjusted using a simple search technique, and together with the new value of P, the whole procedure with the interface at r_n is repeated until suitable convergence of ε_z and P is obtained. The interface is then allowed to move out to r_{n+1} and so on until the outside of the liner is reached or the bore strain has reached some preselected value.

24. Initially, with the interface at the bore, equations (40) and (39) are used instead of (36) and (35); in this case it is unnecessary to determine a value for P as it does not enter into the equations. The method is summarized in the simplified flow chart shown in Fig.2.

RESULTS AND DISCUSSION

25. Numerical results were obtained for a pressure vessel assembly with a liner made of 300 M, which is a vacuum arc remelted steel of basically SAE 4340 composition plus about $1\frac{1}{2}\%$ silicon. A typical stress-strain curve for this material may be represented by the equation

$$\bar{\sigma} = 100(1 + 200\bar{\varepsilon}_p)^{0.09}$$

where $\bar{\sigma}$ is given in tonf/in^2.

26. The results were obtained for a vessel having $1\frac{3}{4}$ inch diameter bore, an external liner diameter of 5.28 inches and an overall diameter of 17 inches. These dimensions were chosen as being typical of the type of equipment in use. For this size of container it would be practical to use a container material having a yield strength of 55-60 tonf/in^2 while still retaining a reasonable ductility. This means that the container bore pressure is limited to about 30 tonf/in^2 if yielding of the container is to be prevented.

27. Thus in Figs 3-5, which show the effect of various assumed coefficients of friction, liner length and tapers, results are presented to show the internal pressure required to produce a container bore pressure of 30 tonf/in^2 as well as that to produce an equivalent bore strain of $\frac{1}{2}\%$ and 1%.

28. Figure 3 shows the variation of these internal pressures with the coefficient of friction for a taper of 2° included angle and an effective liner length of 7 inches. For coefficients of friction greater than about 0.2, the internal pressure required to produce a container bore pressure of 30 tonf/in^2 increases quite slowly with the friction, whilst the pressure required to produce a given equivalent bore strain decreases slowly. In this region, a container bore pressure of 30 tonf/in^2 is always associated with an equivalent bore strain of greater than 1% and so it is the bore strain which really limits the usable internal pressure. For coefficients of friction between 0.1 and 0.2 the same trend of pressure versus friction is observed, but the rate of change is greater. At the lower end of this range, a container bore pressure of 30 tonf/in^2 is produced at an equivalent bore strain of less than $\frac{1}{2}\%$. For lower values of the coefficients of friction, the pressure to produce a given bore strain increases very rapidly for decreasing friction, but the pressure to produce a container bore pressure of 30 tonf/in^2 decreases rapidly. Thus in this region it is the container strength which imposes a limitation on the design, rather than the bore strain.

29. Figure 4 shows the variation of the internal pressure with effective liner length for the same taper as in Fig.3 and a coefficient of friction of 0.2. The variation of pressure with length is of the same type as that with friction. This is to be expected since, considering equation (23) for small tapers and a fixed radial geometry, the restraint imposed on the liner due to the taper is inversely proportional to the product of the coefficient of friction and the effective length of the liner. For the conditions shown, it is the bore strain which limits the usable pressure for any effective liner length greater than about 6 inches.

30. Figure 5 shows that the internal pressures are not very sensitive to changes in the angle of the taper. The pressure for the limiting container bore pressure increases by only about 3% as the included taper angle increases from 2° to 8°, and the pressures for limiting bore strains decrease by even less over the same range. Clearly the limiting value of taper will be determined more by manufacturing limitations on the diameter tolerances rather than theoretical design considerations.

31. For example, for an included taper angle of 2° the relative vertical displacement between the liner and container is about ±0.115 inches for diameter tolerances of ±0.001 inches on each, and double that for a 1° taper.

32. A further point of interest which becomes apparent from the numerical results is that at some value of internal pressure, the pressure on the outside of the liner begins to increase faster than the inside pressure. Once this pressure has been reached the radial bore strain can no longer increase and the equivalent bore strain decreases, though the present calculations are not valid in this region. Figure 6 shows the variations in this maximum equivalent bore strain value with coefficient of friction; the associated values of internal pressure are also shown. Both the strain and pressure increase very rapidly with friction for coefficients of friction less than 0.2 and then tend toward a more gradual, almost linear, increase. This observation is somewhat academic, however, since even for a coefficient of friction of 0.5, the container bore pressure is 45 tonf/in^2 at the limiting strain; this would require a container material almost as strong as the liner. For smaller coefficients of friction the container material would have to be even stronger.

33. It is apparent from these results that it is the product of the coefficient of friction and the effective length of the liner that has the greatest effect on the usable pressure in any tapered pressure vessel assembly design once the diameter ratios and materials have been chosen. Typical values for the coefficients of friction between two metal surfaces are often given as 0.15 and 0.6, but these may be altered quite significantly by placing lead foil between the liner and container, or interspersing graphite or molybdenum disulphide powder between the two. Ideally, before finalizing a design, the coefficients of friction between pieces of liner and container material using the various possible lubricants should be determined; the pieces should have the same hardness and surface finish as in the proposed design and the range of normal pressures studied should be comparable to those expected in use. The effective length of the liner is also a difficult quantity to establish. It will change as the pressure is applied due to the plunger motion and may also change while the pressure remains constant as in the extrusion process where the plunger is continually moving downward. Thus it is necessary to take a range of length that might be used. A reasonable estimate for any one effective length might be the length of the pressure chamber at that point, or perhaps 10% more.

34. The other factors to be decided upon before making any design decision are the criteria to be used in assessing the design.

35. Those used in the present paper are the pressure required to cause yielding of the container and the pressures to produce a given equivalent bore strain. The first is dictated by the method of analysis, though it would be possible to extend the analysis to allow for plastic expansion of the container. The second is more arbitrary, though an equivalent bore strain of greater than 1% will generally correspond to a bore expansion larger than that acceptable for leak-free operation of the system. The permissible bore strain will be dependent on the type of sealing devices employed and the ductility of the liner material.

36. No attempt has been made to demonstrate the effect of different diameter ratios since it is generally accepted that a ratio of about 3 is "optimum"; less than that causes a significant drop in the pressure capability while anything greater produces only small improvements for large increases in physical size, cost and required handling capacity.

CONCLUSIONS

37. A numerical technique for calculating the usable internal pressure in a tapered pressure vessel assembly has been developed which accounts for the elastic-plastic expansion of a work-hardening liner material in an elastic container. It has been shown that the product of the coefficient of friction between liner and container and the effective length of liner is the most important design parameter, once the material and diameter ratios have been chosen. For small values of this product, the usable pressure is limited by the strength of the container, while for large values it is limited by the liner bore strain.

Acknowledgements

One of the authors (R.L.H.) held a Commonwealth Scholarship during the course of this study, for which he is grateful. The authors also thank the National Aeronautical Establishment and the National Research Council of Canada for the computing facilities used during much of this work.

REFERENCES

1. BRIDGMAN, P.W. Studies in Large Plastic Flow and Fracture. McGraw-Hill, London and New York, 1952.
2. BERESNEV, B.I., VERESCHAGIN, L.F., RYABININ, Yu. N. and LIVSHITS, L.D. Some Problems of Large Plastic Deformation of Metals at High Pressures. Acad. Nauk. SSSR, 1960. English translation Macmillan, New York, 1963.
3. PUGH, H.Ll. (editor). Mechanical Behaviour of Materials under Pressure. Elsevier, Amsterdam, London and New York, 1970, p.391.
4. ALEXANDER, J.M. and GUNASEKERA, J.S. Proc. Roy. Soc. London, 1972, A.326, 361.

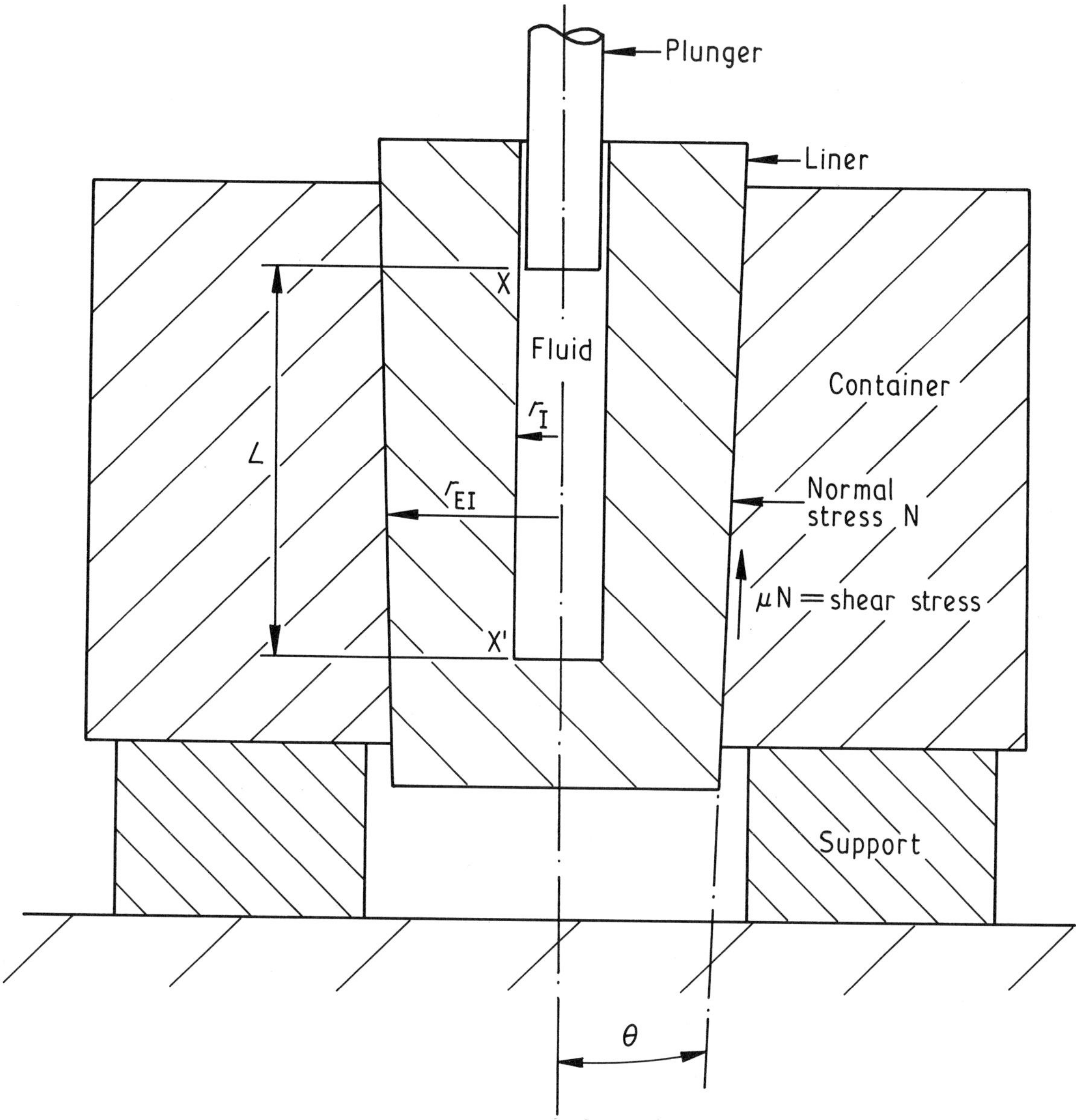

Fig. 1: Tapered pressure vessel assembly, showing forces acting

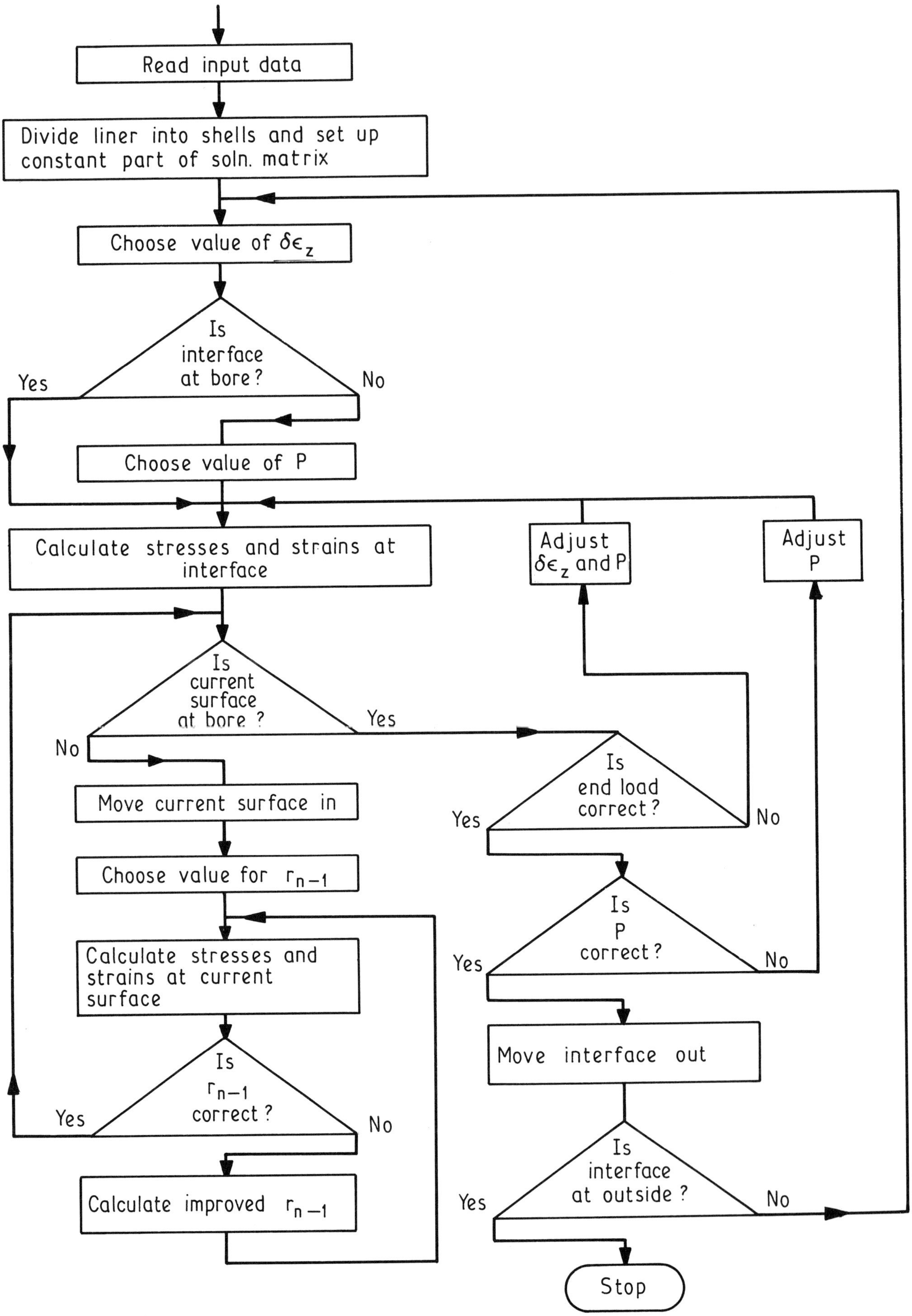

Fig. 2: Simplified flow chart of solution algorithm

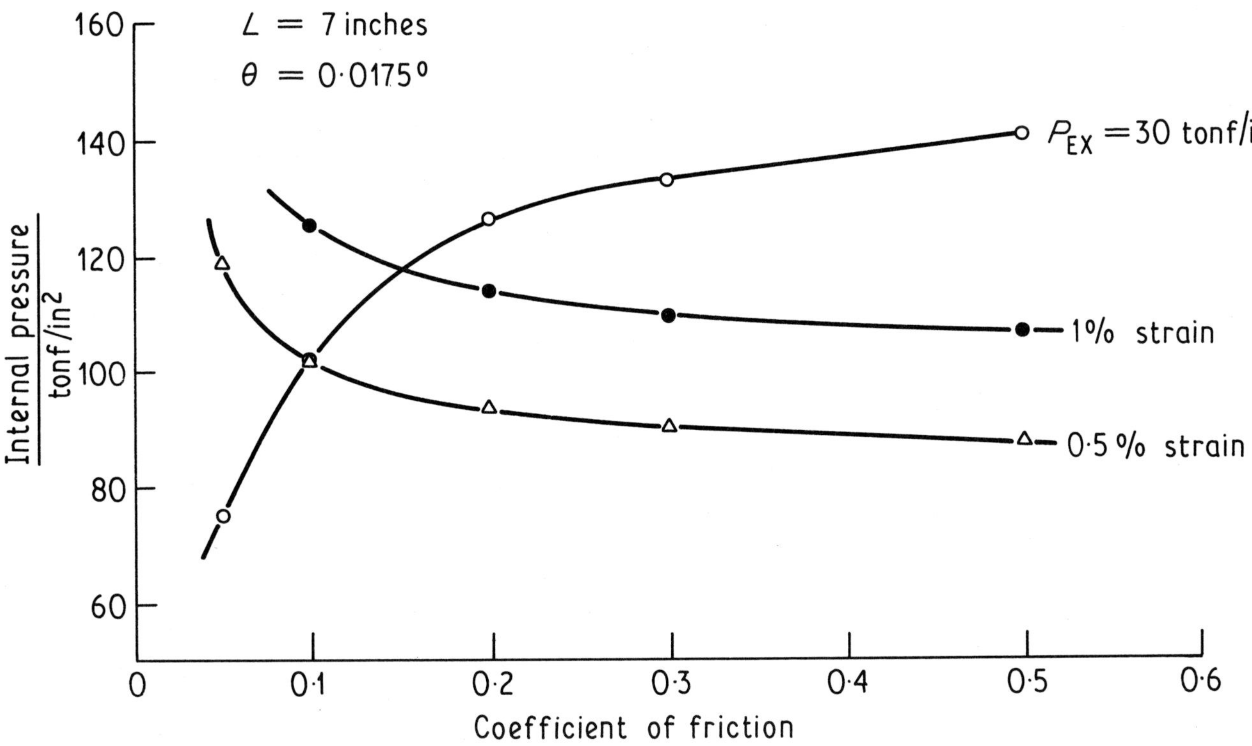

Fig. 3: Variation of internal pressure with coefficient of friction

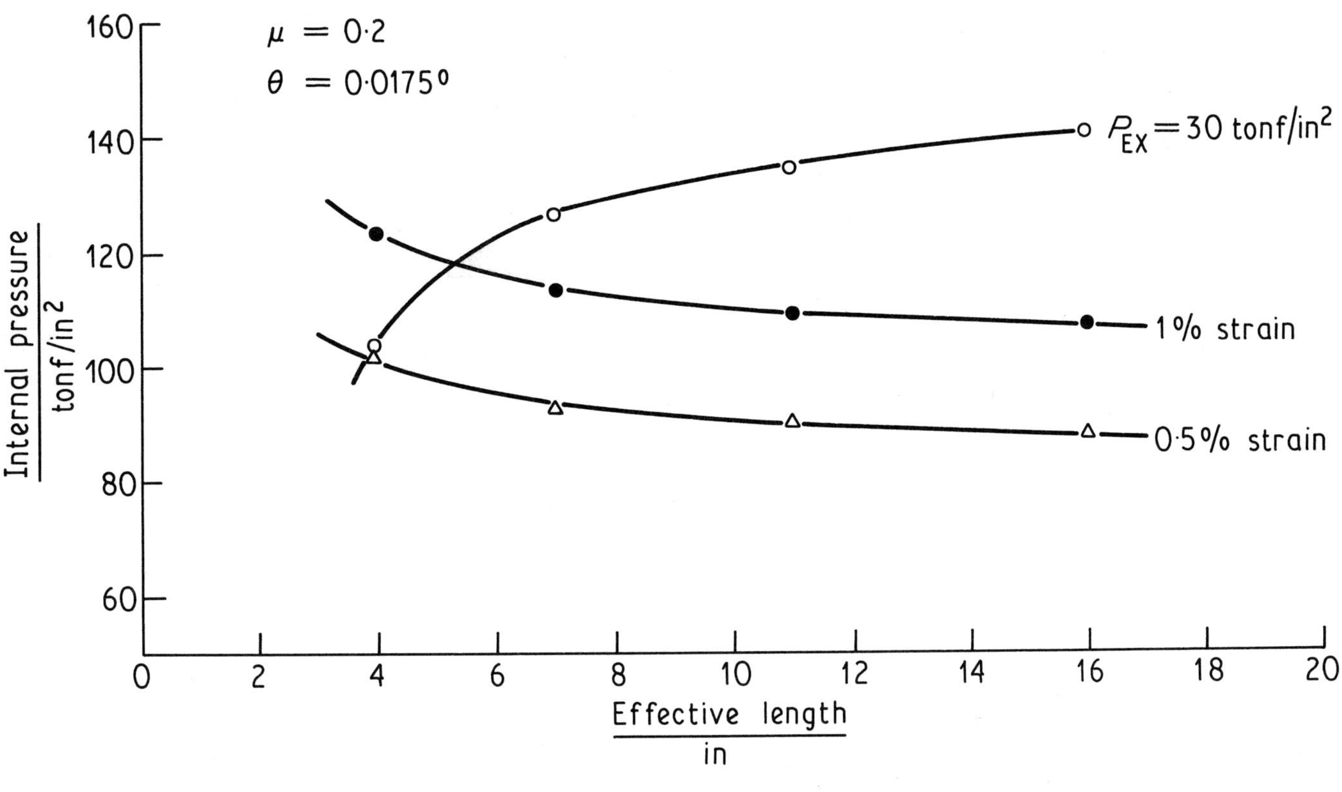

Fig. 4: Variation of internal pressure with effective liner length

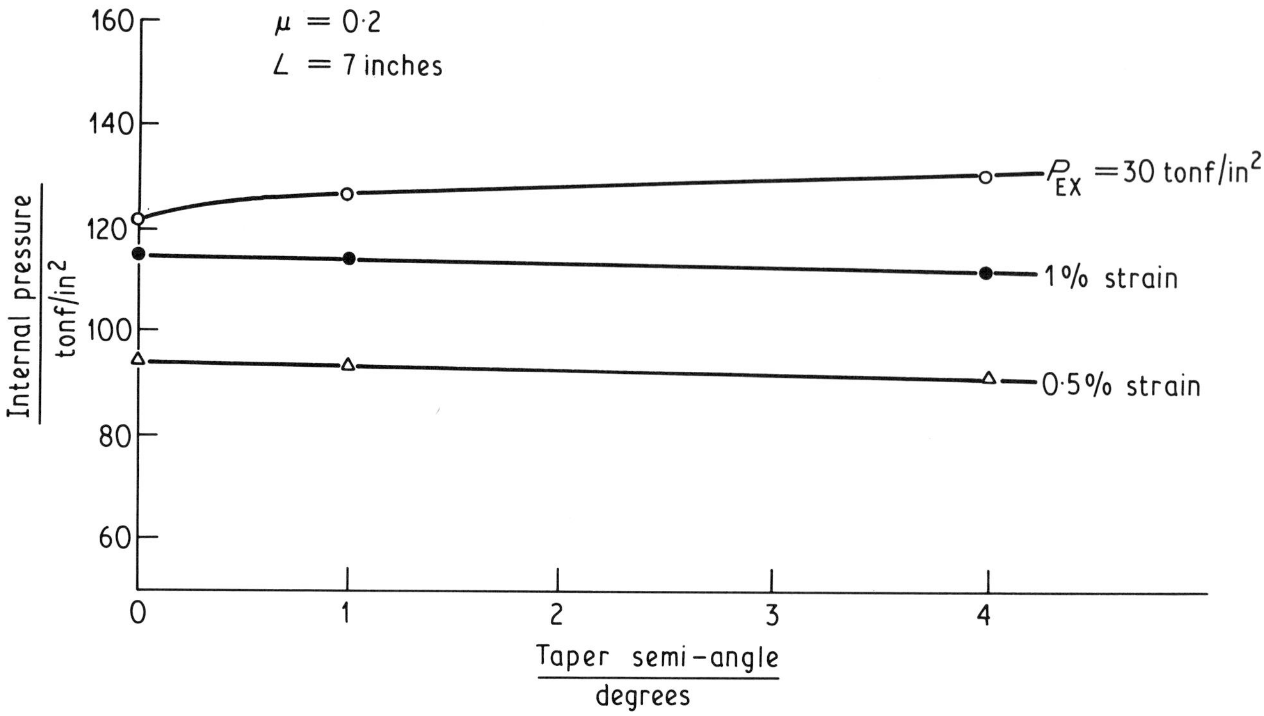

Fig. 5: Variation of internal pressure with taper angle

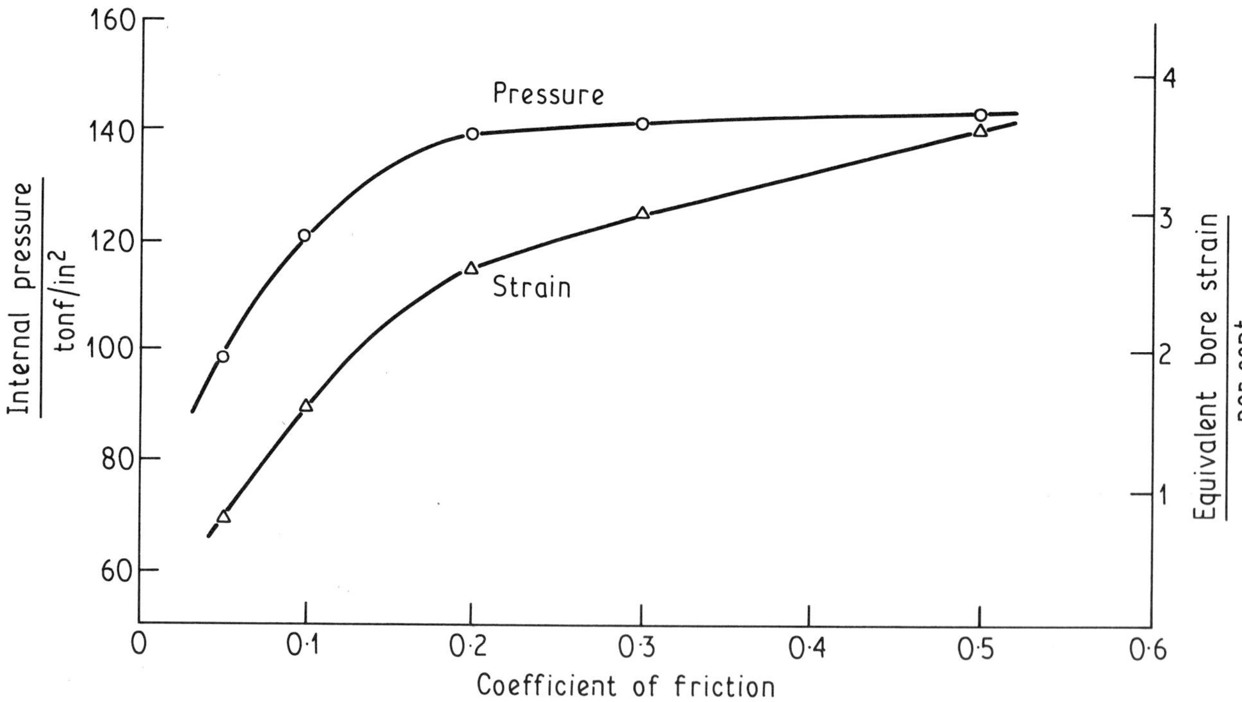

Fig. 6: Variation of limiting equivalent bore strain with coefficient of friction

THE ELASTOHYDRODYNAMIC LUBRICATION OF HIGH PRESSURE PLAIN GLAND CLEARANCE SEALS: THE EFFECT OF SEAL GEOMETRY AND RAM VELOCITY ON SEAL LEAKAGE AND RAM STABILITY

M. J. GRAY, BSc
Research Student, Department of Mechanical Engineering, The Queen's University of Belfast
W. J. SKELTON, BSc, PhD, CEng, MIMechE, Wh.Fell
Lecturer, Department of Mechanical Engineering, The Queen's University of Belfast

SYNOPSIS It is of considerable advantage in high pressure reciprocating pumps to use a fully lubricated seal to minimise wear and heating caused by friction. The simplest of these is the plain gland clearance seal, which is basically a ram fitted into a thick-walled cylinder. This paper describes an analysis of the characteristics of such seals, and in particular the effect of ram velocity on seal leakage, pressure distribution and lubrication. Seal pressures up to 700 MN/m^2 and ram velocities in the range 0 - 2 m/s have been investigated. The analysis presented primarily relates to a synthetic lubricating oil. Some preliminary considerations of special problems related to the analysis of a high pressure water seal are also presented.

1.0 INTRODUCTION

The successful operation of high pressure pumps and intensifiers often depends on forming a satisfactory fluid seal between the moving components, or at least controlling leakage to acceptable levels. Soft rubber or polymer seals, normally backed by metal anti-extrusion or mitre rings, have been widely used to effect a tight seal at high pressures, (1), (2), (3), (4)*, but intimate contact and friction limit both the speed of operation and the life of such seals. This limitation on the speed of operation has led to the development of many high pressure systems which incorporate relatively large slow moving pumps. In many cases it would be desirable to replace these large pumps with much smaller high speed units, however, the economic incentive for such a change often largely hinges on the satisfactory development of a fully lubricated metallic long life (> 10^7 cycles) seal.

To date the most successful metallic seals utilize either multiple segmented rings to form a stationary gland or alternatively an assembly of rings on the plunger to form a moving seal. In general these seals require the use of fairly exotic materials and technology and are therefore expensive.

Several investigators have examined the possibility of using a much simpler one piece plain gland clearance seal where the moving components are separated by a film of fluid, and the operation of such seals at low pressures is well documented (5), (6), (7). However, at high pressures there is significant elastic deflection of the seal gland and ram and also considerable variation of fluid properties, particularly oil viscosity. As a consequence the analysis of even simple seals is extremely complex and involves the use of both simplifying assumptions and numerical computer techniques. Several simple ram/cylindrical gland assemblies have been treated (8), (9), (10), (11), but in general there has been little consideration of the effect of ram reciprocation on seal performance.

This paper presents the results of a theoretical investigation into the effect of ram reciprocation (ram velocities in the range - 2 m/s to + 2 m/s) and pressure, (0 - 700 MN/m^2), on seal pressure distribution, leakage and ram stability. The range of conditions considered might well be found in a high speed/high pressure pump.

2.0 NOTATION

h	clearance height (m)
h_i	initial seal clearance (m)
P	pressure (N/m^2)
x	axial co-ordinate of seal (m)
$\partial P/\partial x$	seal pressure gradient (N/m^3)
B	clearance circumference (m)
ρ	fluid density (kg/m^3)
μ	absolute viscosity (Ns/m^2)
u	ram velocity (m/s) positive out of pressure chamber negative into pressure chamber
v	average fluid velocity (m/s)
P_{max}	pressure in high pressure chamber and end pressure on ram (N/m^2)
D	ram diameter, seal bore (m)
K	diameter ratio of seal cylinder
E	elastic modulus (N/m^2)
ν	Poisson's ratio
R_c	clearance ratio, defined as ratio between the initial clearance and the maximum diametral expansion of ram

* References are given in Appendix 1

NOTATION (contd.)

R_e Reynold's number $(\frac{\rho v h}{\mu})$

d_r radial deflection of ram (m)

d_s radial deflection of seal bore (m)

W mass flow rate (kg/s)

$\left.\begin{matrix}G_o\\G_1\end{matrix}\right\}$ coefficients in the isothermal pressure/viscosity equation

$\left.\begin{matrix}A_D\\B_D\end{matrix}\right\}$ coefficients in the isothermal pressure/density equation

A bar above a quantity represents non-dimensionality

3. THEORETICAL ANALYSIS

When a plain gland seal is pressurised an axial pressure distribution results causing fluid to flow from the seal entrance (high pressure end) to the seal exhaust (ambient pressure end). This distribution causes elastic distortion of both the seal-block and ram. In addition, the fluid properties (density and viscosity) will vary along the clearance. In turn, the pressure distribution is affected by these variables. Thus, the solution of even simple problems can generally not be stated in a closed form and some iterative numerical technique is necessary.

3.1 Iterative Solution

The flow of a Newtonian fluid under laminar conditions at any point in a pressure distribution can be quantified by equation 1 (ref. 12).

$$W = -B\rho\left\{\frac{1}{12\,\mu}\frac{\partial P}{\partial x}h^3 + \frac{uh}{2}\right\} \quad \ldots.(1)$$

The application of this equation where density, viscosity and clearance height are functions of pressure usually requires a numerical solution. In the case of the clearance seal the fluid flow can be treated as uni-directional and consequently the two node elements discussed by Booker and Heubner (13) were used to model the clearance (20 elements).(FIG 1)

The basic computation technique is shown by the flow chart outlined in Fig. 2. Initially a linear pressure distribution is assumed, and used to calculate clearance, density and viscosity at each node. Element density and viscosity are obtained by averaging the relevant nodal values. The matrices describing the properties of the line of elements can then be formed and solved (using the system boundary conditions). This yields a pressure distribution consistent with continuity of mass flow, but generally different from the initial pressure distribution. The entire calculation is then repeated using a new initiating pressure distribution derived by averaging the output and initial distributions from the previous cycle. The output distributions quite rapidly converge to a solution, the maximum discrepancy between initiating and output solutions being less than 0.1% after five or six cycles. The program performs this operation for a range of boundary conditions.

The methods for obtaining the required variables from the pressure are outlined in the sub-sections which follow.

3.2 Elastic Distortion of Seal and Ram

The seal was treated as an open-ended thick-walled cylinder under a variable internal pressure where it was assumed that the deflection at any point is a function of the local pressure only. Using this simplifying assumption the radial bore deflection at any point can be calculated as:-

$$d_s = \frac{P.D}{2E}\left\{\frac{(K^2+1)}{(K^2-1)} + \nu\right\} \quad \ldots.(2)$$

In calculating the ram deflection account was taken of both the local pressure acting on the circumference and the pressure end load. Thus, the radial deflection of the ram at any point can be calculated as:-

$$d_r = \frac{D}{2E}\left\{\nu\,(P + P_{max}) - P\right\} \quad \ldots.(3)$$

When the seal is pressurised the clearance at any point is then given by:-

$$h = h_i + d_s - d_r \quad \ldots.(4)$$

3.3 Viscosity and Density Relationships

The oil data used was taken from the A.S.M.E. lubricant report (14), the figures for oil 1-A at 38°C being chosen. The selection of this fairly high temperature was in anticipation of possible heating effects in pumps using this type of clearance seal.

Equation (5) given by Dowson and Whitaker (15) was used to model the isothermal variation of viscosity with pressure.

$$\ln\bar{\mu} = \frac{G_o\,\bar{P}}{1 + G_1\,\bar{P}} \quad \ldots.(5)$$

Using this approach it was possible to predict the viscosity over the pressure range of interest to within 5% and this was considered acceptable.

Equation (6) from ref. (15) was used to model the isothermal variation of density with pressure.

$$\bar{\rho} = 1 + \frac{A_D\,\bar{P}}{1 + B_D\,\bar{P}} \quad \ldots.(6)$$

In this case the fit with A.S.M.E. data was extremely good.

A preliminary analysis using water has also been conducted. The water density data was obtained from Burnham et. al. (16) and the viscosity data from Bruges and Gibson (17). In this instance a linear interpolation technique was used to model the data.

3.4 Ram Stability Analysis

A ram will not usually run concentrically in a seal. It is, therefore, pertinent to investigate the effects of eccentricity on pressure distributions which may produce a net lateral force on the ram. This has been done by comparing the pressure distributions associated with different clearances for the same seal, the effect of eccentricity being to reduce the clearance on one side of the ram and to increase it correspondingly on the opposite side. By interpolating the pressure differences and integrating over the ram surface, it is possible to estimate the side force.

This form of analysis assumes that the flow remains one-dimensional. Circumferential pressure gradients do exist but these are small compared to those in the axial direction; provided the degree of eccentricity is small. It is considered therefore that this approach is reasonable and that the results obtained are of significance in indicating the main trends.

3.5 Flow Criteria

The fluid finite elements used are based on equation (1) which is valid only if the fluid is Newtonian and the flow laminar. In the present analysis it was assumed that the fluid was Newtonian.

The transition region from laminar to turbulent flow is usually expressed by a critical value of Reynold's Number. Massey (18) gives a R_e value of about 1000 for this transition for flow between parallel plates. Tao and Donovan (6) derived R_e values in the range 750 - 1250 from a series of experiments on fine clearance water seals. These numbers were based on either the radial clearance (6) or the separation between plates (18).

To the authors' knowledge there has been no detailed study of flow criteria for the pressure conditions and clearance geometry presented in this paper. Therefore, in the absence of more relevant data, a transition value of 1000 has been assumed.

4. RESULTS AND DISCUSSION

The results presented mainly relate to oil though a brief consideration of water seals is outlined in sub-section 4.5.

4.1 Pressure Distribution and Clearance Shape

Fig. 1 (a) represents an idealized plain gland seal which, when pressurized, deflects in the manner shown by Fig. 1 (b). The ram is compressed by hydrostatic pressure at the entrance and expanded at the exhaust by the pressure end load. The seal block bore expands at all points other than the exhaust where the deflection is always zero; this is implied by the local pressure assumption in combination the simple elastic theory used. This basic pattern is confirmed by the developing clearance shapes of Fig. 4 which are caused by the axial pressure distributions of Fig. 3 (c). These plots relate to a seal of 0.05 m bore and length.

As the pressure increases the clearance at the exhaust closes. Thus, if the clearance ratio is equal to or less than unity the seal will apparently seize before the maximum pressure is achieved. This represents a significant limitation of the present analysis since clearly the whole pressure distribution will influence seal block deflection at the exhaust where one would anticipate expansion. Preliminary results from a more realistic finite element analysis show that this is so and consequently it is possible to analyse cases where the initial clearances are less than the maximum ram expansion.

The distributions of Fig. 3 show a marked tendency to severe pressure drop near the exhaust at higher pressures. This is probably a result of the throttling effect at the exhaust. The distribution shape at zero ram velocity appears to be independent of seal diameter, though the effect of ram velocity increases in significance as the diameter decreases (Figs. 3 (a) and 3 (b)). This size effect will be discussed in more detail in sub-section 4.3.

4.2 Seal Leakage

The leakage curves of Fig. 5 are essentially similar in shape. Initially the leakage rate increases fairly linearly with increasing pressure but at pressures in excess of about 200 MN/m^2 the rate of increase begins to fall off. In some cases the curves show a peak leakage value at a pressure well below the maximum.

It is possible that the tendency of the leakage curves to maximize is a result of the throttling at the exhaust already mentioned. However, the severe throttling effect is peculiar to the simple elastic analysis and it is conceivable that the more rigorous finite element analysis now underway, may well predict leakage curves with significantly different characteristics. The approximately linear increase in leakage rate at lower pressures may be caused by the exponential increase in oil viscosity compensating for the combined action of increasing clearances and pressure gradients.

It is interesting to note that the flow for a 0.0125 m bore seal at a ram velocity of - 2 m/s is negative for pressures up to about 140 MN/m^2 i.e. the ram is pumping oil into the pressure chamber against the pressure gradient. The practical implication of this observation is that for some operating conditions it would be necessary to arrange for an oil supply at the exhaust end of the seal in order to prevent air being drawn into the system.

The magnitude of seal leakage is very much reduced when the diameter is decreased while at the same time the relative importance of ram velocity increases. This is compatible with the increased effect of ram velocity on pressure distributions shown by Fig. 3. Fig. 6 shows the leakage curves for three seals of bores 0.0125, 0.025 and 0.05 m with a clearance ratio of 1.2 and at zero ram velocity. These curves emphasize the strong dependence of leakage rate on seal diameter. Indeed the leakage curve for the smallest seal (0.0125 m bore) is almost negligible compared to that for the largest seal (0.05 m bore).

4.3 Size Effects

The relationship between seal diameter and leakage can be explored by recourse to equation (1).

Leakage is caused by two mechanisms; (a) pressure gradient and (b) ram (shear) velocity. For pressure alone the leakage can be expressed by:-

$$W_p \quad \alpha \quad \frac{\rho B h^3}{\mu} \quad \frac{\partial P}{\partial x} \qquad \ldots .(7)$$

Thus, the leakage at zero ram velocity is proportional to the fourth power of the seal bore (h^3 and B) and inversely proportional to the seal length. This trend is borne out by Fig. 6 where the largest seal (0.05 m diameter) has 256 times the leakage of the smallest (0.0125 m diameter). Also, if the leakages for seals of different bores and identical pressures at zero ram velocity are compared (see Table 1), then it is apparent that the trends indicated by equation (7) are followed exactly. The leakage caused by ram velocity alone can be expressed by:-

$$W_u \quad \alpha \quad \rho \; B \; h \; u \qquad \ldots .(8)$$

Combining equations (7) and (8) and neglecting any interaction, then:-

$$\frac{W_u}{W_p} \quad \alpha \quad \frac{u}{h^2} \; \frac{\partial x}{\partial P} \; \frac{1}{\mu} \qquad \ldots .(9)$$

For comparing seals of different size at the same pressure it is reasonable to assume that $\frac{1}{\mu}$ is characteristic of the pressure and fluid, while $\frac{\partial P}{\partial x}$ is an inverse function of the seal length. Equation (9) can therefore be reduced to:-

$$\frac{W_u}{W_p} \quad \alpha \quad \frac{Lu}{h^2} \qquad \ldots .(10)$$

Since, by definition, the clearances are proportional to the seal bore, then the magnitude of $\frac{W_u}{W_p}$ is directly proportional to seal length and will increase rapidly as bore size is reduced. It must be emphasized that the formation of equation (10) assumes that there is no interaction between the ram velocity and the geometry of the clearance. However, at the smallest seal bore it can be seen that the change in leakage caused by equal inward and outward ram velocity is not symmetrical about the zero velocity value as would be implied by equation (10). This is directly traceable to the increased effect of ram velocity on the pressure distributions of small bore seals as shown in Fig. 3.

This naive approach to the comparison of seals is not applicable to the calculation of absolute seal leakage or to determining the shape of the leakage/pressure curve.

4.4 Ram Recentring Force

When the initial clearance is reduced the general level of the pressure distribution rises and vice versa if the clearance is increased (cf. a hydrodynamic bearing). Thus, slight eccentricity of the ram will produce pressure differences which will try to recentre it. The pressure difference distributions shown in Fig. 7 are virtually independent of the bore size at zero ram velocity but, as may be expected, there is considerable reduction in the recentring pressure for small bore seals when the ram moves rapidly inwards.

In a high speed pump it is likely that the piston will be driven by a crank or cam and as a consequence it will be subjected to a lateral disturbing force derived from the thrust. The ram stability will be related to the ratio of the lateral recentring force (obtained as outlined in sub-section 3.4) to the thrust, and this ratio is plotted as a function of seal bore size in Fig. 8. These curves are based on an eccentricity to initial clearance ratio of 0.083. The curves indicate that a pump with a ram diameter of 0.0125 m and 0.04 m stroke operating at 1000 R.P.M. at 700 MN/m^2 (these dimensions might be considered reasonable for a large delivery high speed pump) would be vulnerable to ram instability during the delivery stroke.

4.5 A Water Seal and Considerations of Flow

Fig. 9 is the leakage curve at zero ram velocity for a small bore seal using pure water. The leakage rate is very high, giving a maximum Reynolds Number much greater than the transition value discussed in sub-section 3.5. This is in contrast to the maximum values for the oil seals shown in Figs. 5a, b and c, which are well below the transition region. This implies that the use of the Reynolds type flow equation (Eqn. 1) is valid for oils but its use with water is questionable. It is remarkable that in many analyses for water seals, the Reynolds equation seems to be assumed without consideration of the possibility of turbulent flow. Clearly, an analysis which would cater for this possibility would be considerably more difficult than the laminar flow case.

Nevertheless, it may be relevant to speculate on other approaches to the problem. The Bernoulli equation has been examined but this leads to the unreasonable conclusions that the leakage is a function of the overall pressure drop and the exhaust clearance only, but is independent of the seal length. A friction factor approach similar to that used for pipe flow would seem to be more realistic, but defining the relative surface roughness, particularly if the clearances are comparable to asperity heights, and applying this to a geometry far removed from the pipe flow problem is beyond the scope of the present analysis.

5. CONCLUSIONS

The seals were investigated with their application to pumps in mind. Clearly from leakage considerations, it is advantageous to use a small bore seal with a fast cyclic speed. There is, however, a danger that at high speed the flow could be completely reversed and the seal mechanism break down. Furthermore, the stability of these small seals is poor, and side loading would tend to attach the ram to the side wall. Thus, a direct drive from crank or cam to piston could result in excessive friction, wear and possibly jamming. A cross-head arrangement would be desirable to absorb any side loading.

Several shortcomings of the analysis itself have been pointed out in the discussion. A more sophisticated finite element program is being developed to overcome the limitations of the local pressure assumption. It is hoped to publish results from this work in the near future.

Water pumps have recognised importance in industrial applications, principally jet cutting and it would certainly be advantageous to perform similar analyses on water seals. Flow conditions, however, preclude the use of the simple laminar flow approach and work is being directed to their analysis with empirical turbulent flow equations.

APPENDIX 1

References

1. Crossland, B., Skelton, W. J. and Wilson, W. R. D. 'High Pressure Intensifiers: Development of a Unit for pressures of 200,000 lbf/in^2 and possible improvements in design', Proc. Institution of Mechanical Engineers 1967-1968, 182 Pt. 3C, 175-179.

2. Crossland, B. and Logan, J. G. 'Development of Equipment for Jet Cutting', 1st International Symposium on Jet Cutting Technology, B.H.R.A. 1972.

3. Tardiff, H. P. 'High Pressure Dynamic Seals', Canadian Armament Research and Development Establishment, Technical Report 533/65, October 1965.

4. Manning, W. R. D. and Labrow, S. 'High Pressure Engineering', C.R.C. Press Cleveland, 1971.

5. Bowen, J. C. 'An Approach to Pulseless Packingless Pumping', 1st International Symposium on Jet Cutting Technology, B.H.R.A. 1972.

6. Tao, L. N. and Donovan, W. F. 'Through Flow in Concentric and Eccentric Annuli of Fine Clearance with and without Relative Motion of the Boundaries', Trans. A.S.M.E., Nov. 1955, 1291-1301.

7. Exline, P. G. 'Leakage in Capillary Seals of Hydraulic Valves and Pumps', Product Engineering, April 1946, 290-296.

8. Harris, H. D. 'An Elastohydrodynamic Analysis of the Sleeve Type High Pressure Seal', Journal of Lub. Tech., Oct. 1972, 335-340.

9. Wang, N. M. and Kamal, M. M. 'Elastohydrodynamic Effects in a Clearance Seal', Journal of Lub. Tech., April 1970, 310-313.

10. Kamal, M. M. 'A High Pressure Clearance Seal', Journal of Lub. Tech., April 1968, 412-416.

11. Huhs, E. 'The Flow of Viscous Fluids in Clearance Seals at very High Pressures', 5th International Conference on Fluid Sealing, B.H.R.A., 1971 Paper E3.

12. Fuller, D. D. 'Theory and Practice of Lubrication for Engineers', Wiley, New York, 1956.

13. Booker, J. F. and Huebner, K. H. 'Application of Finite Element Methods to Lubrication: An Engineering Approach', Journal of Lub. Tech., Oct. 1972, 313-323.

14. A.S.M.E. Research Committee on Lubrication 'Viscosity and Density of Over 40 Known Lubricants', New York, 1953, Vol. 2.

15. Dowson, D. and Whitaker, A. V. 'A Numerical Procedure for the Solution of the Elastohydrodynamic problem of Rolling and Sliding Contacts Lubricated by a Newtonian Fluid', Proc. Institution of Mechanical Engineers, 1965-1966, Vol. 180, 3B Paper No. 4.

16. Burnham, C. W., Holloway, J. R. and Davis, N. F., 'The Specific Volume of Water in the Range 1000 to 8900 bars, 20°C to 900°C', Amer. Journal of Sci., 1969, Schairer Vol. 267-A, 70-95.

17. Bruges, E. A. and Gibson, M. R. 'Dynamic Viscosity of Compressed Water to 10 kilobar and Steam to 1500°C, Journal of Mech. Eng. Sci., 1969, Vol. 11, No. 2, 189-205.

PRESSURE MN/m^2	BORE = 0.0125 m			0.025 m			0.050 m		
	u = 2 m/s	0 m/s	- 2 m/s	2 m/s	0 m/s	- 2 m/s	2 m/s	0 m/s	- 2 m/s
0	.000270 .000270	0 0	- .000270 - .000270	.00108 .00108	0 0	- .00108 - .00108	.00432 .00432	0 0	- .00432 - .00432
100	.000690 .000968	.000286 .00571	- .000081 .000194	.00614 .0107	.00457 .00914	.00304 .00760	.0793 .152	.0731 .146	.0670 .140
200	.00112 .00168	.000570 .00114	.000093 .000642	.0112 .0203	.00912 .0182	.00709 .0162	.154 .300	.145 .290	.138 .283
300	.00148 .00223	.000776 .00155	.000195 .000937	.0154 .0274	.0124 .0248	.00989 .0223	.209 .407	.119 .397	.188 .387
400	.00175 .00261	.000897 .00179	.000226 .00107	.0175 .0318	.0143 .0287	.0114 .0257	.242 .474	.230 .459	.217 .447
500	.00197 .00285	.000952 .00190	.000206 .00109	.0189 .0341	.0152 .0305	.0118 .0270	.258 .502	.244 .488	.230 .474
600	.00215 .00301	.000963 .00193	.000157 .00103	.0196 .0349	.0154 .0308	.0116 .0269	.263 .509	.246 .492	.231 .477
700	.00232 .00312	.000945 .00189	.000096 .000921	.0198 .0347	.0151 .0302	.0110 .0260	.260 .502	.242 .484	.225 .467

TOP FIGURE .05 m LENGTH SEAL, BOTTOM FIGURE .025 m LENGTH

TABLE 1 LEAKAGE (kg/s) OF OIL SEALS FOR VARIOUS SIZES AND OPERATING CONDITIONS: R_c = 1.2 IN ALL CASES

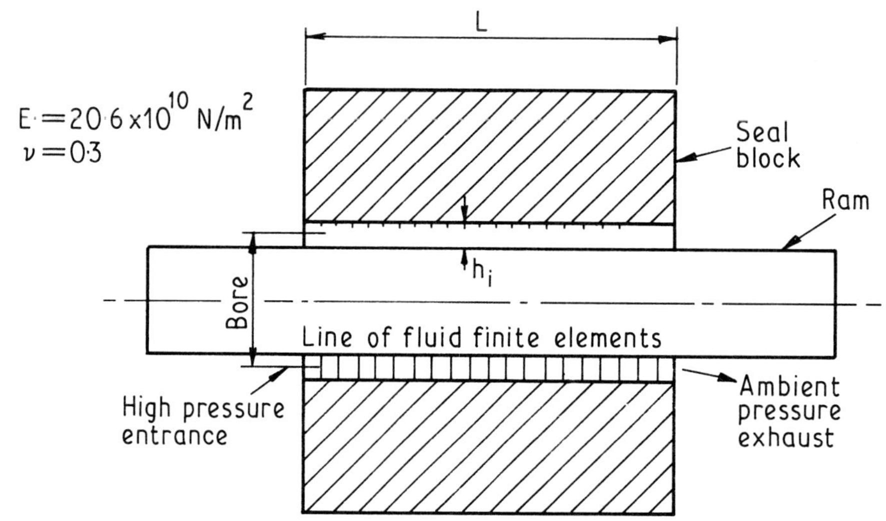

a Unpressurised

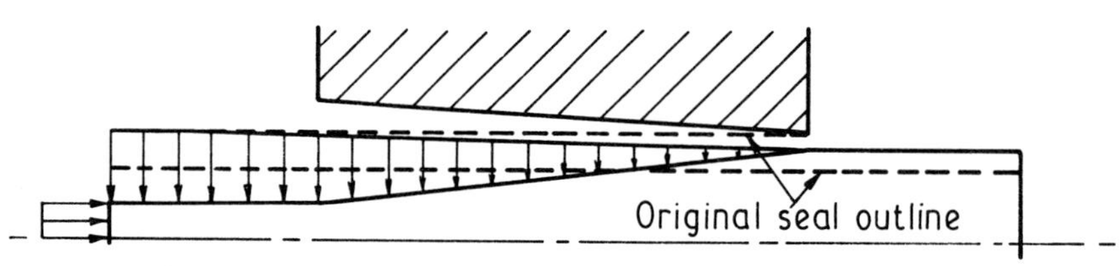

b Pressurised (linear pressure distribution)

Fig. 1: Basic Geometry of a plain gland clearance seal

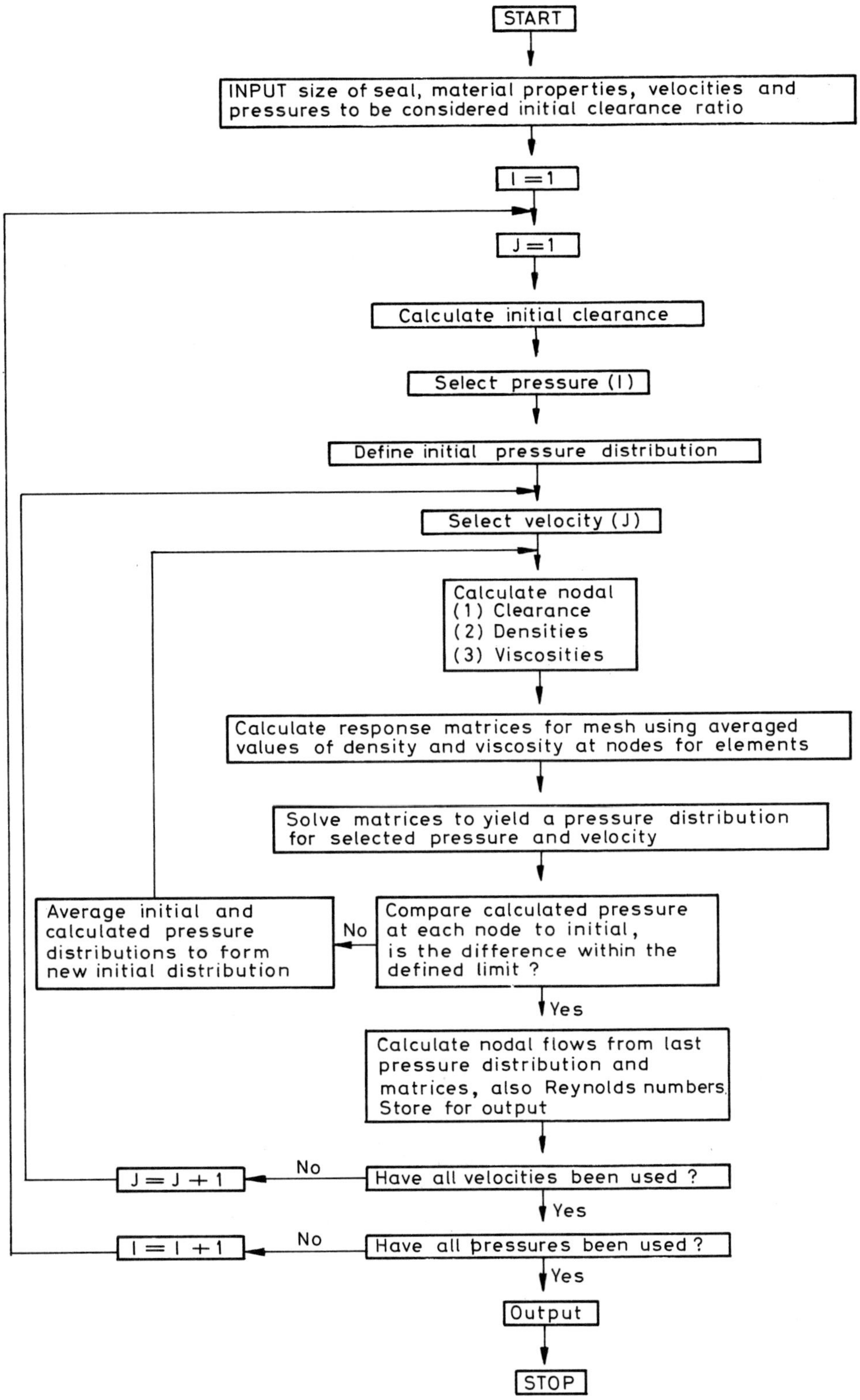

Fig. 2: Simplified block flow chart for iterative solution of plain gland clearance seal

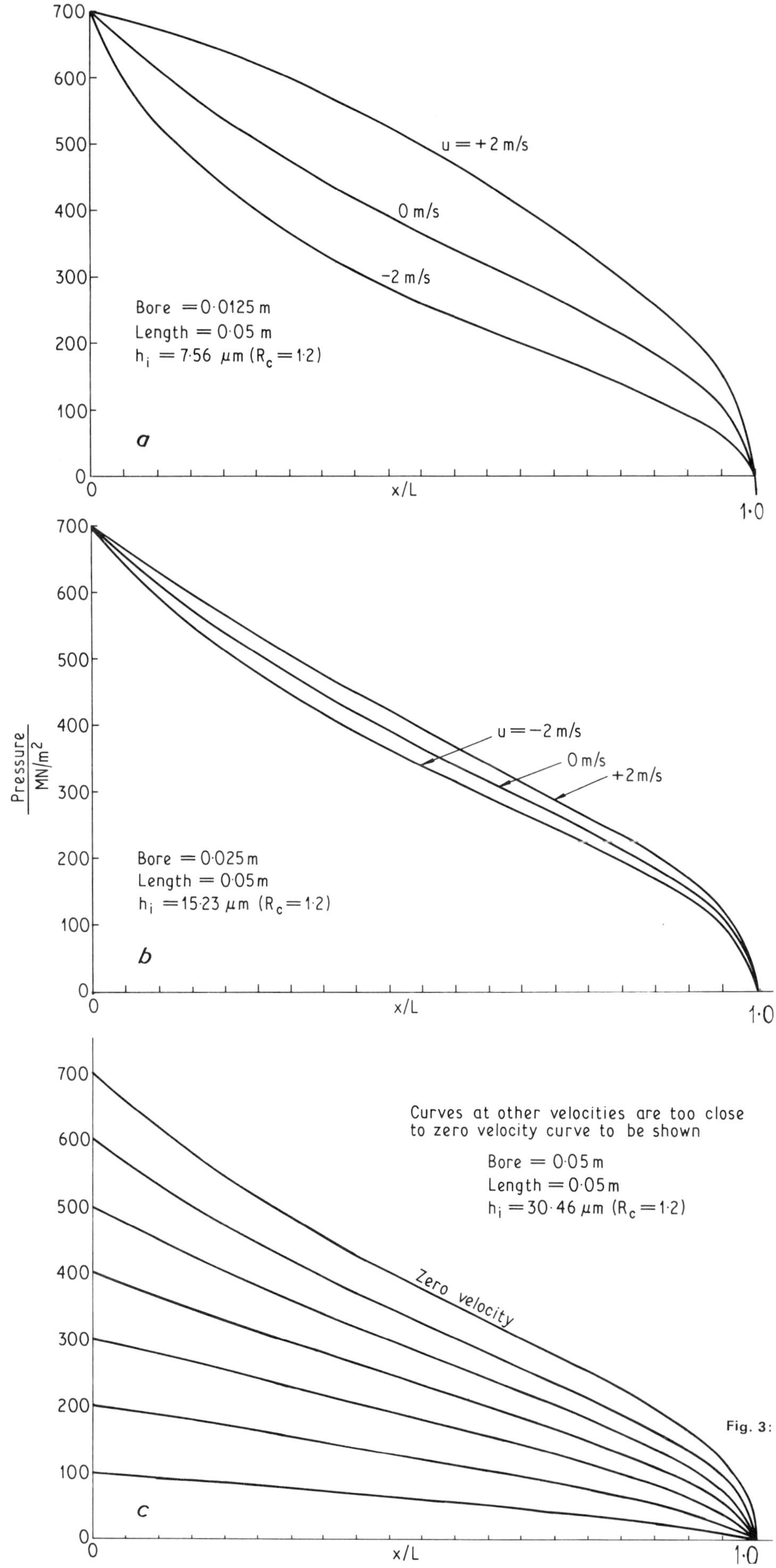

Fig. 3: **Axial pressure distributions at 700 MN/m² showing effect of ram reciprocation**
(a) bore = 0.0125m; length = 0.05m; h_i = 7.61 μm (R_C = 1.2)
(b) bore = 0.025m; length = 0.05m; h_i = 15.23 μm (R_C = 1.2)
(c) bore = 0.05m; length = 0.05m; h_i = 30.46 μm (R_C = 1.2)

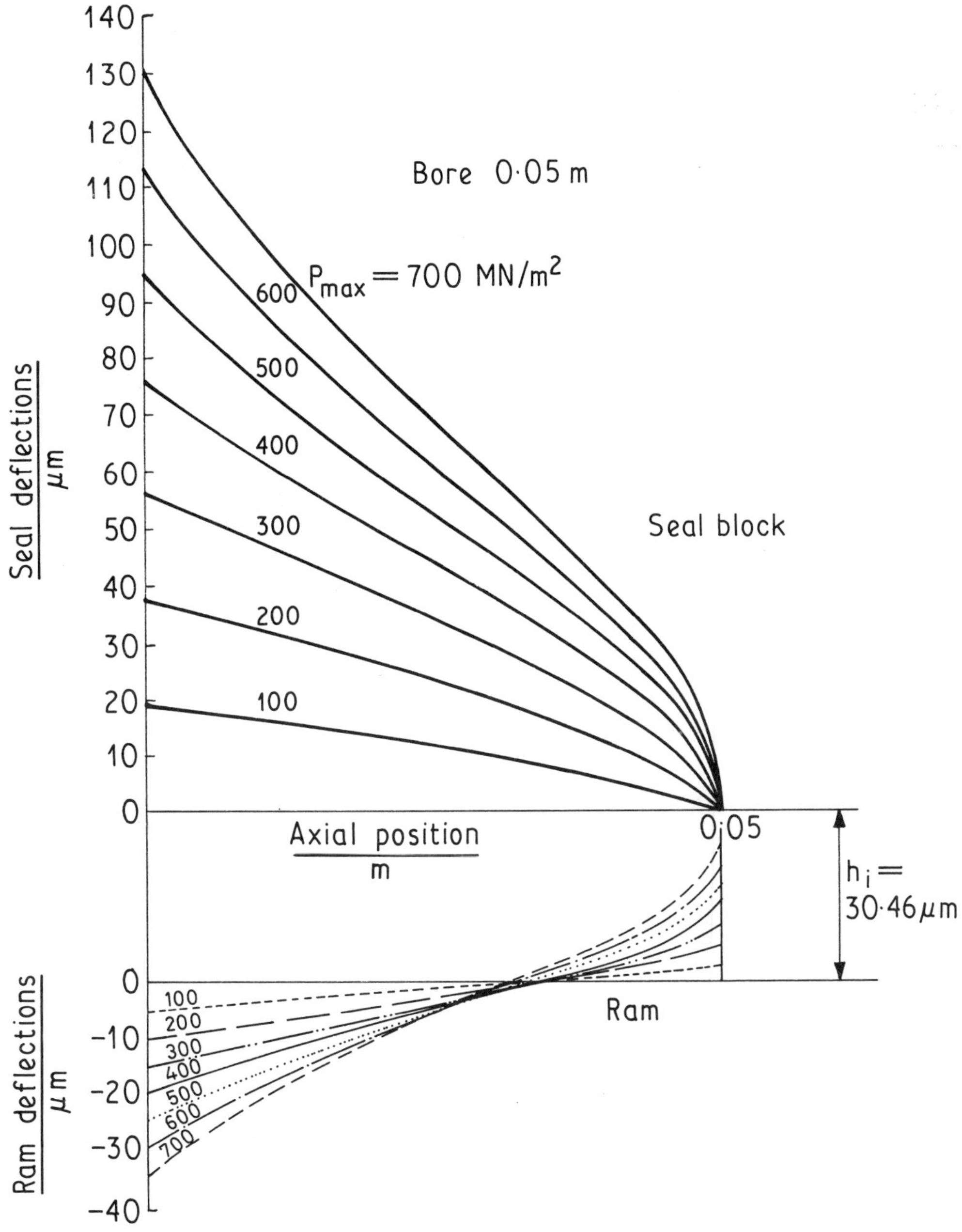

Fig. 4: Ram and seal block clearance deflections associated with the pressure distributions of Fig. 3(c)

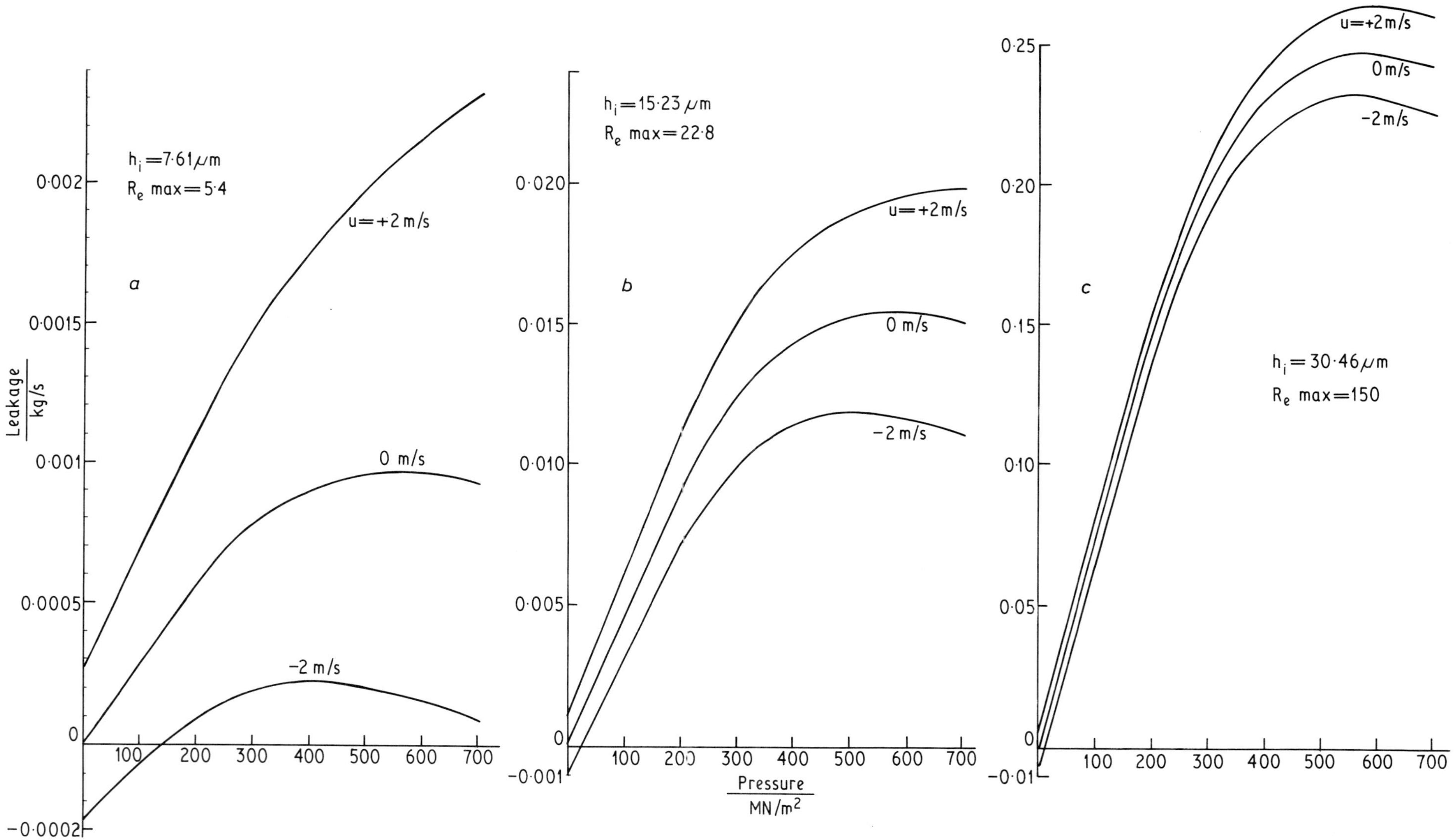

Fig. 5: Seal leakage rates for pressures up to 700 MN/m² showing effect of ram reciprocation
(a) bore = 0.0125m; length = 0.05m; h_i = 7.61μm (R_C = 1.2)
(b) bore = 0.025m; length = 0.05m; h_i = 15.23μm (R_C = 1.2)
(c) bore = 0.05m; length = 0.05m; h_i = 30.46μm (R_C = 1.2)

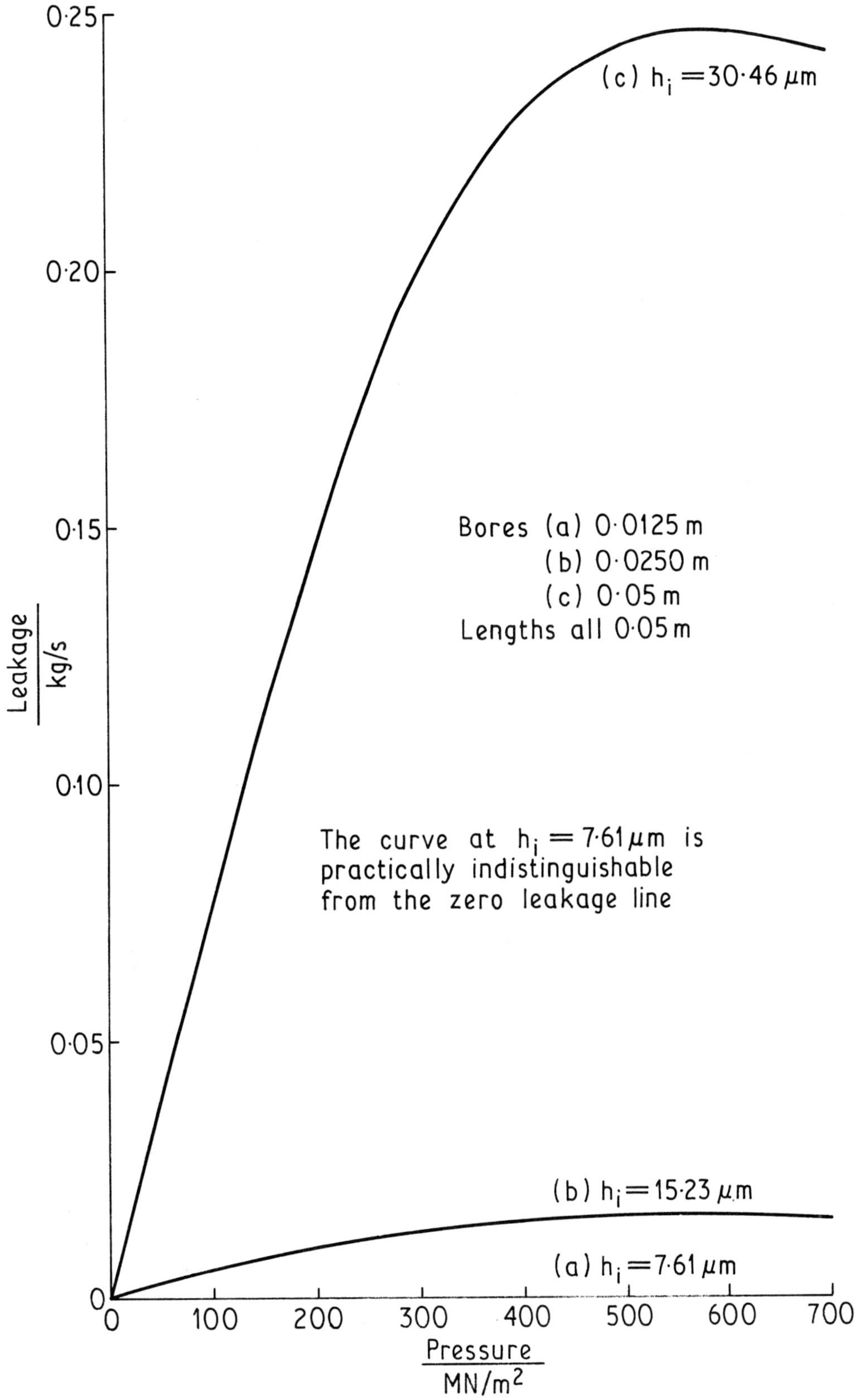

Fig. 6: Comparison between the leakage of three seals at zero ram velocity

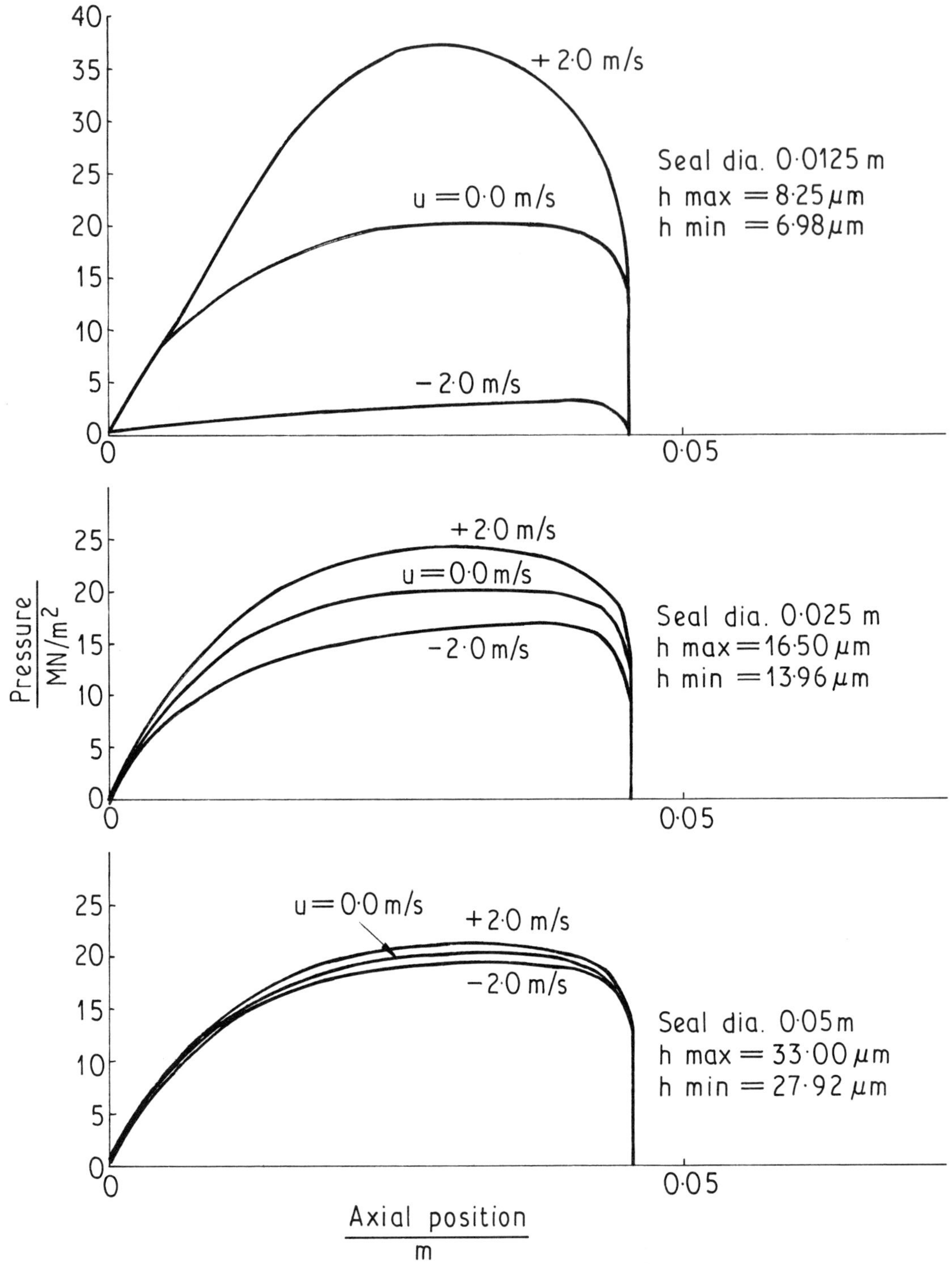

Fig. 7: Pressure differences between distributions (at 700 MN/m^2) for two initial clearances, and three seal bores

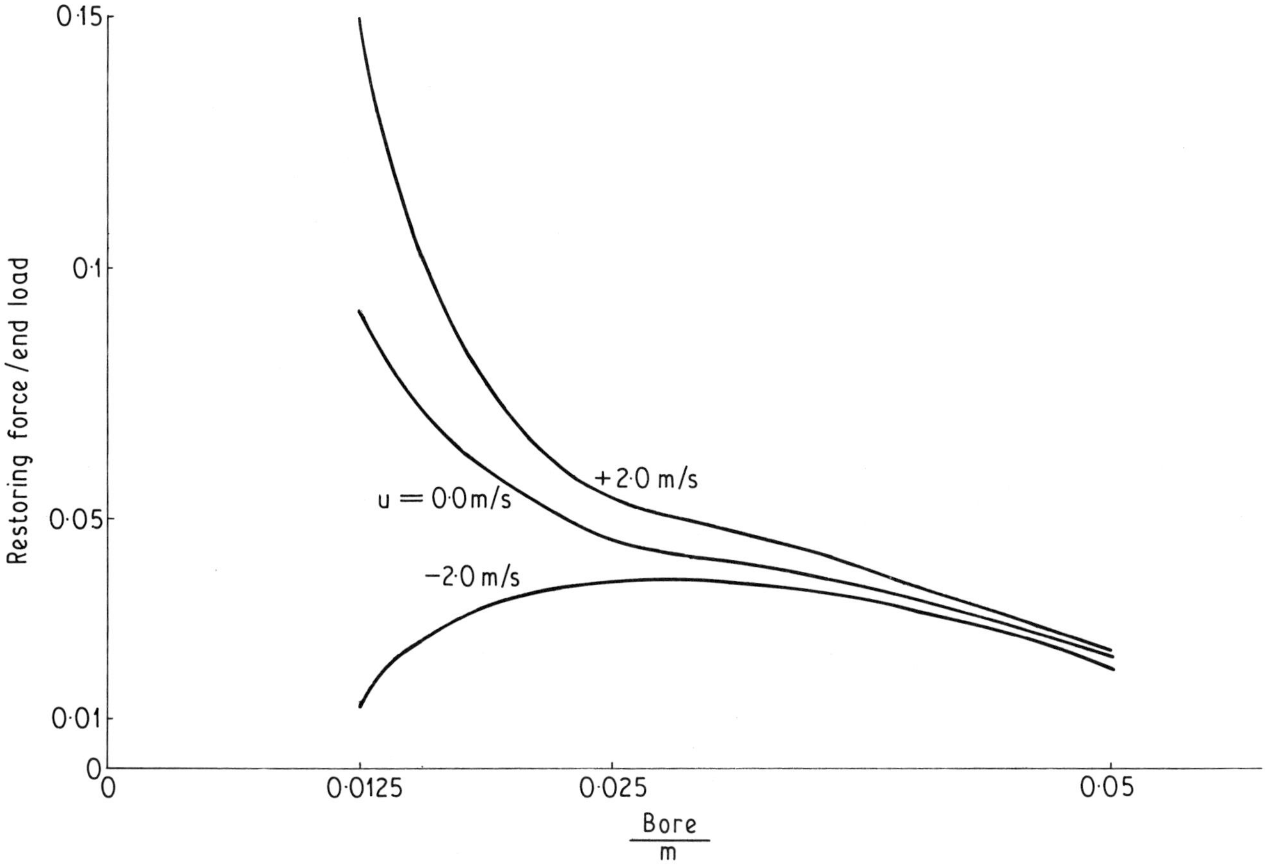

Fig. 8: Recentring force as a fraction of the end load at 700 MN/m^2 seal length .05m in all cases

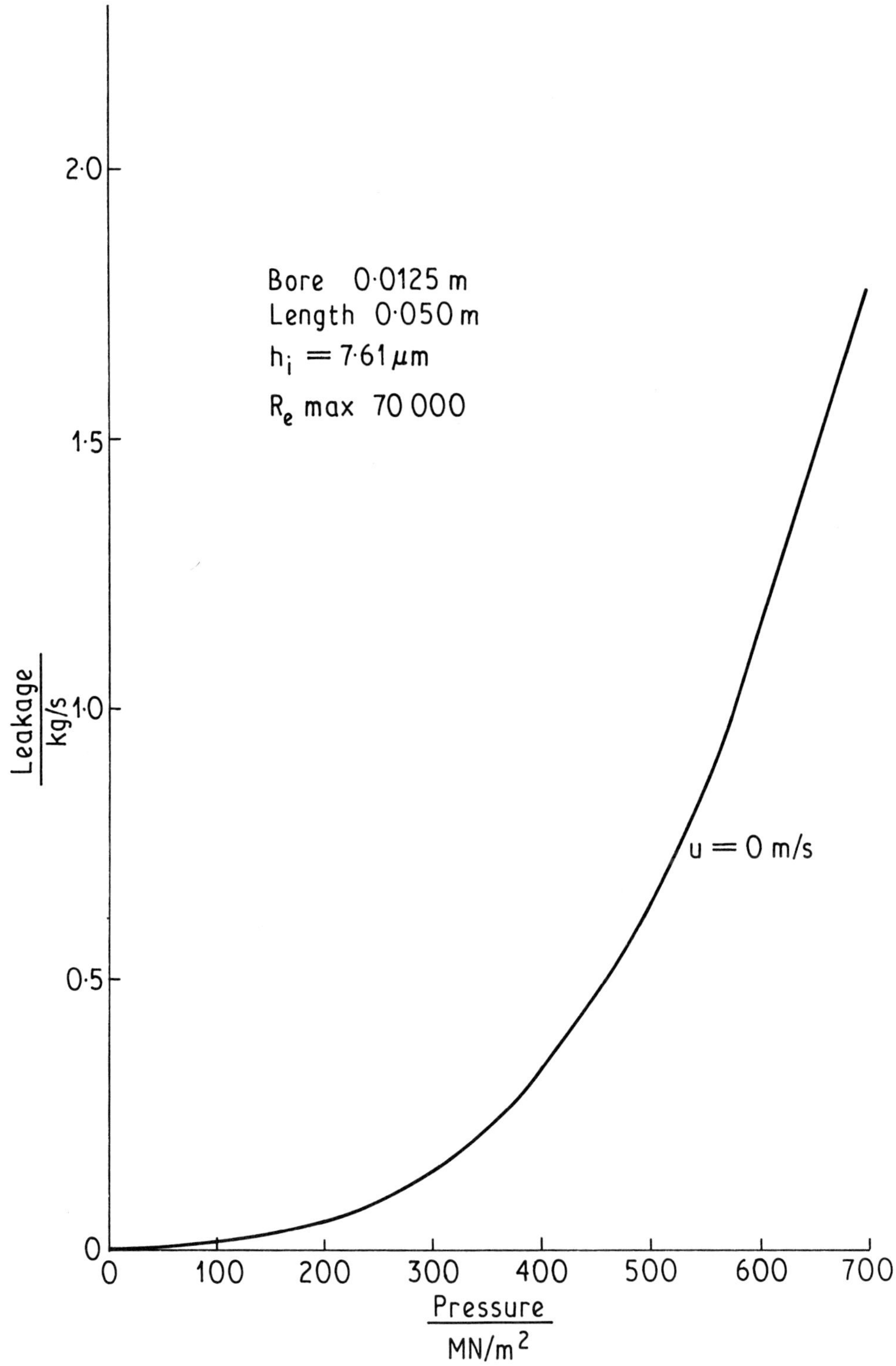

Fig. 9: Leakage of a water seal (40°C) at zero ram velocity

SOME INVESTIGATIONS ON THE BEHAVIOUR OF HIGH PRESSURE PACKINGS USED IN SECONDARY COMPRESSORS FOR LOW DENSITY POLYETHYLENE PRODUCTION

A. TRAVERSARI, Dr.Ing.
NUOVO PIGNONE Florence (Italy)
E. GIACOMELLI, Dr.Ing.
NUOVO PIGNONE Florence (Italy)

SYNOPSIS The life of packings is an important factor in availability of secondary compressors in polyethylene plants. To achieve continual improvement of these parts it has been considered useful to investigate their behaviour on the basis of pressure measurements, at the sealing elements, on machines in service and of a series of observations on packings disassembled after long periods of operations (of the order of two years).

1. INTRODUCTION

The polymerization of ethylene for low density polyethylene production is achieved industrially by bringing the gas to a pressure ranging between 1500 and 3500 atm. This pressure is reached by the use of a special type of reciprocating compressor. The cylinders of these compressors and in particular the sealing elements are subjected to extremely severe working conditions.

Generally speaking two types of seals have been used up to the present for this service:

- Piston rings against a hard metal liner;
- Packing on a hard metal plunger or on one in steel coated with hard metal.

The first system, though giving quite good results, tends more to be affected by operating conditions of the plant, owing to the type of lubrication obtained by injecting oil on the gas suction stream. The quantity, the type of low polymers and the type of catalyst carrier in fact have a considerable influence on the lubrication conditions of the system with liner and piston rings.

The packed plunger solution is on the other hand more independent of these factors since the lubricant is injected directly onto the sealing elements with the following advantages:

a) All the lubricant injected into the cylinder lubricates the sealing elements, while in the piston ring system most of the oil is carried away by the gas;

b) The lubricant can be injected between the sealing elements in the most suitable place, while in the piston ring system, the first rings only are adequately lubricated;

c) Less dilution of the lubricant by the ethylene, since contact between the two substances is less close than with injection into the gas flow;

d) Less pollution of the gas by the lubricant for the reason mentioned in point "a".

What has been said above, together with the considerably increased capacity per line of the new plants, has for some years led to the trend to prefer the packing system to the piston ring one, even on first stage cylinders, that is for pressures up to 1500 atm.

This is also a consequence of the fact that for tungsten carbide liners, manufacturing problems are greater than those for plungers.

It is therefore necessary that the packings as well as the other critical parts of the compressor, have an efficiency that ensures high continuity of operation of the machine and consequently of the plant.

The results obtained in many plants are completely satisfactory. Nevertheless the need of further improvement in present efficiency and of knowing the reasons for certain wear patterns and behaviour connected with the various plants, have led to a more thorough examination of the actual operating conditions of some packings.

This paper shows a series of surveys carried out on 1st and 2nd stage packings of operating machines.

These surveys are examined in the light of the wear patterns encountered on the sealing elements of some packings disassembled after a period of normal service (10,000 - 20,000 hours).

2. TYPICAL HYPER PACKING ARRANGEMENT

Packings consisting of one or two split type breaker rings and five radial tangential type sealing rings are considered (fig.1).

The rings are in special bronze alloys, usually with high lead content, so as to guarantee sufficient strength, low friction coefficient and high thermal conductivity, for a quick dissipation through the packing cups, of the friction heat.

The hardness of these rings usually varies from 55 to 80 Brinell (measured with a 10 mm ball and 500 kg load).

The rod on which the sealing elements slide is

in solid tungsten carbide, with a very high surface finish (1 to 2 microinches CLA). Thus the plunger has exceptional characteristics of hardness, wear resistance, anti-seizing properties, low friction coefficient and very high thermal conductivity.

The synthetic type lube oil of the cylinders has lower lubricating properties than oils used for normal services, since for ethylene polymerization, pollution of the final product must be reduced to a minimum.

Packing performance is greatly influenced by the above parameters and by the efficiency of the breaker rings, whose action is very important and requires special consideration.

We shall therefore consider their role by analysing the operating conditions of a packing.

Let us take the hypothesis that at any moment the pressure inside the cylinder can be considered as consisting of a constant part, equal to suction pressure, and a fluctuating part, equal to the difference between discharge and suction.

Thus we have the static part of the pressure, which tends to be distributed in such a way as to overload the last rings (frame side).

The last of all has to handle almost the whole load (1), as it has been experienced in some packings operating at constant pressure on ammonia synthesis compressors.

The variable part of the pressure is characterized by an increase due, to the polytropic compression, by a steeply decreasing section as a result of the expansion of the gas remaining in the clearance volume, as well as, of course, by two sections with almost constant pressure, corresponding to discharge and suction.

During the compression stroke, the pressure increases quickly inside the cylinder. The pressure breakers try to oppose themselves to the sudden propagation of the pressure wave towards the sealing elements.
rings.

The most important function of the breaker ring is therefore to retard the "backflow" from the packing towards the cylinder when the plunger begins its back stroke.

If this retarding action fails, the pressure immediately upstream of the first sealing element will tend to fall suddenly.

In fact it will follow the pressure in the compression chamber, which, owing to the steep slope of the expansion curve, will very quickly return to suction value. The resultant of the forces acting on the first sealing couple is suddenly inverted, resulting in an expolding action on the radial ring and especially on the tangential one, which makes a stronger sealing action. The violence of this phenomenon is connected to the efficiency of the breaker rings and has a bad effect on the packing, eventually causing one of the following troubles:

a) Breakage of the dowel pin between the radial and tangential rings;

b) Breakage of the lips of the tangential cut rings;

c) Damage to the garter springs of the sealing elements.

After a certain time it may happen that the first sealing pair no longer performs its function, so that the troubles described above, occur to the second pair and the process of progressive damage of the various rings of the packing, continues more or less quickly.

3. TEST METHOD

To simplify the surveys, the measurements were taken at the lubricating oil injection quills, in three positions, and also in the compression chamber.

The measurements were carried out by the application of strain gauges in the axial and circumferential directions on the outer surface of the parts concerned (fig.1).

The surveys were carried out with the use of a carrier amplifier; the output signal was shown visually either on an oscilloscope or on a galvanometric recorder with photosensitive paper on which also the positions of the dead centres, transmitted by a special pickup on the compressor crank shaft, were indicated.

In this way indicated diagrams of the pressure in the cylinder and in the small chambers in which the oil is injected, were obtained with a good approximation.

The first series of tests was carried out on a second stage cylinder of a secondary compressor, driven by a steam turbine (fig.2) and operating in a polyethylene plant with a tubular reactor.

The packing arrangement consisted of two three piece breaker rings with a diametral clearance on the rod of 0.08 mm and five radial tangential pairs, having, like the breaker rings, an axial clearance of 0.15 mm with respect to the packing grooves.

Fig.3 shows the diagrams obtained. From these diagrams the plots of fig.4 and 5 were drawn.

These give the curve of the pressure along the packing, for the compression and suction strokes respectively, at certain crank angles.

In this way it is possible to distinguish more clearly the work performed by each ring and thus those most affected by the variable pressure or by an almost static pressure.

The pressure values, at the rings not concerned in the measurements, were obtained by taking into account the formulas for labyrinth seals (2).

A second series of tests was carried out on a first and a second stage cylinder of a compressor having a capacity of 53,000 kg/h and operating in a plant with vessel reactor.

Both packings had a three piece breaker ring, with small circumferential clearance and five pairs of radial tangential seals. All the sealing elements had an axial clearance of about 0.15 mm with respect to the housing groove.

Figure 6 shows the indicated diagrams of the pressures in both cylinders as a function of the crank angle.

The distribution of the pressures along the

packing during the suction and discharge strokes are shown in fig.7 and 8 for the first stage and in fig.9 and 10 for the second stage cylinder.

4. ANALYSIS OF THE RESULTS

Let us first consider the behaviour of the second stage packing, whose indicated diagrams are shown in fig.3.

This packing was operating with a suction pressure of 800 ats and a discharge pressure of about 2500 ats.

It can be seen that the curves of the pressures after the first and after the second breaker ring, tend to follow, with some phase difference, both the compression and the expansion polytropic curves.

The retard and the differential pressure that these rings bear appear to be insufficient. Both create a drop of about 800 - 900 atmospheres, which is equivalent to about 55% of the total pressure fluctuation between suction and discharge both during expansion of the gas and during the compression phase.

The difference in the behaviour of the two breaker rings for the whole forward stroke is probably due to a difference between the diametral clearances, so that there is a higher pressure drop on the second ring.

The first two pairs of seals have a clear effect on retard and sealing. For a period longer than that of the discharge these elements together support a differential pressure of about 1300 atmospheres, which is 70% of the total pressure fluctuation in the cylinder. The next three rings together support a differential pressure of the same order.

The observations just made are also clear from fig.4 and 5 from which it can be seen that practically the first three sealing elements only are affected by the pressure fluctuation of the cylinder, while the last two are subjected to an almost steady pressure.

The pressure variation occurring at the first sealing pair, is propagated to the following elements in the proportion of 70% at the second pair, 30% at the third and in a negligible amount at the two final sealing elements.

As regards the steady part of the pressure it can be observed that, at the last lube oil injection quill,it has undergone no significant change and therefore more than 60% should be supported by the last sealing element on the basis of the calculation using the hypothesis previously chosen.

Now let us consider the diagrams of figure 6.

It can be seen that the pressure pulsations, upstream of the suction valves and downstream of the delivery valves, have a considerable influence on the curves of the pressures in the cylinder.

The pressure pulsation at the discharge of the first stage has a frequency corresponding to the compressor rpm, while in the first stage suction and second stage suction and discharge the frequency is double since there are connections between the pipes quite near to the measurement points.

The behaviour of the breaker rings appear to be good since they effect a retarding action, especially evident during the compression period, and considerable sealing in both stages, during compression and expansion.

The breaker ring alone of each packing, in fact, support more than 80% of the pressure fluctuation of the gas which is about 1000 ats in the second stage and 700 ats in the first.

The breaker ring is more efficient in the first stage. This can be seen from the less steep slope of the corresponding pressure curves and it is probably due to the higher variability of the specific volume of the gas, which is of the order of 5% in the second stage and 16% in the first.

This fact may be partly explained by the difference existing between the polytropic coefficient of the first and second stage.

It has also to be remembered that the nearer the physical conditions of a gas are to those of a liquid, the more difficult is the task of the breaker ring and the less its effect.

Figures 7 and 8 show that the fluctuating part of the pressure concerns only the first three seal rings, of which the second and third support a differential pressure of the order of 50% and 30% respectively compared to the first.

The steady part of the pressure is practically supported by the last two sealing elements.

The same remarks apply also to figures 9 and 10 for the second stage, with the only difference that the differential pressure supported by the last two sealing pairs is much higher since the suction pressure is higher.

We had the opportunity to inspect some sets of packings after operating periods of 10,000 to 20,000 hours.

Figure 11 shows the radial thickness found after disassembly on a typical first stage and second stage packing which gives a diagram of the wear of each element. This wear tends to be greatest on the first and last elements.

The development of the curve is fairly similar for both radial and tangential rings, and for the first or second stage.

The difference concerns the value of the wear, which is higher in tangential cut rings as compared with radial cut ones, and the behaviour of the extremity sealing elements.

On the first stage it is seen that the first pair did not have the chance of wearing completely. In fact it was found that the dowel had broken, probably in consequence of the "backflow".

After that breakage most of the pressure loaded the second pair, causing an accentuaded wear. The wear of the tangential ring was higher than the amount allowed by the butt gap. This is found quite often as a consequence of non-uniform wear of the rings.

On the second stage, the "backflow" caused, first the breakage of the springs (still of the coil type) of both the rings of the first pair, making it inefficient after limited wear, and later the breakage of the dowel of the second pair. The greatest work in supporting the variable pressure was then carried out by the third sealing element.

The maximum wear on the frame side elements of both stages, is due to the value of the constant pressure to which they are subjected.

Wear of the radial ring of the last sealing couple of the first stage packing is something of an exception. This situation, encountered several times in other compressors, can be explained as follows.

The radial rings when subjected to an almost steady pressure, tend to establish themselves in their position without effecting an appreciable sealing action towards the rod, but simply creating a barrier to the pressure at the cuts of the tangential rings.

when subjected to variable pressure, the first radial rings are forced to exercise a sealing action to the gas that tends to re-enter the cylinder during the suction phase.

The sealing effect is not and must not be complete, since there are radial cuts, which allow the passage of the gas.

However this temporary and limited task can produce appreciable wear effects. In these cases the curve of the wear of the radial rings is maximum on the pressure side and tends to decrease towards the low pressure zone. From dimensional checks and from the surveys given above, a general wear pattern can be deduced connected with the distribution of the pressure along the packing.

The steady part of the pressure causes a type of wear represented by the curve "a" in figure 11 in which the maximum values are reached on the frame side sealing pair.

The fluctuating part of the pressure on the contrary causes a wear with an opposed development, curve "b" of figure 11, where the highest values are found on the first sealing pair.

The resultant wear will thus be a curve as "c" with its maximum values at the ends of the packing.

The maximum absolute value tends to be displaced towards the first ring (pressure side) or towards the last (frame side) according to whether there is a prevalence of the fluctuating part or of the steady one.

From what has already been said it is therefore confirmed that the practical wear pattern is usually as in the curve "d" of figure 11. The phenomenon of "backflow", sooner or later makes some sealing element inefficient and its wear will be much lower than that of the other elements.

5. CONCLUSIONS

Analysing the results obtained, it can be seen that the surveys of the pressures at the sealing elements are in agreement with the wear patterns in many packings. It is also found that a correct sizing of the passage area of the breaker rings is necessary for a good performance of a packing. In this way the dangers arising from exploding action due to the "backflow" are greatly reduced or eliminated.

On the other hand the already long life obtained with the packings when the plunger is driven by a perfectly aligned crank mechanism (3), does not allow short term results to be obtained.

It can be seen that it is very useful to set up a mathematical model to simulate the operation of a packing. The use of a computer can help the designer in his search for an optimum solution and in increasing the efficiency of existing packings.

6. ACKNOWLEDGEMENTS

The authors wish to thank NUOVO PIGNONE for permission to publish information on the design of **packings of its manufacture.**

REFERENCES

1. ENGLISCH C. – KOLBENRINGE – ERSTER BAND – p.59 Theorie Herstellung Und Bemessung Springer Verlag Wien 1958

2. COSIMI L. – Il compressore a pistone a secco con tenuta a labirinti – Il calore – 1961 – N°3 p.91

3. ANDRENELLI A. – Special Features in Reciprocating Compressors for Polyethylene Production. Proc. Instn. Mech. Engr. 1969-70 Vol. 184 P_t3R

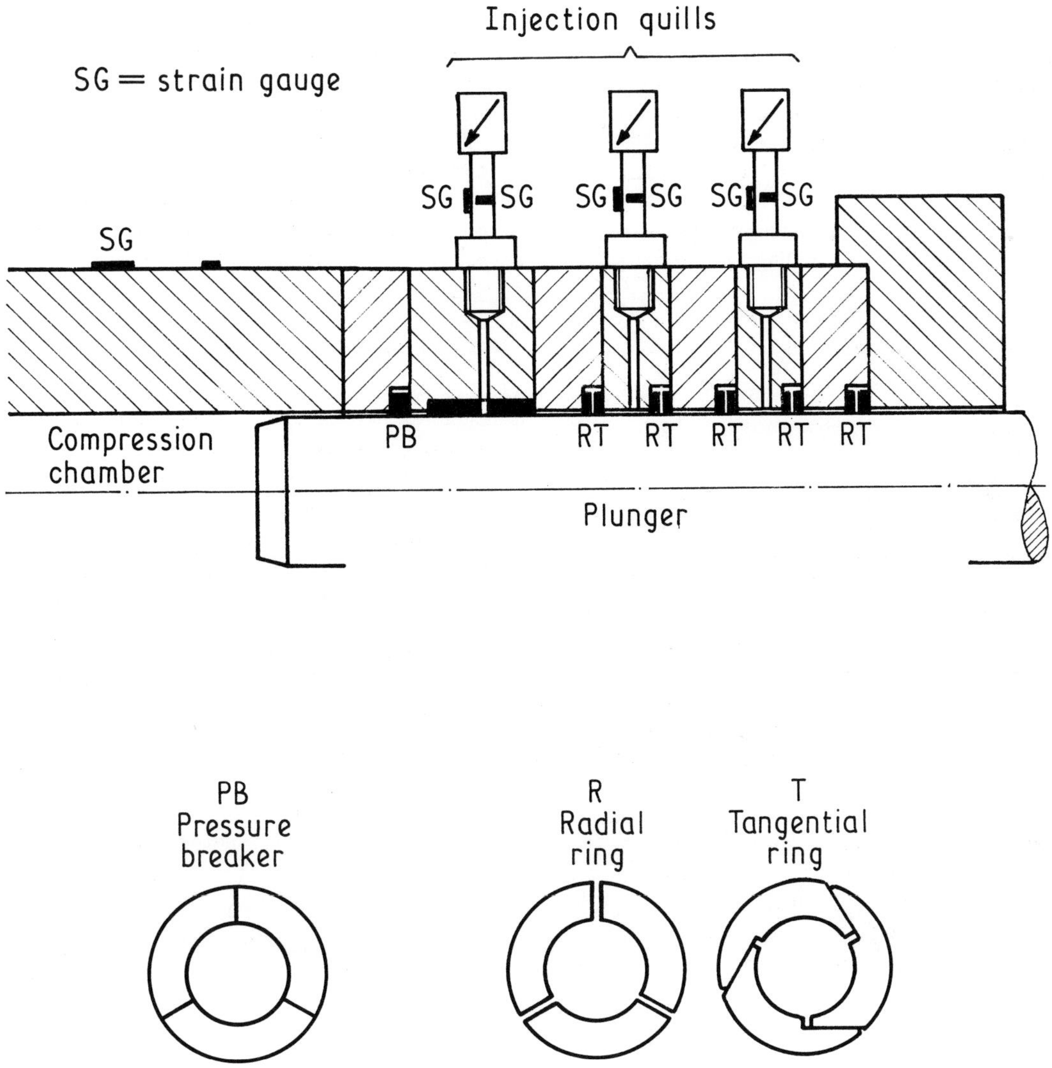

Fig. 1: Packing arrangement and survey technique

Fig. 2: Polyethylene compressor driven by steam turbine on tubular reactor plant

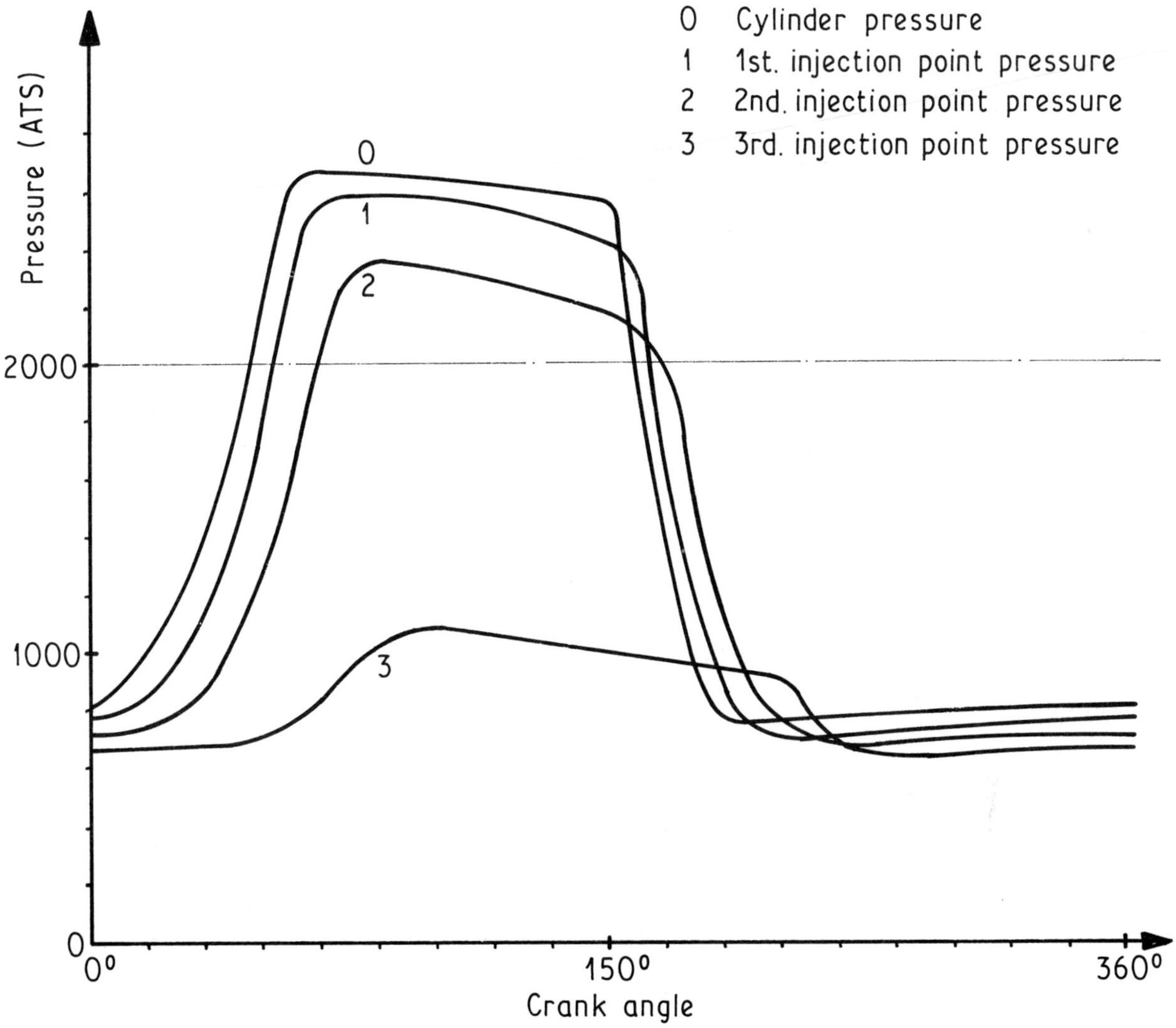

Fig. 3: Survey on a 2nd stage cylinder (first series of test)

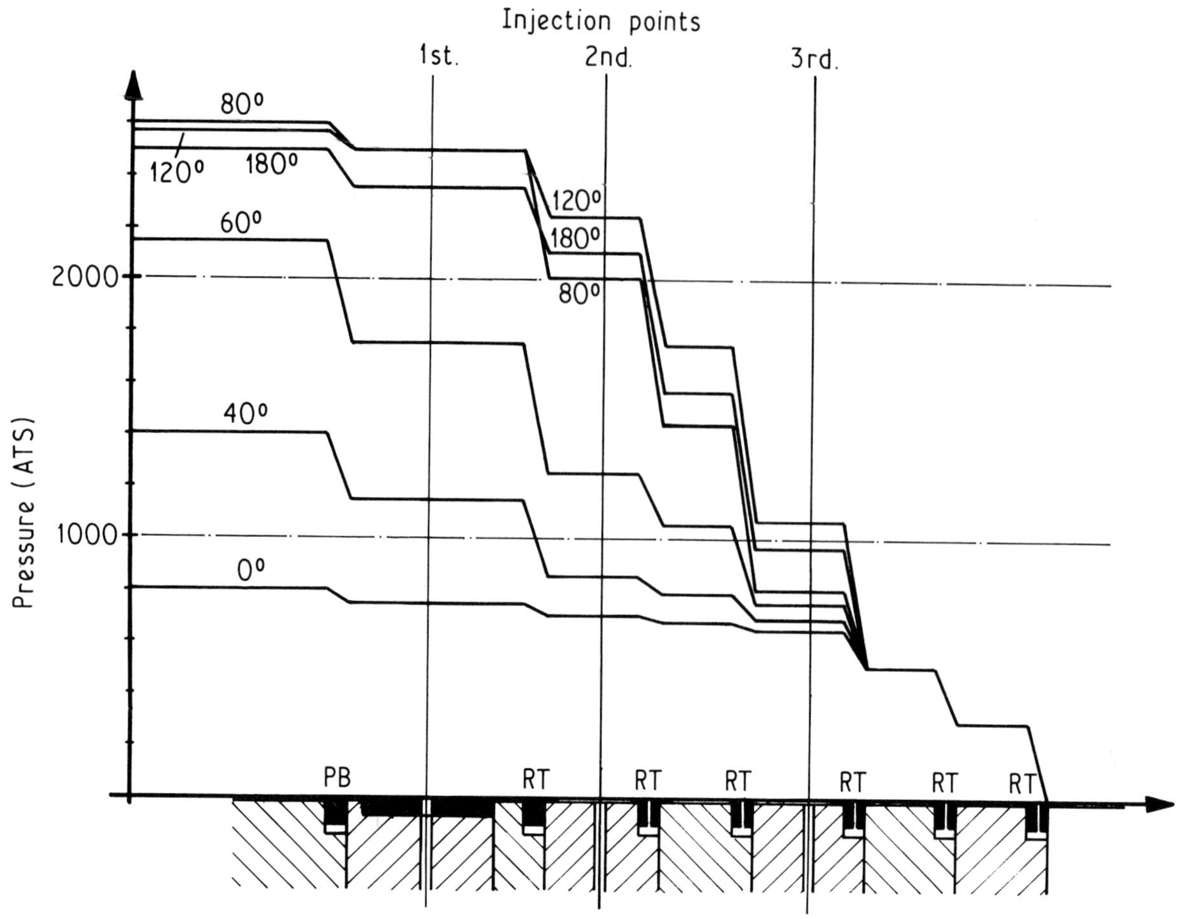

Fig. 4: Pressure distribution on the 2nd stage packing during discharge stroke (first series of test)

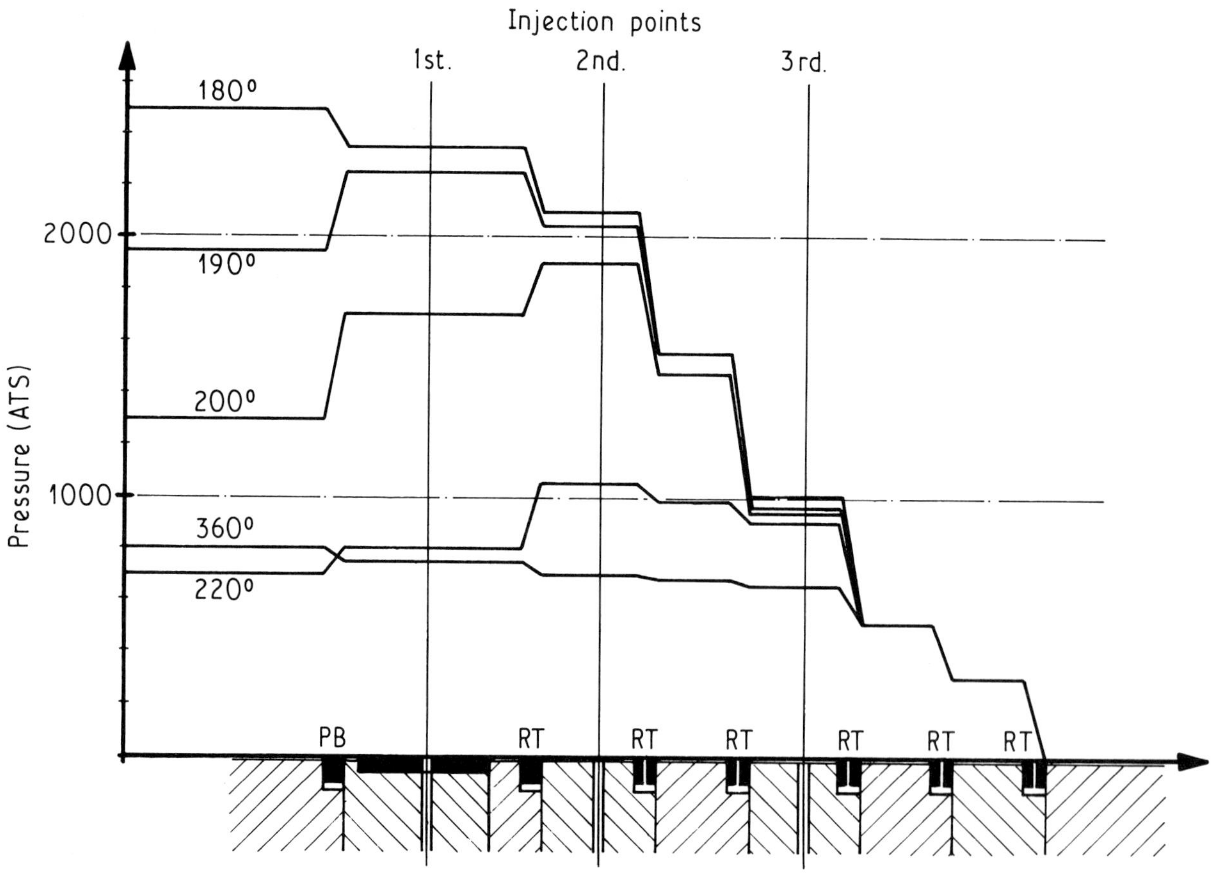

Fig. 5: Pressure distribution on the 2nd stage packing during suction stroke (first series of test)

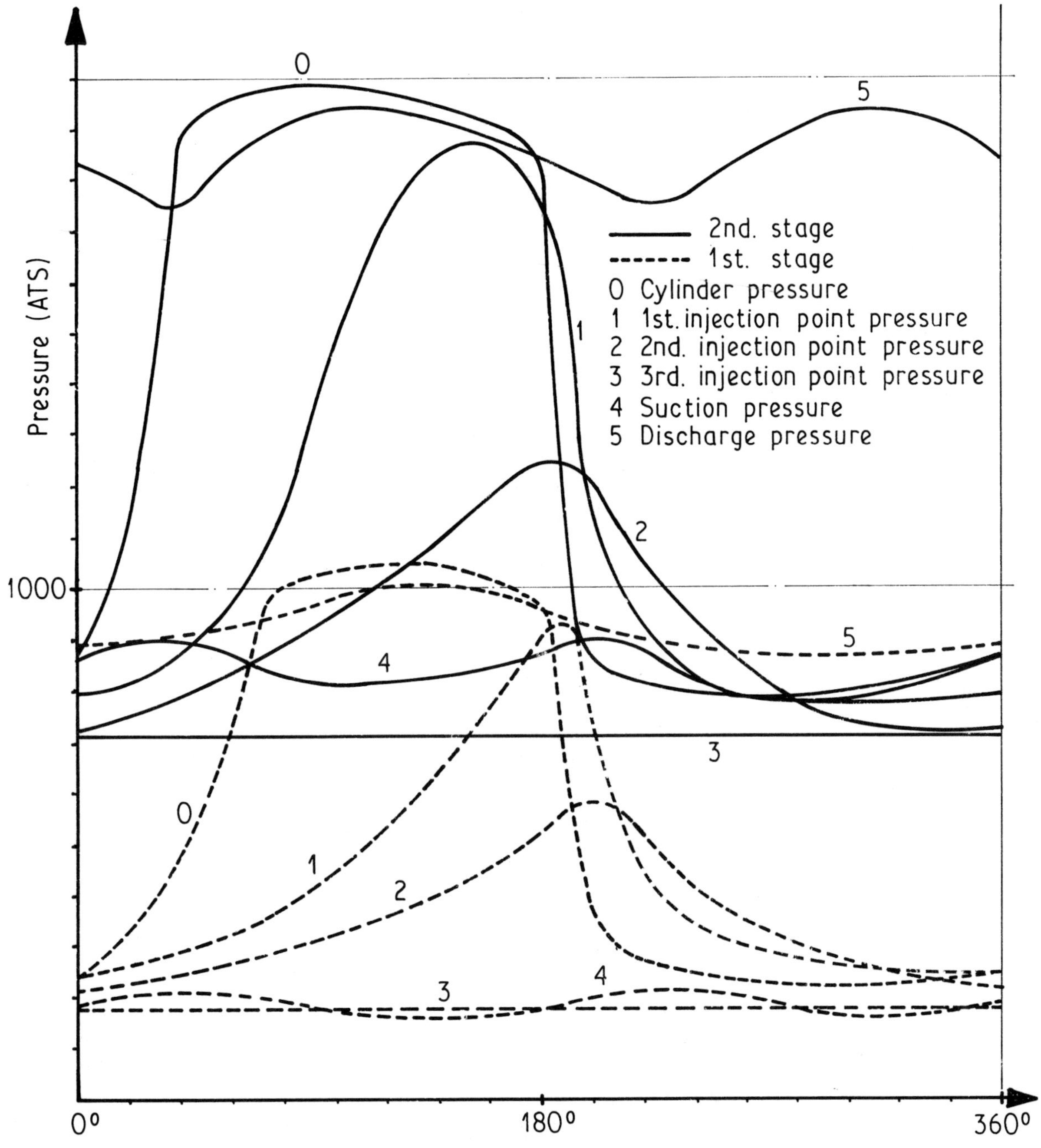

Fig. 6: Survey on a 1st and on a 2nd stage cylinder (second series of test)

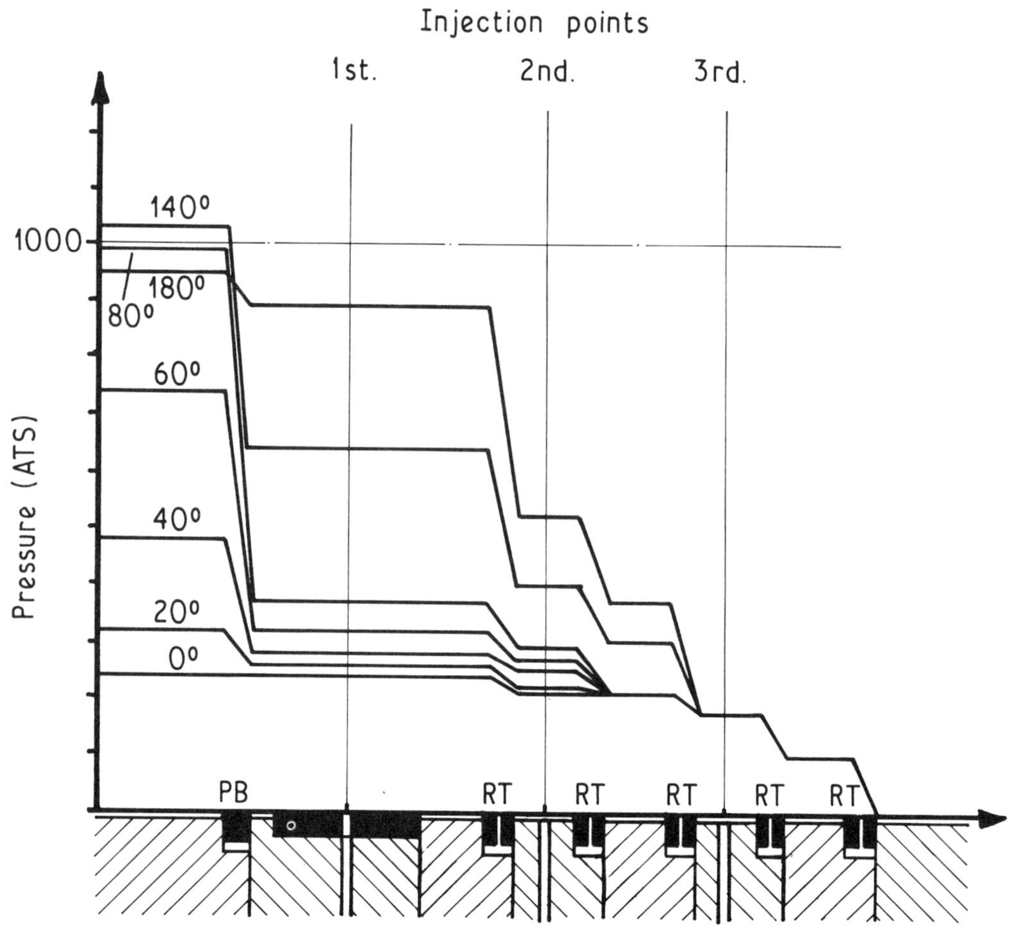

Fig. 7: Pressure distribution on the 1st stage packing during discharge stroke (second series of test)

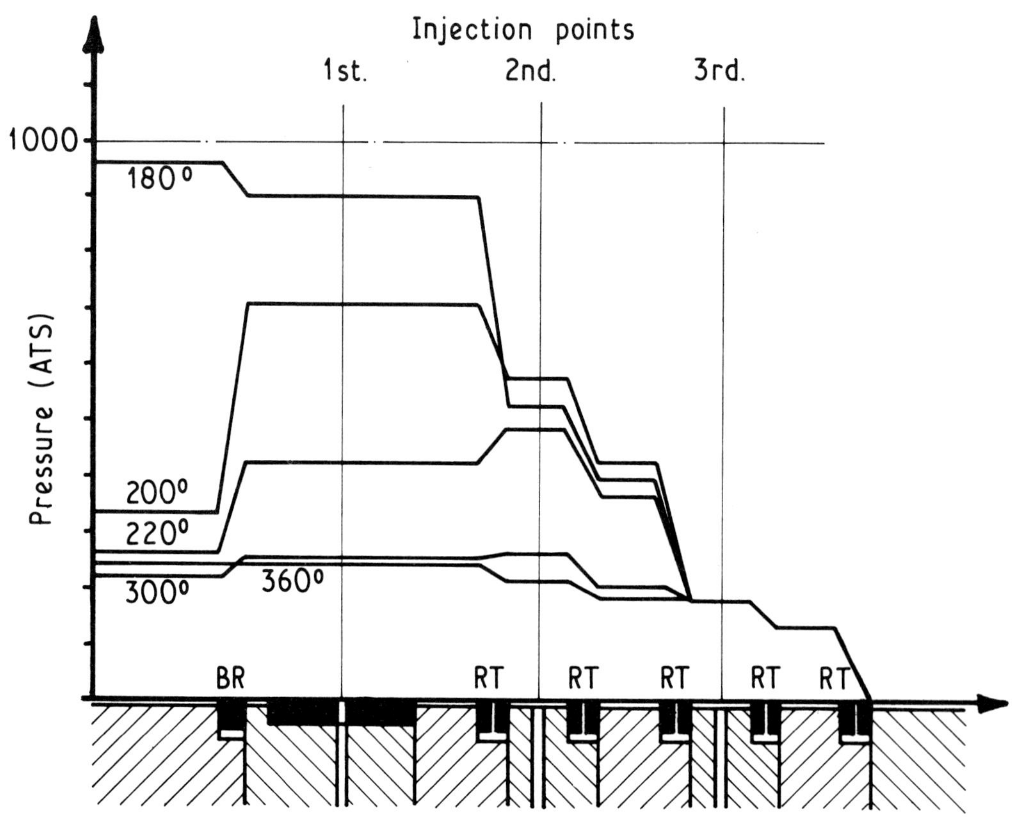

Fig. 8: Pressure distribution on the 1st stage packing during suction stroke (second series of test)

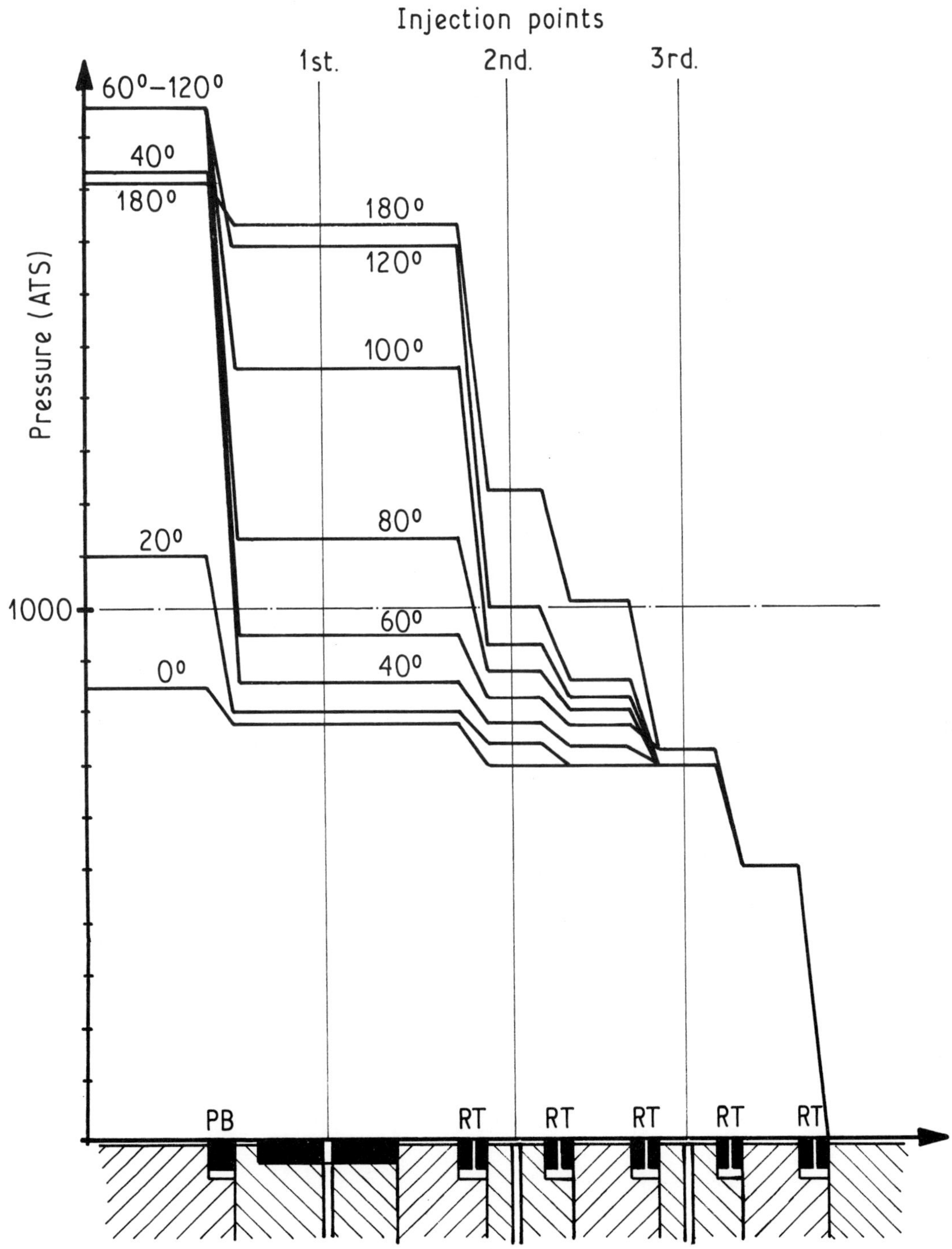

Fig. 9: Pressure distribution on the 2nd stage packing during discharge stroke (second series of test)

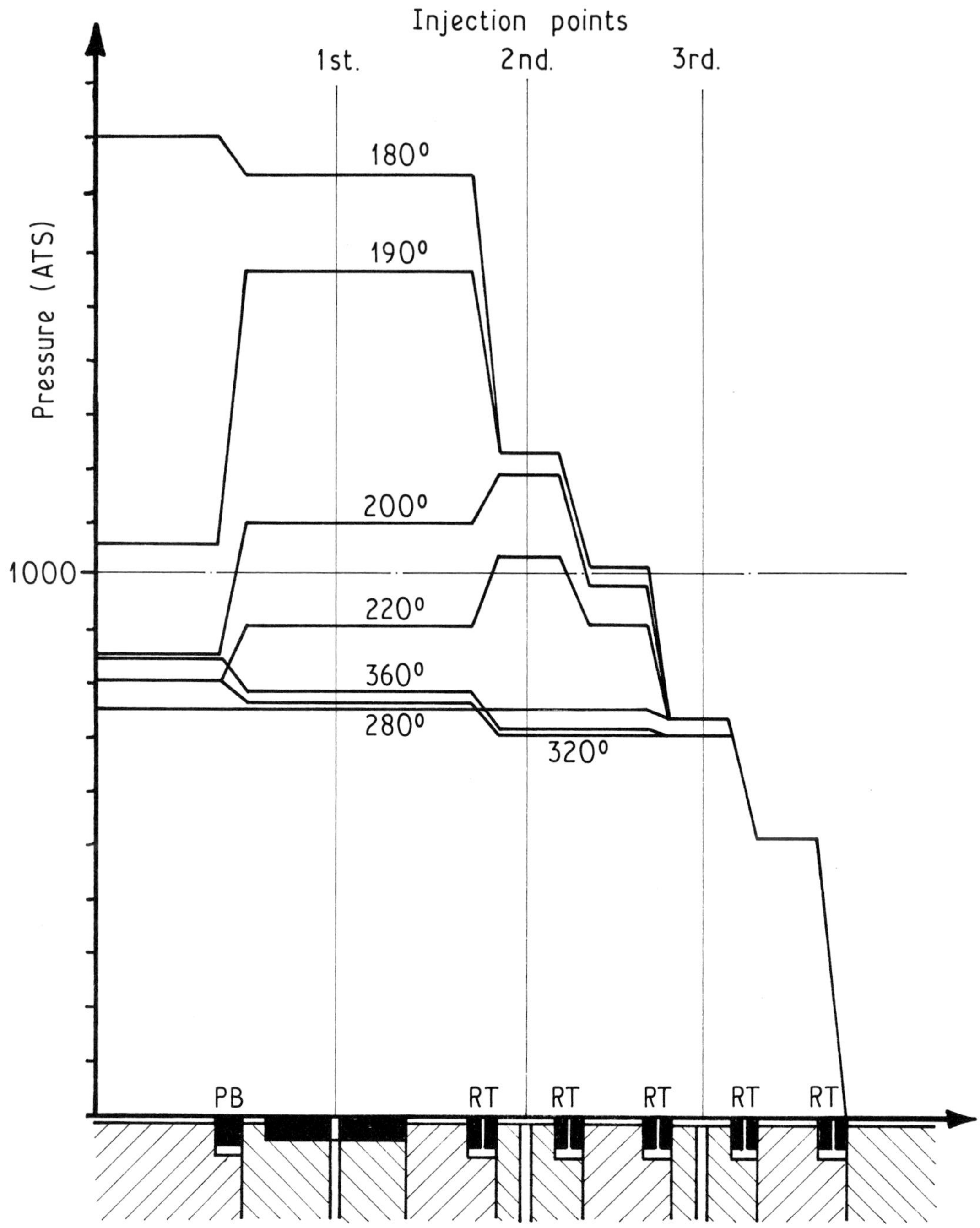

Fig. 10: Pressure distribution on the 2nd stage packing during suction stroke (second series of test)

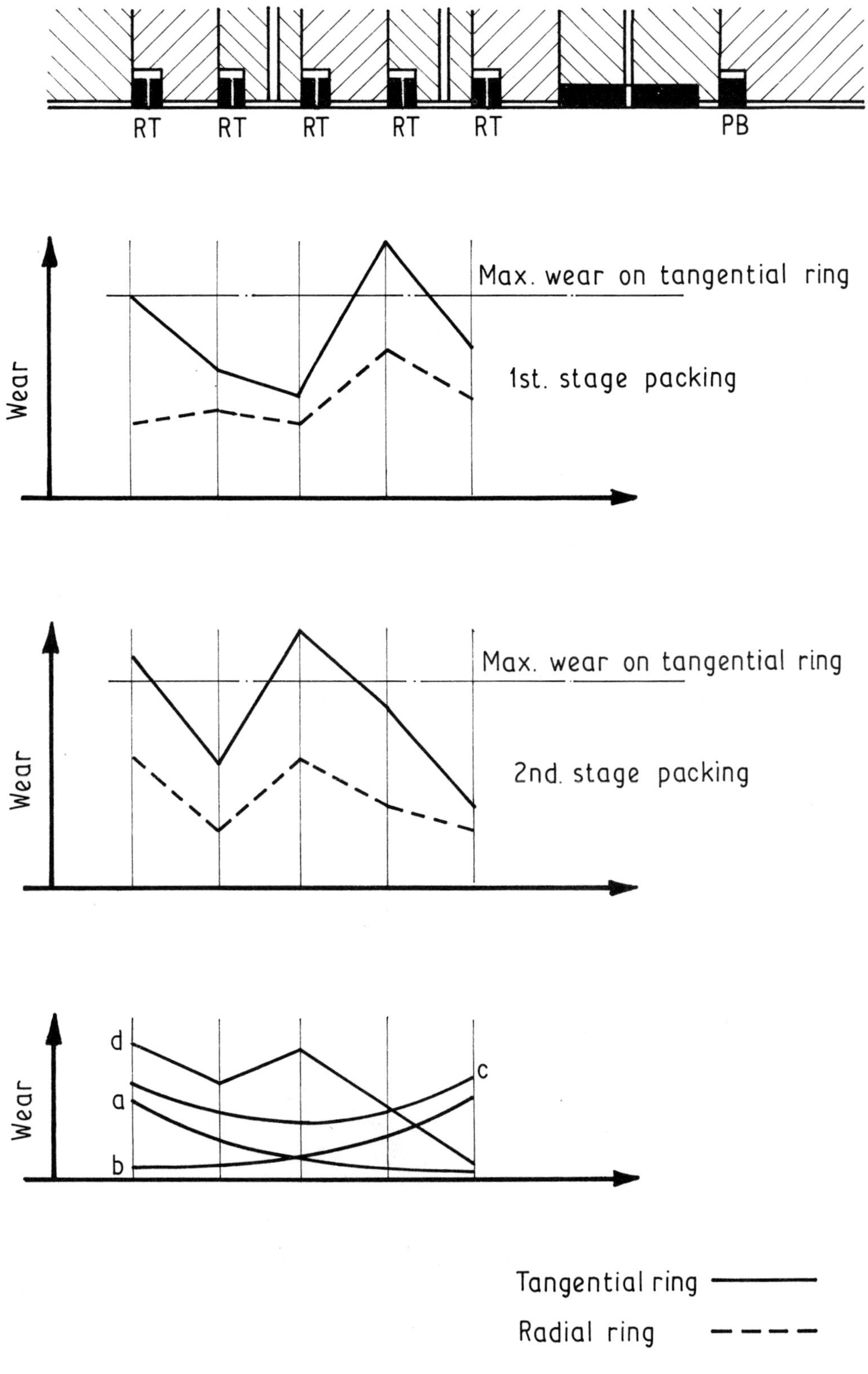

Fig. 11: Wear distribution on a 1st and 2nd stage packing

THE DEVELOPMENT AND PERFORMANCE OF A 340 MPa (50,000 lbf/in²) PUMPING SYSTEM

C. G. LUDLOW[1], R. PAKSERESHT[2] and B. CROSSLAND[3]
[1] Senior Research Officer, Dept. of Mech. Eng., The Queen's University of Belfast
[2] Research Student, Dept. of Mech. Eng., The Queen's University of Belfast
[3] Professor and Head of the Dept. of Mech. Eng., The Queen's University of Belfast

SUMMARY The paper describes the design and performance of a double acting jet cutting pump for pressures up to 340 MPa, which uses a water/oil emulsion as the cutting fluid, though an alternative stainless steel system has been designed for pumping pure water. Three types of accumulators have been tested: a simple fixed volume accumulator, a high pressure air bag accumulator and a detensifier. These accumulators have been tested with several different nozzle sizes, and the pressure pulsations close to the nozzle have been measured with a Belfast strain gauge pressure transducer. The results of these tests are presented and discussed.

Using the accumulator systems tests have been carried out with varying levels of pressure pulsation to investigate the maximum cutting speed for through cuts in neoprene rubber. These tests have indicated for pressure pulsations varying from 10% up to nearly 100% of the output pressure, that there appears to be no significant increase in the maximum cutting speed for pulsation levels of 10-20%, and no significant change in the texture of the cut surface could be noted. However, this may not be representative of all materials, but if it is then it would appear that a simple double ended intensifier with a relatively small fixed volume accumulator may be adequate for most requirements.

INTRODUCTION

1. The erosion of turbine blades by high velocity water droplets has been well known for half a century, and research workers in the field of high pressures have observed and sometimes suffered from the penetrative power of fluid jets escaping from high pressure systems. Nevertheless the possibility of using high velocity water jets for cutting is of relatively recent origin. Probably the Russians were the first to recognise the possibility of high velocity jet cutting in the late fifties, but the main developments in the field have occurred in the last decade. The literature in the field is extensive and reference is made to a bibliography of literature in the subject (1) and to the proceedings of two international symposia (2) and (3).

2. Essentially there are two possible modes of operation of jet cutting systems: pulse jets and continuous jets. Though both systems have possible fields of application we are here concerned with continuous flow systems which are already beginning to find their field of application in continuous cutting of non-metallic materials, such as asbestos sheet, wood, cardboard, expanded rubber and polystyrene foam, hard rubbers, explosive, frozen food, concrete, rocks, etc..

3. For continuous flow systems pressures from as low as 100 MPa* up to 1 or 2 GPa are required, depending on the duty to be performed. Jet diameters of from 100 up to 400 μm are commonly referred to in the literature on jet cutting.

* $1\ N/m^2 = 1$ Pa, 1 GPa = 1000 MPa = 145038 lbf/in² (or psi)

These requirements lead to a hydraulic output of 5.6 KW for a 400 μm jet at 100 MPa and 178 KW for 1 GPa. For some applications multiple jets may be required to allow simultaneous multiple cuts to be made.

4. Ideally the cutting fluid should be pure water especially for the cutting of food, and from the viewpoint that because of the filtration problems most systems are total loss systems. However, this poses some problems as it is normally considered desirable to lubricate the ram seal, and it also imposes severe limitations on the choice of the materials of construction because of corrosion and strength problems. Consequently, in many potentially practical applications an oil/water emulsion is used as the cutting fluid with high pressure packed glands or soft metal lubricated rams for high pressure, using, for instance, lead or silver or indium, which have been developed by Crossland and Logan (4). Some pure water systems operating at relatively low pressures have been developed without oil lubrication of the ram seal element, such as the tungsten carbide seal mentioned by Walstad and Noecker (5). An alternative is to use a forced oil lubricated ram seal with some small consequential contamination of the water being pumped, this is basically the system described by Bowen (6).

5. At the pressures generally required for jet cutting reciprocating pumps are employed which are usually hydraulically driven at relatively low cyclic speeds. Consequently, there is the danger of some fluctuation in the delivery depending on the number of pump cylinders employed and on their phasing, and this will lead to a

fluctuation of output pressure, and a consequential fluctuation of the velocity of the jet produced. A fluctuation of the velocity of the jet could lead to intermittent cutting under some condition, and in other circumstances it would lead to a cyclic variation in the texture of the cut surface which might, for some purposes, be considered unacceptable. No evidence of the effect of pressure fluctuation on the cutting performance of jets has come to the attention of the authors. However, that it is considered important to reduce the pressure fluctuations is obvious from the various efforts to produce pulse-less flow. One approach to this problem is the so-called pip-less pumping system which is applicable to two individually hydraulically driven pumps delivering to the same high pressure manifold. While one of these pumps is delivering the other is recharged and then pressurised to a pressure slightly below the manifold pressure, so that as soon as the first pump completes its delivery stroke the second pump can start its delivery stroke. This leads to a rather complex and expensive low pressure system. An alternative system is two simplex intensifiers which are controlled by a cam driven through an overload clutch with an adjustable stop to obtain the required precompression of fluid before the particular cylinder starts to deliver. This unit is described by Bowen (6) and shown diagrammatically in fig. 1. It is claimed that this system reduces the pressure fluctuations to ± ½% at an output pressure of 200 MPa. Yet a further alternative is to use a simple accumulator between the pump and the nozzle as described by Walstad and Noecker (5). The volume of such an accumulator will depend on the size of the intensifier, the output flow required and the pressure variation which can be tolerated.

6. One of the authors has had a considerable interest in the design and development of high pressure intensifiers for many years, so it was a natural extension of this interest to consider the design of jet cutting intensifiers. Two units have been designed one for 340 MPa and the other 680 MPa both working with oil/water emulsions, though pure water systems are under development. The present paper is concerned with the 340 MPa unit and various methods of reducing the pressure pulsations. This has allowed experiments to be carried out to examine the effect of pressure pulsations on jet cutting; which is also briefly reported.

DESIGN OF THE 340 MPa PUMPING SYSTEM

7. The design of the Belfast intensifier for 1.4 GPa which is described by Crossland, Skelton and Wilson (7) and Crossland and Logan (4) was based on the availability of 70 MPa low pressure pumping sets. However, this operating pressure has posed problems of the availability of suitable pumps and control valves, and additionally the capital cost of such equipment is high. It was consequently decided to reconsider the whole design philosophy of intensifiers. From the viewpoint of keeping the pressure fluctuations small it might have been desirable to use a small high frequency intensifier such as that described by Crossland, Jagger and Agnew (8) and shown in fig. 2. In this pump the cylinders of a conventional Towler in-line pump are each coupled hydraulically to individual high speed intensifiers running at the pump speed of 1440 c/min. However, an essential characteristic of this pump is that the fluid being pumped must be the same as that used in the in-line pump, and it was found that a limitation on the fluid which could be used was imposed by cavitation erosion problems on the low pressure piston of the intensifier, and this would certainly prohibit the use of oil/water emulsions. No simple solution could be found to this problem so it was concluded that a basically conventional low speed intensifier pump was likely to be the best solution at this stage.

8. When deciding on the operating pressure for the low pressure hydraulics it was decided to limit the pressure to 17 MPa as there was a large range of pumps and valves available in this pressure range. Instead of producing the low pressure jack, which is an integral part of the design of the Belfast 1.4 GPa intensifier it was decided that it would be cheaper to use a commercially available double acting ram, and fit to it two fairly conventional high pressure cylinders. The general layout is shown in fig. 3 and a photograph is shown in fig. 4.

9. The liners of the high pressure cylinders and the high pressure components such as the ball valve assemblies are made of NCMV steel, and the cylinders and head are heavily overstrained to get the maximum residual compressive stresses at the bore. At 340 MPa it is possible to keep the hoop stress wholly compressive, which should give a greatly enhanced fatigue performance.

10. Two main types of ram seals are being investigated: a seal fixed in the high pressure cylinder and a seal at the head of the reciprocating ram. Solid lubricated seals using lead or indium or graphite, seals using some form of elastomer, etc. are being investigated currently, while for the pure water system other seals such as a clearance controlled tungsten carbide seal will also be examined. However, experience so far indicates at an output pressure of 340 MPa that seals do not pose a serious problem.

11. The inlet and outlet valves are based on the conventional ball valve assemblies which have been well proven on the Belfast 1.4 GPa intensifiers. The main difference is that the balls are more closely guided than in the previous design. These ball valves have given very little trouble.

12. Three basic types of accumulators have been examined:

(a) A fixed volume accumulator between pump and nozzle.

(b) A high pressure gas bag accumulator shown in fig. 5 in which a conventional rubber air bag is fitted in a high pressure container. The charging pressure of the air bag can be varied using a small diaphragm gas compressor, which can deliver at pressures up to 100 MPa.

(c) A detensifier in which a high pressure ram is connected to a low pressure piston in an oil filled cylinder, which is in turn connected to a conventional low pressure air bag accumulator. Fig. 6 shows a general arrangement of this unit.

13. The low pressure hydraulics consists of a 25 KW power pack with a conventional gear pump, adjustable relief valve, a solenoid operated change-over valve, safety valve and oil reservoir fitted with a cooler on the return oil line from

the relief valve. The solenoid operated change over valve is triggered by microswitches controlled by the motion of the ram of the jack.

14. The performance of the high pressure pump has been examined at various flow rates which were dictated by nozzles of several sizes. Fig. 7 shows the arrangement of the nozzle which has a tungsten carbide or tool steel nozzle element. A strain gauge transducer described by Birks (9) was fitted adjoining the nozzle, which together with a recorder allowed the pressure fluctuation to be measured. For cutting experiments the nozzle was used in conjunction with a hydraulically driven linear table, the speed of which could be varied over a wide range.

RESULTS OF TESTS ON THE PERFORMANCE OF THE PUMPING SYSTEM

15. Compressibility data on the 5% cutting oil/water emulsion used for the work reported in this paper is presented in fig. 8.

16. Needless to say without any reservoir between the pump and nozzle the pressure at the nozzle approaches the priming pressure in between the delivery stroke as can be seen in fig. 9. The pressure drops for a 0.202 mm diameter nozzle at various pressures and for four sizes of fixed volume accumulator are shown in fig. 10, while typical pressure waveforms are shown in fig. 9.

17. Data have also been obtained for the high pressure air bag accumulator shown in fig. 5, for two initial air bag pressures and the results are shown in fig. 11. Similar data for the detensifier for two air bag pressures are given in fig. 12. Typical pressure waveforms for the air bag accumulator and the detensifier are shown in fig. 9.

RESULTS OF CUTTING EXPERIMENTS

18. In these tests the level of pressure pulsations was varied by using the detensifier and varying the air bag pressure, while keeping the maximum pressure at 290 MPa. At each condition the feed rate was widely varied using 12 mm thick neoprene as the test material. The maximum feed rate for a through cut was established and the texture of the cut surface examined.

19. Fig. 13 shows the variation of maximum feed-rate with pressure fluctuation for a through cut. At higher feed rates the through cut was not continuous. Fig. 14 shows typical examples of the cut surfaces.

DISCUSSION OF RESULTS

20. The compressibility data for 5% cutting oil/water emulsion at 20°C shown in fig. 8 is compared with the pure water data given by Bridgman (10), and it will be noted that there is reasonable agreement.

21. For fixed volume accumulators the data given in fig. 10 seems reasonable; for large volumes, or higher pressure where the compressibility is greater, the pressure fluctuations are less. The analysis of such a problem could be extremely complex, but a simple minded analysis is possible which can provide a useful design procedure. The basic assumptions of this simple minded analysis are illustrated by fig. 15 in which the pressure at the nozzle is given by ABCD. At B the change-over valve operates and delivery from the cylinder involved ceases. Consequently, the pressure in the delivery manifold begins to fall-off and, if delivery did not start from the second cylinder, it would fall to the priming pressure. When the change-over valve operates there is a time delay, α, before the pressure in the second cylinder begins to rise and a further time delay while the pressure increases to the manifold pressure, when delivery from the second cylinder will commence.

22. The time from the change-over valve being actuated until the second cylinder begins to deliver is t_1. This time is made up of a dwell period α and the time for the pressure in the second cylinder to reach the manifold pressure at point C in fig. 15, which may be assumed to be linearly related to the compressibility at pressure P_L. So the time t_1 can be expressed as

$$t_1 = \alpha + \beta \times \text{compressibility at pressure } P_L \qquad \text{........(1)}$$

23. During the period from B to C the flow rate through the nozzle will, at any instant, be proportional to the square root of the current pressure. However, the precise form of the change of pressure with time is not readily determined, so it is assumed that the flow rate varies linearly with time. Consequently it may be assumed that the pressure acting over this time interval will be constant at P^1, where P^1 is given by

$$\sqrt{P^1} = \frac{1}{2}\left\{\sqrt{P_m} + \sqrt{P_L}\right\} \qquad \text{......(2)}$$

The delivery through the nozzle during this time will be

$$\frac{\pi}{4}\, D_N^2\, C_D \sqrt{\frac{2P^1}{\rho}} \cdot t_1 = V_1 \qquad \text{......(3)}$$

where D_N is the diameter of the nozzle

C_D is the coefficient of discharge

ρ is the density of Pressure P^1

24. If there is a fixed volume accumulator with a volume V_A then the volume of oil which will be released as the pressure sinks from P_m to P_L will be, ignoring the elastic distortion of the accumulator,

$$V_A\,\{\text{\% compressibility at } P_m - \text{\% compressibility at } P_L\} = V_2 \qquad \text{......(4)}$$

At point C the volumes V_1 and V_2 are equal.

25. The procedure adopted has been to deduce α and β from the experimental data for V_A = 185 ccs and for pressures of 276 and 68.9 MPa, which is given in fig. 10. Then using a trial and error procedure the curve for this particular reservoir has been deduced and this is plotted in fig. 10. Using the same values of α and β the curve for V_A = 1750 ccs has been deduced and this is also shown in fig. 10. The form of curve deduced and its magnitude seems reasonable. Perhaps it would have been more reasonable to have deduced α and β from the experimental data for V_A = 1750 ccs, and then to have deduced the other curves, as the pressure fluctuation at V_A = 185 cc are large and the procedure described is more likely to be in error as the percentage pressure fluctuation increases.

26. An initial attempt has been made to apply the procedure indicated above to the high pressure air bag accumulator, but a more detailed treatment is required as the simple gas laws become more inaccurate at the pressures involved. The simplified treatment involves the gas trapped in the air bag being treated as a perfect gas, so that for a given pressure the volume of oil and gas can be calculated assuming isothermal conditions. Under dynamic conditions the pressure fluctuations in the air bag occur isentropically, and it is possible to calculate the volume change and also the volume change due to expansion of the oil. However, such an analysis underestimates the pressure fluctuations by as much as 50%. The experimental data nevertheless shows that cyclic pressure fluctuations can be very considerably reduced with a high pressure air bag accumulator. There are two problems in the use of such an accumulator; firstly if an initial gas pressure in excess of the normal gas bottle pressure is required then a diaphragm gas compressor will be needed. Secondly air bag accumulators operating at such high pressures have not been previously envisaged and they are consequently not well proven. It would be necessary to be careful in the selection of the synthetic rubber employed for the bag as some synthetic rubbers reach their glass transition point at the pressures employed in this system.

27. The detensifier has produced some rather curious results as can be seen in fig. 12. The detensification ratio is 12.6 and at an initial gas bag pressure of 4.8 MPa this requires a high pressure in excess of 61 MPa before the piston of the detensifier begins to move. At about 61 MPa it can be noted that the cyclic pressure fluctuations fall to a very low level. Similarly for an initial gas bag pressure of 11 MPa the cyclic pressure fluctuation falls to a very low level above about 139 MPa. The question is why does the performance deteriorate so rapidly above this pressure? The reason is that the only available gas bag was a 1 gallon size (i.e. 4544 ccs), and the volume of oil trapped on the low pressure side of the detensifier is only 381 ccs. When the low pressure piston has reached the end of its travel the gas bag pressure will only have risen from its initial pressure of 4.8 MPa to 5.24 MPa or from 11 MPa to 12 MPa. In the case of an initial gas bag pressure of 11 MPa the low pressure piston will reach the end of its travel at a high pressure of 151 MPa, so if the high pressure is set at 200 MPa a cyclic pressure of at least 49 MPa or 24½% will be needed before the detensifier operates and at 250 MPa it would increase to 99 MPa or 40%, which are reasonably in accord with fig. 12. The operation of the detensifier would be greatly improved by using a pint size air bag. Nevertheless, the present unit has demonstrated the potential of a detensifier, though it is a relatively expensive unit.

28. The cutting tests on 12 mm neoprene sheet at various cyclic pressure fluctuations show the effect on the maximum feed rates for through cutting. It indicates that the effect on the maximum feed rate if the cyclic pressure fluctuation is kept within 20% is not very large. The data suggests that the maximum feed rate if the cyclic pressure fluctuation was completely eliminated would be 18 cm/s compared with 16 cm/s for a cyclic pressure fluctuation of 12½%.

CONCLUSIONS

29. The experimental data on cutting of neoprene which is presently being extended to non-homogeneous materials indicates that there is little to be gained either in feed rate or texture of the cut surface by reducing the pressure pulsations below 10-20%. If this is true for other materials besides neoprene, then it is probably economic to use a fixed volume accumulator as its size would not be excessive. With this pump set and intensifier pump an accumulator volume of 1750 ccs would appear adequate, and this would require a 75 mm bore x 400 mm long vessel. A simple minded analytical procedure has been suggested which allows the size of a fixed volume accumulator to be established for a given LP pumping unit and high pressure nozzle diameter.

30. Both the high pressure air bag accumulator and the detensifier give a satisfactory performance, though in the case of the detensifier the performance would be extended considerably by using a small size of air bag on the low pressure side of the unit. However, neither the reliability of the air bag in the high pressure air bag accumulator nor the life of the ram seals in the detensifier have as yet been established.

31. Both development and analytical work on the system are proceeding.

ACKNOWLEDGEMENTS

32. The work reported here was carried out in the Department of Mechanical Engineering of The Queen's University of Belfast with the aid of a generous grant from the Science Research Council. Thanks are due to Dalavan-Watson Ltd. who have provided the nozzles used in this work. The assistance of Mr. R. H. Agnew, Superintendant of Workshop, and Mr. Johnston Harvey, Superintendant of Laboratories and their staff is gratefully acknowledged.

BIBLIOGRAPHY

1. Richardson, C. A. and Thronton, W. A. 'A Bibliography of Jet Cutting Technology' published BHRA Fluid Engineering 1973

2. Proc. First Int. Sym. on Jet Cutting Technology, BHRA Fluid Engineering 1972

3. Proc. Second Int. Sym. on Jet Cutting Technology BHRA Fluid Engineering, 1974

4. Crossland, B. and Logan, J. G. 'Development of Equipment for Jet Cutting', Proc. First Int. Sym. on Jet Cutting Technology, 1972, C1, pp 1-12

5. Walstad, O. M. and Noecker, P. W. 'Development of High Pressure Pumps and Associated Equipment for Fluid Jet Cutting', Proc. First Int. Sym. on Jet Cutting Technology, 1972, C3, pp 21-28

6. Bowen, J. C. 'An Approach to Pulseless Packingless Pumping', Proc. First Int. Sym. on Jet Cutting Technology, 1972, C4, pp 29-36

7. Crossland, B., Skelton, W. J. and Wilson, W. R. D. 'High Pressure Intensifiers: Development of a Unit for Pressures of 200000 lbf/in^2 and Possible Improvements in Design', Proc. Inst. Mech. Engrs., 1967, Vol. 182, pt 3C, pp 175 - 179

BIBLIOGRAPHY (contd.)

8. Crossland, B., Jagger, R. and Agnew, R. H. 'Development of a 50000 psi Large Flow Pump', Proc. 27th Fluid Power Convention, Chicago, 1971, pp 105 - 116

9. Birks, A. W. 'Transducers and Fluids for High Pressure (1.4 GPa) Systems', to be presented at the Second Int. Conf. on High Pressure Engineering, 1975

10. Bridgman, P. W. 'Water, in the Liquid and Five Solid Forms, Under Pressure', Proc. Am. Acad. Arts Sci., 1911, Vol. 47, pp 441 - 558

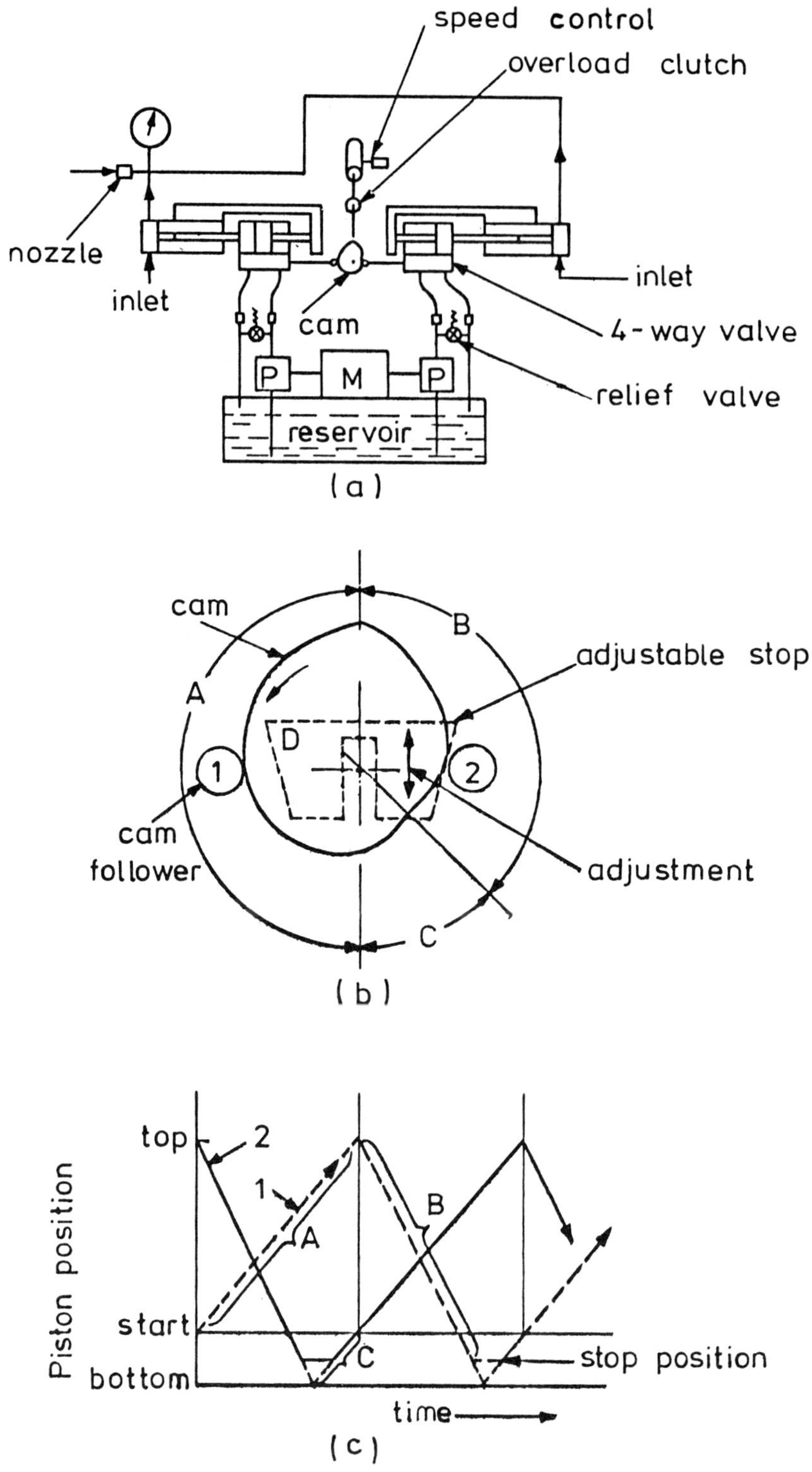

Fig. 1: Pulseless pump described by Bowen (Ref.6)

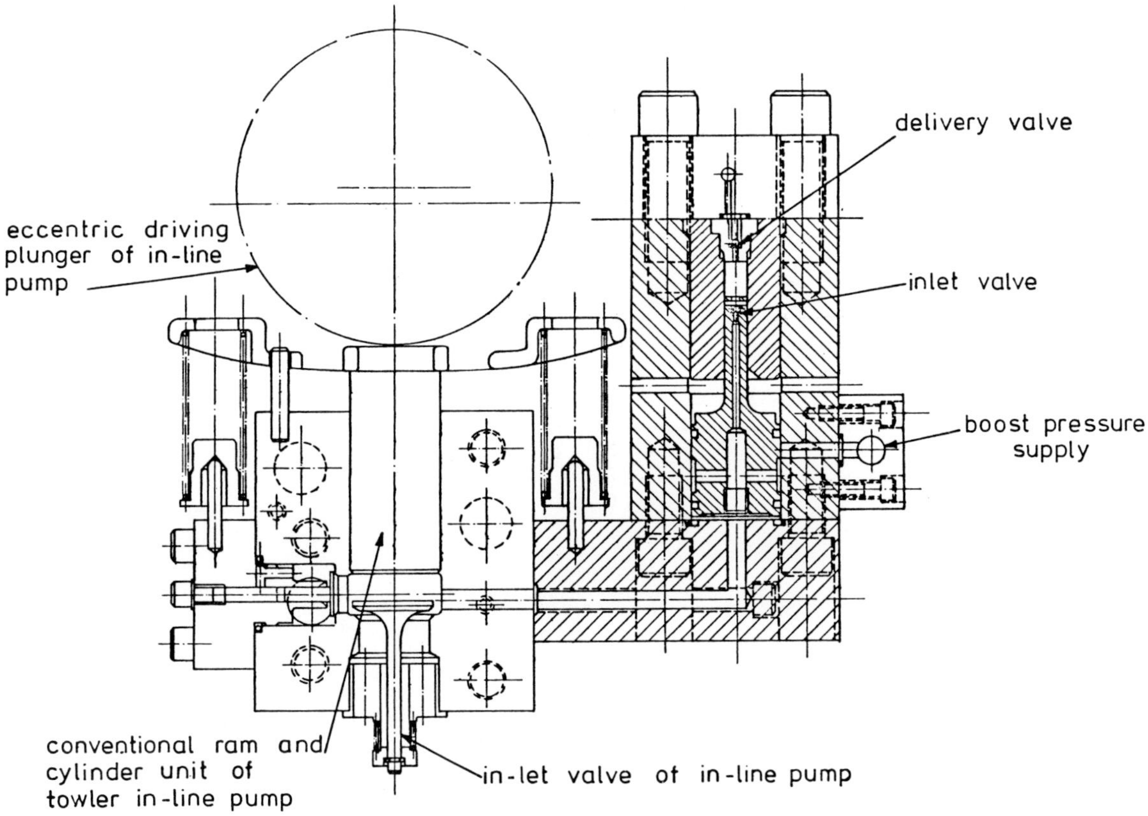

Fig. 2: 340MPa (50000 lbf/in²) large flow pump (Ref. 8)

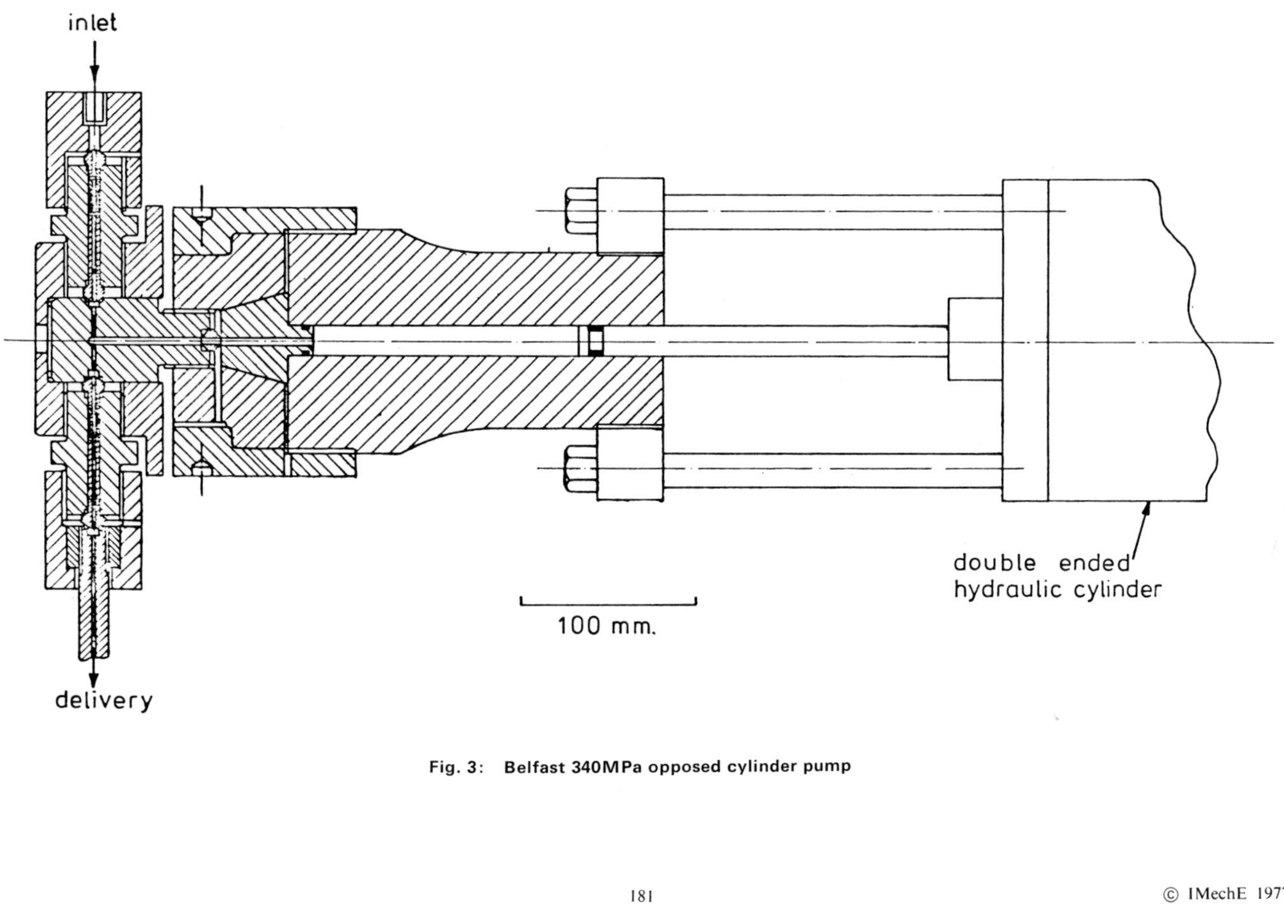

Fig. 3: Belfast 340MPa opposed cylinder pump

Fig. 4: Belfast 340MPa opposed cylinder pump mounted on a supporting frame

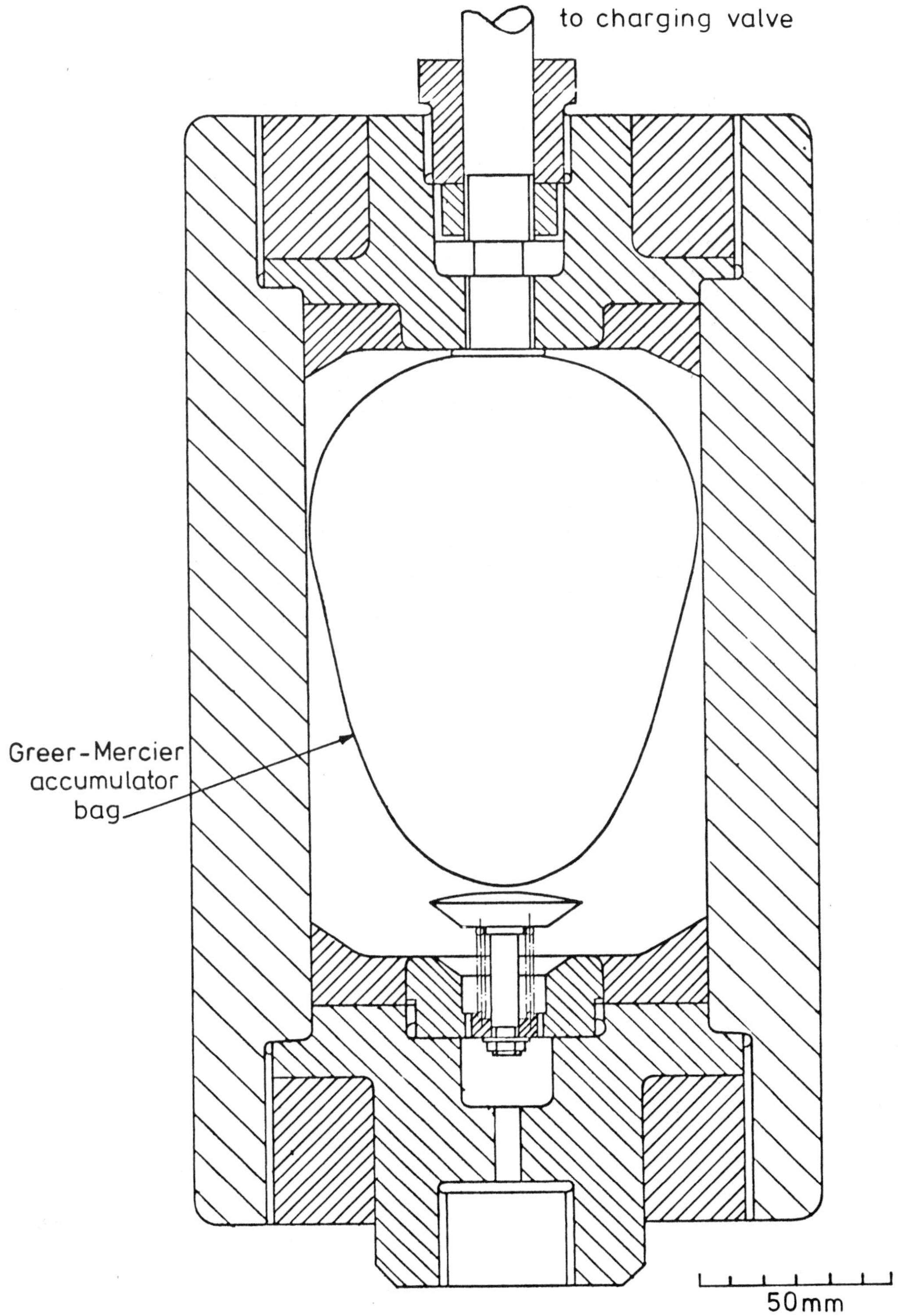

Fig. 5: High pressure air bag accumulator

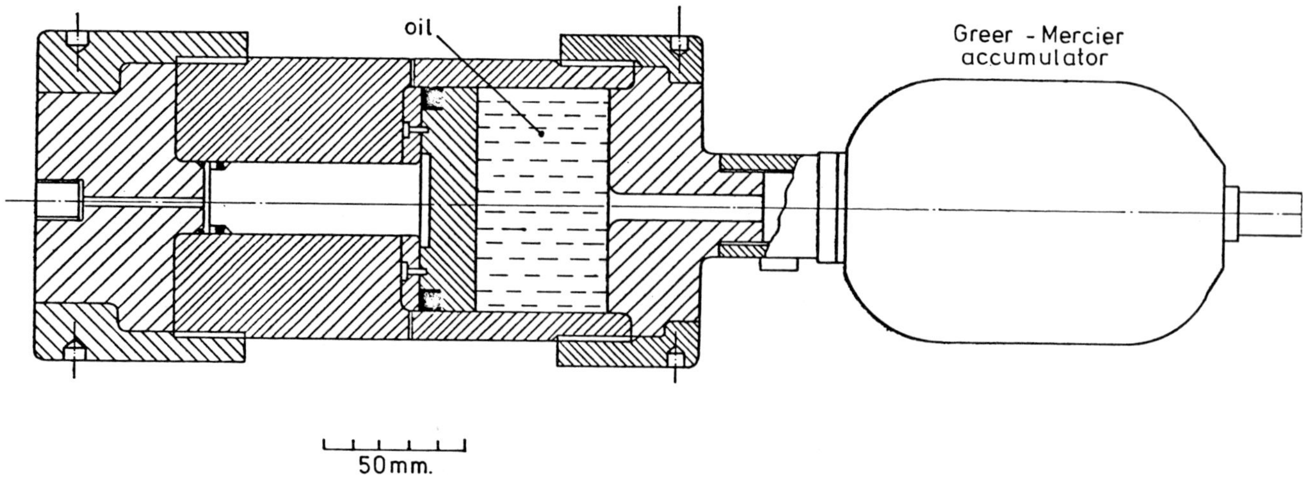

Fig. 6: Detensifier with LP air bag accumulator

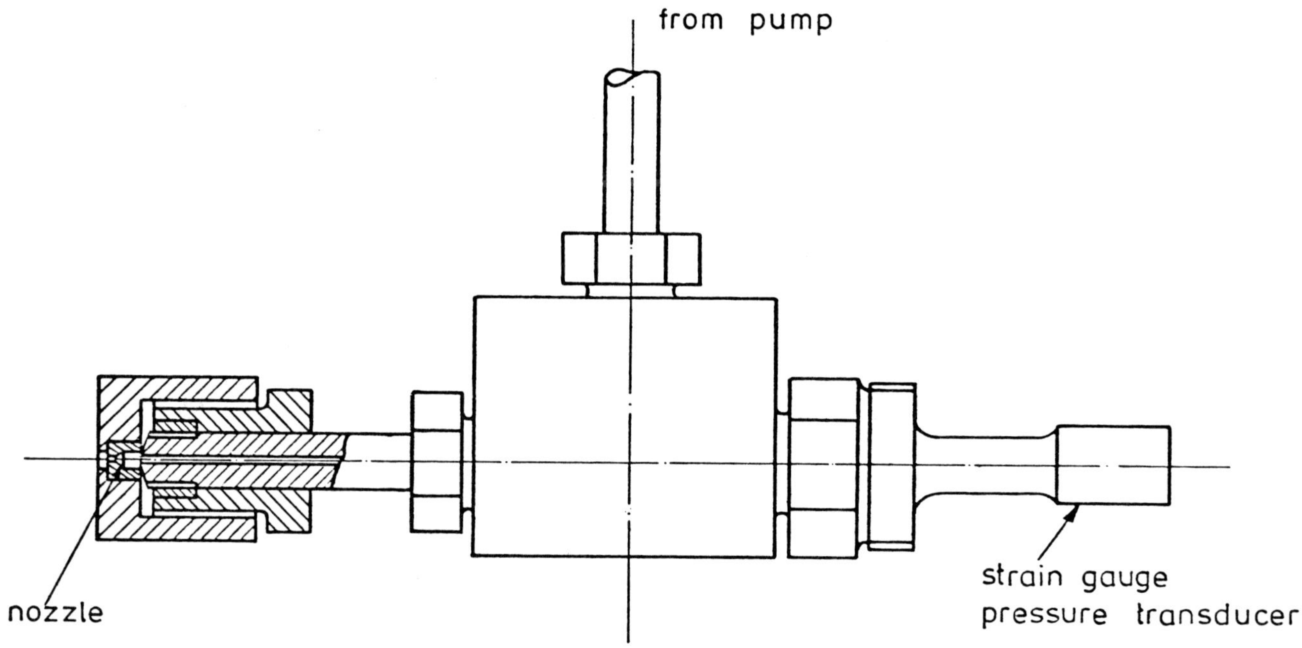

Fig. 7: Arrangement of nozzle with strain gauge transducer

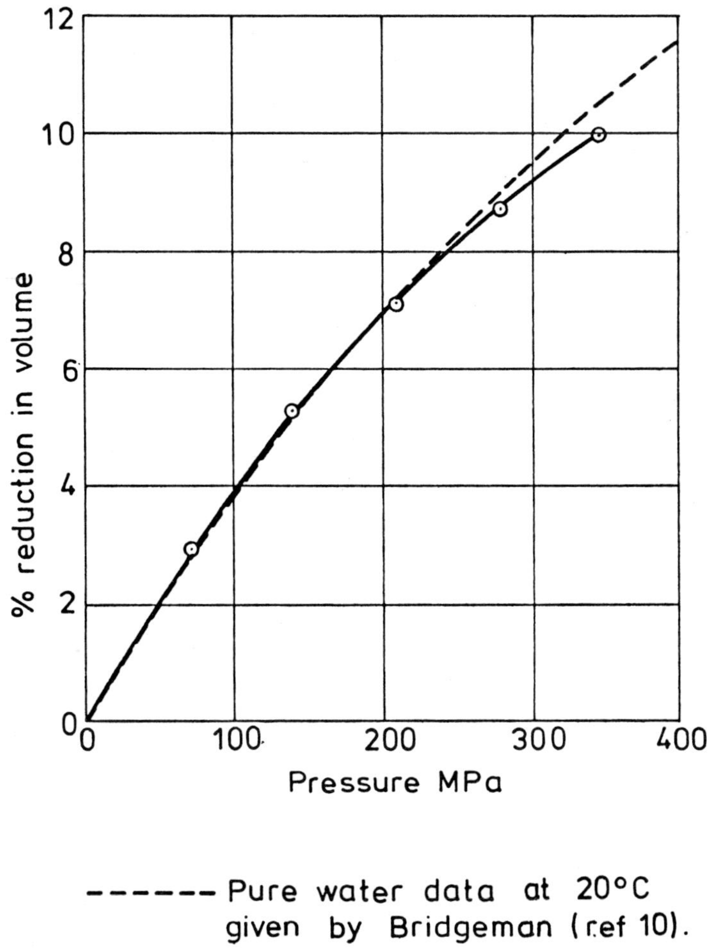

Fig. 8: Compressibility of 5% cutting oil/water emulsion at 20°C

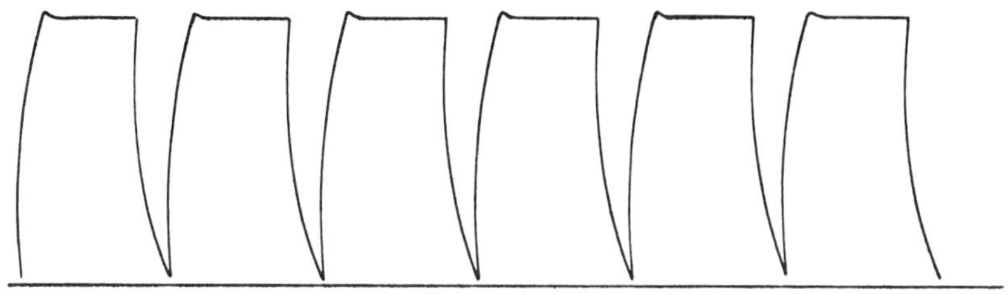

(a) Pressure trace no accumulator.

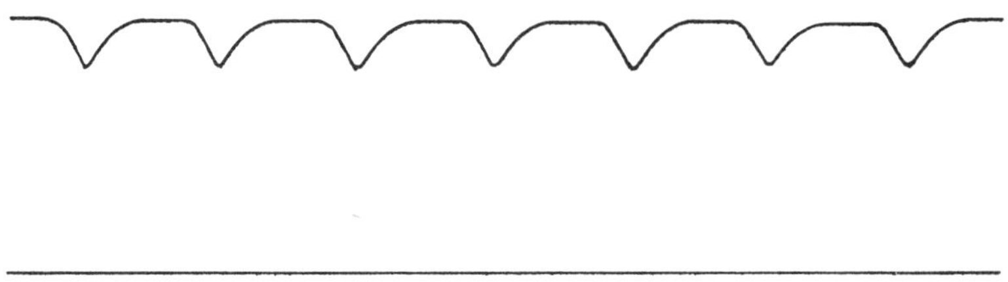

(b) Pressure trace fixed volume accumulator of 640 cm^3

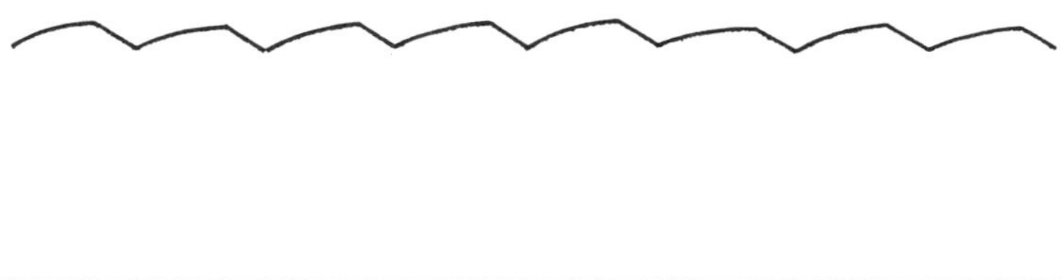

(c) Pressure trace for high pressure air bag accumulator. (initial bag pressure 10 MPa)

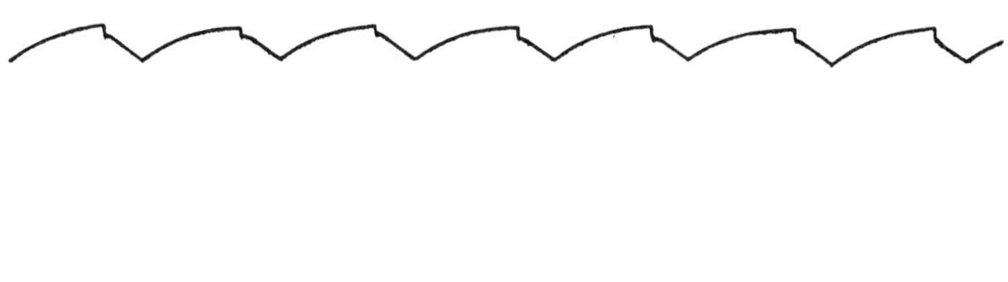

(d) Pressure trace for detensifier with an initial air bag pressure of 20 MPa.

Fig. 9: Pressure waveforms at 290 MPa with 0.2mm diameter nozzle with various accumulators

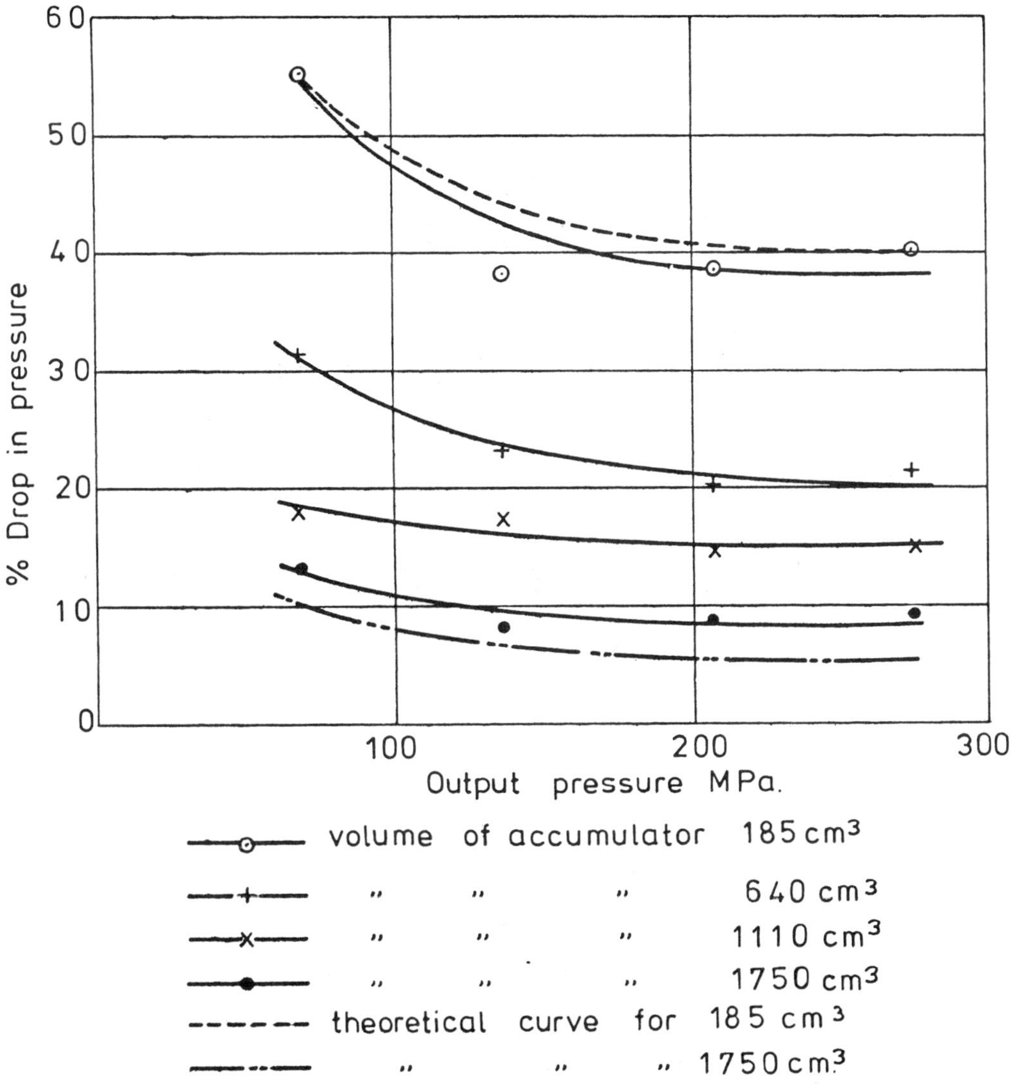

Fig. 10: Pressure drop with fixed volume accumulator for a nozzle diameter of 0.2 mm

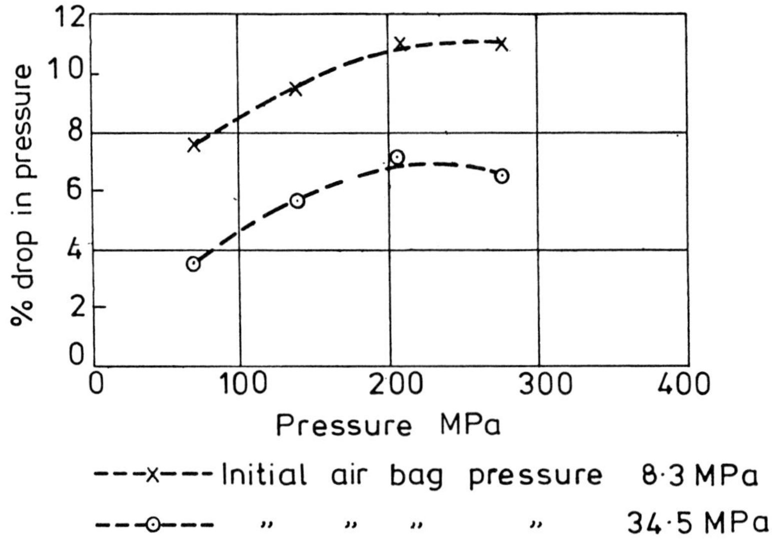

Fig. 11 : Pressure drop with High pressure air bag accumulator diameter.0.2 mm

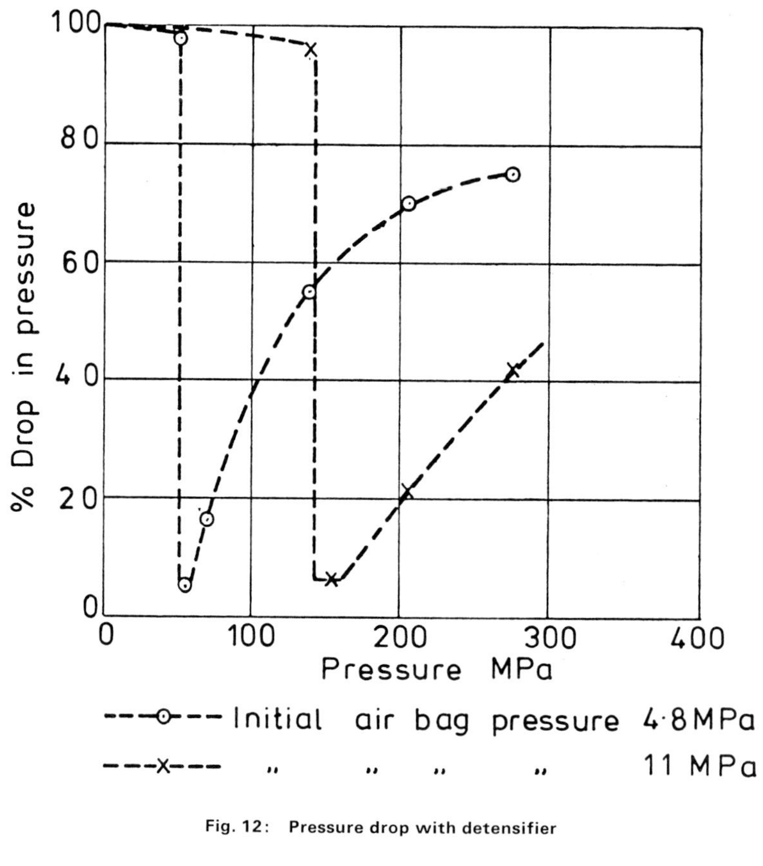

Fig. 12: Pressure drop with detensifier

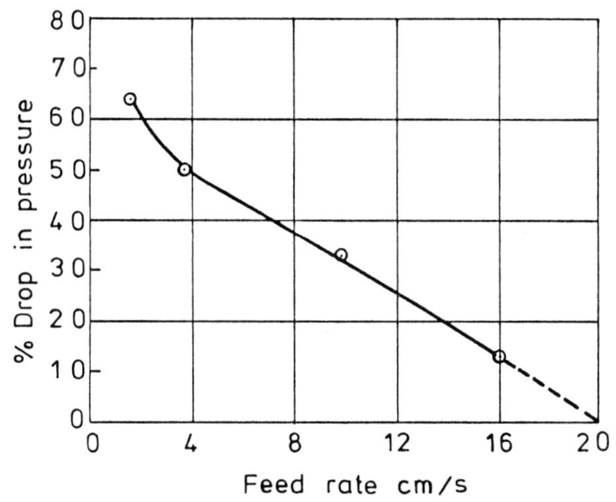

Fig. 13: Maximum feed rate for a through cut in 12 mm thick neoprene as the level of pressure fluctuation is varied for a peak pressure of 290 MPa and a nozzle diameter of 0.2 mm

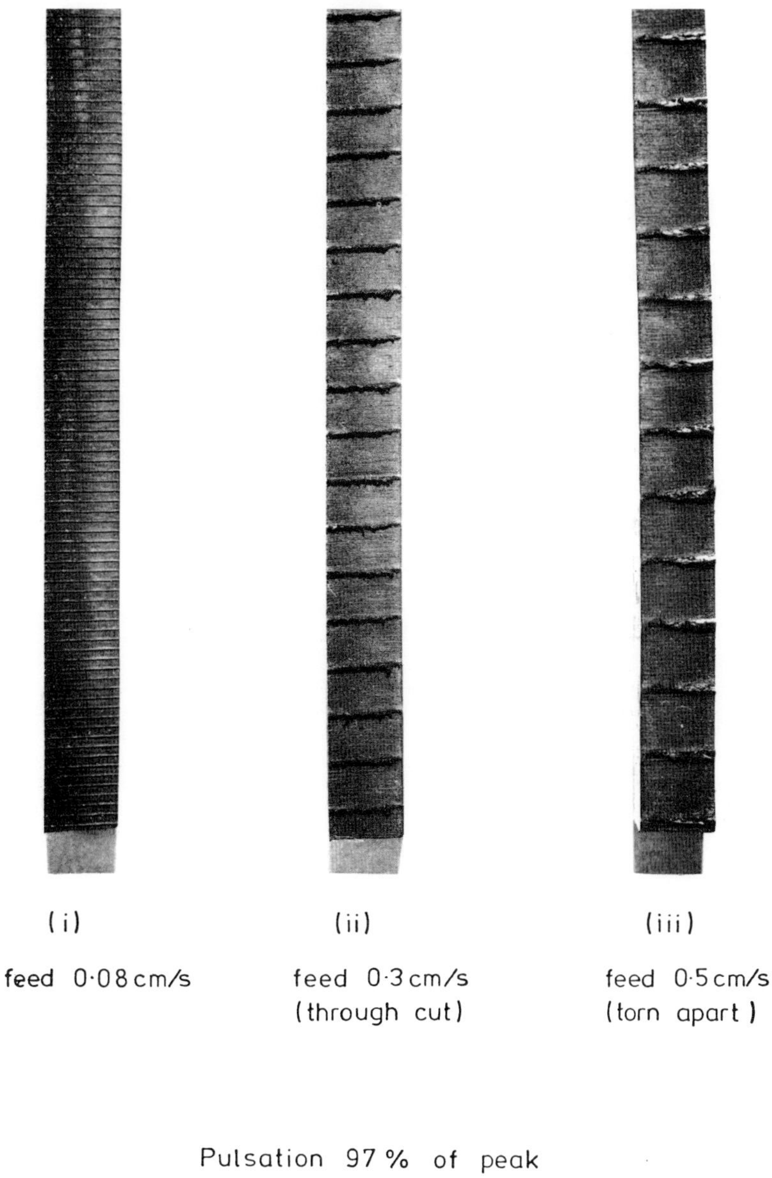

Fig. 14(a): Cuts in 12 mm neoprene at a peak pressure of 290 MPa

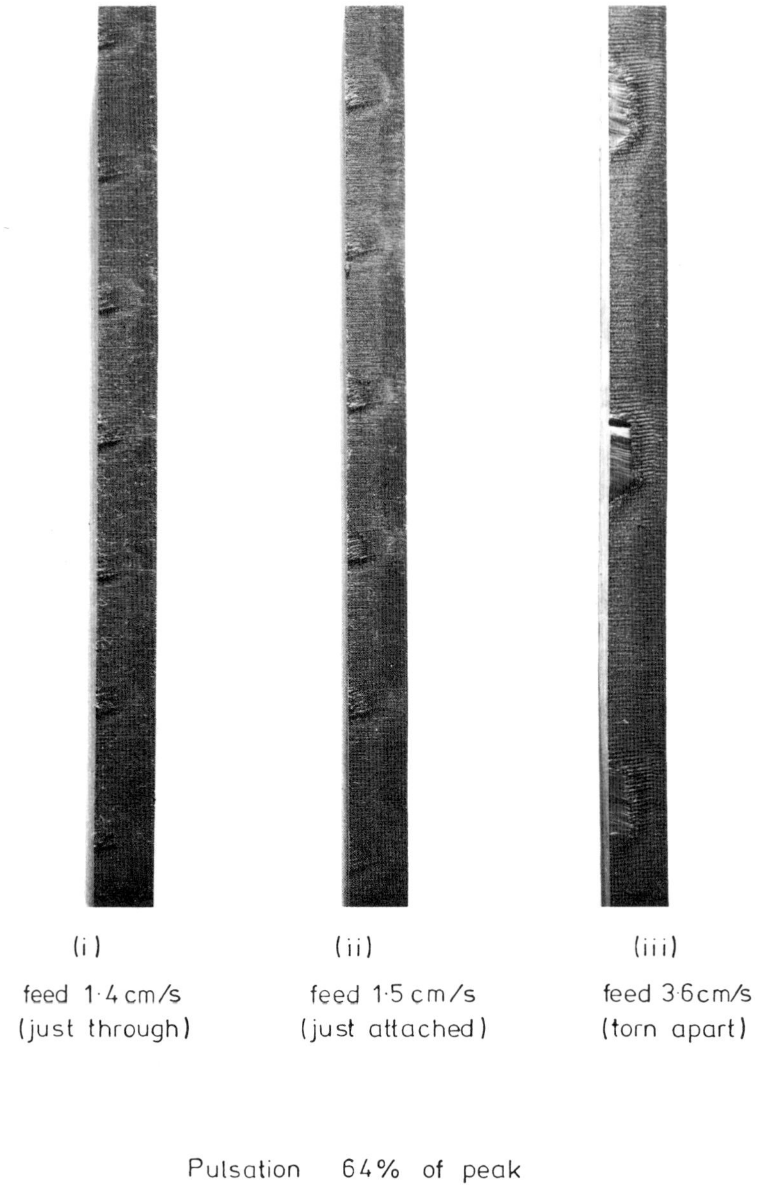

Fig. 14(b): Cuts in 12 mm neoprene at a peak pressure of 290 MPa

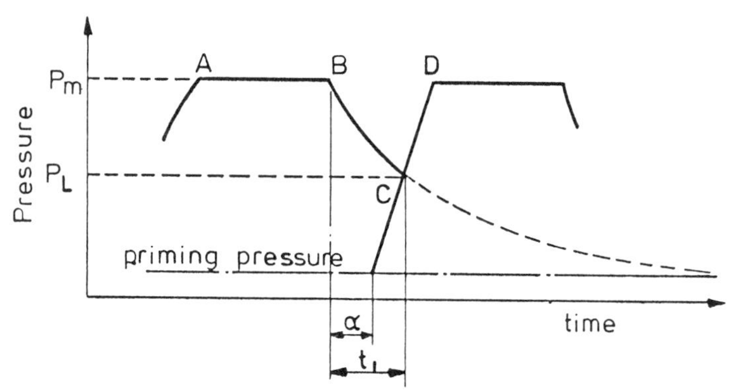

Fig. 15: A simplified model of action in an opposed cylinder pump

SOME PERSPECTIVE ELABORATIONS IN THE FIELD OF HYDRAULIC EXTRUSION

YU. S. KONYAEV, Doctor
Institute for High Pressure Physics, USSR Academy of Sciences

SYNOPSIS The possibility of making use of high pressure /to I00 kbar/ is explained and the methods of hydroextrusion and high temperature /to I.500°C/ are applied at gas extrusion of materials. Some perspective tendencies in the development of the method of pressing materials by liquids of high pressure are mentioned.

At present, hydraulic extrusion as a method of pressworking of metals has gained a wide range of application. It is used for making various products from non-ferrous metals and steel. However the advantages inherent in pressworking of metals which can raise physico-mechanical and operative properties of material (strengh and plasticity, density, magnetic properties, etc.) are not made use of.

From this point of view hydraulic extrusion as a method of pressworking of metals is very promising.

It is known that at present many difficulties are caused by deformation of intermetal combinations and alloys based on them (having the properties of superconductivity, high heat resistance, semi-conductivity) which are of importance for modern technology. There seems that pressure exerted by the available plants (20-30 kilobars) are not adequate for the required straining of these materials.

This is why the Institute for High Pressure Physics of the USSR Academy of Sciences has undertaken a development of equipment and processes necessary for hydraulic extrusion with the applied pressure up to I00 kilobars. The equipment designed and made allows extrusion of materials applying pressure of up I00 kilobars and counter-pressure of up to 25 kilobars. The design was based on a two-stage diagram according to which the entire arrangement for hydraulic extrusion is made as a self-contained unit and assembled beyond the press, then it is loased into the first-stage container wherein the pressure is created by piston.

The pressure in the first-stage container not only supports the loaded components of the second stage but also acts as counter-pressure. The above-described diagram was used for designing and making an experimental plant for which purpose we modified a universal hydraulic press of 630 ton capacity manufactured in series production. In general, such devices can be attached both to the upright and horisontal hydraulig presses of any type and capacity.

. Used as working medium in the first-stage container was kerosine, the second-stage container was filled with lead or indium.

This plant was used for deforming a considerable amount of materials of various compositions and purposes. Among these there were hardened steels with from 0.35 to I8% carbon, grey cast iron, some heat-resistant alloys and intermetal alloys, chromium, tungsten and tungsten alloys, There were also materials for constant magnets as well as a number of intermetalloids, semiconductors. We have primarely been concerned with the problem wether it is possible to subject these and similar materials to cold plastic deformation. The level of counter-pressure in our experiments depended upon the type of material, its purity, methods of its manufactures. Counter-pressures of 20-25 kilobars are generally considered adequate for deforming a wide range of materials, including some intermetalloids.

However those engaged in research in hydraulic extrusion are keenly interested in employing extrusion under elevated temperatures. The researchers expected the gains to be yielded from a combined influence produced by high hydrostatic pressures and elevated temperatures upon the structure and properties of steel. Another expectation is that of carrying out thermal and mechanical treatment of structural steel under high hydrostatic pressures.

The main cause of failure when attemting at hot hydraulic extrusion is, first of all, the fact that used as working medium were the liquids which change their properties with a rise in temperature under pressure. This means changes in the conditions under which the process occurs. In this connection, when designing a plant for hydraulic extrusion under elevated temperature, the authors of this work, first of all, undertake selection of high-pressure working medium. This medium was expected to meet the following requirements:

a) to be able to trensfer high pressure without losses in hydraulic static proper-

ties and to be able of plastic flow through the aperture of the die; b) to retain constant properties in a wide range of temperature under varying high pressures.
A study of literary sources led us to a conclusion that none of the avilable liquids or molten salts or metals can meet these requirements. In this connection a research was undertaken into placticability of application of gaseous medium for pressworking of metals. A study of cold-gas extrusion and comparison of properties of materials extruded in gaseous and liquid media showed that gaseous extrusion does not involve any additional processing difficulties as hydraulic extrusion does. This means that gaseous medium can, in the principle, be used for deforming metals under elevated temperatures.
It was also found out that the gas-extrusion effort is less than that needed in hydraulic extrusion on the average by 2500-3000 kg per sq cm.
Further work in this field resulted in construction of a plant allowing reliable deformation of various materials under pressure of gas of up to IO kilobars in the adjustable range of temperatures of 20 to IOOO°C with heating rates of up to IOO degrees per second. The high pressure chambers were made as two designs differing mainly in the method for heating the blank: one method is to make electrical current pass through the medium, another employs transfer heet from the resistance. In the two cases the main units of the chambers are a container unit, a die holder unit and a hydraulic batcher unit.
Heating was achieved by the use of A.C. allowing high power.
A series of experiments made with a plant for high-temperature gas-extrusion allowed a determination of optimum temperature condition, distribution of temperature over the length of blank and of dependence between the heating rate and the applied power. This, in turn, made it possible to carry out thermal and mechanical treatment using gas-extrusion and to compare the achieved results with those achievable in thermal and mechnaical treatment by rolling under similar conditions.
As a result of treatment wich involves austenisation at 900°C, deformation by gas-extrusion at 500°C to the 20% stage hardening into martensite of steel, grade 45XH3MC, in Ø IO mm rods, the following level of properties can be achieved: the limit of strength 250-260 kg per sq mm, the yield point 215-220 kg per sq mm, elongation II-I3%, contraction 47-53%.
The comparison of properties of steel treated for equal strength by thermal and mechanical method of rolling and of gas-extrusion testifies to the fact that to obtain equal strength under thermal and mechnaical treatment of rolling, a considerable degree of deformation of austenite (75%) was required under rather low temperature (500°C) which naturally called for much power consumption. Besides, the plasticity was lower as compared with treatment by gas-extrusion: relative elongation by 3.5 times, contraction - 2.8 times.
Carrying out thermal and mecha nical treatment by high-temperature gas extrusion is only one trend in application of the suggested plant.
Though a high-temperature gas-extrusion technique is a rather recent development one can say even now that it is a good novelty in pressworking of metals.
While posessing all the advantages of hydrostatic extrusion, employment of gas as a working medium allows this technique to be carried out under very high temperature sparing the tools which show thereby rather good durability.
The hydraulic extrusion devices known already in this and other countries employ the so-called press principle. This is primarily rooted in the fact that by the time hydraulic extrusion was developed, conventional pressworking had already gained wide application. The press principle of hydraulic extrusion can easily be put to life (only small changes are required in the presses) and allows high pressures of liquid to be obtained wich depend only on the strength of the material a container is made from. However to create a reliable and durable sealing for the movable ram is very difficult. Besides, if good sir-tightness of the container is to be provided, its internal surface should be of a very high finish and allows its work only in the elastic range without any plastic deformation. Both the diameter and the length of the container and, consequently, that that of the workpieces cannot exceed certain limits since they depend upon the design features of the press (the axial effort and the amount of stroke).
This is why, as soon as the described technique came into being, our Institute carries out hydraulic extrusion according to the pump principle. With this aim in view we have created a high-pressure hydraulic compressor. It is intended for exerting hydrostatic pressure of up to I6000 bars in practically any volume. The compressor in conjunction with a plant. Device or apparatus can be used for research in the above-described range of pressures.
The main unit of the compressor is its head wich is to be reliable if the entire machine is expected to perform well. The hydraulic compressor can be used in any industrial branch when high hydrostatic pressures are necessary, for instance for pressure testing and sir-tightness testing of vessels and instruments, for increasing the strength and plasticity of steels and alloys, for chemical polymerization, for powder metallurgy when applying isistatic compaction of powder materials, for pressworking of metals (hydrostatic stamping, drawing, shaping, bending, forge drawing), for manufacturing large-size products under impulse loads,etc.
Using a hydraulic compressor for treating the materials by hydraulic extrusion allows a number of new processes to be applied, in particular machining difficult-to-deform metals and alloys, to save much industrial equipment for other jobs, such as presses, to save investment in this branch,

to spare floor space, to render hydraulic extrusion more versatile and universal since it provides the possibility for making products of various shapes and length without much expenditure on equipment.

To conclude, one can state that the prospects for the method of hydraulic extrusion is advantageous over the other techniques since it allows less abrasion of working tools and their longer life, consist in wide application of high pressure generators of continuous operation, commercial application of high-temperature gas extrusion and carrying out differential hydraulic extrusion at the pressure of up to IOO kilobars.

DESIGN AND PERFORMANCE OF WATER JET SYSTEMS FOR SLOTTING OF CONCRETE

A. G. NORSWORTHY, Research Assistant, U. H. MOHAUPT, Assistant Professor and D. J. BURNS, Professor and Chairman
Department of Mechanical Engineering, University of Waterloo

Pulsed and continuous water jet systems are being studied as alternatives to the pneumatic hammer and diamond saw respectively for demolishing or modifying concrete structures. After a brief review of pulse jet designs, this paper concentrates on the use of continuous water jets for the precision slotting of concrete. It considers the effects of jet pressure, nozzle diameter, traverse velocity and multiple passes on slotting efficiency and concludes with a discussion of the potential of hybrid systems consisting of a continuous water jet and mechanical cutter.

Notation

C characteristic speed
C_d nozzle discharge coefficient
C_1, C_2, C_3, C_4 and C^* constants
d_o initial jet diameter approximated by nozzle diameter
f porosity
g surface averaged grain diameter
h slot depth
h_1 slot depth after first pass
h_2 increment in slot depth produced by second pass
k permeability
n fluid viscosity
p_i threshold pressure at which jet does little or no cutting
p_o nozzle pressure
p_s pressure at cutting surface
p_v vapour pressure
q jet flow rate
s Plotdjakonoph coefficient
V nozzle traverse velocity
V_w jet velocity
$\dot{W}$ power
W_w Weber's number of the water jet
W_t Weber's number of relative traversing velocity
$\alpha, \beta, \gamma, \delta$ constants
θ angle between jet axis and traverse direction
θ_o initial angle between jet axis and traverse direction
μ_r coefficient of internal friction
μ_w Coulomb friction coefficient between water and rock (concrete)
ρ_1 density of water
ρ density of fluid in jet
σ surface tension of water in air
τ shear stress
τ_o shear strength
ϕ strength coefficient

1. INTRODUCTION

Table 1, extracted from McCurrich and Brown,[1] summarizes the major limitations of existing techniques for modifying or demolishing concrete structures and slabs, e.g. road surfaces. If the concrete is reinforced some of these techniques have to be combined with oxy-acetylene cutting. These limitations and noise control legislation are forcing industry to look for alternatives.

TABLE 1

Method	Limitations
1. Pneumatic hammer	Slow, creates noise, vibration and dust
2. Rotary percussive drill	Only for holes. Noise and dust
3. Heavy impact, e.g. tractor-mounted drop hammer	For large work requires heavy equipment
4. Explosives	Lack of precision. Difficult in occupied areas, dust
5. Hydraulic bursters in pre-drilled holes	Slow. Problems with reinforcement
6. Thermic lance and flame jet	Expensive. Fumes and fire hazard
7. Plasma arc	Expensive. Fumes and fire hazard
8. Diamond saw	Expensive. Difficult on thick sections and where flat surfaces are not available

A possible alternative to the larger pneumatic hammers is the so-called pulsed water jet which demolishes concrete with high velocity slugs of water (2–4). Pulsed water jet machines can be classified as pressure extrusion, impact extrusion, annular cumulation or cumulation devices.

A pressure extrusion device is essentially a snap-acting intensifier. Energy from the primary power source is stored by compressing a gas to a low pressure, e.g. 20·7 MPa (3000 lbf/in^2). A trigger mechanism suddenly releases this gas to drive directly or indirectly a low pressure piston. This piston is connected directly to a ram of much smaller diameter which compresses the water and extrudes it through the nozzle.

Pressure extrusion and impact extrusion devices are similar except in the latter the low pressure piston is allowed to accelerate and store considerable energy before impacting directly or indirectly with the high pressure ram. Both types have been used to fracture rock and/or concrete at water pressures up to 1380 MPa (200 000 lbf/in^2).[4] However, in almost all cases the ancillary equipment has been very large and the firing rates have been too low to be economic.

An annular cumulation device operates with a hollow cylindrical charge of water.[5] This annular charge is impacted by a high velocity piston. As the piston advances it is preceded by a shock wave. This shock wave pressurizes the water, which collapses inward on itself to form a hypersonic jet, which travels along the polar axis of the original water charge. A special nozzle is not required since the jet is self-focusing.

A major difficulty is the generation of the hollow cylindrical charge of water. If rapid cycling is not required this problem can be solved by using a preformed tube of gel as the charge. Cooley and Clipp[5] have shown that a small hand-held device using preformed charges is capable of piercing 13 mm ($\frac{1}{2}$ inch) thick armour plate with one shot.

A cumulation effect can be achieved with a 'solid' cylindrical charge of water by extruding it through a Voitsekhovsky-type nozzle.[6] This type of nozzle has a cross-sectional area which decreases exponentially with length. It is separated from the water charge by a bursting disc. When a high velocity, low pressure piston impacts the water and breaks the disc, the nozzle shape causes unsteady acceleration of the water. In particular there is an accumulation of kinetic energy in the leading part of the jet and pressures of the order of 6·9 GPa (10^6 lbf/in^2) have been reported.[6] This approach would be more attractive if techniques could be found to reduce the high cost and increase the fatigue life of Voitsekhovsky-type nozzles. The nozzle used by Cooley[6] cost several thousand dollars and had a fatigue life of 100–200 cycles with an internal pressure of 15 000 atmospheres.

Although cumulation devices can produce much higher impact pressures than extrusion devices, the shock waves resulting from impact effects may not be as important as the penetration effect of the jet and subsequent pressurization of the hole. The duration and magnitude of the hole pressure rather than dynamic effects may well decide the extent of crack growth. This opinion is based on our experience in working with the Institute of Gas Technology, Chicago, on the design and development of a pressure extrusion machine for breaking concrete slabs, e.g. road surfaces. This machine is for operation at pressures up to 690 MPa (100 000 lbf/in^2) with a water slug of 328 cm^3 (20 in^3) being fired every seven seconds through a nozzle of the order of 1–2 mm. Its size and power requirements are such that it can be mounted on and driven by the hydraulics of a large backhoe. One shot from this device will penetrate as much as 5 inches and split free-standing blocks. Spalling may occur if the slug hits hard aggregate such as granite or quartz.

It is unlikely that pulsed water jets will be suitable for precision cutting of concrete because of the random nature of the crack growth and local spalling. However, they may be very useful when cutting wide slots (trenches) in concrete slabs if the boundaries of the trench have been precut with a diamond saw or its equivalent. Also shallow precutting with a saw may serve to control the direction of the cracks generated by a pulse-jet slug.

Flail and diamond saws are currently used for 'precision' slotting of concrete. For example, flails are used to cut anti-skid slots on road surfaces. Diamond saws are used to cut expansion slots and slots for traffic sensors. Saws are also used to define the perimeter of a section that must be replaced or removed to gain access to underground cables and pipes; pneumatic hammers can then be used to break up the section of interest.

Olsen[7] and the present authors[8,9] have investigated the use of continuous water jets for the 'precision' slotting of concrete. Both studies have considered the effect of jet pressure, nozzle diameter, stand-off distance, traverse velocity and multiple passes on slotting efficiency.

Olsen used a 37 kw (50 hp) intensifier capable of producing 95 cm^3/sec (1·5 gpm) at 413 MPa (60 000 lbf/in^2) to feed single nozzles of 0·152 mm (0·006 in) or 0·305 mm (0·012 in) diameter at traverse speeds of 25 mm/sec to 25 m/sec.

At pressures less than 345 MPa (50 000 lbf/in^2) the jet easily removed the sand and cement and some small aggregate but did not cut the larger aggregate. At pressures above 345 MPa the granite aggregate was cut but some pieces of quartz aggregate were not cut at the maximum pressure of 413 MPa (60 000 lbf/in^2). This supports the results of McCurrich and Brown[1] and suggests that pressures less than 345 MPa are only suitable for such tasks as roughening concrete, cutting grout and cleaning used bricks.

Two methods were used to form a slot (kerf) significantly wider than the water jet. In one method the nozzle was oscillated side to side in a direction perpendicular to the traverse direction. Olsen suggests that this method would be suitable for producing shallow grooves of 5 to 10 mm for skid control on highways. Up to 80 such grooves could be produced for a power input of 600 kw with 0·15 mm nozzles oscillating simultaneously and moving forward (traversing) at 1 to 2 km per hour.

In the other method the nozzle makes multiple passes on parallel but not coincident paths. This is more efficient than the oscillating nozzle method when cutting wide deep slots. Olsen reports that parallel cuts can be several millimeters apart and still lead to break off of the material between adjacent cuts. He estimates that multiple nozzles of 0·2 to 0·3 mm diameter and a power input of 600 kw would suffice to advance two 50 mm deep slots at 1 to 2 km/hr.

These estimates are included only to illustrate the general capabilities of continuous jet systems. Olsen did not publish a detailed economic analysis. Some support for Olsen's observations can be found in our paper presented at the same symposium[8] and more recent work.[9] This data was obtained using the equipment shown in Figs. 1–3.

2. EQUIPMENT FOR CONTINUOUS JET CUTTING

Figure 1 shows one end of our double-acting fluid support intensifier capable of producing 60 cm^3/sec (1 gpm) at 690 MPa (100 000 lbf/in^2). This intensifier was designed to evaluate the fluid support concept[10] and to provide an output well beyond the range of commercial water intensifiers, i.e. 207 MPa to 483 MPa.

Vessel B and its end plug A are only subjected to cyclic pressure when the fluid support space between vessels B and D is charged and discharged. The inner vessel D and its cylinder head C are held in compression by the fluid support pressure to minimize the risk of fatigue failure. Vessel D is made from two H 111 tool steel cylinders which are a push-fit; it is subjected to ranges of shear stress up to 514 MN/m^2 at 10 to 20 cycles/minute.

The hollow ram, F which has an OD of 3·81 cm and a

stroke of 15·25 cm is separated from the bore of vessel D by a core bar E. This core bar also stops the SCN neoprene seal from moving up the cylinder. This arrangement of ram, core bar and seal has been used by Crossland *et al.*[11] at pressures up to 13·8 k bar and has operated very successfully in our intensifier in five years of laboratory use.

The ram F is driven by an actuator, Q, which operates at a maximum pressure of 207 bar and is rated at 578 000 N. As the actuator ram, M, drives the ram F to pressurize the water in vessel D it also serves to recharge the equivalent of vessel D in the other end of the intensifier. A small external pump supplies water to the recharging chamber K. The movement of the ram M within the chamber K pressurizes this water to 69 bar and recharges the other end of the intensifier, via the equivalents of port H and the ball valve in the top of the hollow ram F shown in Fig. 1.

Limit switches within the recharging chambers control the reversal of the double-acting actuator. The support spaces between the inner and outer vessels also serve as an accumulator of about 1280 cm^3 to minimize fluctuations in nozzle pressure when the actuator reverses.

Figure 2 shows the nozzle unit used to support the synthetic sapphire nozzles, which had inside diameters of 0·178 mm (0·007 in) or 0·381 mm (0·015 in). This arrangement has worked very well; it may be possible to improve the efficiency of these nozzles by varying the geometry as suggested by Chadwick *et al.*[12]

The jet discharging from the nozzle unit is obscured by a conical envelope of mist. Laser photographs of the jet alongside a reference wire[13] show that the jet remains parallel at small standoff distances, e.g. 0·89 mm. At somewhat longer standoff distances jet divergence due to air drag becomes very noticeable. For some applications this is undesirable; it can be minimized by adding very small amounts of long chain polymers to the water.[14] In the present tests a mixture by volume of 99% water and 1% oil, normally used as a coolant on machine tools, was the cutting fluid. This oil was added as a lubricant; tests have not been run to ascertain whether the oil improves the coherence of the jet.

Figure 3 shows the rotary coupling[15] which has been developed for moving the nozzle unit with respect to the intensifier. The retaining screw is tightened to form a seal between the cone, shaft and body. When the coupling is pressurized the shaft and body tend to separate from the cone and act as a controlled clearance seal. If the screw is adjusted correctly the leakage is very small and the torque required to rotate the shaft is in the range 2 to 6 Nm. Other forms of this rotary coupling are being studied,[9] to date none have proved as reliable as the metal-metal seal shown in Fig. 3.

The equipment shown in Figs. 1–3 has been used to establish which nozzle arrangements are likely to be suitable for the precision slotting of concrete. The data presented herein was obtained with concrete blocks, 15 cm × 15 cm, having compressive strengths between 20·7 and 30 MPa. The aggregate was by volume: 65% dolomite, 15% limestone, 10% granite and 10% miscellaneous.

The slot depth was measured by inserting pieces of shim of 0·051 to 0·102 mm thickness; depth scales were scribed on the sides of these shims. At least twenty readings were taken along each 15 cm slot.

3. ANALYSIS OF SLOTTING DATA

Figure 4 compares previously published data[8] for a 0·178 mm nozzle with the results to date of tests with a 0·381 mm nozzle. Figure 4 shows average slot depths; variations due to the inhomogeneous nature of concrete are shown on Figs. 5 and 6.

Figure 5 also illustrates the influence of traverse velocity on slot depth. It shows that high traverse velocities are desirable if slot depth times traverse velocity is used as a measure of slotting efficiency.

The data presented in Figs. 4–6 was obtained with pressures generally within the range of water intensifiers now being offered commercially. Three cutting theories have been used to estimate the potential of operating at higher pressures and to correlate the data shown in Figs. 4–6.

Crow *et al.*[16] developed a theory for hydraulic rock cutting based on cavitation drag. The jet curves as it proceeds through any material and exerts a local average surface pressure, p_s. Crow *et al.* suggest that the state of flow is such that cavitation bubbles form behind protruding grains so that the downstream faces generally see the vapour pressure p_v of the water. Since the upstream face bears the surface pressure; p_s, there is a shear stress, τ, where

$$\tau = \mu_w(p_s - p_v) \qquad (1)$$

and μ_w is the Coulomb friction coefficient between water and rock for cavitation conditions.

This shear stress, τ, is related to the inherent shear strength of the rock, τ_o, by the Mohr-Coulomb failure criterion

$$\tau = \tau_o + \mu_r(p_s - p) \qquad (2)$$

where μ_r is the coefficient of internal friction and p is a pore pressure arising from subsurface permeation.

These and the flow equations are combined with Darcy's law for flow through a porous medium[16] and integrated with respect to the local incident angle of the jet stream, θ, to obtain an equation for the slot depth, h, where

$$h = \frac{2d_o p_o \mu_w}{\tau_o} \int_0^{\theta_o} \frac{e^{\mu_w(\theta - \theta_o)} \sin\theta \, d\theta}{1 + (V/C)\sin\theta} \qquad (3)$$

where the so called intrinsic velocity C is

$$C = \frac{k\tau_o}{nf\mu_r g} \qquad (4)$$

and the other parameters are as defined in the notation.

Crow *et al.* used equation (3) to correlate slotting data for Wilkeson sandstone. The quantities k, f, τ_o, μ_r and g were determined in ancillary tests; these gave $\tau_o = 11$ MPa and C = 33·8 cm/sec. Values for μ_w were obtained by curve-fitting; the best fit was obtained with $\mu_w = 2{\cdot}0$.

Equation (3) shows that for a particular traverse velocity and pressure the slot depth increases linearly with increase in nozzle diameter. Therefore, the data on Fig. 6 for traverse velocities of 5·08 and 22·0 cm/sec can be scaled by 0·381/0·178 to estimate the effect of operating with the larger, i.e., 0·381 mm nozzle. These scaled curves are shown on Fig. 7 together with a curve for a low traverse velocity, 0·25 cm/sec, calculated directly from equation (3) using values of τ_o, C and μ_w derived by fitting equation (3) to the data on Fig. 6. Since the quantities k, f, τ_o, μ_r and g are not easily definable for concrete, Crow *et al's* experience with Wilkeson sandstone was used as a guide when choosing combinations of τ_o and C. The best fit was obtained with τ_o = 11·8 MPa, C = 25·4 cm/sec and μ_w = 0·6. However, this is a very doubtful procedure and the 0·25 cm/sec curve on Fig. 7 must be regarded as a very rough estimate.

Mohaupt and Burns[13] used an energy balance to predict slot depth for homogeneous materials which deform plastically before fracturing at a constant fracture energy. This energy balance gives

$$h_1 = \frac{p_o - p_i}{(C_1/d_o) + (C_2 d_o + C_3)(V/q)} \tag{5}$$

where p_i is the threshold pressure below which the jet will not cut the concrete

q is the flow rate

and C_1, C_2 and C_3 are constants to be determined from actual slotting data.

If the material is brittle the energy absorbed by the plastic zone ahead of the cut will be very small and equation (5) reduces to

$$h_1 = \frac{p_o - p_1}{(C_1/d_o) + C_2 d_o (V/q)} \tag{6}$$

Figure 8 shows the two data points used to evaluate C_1 and C_2 and the predictions of equation (6) for a threshold pressure of 13·8 MPa. This equation was also used to estimate the potential of operating with a larger nozzle and higher pressures, cf Fig. 7.

Kinoshita *et al.*[17] used dimensional analysis to obtain an equation for slot depth in rock cutting

$$h = \frac{C_4 d_o W_w^{\alpha}}{W_t^{B} (1/d_o)^{\gamma} \phi^{\delta}} \tag{7}$$

where

$W_w = \dfrac{V_w^2 \rho_1 d_o}{\sigma}$, Weber's number of the water jet

$W_t = \dfrac{V^2 \rho_1 d_o}{\sigma}$, Weber's number of relative traversing velocity

$\phi = \dfrac{S d_o}{\sigma}$, strength coefficient

and the parameters ρ_1, S, V_w and σ are defined in the notation.

The constants α, B, γ, δ and C_4 in equation (7) must be derived from experimental data. The data presented herein is for one type of concrete. Therefore, the strength coefficient, ϕ, cannot be evaluated and it must be lumped with the constant C as a constant C*. This leaves four unknowns, C*, α, B and γ which were evaluated using the three data points shown on Fig. 9 and data obtained with a 0·381 mm nozzle operating at 345 MPa pressure and 5·08 cm/sec. Figures 9 and 7 show the prediction of equation (7) for the 0·178 and 0·381 mm nozzles respectively.

Figure 7 compares the predictions of the three theories for slotting with a 0·381 mm nozzle at pressures up to 690 MPa. Figure 7 suggests that a 0·381 mm nozzle operating at 690 MPa and a traverse velocity of the order of 0·25 cm/sec will cut a slot about 3 cm deep. Figure 10 shows that a similar slot depth can almost certainly be obtained at much lower pressures, 276 MPa, with three nozzles operating in tandem.

The power, $\dot{W}$, required to drive a single nozzle is

$$\dot{W} = C_D \frac{\pi}{4} d_o^2 p_o^{3/2} \sqrt{\frac{2}{\rho}} \tag{8}$$

This equation shows that three nozzles operating at 276 MPa require about 25% less power than one nozzle operating at 690 MPa. This and the increase in equipment reliability to be expected at lower pressures shows that there is much to be gained by operating several nozzles in tandem.

Figure 10 also shows previously published data for other standoff distances, pressures and nozzle diameters.[7,8] Momentum loss to the side walls of the kerf probably explains the reduction in the incremental depth per pass observed during the second and third passes. The cutting theory proposed by Mohaupt and Burns, equation (6), predicts this reduction. The constant C_1 is derived by considering friction losses.[13] Therefore, for a second pass the equations must be written as

$$\frac{C_1}{d_o}(h_1 + h_2) + C_2 d_o \frac{V}{q} h_2 = p_o - p_1 \tag{9}$$

where h_2 is the increment in slot depth produced by the second pass and h_1 is the slot depth produced by the first pass.

To facilitate comparison of equations (9) and (6) rewrite equation (9) as

$$h_2 = \frac{p_o - p_1 - (C_1 h_1/d_o)}{(C_1/d_o) + C_2 d_o (V/q)} \tag{10}$$

The extra term, $(C_1 h_1/d_o)$, in equation (10) should account for the difference between h_2 and h_1. Tests are required to verify this experimentally.

4. HYBRID SYSTEMS

This and previous studies by Olsen and the present authors,[8,9] show that jet slotting of concrete is possible at pressures within the limits imposed by fatigue and seal wear of existing continuous intensifiers. A preliminary economic analysis[9] for cutting anti-skid slots shows that continuous water jets compare favourably with flails. However, the aforementioned studies have shown that pressures in excess of 345 MN/m^2 (50 000 psi) are required to cut hard aggregate such as granite and quartz. Although a few uncut pieces of aggregate will not affect the performance of anti-skid slots, it is pertinent to ask if the efficiency of this and other concrete slotting operations would be improved by using a hybrid system consisting of a water jet and a mechanical cutter.

This approach has already been used by Maurer, Heilhecker and Love[18] to improve the efficiency of well drilling at depths up to 3900 m (12 800 ft). A series of full-scale field trials have shown that drilling bits, fitted with nozzles operating with drilling mud at pressures of 69 MPa (10 000) to 103·5 MPa (15 000 psi) can drill two to three times faster than conventional bits.

The high pressure bits use diamonds, drag blades or roller cones to assist in disintegrating the rock and to allow for drilling through hard materials not cut by the jets. Sintered tungsten carbide nozzles are used to reduce erosion by the sand-laden drilling mud. For example, six 3·3 mm (0·13 inch) diameter nozzles were used in a 22·2 cm (8¾ inch) diameter bit operating at 96·6 MPa (14 000 psi). To achieve this pressure, the 5·86 kg/l (10 lb/gal) mud was pumped at 1238 l/min (327 gal/min) with a power input of 2104 kw (2820 hp).

Studies are also under way to use continuous water jets to augment the performance of tunnel boring machines which use disc cutters to disintegrate the rock into small chips a few inches across. The rate of advance and life of disc cutters is limited by abrasion and the enormous thrust required to force the cutters into the rock face. If the disc cutter is preceded by one or more water jets, the thrust requirements are reduced because the rock is less able to resist the wedge effect of the disc in the water jet slot. Crow[19] has shown that maximum efficiency is obtained when several jets in tandem precede the disc cutter. For example, he has calculated that a 6·1 m (20 ft) diameter tunnel could be advanced in Colorado red granite at 3 m/hr (10 ft/hr) if the boring machine had 20 discs and 149 nozzles of 0·25 mm (0·01 inch) diameter operating at 257 MPa (37 200 psi). This implies a water supply of 326 l/min (86 gal/min) and a power input of 1395 kw (1870 hp).

The advances which have been made in well drilling and the studies in progress on tunnel boring suggest that a hybrid system should be considered for slotting and demolition of concrete structures and slabs. A mechanical cutter-water jet system might be very useful for deep slotting, particularly when the concrete is reinforced.

References

1. McCurrich, L. H. and Browne, R. D. 'Application of water jet cutting technology to cement grouts and concrete.' Proc. 1st Intnl. Symposium on Jet cutting Technology, Coventry, BHRA, April 1972.
2. Bresee, J. C., Cristy, G. A. and McClain, W. C. 'Some comparisons of continuous and pulsed water jets for evacuation.' Ibid B9–125.
3. Moodie, K. and Taylor, G. 'The fracturing of rocks by pulsed water jets.' Proc. 2nd Intnl. Symposium on Jet cutting Technology, Cambridge, BHRA April, 1974.
4. Singh, M. M., Finlayson, L. A. and Huck, P. J. 'Rock breakage by high pressure water jets.' Ibid 1, paper B8–113.
5. Cooley, W. C. and Clipp, L. L. 'High Pressure Water Jets for Undersea Rock Evacuations.' ASME paper 69–WA–UnT–7, Jnl. of Engineering for Industry, 1970.
6. Cooley, W. C. 'Rock breakage by pulsed high pressure water jets.' Ibid 1, paper B7–101.
7. Olsen, J. H. 'Jet slotting of concrete.' Ibid ref 3, paper G1.
8. Norsworthy, A., Mohaupt, U. H. and Burns, D. J. 'Concrete Slotting with continuous water jets at pressures up to 483 MPa (70 ksi).' Ibid ref. 3, paper G3.
9. Norsworthy, A. G. 'A study of jet cutting applied to concrete.' March, 1975, M.A.Sc. thesis, University of Waterloo.
10. Mohaupt, U. H. and Burns, D. J. 'Machining with continuous fluid jets at pressures of 2 to 7 k bar.' Ibid ref. 1, paper G4–37.
11. Crossland, B., Skelton, W. J. and Wilson, W. R. D. 'High Pressure Intensifiers: Development of a unit for pressures of 200 000 lb/in² and possible improvements in design.' Proc. Instn. Mech. Engrs., Vol 182, Part 3C, p. 11, 1967–68.
12. Chadwick, R. F., Kurko, M. C. and Corrivenn, J. A. 'Nozzle for producing fluid cutting jet.' US Patent 3 756 106, Sept. 4, 1973.
13. Mohaupt, U. H. and Burns, D. J. 'Machining unreinforced polymers with high velocity water jets.' Experimental Mechanics, p. 152–157, April, 1974.
14. Franz, N. C. 'Fluid additives for improving high velocity jet cutting.' Ibid. ref. 1, A7–93.
15. Mohaupt, U. H. 'High Pressure Rotary Coupling.' US Patent 3 877 732, April 15, 1975.
16. Crow, S. C., Lade, P. V. and Hurlburt, G. H. 'The mechanics of hydraulic rock cutting.' Ibid ref. 3, B1–1.
17. Kinoshita, T., Hoshino, K. and Takagi, K. 'Rock breaking with continuous high speed water jet streams.' Ibid ref. 1, B2–13.
18. Maurer, W. C., Heilhecker, J. K. and Love, W. W. 'High pressure drilling.' Jnl. of Petroleum Technology, p. 851, July, 1973.
19. Crow, S. C. 'Design principles for jet tunneling machines.' University of California, Los Angeles, Report ENG 7471, October, 1974.

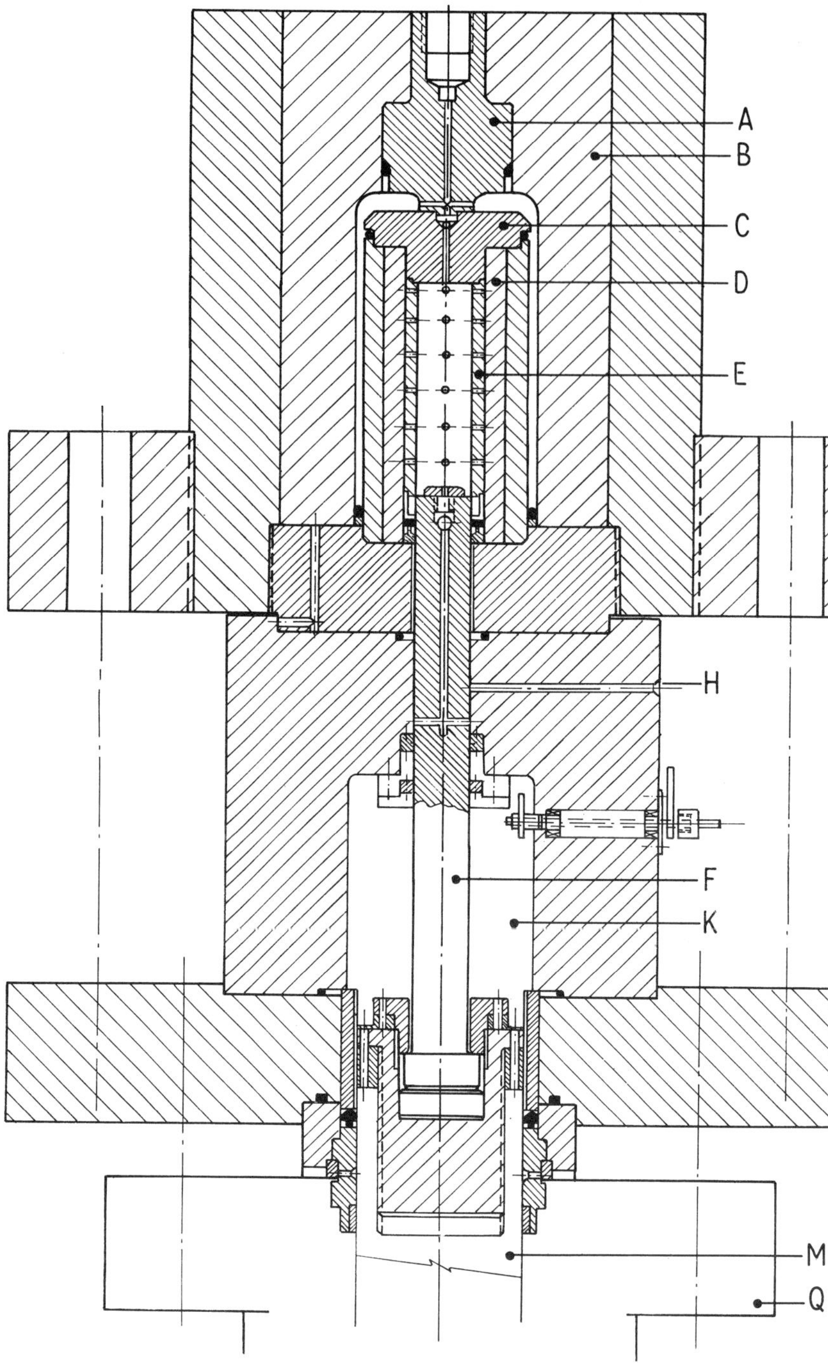

Fig. 1: Fluid support intensifier

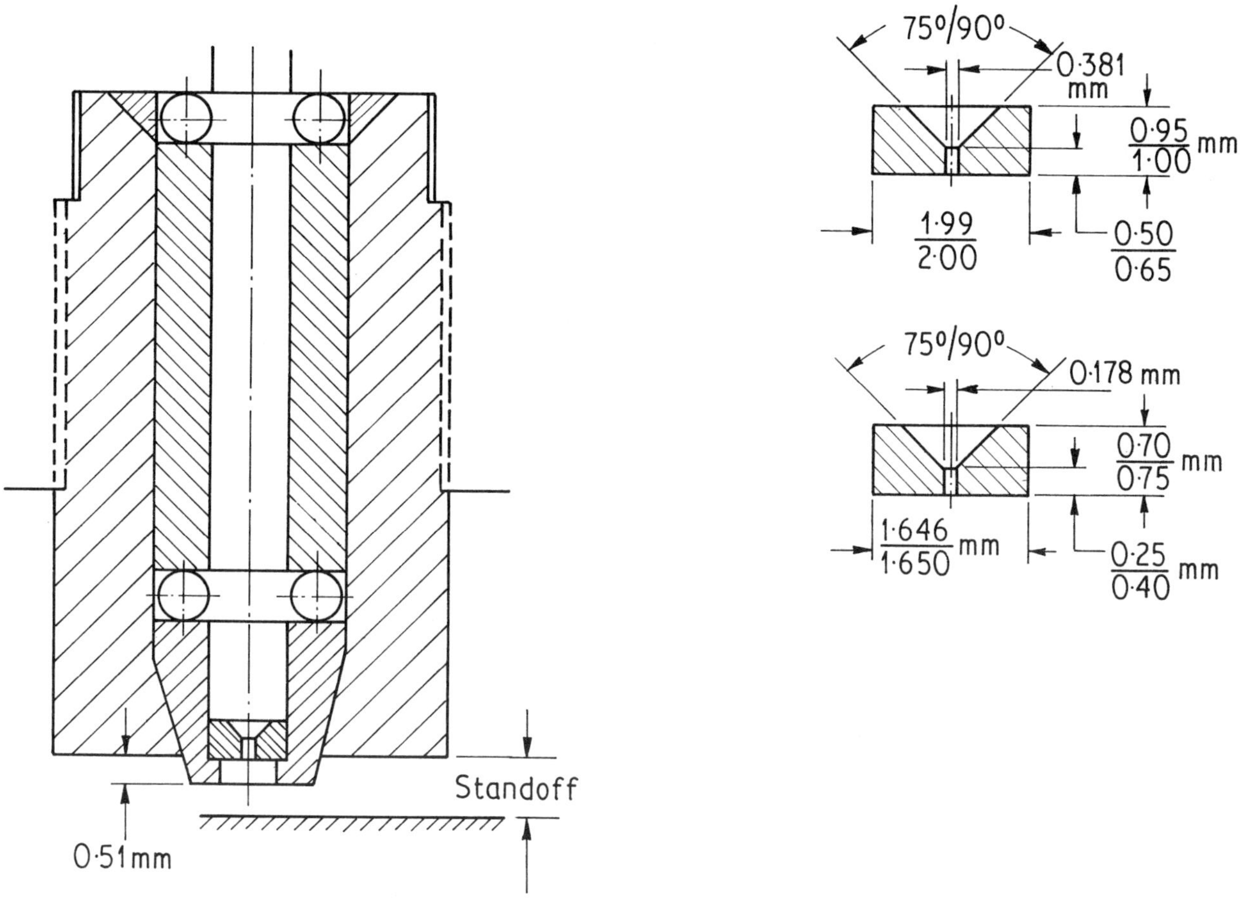

Fig. 2: Nozzle unit and sapphire nozzles

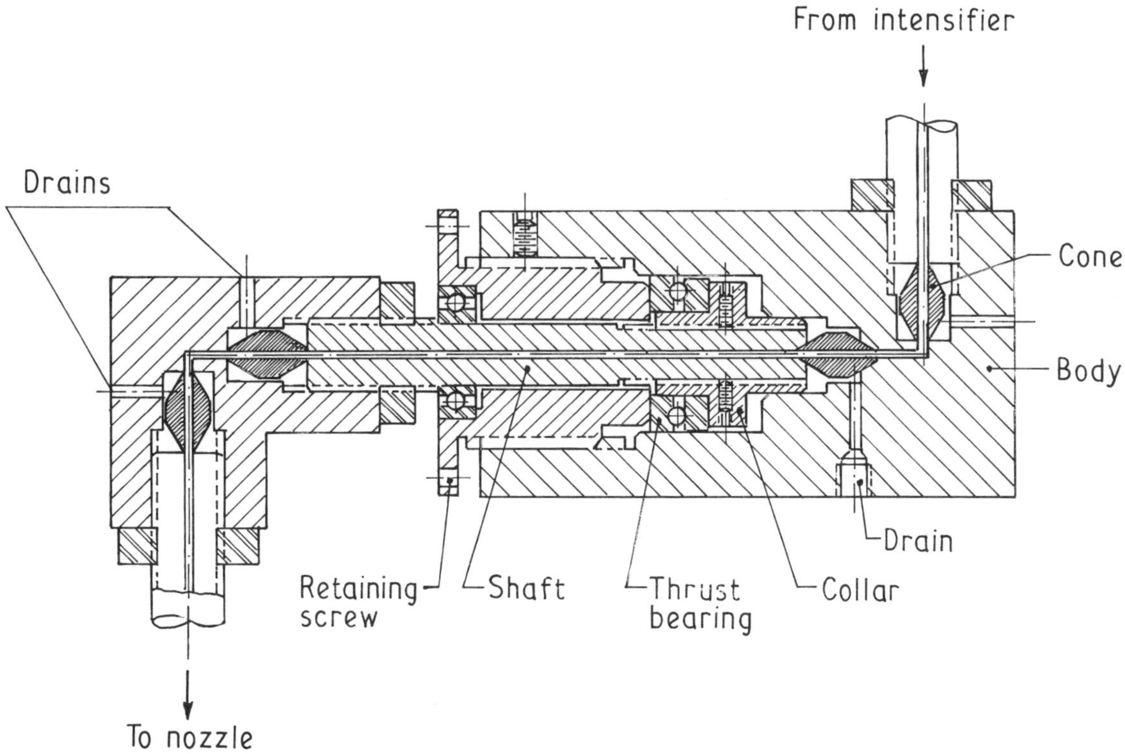

Fig. 3: Rotary coupling

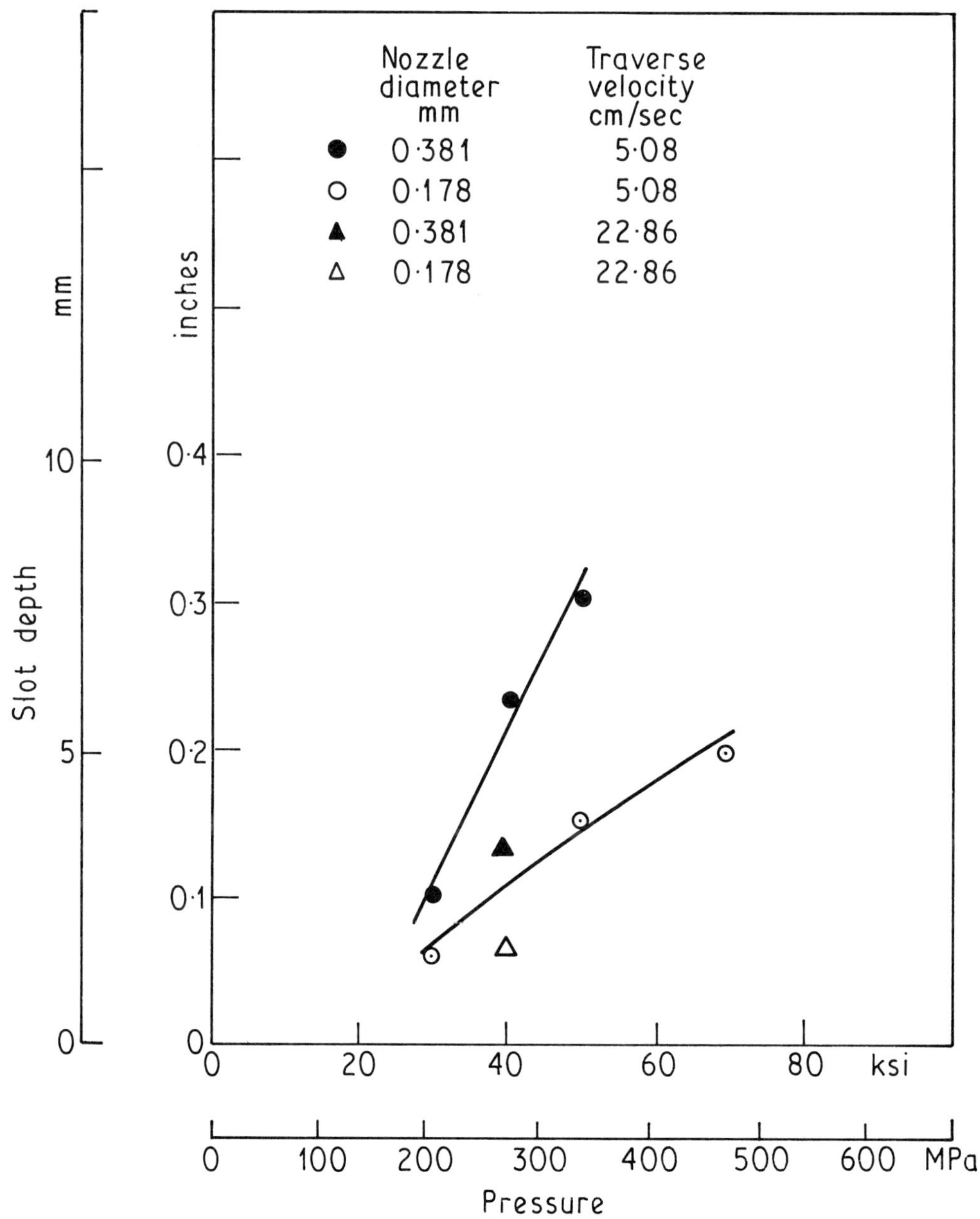

Fig. 4: Single pass with a standoff of 1.25 cm

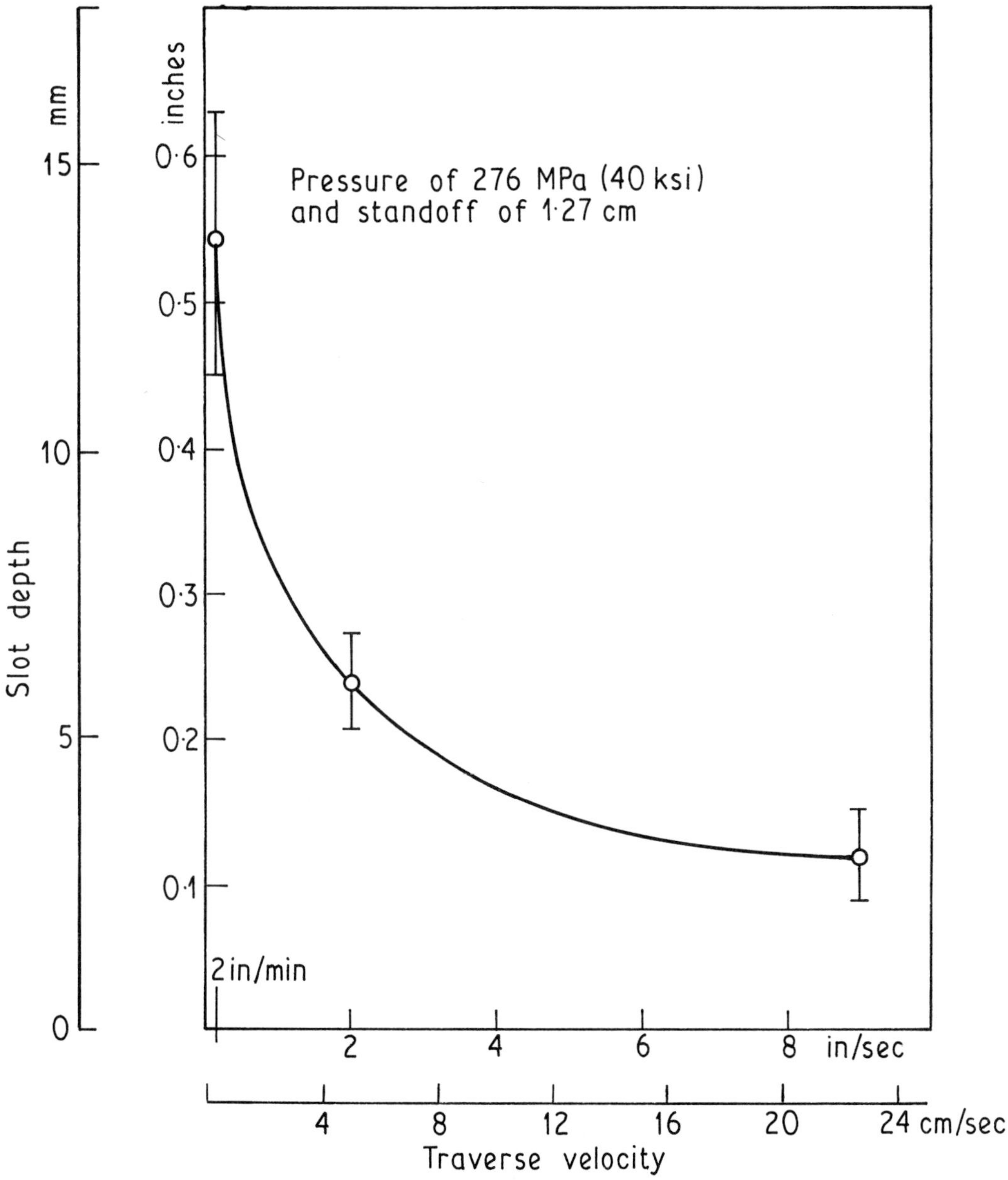

Fig. 5: Single pass with a 0.381 mm nozzle

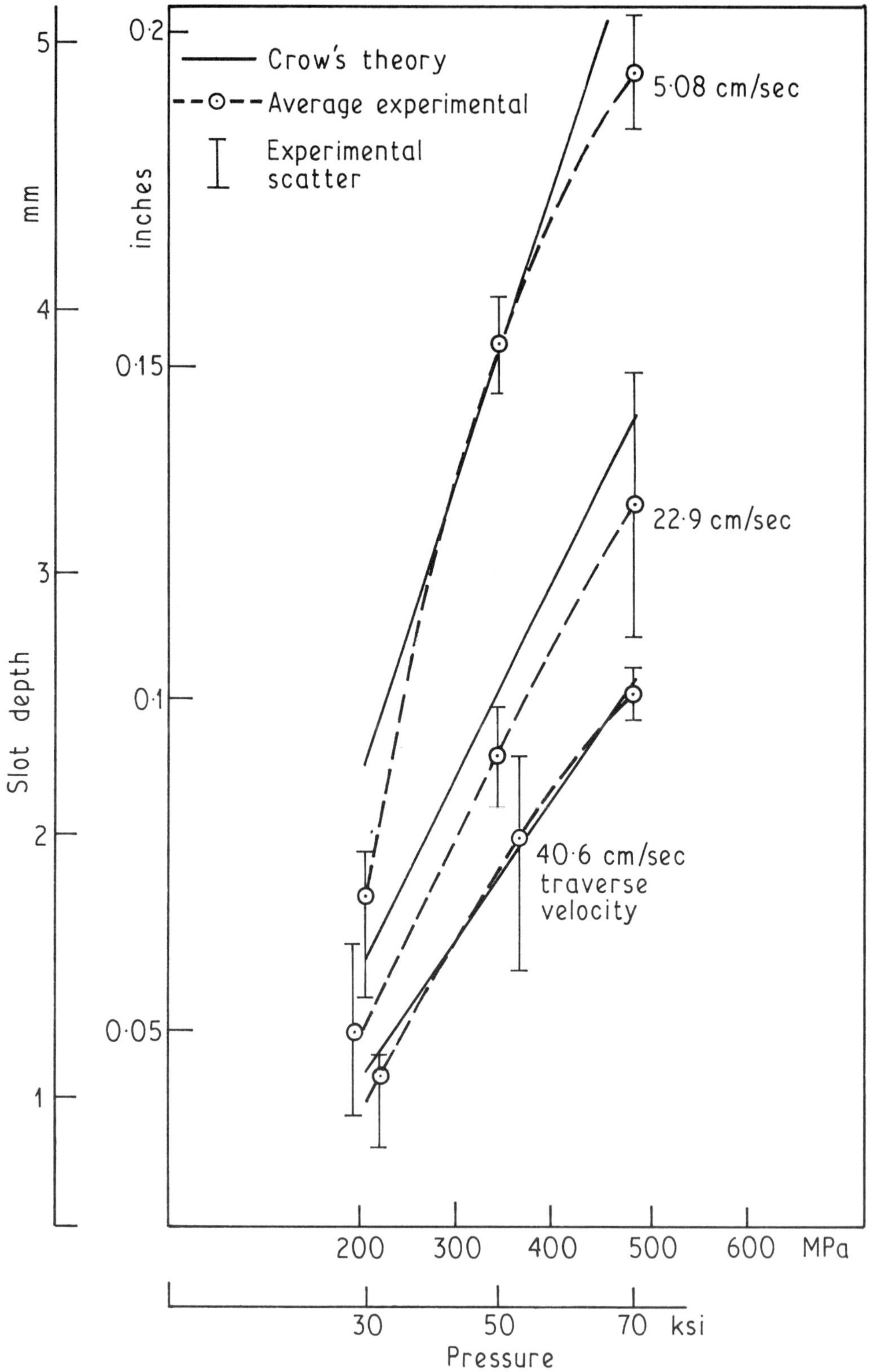

Fig. 6: Single pass with a 0.178 mm nozzle

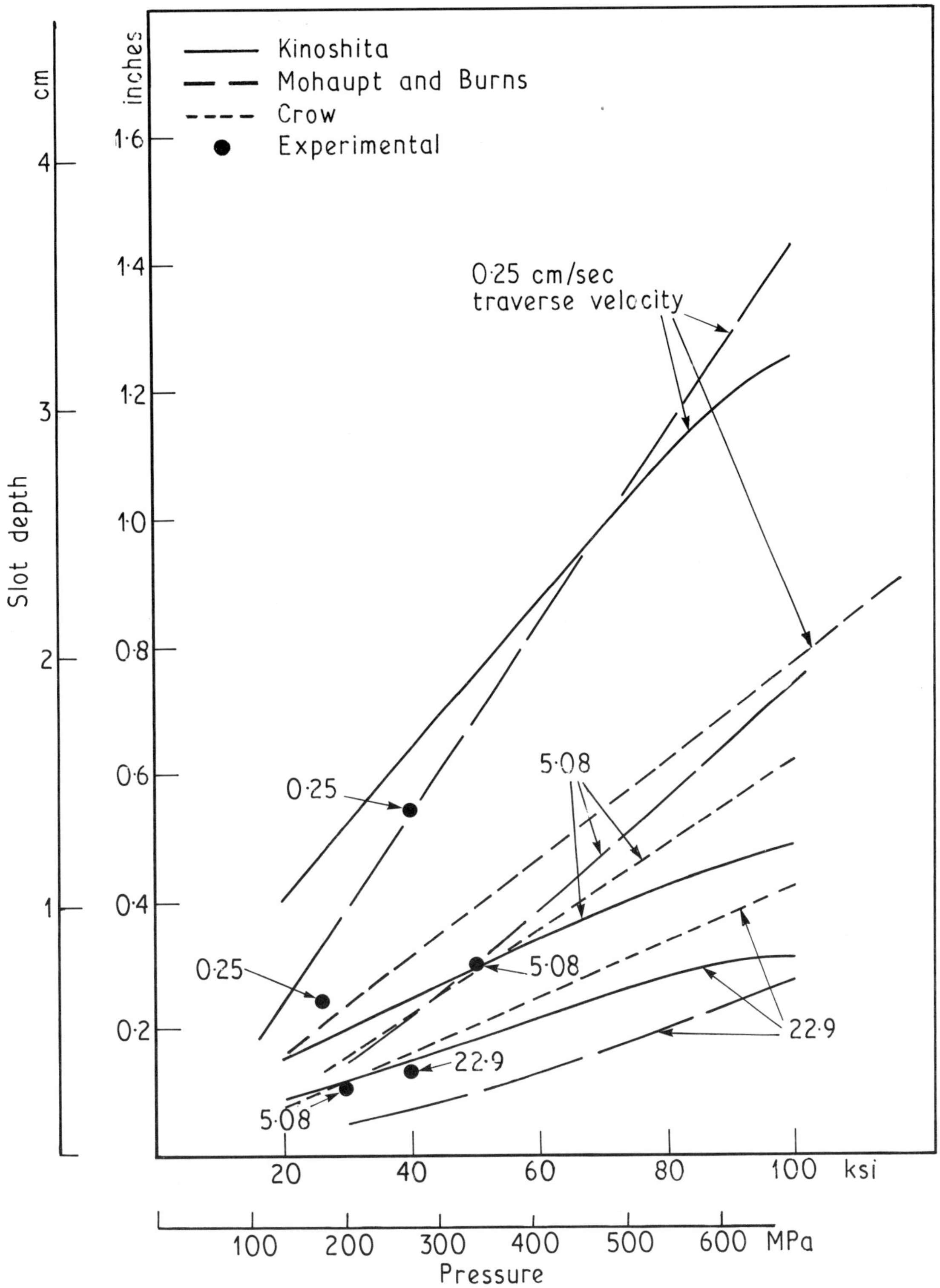

Fig. 7: Predictions for single pass with 0.381 mm nozzle

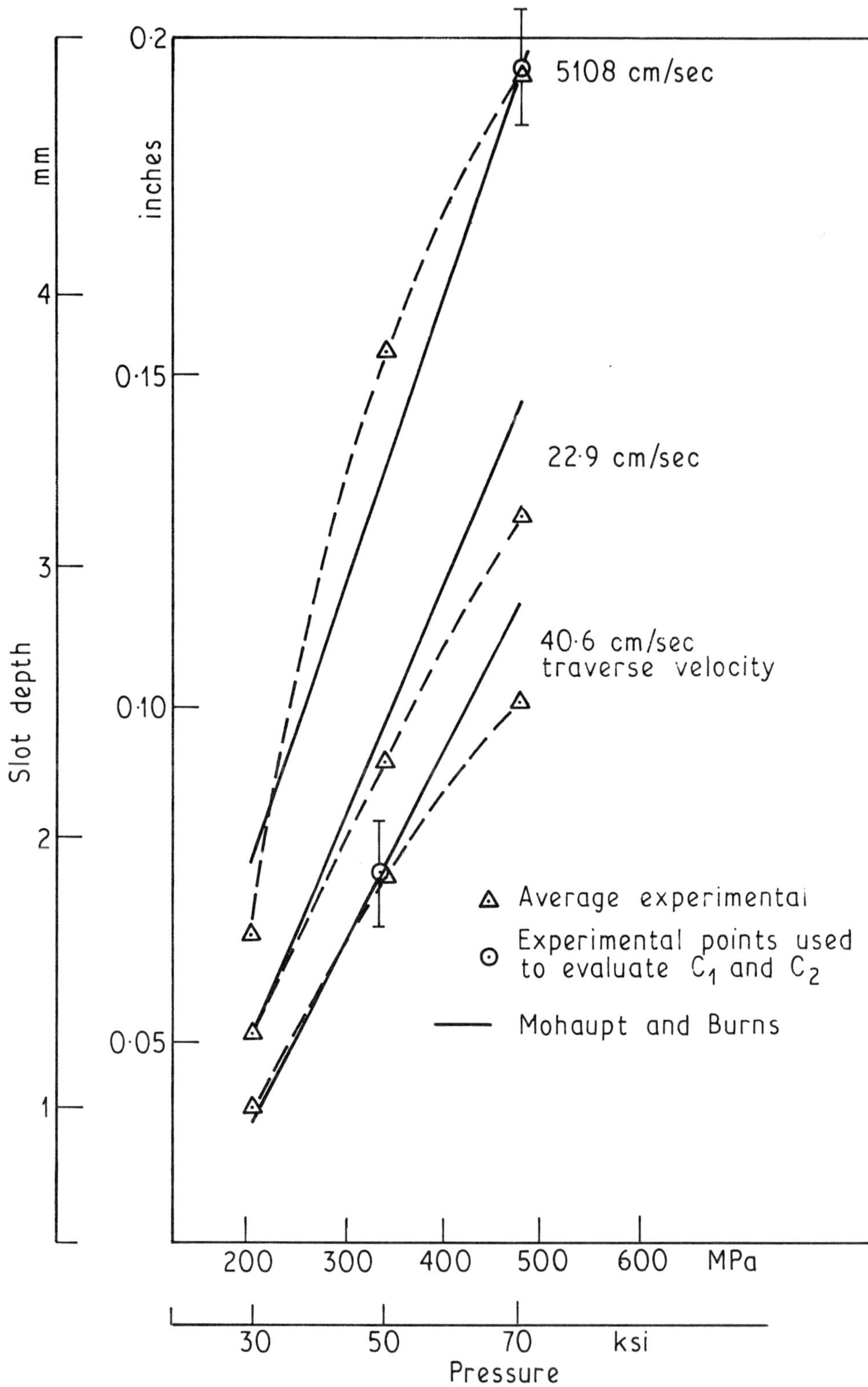

Fig. 8: Single pass with a 0.178 mm nozzle

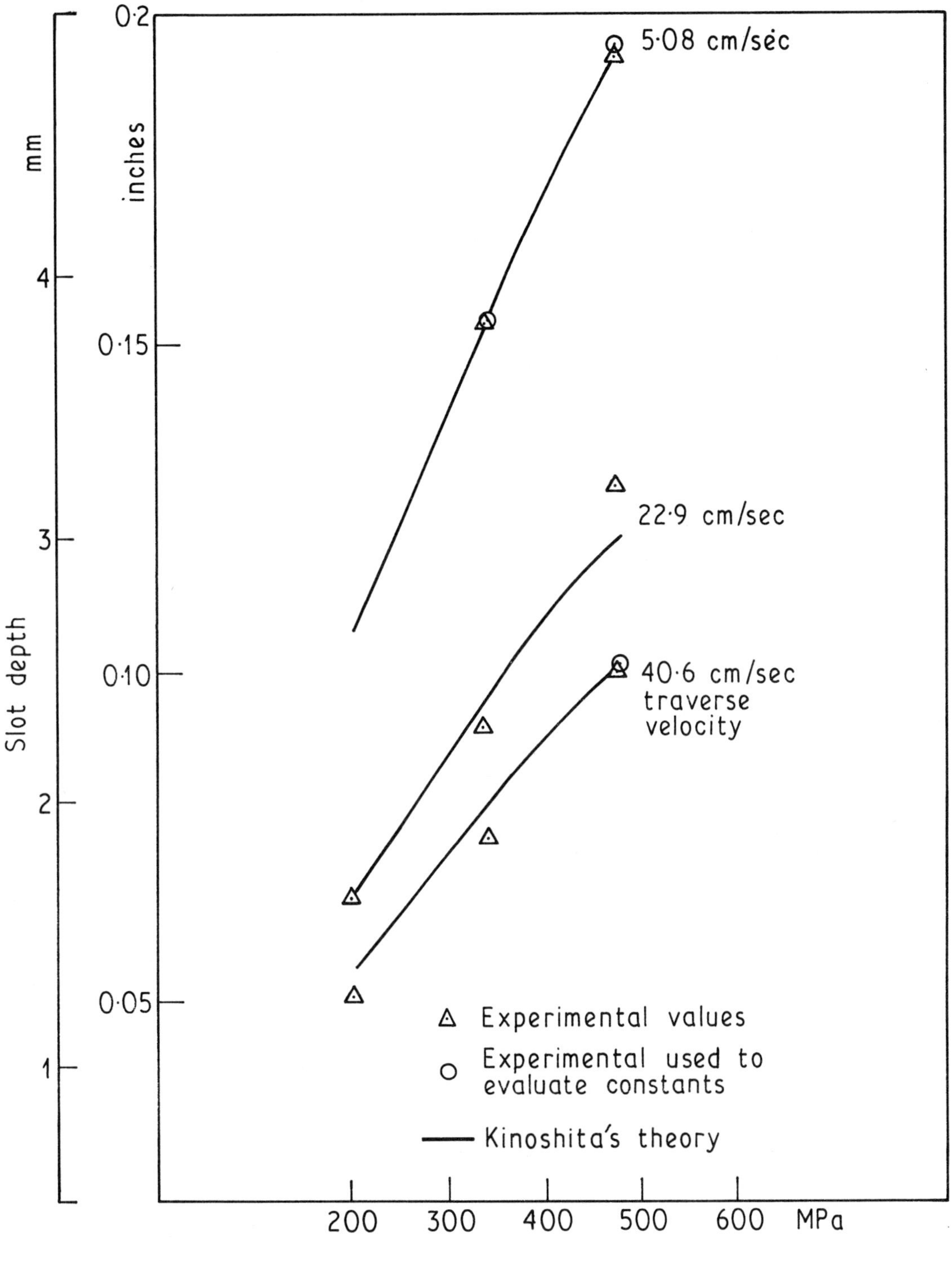

Fig. 9: Single pass with a 0.178 mm nozzle

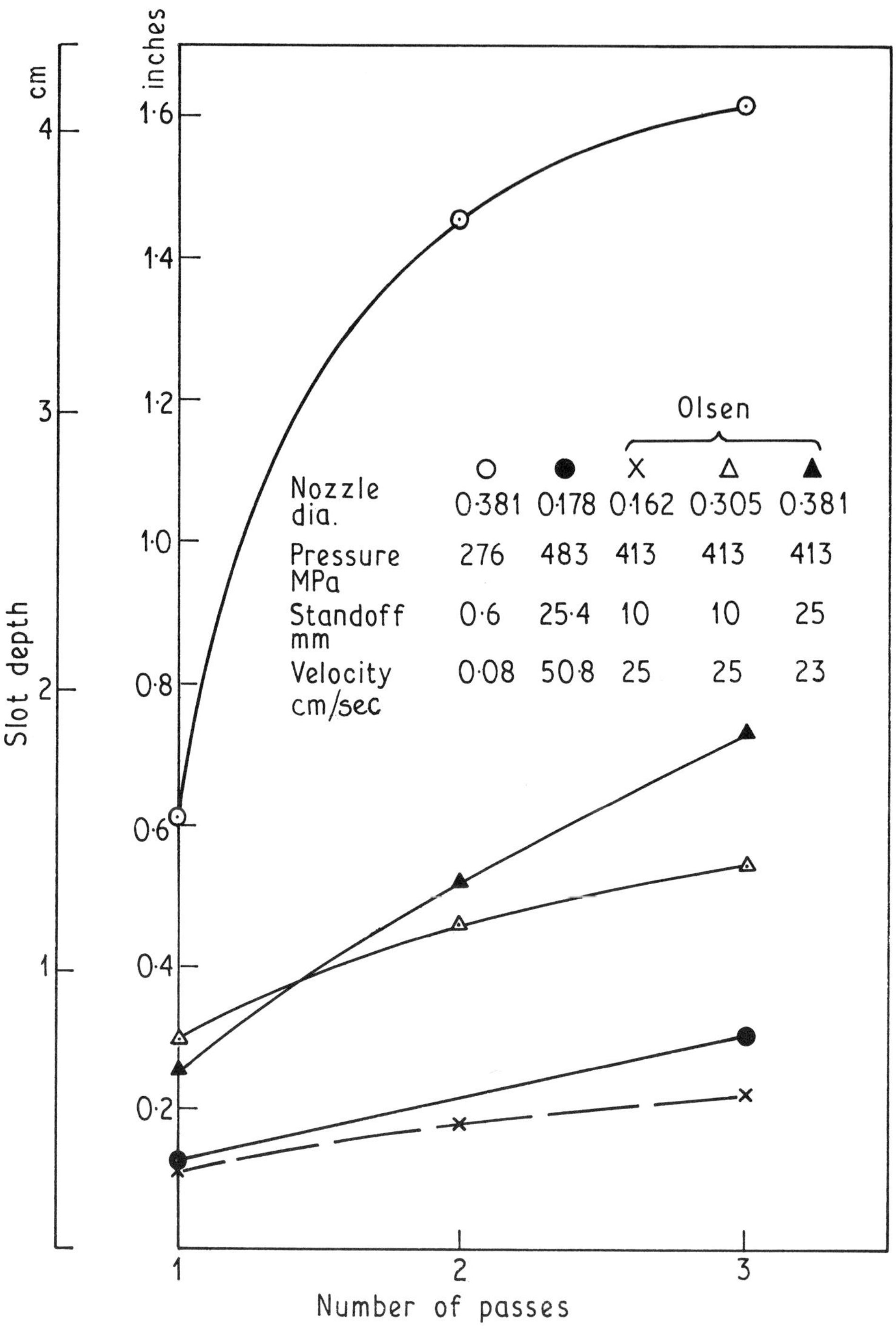

Fig. 10: Effect of multiple passes

EXPERIENCE WITH AN INTERNALLY-HEATED GAS-MEDIUM DEFORMATION APPARATUS TO 500 MPa (5 kbar)

M. S. PATERSON, ScD
Research School of Earth Sciences, Australian National University, Canberra

SYNOPSIS The requirement for higher temperatures, more flexibility and longer-term reliability for creep experiments has led to modifications in design of an apparatus for deforming rocks and minerals, especially in the furnace design and specimen arrangements. Temperatures of at least 1300° C are now available at 300 to 500 MPa confining pressure. The effectiveness of the internal furnace depends on minor details in design, the choice of suitable ceramics and the careful control of dimensions in combating heat loss by convection.

INTRODUCTION

The apparatus concerned is used primarily for studies on the plastic flow of minerals and rocks. Its main advantages for this work lie in its internal furnace, internal load cell and gas pressure medium. Although the pressure vessel and pumping system were designed for and initially tested to 1000 MPa (10 kbar), the apparatus has rarely been used above about 500 MPa and most use has been at 300 MPa. On the other hand, the direction of research has called for higher temperatures than originally aimed at, especially in order to reach the high temperature creep range of olivine and olivine-rich rocks. The melting point of pure magnesium olivine (Mg_2SiO_4) is about 1900°C and the work of Kohlstedt and Goetze (ref. 1) at atmospheric pressure has shown the desirability of reaching temperatures as high as 1600°C in order to produce on the laboratory time-scale the type of behaviour that is thought to be characteristic of that in the earth at lower temperatures on the geological time-scale. It is therefore of considerable geophysical interest to carry out experiments up to such temperatures, although in the rocks themselves a temperature limit of around 1200 to 1400°C is set by the partial or incongruent melting that occurs due to the presence of other phases. At the same time, it is important to be able to work at elevated pressures, not only to gain information on the relatively small effects of pressure on the processes of crystal plasticity and diffusion within individual crystals but also, in the case of the rocks, to avoid confusing complications from brittle phenomena that are characteristic only of behaviour at atmospheric or very low pressures.

A general description of the apparatus has been given in an earlier paper (ref. 2). In brief, it is based on a pressure vessel of 63.5 mm inside diameter, with axis arranged vertically. To deform the specimen, a piston of 31.8 mm diameter is moved through the upper closure of the cylinder by means of a hydraulic jack, fed from a variable-speed screw press which provides for strain rates in the range of 10^{-3} to 10^{-8} sec^{-1}. The lower end of the specimen bears against an anvil attached to the internal load cell mounted on the lower closure of the vessel. Only this anvil and a detachable extension piece attached to the upper piston enter the hot zone of the furnace where the specimen is situated and so only these parts need be made of heat-resistant materials (now normally sintered titanium carbide or alumina). The pressure medium, argon, is pumped into the vessel through high-pressure tubing from an external pumping system.

The aim of the present paper is to review interesting aspects of the operating experience with the apparatus and to describe some modifications that have improved the performance and extended the range of application, especially to higher temperature and with more diverse specimen arrangements.

SPECIMEN ARRANGEMENTS

The original high temperature arrangement allowed for only one specimen diameter, viz. 10 mm, and specimens of smaller diameter were sleeved out with copper to this diameter. The specimen was attached directly to the piston extension piece at one end and to an endpiece of similar bulk at the other end by means of the thin-walled (0.25 mm) copper jacket used to exclude the pressure fluid from pores or cracks in the specimen (illustrated in ref. 2); the advantage of this arrangement was that it could be completely assembled on the bench and introduced into the apparatus as a unit with the piston, and also independent control of pore pressure was possible through a hollow piston. However, to provide for several specimen sizes in this way is not convenient, partly because a much smaller specimen would tend to get damaged during assembly because of the rather massive pieces attached to it.

A new specimen arrangement now in use for compression tests is shown in Fig. 1. The endpieces to which the jacket is sealed are now minimal in mass and complexity and are therefore cheaper to replace and to provide in a range of sizes, a matter of importance when using the more expensive materials required at the highest temperatures. Since the upper piston extension and anvil are of standard sizes, the distance between the end faces of the specimen assembly, as well as its terminal diameters, are arranged to be approximately the same regardless of actual specimen diameter and length by dimensioning the endpieces suitably. The jacket is sealed to the endpiece as before by pushing on a ring with interference fit so that the wall thickness of the jacket is reduced by roughly one half under the ring, the aim being to ensure that all leak paths along scratches are blocked by the extruding jacket material. However, this procedure is not always successful and it is found that complete sealing is more easily achieved if the bore of the ends of the jackets is lightly polished (a domestic brass-polishing compound is suitable for this purpose).

When diamond drills are used to core the specimens from blocks or crystals, convenient sizes for specimens are 10, 7 and 5 mm diameter, with length twice the diameter, the actual size chosen depending on strength, grainsize, scarcity of material and suchlike considerations. However, in some cases, such as when polished faces are required or when a fine wire saw is used for cutting in order to conserve valuable material, it is convenient to use specimens of square cross-section; if the side of the square section is made equal to about 0.7 of one of the above diameters, that is, 7, 5 or 3.5 mm, then the specimen can be used with the same endpieces and jackets as for the round specimens since the jacket will normally collapse around the specimen without puncturing when the confining pressure is applied.

The choice of material for the endpieces is determined by the requirements that it be stronger in compression than the specimen and chemically inert with respect to it at the temperature of interest. Also it is desirable that the material has enough tensile strength to allow its use also for the sealing rings to ensure that they are matched in thermal expansion and compressibility to the endpieces. Above the range of highspeed toolsteel, special heat-resistant alloys, especially molybdenum-based alloys, are sometimes useful but they are not strong enough for many silicate rocks. Sintered carbides can provide adequate strength in such applications but they are unsuitable if water is present. An optimum combination of strength and resistance to corrosion is often given by dense alumina, although in this case sealing rings of a Mo-based alloy are required.

In general, copper can be used for jackets up to 1000°C and nickel beyond. If sulphides are present, gold can be substituted for copper but up to about 500°C a lining of gold foil of a few hundredths of a millimeter in thickness can be wrapped around the specimen to protect a copper jacket. The use of nickel jackets is limited to less than 1300°C when Mo-based rings are used on account of a Ni-Mo entectic being formed at about that temperature. Platinum jackets can be resorted to for higher temperatures.

FURNACE

Although the basic layout of the furnace has remained as previously described (ref. 2), many details have been changed in attempts to improve the efficiency and to obtain higher temperatures without raising the temperatures at the wall and upper closure of the pressure vessel. The design currently in use is shown, slightly schematically, in Fig. 2.

Vigorous convection of the dense gas tends to lead to overheating of the upper part of the furnace in this vertical configuration but by appropriately reducing the relative power supplied to the upper of the two windings, this tendency is readily counteracted and the hot zone arranged to be symmetrically disposed about the centre of the furnace. Moreover, by separating the two windings by an empirically chosen gap at the centre, the temperature gradient can be flattened over an appreciable length at the peak temperature. A relatively large free space of about 16 mm diameter has been provided in the hot zone of the furnace as a specimen chamber. This provision allows for large deformation of the specimen without damage to the furnace, including unexpectedly large lateral displacements of the specimen endpieces arising from anisotropic behaviour or the development of major shear zones (in extreme cases, the furnace may have to be disassembled to release the specimen but this is preferable to furnace parts being broken). The free space around the specimen is, of course, disadvantageous in respect to convection but the measures just described for controlling the temperature gradient have nevertheless been effective and so the procedure sometimes adopted of filling the space with a loose powder has been avoided, together with the accompanying inconvenience in inserting or removing the specimen and the possibility of the powder fouling moving parts elsewhere or putting unexpected constraints on the specimen.

It has been found that a very important factor in improving furnace efficiency is to reduce the diametral clearance between the bore of the furnace core A (Fig. 2) and the piston extension or anvil to a minimum (say 0.2 mm) which still allows for slight lateral movement of the piston under eccentric loading, it being insufficient to rely mainly on a seal ring of very small clearance at each end of the furnace for controlling convective heat transfer. In order to maintain the same free space around the specimen as previously, this requirement has led to the stepping of the internal bore of the furnace core as shown in Fig. 2. At the same time, in order to avoid the increased likelihood of accidental fracture of the lands between the winding grooves because of the smaller clearances, the winding is now placed on the outside of the core instead of inside as previously. The cores with windings are then encased completely on the outside to minimize convective heat loss in the radial direction. The material at present being used for the cores and the surrounding sleeves is dense alumina.

The dimensions of the furnace parts are adjusted so that the mating end surfaces of the end plates C, the flanges of the core tubes A, the sleeves B surrounding the core tubes, and the centre piece D are all in intimate contact. To maintain this contact throughout temperature cycling, another very important factor in achieving high efficiency in the furnace, the endplates C are springloaded. The gas seals E on the piston extension and anvil at their entrances to the furnace are also spring loaded against horizontal faces, allowing sealing to the maintained even though small lateral movements of the seal rings may arise in case of eccentric specimen behaviour; the diametrical clearance of these rings is minimal consistent with avoiding friction when hot (generally around 0.02 mm.)

The third important factor in achieving high furnace efficiency is the effectiveness of the insulation around the furnace core assembly. In choosing insulation for a furnace operating in gas at high pressure, several factors have to be taken into account:

(1) Because of the high density of the gas, convective movement of the gas through the pores of a porous insulator leads to much greater heat transfer than at atmospheric pressure (cf. measurements of Bett, Saville and Brown, ref. 3). Therefore, pores, if present, have to be small or otherwise of such configuration as to give rise to only a small permeability. The conventional insulating materials, often loose powders, used in furnaces at atmospheric pressure are likely to be unsuitable on these grounds.

(2) On the other hand, unless a fully dense insulation is used, the permeability presumably should not be so small as to lead to entrapment of gas in pores during pressure release, otherwise fracturing by decrepitation will tend to occur.

(3) Because of the rather massive character of the insulation to which the above considerations lead, the intrinsic thermal conductivity of the solid material itself is of great importance and should be chosen to be as low as possible.

An alternative approach, namely, to depend on the low thermal conductivity of the gas itself and to control the configuration of the gas-filled space so as virtually to eliminate convection by using a lamination of metal foils, has been used in an interesting design by Malbrunot, Meunier and Vidal (ref. 4). However, in the present apparatus, solid insulation has been relied upon.

The insulation consists of ground, concentric sleeves F fitted between the fired pyrophyllite end plates C, with radial clearances of about 0.02-0.05 mm between the sleeves. A dense mullite containing 15 percent glass (MV30 of McDanel Porcelain Co., Beaver Falls, Pa. USA) and fired pyrophyllite have been the most satisfactory materials used so far for the sleeves. However, a better material is still sought for two reasons, namely, to reduce the incidence of cracking and to provide a still lower thermal conductivity. The cracking of an insulation sleeve often leads to serious deterioration of furnace efficiency, probably due, in part, to increased convective loss through the crack and, in part, perhaps more importantly, to a distortion of the sleeve that leads to increased friction preventing the spring loading from keeping the parts of the furnace mated together and so to serious convective losses through the gaps. The cracking is probably due mainly to thermal stress associated with the temperature gradient within the sleeve, although some may have been due to stress transmitted from inner, hotter sleeves due to inadequate clearances; therefore the usual requirements for minimizing failure from thermal stresses, that is, low thermal expansion and elastic stiffness combined with high strength, are relevant as well as the low thermal conductivity, while care to raise the temperature gradually is also important. Measurements of power consumption indicate that fired pyrophyllite is a more effective insulator than the mullite, in keeping with their relative conductivity values at atmospheric pressure, roughly 1.3 and 2 watts per metre kelvin (0.003 and 0.005 c.g.s. units), respectively; however, the pyrophyllite appears to be somewhat more prone to cracking. It is intended in the future to test the suitability of a very fine textured, low density zirconium oxide as an alternative insulation material.

The procedure of monitoring the furnace temperature with thermocouples placed in the insulation near the furnace windings has been abandoned because the relationship between their readings and the specimen temperature proved to be very variable, presumably because slight changes in gas convection patterns from one heating cycle to another affect the local temperature gradients seriously. Therefore, the specimen temperature is now usually monitored by a thermocouple situated on the centreline of the piston extension at the upper end of the specimen endpiece, the temperature at this thermocouple being set at an appropriate value slightly lower (of the order of 10°C) than the desired specimen temperature; this differential is determined in calibration runs in which the axial temperature distribution in the specimen and vicinity is determined. The thermocouple, a metal-sheathed ceramic-insulated assembly of 1 mm outside diameter, is introduced along the hollow piston and through a pressure seal at the junction of the piston and piston extension. It is also usual to introduce similarly a second thermocouple with its junction about 10 mm higher than that of the first in order to monitor the temperature gradient; the difference between the two thermocouple readings is set at a previously calibrated value (also of the order of 10-20°), corresponding to minimum gradient in the specimen, by adjusting the relative currents in the furnace windings. One of the thermocouples is connected to a proportional temperature controller for long-term temperature stability for creep tests, although the previous expedient of supplying the furnace from a voltage stabilizer and depending on occasional manual adjustment is adequate for short runs.

The temperature in the gas space in the vicinity of the upper closure of the pressure vessel and the temperature just inside the furnace tube wall opposite the hot zone are monitored with two other thermocouples. It has been found desirable to limit the temperature in the top gas space to less than 200°C in order to avoid excessive absorption of gas in the O-ring seals. The furnace wall thermocouple indicates maximum temperatures between about 100 and 200°C or so, depending on the nature of the insulation and the condition of the furnace (the maximum temperature in the wall of the pressure vessel is probably up to 100° lower than this because of the temperature differentials at the intervening interfaces). As the furnace condition deteriorates and power consumption rises for a given specimen temperature, it has often been noted that the temperature in the top gas space rises while that at the furnace wall decreases.

With a furnace in good condition and with pyrophyllite insulation, a specimen temperature of 1200°C can be readily reached with top gas space and furnace wall thermocouples both reading substantially less than 200°C, and with a total power consumption of about 1500 watts. Under these conditions the temperature in the hot zone is constant over a length of 20 mm to within a few degrees (say 1-2° with very careful adjustment or 5-10° without taking great care) and to within about 20° or so over a length of 40 mm (which is the total length of the specimen and end-piece assembly currently in use, Fig. 1). Both platinum -40% rhodium and molybdenum windings of about 0.5 mm diameter have been found satisfactory for these conditions using the same diameter for upper and lower windings. The previously used 1 mm diameter Nichrome windings are only useful below about 1100°C since the windings run at a much higher temperature than the specimen itself when no cement is used to cover the windings. The temperature limit of the present furnace design has not been determined, but 1300°C has been reached with platinum-rhodium windings and it is hoped to reach substantially higher temperatures with molybdenum windings.

GENERAL OPERATING EXPERIENCE

The high pressure gas seals using polyurethane O-rings have continued to prove satisfactory in general although those subjected to elevated temperature have a more limited life because of greater gas absorption and subsequent swelling (at the highest temperatures, new rings may be required each run for the piston junction seal and for the piston seal in the upper closure). The type of seal used on the moving pistons (ref. 5) is illustrated in Fig. 3. However, in two situations difficulties have been experienced with this type of seal. Firstly, where it is used on the intensifier piston, which has a long stroke (380 mm) and is moved considerably faster than the pistons in the main pressure vessel, the O-ring tends to roll on part of its circumference, giving rise to a spiral leak path if there is slight prior damage or a moulding groove on the ring. As a result, this O-ring has to be replaced relatively frequently (ranging from a few runs to many tens of runs). The use of an O-ring of square section does not avoid the problem. The other situation in which the seal design of Fig. 3 has proved unsatisfactory is in a second pumping system, on the piston of a reciprocating gas pump that has been built on the compressed air drive unit of a commercial gas pump to give pressures of 300 to 500 MPa, above the commercially available range; in this case the O-ring disintegrates so quickly under the rapid stroking that it has to be replaced almost every run and a radical alternative is probably needed.

The condensation of water in the bore of the pressure vessel opposite the furnace continues to be a slight problem in spite of passing the gas through a drier and flushing the apparatus at the beginning of each run. It is now thought possible that adsorbed water on ceramic parts in the furnace may be a source and that the problem is aggravated by small clearance between the vessel wall and the tube housing the furnace, but more study is required. Meanwhile, corrosion pitting as severe as that which led to the previous failure (ref. 2) has been successfully avoided. The two pressure vessels now in use have also been tempered somewhat more than the one that failed (ref. 2), their hardness being in the range 46 to 48 Rockwell C; a hardness even a little lower could probably be used with advantage because the permanent growth of the internal diameter of these vessels upon initial pressurizing to 1050 MPa before final grinding was, at most, about 0.05 to 0.08 mm in the central region and less than 0.02 mm near the closures.

Internal load cells of the design previously described (ref. 2), a steel tube with electrical resistance strain gauges attached, have proved quite reliable provided only thin layers of the Allen PBX cement are used. Slight zero drift may occur initially but this is usually negligible after about half an hour. However, a slight drop in load is often indicated when the gas pump operates to restore the confining pressure to the set value if it has fallen slightly due to leakage. This effect can be attributed to stretching of the pressure vessel with slight change in pressure and is a genuine load fluctuation, not an artifact of the gauge. It is difficult to avoid completely without perfect gas sealing but it can be minimized by minimizing the differential between "on" and "off" positions of the pressure controller.

The apparatus has been used so far mainly for constant strain-rate experiments (cf. Fig. 4). However, some creep tests have been done using on-off contacts on the load recorder to interrupt the screwpress as necessary to maintain constant load, the screwpress being set to run slightly faster than necessary to maintain the load. It is intended to use the creep mode of testing more at the highest temperature and one of the aims in the developments described in this paper has been to improve reliability so as to make long-term creep tests more feasible.

REFERENCES

1. KOHLSTEDT, D.L. and GOETZE, C. Low-stress high-temperature creep in olivine single crystals. J. Geophys. Res., 1974, 79, 2045-2051.

2. PATERSON, M.S. A high-pressure, high-temperature apparatus for rock deformation, Int. J. Rock Mech. Min. Sci., 1970, 7, 517-526.

3. BETT, K.E., SAVILLE, G. and BROWN, M. Internal insulation of high-pressure vessels. Engineering Solids under Pressure, 3rd Int. Conf. on High Pressure, Aviemore, 1970 (Ed. H. Ll. D. Pugh). Institution Mech. Engrs, London, 1971, p. 38-42.

4. MALBRUNOT, P., MEUNIER P.A. and VIDAL, D. Design of furnaces for internal heating in high pressure gases. Chem. Eng. World, 1972, 7, 55-58.

5. PATERSON, M.S. O-ring piston seals for high pressure. J. Scient. Instruments, 1962, 39, 173-175.

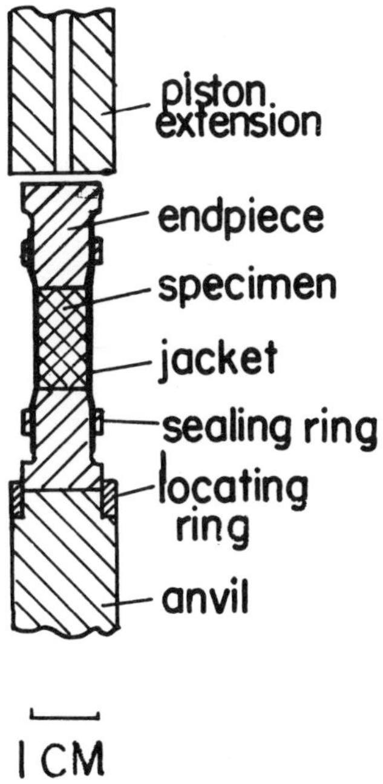

Fig. 1:

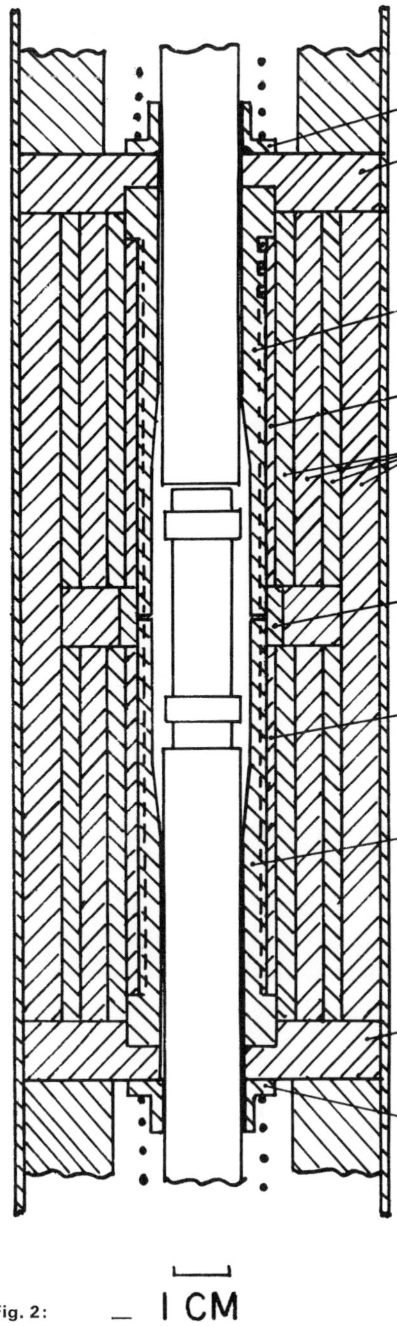

Fig. 2:

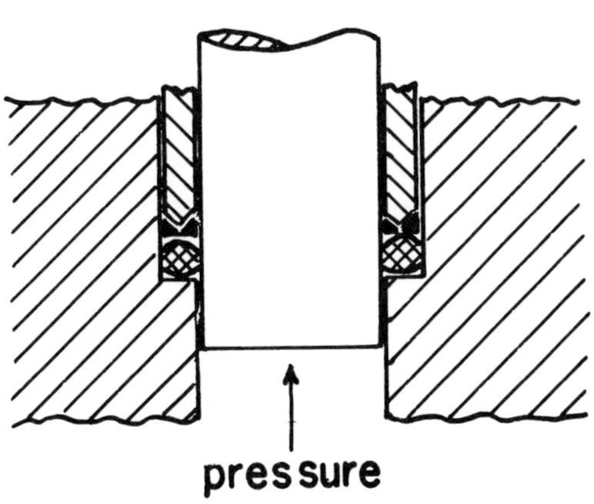

Fig. 3:

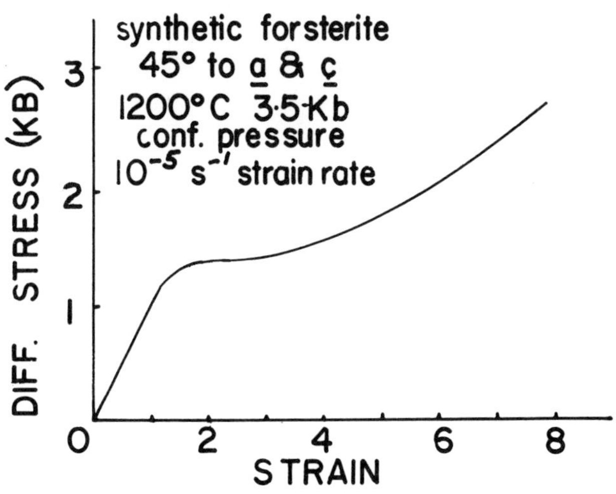

Fig. 4:

MEASUREMENT

TRANSDUCERS FOR VERY HIGH PRESSURES

V. E. BEAN, PhD
National Bureau of Standards, Washington, DC

SYNOPSIS Two new pressure transducers are currently being developed at NBS. In one the sensor is a parallel plate capacitor; in the other it is an ultrasonic delay line. Both have properties superior to the manganin gage and are being developed to be used as transfer standards in our calibration service.

INTRODUCTION

1. The ideal high pressure transducer would be stable, rugged, sensitive, and be free from hysteresis and temperature effects. Since no transducer is completely satisfactory in all these attributes, there exists an element of compromise in the selection of a transducer for a particular application. In the pressure range above 700 MPa the choice is virtually limited presently to the manganin resistance gage.

2. Manganin is an alloy of copper, manganese, and nickel tailored so that the temperature coefficient of resistance is zero near room temperature. It was originally developed for use in precision resistors. Since its resistance increases almost linearly with pressure, it has been widely used as a pressure gage. The art of building and using manganin gages has been discussed by Babb (Ref. 1). Two limitations of the manganin gage have led us to consider the development of other transducers. First, the resolution is limited in practice to about 10^5 Pa assuming one uses the finest commercially available Mueller bridge. And second, to obtain that resolution, a 4 lead resistance measurement technique must be used and there are pressure vessels where 4 well insulated electrical feed throughs are just not available.

3. There are two transducers presently under development at the National Bureau of Standards. In one the sensor is a simple parallel plate capacitor; in the other it is an ultrasonic delay line. We will first consider the capacitance gage.

CAPACITANCE PRESSURE GAGE

4. The change of capacitance with pressure has been shown to be very reproducible for many solid dielectrics by Andeen, Fontanella, and Schuele (Ref. 2). For developmental purposes in our laboratory the sensor is a dielectric disk 25 mm in diameter by 1 mm thick with vapor deposited electrodes on both faces. The high electrode has a guard ring separated from the electrode by a 0.01 mm gap to minimize fringing field effects. The capacitance is proportional to the product of the dielectric constant and the electrode area divided by the electrode separation all of which are functions of temperature and pressure. The pressure coefficient of the two geometric factors, the area and the separation, is simply the compressibility of the dielectric material. The pressure coefficient of the dielectric constant (and in this case, the term "constant" is certainly a misnomer) may be either positive or negative and at the present time there is no way to reliably predict it from theory. Anisotropic crystals may have two dielectric constants, each with its own pressure coefficient independent of the other, and all a function of crystallographic orientation.

5. The heart of this development program is finding the optimum dielectric material. The ideal material would have a huge pressure coefficient, a vanishingly small temperature coefficient and a zero loss tangent. It would be hard and non-porous, easy to polish to obtain a mirror-like surface; it would be nonhygroscopic, nonpoisonous, nonbrittle, have low vapor pressure, be chemically inert, free from polymorphism and inexpensive. Many materials have been studied; the temperature and pressure coefficients of some are plotted in Figure 1.

6. The choice of material is heavily dependent upon the scheme employed to minimize temperature effects. Since the temperature effects are large compared to the pressure effects, simply selecting any single material from the plot will result in a gage that is unacceptably temperature dependent. If there existed two materials with temperature coefficients equal in magnitude but opposite in sign, then one could make a capacitor from each and use them in parallel combination as the sensor thus effectively cancelling the temperature effects. Such a pair of materials is at present unknown.

7. The most promising temperature compensation scheme is the work of Dr. J. H. Colwell in our laboratory. His capacitance bridge measures the ratio of the capacitance of the sensor to that of a standard capacitor,

$$R = \frac{C}{C_{std}} . \tag{1}$$

The temperature coefficient of the ratio is

$$\frac{1}{R}\frac{dR}{dT} = \frac{1}{C}\frac{dC}{dT} - \frac{1}{C_{std}}\frac{dC_{std}}{dT} \tag{2}$$

The way to minimize the temperature problem is to make the standard of a material having the same temperature coefficient but a pressure coefficient widely different from that of the sensor and put both capacitors inside the pressure vessel in intimate thermal contact. In this case the two temperature coefficients need to be equal in magnitude and of the same sign to cancel. Such combinations do exist, for example, calcite cut parallel to the c axis and calcite cut perpendicular to the c axis, or one of the calcites and potassium chloride. This approach is the one presently being pursued.

8. An automatic capacitance ratio bridge has been designed and is now being constructed at NBS that has a resolution 1 part in 10^8 and a range of ±10% of the capacitance standard. It will greatly speed the research and will eventually be incorporated into the final pressure measurement system.

9. The capacitance gage is expected to have a resolution of 700 Pa. The pressure range is limited only by the pressure vessel used to house the capacitor. The biggest disadvantage is that three well shielded electrical feed throughs into the pressure vessel are required.

ULTRASONIC PRESSURE GAGE

10. The ultrasonic pressure gage was first introduced by Heydemann (Ref. 3). Its operation is dependent upon the change of the speed of sound in the sensor with pressure. The sensor is simply a right circular cylinder of quartz with plane and parallel ends and having an AC cut quartz electric-acoustical transducer bonded to one end with epoxy. When the AC transducer is excited by an electrical signal having a frequency near its resonance (nominally 10 MHz), it generates shear waves in the cylinder that are reflected at the other end and converted back to an electrical signal upon reaching the transducer. A gated pulse of 10 MHz carrier that is narrower than the round trip transit time will result in a train of echoes as seen in Figure 2.

11. The sensor is placed in one arm of an ultrasonic interferometer (Fig. 3). The output of the continuous wave oscillator is gated by the rf switch to provide the pulses which are amplified, routed via the circulator to the sensor, reflected, and returned to the circulator. The output of the circulator is the train of echoes (Fig. 2). After amplification, the signal is phase sensitive detected by the mixer using the original continuous wave as the reference. The output of the mixer is a series of dc pulses, one pulse for each echo in the train, with amplitudes proportional to the phase differences between the reference and each echo. With this output displayed on an oscilloscope, the echo to be used is selected and then the oscillator frequency is critically adjusted until the signal from the chosen echo is zero. If this null frequency is close to the resonance frequency of the AC transducer, then any phase change at reflection will remain essentially constant (Ref. 4) so that the null frequency may be expressed as:

$$f_n = \frac{c(P,T)}{2p\ L(P,T)} \left(n + \frac{1}{2}\right) \qquad (3)$$

where c(P,T) is the speed of sound in the cylinder at the ambient pressure and temperature, L is the cylinder length at that pressure and temperature and p and n are integers corresponding to the number of the echo used and the number of wavelengths corresponding to L respectively. The null condition is a function of pressure and temperature and the frequency change required to maintain the null is a direct measure of the change in sensor environment.

12. The interferometer has been automated so that the frequency will track the changes required by the sensor. The tracking interferometer has a stability of a few parts in 10^7 in frequency and is described in the literature (Ref. 5).

13. Once again we are faced with the problem of finding the optimum sensor material. At present the materials being used are AT and BT cut quartz. Single crystal alpha quartz is an anisotropic material with both positive and negative temperature coefficients depending upon crystallographic orientation. The sketch of a natural quartz crystal in Figure 4 (Ref. 6) shows the orientation of the Y-cut which has a temperature coefficient of 95 ppm K^{-1}. The Z-cut which contains the X axis and is normal to the Z axis has a temperature coefficient of -83 ppm K^{-1}. The temperature coefficient for any angle of cut rotated about the X axis is given by:

$$\frac{1}{f}\frac{df}{dT} = \alpha + \frac{1}{2C}\frac{dC}{dT} \qquad (4)$$

and

$$C = m^2C_{66} + n^2C_{44}\ 2mmC_{14} \qquad (5)$$

where α is the linear thermal expansion, C_{ij} are the elastic constants and m and n are the direction cosines. The anisotrophy in the thermal expansion is small and has been ignored. The temperature coefficient calculated from this equation using the data of Bechmann, Ballato and Lakaszek (Ref. 7) has been plotted as a function of orientation about the X-axis (Fig. 5).

14. The pressure coefficient also is a function of orientation and can be calculated from the expression

$$\frac{1}{f}\frac{df}{dP} = -\frac{K}{6} + \frac{1}{2C}\frac{dC}{dP} \qquad (6)$$

where K is the compressibility which has a small and ignored anisotropy. Using the data of McSkimin, Andreatch and Thurston (Ref. 8) the pressure coefficient has been calculated and is also plotted as a function of orientation about the X-axis (Fig. 5). The units on the right side of the plot pertain to the pressure coefficient. The AT cut (35°18') has a zero temperature coefficient and a pressure coefficient of 1.12×10^{-11} Pa^{-1}; for the BT cut (-49°13') the values are zero and $-1.57 \times 10^{-11} Pa^{-1}$ respectively.

15. In our present work we are using both AT and BT cut quartz cylinders which are 15.2 mm long by 8.9 mm in diameter. It appears that the ultrasonic transducer will have a resolution of 30 kPa. It requires only one insulated lead into the pressure vessel.

16. Both of these new transducers have properties superior to the manganin gage and are being developed to be used as transfer standards in our calibration service. The ultrasonic gage is to be used as the transfer standard for our 2.5 GPa controlled clearance piston gage (Ref. 9) and will make that unique device available as a primary calibration standard.

REFERENCES

1. BABB, S.E., Jr. "The Measurement of Hydrostatic Pressure", Mechanical Behavior of Materials Under Pressure, H. Ll. O. Pugh editor, 1970 (Elsevier Publishing Co., London).

2. ANDEEN, C., FONTANELLA, J., and SCHUELE, D. "A Capacitive Gauge for the Accurate Measurement of High Pressure", Rev. Sci. Instr., 1971, 42, 495-6.

3. HEYDEMANN, P.L.M. "An Ultrasonic Pressure Gage", J. Basic. Eng. 1967, 89, 551-3.

4. McSKIMIN, H.J. and ANDREATCH, P. "Analysis of the Pulse Superposition Method for Measuring Ultrasonic Wave Velocities as a Function of Temperature and Pressure", J. Acous. Soc. Am. 1962, 34, 609-15.

5. ANGEL, W.T. and BEAN, V.E. "A Tracking, Pulsed Ultrasonic Interferometer", Rev. Sci. Instr., 1975, 46, 533-5.

6. LACK, F.R., WILLARD, G.W. and FAIR, I.E., "Some Improvements in Quartz Crystal Circuit Elements" Bell System Technical Journal, 1934, 13, 453.

7. BECHMANN, R., BALLATO, A.D. and LAKASZEK, T.J. "Higher Order Temperature Coefficients of the Elastic Stiffness and Compliances of Alpha-Quartz", Proc. IRE 1962, 50, 1812-22.

8. McSKIMIN, H.J., ANDREATCH, P. AND THURSTON, R.N. "Elastic Moduli of Quartz versus Hydrostatic Pressure at 25° and -195.8 °C", J. App. Phys. 1965, 36, 1624-32.

9. JOHNSON, D.P. and HEYDEMANN, P.L.M. "Dead Weight Piston Gauge for Pressures to 26 Kilobars", Rev. Sci. Instr., 1967, 38, 1294-1300.

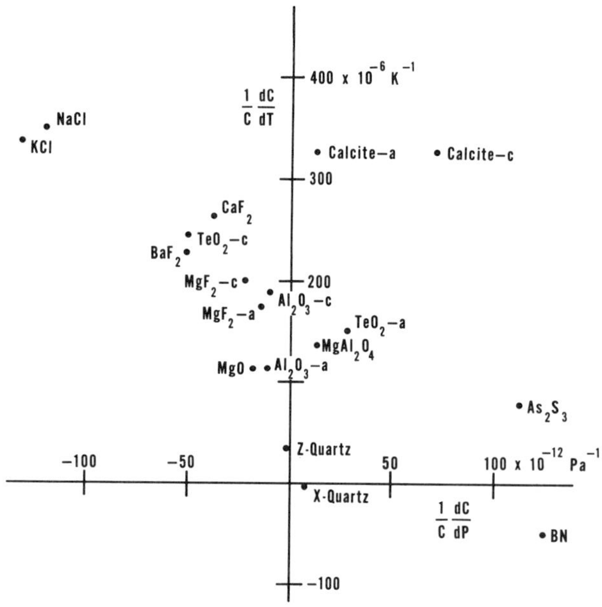

Fig. 1: Temperature and pressure coefficients of capacitance for selected materials. Measurements made parallel to the c and a axes are indicated

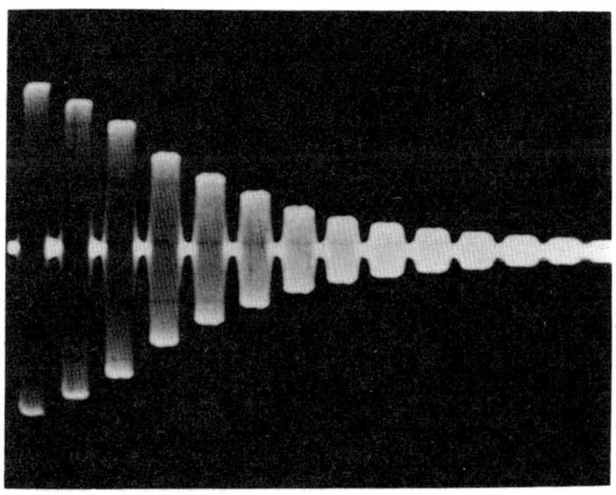

Fig. 2: Train of echoes from the ultrasonic sensor

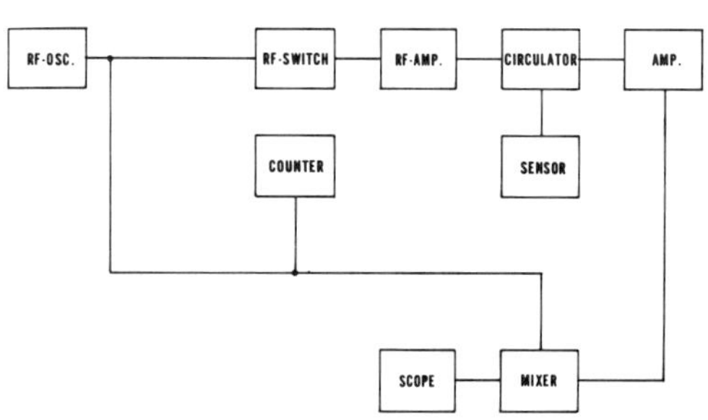

Fig. 3: Ultrasonic interferometer

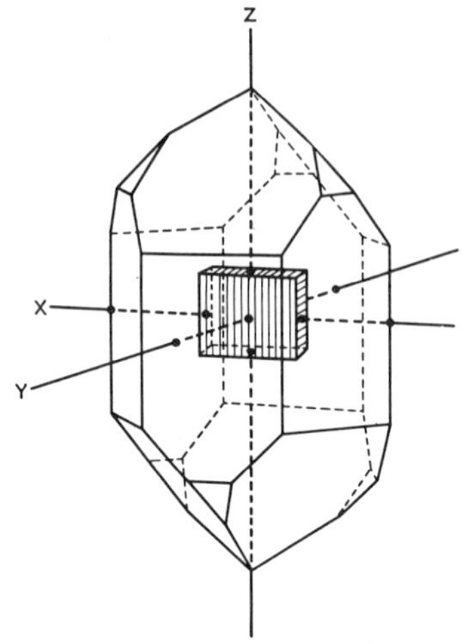

Fig. 4: Quartz crystal showing the Y-cut

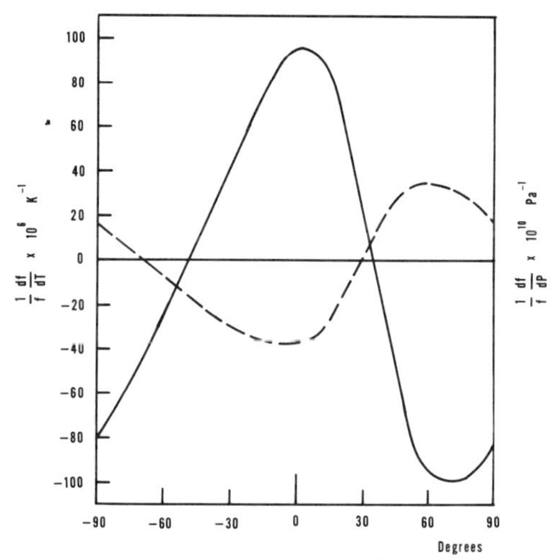

Fig. 5: Temperature and pressure coefficients of frequency of quartz as a function of angle of rotation about the x axis. The temperature coefficient is the solid line; the dashed line is the pressure coefficient

ꞮRANSDUCERS AND FLUIDS FOR HIGH ꟼRESSURE (1·4 GPa) SYSTEMS

. W. BIRKS, Senior Research Officer
epartment of Mechanical Engineering, The Queen's University of Belfast

ʏNOPSIS The development of high pressure apparatus for applications such as the testing of geological ɔecimens under high hydrostatic pressures, jet cutting and isostatic compaction has necessitated developɘnt of pressure, differential pressure and force transducers suitable for systems operating at pressures of ɔ to 1.4 GPa. It has also proved necessary to investigate the pressure transmission characteristics, ɐrticularly in small bore piped systems, of fluidswhich are used in high pressure equipment.

. The measurement of pressures of up to 1.4 GPa

The development of high pressure intensifier ystems capable of producing pressures of up to 1.4 ᴾa necessitated means of measuring and controlling ɹch pressures. Usually, the intensifier is ɔerated from a relatively low pressure supply and ı many cases it is necessary to be able to control ɿe high pressure with precision. Consequently some ɔrm of pressure transducer is needed which must be ɘlatively insensitive to temperature as inevitably ɿe compression of the pressure transmitting fluid ɐises its temperature.

. Manganin wire pressure sensors

The use of the pressure sensitivity of anganin[1] wire as a secondary pressure standard was roposed by Lisell in 1903 and employed extensively y Bridgman[2] and many other experimenters. Manganin as been used for measuring steady state pressures nd for the measurement of dynamic pressure pulses f very short time duration[3]. In the first case ɔils of wire are generally used, in the latter case thin foil resistance element supported on an epoxy esin backing, and very similar to the conventional ɔil strain gauge, is used.

The design of a typical Manganin sensor is hown in Fig. 1. In the form shown the coil is oused in a bellows which contains an electrically on-conducting fluid (e.g. Plexol 201 or petroleum pirit) this allowing the sensor to be used in ystems containing conducting fluids such as brake luid. In the present design the bellows are etachable to allow access to the coil assembly, eals and electrical lead-throughs. Once assembled, owever, it is vacuum filled with the non-conducting luid. If the transducer is to be used in systems n which only non-conducting fluids are used then bviously the bellows assembly is not required. lectrical leads are brought out via tapered high trength steel pins in precision ground ceramic ones which are lapped into tapered holes in the nd plug. Normally a temperature compensating coil s incorporated in the cover plate of the sensor. his provides a degree of compensation for ambient emperature changes but cannot compensate for hanges in the fluid temperature other that is than low changes over a long period of time. Other rrangements have been made, using additional leadıroughs, for three or four wire resistance measurements to be employed or in cases where two separate active elements are used one for control purposes and one for precise measurement of pressure.

The whole unit is housed in a high-tensile steel cylinder with an outer sleeve of low strength, high fracture-resistant steel which is intended to contain the inner component in the event of failure. The coils are wound loosely using selected double-silk-covered Manganin wire, a small ceramic former providing support. Bifilar winding is used to give non-inductive properties. To remove the winding stresses, each coil undergoes a fairly long course of heat treatment which has the effect of quickening the natural ageing process and leaves the wire in a very stable condition. The wire is selected on the basis of its low temperature coefficient of resistivity, typically 5 x $10^{-6}/^{o}C$ at 20°C. Both the active and temperature compensating coils are adjusted to a resistance of 100 Ω within a tolerance of ±0.1%. Other values can be employed if desired.

A considerable programme of testing has been carried out up to pressures of 1.43 GPa including sudden release of pressure. This did not appear to cause any damage to the coil and no distortion of the bellows was observed after ten releases of pressure from 1.43 GPa in 5s and five releases in 1s. There was, however, a tendency for small particles of heat-treatment scale to flake off the bore of the holes through which the terminals pass and cause short circuiting of the coils. This was overcome by polishing any troublesome scale out of the holes following up by ultrasonic cleaning.

Calibration of the Manganin coils is normally carried out at pressures up to 689 MPa (100000 lbf in^{-2}) using a Budenberg deadweight piston gauge and at pressures up to 1.38 GPa (200000 lbf in^{-2}) using a Harwood controlled-clearance piston gauge[4, 5]. It has been found that the zero pressure resistance value changes during the first few pressure cycles (Fig. 2)apoint also noted by Bridgman[2] and all gauges are therefore given five cycles of pressure to 1.38 GPa before final calibration to 'condition' the wire.

The pressure sensitivity for 38 s.w.g. Manganin is typically 2.39 x 10^{-8} ohm/KPa, the non-linearity appears to be less than 0.2% of full range but according to Dadson[6] this is probably less than the error in the piston gauge itself.

Other gauges of wire exhibit slightly different pressure sensitivities but the differences are not consistent with wire diameter and could be accounted for by very small differences in the composition of the wire.

A temperature coefficient of $\pm5 \times 10^{-6}/^{\circ}C$ at $20^{\circ}C$ is about the best that could be expected in Manganin wire (Fig. 3). This may seem small but it is significant when considered in terms of the pressure-resistance coefficient. The temperature changes due to compression or expansion of the pressure transmitting fluid are appreciable and it takes a considerable time before temperature equilibrium is achieved. For example, if the pressure is reduced to atmospheric from 689 MPa the resistance is still changing by a small but significant amount after 30 min. The time required for the resistance to stabilise is dependent upon both the rate of change of pressure and the range through which the pressure changes. Gold-chrome wire[7] has been used as an alternative because of its low temperature coefficient but its pressure sensitivity is only about one third that of Manganin. Another wire, similar in composition to Manganin, is Zeranin (Isabellen-Hutte KG), well known to manufacturers of precision high stability wire-wound resistances. This wire has a very low temperature coefficient of resistivity, this being virtually zero between $0^{\circ}C$ and $70^{\circ}C$ (Fig. 3) and has been found to have a linearity comparable to that of Manganin in its pressure/resistance relation.

The pressure sensitivity for Zeranin is about two-thirds that of Manganin but, because of its low temperature coefficient, stable resistance values are attained after similar pressure changes in 3 or 4 min, compared with 25 to 30 min required with Manganin. Calibration to 689 MPa has been established using a Budenberg deadweight piston gauge and is presently being continued to 1.38 GPa on a Harwood controlled-clearance piston gauge.

Other resistance wires also having extremely low temperature coefficients have been developed for the production of high stability resistors. Examples of these are Evanohm (USA) and Telprecist (GB) but tests have yet to be carried out on these materials. The results of such tests should prove interesting.

3. Instrumentation for use with Manganin sensors

With Manganin sensors it is necessary to measure the resistance in much the same way as with a resistance thermometer and similar methods are applicable. Self heating (I^2R) of the wire must be minimised and thermal effects on the leads must be considered hence the bridge measuring system must operate at low excitation and incorporate lead compensation. Precision Wheatstone bridge measurements may be used with either 3 or 4 lead configuration, generally using a low level d.c. source. However, other precision bridges such as Carey-Foster type may be used, or low frequency a.c. ratio bridge methods may be preferred. All these methods are capable of high precision and excellent resolution but nevertheless they are time consuming and necessitate referring the values of resistance to a calibration curve in order to determine pressure. The accuracy of pressure measurement can be to with in ±0.1% but in general it is more convenient to have a bridge which is direct reading in pressure units and which has a sensitivity factor adjustment.

It is possible to design such a direct reading bridge[8], but it is important to realise that the circuitry used in making the conversion from resistance to pressure assumes a perfectly linear relationship between these two parameters. Such a bridge may offer practical advantages but can neve achieve an accuracy of better than ±0.5%.

The bridge methods referrec to are all 'null-balance' methods, an alternative is to use an 'unbalanced' bridge configuration, a technique well known to users of resistance strain gauges. Here the e.m.f. developed across a Wheatstone bridge du to a change in resistance of one (or more) arms is a direct indication of change in resistance or str (in our case, pressure). However, non-linearity o the single active arm unbalanced bridge must be cor sidered, particularly where the change in resistan is relatively large as is the case with a Manganin sensor. Furthermore, the output voltage is a function of the input voltage, that is the bridge excitation, and a highly stable regulated power su must be used. The bridge must also operate into a very high impedance to reduce loading effects to a minimum. Non-linearity may be reduced by using hi value ratio arms but this is at the expense of sensitivity, or some improvement can be effected b using a constant current source instead of the more usual constant voltage. Compromise can result in a bridge system in which the non-linearity does not exceed 1% for the resistance changes encountered.

Provided only slowly changing system pressures are to be measured and controlled, Manganin or Zeranin pressure sensors are reasonably satisfactor However, where rapid pressure changes are involved such as for example, in isostatic compaction where quick cycle times are desirable, the method is unsatisfactory. Thermal errors due to pressurising and de-pressurising the fluid are considerable and some alternative method of measuring pressure is necessary. For such purposes a resistance strain gauge pressure transducer has been developed.

4. Strain gauge pressure transducer

The strain gauge transducer is shown in Fig. 4 The pressure sensing element is a thick-walled cylinder with a K ratio of 3 to which the temperatu compensated foil type electrical resistance strain gauges are bonded. A high strength nickel-chrome-molybdenum-vanadium steel is used for this member which is autofrettaged at a pressure of 1.5 GPa (215000 lbf in^{-2}) and then subjected to a low tempe ature heat treatment to stabilise the material. A mild steel protective cover is provided which incor porates the electrical connector at one end, provid environmental protection for the strain gauges and wires and also acts as a safety shield. Oblique bleed holes are provided in the mild steel shield i case of fracture of the pressure element and these are plugged with a soft silicone rubber to prevent ingress of dirt and moisture. The strain gauges ar arranged to give a full bridge configuration with t active and two complementary gauges, the bridge resistance being 350 ohms. With an excitation of 1 v d.c. the nominal output at full range pressure (1.38 GPa, 200000 lbf in^{-2}) is 15 mV.

To test the effectiveness of the shield around the sensing element, a deliberate burst test was carried out with the transducer connected to a fairly large volume system. The fluid was discharged safely through the oblique ports and no damage to the electrical connector was observed.

Initial calibration tests to 1.38 GPa showed a slight zero shift in the output for the first three applications of pressure after which no further shift was observed. This effect is well known to users of bonded resistance strain gauges although the causes are open to debate. It is thought to be due to either, or both, redistribution of residual strains and initial straining of the gauge foil, backing and bonding resins. Conditioning is always performed and in the case of the pressure transducer this takes the form of applying five cycles of zero to full rated pressure, after which final calibration is carried out.

Calibration is carried out on a Harwood controlled clearance piston gauge to 1.38 GPa (200000 lbf in^{-2}), the transducer output being measured by means of a precision digital voltmeter checked against a standard cell. For the purposes of calibration the transducer supply is adjusted to 10.00 v d.c. The calibration data (mV against pressure) is processed in a similar manner to the manganin sensor data, i.e. a best straight line fit passing through zero and the sensitivity determined from the slope. A typical calibration shows the combined non-linearity and hysterisis to be well within ±0.5% of full range. No zero shift was observed when the pressure was released suddenly and from this aspect the transducer would appear very satisfactory both for measuring and control purposes.

The effects of changes in ambient temperature on the transducer have been investigated. A shift between + 20°C and 50°C produces a change in the zero pressure of 0.03% per °C of the full range output at 1.38 GPa. The sensitivity between 0 and 689 MPa is changed by 0.05% per °C for the same range of temperature. However, this is for half range pressure and will rather better for the full range condition but tests have yet to be concluded to verify this point.

Work is presently being carried out on transducers featuring strain gauges which are more carefully matched to the coefficient of expansion of the high strength steel and which also incorporate improved wiring to the gauges.

5. The measurement of small differential pressures at high ambient pressures to 1 GPa

There are occasions when there is a requirement to measure a small pressure difference in the presence of a high ambient or background pressure, for example, in the measurement of flow in a high pressure system using an orifice plate or in measuring the pressure across a piston in a high rate of shear viscometer. Commercially available transducers utilising capsules, bellows or diaphragm sensing elements with resistance strain gauges (either bonded to a diaphragm or unbonded and acting as direct displacement sensors) or variable inductance pick offs (differential transformer) are limited to ambient pressures of up to 138 MPa (20000 lbf in^{-2}). Pressure gauges of the Bourdon tube type are unsuitable because of insufficient resolution. Instruments based on resistance changes of manganin can be used in static or very near static situations but transient temperature changes in the pressurising fluid also alter the resistance of the sensing element, making this method unsuitable for dynamic situations.

The high rate of shear viscometer developed by Boyle[9] was designed for measurements to be made at ambient pressures of up to 1.0 GPa (145000 lbf in^{-2}). This device utilizes a system in which the test fluid is driven by a relatively small pressure difference through a capillary tube connecting two reservoirs, both of which are at high pressure. The fluid flow rate depends upon the tube geometry and the small pressure difference between the two reservoirs.

The differential pressure sensor comprises two basic components, the pressure vessel which is subjected to the high ambient pressure and the internal assembly which comprises the active sensing diaphragm together with the displacement pick-off and ancillary components. Figure 5 shows the general assembly of the device, the pressure vessel being of conventional design featuring heavy screwed end caps which retain the end plugs. Cylinder materials are a nickel-chrome-molybdenum-vanadium high strength steel (NCMV) as are the end caps and plugs. The K ratio for this cylinder is 3 and the high pressure pipe connections are made on to the end plugs as shown.

The pressure sensing element (Fig. 6) is a low hysterisis stainless steel diaphragm machined integral with its mounting which is, in turn, supported from a distance piece which forms a pressure transfer port and facilitates assembly of the high pressure seals comprising neoprene O ring and aluminium bronze chamfer ring. In this way the sensing diaphragm is not subjected to the stresses imposed on the containment vessel by the high ambient pressure. If this were the case one would expect very considerable errors due to the stiffening effect upon the diaphragm of these radial stresses. The diaphragm displacement is sensed by means of a variable inductance transducer and the magnetic core of this transducer is attached to the diaphragm at its centre. A low stiffness stainless steel bellows encloses the displacement transducer and the assembly is filled with Plesol 201*. This is a fluid which has good electrical insulating properties, is non-hygroscopic and has a high solidification pressure. Plexol 201 is based on di-ethyl-hexyl sebacate and has found to have no long term detrimental effects on the transducer windings (insulation) nor on the epoxy resin used to encapsulate them.

The half bridge type variable inductance displacement transducer chosen for this particular application was the Type NH1 - 100/B manufactured by Vibrometer. This particular transducer was originally designed as a needle lift transducer for diesel engine injectors but has the advantages of small physical size and high sensitivity for deflections of ±1.5 mm. The metal housing of the NH1 - 100/B has been pierced by drilling two small diameter holes. This prevents collapse of the casing by ensuring that fluid reaches the interior, however considerable care is required when positioning and drilling these holes if damage to the windings is to be avoided.

Electrical connections are made via small hermetic metal to glass seals in the bellows end cap and thence by flexible p.t.f.e. insulated leads to ceramic cone type high pressure lead throughs. The internal electrical connections are insulated with an epoxy resin where they are in contact with the test fluid.

---

* Plexol 201 is a registered trade name for a synthetic di-ester lubricant supplied by Rohm & Haas (U.K.) Ltd.

6. Operation and calibration

The associated instrumentation used comprises a Vibrometer Carrier Frequency Bridge Type 100 - TRI/A. For convenience a digital voltmeter was used to monitor the output from the bridge amplifier. The bridge operates at a supply frequency of 100 kHz hence some care is necessary in screening the input and signal leads particularly those within the differential pressure sensor assembly.

Calibration of the differential pressure sensor was performed at atmospheric ambient pressure against a dead weight piston gauge, i.e. one pressure port open to atmosphere. Following this test the unit was subjected to high ambient pressure with no differential across the diaphragm to establish the effects on the electrical output. Ideally, the calibration would need to be carried out with a true differential pressure applied in the presence of the high ambient pressure, possibly using the apparatus described by Dadson & Daborn[10], Variations in the variable inductance transducer zero displacement output were very small at ambient pressures of up to 1 GPa. This would be expected as symmetrical dimensional changes, or equal changes in the resistance of both windings, do not produce an electrical unbalance. Pressure effects on the magnetic properties of the moving core would also be expected to produce only very small changes in sensitivity.

The bridge output - pressure relationship for a diaphragm of 0.64 mm (0.025 in) thickness is linear and the results lie within a ±1% band. A 0.3 mm thick diaphragm exhibited some degree of non-linearity, possibly due to increased edge effect, however the repeatability was good.

The maximum deflection of the diaphragm having a thickness of 0.64 mm (0.025 in) and diameter 15 mm (0.584) at a pressure difference of 7.0 MPa is 0.199 mm . With the 100 TRI/A Carrier Frequency Bridge the sensitivity is about 0.65 mV/μm and the resolution 1μm, hence the displacement transducer can be used to detect pressure changes of about 35 kPa (5.0 lbf in^{-2}) in the presence of the high ambient pressure. The theoretical value derived from thin plate theory[11, 12] for the diaphragm displacement, at 7.0 MPa, of 0.199 mm is in reasonable agreement with the measured value of 0.184 mm.

7. Resistance strain gauge loadcells for use under high hydrostatic pressures

In investigations into the properties of materials under hydrostatic pressures it is sometimes necessary to measure load (or torque) applied to a specimen, the measurement being made inside the high pressure container. One example is in a machine designed to investigate the properties of rock samples whilst subjected to hydrostatic pressures of up to 689 MPa, where it becomes necessary to measure the compressive load imparted to the rock specimen. In this particular machine the compressive end load is applied to the specimen by means of a ram, the rod of which passes through high pressure seals. A direct measure of the load from the ram pressure and area would give rise to errors due to the frictional losses at the sliding seals. Moreover, the frictional losses could be expected to change as the hydrostatic pressure changes and also as wear occurs, hence the desirability of measuring load by means of an internal load cell.

Another example is in a machine designed for an investigation into material properties under hydrostatic pressures of up to 300 MPa. In this machine it was nedessary to measure the tension force applied to the test specimen.

Whilst the basic design of load cells using bonded electrical resistance strain gauges presents no real difficulties, other than in the limitations of space available inside a high pressure containment, the bonding techniques do present some practical difficulties. Not only must the adhesive meet all the more usual requirements for strain gauge techniques but it must withstand the chemical action of the high pressure fluid, preferably for long periods, without effect on its ability to transmit the mechanical strain to the gauge itself.

The effects of a high hydrostatic pressure environment on the gauge sensitivity, and consequently upon the calibration must also be considered.

8. Loadcell design, general considerations

The design of loadcells utilising resistance strain gauges is well documented in the literature, there being a great variety of configurations ranging from simple cantilever beams to complex toroidal types. It is most desirable that the loadcell shall be relatively insensitive to off-centre loading and this is frequently achieved by averaging the load using a number of seperate load (cell) elements. Such means may be applicable to the more common industrial weighing applications but are not easily adapted to measurements in the confined space available in most high pressure apparatus. It is, therefore more practical to employ multiple strain gauges attached to the elastic load sensing member and disposed in a manner which in itself allows for some degree of averaging. Examples of this technique are illustrated in the compression loadcell which has three longitudinal strain gauges radially spaced at 120° (Fig. 7) and the tension type loadcell (Fig. 10) which incorporates a similar strain gauge configuration.

The materials used are selected on the basis of high strength and low hysterisis, typical materia being EN 26, NCMV steels, Nispan C and berylium copper.

The effects of temperature on the loadcell may require compensation and the normal methods employed with strain gauge techniques for changes due to Young's Modulus (i.e. a change in sensitivity may be adopted. However, experience to date has been limited to apparatus operating at or near ambient temperature (c.20°C) and compensation has not proved necessary. However, the thermal effects even over small ranges of temperature, due to the resistance-temperature coefficient of the strain gauge element and mis-match between the linear coefficient of expansion of gauge element and loadcell material, must be duly compensated for. Fortunately, the introduction in recent years of the "temperature compensated" foil strain gauge has greatly simplified this particular problem. Strain gauges of this type feature a foil resistance element bonded to a very thin epoxy or polyimide resin backing, the characteristics of the foil being adjusted by metallurgical and heat treatment processes to closely match the particular structura material used. A very high degree of thermal match can be achieved and consequently errors from this source practically elliminated.

For strain gauge loadcells subjected to high hydrostatic pressures, the main area of concern is in the effects of pressure on the strain gauges, that is on the foil (or wire) of the resistance element and the backing material (and adhesive). The pressure effect will be to produce a change in resistance (and therefore a bridge output signal eventually) due to the compressibility of the elastic material of the loadcell and also the resin backing and adhesive layer. There may also be an effect on gauge factor which would directly affect the sensitivity of the loadcell. Provided these effects are small and reproducible then, for loadcell considerations, they could be allowed for in the calibration.

9. Effect of pressure on resistance strain gauges

Much of the literature dealing with the effects of hydrostatic pressure on the behaviour of resistance strain gauges has little relevance to modern self-temperature compensated foil type gauges. However, several papers are of interest, the work of Clough et al[13] on SR4 strain gauges (wire type) in castor oil at pressures up to 10000 lbf in^2 (69 MPa) confirmed the pressure effects and gave linear correction factors for grey iron and modular iron. Milligan[14] investigated the effects on bakelite and epoxy backed gauges at pressures up to 140000 lbf in^2 (980 MPa) but the gauges used are now obsolete. Pugh and Gunn[15] review this and other work and investigate the effects on epoxy backed Saunders Roe foil type gauges. These gauges were of the uncompensated foil type and are now obsolete. Temperature compensation with this type of gauge is provided by the 'dummy' gauges in the usual manner. However, the basic design of loadcell (the 'top hat' type) is of considerable interest since it forms a very compact unit. They reported that the loadcell calibration performed at atmospheric pressure agreed with that carried out under pressures of up to 950 MPa to within 1½%.

Montague[16] and Thomson[17] investigated the effects on foil gauges but again these were of the uncompensated type. Hanafee[18] deals with unmounted gauges at pressure of up to 1 GPa, and investigated the pressure effect on a very low range loadcell in Ref. 19. In both instances the conclusion is that there is a small but significant change in gauge factor. Brace[20] observed the behaviour of Constantan foil epoxy backed B.L.H. series FA gauges bonded to a variety of materials whose linear compressibility was precisely known, at pressures of up to 1 GPa. This work is of particular interest both in that the gauges are of the compensated foil type and the bonding techniques are very similar to those developed in our own work. Gauges attached to Armco iron exhibited a linear pressure effect (strain) of $+0.54 \times 10^{-7}$ bar^{-1}. In this investigation a single active gauge was used since the aim was to measure true strain, for a loadcell application a half bridge configuration may be used with the active gauge(s) in one arm and the compensating gauge(s) in the adjacent arm. By this means some degree of compensation for pressure effect (as for temperature effects) can be achieved, but in practice this compensation is never perfect since it is virtually impossible to obtain exactly the same bonding (thickness of adhesive) throughout.

10. Bonding techniques and adhesives

Many adhesives have been employed for bonding resistance strain gauges to components, ranging from solvent systems such as Durofix to high temperature ceramic cements. Normally the choice of adhesive is determined by the temperature at which the gauges are required to operate, by the backing material and component material, strain range, long or short term measurement (creep effects) and resistance to the environment, say water or oil contamination. Certain of the adhesives require a relatively rough surface into which they key, for example the bakelite type cements and ceramic high temperature cements. Others require a very smooth polished surface to establish a chemical bond, for example the Eastman 910. In the case of strain gauges subjected to high hydrostatic fluid pressures the adhesive must also be resistant to attack by the fluid, a factor which may differ greatly under high pressure from the normal ambient pressure effects.

The epoxy and poly-imide based resins offer very high mechanical bond strength together with good resistance to solvent attack and are compatible with the backing materials (often similar resins) used in present day foil strain gauges. These adhesives are an obvious first choice, generally however they do necessitate a hot cure/post cure process. Another type of adhesive is the so called 'impact' type of which the Eastman 910 is perhaps the most widely used. This adhesive polymerises at room temperature under the effect of clamping pressure and gives an exceptionally high bond strength. It is very popular for quick and reliable strain measurements but it is not recommended for transducer applications where very long term stability is demanded. This is due to it being slightly hygroscopic and even for normal short term tests some form of protection is required, this normally being a coating of, for example, silicone rubber.

A considerable amount of work was carried out on the compressive loadcell used in the compression testing machine (Fig. 7) before a satisfactory bonding technique was evolved[21]. Initially the fluid used in this particular machine was the synthetic diester Plexol 201, i.e. di-ethylhexylsebacate but because no adhesive could be found that would withstand this fluid, nor indeed a strain gauge backing either, castor oil was substituted. In the original loadcells Tinsley L.S.G. series foil gauges were used, LSG1/5/CN/E 120 ohm 1 inch linear type, these being an epoxy resin backed gauge. (The type number was changed to 26/120/E/C at a later date). Some trouble was experienced with these gauges due to the foil separating from the backing and later work used the Micro Measurements EA Series gauges EA-06-120, these having a polyimide resin backing.

Eastman 910 adhesive used with MM gauges proved unsatisfactory, the bond failed after one application of pressure to 689 MPa (castor oil). Subsequent work on specimens has further proved this point, somewhat surprising in view of the exceptionally high bond strength normally attained. It is possible that the very thin glue line thickness with this adhesive does not give sufficient 'flexilibity' to allow for the compressibility effect in the gauge backing. The bond was observed to fail between gauge backing rather than at the metal interface.

Polyester type adhesives were also found to be unsatisfactory, even when used with polyester backed gauges, moreover the short storage life of these adhesives is considered a further disadvantage even for more usual applications.

Epoxy resins have proved to be the most

suitable adhesives and in particular the Araldite Strain Gauge Cement used with hardener HY 956 or alternatively HY 951. This resin was developed fifteen or so years ago and incorporates a filler which unfortunately makes it extremely viscous. With the older pattern strain gauges which had relatively thick backings (typically 0.05 mm) this presented no great problems but with the more recent gauges of which the MM.EA series are typical the backing is about one tenth the thickness. Whilst highly desirable from the true strain measurement aspect this thin backing does present some difficulties when the filled and viscous resin is used. Great care is required in clamping the gauges and in avoiding excessive build up of the adhesive layer and in particular the avoidance of blisters beneath the gauge element. The technique evolved and which has given excellent results is detailed in Appendix 1.

11. Calibration and performance of the loadcells

A typical calibration for the compressive loadcell is shown in Fig. 8 the calibration being performed in an Amsler testing machine. The loadcells were load cycled ten times to full range and no zero shift was observed after the third load cycle. Off-centre loading effects were assessed by carefully off-setting the loading bar, which was in contact with the anvil (core bar) of the loadcell, by 3.0 mm. This produced no significant effect on the calibration which showed non-linearity and hysteresis to be within ±0.5% of full range output.

In order to determine the effects of hydrostatic pressure on the loadcell characteristics, the high pressure containment of the compressive testing machine was utilised, an intensifier being used to generate the high fluid pressure. No axial load was applied and the bridge output was recorded at increments of pressure up to a maximum of 689 MPa. The output due to hydrostatic pressure only is given on the same graph as the load calibration (Fig. 8) so that the true axial load can be determined for any combination of axial load and confining pressure.

Four loadcells of this design have been in use over a period of several years, including the initial series of tests using Plexol 201. Of the loadcells subsequently re-gauged and used with castor oil one completed twenty tests at up to full range load (200 kN) and pressures to 680 MPa prior to failure. The failure was due to a hair-line crack propagating through the Araldite adhesive adjacent to the gauge foil terminal. Another loadcell completed 56 tests at similar load and pressure conditions and re-calibration on the Amsler machine gave results 0.7% lower at full range than the original calibration. There was no significant change in the pressure sensitivity. The present loadcell has completed at least 150 tests and although again frequent re-calibration has shown a small and continuous decrease in output, amounting in total to several percent, the unit functions satisfactorily. However, it is essential to maintain a frequent check by way of re-calibration but use has shown that impending failure is usually preceeded by a marked change in the output due to pressure.

It is interesting to note the effects of gauge size on the pressure sensitivity. The gauges used on the loadcell calibration shown in Fig. 8 were 26 mm x 6 mm whereas the MM gauges subsequently used (Fig. 9) were 6 mm x 3 mm. These latter gauges exhibit about one half the sensitivity to pressure although the gauge resistance is the same in both instances and the bonding method common to both. The effect is probably related to the ratio of gauge backing thickness to adhesive layer thickness rather than surface area in itself, however it would appear desirable to use small gauges even where overall size limitations are not critical.

The small tension type loadcell detailed in Fig. 10 illustrates the degree of compactness often demanded. In this instance the gauges are 3mm x 2 mm MM.EA Series using the Araldite bonding as before, the pressure effect on the output is very similar at 300 MPa to that noted for the 6 mm x 3 mm gauges.

12. Further Work

The calibration of strain gauge loadcells at ambient rather than at the high hydrostatic pressures at which they are to be used is not completely satisfactory. Whilst the individual load and pressure calibrations show the relative output due to these effects they do not take into consideration a change in sensitivity due to pressure effects on the gauge factor. Hence it would be most desirable to have some means of calibrating under hydrostatic pressure conditions.

The apparatus described by Pugh and Gunn[15] provides means for such calibration, and this and other possible methods are being investigated. Experience to date has been mainly with castor oil, but this is not a suitable fluid for many higher pressure applications. Tests are proceeding with adhesives based on polyimide resins for such applications.

13. Fluids for use in high pressure apparatus

The selection of a fluid for use in equipment and apparatus operating under high hydrostatic pressures is influenced by several factors, perhaps the most important of which is the ability of the fluid to transmit the pressures without solidifying. Other vital considerations are the effects of the fluid on the constructional materials, i.e. sealing rings, both metal and elastomer and the temperatures at which the fluid may be used. It is important to note that the effects of fluids at high pressures of up to 1.4 GPa, even at room temperature, can be greatly different to the effects at the more usual hydraulic system pressures in everyday use. This is particularly so in the case of elastomer materials used in seals.

When electrical measurements must be made in the presence of the working fluid, for example where the resistance of a test sample subjected to hydrostatic pressure is required or when electrical input transducers are enclosed within the apparatus, then other factors influence the choice of fluid. These are the electrical conductivity of the fluid (it must obviously have good insulation properties) and its effects upon the materials used in the electrical devices. For example lead out wires must be insulated, typically by means of a varnish, and the fluid must not attack this. Similarly, where electrical resistance strain gauges are employed in loadcells the fluid must not attack the backing materials used in the gauge construction, nor the adhesives used in bonding the gauge to the material surface. It is also highly desirable that the fluid is non-hygroscopic since this will greatly affect the electrical properties over a period of time.

There appears to be very little information available relating to the practical choice of a fluid particularly as concerns ability to transmit pressure without solidifying or "freezing" and long term effects on varnishes and resins. Reference (22) gives some guidance as to choice of fluids for hydrostatic compaction work.

The effects of various hydraulic fluids on elastomer seals e.g. O rings, is well covered in the manufacturer's literature. In general this data only applies to the more usual operating conditions met with in everyday hydraulic equipment. Consequently a series of tests have been carried out using simple apparatus, to determine the characteristics of a number of fluids, at room temperature, up to pressures of 1.4 GPa.

14. A simple means of determining fluid characteristics

The absolute viscosity of fluids has been investigated by Wilson [23], who used a falling plate viscometer to pressures of about 1 GPa, and later by Boyle [24] using a high rate of shear (capillary tube) viscometer to similar pressures. Scaife [25] investigated the electrical properties of one particular fluid used by both Wilson and Boyle, namely the synthetic diester, bis diethyl hexyl sebacate. However, both the falling plate viscometer and the high rate of shear device are complex and require considerable back-up facilities. The methods are therefore costly and time consuming.

In practice what is required is a simple means of determining the pressure freezing point and/or the pressure drop along the relatively small bore high pressure tubing used in a system. A great many systems use ¾" O.D x 1/16" I.D. composite tube (inch sizes are quoted as this tube is of U.S.A. origin) hence this is the tube chosen for the freezing rig work.

The freezing rig, Fig. 11, consists of pressure generating apparatus, in this case a standard 'Belfast' intensifier used in conjunction with a motor driven hydraulic pump and hand priming pump. A light mineral oil is used in the low pressure system and a fluid having a high solidification pressure or freezing point is used in the high pressure system. This fluid is a mixture of Plexol 201 (basically bisdiethylhexylsebacate) and petroleum.

The high pressure output from the intensifier is taken to a cross piece which connects a relief valve, a manganin pressure sensor and the input side of a fluid isolator. On the down line side of the isolator is the test section comprising 1 m length of (1.5 mm) 1/16" inch bore high pressure composite tube. The pressure difference along this length of tube is obtained from two resistance strain gauge pressure transducers previously described.

A filling and shut off valve is provided at one end of the test section, filling being performed by evacuating the isolator and test section of tube via two filling flasks, one acting as a trap, the other as a fluid reservoir. Venting the vacuum line to atmosphere results in fluid being drawn into the system, after which the shut-off valve is closed and pressure applied.

The fluid isolator used is a modified manganin sensor body where the encapsulating bellows normally enclosing the manganin coil is used as a separator. This proved completely satisfactory for all but one fluid tested where the compressibility was unduly great (40%), the bellows being limited to about a 30% volumetric change. A sliding piston isolator was devised for this particular test, the seals being standard SCN (Walker & Co.) rings the piston sliding in the manganin sensor body. This isolator has since proved completely reliable over a number of tests to 1.5 GPa.

The manganin pressure sensor was incorporated firstly to act as a transfer standard for pressure measurement and secondly to provide a safeguard against exceeding an input or driving pressure in excesssof the autofrettage pressure for the strain gauge transducers.

With some of the fluids there is no well defined freezing point merely a gradual increase in pressure difference being observed. Others, such as the castor oil, exhibit a well defined freezing point and in this particular instance a marked time dependence also.

15.

From this data, and based on a 2% pressure drop along the 1 m test section, the following maximum operating pressures for the various fluids are suggested,

Fluid	Pressure limit MPa	Pressure limit (lbf/in^{-2})
Shell Tellus 27	520	82000
Shell Tellus 27 + 10% 120° BP Petroleum V/v	565	90000
First Cut Castor Oil	550	80000
Castor Oil + 10% methyl Alcohol V/v	1030	150000
Castor Oil + 20% methyl Alcohol		
Silicone fluid MS200 3 c.s.	980	142000
Lockheed H.D. Brake Fluid	>1380	>200000
Mobil 46 Brake Fluid	>1380	>200000
Ucon LB series Fluid 650X	690	100000
Ucon LB series Fluid 70X	1170	170000
Plexol 201	1310	190000
Plexol 201 + 10% 120° BP Petroleum V/v	1310	190000
Plexol 201 + 40% 120° BP Petroleum V/v	1380	200000
Plexol 201 + 80% 120° BP Petroleum V/v	>1380	>200000
Di-2-ethyl hexyl sebacate	1310	190000
Plexol 244	1380	200000
Plexol 262	>1380	>200000
Plexol 273	>1380	>200000

These limits are for an operating temperature of around 20°C, the limit is markedly affected by temperature increasing as the temperature increases as would be expected. The size of pressure vessel, length of high pressure tubing and its internal diameter will all contribute to the practical limit of pressure at which a particular fluid may be used

Acknowledgements

The author wishes to acknowledge the help and co-operation of all who have assisted in the design and testing of the various systems. Particular thanks are due to Professor B. Crossland for his support and guidance and to Mr. C. Ludlow for his contribution to the designs. The laboratory and workshop staff in the Department of Mechanical Engineering have also provided much assistance, particularly Mr. J. Maginnes who assisted throughout with the high pressure equipment.

References

1. 'Manganin' is a trade name of Isabellen Hutte KG. Data booklet is published under this title by the manufacturer.

2. Bridgman, P. W. The measurement of hydrostatic pressures up to 20000 kgf/cm^2. Proc. Am. Acad. Arts. Sci. 47 1911 p.321-341.

3. Lichtenberger, A. Contribution to the investigation of thin piezoresistive Manganin pressure gauges. Application to shock waves. Report 16-73 I.S.L. St. Louis (France) 17.5.73.

4. Harwood Engineering Co. Inc. Handbook for the Harwood Controlled Clearance Piston Gauge Harwood Eng. Inc. Walpole Mass. U.S.A.

5. Heydemann, P.L.M. Piston Gauges. Thermodynamic Measurements Chapter III a-3, N.B.S. Washington DC. (To be published).

6. Dadson, R. S., Greig, R. C. P., and Horner, R. Developments in the accurate measurement of high pressure. Metrologia 1, 1965, p. 55.

7. Darling, H. E., Newhall, D. H. A high pressure wire gauge using gold chrome wire Trans. A.S.M.E. 75, 1953, p. 311.

8. Gall, C. A., Birks, A. W. A direct reading bridge for the measurement of hydrostatic pressure by means of Manganin sensors. Strain 3, No. 4, 1967, p. 27.

9. Boyle, W. P., High Pressure Viscometry. Ph.D. Thesis. Queen's University, Belfast, 1970.

10. Dadson, R. S. & Daborn, J. E. Differential Pressure Measurement, a new primary standard Physics Bulletin, 24, 1973.

11. Neubert, H. K. P. "Instrument transducers", Oxford, p. 40, Thin Plates and Diaphragms summarised from Timoshenko Strength of Materials II.

12. Roark. Formulas for stress and strain. McGraw-Hill 4th Edn. Ch. 10, p. 210.

13. Clough, W. R., Shank, M. E.., Zaid, M. Behaviour of the SR4 wire resistance strain gauge on certain materials in the presence of hydrostatic pressure. Proc. Soc. Exp. Stress Anal. X2. 167-176, 1953.

14. Milligan, R. V. The effects of high pressure on foil strain gauges. Soc. Ex. Stress Anal. Paper 803, 1963.

15. Pugh, H. Ll. D, Gunn, D. A. A strain gauge load cell for use under high hydrostatic pressures. NEL report No. 143 East Kilbride, Glasgow, March, 1964.

16. Montague, P. The effect of hydrostatic pressure on the behaviour of foil strain gauges. Strain 2 2, 1966, p. 17-22.

17. Thomson, K. F. Hydrostatic pressures on foil strain gauges. Strain 2 3, 1966.

18. Hanafee, J. E. Pressure dependance of resistance strain gauges. Rev. Sci. Inst. 41, 885 (N), 1970.

19. Hanafee, J. E. and Radcliffe, S. V. Effect of high pressure on a strain gauge load cell. Rev. Sci. Inst. 38, 1967, p. 329.

20. Brace, W. F. Effect of Pressure on Electric resistance Strain Gauges. Exp. Mechanics, 4 (7) p. 212-216, 1964.

21. Birks, A. W., Ludlow, C. G. A strain gauge loadcell for the measurement of compressive loads under conditions of high hydrostatic pressure. Strain 5, No. 4, October 1969.

22. Boyer, C. B., Peterson, J. H., Orcutt, F. Equipment for hydrostatic pressing of metal powders. Proc. 27th Nat. Conf. on Fluid Power, Chicago, October 1971.

23. Wilson, W. R. D. Viscometry of lubricants under high pressure and temperatures. Ph.D. Thesis, Queen's University, Belfast, 1967.

24. Boyle, W. P. High Pressure Viscometry. Ph.D. Thesis, Queen's University, Belfast, 1970.

25. Scaife, W. G. The relative permittivity of bis (diethyl hexyl) sebacate as a function of pressure and temperature. High temperature High Pressure, 2, 1970 p. 453-460.

Appendix 1

Procedure for bonding foil strain gauges using Araldite Strain Gauge Cement and Hardener HY941 or HY956

Refer to Araldite Strain Gauge Cement Instruction Sheet No. A13a C.I.B.A. Ltd., Duxford, Cambridge

1. Initial surface finish should be a good machine finish, preferably ground and free from deep scores.

2. Vapour blast using a 120/220 mesh grit, alternatively use a similar dry (sand) blast.

3. Thoroughly clean, preferably by ultrasonic means, using acetone. Avoid scratches on the surface where the gauge is to be fixed. Do not use trichlorethylene or carbontetrachloride. Use reagent grade acetone not industrial (thinners) grade.

4. Mark out gauge position index lines using a dry ball point pen or hard pencil. Do not use a scriber. Position gauges using Mylar transfer tape.

5. Wash off using acetone immediately prior to fixing gauges. Keep components in a dessicator with some silicagel to avoid atmospheric corrosion.

6. Mix resin and hardener as recommended on the Araldite data sheet. Apply to surface to give a thin layer of about 0.1 mm thickness. The components may be pre-heated to 50-60°C to facilitate spreading. Reposition gauges by hinging tape back into original position. Place Mylar or polythene sheet over gauges and clamp.

7. Cure adhesive at 60°C for 5 hours.

8. Wire up gauges and terminals using a fine enamel insulated copper wire, remove traces of solder flux. Check resistance and insulation from gauge to component.

9. Postcure 60°C for 15 hours or 80°C for 5 hours.

<u>Note</u>: No gauge grid protection is used, hence particular care is required in handling and assembly. Protect gauges with Mylar adhesive tape prior to final installation in testing machine.

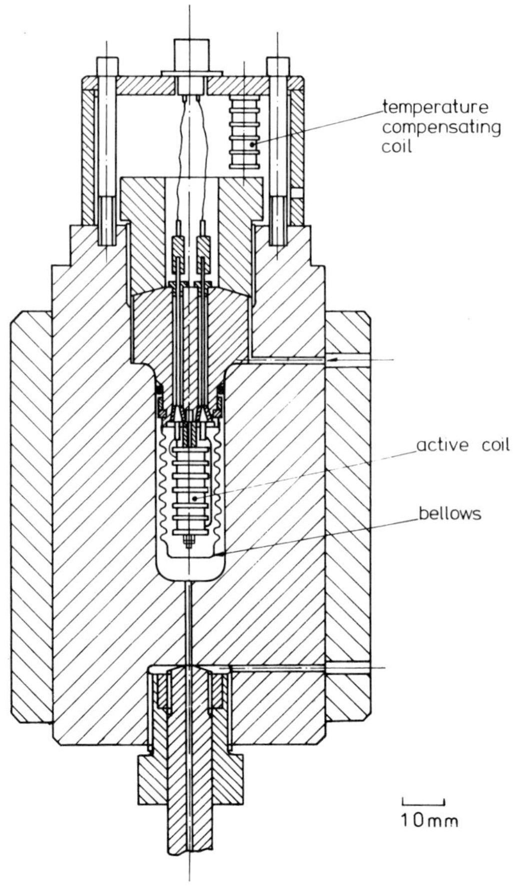

Fig. 1: 1.4 GPa (2000,00 lbf/in⁻²) manganin pressure transducer

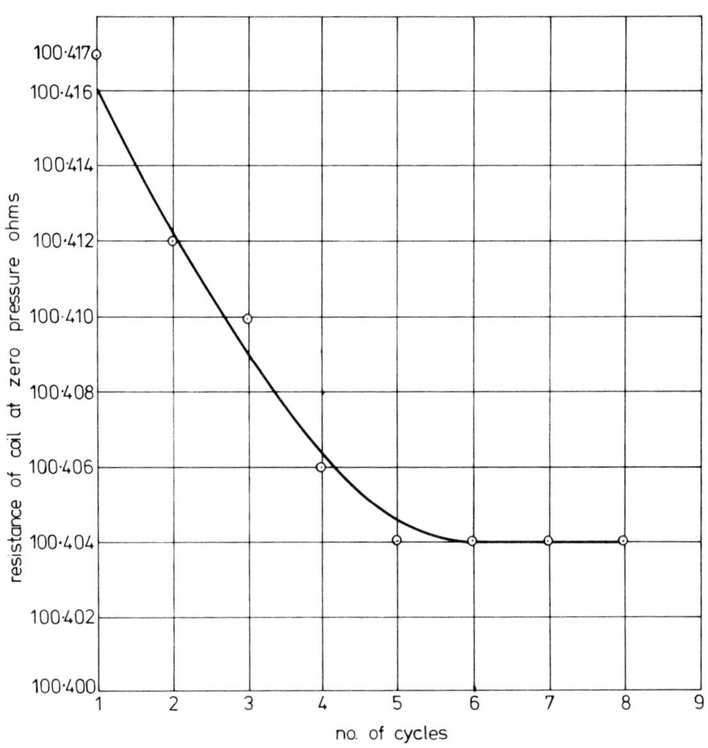

Fig. 2: **Effect of pressure cycling to 690MPa (100,000 lbf/in) on the zero pressure resistance of a manganin coil.**
Procedure followed — zero pressure to 690MPa for 5 mins. 690MPa to zero pressure wait for 5 mins. and take reading

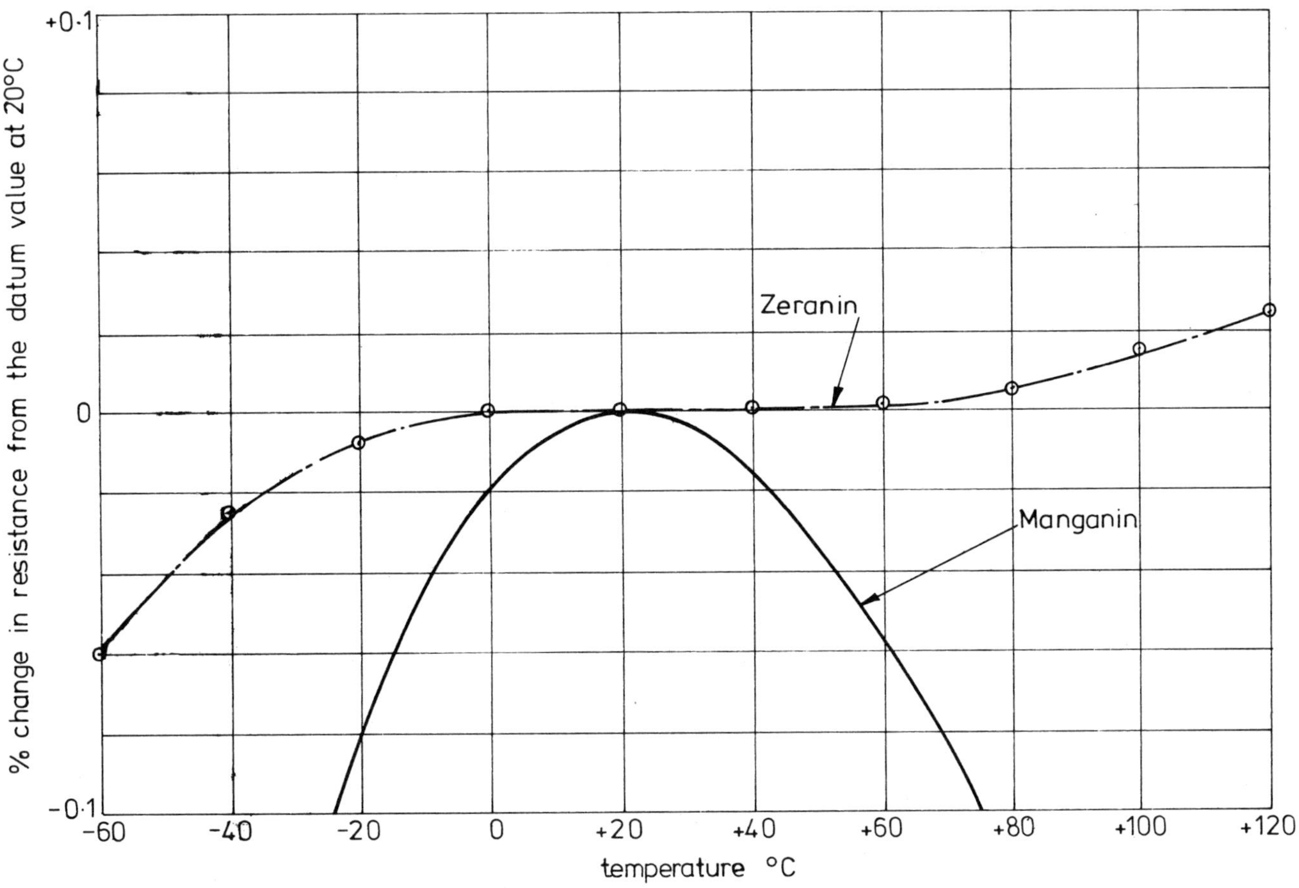

Fig. 3: Temperature coefficient of resistance for manganin and zeranin (Isabellenhutte-Heusler KG)

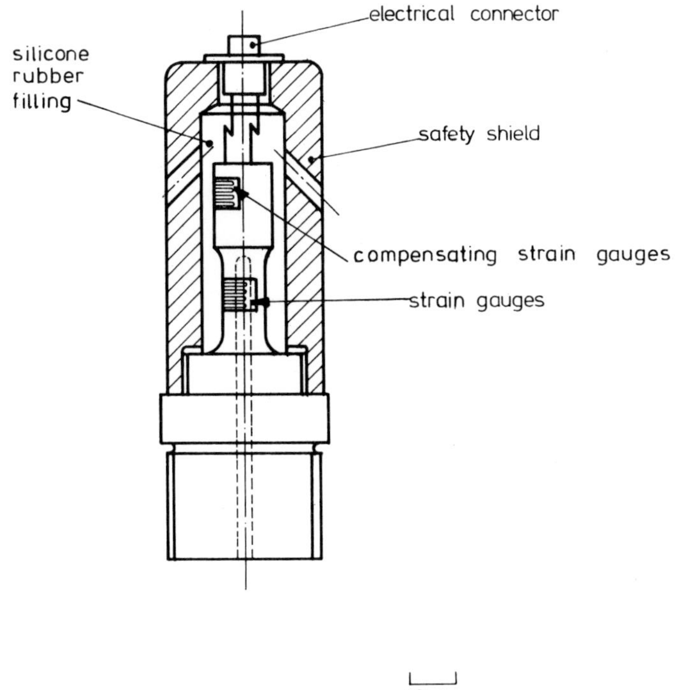

Fig. 4: 1.4 MPa (200,000 lbf/in^{-2}) strain gauge pressure transducer

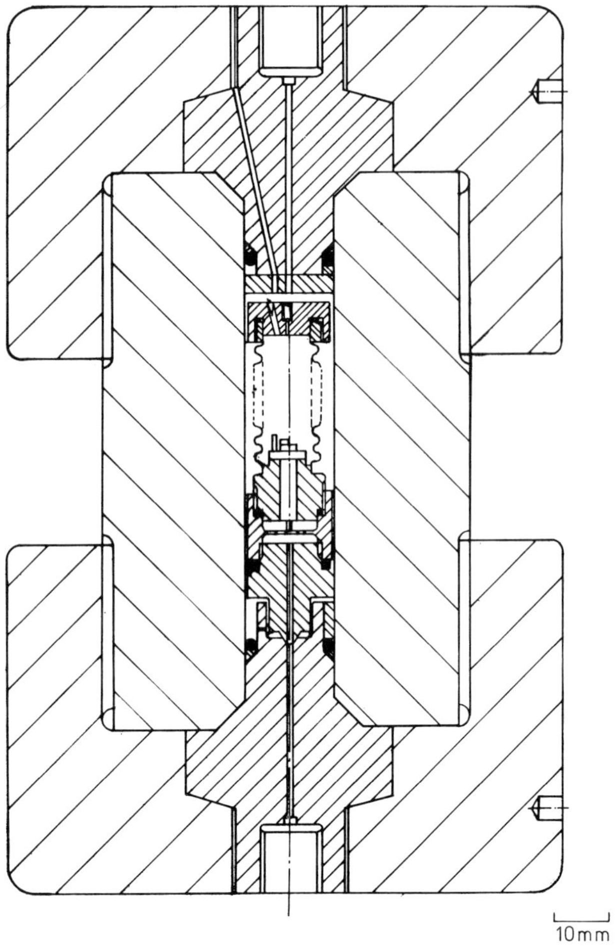

Fig. 5: Differential pressure sensor in vessel

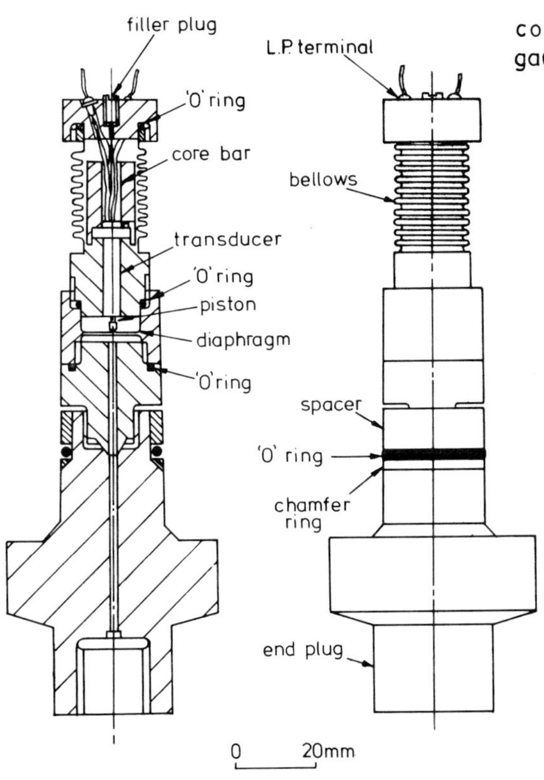

Fig. 6: High pressure differential pressure sensor

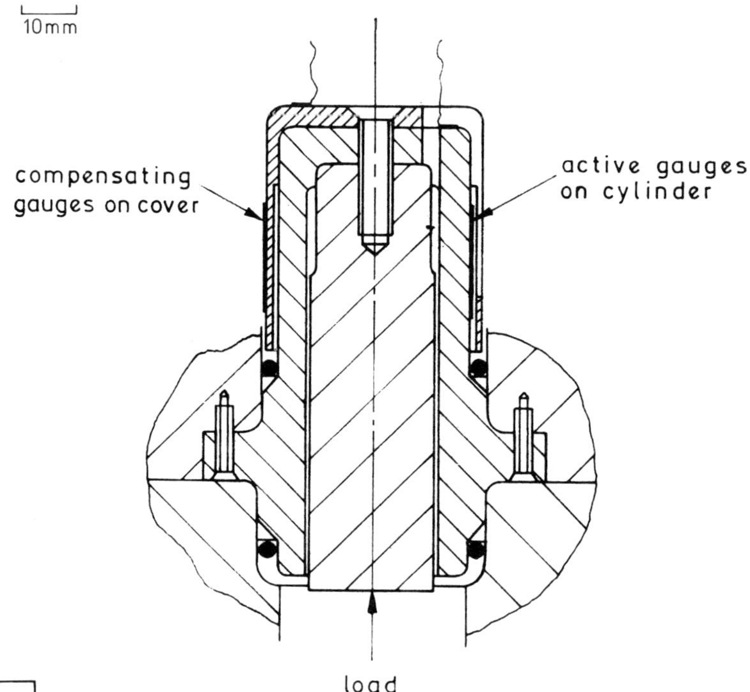

Fig. 7: Load cell

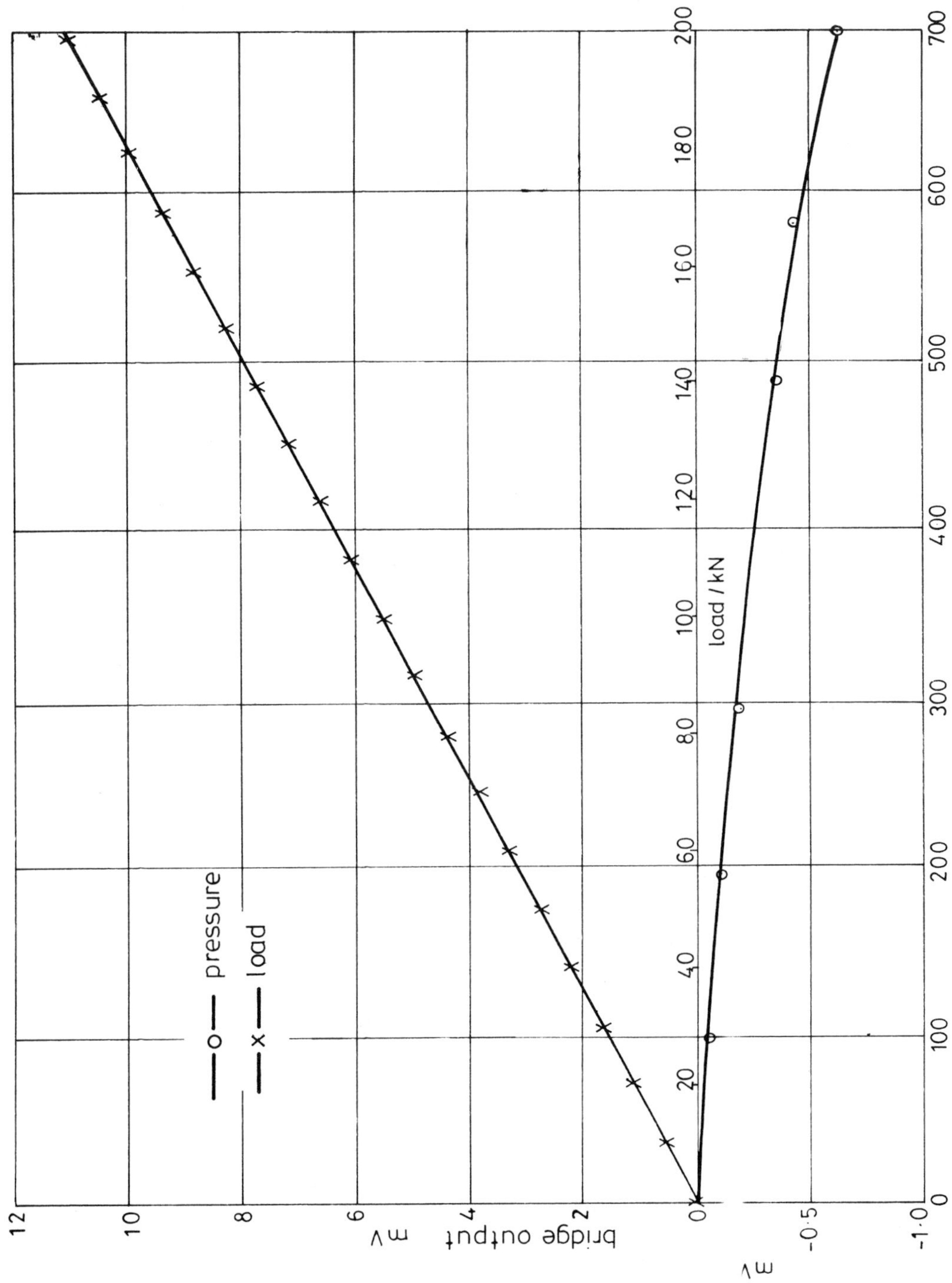

Fig. 8: **Loadcell calibration (tinsley-telcon gauges 26/120/E/C)**

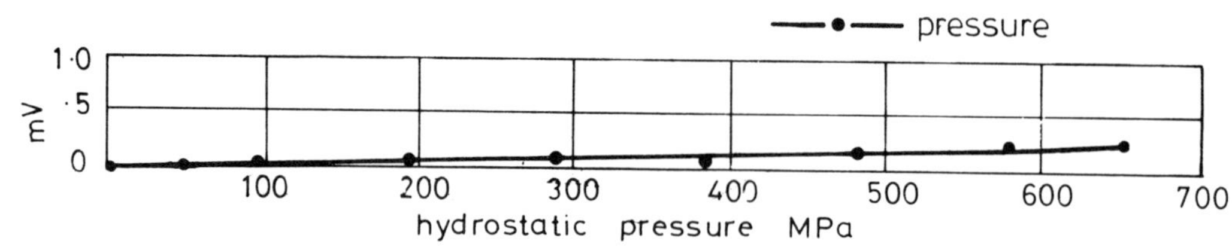

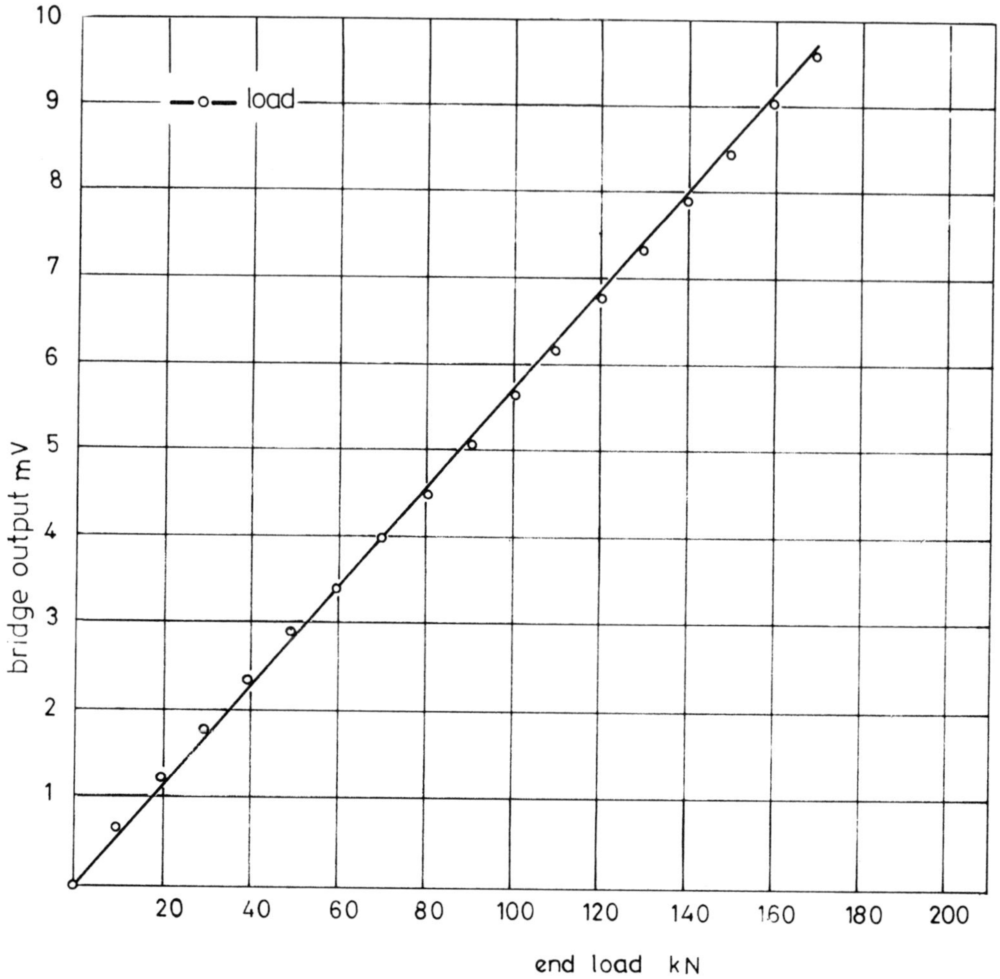

Fig. 9: Loadcell calibration (MM gauges EA-06-120)

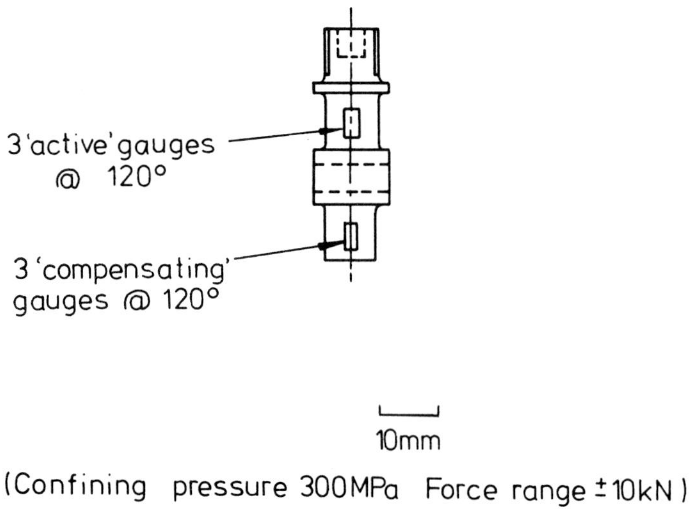

Fig. 10: Strain gauges loadcell for tension-compression machine (Confining pressure 300 MPa force range ± 19KN)

ITEM	CICo PART No.
1 Motor driven pump, M	Enerpac BPN
2 Belfast intensifier	HP 8600
3 Priming pump (hand operated)	Enerpac P-39
4 Shut off valve	HP 8601
5 Cross connector	X-12-1C
6 Manganin Pressure sensor	HP 8606
7 Fluid isolator	HP 8606 MOD
8 Tee connector	T-12-1C
9 Pressure transducer	HP 8615
10 Pressure transducer	HP 8615
11 Shut off valve (also filling)	HP 8608F
12 1/16" I.D. × 3/4" O.D. HP pipe	P-12-1

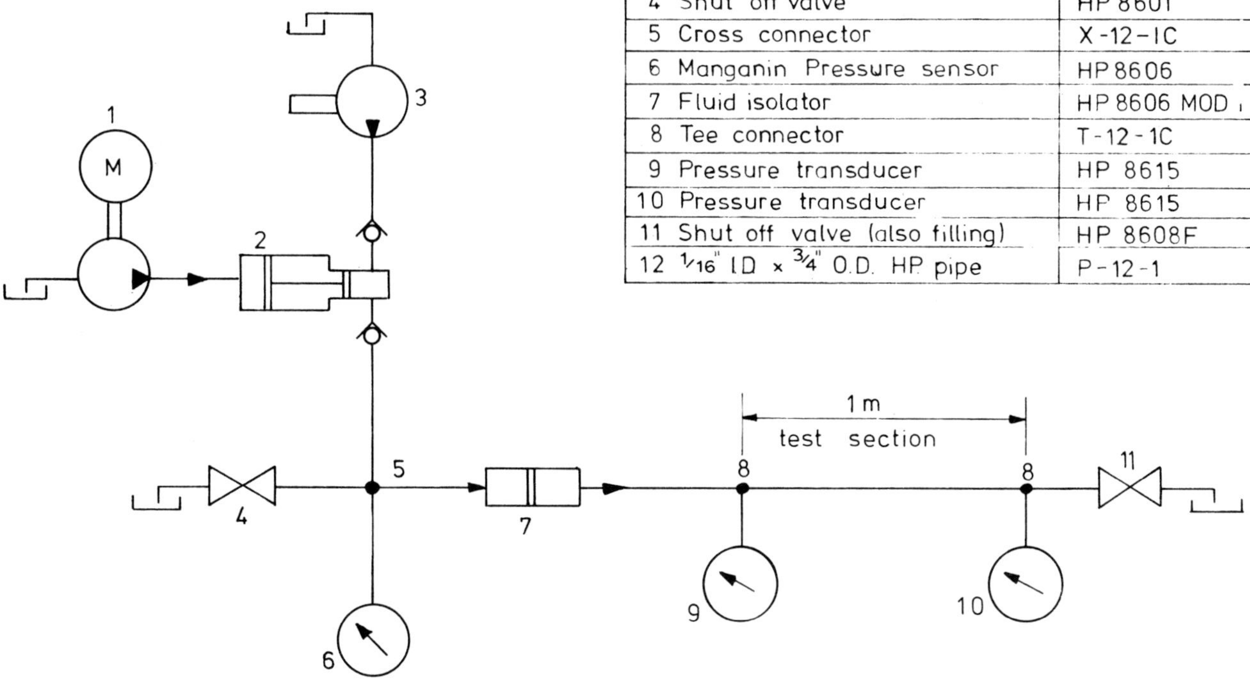

Fig. 11: **Hydraulic circuit for freezing rig**

SOME CONSIDERATIONS OF PISTON GAGES

D. H. NEWHALL, BSE(ME), MSE(EM), FIMechE
Harwood Engineering Company, Inc., Walpole, Massachusetts, USA
L. H. ABBOT, BS
Harwood Engineering Company, Inc., Walpole, Massachusetts, USA

SYNOPSIS Part A of this paper lists the instrumental corrections applicable to controlled clearance piston gages, discusses the respective uncertainties, and tabulates these values for two extreme examples. In Part B the spherical piston gage is described and two major advantages are discussed. Three additional applications of the spherical piston are noted.

Part A. INSTRUMENTAL UNCERTAINTIES OF THE CONTROLLED CLEARANCE PISTON GAGE

INTRODUCTION

1. The controlled clearance free piston gage (1) (2) (3) (4) (5) provides probably the most accurate pressure measurements over a wide range. While it was originally conceived and developed to evaluate high pressure it has been found equally useful in low ranges, overlapping to some extent the usefulness of mercury manometers (3). The question has arisen, exactly how accurate is this instrument.

1.1 The measurement of pressure with a free piston gage requires the knowledge of the force on the piston and of the effective area of the piston. Force measurements to within two or three parts per million are possible, with some laboratories claiming nearly another order of magnitude (6). Since the force is measurable with a far greater certainty than the area, the degree of accuracy is limited principally by the evaluation of the effective area. Corrections in force and fluid heads (7) (5), common to all types of free piston gage, are not of concern in this paper, which addresses itself to the uncertainties related to the effective area of the controlled clearance type of gage in particular.

1.2 There have been developed, to date, three main types of free piston gages. The most familiar is the simple piston gage, frequently attributed to Amagat(8). Here a cylindrical piston of known diameter with vertical movement in a closely fitting cylinder is loaded with a known dead weight and supported by hydrostatic pressure. By definition the known weight and the effective area of the piston determine the pressure. Because of the expansion of the cylinder and the contraction of the piston the leakage between the piston and cylinder increases with pressure, and this gage becomes impractical at higher pressures. The re-entrant type developed by Bridgman (9) shown in Fig. 1, presents the opposite problem, since the cylinder has a tendency to seize the piston as pressure increases. In either case, the unpridictability of the deformations of the piston and cylinder causes considerable uncertainty in the determination of the effective area. As the name implies, the principal virtue of the controlled clearance gage, Fig. 2, is that the amount of clearance can be established with enhanced precision at the will of the operator. To show the significance of this, we reproduce a graph, Fig. 3, due to Johnson et al (10) of NBS, comparing the pressure change in effective area of a working standard simple piston gage and that of a controlled clearance gage.

INSTRUMENTAL CORRECTIONS, CONTROLLED CLEARANCE GAGE

2. The pertinent parts of a controlled clearance piston gage are shown schematically in Fig. 2. Herein P_m is the pressure to be measured, developed by some sort of generator and applied to the free piston a supporting the deadweight W, as well as to a test gage or other subject under investigation. c is the high pressure "jacketed cylinder," within the "jacketing cylinder" d and sealed by packing rings b. P_j is the "jacket pressure," developed independently of P_m and adjusted so as to constrict the jacketed cylinder very nearly to the extreme of making contact with the piston, sufficient leeway being accorded that the piston may rotate freely while the leakage of pressure fluid is almost reduced to vanishing point.

The fundamental relation of a piston gage is:

$$P_m = \frac{W}{A_e} \quad (1)$$

wherein P_m is the measured pressure, balanced by the dead weight load W, and acting over an effective area A_e. If a stress upon a piston of nominal area A_o causes a small change in its diameter d_o, and a diametral strain e, the changes are related as follows:

$$\frac{\Delta A}{A_o} = \frac{2\Delta d}{d_o} = 2e \quad (2)$$

The deformations of the piston and jacketed cylinder in Fig. 2 are shown with considerable exaggeration. The exact deformation within the axial distance between that portion of the piston under true hydrostatic pressure and that portion under simple axial load is quite complex (10) (11). In one form or another, this problem exists for all free piston gages and is the subject of considerable discussion in the major works in this field. There are two points in general agreement - the design should provide that the pressure gradient is very short*, probably less than a diameter of the piston, and that this gradient should be located several diameters from either end of the piston or the cylinder. It will be noted that this condition is assured by making a slight constriction in the bore of the jacketed cylinder, short in length and about midway along the jacketed portion. Without this constriction the gradient would move up or down with increasing or decreasing measured pressure, and variable end effects would follow. These would invalidate the evaluation by means of elastic theory of the clearance changes due to changes of jacket pressure. Moreover, the wedging of the packings bounding the jacket space causes local contractions of the jacketed cylinder which might result in a spurious gradient at one of these points, resulting in loss of clearance control. The built-in constriction precludes this possibility. From the theory of elasticity, the diametral strain at the bottom of the piston, where it is under hydrostatic pressure P_m, is:

$$e_{P_m} = \frac{P_m}{E_p}\left[1-2\mu_p\right] \tag{3}$$

and at the top, where it is subjected to longitudinal load only, the diametral strain is, by Hooke's Law, for simple stress:

$$e_W = \mu_p \frac{P_m}{E_p} \tag{4}$$

The effective strain, being the mean, is:

$$e_p = \frac{1}{2}\left(e_{P_m} + e_W\right) = -\frac{P_m}{2E_p}\left(1-3\mu_p\right) \tag{5}$$

and:

$$\left(\frac{\Delta A}{A_o}\right)_{P_m} = 2e_p = -\frac{P_m}{E_p}\left(1-3\mu_p\right) \tag{6}$$

2.1 The area of the piston is dependent on its temperature, which is a function of the work done as it is rotated against the viscous drag of the pressure fluid, and of the environment. Many factors apply here, - particularly the speed of rotation, the viscosity which is strongly affected by the pressure of the fluid (10) (14), the clearance between piston and cylinder, and the thermal conductivities of the fluid and of the bulk of the surrounding metallic structure. It would be hopeless to try to evaluate all these effects. The obvious thing to do is to reduce the speed of rotation, but not to the point where the film breaks down, resulting in a metal-to-metal contact, and to place a thermocouple, thermistor or other temperature-sensitive element as close as possible to the piston Lacking such an element it is possible to make an order-of-magnitude estimate of the difference in temperature between the piston and the ambient air from a knowledge of the power input and the geometry and masses of the jacketed and jacketing cylinders. For instance a 15 W motor rotating the piston at 4 revolutions per second to supply a torque of 0.075 m X 1 N is supplying about 6000 J/hour, enough to raise the temperature of the cylinders about 0.7^o K in one hour. A very rough calculation indicates that about 10 hours would be required before temperature equilibrium with ambient air is attained, and at the end of a working period of 3 hours the temperature differential would be of the order of 2° K. With bench or console models rotation is by occasional manual impulse and the temperature effect is inconsequential. If α_p is the linear thermal coefficient of expansion of the piston, the effect on the area is $2\alpha_p$ and if T^oC is the temperature, ^oC, we may write:

$$\left(\frac{\Delta A}{A_o}\right)_T = 2\alpha_p\left(T-20^\circ C\right) \tag{7}$$

The area of the piston under load is:

$$A_p = A_o\left[1+\left(\frac{\Delta A}{A_o}\right)_{P_m}+\left(\frac{\Delta A}{A_o}\right)_T\right] \tag{8}$$

If the jacket pressure were increased to the point where the cylinder and piston were just in contact without clearance ("stall pressure") the piston and the bore of the cylinder would be of equal area, A_p. If now the jacket pressure be decreased an amount ΔP_j the bore of the cylinder would be slightly greater in area than the piston and would now be:

$$A_c = A_p\left[1+\left(\frac{\Delta A}{A_o}\right)_{\Delta P_j}\right] \tag{9}$$

From the Lamé equations and Hooke's law of deformation in combined stress, if a cylinder of wall ratio ω=o.d./i.d., is strained diametrally at the bore because of external pressure ΔP_j, (15), then:

$$\frac{\Delta d_c}{d_c} = e_c = +\frac{2\Delta P_j\,\omega^2}{E_c(\omega^2-1)} \tag{10}$$

and:

$$\left(\frac{\Delta A}{A_o}\right)_{\Delta P_j} = \frac{4\Delta P_j\,\omega^2}{E_c(\omega^2-1)} \tag{11}$$

where e_c is the diametral strain, E_c is Young's modulus of the material of the cylinder and ΔP_j is the amount of relaxation from stall pressure. When the clearance is small, the effective area may be taken as the arithmetic mean of the cross sectional areas of the piston and cylinder:**

---

*In order further to condense the pressure gradient a modification has been designed (12) wherein the cylindrical piston terminates at its lower end in a sphere whose major diameter is slightly greater than that of the piston: as a result the gradient is reduced effectively to a line of contact (13). Aside from the fact that an ideal pressure gradient is approximated, the axes of the piston and of the cylinder may diverge very slightly without causing binding. This concept is discussed in Part B of this paper.

$$A_e = \frac{A_p + A_c}{2} \tag{12}$$

$$= A_o\left[1 + \left(\frac{\Delta A}{A_o}\right)_{P_m} + \left(\frac{\Delta A}{A_o}\right)_T + \frac{1}{2}\left(\frac{\Delta A}{A_o}\right)_{\Delta P_j}\right]$$

and substituting:

$$A_e = A_o\left[1 - (1 - 3\mu_p)\frac{P_m}{E_p} + 2\alpha_p(T - 20°C) + \frac{2\omega^2 \Delta P_j}{(\omega^2 - 1)E_c}\right] \tag{13}$$

Equation (13), expressing the effective area in terms of observed readings and of physical properties of the materials of construction of the piston and cylinder, shows three corrections which are to be applied to the nominal area. It will be noted that each of these corrections is linear. Fig. 4, illustrates the relative proportions of these corrections.

UNCERTAINTIES IN CORRECTIONS

3. Each of the parameters in equation (13) is subject to uncertainty, affecting the overall uncertainty in the effective area, which can be expressed as a partial differential. If we let ϵ represent the uncertainty, or possible error, we may write, in general,

$$\epsilon\,(A_e)_{x,y,----} = Y_{x,y,----}\,\epsilon_{x,y,----} \tag{14}$$

where x, y, -----are parameters in equation (13) and Y is one of the three corrections and is a function of x, y, ---. The overall uncertainty in the effective area is then:

$$\epsilon\,(A_e) = \Sigma\, Y_{x,y,----}\,\epsilon_{x,y,----} \tag{15}$$

The individual uncertainties are tabulated in Table I for the general case, and for two examples of the controlled clearance gage in Table II.

DISCUSSION

4. The parameters representing material properties are given for room temperature and atmospheric pressure. These properties are functions of both variables. As to temperature the functions are sufficiently known and the information is available. The effect of pressure is small. For Young's modulus and Poisson's ratio of the piston and cylinder materials determinations of the pressure effect have not, to our knowledge, been made; but Bridgman (15) tabulates percentage change of rigidity and of incompressibility under 1000 MN/m^2 (knowing any two of these properties the other two may be derived), and it appears that the effect of pressure on the values given would be negligible. The effect of pressure on thermal expansion is larger, but the temperature difference is so small that this error, too, need not concern us.

4.1 It is probable that the accuracies attainable with the various ranges of controlled clearance gage are sufficient for all current purposes. The time may come, however, when this is no longer true. Referring to Table II, it will be evident that for the lower pressure ranges the prime subject for improvement is the nominal area. It is, of course, obvious that for any range, it is advantageous to make the diameter of the piston as great as possible, c.f. Table I in reference (2). Here we are dependent upon the national standardizing laboratories and we understand that the tolerance quoted here may be improved by a factor of 5. The spherical piston, subject of Part B of this paper, should offer an advantage in this respect, since it is free from any question as to taper of the piston. As for the higher pressure ranges, in addition to further restriction of the nominal area, it is evident that more accurate values for the elastic properties will be required: by the use of resonance methods it is hopeful that these can be achieved. And if better knowledge of the temperature of the piston is required: it could be attained in the manner already suggested.

4.2 It is a pleasure to acknowledge the helpful discussions with Dr. D. P. Johnson and pertinent suggestions by Dr. J. M. Alexander in the preparation of this paper.

REFERENCES PART A

1. U. S. Patent No. 2,796,229

2. Johnson, D. P. and D. H. Newhall, Trans. ASME, 75, 301-310, (1953)

3. Brombacher, W. G., Inst. Cont. Syst. 40, 87, (Sept., 1967)

**Rigorously, the effective area is:

$$\pi\left[r_2 r_1 + \frac{(r_2 - r_1)^2}{6}\right]$$

where r_2 and r_1 are respectively the radii of the bore of the cylinder and of the piston. This was established by Myers and Jessup (13) who pointed out also that the difference between the true area and the mean area,

$$\pi\left[\frac{r_2^2 + r_1^2}{2}\right],$$

is very small. If the radii are respectively 2.54254 and 2.54000 mm (0.1000 and 0.1001 in) corresponding closely to a Harwood 300 DWT for 1.3 MN/m^2 (9000 psi) range, the difference between areas is about 1 part in 3,000,000. If the jacket pressure were applied until the piston and cylinder were in contact and then backed off until the clearance was 2.54000 mm the ΔP_j would be 2.0868 MN/m^2. In actual practice the ΔP_j would be less than one-tenth as much, with the clearance in accordance, and the agreement between the true and the mean values of effective area would be far better than the figure above, say one part in 30,000,000.

4. Lloyd, E. C., idem. p. 105

5. Heydemann, P. L. M. and B. E. Welch, Experimental Thermodynamics Vol II, Chapter 4, Part 3, B. LeNeindre and B. Vodar, Eds., Butterworths and Co., London, 1975

6. Person communication, D. P. Johnson

7. Cross, J. L., Nat. Bur. Stds. Monograph No. 65, (1964)

8. Amagat, E. H., Ann. Chem. Phys., Ser. 6, 29, 68 and 505, (1893)

9. Bridgman, P. W., Proc. Am. Acad. Arts Sci., 47, 321-343, (1911)

10. Johnson, D. P., J. L. Cross, J. D. Hill, and A. H. Bowman, Ind. Eng. Chem., 49, 2046-2050, (Dec. 1957)

11. Bennett, C. O. and B. Vodar, High Pressure Measurement, A. A. Giardini and E. C. Lloyd, Eds., Butterworths, Washington, 365-379, 1963

12. U. S. Patent No. 3,360,071

13. Meyers, C. H. and R. S. Jessup, Nat. Bur. Stds. Research Paper No. 324, June, 1931

14. Bridgman, P. W., Proc. Am. Acad. Arts Sci. 61, 57-99, (1926)

15. Timoshenko, S. P., Strength Of Materials, Vol. II, 3rd Ed. Van Nostrand, New York, 1958

16. Bridgman, P. W., Physics Of High Pressure, G. Bell and Sons, London, 1952, re-publ. by Dover, N. Y., 1970, p. 386

Part B. ADVANTAGES OF SPHERICAL PISTONS IN FREE-PISTON GAGING DEVICES

INTRODUCTION

1. The controlled clearance free piston gage, subject of Part A of this paper, was introduced by Johnson and Newhall (1) (2) in 1952 and has established its place firmly as a pressure standard for ranges above the mercury manometer level. In the succeeding years it has undergone very few changes, the most important having to do with mutual alignment of the piston, its containing cylinder and any guide bearings relating to the supported weights. The present arrangement is considered satisfactory, but the concept of the spherical piston (3) appeared as an alternate solution of the alignment problem.

PRINCIPLE

2. The spherical piston may take many forms, but basically it consists in a cylindrical column terminating at one or both ends in a spherical surface of slightly greater diameter, which diameter determines the nominal piston area. Since the largest portion of the piston is spherical it can rotate about a horizontal axis to accommodate any likely trivial mis-alignment.

ADVANTAGES

3. A spherical piston offers two important advantages over a plain cylindrical piston. First because of its ability to rotate about a horizontal axis it is inherently free from binding. Second, whereas in any piston gage an assumption has to be made regarding the precise location and value of the effective area, owing to its geometry the spherical piston significantly reduces this uncertainty.

Freedom from binding

3.1 A spherical piston gage is shown in Fig. 1 where S is a hard sphere, closely fitting the containing cylinder M and supporting the weights W through the coupling C and the support H against the thrust of the pressure to be measured. The guide bearing B eliminates friction axially and rotationally. The specification of a perfectly functioning piston in a free piston gage is that it be acted on only by a vertically and axially applied load and its equal and opposite hydrostatic resistance, with a leakage film preventing friction between it and its containing cylinder, and with no perceptible friction with respect to axial and/or rotational motion at any guide bearing. With neither a conventional nor a spherical piston can precisely axial loading be expected, since tolerances must be allowed for the loading or removal of the weights and for machining exigencies. Rotation of the piston tends to maintain an unbroken leakage film around the piston, but the clearance is small between a conventional straight piston and its cylinder except at the mid-region where the clearance is very much less. It is evident that the location and nature of a guide bearing is very critical. The spherical piston makes only line contact (except for the liquid film) with the cylinder and because of the reduced diameter of the column can incline slightly without making other contact with the cylinder. There is an appreciable tolerance, - 0.09 mm in the Harwood prototype, - hence the piston cannot bind. It is obvious that the greater the distance between the great circle of the sphere and the guide bearing in relation to the diameter of the sphere, the better the design. The fabrication problems in systems involving co-axial tandem hydraulic pistons, cylinders and bearings are greatly eased with the use of the spherical pistons.

Precision of area

3.2 With respect to the effect of load on the area of the piston, the same considerations apply as for a conventional piston: above the great circle the pressure is uni-axial, while below it the pressure is hydrostatic. The spherical piston has an advantage since the pressure gradient is located specifically at the great circle and there is greater assurance as to the influence of end effects of the cylinder.

APPLICATION

4. A prototype gage as shown in Fig. 1 has been constructed.[2] This was a bench model capable of reading 5.5 MN/m^2 (800 psi), with a sphere of 14.2875 mm (9/16 in) diameter. The cylinder M was jacketed for controlled clearance. The guide bearing B provided for manual rotation of the floating system. Unfortunately the instrument and most of the records

relating to it were destroyed by fire before evaluation was complete, and replacement has not yet been feasible: the limited results, however, were encouraging.

Load cell

4.1 A second design is for a load cell as shown in Fig. 6. In this case the generally cylindrical piston is much larger (about 90 mm in diameter by 250 mm long) than that of the dead weight tester just described. Whereas that piston was a true sphere, for the load cell the sphericity consists of an essentially spherical zone about one-quarter-way up from the lower face of the piston and having a radius of curvature slightly (about 10 percent) greater than the radius of the cylinder. To accommodate the inevitable eccentricity in the application of the load a portion of the top surface of the piston is made spherical and resting on it is a matching load-bearing member. Provision is made to force oil under pressure between the two mating surfaces. Guide zones are provided at each end of the piston. Here the clearances are small but larger than that at the spherical surface and several circumferential grooves are turned to promote even distribution of the leaking pressure fluid as a lubricant. For the best accuracy a jacketing cylinder provides controlled clearance.

Intensifier

4.2 The third design, Fig. 7, is for an intensifier having an accurately known pressure ratio. There are situations where, for lack of space, unfriendly environment, etc., it is impossible to place a suitable pressure measuring device within a region at high pressure. With such a problem it is customary to measure in conventional fashion the pressure in the low-pressure side of an intensifier, multiplying this by the piston ratio and making an allowance for the estimated friction. By using a sphere at both ends of the piston, the uncertainty due to friction in the intensifier is minimized if not effectively eliminated. The design shows the two separate cylinders, in which the piston moves disposed vertically and supported by platens held together by tie rods. A belt drive is provided to rotate the piston and eliminate bearing friction. Here again the controlled clearance feature is applicable.

Differential piston

4.3 The fourth design Fig. 8 shows a differential piston in a dead weight tester, having spheres of slightly different diameter at the two ends and having the load hung from a bearing fixed at about the mid-point of the piston. As in the previous case the two cylinders are supported by platens held by tie-rods, piston rotation is provided, and for highest accuracy controlled clearance is used.

CONCLUSIONS

5. Two significant advantages of a spherical piston over the conventional straight piston have been cited. Furthermore, it appears that the designs for a spherical piston are simpler, and presumably less expensive. Per contra, no disadvantage compared with a conventional piston has been recognized.

5.1 It is regrettable that we have not had the opportunity to accummulate more experience with the spherical gages. Rather than wait indefinitely for this, it has seemed preferable to disclose the information that is presently available.

REFERENCES PART B

1. U. S. Patent No. 2,796,229

2. Johnson, D. P. and D. H. Newhall, Trans. ASME, 75, 301-310, (1953)

3. U. S. Patent No. 3,630,071

TABLE I

INDIVIDUAL UNCERTAINTIES

CORRECTIONS FOR EFFECTIVE AREA

1. $\epsilon(A_e)_{\mu_p} = \frac{3P_m A_o}{E_p} \epsilon_{\mu_p}$

2. $\epsilon(A_e)_{E_p} = \frac{P_m A_o}{E_p^2}(1-3\mu_p)\epsilon_{E_p}$

3. $\epsilon(A_e)_{\alpha_p} = 2A_o(T-20^\circ C)\epsilon_{\alpha_p}$

4. $\epsilon(A_e)_{E_c} = \frac{4\omega^2}{\omega^2-1}\frac{A_o}{E_c^2}\Delta P_j \epsilon_{E_c}$

5. $\epsilon(A_e)_{P_m} = \frac{-(1-3\mu_p)A_o}{E_p}\epsilon_{P_m}$

6. $\epsilon(A_e)_{\Delta P_j} = \frac{4\omega^2}{\omega^2-1}\frac{A_o}{E_c}\epsilon_{\Delta P_j}$

7. $\epsilon(A_e)_{\Delta T} = 2\alpha_p A_o \epsilon_T$

8. $\epsilon(A_e)_{\omega} = \frac{-4A_o \Delta P_j}{E_s}\frac{\omega}{(\omega^2-1)^2}\epsilon_\omega$

9. $\epsilon(A_e)_{A_o} = \left[1-(1-3\mu_p)\frac{P_m}{E_p} + 2\alpha_p(T-20^\circ C) + \frac{2\omega^2}{\omega^2-1}\frac{\Delta P_j}{E_c}\right]\epsilon_{A_o}$

TABLE II

UNCERTAINTY OF EFFECTIVE AREA

PARAMETER		SYMBOL	VALUE	UNCERTAINTY OF VALUE	ppm DWT 200	ppm DWT 1000
1. Poisson's ratio, piston		μ_p	0.21	±0.002	0.01_3	$13._0$
2. Young's modulus, piston	MN/m^2	E_p	632,000	±6,200	0.007_9	$7._9$
3. Thermal expansion, piston	per °C	α_p	4.01×10^{-6}	$\pm0.5\times10^{-6}$	0.01_0	$2._0$
4. Young's modulus cylinder	MN/m^2	E_c	208,000	±3,400	2.5	$22._1$
5. Measured pressure	MN/m^2					
	DWT 200	P_m	1.38	±0.00014	0.00008_1	
	DWT 1000	P_m	1,380	±1.4		$+0.8_1$
6. Differential jacket pressure	MN/m^2					
	DWT 200	ΔP_j	5.52	±0.069	0.1_9	
	DWT 1000	ΔP_j	68.96	±3.4		0.1_3
7. Differential temperature	°C					
	DWT 200	T-20°C	0.01	±0.5	$4._0$	
	DWT 1000	T-20°C	2.0	±0.5		$4._0$
8. Wall ratio						
	DWT 200	ω	1.813	±0.00045	0.01_7	
	DWT 1000	ω	12.824	±0.014		0.00_9
9. Nominal area	mm^2					
	DWT 200	A_o	645.4	±0.0058	$9._0$	
	DWT 1000	A_o	3.227	±0.00034		$340._0$
Total Uncertainty of Effective Area,				ppm	$\pm15._7$	$\pm400._0$
" " " " "				percentage	$\pm .001_7$	$\pm 0.04_0$

PART A

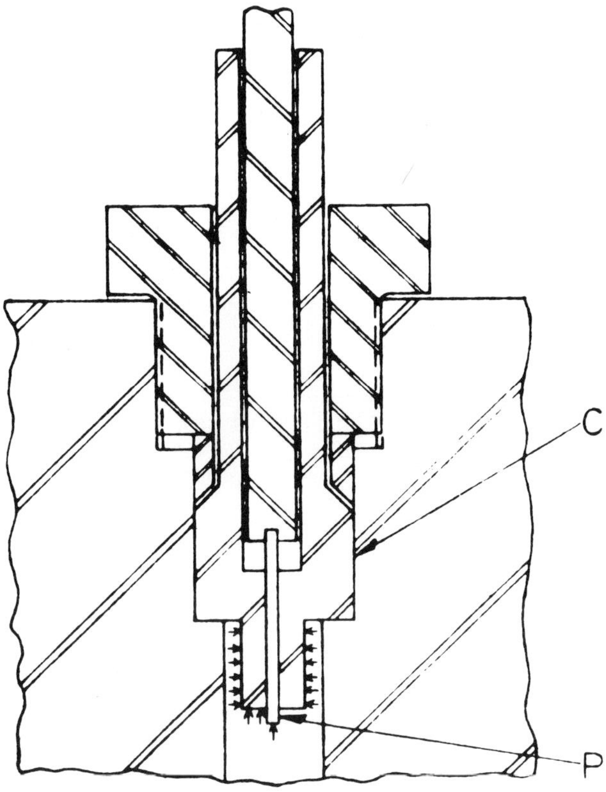

Fig. 1: Bridgman gage, re-entrant cylinder P represents the piston and C the re-entrant cylinder

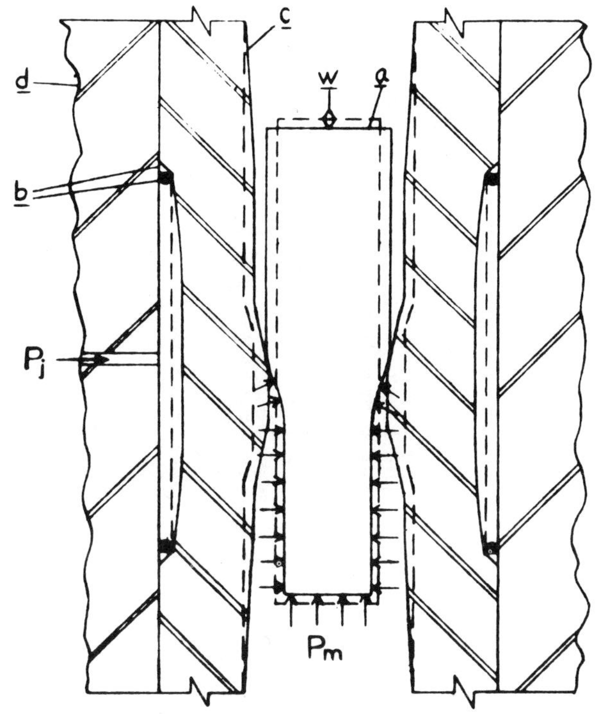

Fig. 2: Controlled-clearance piston gage. Dotted lines show configuration at atmospheric pressure; full lines, under measured pressure, Pm, and jacket pressure, Pj

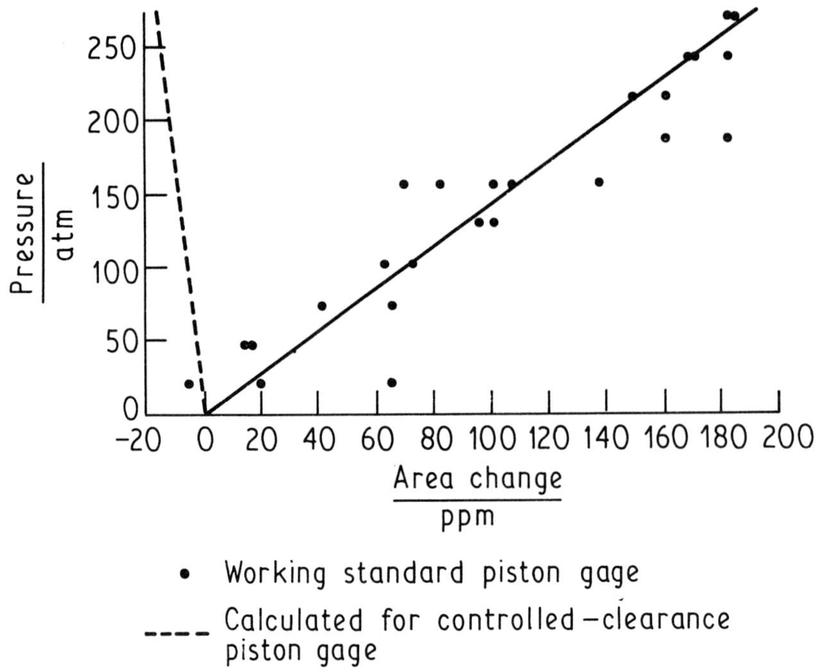

Fig. 3: Change in effective area with pressure

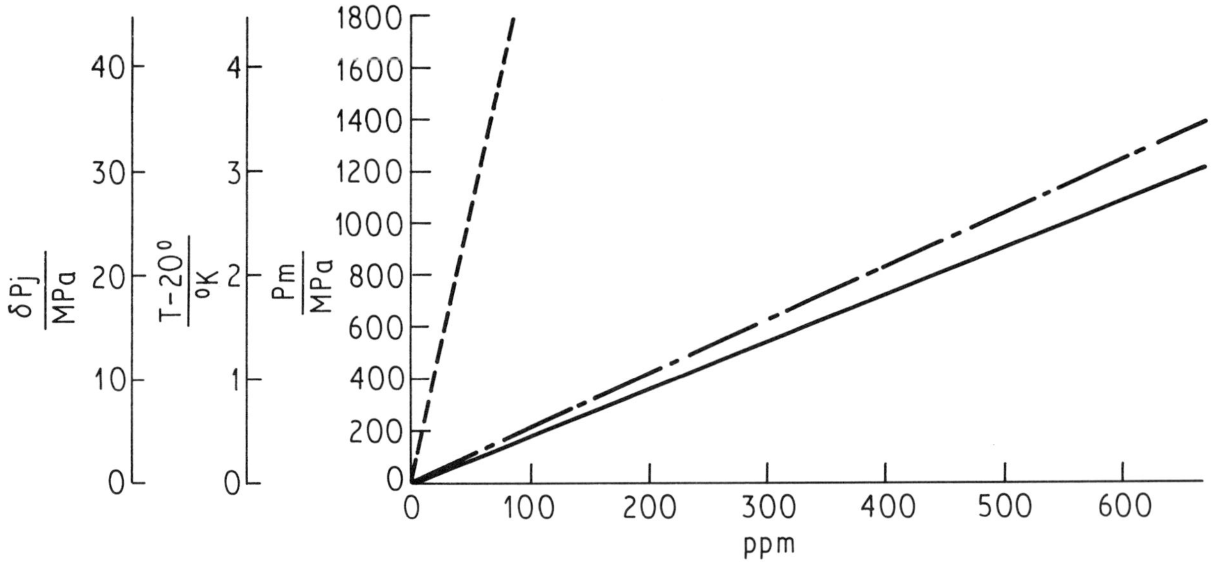

Fig. 4: Relative proportions of corrections
Abscissae: corrections, ppm, (parts per million)
Ordinates:

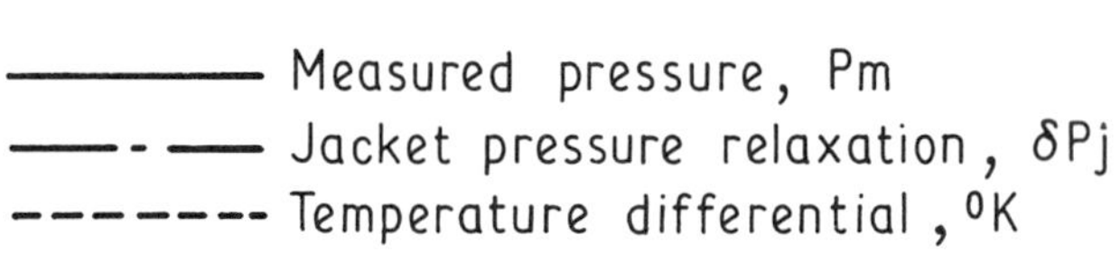

PART B

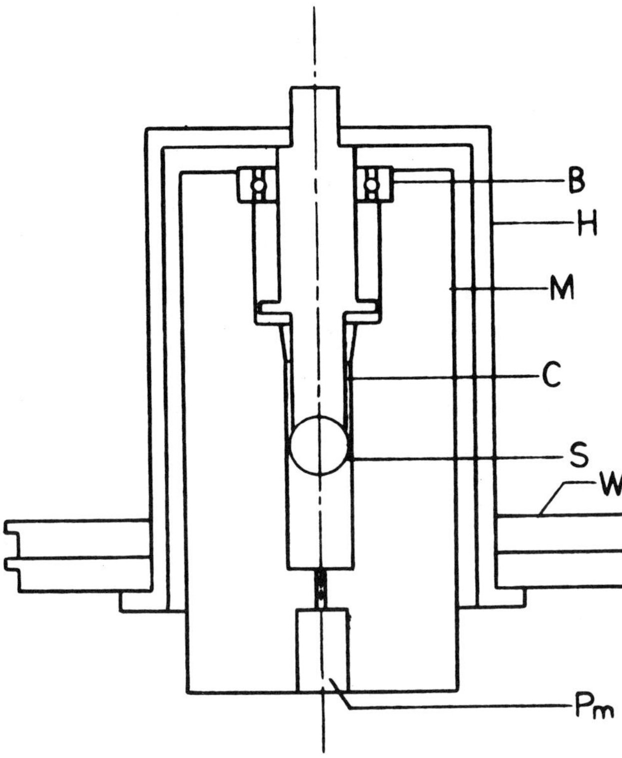

Fig. 1: Spherical piston gage
P_m Pressure to be measured
W Weight
H Support
C Coupling
M Cylinder
B Guide bearing
S Sphere

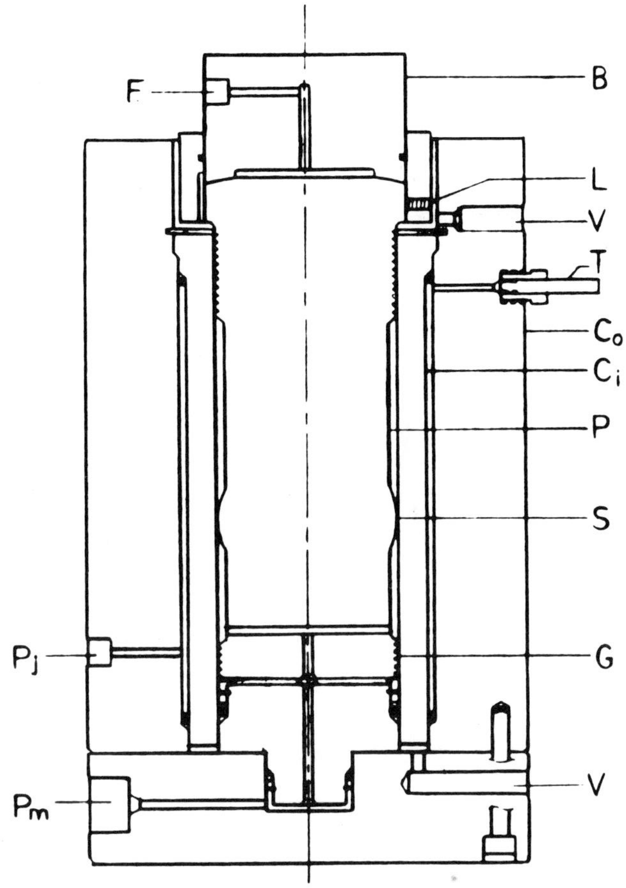

Fig. 2: Load cell
S Spherical zone
P Piston
B Bearing
G Guide zone
P_m Measured pressure
P_j Jacket pressure
F Lubricating fluid
V Vent
T Dead end plug
C_1 Jacketed cylinder
C_0 Jacket cylinder
L Set screw

PART B

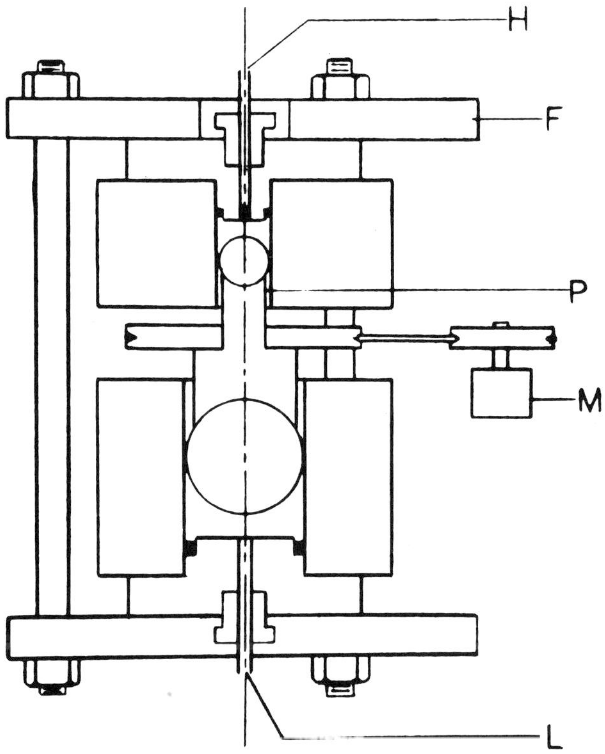

Fig. 3: Intensifier
P Piston
L Low pressure
H High pressure
F Frame
M Motor for rotating piston

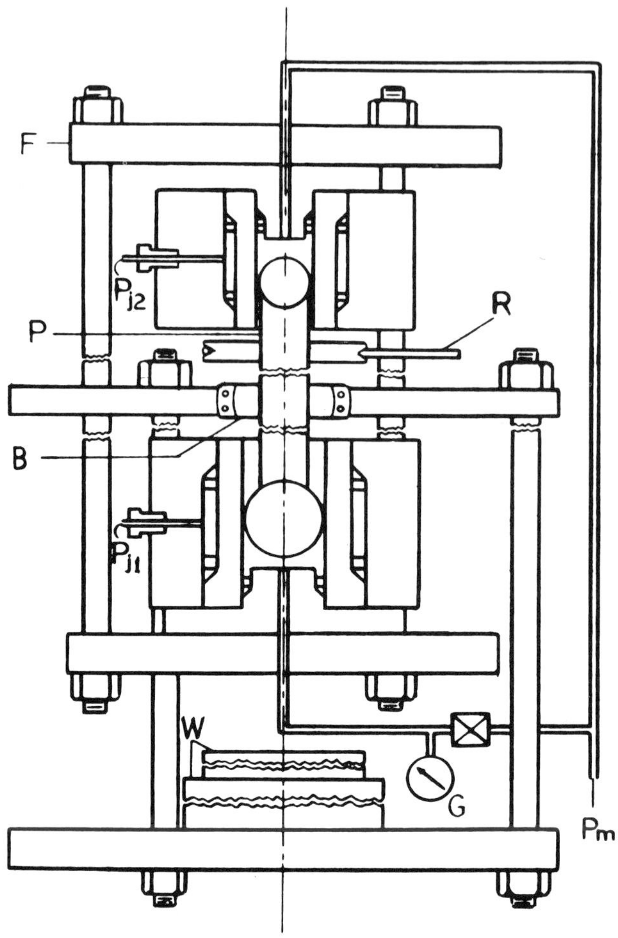

Fig. 4: Differential piston
P_m Measured pressure
P_{j1} Jacket pressure, lower cylinder
P_{j2} Jacket pressure, upper cylinder
R Belt drive
W Weight
G Gage or other test piece
B Bearing supporting weight
F Frame
P Piston

STRESS ANALYSIS OF HIGH PRESSURE VESSELS BY MEANS OF PHOTO-ELASTIC INVESTIGATIONS

A. KUSKE. W. STEINCHEN and J. ZECH
Institut für Maschinewesen, Universität Clausthal, D 3392 Clausthal-Zellerfeld, West-Germany

SYNOPSIS—In recent years, the 'Institut für Maschinenwesen' has had to solve an increasing number of problems concerning high pressure vessels. Mainly the method of photo-elasticity has been used. Numerical methods fail in cases of complicated shapes. Different examples are given: viz. 1. atomic reactor: shape, pressure, the model, stress distribution; 2. receiver: very high stresses and temperature used for metal pressing, optimal shrinkage to minimize stresses; 3. cable covering mould: kind of tool, load, thermal stresses; 4. autoclave: very high pressure and temperature, notch effect at bottom, different types of plugs. The examples presented show that photoelasticity is a useful tool for the solution of pressure problems and has advantages in comparison with theoretical calculations.

INTRODUCTION

In recent years, the 'Institut für Maschinenwesen' has had to solve an increasing number of problems in the development of new types of pressure vessels.

Simple two-dimensional problems can easily be solved numerically (finite elements), but in cases of complicated three-dimensional shapes or unknown distribution of loading forces (e.g. screw thread) the stress distribution can only be found by experimental methods. That is why most of these problems were solved by the photoelastic methods most suited for three-dimensional cases. Mainly the freezing method[1] was applied, where the stresses in a model of the prototype, machined out of an epoxy-resin, are frozen in at 130°C under load. After cutting slices out of the model, the three-dimensional stress distribution in the whole model can be determined. These results are applied to the prototype observing the model laws of similarity. Only in a few cases was two-dimensional photoelasticity used instead of the freezing method.

The vessels mentioned in this report were loaded by internal pressure, shrinkage forces, concentrated forces or by inhomogeneous temperature distributions. Four examples of high pressure vessels are described:

1. an atomic reactor
2. a receiver
3. a cable covering mould
4. an autoclave.

1. *Atomic reactor*

The prototype of the 300 MW atomic reactor to be investigated contains vessels of different shapes. These vessels are loaded by internal pressure—by gas or steam—and by bolt forces in the bolted flange. Because of the flanges and openings in the vessels, the three-dimensional stress distribution is so complicated that it could not be calculated by means of equations for shells.

The photoelastic freezing method which was used makes it possible to determine the influence of the shape of the vessel on the stress distribution and to determine directly the stress concentration-factors. The models of epoxy-resin were cast roughly to shape and then machined to the final shape.

The cover of the steam generator, a critical part of the atomic-reactor, was to be investigated. Horizontal and vertical sections of this cover are shown in Fig. 1. The stress distribution is mainly influenced by the flange and by the tubes penetrating the cover. The internal pressure and the forces of the bolted flange were investigated separately by two different models with two different load factors. This had to be borne in mind in evaluating the stresses in the prototype from the experimental results. The first model was loaded by an internal pressure of 1·0 N/cm². The pressure was kept constant by means of a column of a suitable oil. In order to compensate for the influence of the weight of the oil and the model, the whole model was submerged in a bath of the same oil.

Because of the critical strength of the bolt connection the two flanges had been cemented together for this experiment, but this does not influence the stress distribution outside the flange. Figure 2 shows the first model under internal pressure, Fig. 3 shows a radial slice cut out of this model photographed in polarized light. The dark lines are the isochromatics, from which, for example, the meridian stresses at the outer edge σ_o of the prototype can be calculated by applying optical laws and model laws:

$$\sigma_o = 2650 . n \; |N/cm^2|$$

where n is the fringe order of the isochromatics which can be read from Fig. 3 and the factor 2650 is obtained from the pressure relation, optical value and the thickness of the model slice. The stress at the inner edge σ_i is calculated in a similar manner

$$\sigma_i = (2650 . n - 450) \; |N/cm^2|$$

Here the radial pressure of 450 N/cm² is to be subtracted from the difference of principal stress given by the fringe order n. The maximum fringe order n = 4·5, corresponds to stresses of 12 000 N/cm² in the prototype. The section of zero-moment where $\sigma_o = \sigma_i$ can also be obtained from the fringe pattern in Fig. 3 and this is the most suitable area for the weld.

The lower part of the cover is subjected to the forces acting on the flange. As the bolted flange has no influence on the stress distribution in the area of the pipe ducts, these are not incorporated in the second model, which is shown in Fig. 4.

It is not necessary to make all bolts out of the model material. In those parts of the model where no slices are cut, steel bolts are used. Figure 5 shows the stress distribution in the flange caused by internal pressure (p) and by the bolted flange (s) as well as by both acting together.

2. *Receiver*

Vessels with high internal pressure are usually constructed of several thick tubes shrunk together. Typical examples of these vessels are receivers used for metal forming processes. These receivers are undergoing repetitive internal pressures from 0 N/mm^2 up to 1000 N/mm^2, shrinkage stresses and temperatures of more than 500°C. Simple calculation methods for those compound tubes are well known. But if they contain axially drilled holes in one or several tubes needed for a heating system, the stress distribution can be determined easier by experimental methods because there is a loss of stiffness and stress concentrations at holes. Photoelasticity proved to be very useful for these investigations.

A receiver constructed of 3 tubes shrunk together was investigated. A large number of holes was necessary to accommodate an electrical heating system. The questions to be answered were:

(a) is the loss of stiffness caused by the holes negligibly small?
(b) what is the stress concentration factor at the holes?
(c) which is the most suitable pitch diameter for the holes?

The two-dimensional method of photoelasticity was applied. Several models were necessary to check the influence of the holes on stiffness and stress concentration.

(a) Stiffness

The models consisted of three plane rings (with given interferences) which were shrunk together. The difference of the principal stresses ($\sigma_1 - \sigma_2$) in the model—caused by shrinkage—is read from the isochromatics which are observed in polarized light. The experimental result for the stress distribution coincides with that found by calculation. Holes were then drilled in the model. The total fringe system was not changed visibly by the holes; indicating no significant influence on the stiffness.

(b) Stress concentration

In order to determine the stress concentration factor $K_t = \sigma_{max}/(\sigma_1 - \sigma_2)$ the maximum stress at the holes (σ_{max}) was compared with the stress difference ($\sigma_1 - \sigma_2$) at the corresponding pitch diameter before drilling the holes. The stress concentration factor K_t was found to be between 1·6 and 1·7 for all holes at different pitch diameters.

(c) Pitch diameter

The determination of an optimum pitch diameter for the holes is an important problem. The smaller the pitch diameter the better the heat transfer to the centre of the receiver, where the metal ingot is located. But the smaller the pitch diameter the higher the shrinkage stresses in that part of the receiver and the resulting stress concentration. Therefore a diagram was constructed indicating those pitch diameters which are suitable for drilling. The suffices in the following terms correspond to the equivalent ones of the Smith-diagram (Fig. 6):

$$\sigma_{sh} = \sigma_{min}; \quad \sigma_{sh} + \sigma_p = \sigma_{max};$$
$$\sigma_{sh} + \sigma_p/2 = \sigma_m; \quad \sigma_p/2 = \sigma_a$$

small suffices indicate existing stresses in the investigated specimen due to the load:

σ_{sh} = shrinkage stresses
σ_p = stresses caused by internal pressure.

The stress distribution caused by shrinkage σ_{sh} (Fig. 7) is superimposed on the repetitive stresses caused by internal pressure σ_p (Fig. 8).

The mean stress $\sigma_m = \sigma_{sh} + \sigma_p/2$ is shown in Fig. 9.

Now the allowable alternating stress, σ_A can be obtained from the Smith diagram (see Fig. 10) curve (1) as a function of σ_m.

This curve (1) must be compared with the curve of

$\sigma_a . K_t = \sigma_p/2 . K_t = \sigma_a^*$ (alternating stress including stress concentration at the holes)

(curve (2)). If the condition $\sigma_a^* \leqslant \sigma_A$ is satisfied, holes can be drilled at that pitch diameter (crossed area). Moreover, the condition $\sigma < \sigma_y$ curve (3), (4) (σ is the maximum stress at any point of the receiver, σ_y is the yield stress) should be satisfied at every point to avoid damage to the receiver.

3. *Thermal stresses in a cable covering mould*

The stresses resulting from temperature differences have not been mentioned in the example above. The reason for preheating a receiver or a mould is to avoid thermal stresses generated by the warm ingot and to avoid rapid cooling of the ingot. However, preheating of a vessel causes thermal stresses during the heating period. The magnitude of these stresses depends on the geometry of the vessel and on the amount of heat transmitted per unit time.

Since the calculation of the thermal stresses in cases of complicated geometry is only possible by expensive numerical methods, photoelasticity is chosen in this case because the optical effect is caused by the inhibited strains which cause stresses, while thermal dilatation produces no double refraction. (Other methods, e.g. strain gauges cannot separate the two effects.) Using model laws, the experimental results are transferable to the prototype. Of course, a time factor must be determined if the temperature field is unsteady.

Figure 11 shows the isochromatics of a model of a cable covering mould.[3] Only the thermal stresses are considered in this example. The four quarters of Fig. 11 show the thermal stresses in the mould after different times of heating. The electrical heating system is located in the holes. The front and rear planes of the model are protected against heat loss during the experiment which means that the heat flows only in the plane of the model. The temperature at several diameters was measured by thermo-couples.

Figure 12 shows the plot of thermal stresses against heating time for a point of the model located at the inner edge of the mould. The maximum stresses were reached in the model after 30 min, which means a time of about 15 min for the prototype under the conditions of this experiment. After the thermal stresses have been determined, the stresses caused by shrinkage and internal pressure can be superimposed.

4. *Autoclave*

In high pressure vessels for producing some ceramic and sinter materials of large dimensions, the pressure is for example about 20 000 to 40 000 N/cm^2. The purpose of this investigation is to optimize the closure and to allow even higher pressures to be used.[4] Different types of plugs are used in such cases. Most of them are threaded with different shapes of threads or in form of bayonet-type fittings. It is well known that high stresses would occur in the flat base of a vessel. Actually, the stresses are higher than the elementary theory predicts because of the notch effect which is shown in Fig. 13.

Several different problems of high pressure vessels were solved by photoelastic experiments requiring the application of the freezing method. The solution of this problem with numerical methods, i.e. the finite-element-method would require considerable effort to get approximate results.

In Fig. 14 a vessel is shown with two possible forms of closures, the upper end with a bayonet plug and the lower end with a threaded plug. The photoelastic models for this series of investigations were made of Lekutherm X 30, an epoxy-resin produced by Bayer-Leverkusen. It was found that the most serious problem was that of the unequal distribution of the forces over the threads of the screw. The stress in the first thread is higher than in the others. In order to equalize these forces, different improvements were suggested, first the influence of the length of the screw, i.e. number of threads was investigated (Fig. 15).

Figure 16 shows a better shape of a thread for stress analysis than the thread of DIN 405. In another suggested variation the closure consists of two equal parts (Fig. 16 lower end). The outer part is screwed against the inner one. As the thread does not fit exactly but has a certain clearance, the outer part of the plug is loaded first when the pressure is increased. The thread of the inner part of the plug, however, is not loaded until the pressure exceeds a certain limit. The stress distribution for the split plug is shown in Fig. 17. Figure 18 shows the distribution of isochromatics and the shape of the split plug: the outer part is screwed against the inner one, which is conically threaded.

Another idea was tried for the lower end of the same model (Fig. 14). Here the thread having a certain clearance was deformed by a ring shrunk on the cylinder. The effect of this procedure is a more equal distribution of the force on the threads.

If the cylinder has a bayonet-plug which can be opened or closed rapidly, the problem of stress concentration is even more serious. This cylinder can be closed turning the plug which is a combination between thread and bayonet by 30°. The operating condition is shown in Fig. 19. The maximum stress in a plug containing slits in the axial direction was found to be 40% lower than that in a normal thread plug.

References

1. A. Kuske. 'Taschenbuch der Spannungsoptik.' Stuttgart 1971.
2. A. Kuske and G. Robertson. 'Photoelastic stress analysis.' John Wiley & Sons, London 1974.
3. W. Bock. 'Beitrag zur festigkeitsmäßigen Optimierung mittels Spannungsoptik von Strangpreßwerkzeugen am Beispiel einer Kabelmantelmatrize unter Berücksichtigung von Wärmespannungen.' Diss. TU Clausthal 1975.
4. E. R. Koch. 'Beitrag zur festigkeitsmäßigen Optimierung von Bauteilen am Beispiel eines Hochdruckautoklaven.' Diss. TU Clausthal 1971.

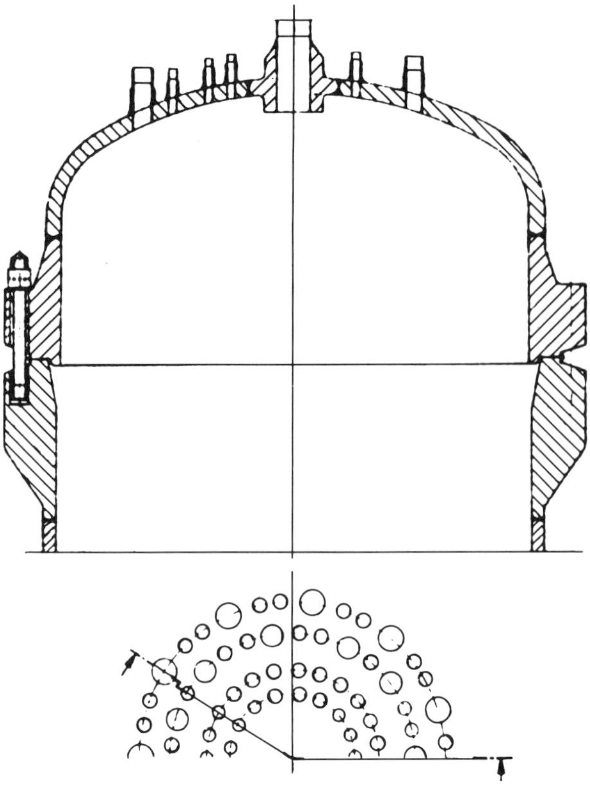

Fig. 1.

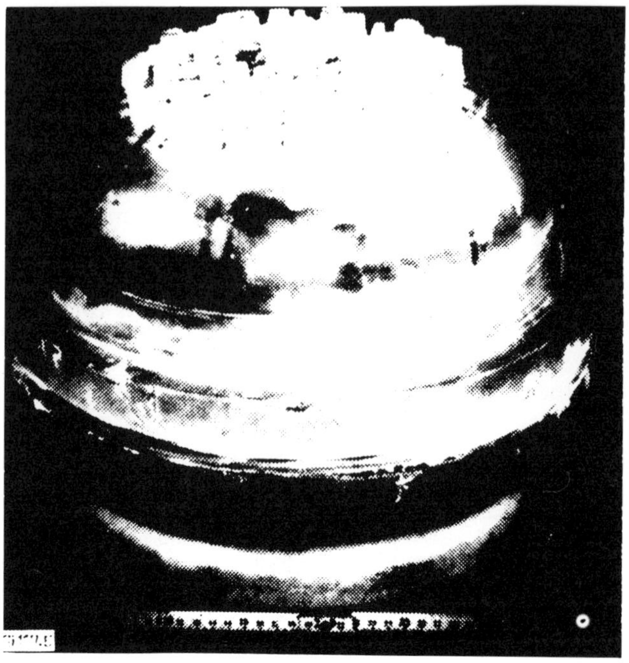

Fig. 2.

Fig. 3.

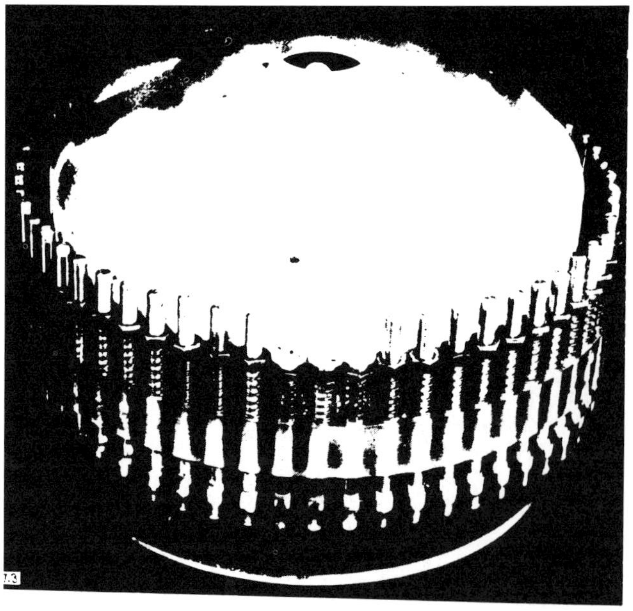

Fig. 4.

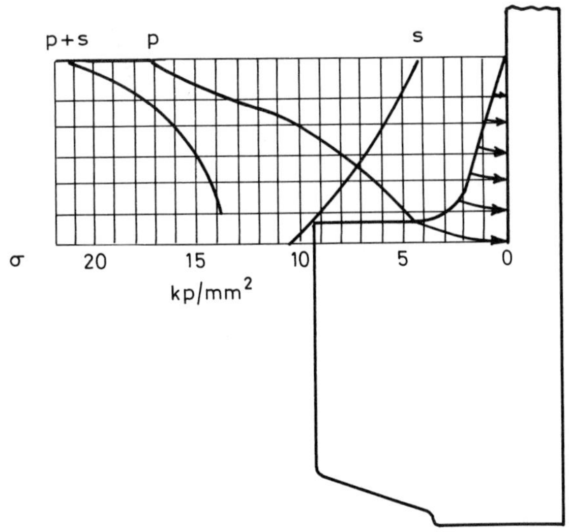

Fig. 5.

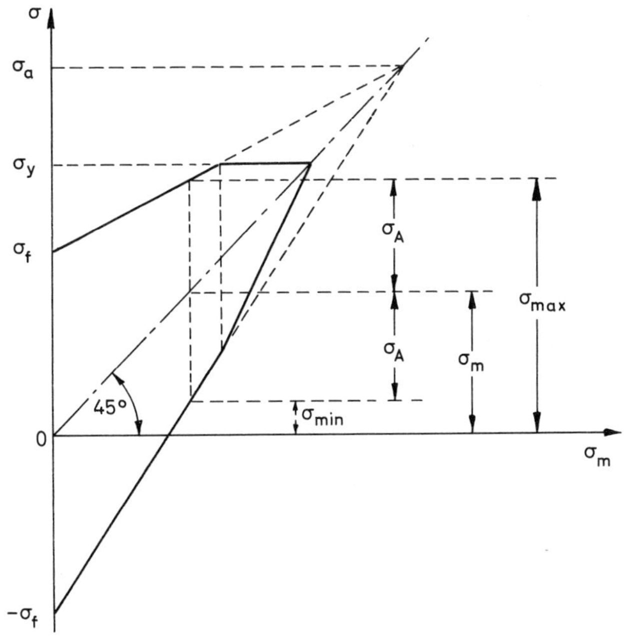

Fig. 6.

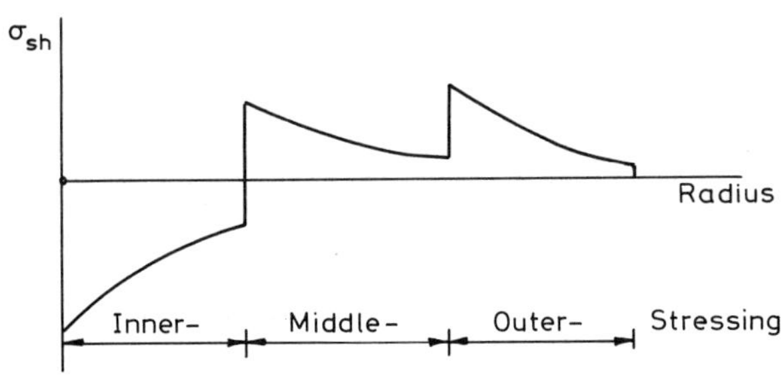

Fig. 7.

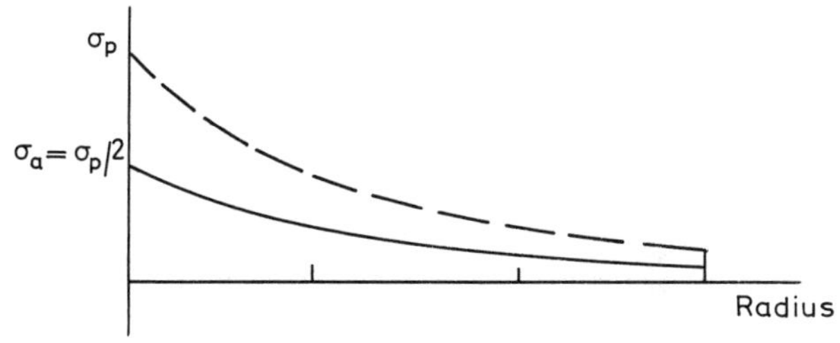

Fig. 8.

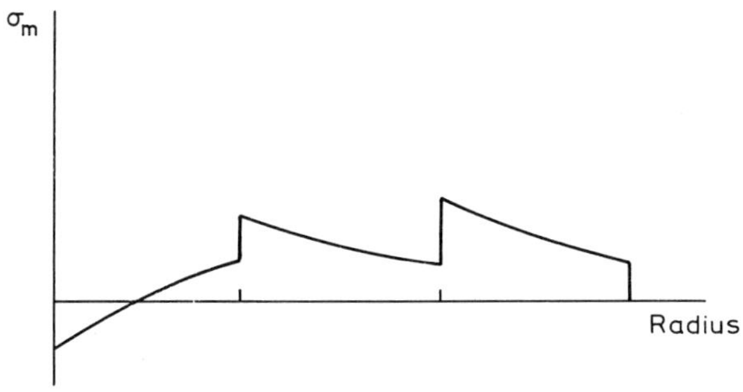

Fig. 9.

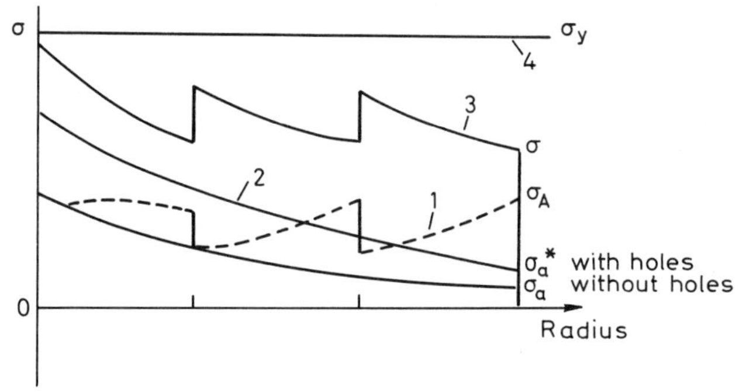

Fig. 10.

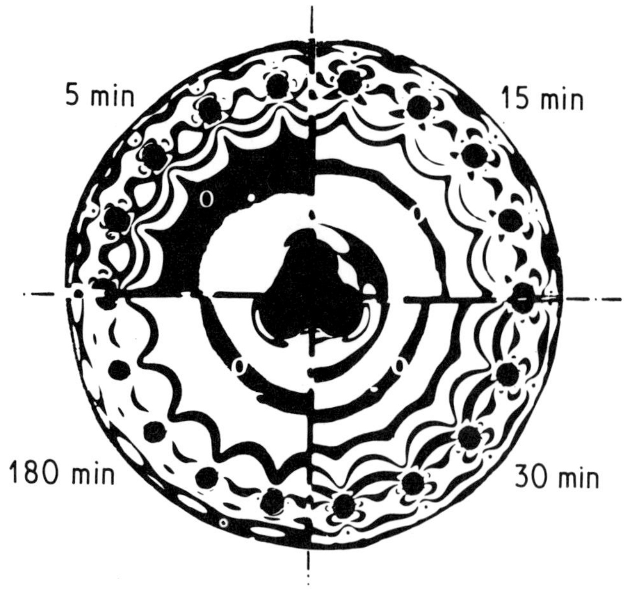

Fig. 11.

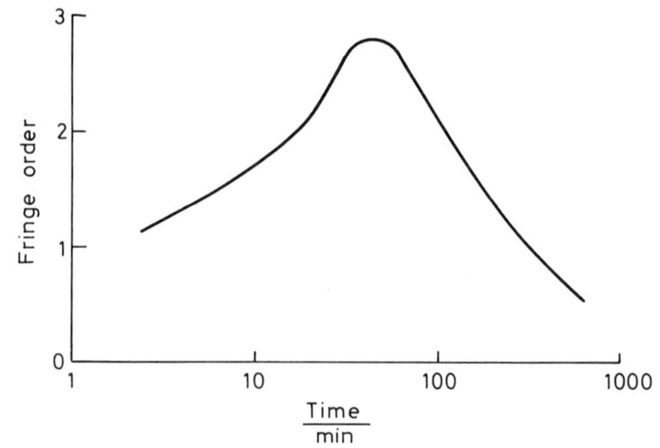

Fig. 12.

Fig. 13.

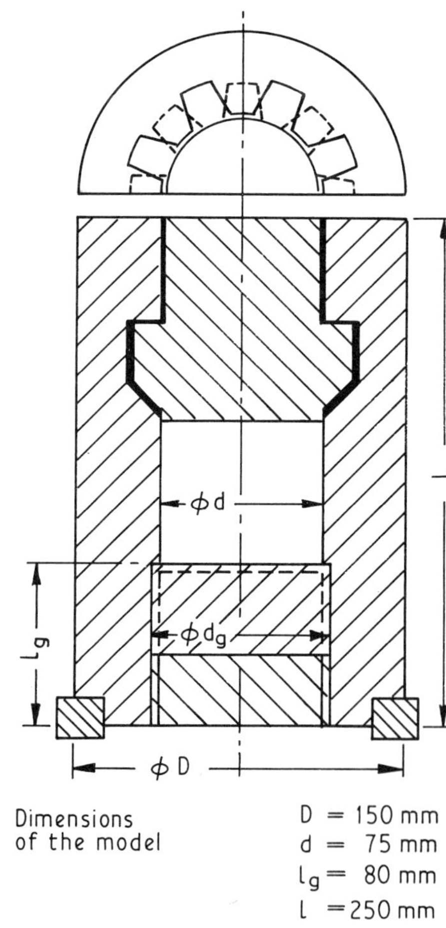

Fig. 14.

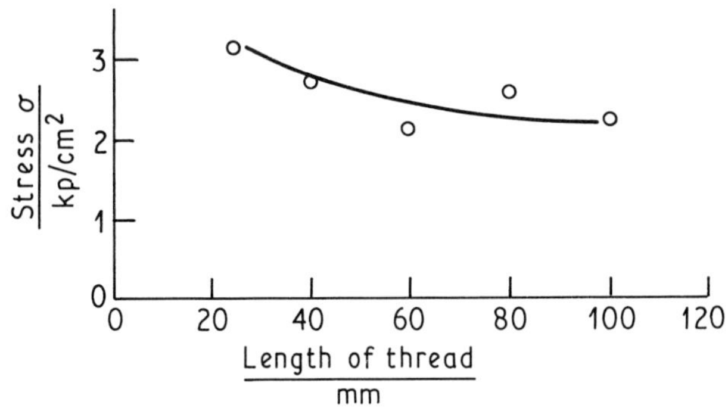

Fig. 15.

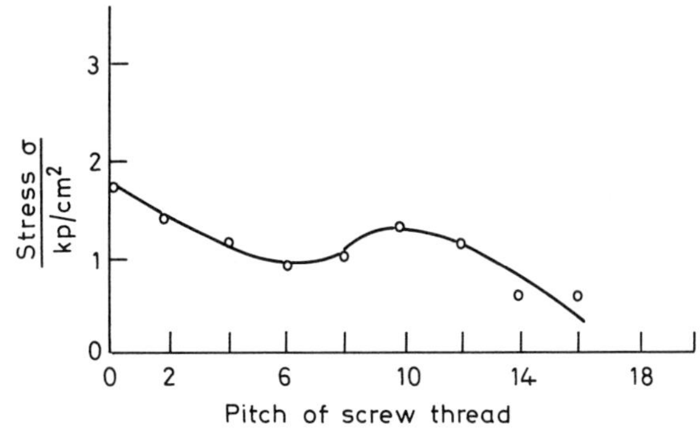

Fig. 17.

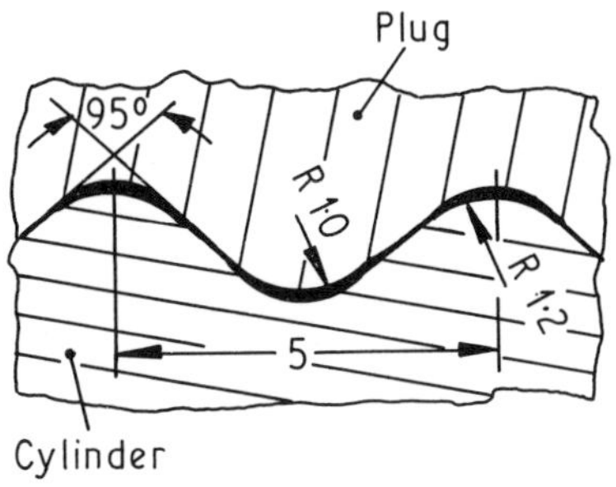

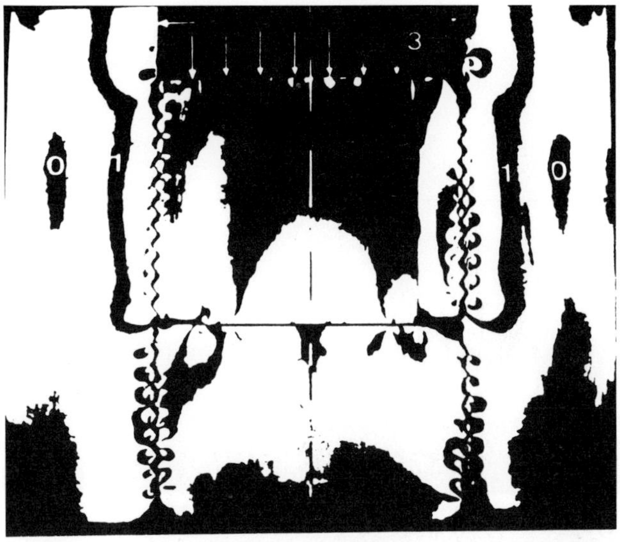

Fig. 18.

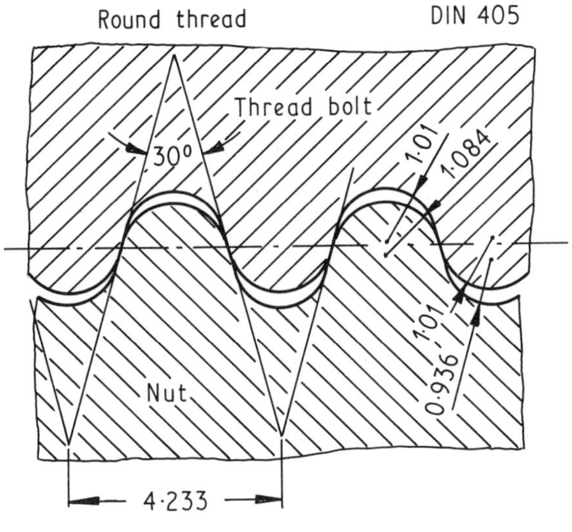

Fig. 16.

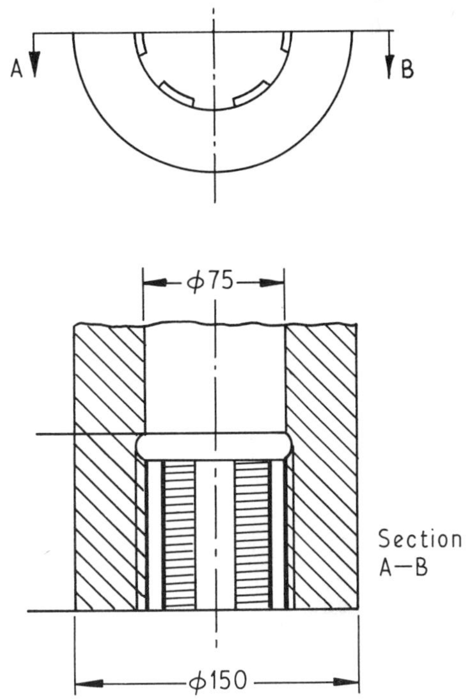

Fig. 19.

FRACTURE TOUGHNESS MEASUREMENT AND ULTRASONIC CRACK MEASUREMENT IN THICK-WALL CYLINDER GEOMETRIES

D. P. KENDALL, MSc, J. H. UNDERWOOD*, MSc and D. C. WINTERS, MSc
US Army Benet Weapons Laboratory, Watervliet, NY, USA
*currently on US Army Fellowship at National Engineering Laboratory, East Kilbride

SYNOPSIS Fracture toughness tests were performed using 'C' shaped specimens cut from thick-walled cylinder forgings of a Ni-Cr-Mo steel. The measured critical K_I value is essentially constant over a range of specimen sizes and is in good agreement with the known K_{Ic} value for the material. A continuous measurement of the crack length was obtained during the fracture toughness tests using an ultrasonic pulse-echo technique. Based on these ultrasonic crack growth measurements and calculated crack growth values from a compliance analysis and on prior work with 'C' specimens, a K_{Ic} test procedure is recommended for the 'C' specimen.

INTRODUCTION AND OBJECTIVES

1. High pressure containers are, more often than not, in the form of thick-walled cylinders. In previous work (ref. 1) a new fracture toughness test method was proposed which is particularly suited to thick-walled cylinder geometries. The method utilizes a 'C' shaped specimen which consists of a portion of a disc cut from a thick-walled cylinder. It is edge notched on the inner radius and pin loaded in tension through holes drilled near the ends. A sketch of typical specimens of this type over a range of diameter ratios is shown in Fig. 1. This specimen has the advantage of providing the maximum possible utilization of available material from a thick-walled cylinder and has an economic advantage due to its ease of machining.

2. This report describes an experimental study of the applicability of the 'C' specimen to a variety of cylinder geometries. 'C' shaped specimens were fabricated from alloy steel forged cylinders and tested using the accepted plane-strain fracture toughness test procedure (refs 2 and 3) except where the procedure could not apply to 'C' specimens. During the fracture toughness tests the crack length was continuously measured using an end-on ultrasonic technique. From the ultrasonic measurements the point on the load-deflection curve which corresponds to the critical crack extension can be determined. Then the procedure for determining K_{Ic} from the load-deflection curve can be established for the 'C' specimen.

3. The specific objectives here are first to demonstrate that 'C' specimen tests give the expected value of fracture toughness over a range of specimen sizes, second, to describe the procedure for measuring crack growth in cylinder geometries using the end-on ultrasonic method third, to outline a K_{Ic} test procedure for the 'C' specimen which can be used for a range of thick-walled cylinder geometries.

TEST SPECIMENS

4. The two groups of specimens listed in Table 1 are from two types of forged steel hollow cylinders. The first group were machined from a single forging of about 0.4 m outside diameter and 11 m in length. The group one specimens were taken from the forging in such a way that the plane of the crack was normal to the tangential direction of the cylinder and so that the specimen crack tip was located at the approximate mid-wall-thickness point of the original forging. The second group were machined from two gun barrels which have outer radii and wall thicknesses of the same values as those listed for the 'C' specimens, see Table 1. The table lists the dimensions of both groups of specimens. The material for all specimens is a vacuum arc remelted, quenched and tempered 4330 steel, modified by added Ni, Cr and V. The measured chemical composition for the group one material is the following in weight per cent: C-0.34, Ni-3.08, Cr-1.15, Mo-0.58, Mn-0.50, Si-0.22, V-0.13, P-0.012, S-0.011. The mechanical properties measured from the forgings are shown in Table 2.

5. Both the British and American standard methods for fracture toughness testing (refs 2 and 3) require that three specimen dimensions a, b, B, be at least $2.5(K_{Ic}/\sigma_y)^2$. The requirement that the specimen thickness, B, be at least this value should in general be easily met with 'C' specimens since it involves only cutting the proper thickness slice perpendicular to the cylinder axis. However the minimum crack length and ligament length requirements, a and b, can only be met if the specimen wall thickness, W, is adequate. An indication of how the crack length of the specimens tested in this study compares with the minimum size requirement is shown by the ratio of these values, $a/2.5(K_5/\sigma_y)^2$, listed in Table 2. The K_5 values are the values at the 5 per cent secant offset load from each test, see Table 1. Without considering at this point how close these K_5 values may be to K_{Ic}, the $a/2.5(K_5/\sigma_y)^2$ values indicate that our specimens range from about the minimum required size to about twice that value.

TEST PROCEDURE

Fracture toughness tests

6. The test specimens were machined to the dimensions in Table 1 and then notched at the midpoint on the bore surface with a power hacksaw. The notches in the group one specimens were sharpened with a 0.4 mm jeweller's saw. The group two notches were cut to an estimated 0.15 mm radius using a broaching procedure with a 'V' shaped blade in a power hacksaw. The latter procedure proved to give straighter initial notch fronts and more uniform crack growth, see Figs 2c and 2d. All specimens were then preloaded in compression to loads corresponding to 65 $MN/m^{3/2}$. We believe that a tensile residual stress is produced just ahead of the notch tip due to the compressive preload and that the residual stress produces faster and more uniform crack initiation along the notch front.

7. Next the specimens were precracked in a Sonntag SF-1U fatigue machine at 30 Hz. The K value corresponding to the maximum fatigue load was never more than 55 $MN/m^{3/2}$. The minimum load was always 10 per cent of the maximum load. In general the notch was extended 5 mm by fatigue. Five of the original fourteen specimens planned for group one developed biplanar fatigue cracks due to lack of proper notch sharpening. The results from these specimens have not been considered. Of the remaining nine group one specimens reported, the specimen with the fatigue crack most affected by an improper notch is specimen 1A, which is shown in Fig. 2a. The small area of biplanar crack growth could have had an effect on the subsequent fracture. The fatigue crack of specimen 3B shown in Fig. 2b is more typical of group one. The group two specimens, with adequate notch sharpness, had uniform fatigue cracks, seen in Figs 2c and 2d, specimens 348A and 490A respectively.

8. The fracture toughness tests were conducted in a hydraulic testing machine that provided an electrical output proportional to the load. This output was used to drive one of the Y axes of the 2-pen X-Y_1-Y_2 recorder. Specimen extension was measured by means of a special, spring-loaded displacement transducer which was fitted into indentations made on the inside surface of the specimen at the mid-thickness and on a line connecting the centres of the loading holes. The output of this transducer was connected to the X axis of the recorder.

9. The fracture toughness tests described above were performed in the same general manner as recommended for the standard fracture toughness specimens. Thus, if the ultrasonic measurements show that the critical initial crack growth for the 'C' specimen occurs at the 5 per cent secant offset load, as it does for the standard bend and compact specimens, then the critical K values obtained in these tests are good measures of K_{Ic}. Typical load-deflection plots from the two groups of tests are shown as Fig. 3 and Fig. 4.

Ultrasonic crack growth measurement

10. Crack growth measurements are performed using a normal probe (lead metaniobate) operating at 5 MHz. The measurements are made from the end-on position, ie facing the tip of the crack (ref. 4). The crack grows directly towards the transducer so that no mechanical drive system has to be used as in Clark's method (refs 5 and 6). The crack signals reflected from the facets at the tip of the crack are very small. It is important therefore, that the transducer signal-to-noise ratio be large, that receiving gain be high and that the oil film coupling be very good. To detect the fatigue cracks grown at the low loads used with the 'C' specimens it is generally necessary to put a 5 to 10 kN load on the specimen. At zero load the probe reads the starter notch and at a load of 5 to 10 KN the probe changes to the crack tips. This method is sensitive to crack length changes of 0.25 mm.

11. The detected output signal of the flaw detector (refer to Fig. 5) is connected to a voltage threshold circuit in the electronic crack depth reader (ref. 7) so that noise signals are not read. The basic timing signal from the oscillator in the flaw detector is used to gate on a high speed counter in the crack depth reader. It counts from 0 to 1000 for each transmitted pulse. The first received signal (echo) after the transmitted pulse is used to 'read' the contents of the counter.

12. The basic count rate is 11.6 MHz. Each count corresponds to 0.25 mm of sound travel in steel. By using this digital method, gates can be used to prevent a reading in undesired 'time' areas, ie the duration of the transmitted pulse. The starter notch and the back wall of the 'C' specimen can be gated out also, allowing reading only on the crack. The readout rate of the crack depth reader is the same as the pulse repetition frequency of the flaw detector, ie from 80 to 2000 readings per second.

13. A Burr-Brown, DAC50-12B-BCD, commercial digital-to-analog convertor is used to convert the 'depth' number (held in the output register of the crack depth reader) to a d.c. voltage. The output voltage accuracy is greater than one part in a thousand. The output operational amplifier drives the X-Y_1-Y_2 recorder input directly.

DISCUSSION OF TEST RESULTS

Fracture tests and crack growth measurements

14. The basic test results are plots of load against specimen deflection and crack growth against specimen deflection (see Figs 3 and 4) which were obtained as described in the preceding section on test procedure. From these plots critical K values were determined for each specimen using the appropriate x/W and a/W ratios to obtain $KB\sqrt{W}/P$ from ref. 8. The critical K values corresponding to the maximum load, K_{max}, and the 5 per cent secant offset load, K_5, were calculated and listed in Table 1. As seen in the table, comparatively little variation in K_{max} or K_5 is apparent over the range of sizes of the group one specimens. The group one K_5 values are in the range 122 to 131 $MN/m^{3/2}$. These values can be compared with the K_{Ic} values from two 25 mm compact tension specimens from the same forging, which are 129 and 133 $MN/m^{3/2}$. The group two results, from two gun tubes of nominally the same alloy, indicate that tube 490 has a significantly lower toughness than the group one forging. No K_{Ic} values are available for the gun tube forgings. However, two independent indications

of this lower toughness are the low value of reduction-in-area for tube 490, see Table 2, and second, the small amount of shear deformation noted for tube 490 specimens, see Fig. 2d. We interpret the fracture test results (particularly those from group one) as an indication that the 'C' specimen can be used for K_{Ic} tests and that the 5 per cent secant offset criteria can be used as with the standard K_{Ic} specimens.

15. As a further verification of the 5 per cent offset criteria for 'C' specimens, consider Fig. 6. It shows the relative secant offset against relative crack growth data from all the specimens for which ultrasonic crack growth results were obtained. The relative crack growth values plotted are the ultrasonic crack growth measurements, Δa, normalized by the initial crack length, a, measured after the test. The relative secant offset, s, is the per cent secant offset in the usual sense, divided by 100; so that a 5 per cent secant offset corresponds to s = 0.05 in Fig. 6. The most significant information in Fig. 6 is that for a value of s = 0.05, which corresponds to the established K_{Ic} failure criteria, the values of measured relative crack growth vary from 0.02, the value upon which the K_{Ic} failure criteria is based, up to 0.05. Specimen 1A should be ignored because of the biplanar fatigue crack mentioned previously and the fact that specimen 1A is under the minimum size requirement for a K_{Ic} test.

16. A measure of the accuracy of the ultrasonic method in measuring the crack length can be seen in the ratio a_u/a listed in Table 2, where a_u is the total length of notch plus fatigue crack measured ultrasonically and a is the same length measured from the fracture surface after each test. This ratio is always found to be near 1.0.

Calculated crack growth values from compliance

17. The relative crack growth parameter, $\Delta a/a$, can also be calculated using a compliance analysis and can then be compared with the ultrasonic measurements. Beginning with Irwin's compliance relation (ref. 9) for plane-strain conditions:

$$\frac{K^2(1 - \nu^2)}{EB} = \tfrac{1}{2}(P/B)^2 \frac{dc}{da} \tag{1}$$

using $dc/da \cong \Delta c/\Delta a$, a good approximation for small changes in crack length, and $\Delta c = c\left[s/(1-s)\right]$ arrive at

$$\frac{\Delta a}{a} = \frac{\frac{s}{1 - s}[cEB]}{2(1 - \nu^2)(a/W)[KB\sqrt{W}/P]^2}. \tag{2}$$

For similar geometries, the two terms in brackets [ ] in equation (2) can be expressed in terms of a/W and x/W. First, it is easy to show that K for the 'C' specimen in the form of a Wilson-Srawley (refs 10 and 11) type function is essentially a constant, denoted Y here for a convenience.

$$Y = \frac{[KB\sqrt{W}/P](b/W)^{3/2}}{(3x/W + 2 + a/W)}. \tag{3}$$

For the group one specimens, for example, Y = 1.37 ± 0.5 per cent. Second, equation (1) re-arranged and written in terms of b rather than a can be used to obtain an expression for c:

$$dc = \frac{-2(1 - \nu^2)[KB\sqrt{W}/P]^2}{EBW} db.$$

Using equation (3) and integrating

$$c = \frac{(1 - \nu^2)Y^2(3x/W + 2 + a/W)^2}{EB(b/W)^2} + c_1 \tag{4}$$

where c_1 can be evaluated from equation (4) by setting a = 0 and $c = c_0 = \{6\pi(x/W + \tfrac{1}{2})^3\}/EB$, the conditions for an uncracked 'C' specimen. The expression for c_0 is from a curved beam analysis (ref. 12) which only approximates the thick-wall geometries considered here, but c_0 does not have a large effect on the total compliance. Equation (4) becomes

$$cEB = Y^2(1 - \nu^2)(W/b)^2(3x/W + 2 + a/W)^2 + 6\pi(x/W + \tfrac{1}{2})^3 - Y^2(1 - \nu^2)(3x/W + 2)^2. \tag{5}$$

Finally, combining equations (2), (3) and (5),

$$\frac{\Delta a}{a} = \frac{s/(1 - s)}{2(a/b)}\left\{1 + \frac{6\pi(b/W)^2(x/W + \tfrac{1}{2})^3}{Y^2(1 - \nu^2)(3x/W + 2 + a/W)^2} - \frac{(3x/W + 2)^2(b/W)^2}{(3x/W + 2 + a/W)^2}\right\}. \tag{6}$$

Equations (5) and (6) are general expressions for compliance and relative crack growth for 'C' specimens under plane strain conditions. For $\nu = 0.3$ the plane strain compliance is about 10 per cent less than the plane stress value, which can be calculated from equation (5) with the $(1 - \nu^2)$ terms omitted. The plane strain and plane stress calculations of relative crack growth are little different. In fact, for the geometries considered here, the sum of the second and third terms inside the brackets of equation (6) is close to zero, so that

$$\frac{\Delta a}{a} \cong \frac{s/(1 - s)}{2(a/b)}.$$

18. The expression for compliance, equation (5), was used to calculate c for comparison with the measured values, which are simply the inverse of the initial slopes of the load-deflection plots. This comparison is shown in Table 2. The good agreement between measured and calculated compliance indicates that the tests were properly performed and that the relative crack growth values from the tests are reliable. The measured values of $\Delta a/a$ at s = 0.05 were taken from Fig. 6 and are compared with the calculations from equation (6) (see Table 2). The significantly higher measured values are due to the fact that the ultrasonic method measures the crack growth at the centre of the crack front (at the points indicated by the 'V' marks on the fracture surfaces, see Fig. 2). The crack often 'tunnels' ahead at the centre of the specimen and lags behind at the surface. It is the average crack growth which affects compliance and which should be compared with the calculated $\Delta a/a$ values. This 'tunnelling' behaviour can be seen on the fracture surfaces of 25 mm compact specimens of a similar alloy as that used here, see Fig. 7. The compact specimens were unloaded at a point somewhat beyond the 5 per cent offset load and were then subjected to fatigue loading. The 'tunnel' crack growth which occurred during the monotonic loading can be easily seen as the narrow dark area between the two lighter areas of fatigue crack growth.

CONCLUSIONS AND RECOMMENDATIONS

19. We conclude from the tests and analysis that (a) the 'C' specimen can be used to measure K_{Ic} over a range of geometries by using the same general procedure as the standard test method, and (b) the 5 per cent secant offset procedure for determining K_{Ic} values results for 'C' specimens in about 2.5 per cent average crack growth and is thus a comparable procedure to that for the standard K_{Ic} specimens.

20. Based on this work and on prior results (refs 8 and 13), we recommend that 'C' specimen fracture toughness tests be performed using the standard test method where possible and with specimens of the following dimensions (refer to Fig. 1): $x/W = 0.5$, $B/W = 0.5$, $0.25 \leq h/W \leq 0.35$ and $1.5 \leq r_2/r_1 \leq 3.0$. For specimens in this range we believe that the K calibration in Table 3 is accurate within 1 per cent.

21. Only when additional comparative tests are performed and when the offset fracture criteria and K calibration are further verified can the 'C' specimen be included in a standard K_{Ic} test method.

LIST OF SYMBOLS

a	=	crack depth
a_u	=	crack depth as measured by ultrasonics
b	=	remaining uncracked ligament; $b = W - a$
B	=	specimen thickness
c	=	elastic compliance
E	=	elastic modulus
h	=	diameter of loading holes in 'C' shaped specimen
K	=	opening mode stress intensity factor
K_5	=	K value corresponding to 5 per cent secant offset
K_{max}	=	K value corresponding to maximum load
K_{Ic}	=	plane strain fracture toughness of material
P	=	applied load
r_1	=	inner radius of 'C' chaped specimen
r_2	=	outer radius of 'C' shaped specimen
s	=	relative secant offset used with load-deflection plot
W	=	specimen depth; for 'C' shaped specimen $W = r_2 - r_1$
x	=	offset from the inner radius of loading holes in 'C' shaped specimen
σ_y	=	yield strength of material
σ_u	=	ultimate strength of material
ν	=	elastic Poisson's ratio

REFERENCES

1. KENDALL, D. P. and HUSSAIN, M. A. A new fracture-toughness test method for thick-walled cylinder material. Experimental Mechanics, Vol. 12, April 1972, p 184-189.

2. Standard method of test for plane-strain fracture toughness of metallic materials. E399, 1973 Annual Book of ASTM Standards, Part 31, American Society for Testing and Materials, Philadelphia, PA, 1973.

3. Methods for plane strain fracture toughness (K_{Ic}) testing. DD3: 1971 British Standards Institution.

4. MILLER, J. J. Ultrasonic measurement of crack depth in thick-walled cylinders. Benet R & E Laboratories Technical Report, WVT-7017, Watervliet, NY, Feb. 1970.

5. CLARK, W. G. Jr. Ultrasonic detection of crack extension in the WOL type fracture toughness specimen. Materials Evaluation, August 1967, p 185.

6. CLARK, W. G. Jr. and CESCHINI, L. J. An ultrasonic crack growth monitor. Materials Evaluation, August 1969, p 180.

7. WINTERS, D. C. Automated crack position and depth measurements in cannon tubes. Benet Weapons Laboratory Technical Report, WVT-7248, Watervliet, NY, Sept. 1972.

8. UNDERWOOD, J. H., SCANLON, R. D. and KENDALL, D. P. K calibration for C-shaped specimens of various geometries. Fracture Analysis, ASTM STP 560, American Society for Testing and Materials, Philadelphia, PA, 1974, p 81-91.

9. IRWIN, G. R. Structural aspects of brittle fracture. Applied Materials Research, Vol. 3, April 1964, p 65-81.

10. WILSON, W. K. Stress intensity factors for deep cracks in bending and compact tension specimens. Engineering Fracture Mechanics, Vol. 2, 1970, p 169-171.

11. SRAWLEY, J. E. and GROSS, B. Stress intensity factors for bend and compact specimens. Engineering Fracture Mechanics, Vol. 4, 1972, p 587-589.

12. TIMOSHENKO, S. Strength of Materials Part I, D. Van Nostrand Co., Princeton, NJ, 1955, p 378-379.

13. UNDERWOOD, J. H. and KENDALL, D. P. K results and comparisons for a proposed standard C-specimen. Benet Weapons Laboratory Report, WVT-74041, Watervliet, NY, Sept. 1974.

Table 1

Specimen dimensions and fracture test results

Specimen	r_2 mm	B mm	W mm	h mm	x/W -	a/W -	KB $\sqrt{W}$/P (ref. 8)	K_{max} $MN/m^{3/2}$	K_5 $MN/m^{3/2}$
Group 1									
1A	80	50.8	39.8	12.7	0.750	0.484	17.5	124	124
3A	120	50.8	59.9	12.7	0.750	0.500	18.3	126	-
3B	120	50.8	59.9	12.7	0.750	0.508	18.8	127	125
4A	133	47.0	66.5	15.9	0.750	0.512	19.0	125	124
4B	133	47.0	66.5	15.9	0.750	0.512	19.0	125	124
5B	146	50.8	72.8	19.0	0.750	0.526	20.0	131	128
6B	159	50.8	79.7	19.0	0.750	0.538	20.8	123	122
7A	192	50.8	95.7	25.4	0.750	0.483	17.4	130	130
7B	192	50.8	95.7	25.4	0.750	0.530	20.2	131	131
Group 2									
348A	119	38.1	49.3	19.0	0.964	0.547	23.6	131	126
348B	119	38.1	49.3	19.0	0.964	0.572	25.7	127	120
490A	123	38.1	53.3	19.0	0.893	0.528	23.1	106	103
490B	123	38.1	53.3	19.0	0.893	0.538	23.9	106	105

Table 2

Specimen material properties and crack growth data

Specimen	σ_y MN/m^2	σ_u MN/m^2	RA %	$\frac{a}{2.5(K_5/\sigma_y)^2}$ -	$\frac{a_u}{a}$ -	c mm/MN		$\frac{\Delta a}{a}$ -	
						Measured	Calculated	Measured	Calculated
1A				0.92	0.92	13.8	14.4	0.071	0.029
3B				1.43	0.97	14.1	15.7	0.049	0.026
5B	1360	1460	38	1.71	0.99	15.7	16.9	0.041	0.025
6B				2.14	0.98	17.8	17.8	0.044	0.023
7B				2.16	0.99	19.8	17.2	0.053	0.024
348A	1190	1310	27	0.96	1.00	31.9	31.8	0.022	0.024
348B				1.10	1.01	37.9	35.5	0.026	0.021
490B	1170	1280	20	1.42	0.97	29.8	31.6	0.040	0.025

Table 3

K calibration for 'C' specimens with x/W = 0.5 and $1.5 \leq r_2/r_1 \leq 3.0$

a/W	$\frac{KB\sqrt{W}/P}{1 + 0.035 \ln(r_2/r_1)}$	a/W	$\frac{KB\sqrt{W}/P}{1 + 0.035 \ln(r_2/r_1)}$
0.45	12.7	0.50	14.9
0.46	13.1	0.51	15.4
0.47	13.5	0.52	15.9
0.48	14.0	0.53	16.4
0.49	14.4	0.54	17.0
		0.55	17.6

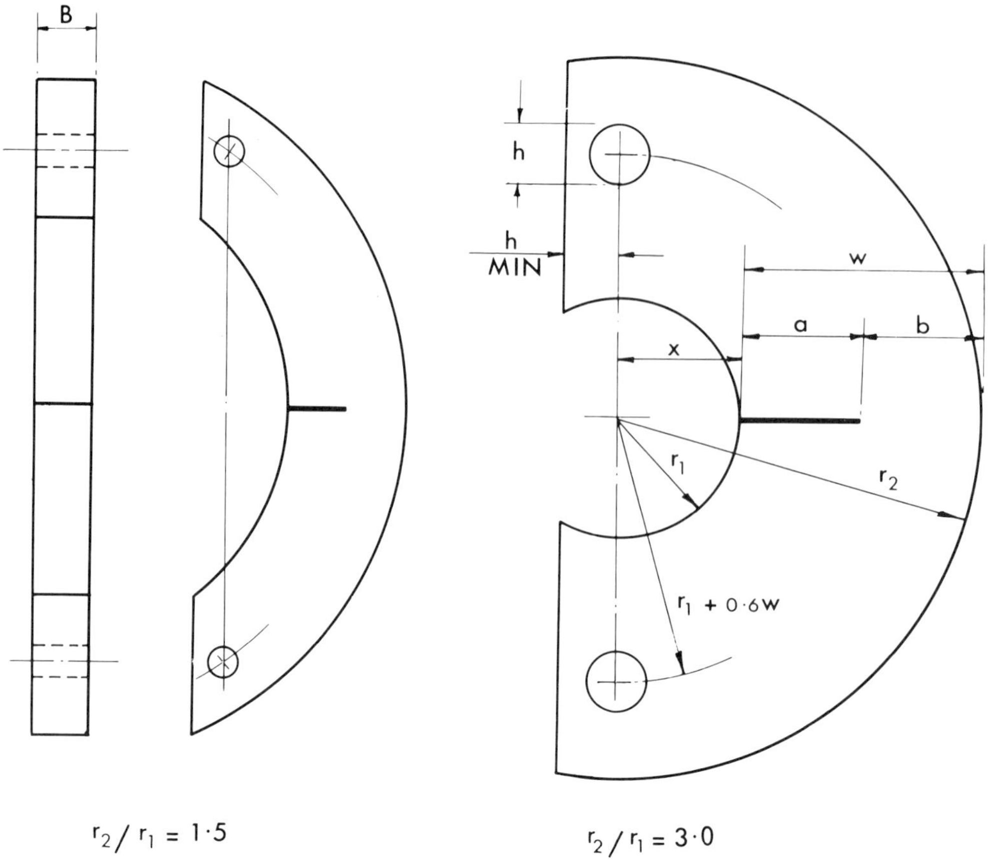

Fig. 1: 'C' specimen sketch and nomenclature

Fig. 2: Fatigue and fracture surfaces
(a) spec 1A (b) spec 3B
(c) spec 348A (d) spec 490A

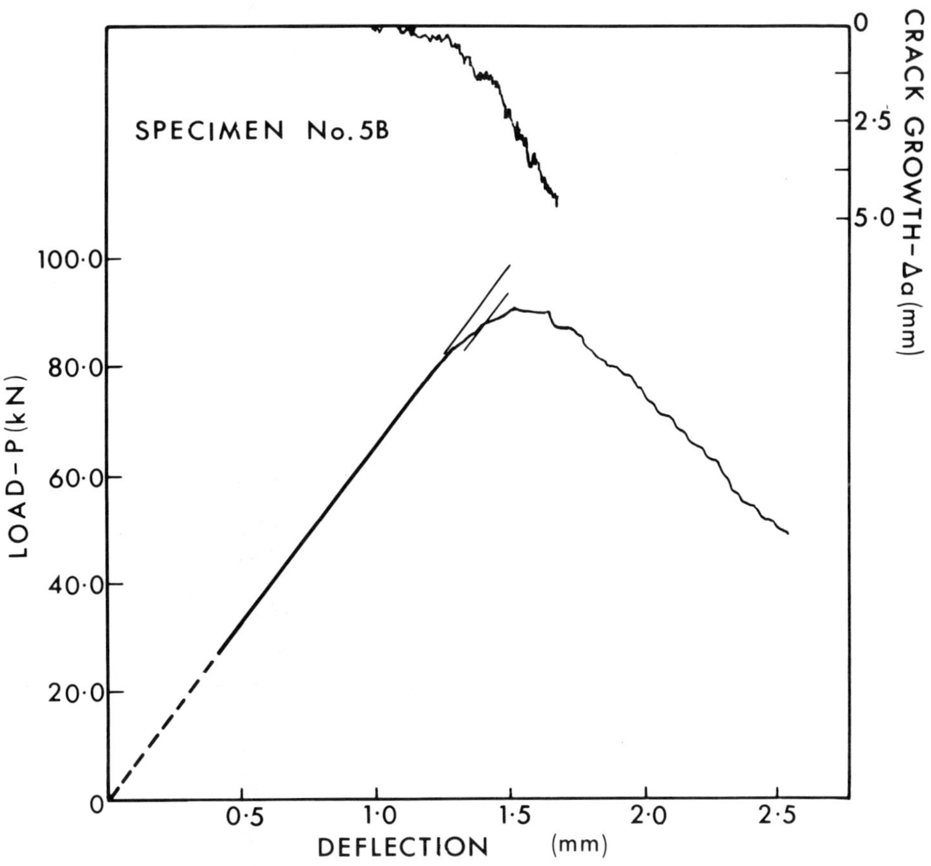

Fig. 3: Load-deflection-crack growth plot for spec 5B

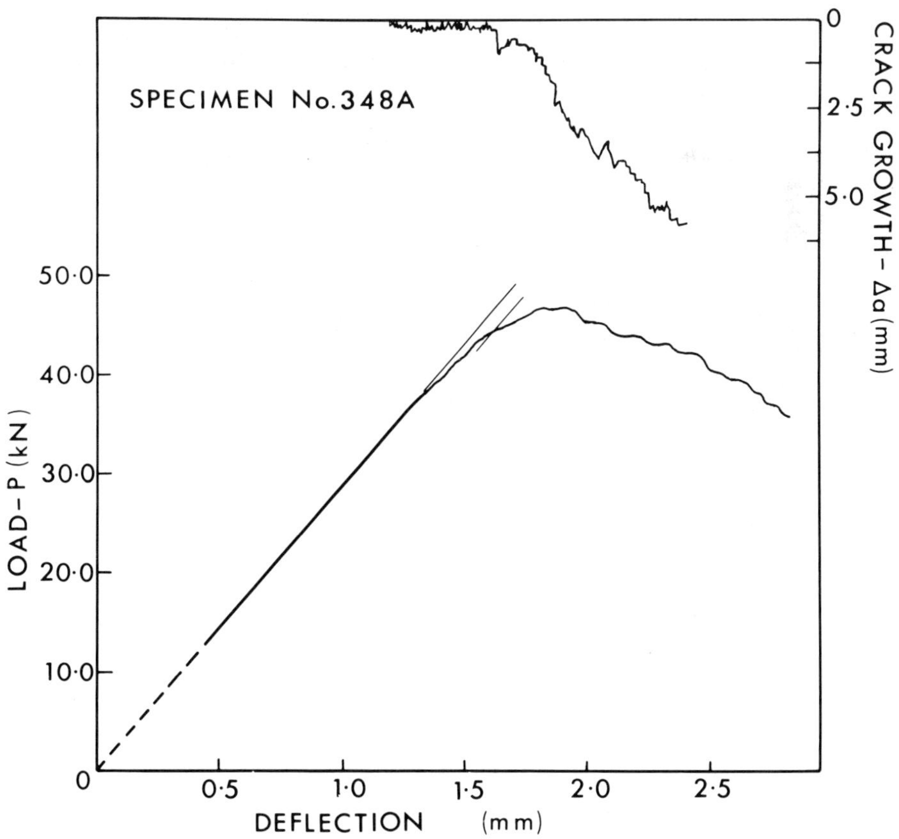

Fig. 4: Load-deflection-crack growth plot for spec 348A

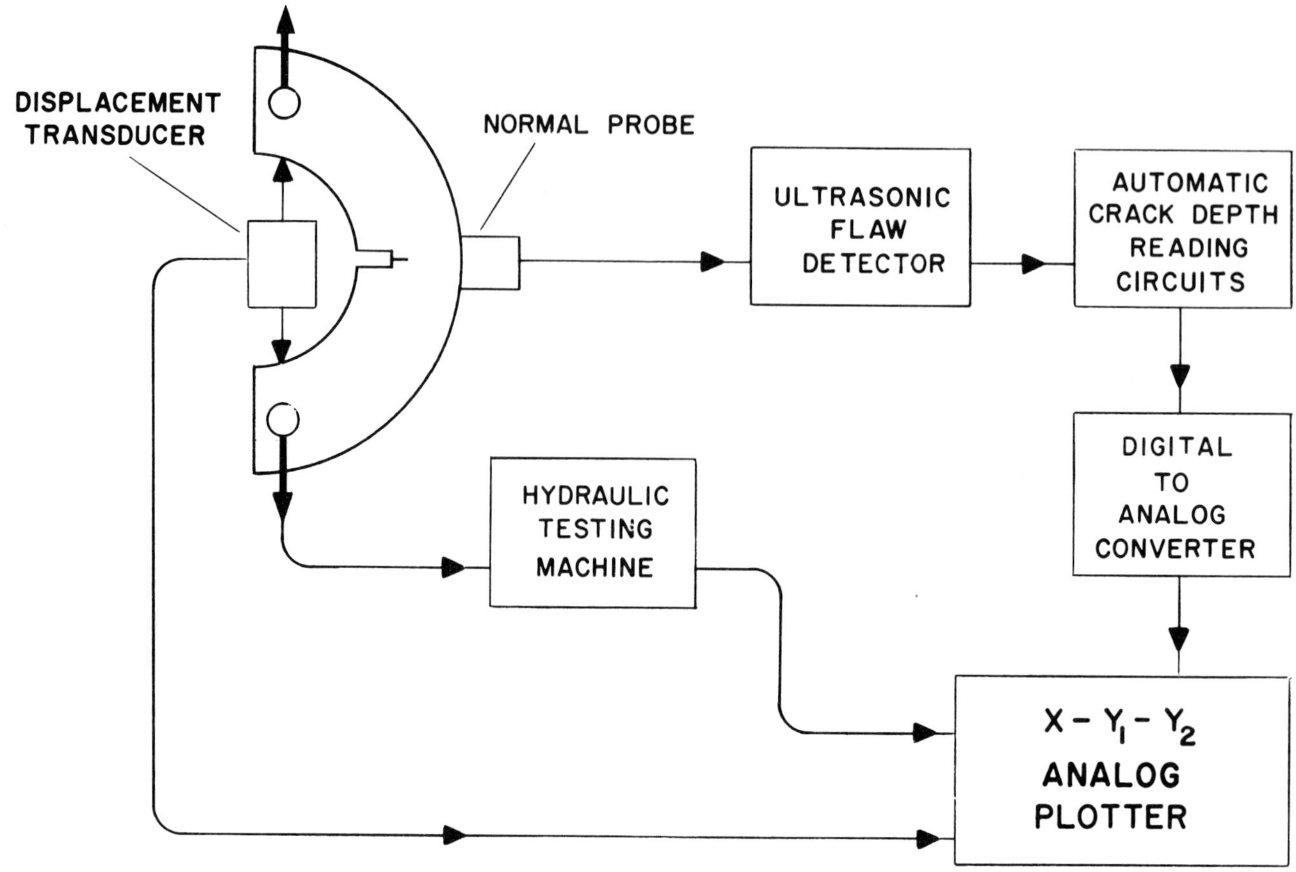

Fig. 5: Diagram of ultrasonic crack growth measurement procedure

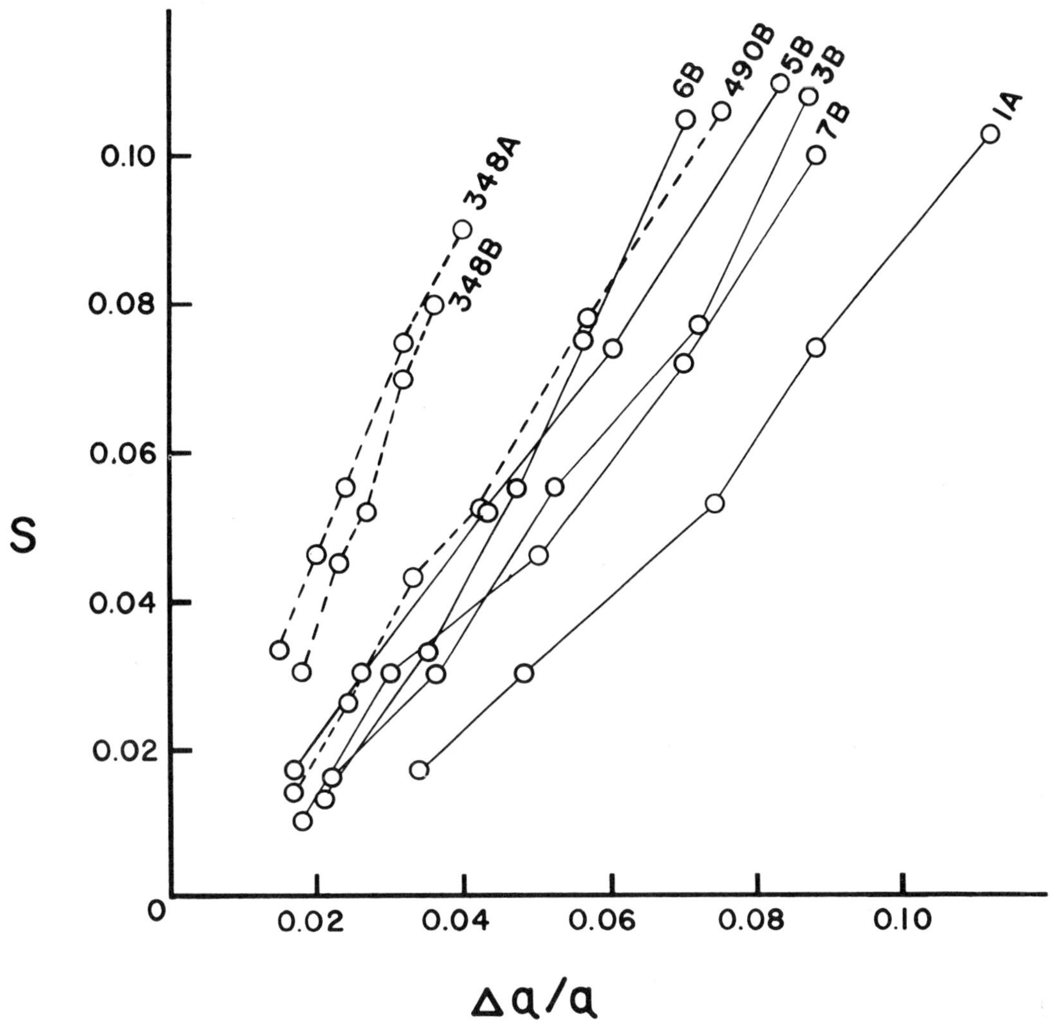

Fig. 6: Relative secant offset against relative crack growth

Fig. 7: Compact specimen fracture surfaces showing 'tunnelling' during crack growth

FATIGUE AND FRACTURE

FATIGUE AND FRACTURE OF THICK-WALLED PRESSURE VESSELS

J. S. C. PARRY, PhD
University of Bristol and
R. W. E. SHANNON, PhD
British Gas Corporation Engineering Research Station, Newcastle-upon-Tyne NE99 1LH

SYNOPSIS A review of published work on the fatigue and fracture of thick-walled pressure vessels including a summary of the experimental work on fatigue is given together with the various factors that can affect fatigue strength. A brief introduction to fracture mechanics is followed by a summary of the methods of estimating the stress intensity factor K_I for cylinders containing cracks. The application of fracture mechanics to the failure of pressure vessels and the propagation of fatigue cracks is discussed together with the estimation of critical crack size and the effects of residual stress.

INTRODUCTION

1. Traditional pressure vessel design is based on the stress analysis of unflawed structures and the use of materials which exhibit ductile behaviour. In addition, a safety factor is applied to allow for the uncertainty of knowledge about the material behaviour and loading. Hence it is not surprising that pressure vessels usually perform their function successfully.

2. However, there have been occasions when catastrophic or fatigue failure has occurred without warning. The investigation of such failures has helped add to the collective experience, and designers have gradually learned how to avoid such situations, although in some cases the causes have not been completely understood.

3. Brittle fracture is by far the most spectacular mode of pressure vessel failure. It is by no means a new phenomenon, however it is one which the advance of technology has tended to highlight. The most dramatic examples have usually involved plate structures, such as bridges and ships hulls, but the problem has by no means been absent in heavy section components of the type used for very high pressure vessels. Examples of the latter which occur in the commercial field are chemical reaction vessels, isostatic compaction chambers, and in the military sphere, high performance guns.

4. Fatigue failures which are caused by pressure fluctuations are more insidious in that a crack may start at a point where there is a stress raiser and will spread slowly, sometimes over long periods of time, until it reaches a critical size when it will initiate a catastrophic brittle fracture. Alternatively, if the toughness of the material is sufficiently high the crack will spread right through the vessel wall causing a leak and if the high pressure fluid is flammable a fire may ensue or even an explosion. High pressure chemical plant, diesel engine injection equipment and isostatic compaction equipment are some of the many fields where the possibility of fatigue failure must be considered.

5. In recent years there has been increasing use of higher strength materials to save weight and cost. For most materials these higher strengths have been achieved at the expense of reduced ductility, resulting in their resistance to brittle fracture being impaired and the danger of catastrophic failure occurring at stress levels well below general yield. Also the fatigue resistance of materials does not necessarily increase in proportion to the tensile strength so that sometimes high strength vessels can be more susceptible to fatigue failure.

HISTORICAL REVIEW

6. Probably the first fatigue failures in high pressure vessels occurred on the early plant producing the first commercial polyethylene during the 1939-1945 war (ref. 1). These failures caused great concern since the plant had been carefully designed, due allowance having been made for the fatigue nature of the loading, yet failures occurred in components which were thought to have a reasonable factor of safety. Clearly there were weakening factors influencing these failures that were not understood and Morrison et al (ref. 2) started to study the problem after the war.

7. In contrast the problems of brittle fracture are much older and started with the construction of the first guns. By the beginning of the nineteenth century brittle fracture of cast guns had become so common as to be sometimes considered as one of the ordinary risks of a soldier's or seaman's life.

8. In 1861 a Captain Rodman (ref. 3) reported an extensive series of experiments aimed at finding ways of producing safe cast iron guns. The most significant development resulting from Rodman's most comprehensive studies was the introduction of bore cooling when casting guns. The process of differential cooling from inside to out induced residual compression in the inner portions of the tube. Thus the magnitude of the net tension stress produced during firing was significantly reduced and this improved the endurance of the gun. This study is one of the earliest recorded examples

of an attempt to increase the reliability of a structure made from a classically brittle material which was actually successful.

9. A few years later Kirkaldy (ref. 4) recognised the main factors which influence the fracturing characteristics of a brittle material. He listed three important ways in which the fracture mode of wrought iron could be changed from fibrous (ductile) to a brittle form of failure without lowering the test temperature. These were (i) by using a re-entrant notch (ii) by increasing the hardness of the iron through heat-treatment (iii) "by applying the strain so suddenly as to render the material liable to snap from having less time to stretch."

10. Towards the end of the 19th century the re-requirement to use higher pressures in guns was met by the introduction of built up and wire wrapped methods of construction; the objective being to improve the distribution of pre-stress, such that firing loads would result in a more uniform tension throughout the wall thickness. However, monoblock vessels were generally preferred because they were easier to built but it was not until the Auto-frettage method of producing beneficial residual stresses was introduced by Malaval (ref. 5) in 1906 that they were able to compete favourably.

11. The commercial use of chemical reactions at high pressures and temperatures began about this time and although the development of the reaction vessels created considerable problems, early manufacturers were able to exploit the techniques pioneered by gun designers. The safety of such structures was normally predicted on the basis of arbitrary rules for materials selection, service experience and the designer's intuition. The only engineering approach used in dealing with brittle fracture was that involving the application of the impact energy - transition temperature type fracture test for material evaluation, although it was generally accepted that laboratory notch tests could not simulate the behaviour of brittle fracture in a large structure.

12. By the mid-1950's a "fracture safe" approach was being adopted and the gap between the measurement of material fracture parameters and their application to specific designs has to some extent been filled by the analyses of linear elastic fracture mechanics (ref. 6) and general yielding fracture mechanics (ref. 7). The former enables the severity of the crack tip environment to be related to the geometry of the component and the global loads applied to it through the use of a single stress field parameter called the stress intensity factor K. Whilst the latter deals with those situations where the conditions at failure differ markedly from linear elastic behaviour, and large amounts of plasticity are associated with the crack tip. It is now well established that for elastic-plastic deformations near the crack tip a singularity still exists and that the crack opening displacement or COD near the crack tip reflects the nature of this singularity.

FATIGUE FAILURE

13. At first sight the fatigue of thick cylinders appears to be a problem in multiaxial combined stress fatigue. The stresses in a plain cylinder are easily determined and the latest stress analysis techniques enable the stresses at stress concentrations to be estimated. Also there are several hypotheses to allow for the effects of combined stress that have experimental evidence to support them. Such techniques can be applied successfully where the fracture surface is not in contact with the pressurising fluid, for example, cracks in the threads of screwed end closures outside the high pressure seal. However, when the high pressure fluid can enter the fatigue crack other weakening factors are involved and the fatigue strength is only a fraction of the predicted value.

14. The fracture surface of such fatigue cracks has a characteristic appearance. Unless any geometrical constraint prevents it, the crack always spreads on an axial radial plane. The crack surface is usually very smooth particularly if the failure has developed during a large number of cycles and the crack seems to spread at roughly the same rate in all directions giving it a semi-elliptical shape (see Fig. 23 in ref. (2)).

Experimental Results

15. The fatigue limits (10^7 cycles) of repeated pressure tests reported by Morrison et al (ref. 8) on cylinders with diameter ratios ranging from W = 1.2 to 3.0 made from a 2½% Ni-Cr-Mo steel EN25T correlated very well when compared on the basis of the maximum shear stress at the bore. Further tests on six other materials tended to confirm this finding and the empirical relationship that for ductile ferrite steels the safe range of bore shear stress was roughly one third of the ultimate tensile stress was proposed. Fig. 1 shows the results of over 100 repeated pressure tests on four ferritic steels in which the shear stress at the bore divided by the ultimate tensile stress is plotted against the number of cycles to cause failure. These results refer to cylinders with bore diameters ranging from 9.5 mm to 25 mm and wall thickness from 2.5 mm to 15 mm. This simple result did not explain why the fatigue strength of a cylinder was only about 60% of what would be expected from the normal fatigue properties of the material but it did provide a reasonable basis for design.

16. This and other work by Parry (ref. 9, 10) revealed a number of factors that affect the fatigue strength of a cylinder, some of which are useful and some perhaps help to explain the abnormal behaviour of these failures. As was mentioned earlier any process which produces compressive stresses at the bore is beneficial, for example autofrettage or even a light honing operation. The obvious explanation is that since repeated pressure fluctuates the stresses from zero to a maximum value, the compressive residual stress will reduce the mean stress of the cycle so increasing the safe range of stress. Burns et al (ref. 11) studied this effect and confirmed the beneficial effects of reducing the mean stress as well as the detrimental effect of increasing it.

17. Nitriding and carburising increase the intrinsic strength of the surface material as well as inducing compressive stresses and these processes can produce very significant increases of fatigue strength, up to 40% having been reported (Fig. 1). However, once the hardened layer is cracked, for instance at a stress concentration or by over-

pressurisation of the plant, the beneficial effects may be lost and the cylinder may be weaker than if it had not been hardened.

18. Surprisingly it was found that protecting the bore with a thin film of rubber or a layer of neoprene produced a large increase of strength (up to 30%). It appears that the high pressure fluid must have some physically damaging effect on the metal. If the metal surface can be protected or if the surface compressive stress is large enough to prevent or discourage the penetration of the fluid into the metal surface there is an appreciable increase in fatigue strength.

19. This was discussed by Frost (ref. 12) who called it the "hydrowedge" effect and suggested that perhaps the maximum hoop tension should be the criterion of failure. To allow for the hydrowedge effect of the pressure acting in any crack or surface defect a tensile stress equal in magnitude to the pressure was added to the hoop stress. The resulting stress is equal to twice the shear stress at the bore so the correlation between cylinders with different diameter ratios is still maintained.

20. Cross-bores Of course any stress concentration or a poor quality surface finish in the bore will reduce the fatigue strength (ref.9), see Fig. 1. Perhaps one of the most useful results is the way a stress concentration in the form of a cross-bore (i.e. a small radial hole through the cylinder wall) can be strengthened by the autofrettage process. For a cylinder with diameter ratio W = 2.25 the higher the autofrettage pressure the greater the increase in fatigue strength until when the pressure is sufficient to yield the whole wall thickness the fatigue strength is nearly as high as that of a plain cylinder without any stress concentration, i.e. a 100% increase in strength (Fig. 2). Unfortunately autofrettaging plain cylinders does not produce so dramatic an effect (up to 25% increase having been reported). The reason for this is not properly understood but is probably connected with the much higher strain gradients in the cross-bored cylinders.

21. The stress raising effect of a cross-bore can be reduced by rounding the corners where the cross-bore and the main bore intersect but practically this can be a difficult operation unless the bore is very large. Another method of improving the strength proposed by Cole (ref. 13) is either to make the hole elliptical or to offset a circular hole so that its axis is no longer radial, the limiting case being when the hole is tangential to the cylinder bore. However, theoretically an optimum position is reached before the limiting case.

22. Directionality and higher tensile strengths

The dangers of trying to increase the strength of a pressure vessel by increasing the tensile strength without careful analysis of the material properties and design have already been mentioned. The fatigue strength of some 13% Cr steel tubing was actually reduced as the tensile strength of the material was increased (ref. 9). The maximum stress (either hoop or shear stress) in a cylinder always acts across a longitudinal plane. A tube or cylinder made from a forging, whether hollow or a solid bar always has the inclusions and impurities elongated in the longitudinal direction. This makes the fatigue strength in the longitudinal direction greater than that in the hoop and radial directions. Thus the hoop and maximum shear stress act in these weaker transverse directions. Boyd (ref. 14) investigated the variation of bending fatigue strength of EN25 steel in both the longitudinal and transverse directions as the tensile strength was altered by the heat treatment and showed that whilst the longitudinal fatigue strength increased with the tensile strength, the transverse fatigue strength remained constant or even decreased. This probably explains the behaviour of the 13% Cr steel tubing and emphasises the importance of checking the transverse fatigue strength of high tensile materials. Although this example acts as a warning the fatigue strength of a vessel can be increased by using higher tensile material and this has been demonstrated by Crossland and Skelton (ref. 15) for cross-bored cylinders. The low transverse fatigue strength of a steel is related to the quantity of inclusions and any of the cleansing processes such as vacuum re-melting, that are used in modern steel making procedures greatly improve the transverse properties.

23. Short life experimental results Fatigue data for lives of less than 10^5 cycles to failure given by Davidson et al (ref. 16), Austen et al (ref. 17 and 18) and Jones and Tomkins (ref. 19) covers a range of diameter ratios as well as the effects of autofrettage, bore rifling and chromium plating. The results from different diameter ratios, W = 1.2 to 2.0 correlates best on the basis of the maximum tensile stress at the bore and it appears that in the limited life region this is the best criterion whereas at the fatigue limit the bore shear is the critical quantity.

Design criteria

24. For design purposes it is desirable to have methods of predicting the fatigue limit and short life behaviour of a pressure vessel. The empirical rule proposed by Morrison et al (ref. 8) that for ductile ferritic steels the safe range of bore shear stress is one third of the tensile strength, is crude but simple to apply.

25. More sophisticated methods have been proposed by Haslam (ref. 20 and 21), Tomkins (ref. 22) and McDiarmid (ref. 23) and they all fit the existing experimental data satisfactorily. Haslam used fracture mechanics to obtain crack growth rates with which to predict the short life behaviour and used Frost's "hydrowedge" with an empirical relation to predict the fatigue limit from the transverse bending fatigue limit. Tomkins used push-pull fatigue data and related the shear strain in the two tests to predict the mortal part of the pressure versus cycles to failure curve, the fatigue limit being when the shear strain range in the cylinder was equal to half that in the push-pull test at the fatigue limit. McDiarmid developed a general criterion of fatigue failure under multiaxial stress based on a critical range of shear stress modified for the effects of the normal stress acting on the maximum shear stress plane and used the torsion and transverse bending fatigue limit to predict the fatigue limit of a pressure vessel.

26. However well these criteria appear to predict the behaviour the only safe procedure when a design involves a new material or constructional technique is to obtain experimental results to confirm the accuracy of the methods used.

FRACTURE BEHAVIOUR

27. The stress intensity approach The local stresses near a flaw depend on the product of the nominal stress σ and the square root of the flaw depth a. The importance of this was first appreciated by Irwin (ref. 24), who coined the term stress intensity to emphasise this fundamental relationship. In general, the stress intensity factor is given by

$$K_I = A\sigma(a)^{\frac{1}{2}} \qquad (1)$$

where A is a factor (of the order $\sqrt{\pi}$ for plane problems) which depends on geometry and loading conditions.

28. Stress fields in the vicinity of a crack tip can be divided into three types corresponding to the three basic modes of crack surface displacement. Mode I is where the crack surfaces move directly apart. Mode II is where the crack surfaces slide over each other in a direction perpendicular to the leading edge of the crack. Mode III is where the crack surfaces slide parallel to the leading edge of the crack. Mode I or the opening mode is of primary interest since it dominates in most cases of engineering importance. The higher the value of K_I the more severe the crack and when a critical value K_C is reached, sufficient energy is being supplied to the crack tip to promote crack extension, usually in an unstable manner.

29. Local crack tip stress fields may be characterised as essentially plane stress or plane strain. For example a plate loaded in uniaxial tension is in a state of plane stress. However, when a crack is introduced, the highly stressed material adjacent to the crack tip is constrained by the less highly stressed surrounding material and stresses are introduced in the through thickness direction at the crack tip. This situation is referred to as plane strain. When the crack tip plastic zone size becomes comparable to the thickness of the plate, yielding in the through thickness direction occurs, allowing relaxation of the through thickness stresses, so that the whole plate approaches a state of plane stress.

30. The load and crack length at instability (and therefore K_C) decrease with increase in thickness, reaching a minimum value when plane strain conditions are approached closely. This is described as the plane strain fracture toughness value K_{IC} and can be considered as a material property like the proof stress in a tensile test. To be certain that plane strain conditions prevail in a test, the thickness of the specimen must be about 50 times the radius of the plane strain plastic zone associated with the crack tip, which is equivalent to 2.5 $(K_I/\sigma_y)^2$, where σ_y is the material yield stress.

In principle the measurement of K_{IC} is simple, however in practice there are a number of experimental and interpretational difficulties which complicate the test procedure but test methods are now well developed and several standards have been issued for guidance (ref. 25 and 26).

31. The COD approach There is still a problem in quantifying the resistance of a material to crack extension when general yielding occurs in the cracked region. If a body containing a crack is loaded, the crack tip opens without extension by an amount δ given approximately (ref. 27) by $\delta = K_I^2/E\sigma_Y$. This movement is associated with the development of the crack tip plastic zone and is called the crack opening displacement. The critical value δ_c at which crack growth starts, provides an alternative method of measuring fracture toughness (ref. 28) and has the advantage that it still has meaning at and beyond general yield.

Stress intensity factors for thick-walled cylinders

32. Cylinders with straight fronted cracks Bowie (ref. 29) solved the plane problem of a system of equal length radial cracks emanating from the surface of a circular hole in an infinite plate under biaxial tension. With the application of fracture mechanics concepts to this solution, stress intensity factors K_I corresponding to long axial cracks in a circular thick-walled cylinder of very large diameter ratio W were computed (ref. 30). More recently data for a single radial crack in the bore surface of cylinders with finite value of W have been derived by Bowie and Freese (ref. 31) and Shannon (ref. 32), the former using a modified mapping collocation technique developed by Bowie and Neal (ref. 33) and the latter a finite element technique. Very good agreement exists between the two methods over a wide range of cylinder geometries and crack depths, see Fig. 3.

33. Values of K_I for cracked cylinders have also been obtained experimentally by Shannon and Austin (ref. 34) using controlled fracture tests, and by Underwood et al (ref. 35) using a compliance test method. A comparison of these experimental results with the theoretical predictions confirms the accuracy of the latter when due account is taken of the errors associated with the experimental techniques (ref. 32).

34. A detailed study of the data in Fig. 3 reveals several important facts about the behaviour of vessels with differing geometries. Since the effect of increasing pressure on K_I is obvious, it only remains to consider the effect of changes in diameter ratio and absolute cylinder dimensions. Cylinders with constant bore size or constant wall thickness t experience a decrease in K_I with increasing W for any particular value of absolute crack depth a. In addition, cylinders with the same W and absolute crack depth also experience a decrease in K_I with increasing bore.

For cylinders having the same diameter ratio, but different cross-sectional dimensions, the value of K_I for a particular value of a/t increases with increasing vessel size. Thus for a critical value of stress-intensity equal to K_{IC} the critical crack depth ratio (a/t)c will reduce for increasing cross-sectional dimensions, but the absolute value of crack depth will increase.

35. The data in Fig. 3 correlates impressively if the shear stress $_\theta\tau_r$ at the bore of an uncracked cylinder and the geometry term $(Rt/a)^{\frac{1}{2}}$ (where R is the bore radius) are used to normalise the stress intensity factor as shown in Fig. 4. If a single curve is assumed to describe the data in Fig. 4, then when cylinders of the same fracture toughness and bore size are stressed, to the same shear stress level they will fail at the same value of a/t i.e. geometrical similarity at fracture.

36. Let us examine the benefits of residual stress systems in thick-walled cylinders. In all cases where the zone of residual compression at the bore is deeper than the critical crack size in a plain cylinder, the fracture resistance will be increased as manifested by a longer critical crack depth. In contrast to this a reduction in critical depth will ensue for those cases where the crack has extended into the original area of residual tension.

37. Kutter (ref. 36) has explored the problem of multiple cracks in an infinite plate and shown that two diametrically opposed cracks represent the most critical geometry. Three or four cracks have lower stress intensities than two cracks, but greater than one crack and, except for the case of one crack, the stress intensity decreases with the number of cracks due to the interaction of the crack tip stress fields. Values of K_I for two diametrically opposed equal depth cracks and a range of cylinder geometries are shown in Fig. 5. A comparison of Fig. 3 and 5 shows that for low values of crack depth and diameter ratio a single crack has a higher stress intensity than two cracks. However, at a large value of crack depth the situation is reversed, with the point at which both cases have a common value occurring at progressively lower values of crack depth with increasing W.

38. Cylinders with semi-elliptical cracks In practice, straight fronted cracks of the type so far considered only occur when a long stress raiser provokes initiation. Natural or fatigue cracks are normally semi-elliptical in shape and have a maximum value of stress intensity at the point of deepest penetration. As yet no analysis of a pressurised thick-walled cylinder containing such a crack has been completed and only empirical solutions exist. In general, these are based on a combination of existing solutions. For example Underwood (ref. 37) assumed that the effect of a given crack shape on the K_I of a pressurised cylinder was the same as the effect of that shape on the K_I of a plate under uniform tension. The shape factor chosen was that proposed by Rice and Levy (ref. 38). This was combined with the analyses of Bowie and Freese (ref. 31) for a straight fronted crack in a pressurised cylinder and an example of the results for one geometry of cylinder are shown in Fig. 6. The accuracy of such a solution is questionable, but it is believed to be within 10% of the true result for shallow cracks with low values of aspect ratio a/2C. However, straight fronted (i.e. constant depth) cracks represent the weakest case for any given crack depth in a cylinder.

Cyclic (sub-critical) crack growth

39. The total useful life of a cyclically loaded component depends upon the rate of growth of flaws from a sub-critical size to the critical condition when failure occurs. Thus the rate of crack growth under service conditions must be known as well as the critical combination of stress and defect size needed to cause failure.

40. Fatigue crack propagation is a localised phenomenon dependent upon the material, environment, and stress conditions at the crack front. Paris and Erdogan (ref. 39) have reviewed most of the existing crack propagation laws and shown that, in general, the data could be adequately represented by

$$\frac{da}{dN} = C(\Delta K_I)^n \qquad (2)$$

where ΔK_I is the maximum range of stress-intensity and C and n are constants which depend on such things as mean load, material and environment, etc.

Using an expression similar to equation 1 to represent the variation of K_I with crack depth in a cylinder, and substituting in equation 2, integration will yield an expression for the number of cycles N for a crack to grow from an initial depth a_o to a final depth a_f.

$$N = \frac{1}{(1-n/2)C\sigma^n\pi^n/2}\left\{a_f^{\,1-n/2} - a_o^{\,1-n/2}\right\} \qquad (3)$$

Haslam (ref. 40) has followed this type of approach with some success, however the method involves a careful choice of the parameter C and the crack limits of initiation and criticality.

Stress corrosion cracking

41. Slow crack growth can also occur under conditions of sustained monotonic loading, particularly in the presence of a corrosive environment and Beacham and Brown (ref. 41) have shown that it is possible for slow crack growth to occur at stress-intensity levels less than K_{IC}. The resistance of materials to stress corrosion is generally assessed by measuring the time required to fracture a cracked specimen under sustained load in a hostile environment. For a number of materials a definite limiting value of K_I is found to exist below which no apparent crack growth occurs. This limit, termed K_{ISCC}, is not however a feature of all materials, but where it exists it provides a useful index of a material's stress corrosion resistance. A single, unified theory to explain stress corrosion, has not yet been developed, mainly because of the large number of factors which appear to influence the mechanism.

FRACTURE TOUGHNESS AND ITS USE IN CYLINDER DESIGN

42. The accurate measurement of plane strain fracture toughness requires that certain specimen dimensions exceed a critical size which is a function of the toughness and strength of the material. For a material which has been fabricated into a thick-walled pressure vessel, the wall thickness and curvature often prevent the use of either of the standard test specimens (ref. 26). However, two new types of specimen, based on the cross-sectional geometry of a cylinder, have been proposed (ref. 42, 43). In one case approximately one-half of a disc cut from a cylinder is used, whilst in the other the entire cylinder cross-section becomes a specimen. Comparative tests, using materials of known fracture toughness have shown good agreement between the new proposed test methods and the ASTM standard (ref. 25).

43. Fracture toughness data for a wide range of materials have been published by several authors (ref. 44-50) and includes data for several common pressure vessel steels tested at ambient temperature (ref. 44-47). The latter illustrate the decrease in toughness which accompanies an increase in strength resulting from differing heat treatments.

When very high strength materials are tested over a

wide range of temperatures only minor variations in fracture toughness occur. However, with medium strength steels toughness decreases with decreasing temperature (ref. 46) with the effect becoming most pronounced in low strength structural steels (ref. 49) where a very steep decrease in K_{IC} occurs over a narrow temperature range.

44. The application of fracture toughness to the design of thick-walled cylinders is straight forward in principle if any of two of three parameters (flaw size, working stress and fracture toughness) are known. The value of the third, necessary for safety can be calculated from a relationship connecting all three as indicated by equation 1. For example, consider a cylinder with a diameter ratio of 2.0 and a bore diameter of 200 mm manufactured from a material having a tensile strength of 1500 MPa. Using the data given in Fig. 3, curves representing the relationship between failure pressure and flaw size can be constructed for a range of values of fracture toughness, see Fig. 7. Thus, knowing the material toughness and the design pressure, the value of crack depth to cause failure can be determined by interpolation from Fig. 7. This is the critical crack depth to cause failure and values less than this should be used as a maximum allowable flaw size for non-destructive testing purposes.

45. Use of the data given in Figs. 3 and 5 for long straight fronted cracks will give a conservative design compared to that given in Fig. 6 for a curved fronted crack. However, such caution can be fully justified in those cases where either the geometry of the initiating flaw is unknown, or a very brittle material has been used for the construction of the vessel.

CONCLUSIONS

46. Although fatigue and fracture of pressure vessels are still very serious problems for the designer there is now sufficient understanding of these phenomena so that usually a rough estimate of the conditions to cause failure can be made. There is still a great deal to be discovered and explained, and in many areas the designer has to extrapolate too far for comfort. However, it appears that most of the recent failures have not been caused by unknown phenomena but by the designer not foreseeing that a dangerous situation existed or being unaware of the appropriate information. Perhaps this is unfair as it is so easy to be wise after the event!

REFERENCES

1. MANNING W.R.D. and LABROW S. High Pressure Engineering, Leonard Hill 1971, p.108-124.

2. MORRISON J.L.M., CROSSLAND B. and PARRY J.S.C. "Fatigue under triaxial stress: development of a testing machine and preliminary results." Proc. Instn. Mech. Engrs. 1956, 170, p. 697.

3. RODMAN T.J. Reports of experiments on the properties of metals for cannon. 1861. Crosby, Boston, Mass.

4. KIRKALDY D. Results of an experimental inquiry into the tensile strength and other properties of various kinds of Wrought Iron. 1862. W.F. Kirkaldy, Strength and properties of materials with descriptions of the system of testing (Sampson, Low, Martson, Searle and Rivington Ltd., London 1891).

5. JACOB L. Résistance et Construction des Bouches à Feu, Doin et Fils, Paris 2nd Ed. 1920, p. 157.

6. IRWIN G.R. Structural aspects of brittle fracture, Appl. Mater. Res. 1964, 3(2), p. 65.

7. WELLS A.A. Unstable crack propagation in metals-cleavage and fast fracture. Cranfield Crack Propagation Symposium, Sept. 1961, Vol. 1, p. 210.

8. MORRISON J.L.M., CROSSLAND B. and PARRY J.S.C. "Strength of thick cylinders subjected to repeated internal pressure." Proc. Instn. Mech. Engrs. 1960, 174, p. 95-117.

9. PARRY J.S.C. "Fatigue of thick cylinders: further practical information." Proc. Instn. Mech. Engrs. 1965-66, 180, p. 387.

10. PARRY J.S.C. "Further results of fatigue under triaxial stress." Proc. Int. Conf. on Fatigue of Metals. (Instn. Mech. Engrs), 1956, p. 132.

11. BURNS D.J. and PARRY J.S.C. "Effect of mean shear stress on the fatigue behaviour of thick-walled cylinders." Proc. Instn. Mech. Engrs. 1967-68, 182 Pt. 3c. Conference on High Pressure Engineering, p. 72.

12. FROST N.E. "A note on the behaviour of fatigue cracks." J. of Mech. and Physics of Solids 1961, 9, p. 143.

13. COLE B.N. "Strategy for cross-bores in high pressure containers." J. Mech. Engrs. Sci. 1969, 11, p. 151.

14. BOYD R. "The fatigue strength of an alloy steel: effect of tempering temperature and directional properties." Proc. Instn. Mech. Engrs. 1965-65, 179, p. 733.

15. CROSSLAND B. and SKELTON W.J. "Effect of hardness on the fatigue strength of thick-walled cross-bore cylinders of EN25." Proc. Instn. Mech. Engrs. 1967-68, 182, Pt. 3c. Conference on High Pressure Engineering, p. 106.

16. DAVIDSON T.E., EISENSTADT R. and REINER A.N. "Fatigue characteristics of open-end thick-walled cylinders under cyclic internal pressure." J. Basic Engrg. Trans. Am. Soc. Mech. Engrs. 1963, 85, (Series D), p. 555.

17. AUSTIN B.A. and CROSSLAND B. "Low-endurance fatigue strength of thick-walled cylinders: development of a testing machine and preliminary results." Proc. Instn. Mech. Engrs. 1965-66, 180, Pt. 1, p. 43.

18. AUSTIN B.A., REINER A.N. and DAVIDSON T.E. "Low cycle fatigue strength of thick-walled pressure vessels." Proc. Instn. Mech. Engrs. 1967-68, 182, Pt. 3c. Conference on High Pressure Engineering, p. 91.

19. JONES P.M. and TOMKINS B. "Low endurance fatigue of very thick walled cylinders." Proc.

Instn. Mech. Engrs. 1967-68, 182, Pt. 3c. Conference on High Pressure Engineering, p. 81.

20. HASLAM G.H. "The fatigue limit of cylinders subjected to repeated internal pressures." High temperatures: High pressures 1969, 1 (6), p. 705.

21. HASLAM G.H. "Estimation of the fatigue life of a thick-walled cylinder subjected to repeated internal pressure." J.Mech. Engng. Sci. 1971, 13, p. 130.

22. TOMKINS B. "Fatigue failure criteria for thick-walled cylindrical pressure vessels." Int. J. Pressure Vessels and Piping, 1973, 1. p.37.

23. McDIARMID D.L. "Fatigue design for multiaxial stress conditions." Inst. Mech. Engrs. Conference on Designing against Fatigue, Oct.1974.

24. IRWIN G.R. "Analysis of stresses and strains near the end of a crack traversing a plate." J. Appl. Mech. 1957, 24, p. 361.

25. ASTM "Tentative method of test for plane-strain fracture toughness of metallic materials." E 399-70T ASTM Standards Part 31, July 1971, p.919.

26. British Standards Institution "Methods for plane strain fracture toughness (K_{IC}) testing DD3, 1971.

27. WELLS A.A. "Notched bar tests, fracture mechanics and the brittle strength of welded structures." Hordrement Lecture, 1964, Br. Weld J. 1965, 12 (1), p. 2.

28. DOBSON M.O. "Practical fracture mechanics for structural steels." Chapman and Hall, London, 1969.

29. BOWIE O.L. "Analysis of an infinite plate containing radial cracks originating at the boundary of an internal circular hole." Jnl. Maths. Phys. 1956, 35, p. 60.

30. PARIS P.C. and SIH G.C. "Stress analysis of cracks." ASTM STP 381. Am. Soc. for Testing and Materials, 1965, p. 30.

31. BOWIE O.L. and FREESE C.E. "Elastic analysis for a radial crack in a circular ring." Eng. Fract. Mech. 1972, 4, p. 315.

32. SHANNON R.W.E. "Stress intensity factors for thick-walled cylinders." Int. J. Pressure vessels and Piping. 1974, 2, p. 19.

33. BOWIE O.L. and NEAL D.M. "A modified mapping-collocation technique for accurate calculation of stress intensity factors." AMMRC TR69-28, Nov. 1969.

34. SHANNON R.W.E. and AUSTIN B.A. "The application of fracture mechanics to failure of thick-walled cylinders." Practical Application of Fracture Mechanics to Pressure Vessel Technology. Conference, London, Inst. Mech. Engrs. 1971, p.109.

35. UNDERWOOD J.H., LASSELLE R.R., SCANLON R.D. and HUSSAIN M.A. "A compliance K calibration for a pressurised thick-walled cylinder with a radial crack." Eng. Fract. Mech. 1972, 4, p. 231.

36. KUTTER H.K. "Stress analysis of a pressurised circular hole with radial cracks in an infinite elastic plate." Int. J. Fract. Mech. 1970, 6, p. 233.

37. UNDERWOOD J.H. "Stress intensity factors for internally pressurised thick-walled cylinders. Stress analysis and growth of cracks." ASTM STP 513. Am. Soc. for Testing and Materials, 1972, p.59.

38. RICE J.R. and LEVY N. "The part-through surface crack in an elastic plate." Technical report AASA NGL 40-002-080/3 to the National Aeronautics and Space Administration 1970.

39. PARIS P.C. and ERDOGAN F. "A critical analysis of crack propagation laws." Trans. ASME 1963, 85, Series D, p. 528.

40. HASLAM G.H. "Fatigue of cylinders subjected to pulsating internal pressure." Trans. ASME 1972, 94, Series B, p. 284.

41. BEACHAM C.D. and BROWN B.F. "A comparison of three pre-cracked specimens for evaluating the susceptibility of high strength steels to stress corrosion cracking." Stress corrosion testing ASTM STP 425, Am. Soc. for Testing and Materials, 1967.

42. KENDALL D.P. and HUSSAIN M.A. "A new fracture toughness test method for thick-walled cylinder material." Proc. Soc. for Experimental Stress Analysis, 1973, 29 (1), p. 184.

43. JONES A.T. "A radially cracked, cylindrical fracture toughness specimen." Eng. Fract. Mech. 1974, 6, p. 435.

44. JONES M.H. and BROWN W.F. Jr. "The influence of crack length and thickness in plane strain fracture toughness tests." Review of Developments in plane strain Fracture Toughness Testing ASTM STP 463, Am. Soc. for Testing and Materials, 1970, p. 63.

45. MAY M.J., PRIEST A.H. and WALKER E.F. "Fracture toughness of high strength steels." The British Iron & Steel Research Assoc., MG/E/302/67.

46. WESSEL E.T. "State of the art of the WOL specimen for K_{IC} fracture toughness testing." Westinghouse Research Laboratories, 67-1D6-BTLFR-P2, May 1967.

47. LOGAN J.G. and CROSSLAND B. "The fracture toughness of EN25 and a 3% Ni-Cr-Mo-V steel at various strength levels together with Charpy impact data." Practical Application of Fracture Mechanics to Pressure Vessel Technology. Conference, London. Inst. Mech. Engrs. 1971, p. 148.

48. STEIGERWALD E.A. "Crack toughness measurements of high-strength steels." Review of Development in plane strain Fracture Toughness Testing. ASTM STP 463. Am. Soc. for Testing and Materials 1970, p. 102.

49. WESSEL E.T. "Linear elastic fracture mechanics for thick-walled welded steel pressure vessels. Material property considerations." Practical Fracture Mechanics for Structural Steel UKAEA Risley, 1969.

50. WALKER E.F. and MAY M.J. "Fracture properties of a Ni-Cr-Mo-V steel." Fracture Toughness of High Strength Materials: Theory and Practice. The Iron and Steel Institute, I.S.I. Pub. 120.

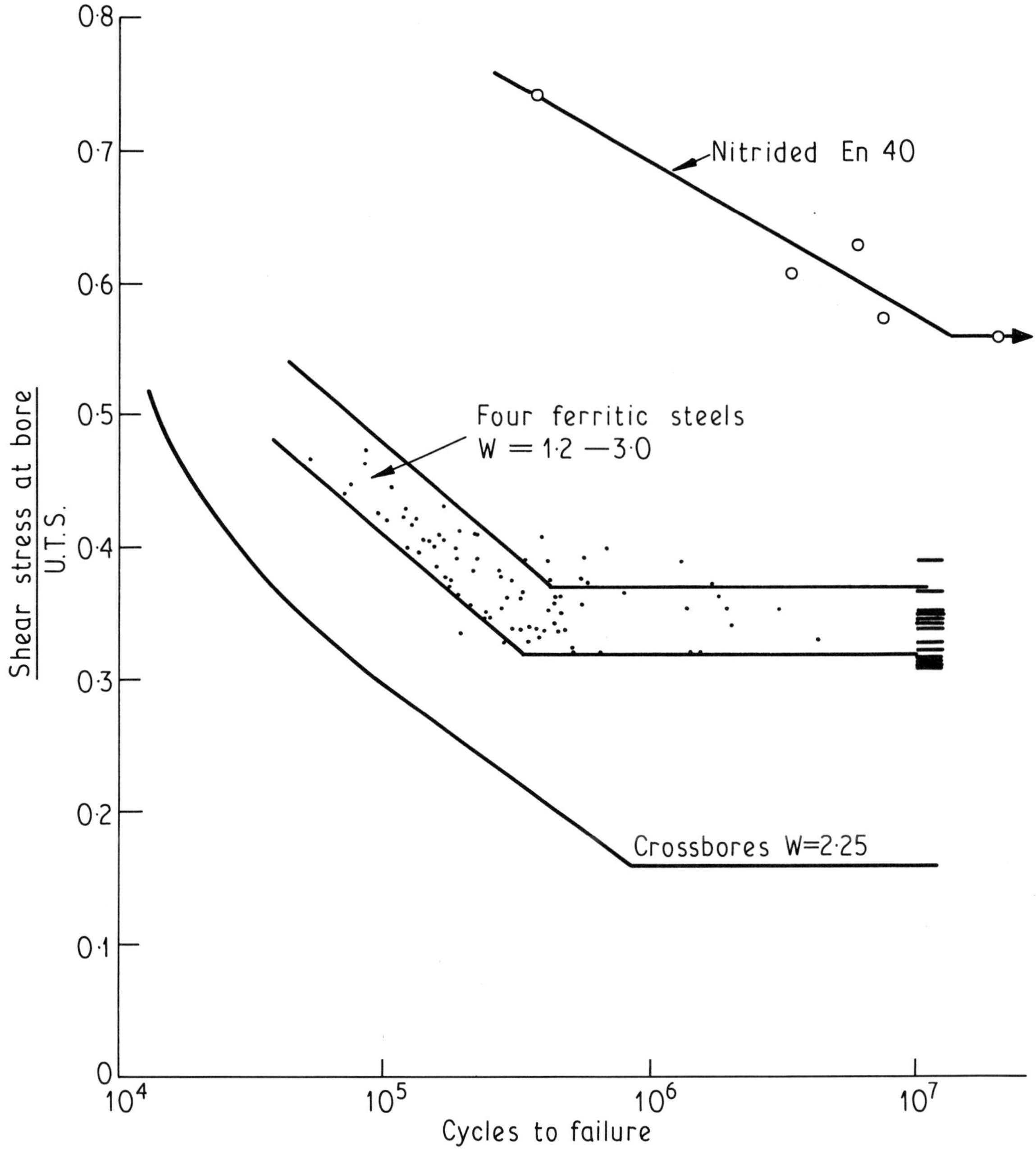

Fig. 1: Repeated pressure fatigue tests on thick-walled cylinders

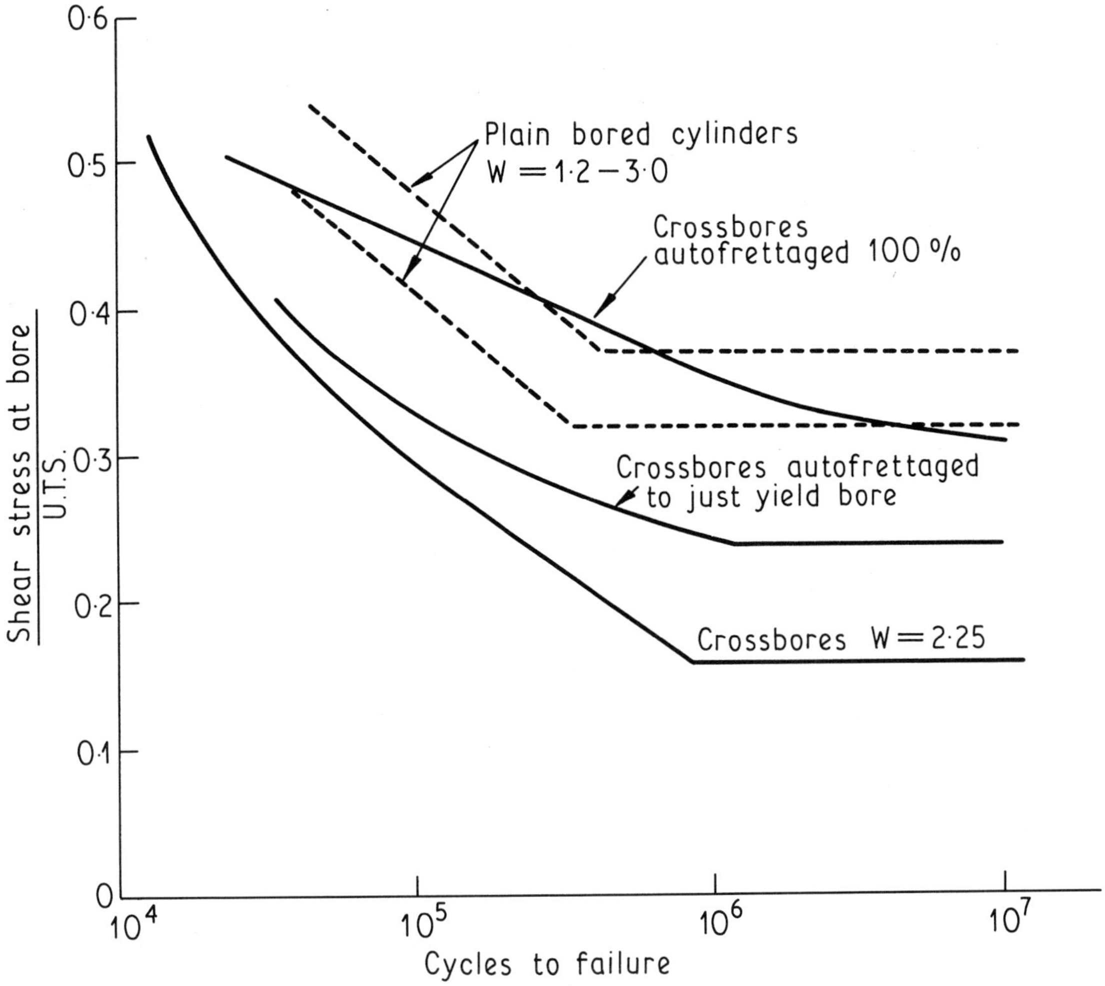

Fig. 2: Repeated pressure fatigue tests on cross-bored cylinders

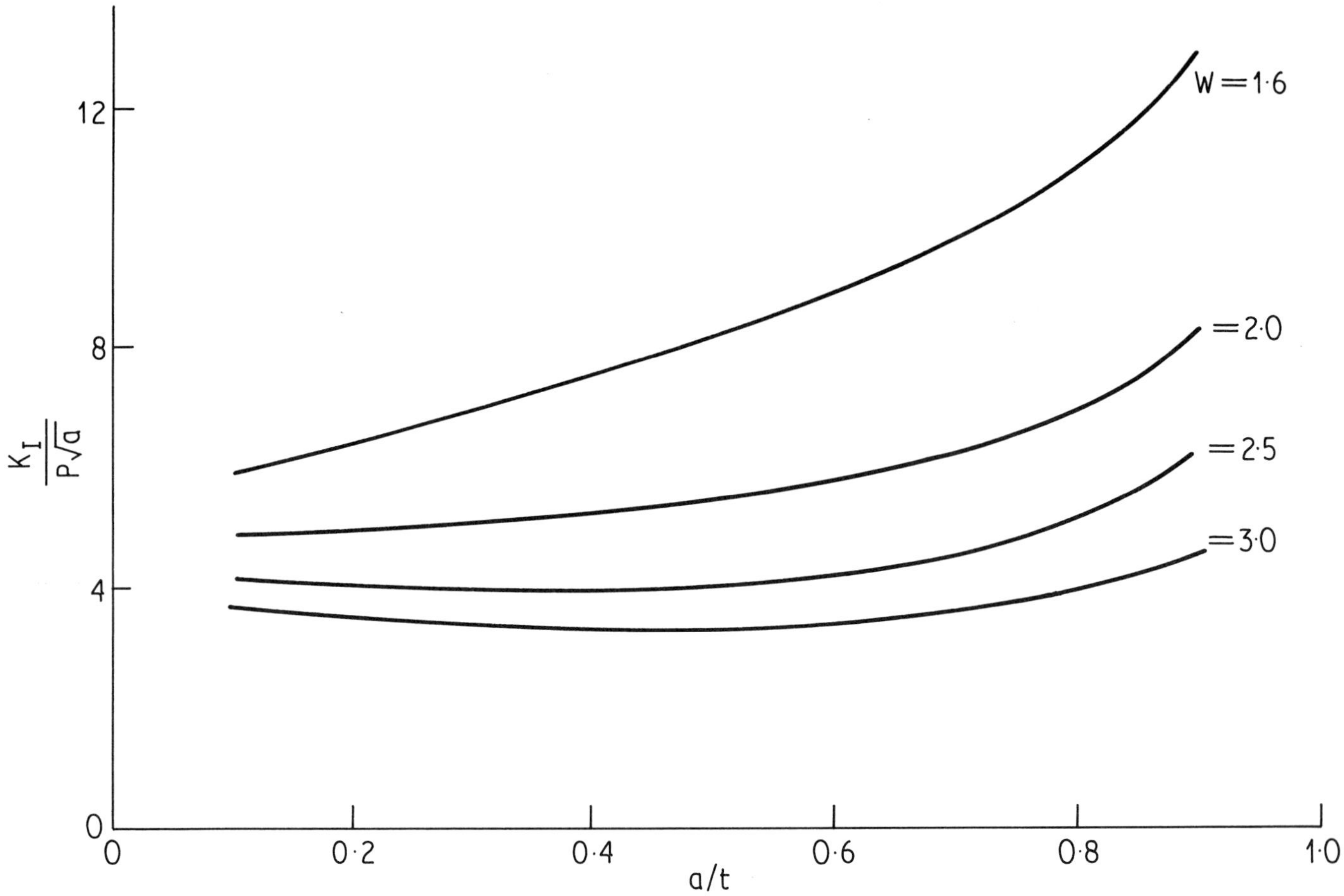

Fig. 3: Stress-intensity factors for a single radial crack in a pressurised cylinder

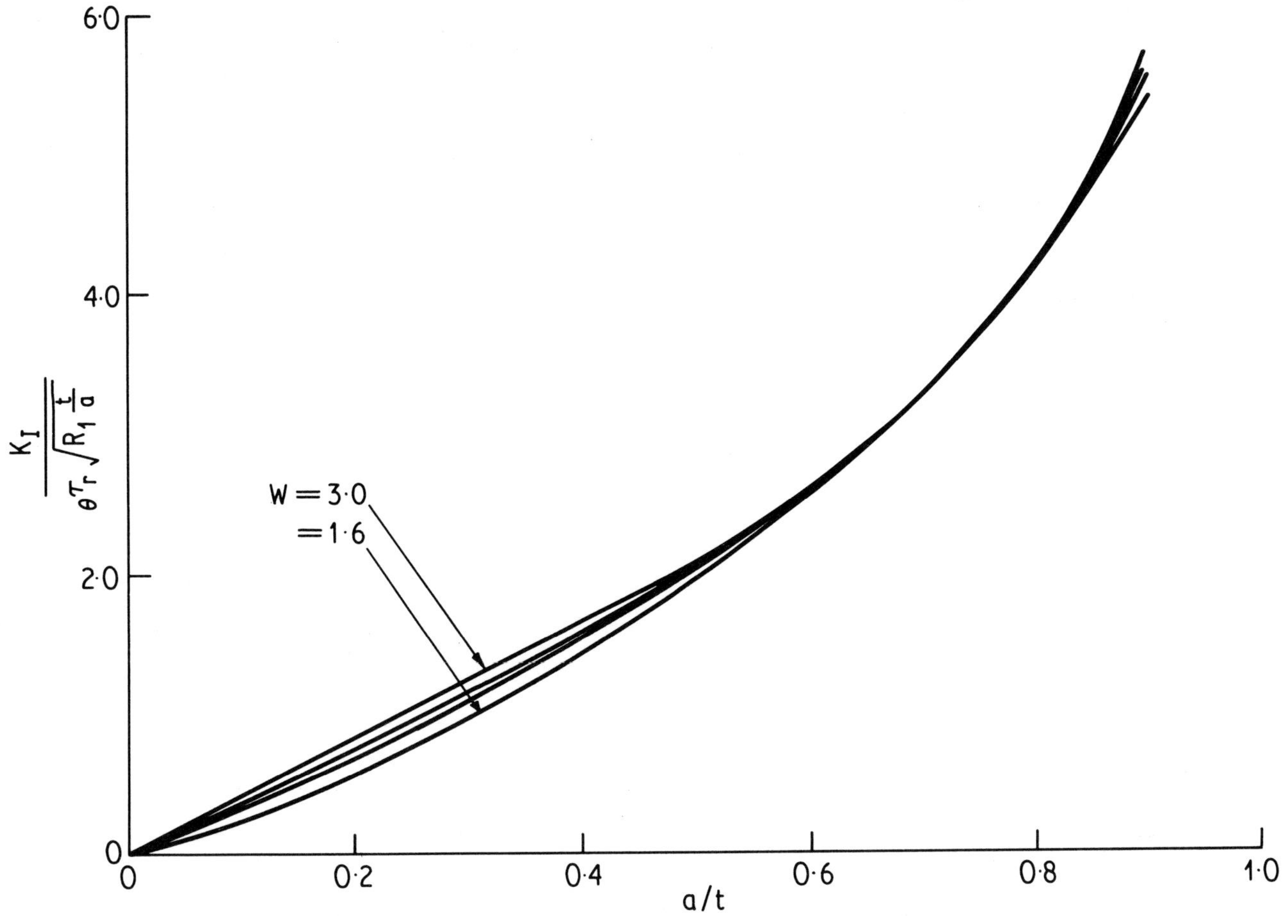

Fig. 4: Stress-intensity factors normalised with a simple stress parameter ${}_\theta\tau_r\sqrt{R_1\frac{t}{a}}$

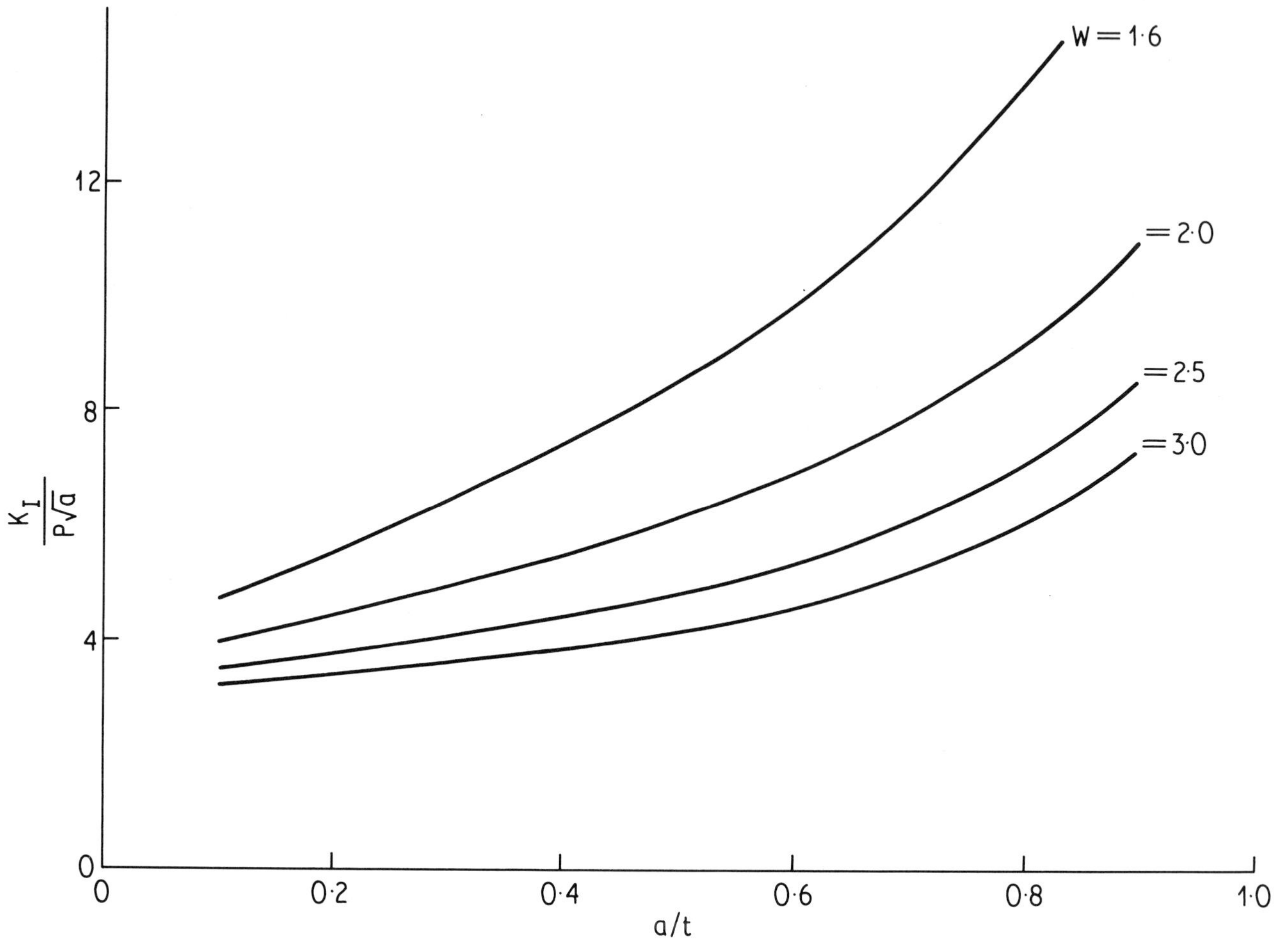

Fig. 5: Stress-intensity factors for two diametrically opposed cracks in a pressurised cylinder

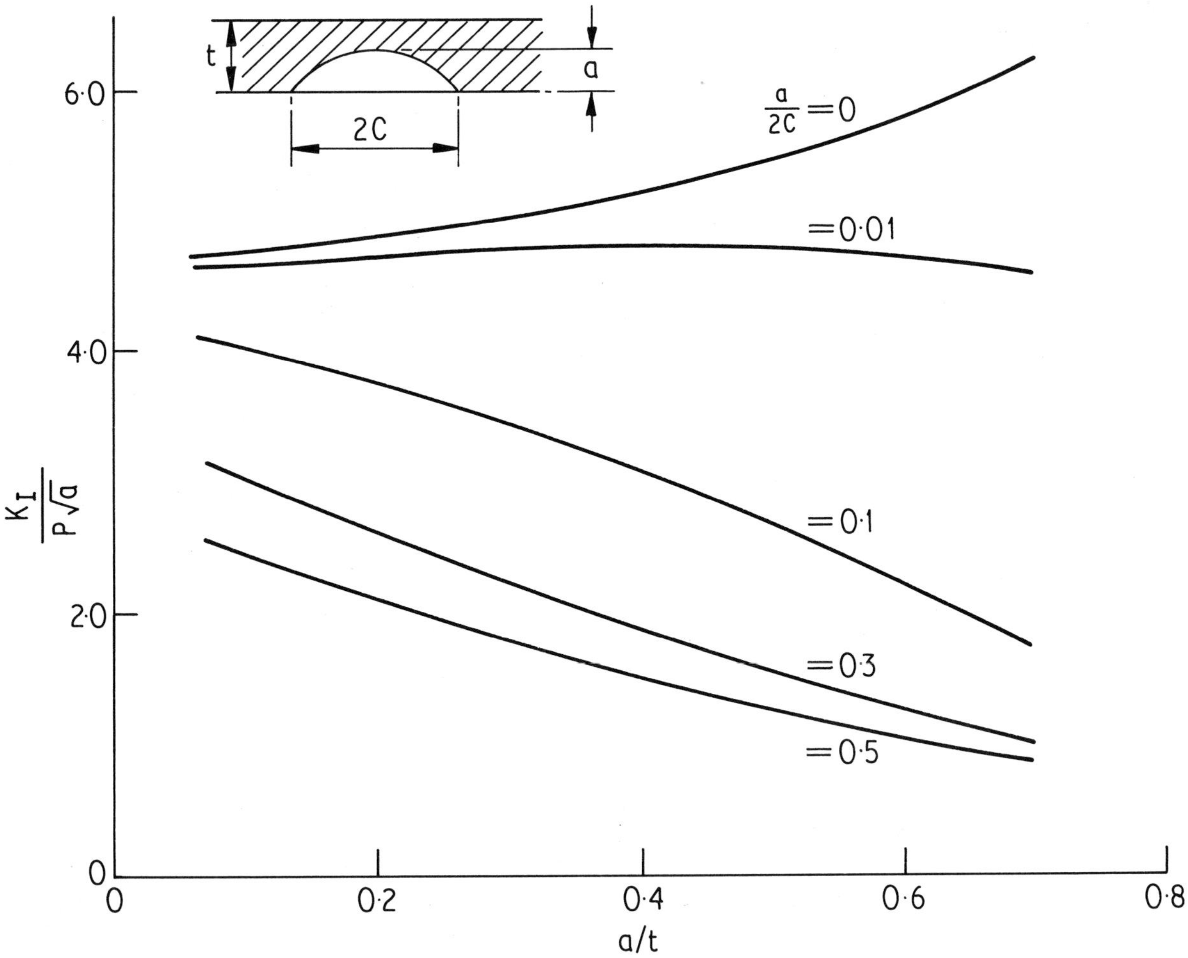

Fig. 6: Stress-intensity factors for a semi-elliptical crack

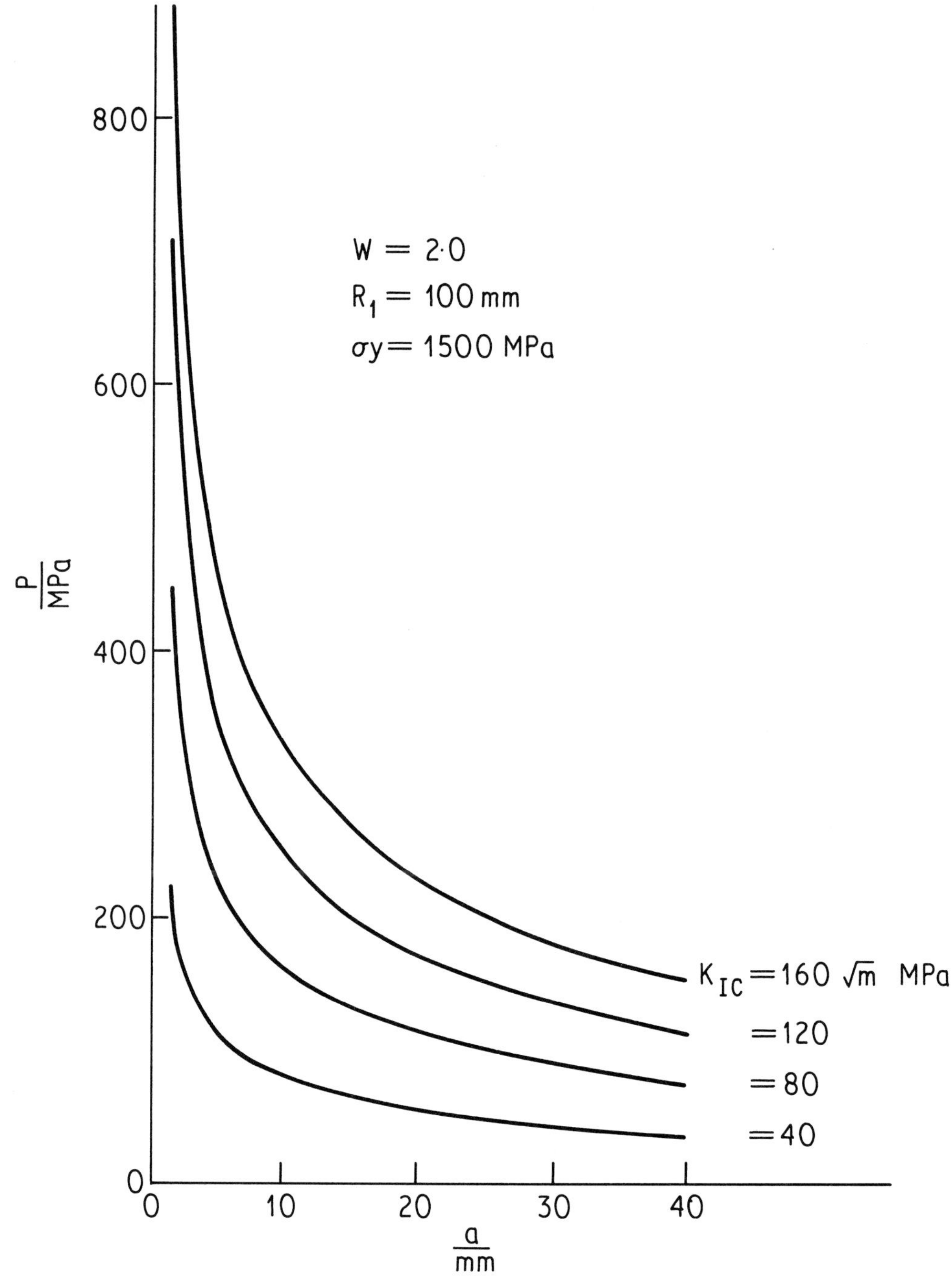

Fig. 7: Design curves for a cylinder with a diameter ratio of 2.0

THE FATIGUE AND BURST PROPERTIES OF HIGH PRESSURE TUBING MADE FROM STAINLESS STEELS

J. ROGAN, BTech, DIC, PhD
Department of Mechanical Engineering, Imperial College of Science and Technology, London

SYNOPSIS The first part of this paper is concerned with the fatigue and burst properties of thick-walled tubing made from 0.13% Carbon 13% Chromium stainless steel. The tube sizes studied range from 10 mm to 32 mm bore diameter with outside to inside diameter ratios varying from 1.8 to 2.6.

All tubes have been tested in the as-received and autofrettaged condition and the fatigue strength has been determined from tests in repeated pressure and at high mean pressures.

The mode of fatigue failure for all tubes was completely stable with no tendency for crack extension along the tubes.

The burst behaviour is typical for ductile materials and a simple formula can be used to predict the burst pressure.

The second part of the paper discusses factors affecting the fatigue strength of small bore tubing made from austenitic stainless steels. The fatigue strength is shown to depend upon the hardness of the cold worked material near the bore of the tubing, and upon the smoothness and freedom from radial fissures at the bore surface.

INTRODUCTION

1. In many high pressure processes used in the petrochemical industry resistance to corrosion attack is a desirable property of the materials used in plant fabrication.

2. One of the most widely used corrosion resistant materials is 0.13% Carbon 13% Chromium Type 410 stainless steel which can be heat treated to a wide range of strength levels, and which is also available as cold-reduced thick-walled tubing in bore sizes from 0.125 in. to 1.25 in. (3.2 to 32mm).

3. This material is usually heat treated to modest strength levels with tensile strengths in the range 98,000 to 120,000 lbf/in^2 (676 to 828 MN/m^2). This gives thick-walled tubing which exhibits excellent ductility and fracture properties, good resistance to corrosion and high fatigue strength.

4. The other widely used corrosion resistant materials are the austenitic stainless steels of which types 304 and 316 are most common for high pressure applications.

5. These can only be hardened by cold mechanical deformation and are used as cold drawn thick-walled tubing and as small cold worked blocks for valve bodies, check valves and similar components.

6. Despite the wide applications found in high pressure processes for thick-walled tubing made from stainless steels, there is no published work upon their fatigue and burst properties apart from some results on free machining EN56 reported by Parry (Ref. 1).

7. This paper presents results for a wide range of tube sizes and discusses those factors which affect fatigue behaviour.

TYPE 410 TUBING 0.13% Carbon 13% Chromium

8. All of the thick-walled 410 tubing used in the present work was produced by one manufacturer. For each tube size tests were done on tubing made at different times from different heats of steel. So the results are fully representative of the variation in properties of tubing made by one manufacturer.

9. They do not include variations in fatigue strength and mode of fracture which can occur when testing nominally identical pieces of high pressure tubing made by different manufacturers using different tube production processes. The production route can have a profound effect upon the mode of fatigue fracture, (see paper by author in this Conference upon the fatigue properties of low-alloy high strength steel tubing), but this has not been investigated in the present work.

Fatigue Strength

10. The fatigue strength of the whole range of 410 tubing has been measured in repeated pressure and at high mean pressures. A typical fatigue curve for 410 tubing is given in figure 1 which also shows a typical curve for non-autofrettaged tubing made from low alloy high strength steel.

11. In the non-autofrettaged condition the fatigue limit is similar for both types of tubing having the same diameter ratio.

12. The fatigue strength of the 410 tubing can be significantly increased by a pressure test designed to induce mild compressive residual stresses at the bore of the tubing. A proper autofrettage produces a still further improvement in fatigue strength as can be seen in figure 1.

13. The benefits of autofrettage are limited by the strength level to which the 410 tubing has been heat treated. Much higher fatigue strengths can be achieved from harder 410 tubing but there may be some reduction in the excellent toughness of the tubing. The effect of increased strength level upon fatigue properties remains to be fully investigated.

14. A very large number of fatigue tests have been performed upon the 410 tubing over the whole range of bore sizes and outside-to-inside diameter ratios, K ratios.

15. The regular nature of the fatigue behaviour makes it possible to derive simple expressions which can adequately predict the fatigue strength AT THE FATIGUE LIMIT in repeated pressures and at high mean pressures.

16. These are valid over the range of bore sizes from 0.4 in. to 1.25 in. (10 to 32 mm), the range of diameter ratios from K = 1.8 to K = 2.6, and the range of static tensile strength levels from 98,000 to 110,000 Lbf/in^2 (676 to 760 MN/m^2).

17. For tests in repeated pressure, the range of bore shear stress at the FATIGUE LIMIT $\Delta\tau = \Delta P.\frac{K^2}{K^2-1}$ is found to be $\Delta\tau = \frac{\sigma_{UTS}}{2.79}$ for as-received tubing.

18. Pressure testing increases the range of shear stress at the fatigue limit by 1.15 while autofrettage produces a further increase to $\Delta\tau = 1.3.\frac{\sigma_{UTS}}{2.79}$.

19. The effect of high mean shear stress (high mean pressure) upon the range of shear stress at the fatigue limit is shown in figure 2. These results cover the range of tube sizes, diameter ratios and strength levels and also show the effect of pressure testing and of autofrettage.

20. The effect of high mean shear stress is least for tubing in the as-received condition since the maximum cyclic pressures in high mean pressure tests causes slight cyclic strain hardening at the bore of the as-received tubes.

21. This strengthening effect due to cyclic strain hardening is particularly noticeable in 410 tubing (compared with low alloy steel tubing) because the 410 has been heat treated to give a fairly low ratio of yield strength to ultimate tensile strength. Thus it shows a distinct cyclic strain hardening capacity.

22. The lines R, T and A drawn on figure 2 are representative for the range of shear stress at the fatigue limit and the effect of high mean shear stress for type 410 tubing in the as-received, pressure tested, and autofrettaged conditions respectively.

23. The fatigue strength in repeated pressure is directly proportional to the static tensile strength and the shear stresses referred to in figure 2 have been modified to take account of the different strength levels to which the various tubes have been heat treated.

24. Thus the shear stresses have been modified by choosing the UTS of one tube as "reference", and then scaling the experimentally determined actual range and mean shear stresses of the other tubes by the factor Actual UTS/Reference UTS. The reference tensile strength was chosen as 108,000 Lbf/in^2.

25. It would now be possible to fit equations to the lines R, T and A in figure 2 and derive expressions for the range of shear stress at the fatigue limit for any mean stress for any type 410 thick-walled tube.

26. Alternatively, the lines R, T and A in figure 2 may be directly used to predict fatigue strength and to design the tube required for any given applications, as shown below.

27. Suppose type 410 tubing is to be used in a higher pressure interstage cooler. The mean interstage pressure is say 20,000 Lbf/in^2 and taking account of vibration stresses, pressure end load and flanged joint loads, margin for burst pressure over operating pressure, etc., the diameter ratio is chosen as K=1.8. The minimum tensile strength allowed in the tube specification is say 98,000 Lbf/in^2.

28. Thus K = 1.8 $K^2/K^2-1 = 1.446$

$\sigma_{UTS} = 98,000$ Lbf/in^2.

The modified shear stress due to a mean pressure of 20,000 Lbf/in^2 is

$$\text{mean } \tau = 1.446 \times \frac{108,000}{98,000} \times 20,000 = 31,870 \text{ Lbf/in}^2.$$

29. A vertical line drawn on figure 2 through this point 31,870 Lbf/in^2 on the mean shear stress axis intersects the line R and predicts a range of shear stress at the fatigue limit of 35,500 Lbf/in^2. This corresponds to a range of pressure of

$$35,500\left(\frac{K^2-1}{K^2}\right) \times \frac{\text{Actual UTS}}{\text{Reference UTS}} = \frac{35,500}{1.446} \times \frac{98,000}{108,000}$$

$$= 22,280 \text{ Lbf/in}^2$$

30. Hence the as-received type 410 tube will withstand indefinitely a range of pressure of 22,280 Lbf/in^2 cycling about a mean pressure of 20,000 Lbf/in^2. These pressure conditions are shown in figure 3. The designer can then compare these with the actual pressure conditions in plant service and will see that fatigue failure will never occur. If at some later date the interstage pressure is uprated or the cyclic pressure in the coolers is increased, the new conditions can be evaluated as above and the new 'margin of safety' calculated.

Mode of Failure

31. The mode of fatigue failure of every specimen tested was of the safe stable type with no tendency whatsoever for brittle axial crack extension following post-failure pressure cycling.

32. Typical fatigue cracks for a range of type 410 tubing are shown in figure 4.

Static Burst Pressure

33. Burst tests were performed on pieces cut from all the samples of 410 tubing. All burst failures were accompanied by gross plastic deformation and the mode of burst also shown in figure 4 is typical of materials with good ductility.

34. The actual burst pressure can be accurately predicted for the whole range of tubing, using a mean diameter formula of the type

$$P_{ULT} = 2.17\sigma_{ULT}\left(\frac{K-1}{K+1}\right)$$

AUSTENITIC STAINLESS STEEL TUBING

35. Small bore tubing made from type 304 Austenitic stainless steel is widely used in high pressure systems. The main areas of use are in catalyst lines from initiator pumps to injection points, and for instrumentation and high pressure lubrication.

36. Type 304, in common with all the non-precipitation-hardening austenitic materials can only be hardened by cold mechanical deformation. Thus in fabricating tubing from 304 steels, great care must be taken to ensure a final cold working treatment which gives adequate hardness (greater than 270 VPN) to the material at the inside diameter of the tube. Also the presence of radial flaws at the bore can seriously reduce fatigue strength, so the material at the inside diameter should be free from cold working cracks and laps and be as smooth as practically possible.

37. Small bore tubing from several different manufacturers has been studied. The structure at the tube inside diameter is very important and figure 5 shows micro photographs of the bore surface of 3 types of 0.125 in.ID x 0.375 in.OD tubing.

38. The tube N and tube T in figures 5(a) and (b) respectively have good bore surface finish. However, the material at the bore of tube N is too soft and this results in a low fatigue strength. Tube S in figure 5(c) shows damaged material at the tube bore, this "crowning" being typical of austenitic material which has been excessively sunk during the final manufacturing stage. Thus although the hardness is high at the bore of tube S its fatigue strength is low because of the bore defects.

39. Microhardness traverses across the tube wall from inside to outside diameters are shown for the three tubes on figure 6, while figure 7 presents the results of repeated pressure fatigue tests.

40. These results, and results from tests on tubing of larger bore size in both 304 and 316, and from components such as valve bodies, show that the best fatigue properties are achieved when the material at the inside diameter is cold worked to a hardness of 300 VPN or greater. Of equal importance is the condition of the material at the bore surface which should be free from laps and fissures and from the 'crowned' or 'coronet' defects which are caused by incorrect cold drawing during the last manufacturing sequence.

41. The fatigue strength can be improved by a modest pressure test and still further improved by a proper autofrettage to 105,000 Lbf/in^2 for the K = 3 tubes.

42. The mode of fatigue failure is always of the safe, stable type.

43. The fatigue properties of austenitic steels will be discussed in more detail in a paper by Mr. P. S. J. Crofton and the author which deals with a wide range of "new" materials which are increasingly being used in high pressure applications.

CONCLUSIONS

44. The fatigue properties of stainless steel tubing have been presented and the factors affecting fatigue strength discussed.

45. Type 410 tubing is shown to have excellent fatigue strength and mode of fracture which, coupled with its resistance to corrosion, make it an ideal material for many high pressure processes.

46. Simple expressions have been derived from which the fatigue strength and the burst strength of a wide range of thick-walled tubing can be predicted.

47. Those factors which influence the fatigue properties of austenitic stainless steel tubing have been briefly discussed.

ACKNOWLEDGEMENTS

Most of this work was supported by a research grant from T.I. Chesterfield Ltd., Stainless Tube Division, which was gratefully received. The author would like to thank Mr. G. Trotman and Mr. T. Thursfield of that company for their contribution to the project.

The work was carried out in the Mechanical Engineering Department of Imperial College and the encouragement of the Head of Department, Professor Sir Hugh Ford, F.R.S., is gratefully acknowledged.

The properties of a wider range of materials is being investigated by Mr. P. S. J. Crofton, a research student with the author, who also made a significant contribution to the present work.

REFERENCE

1. Parry, J. S. C. 'Fatigue of thick cylinders: further practical information' Proc. I. Mech.E., 1965-66, 180, pt. 1, 387.

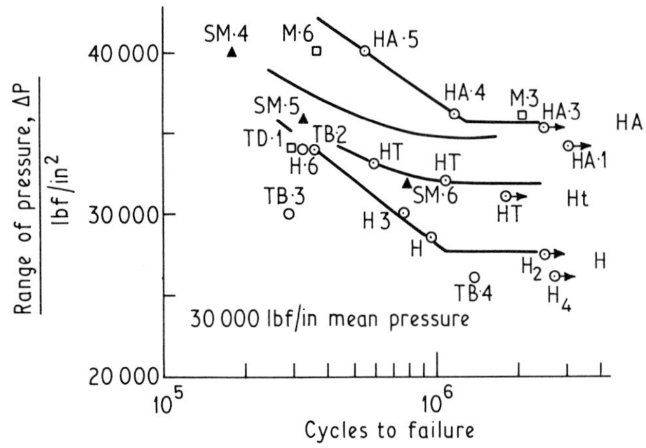

Fig. 1: Fatigue strength of type 410 tubing tested about 30,000 lbf/in² mean pressure. Tube dimensions .9in ID × 2.125in OD K = 2.36
H as-received HT pressure tested HA autofrettaged
Low alloys high strength steel tube results for tubes with K = 2.3 to 2.7 shown for comparison

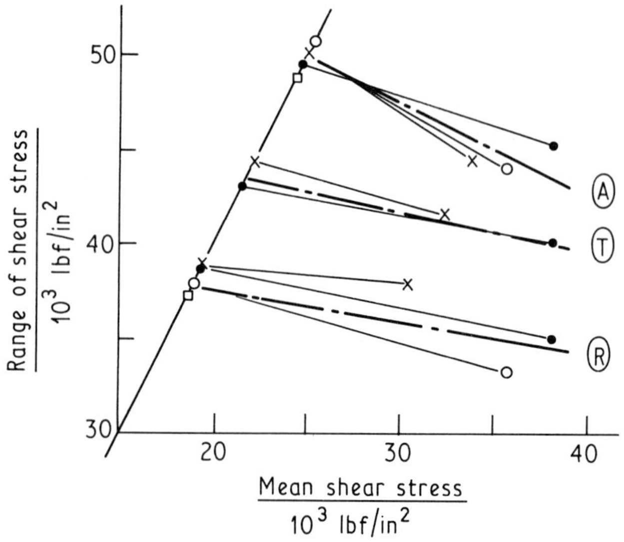

Fig. 2: Effect of mean shear stress upon the range of shear stress at the fatigue limit for 410 tubing
R as-received. T pressure tested.
A autofrettaged.

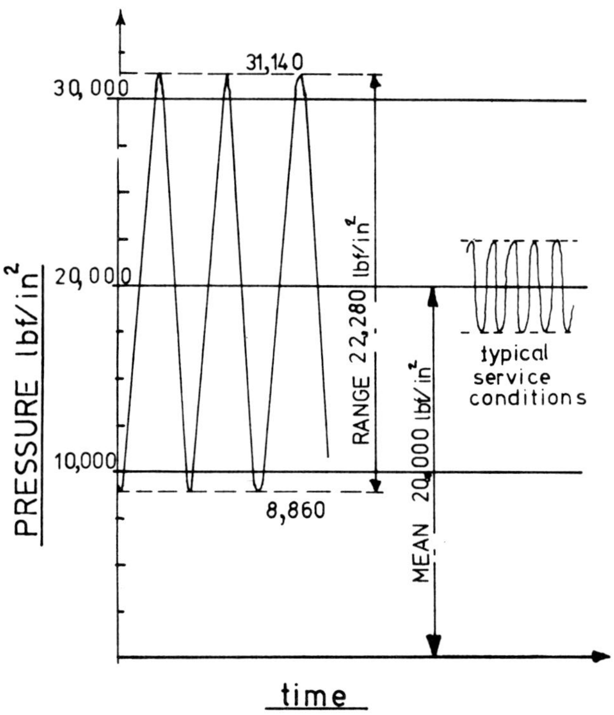

Fig. 3: Pressure conditions at the fatigue limit for K = 1.8 tube used in design example

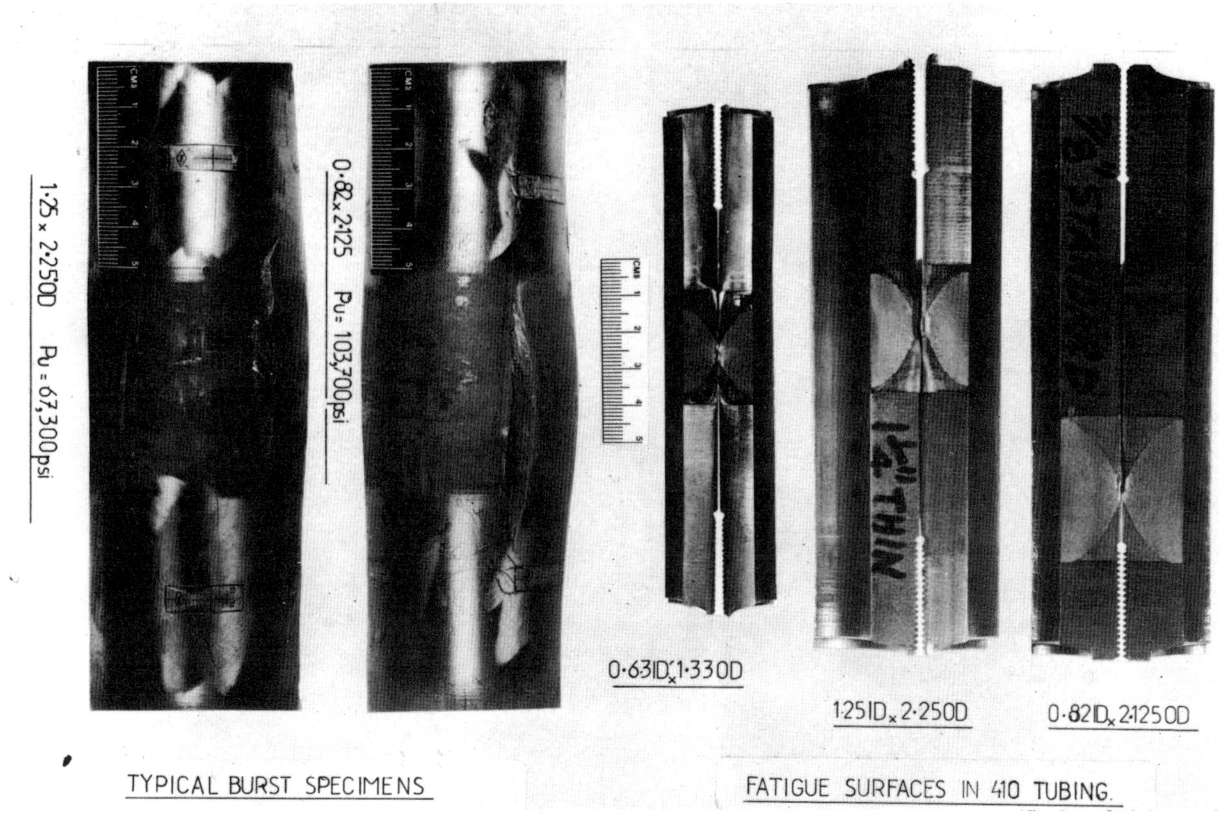

Fig. 4: Mode of fatigue cracking and of burst for several types of 410 tubing

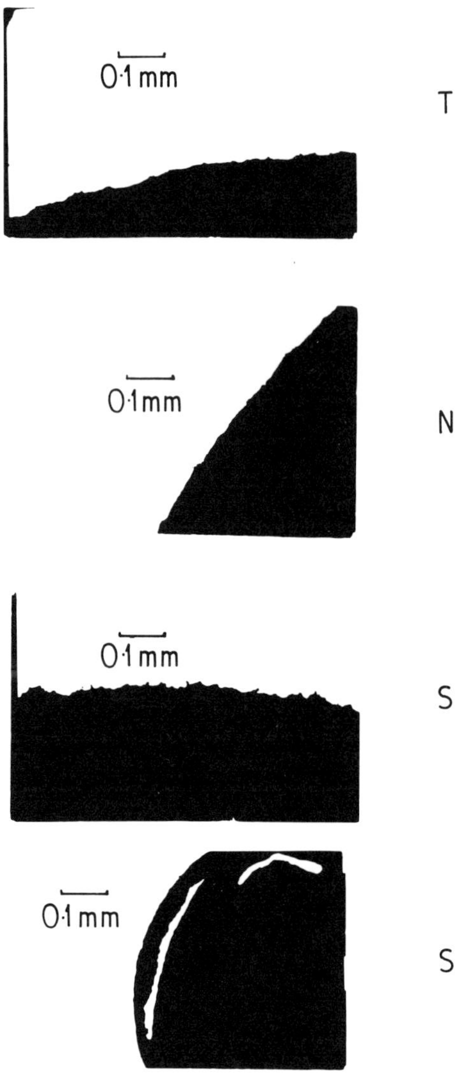

Fig. 5: Microphotographs of bore in three types of austenitic stainless steel tubing

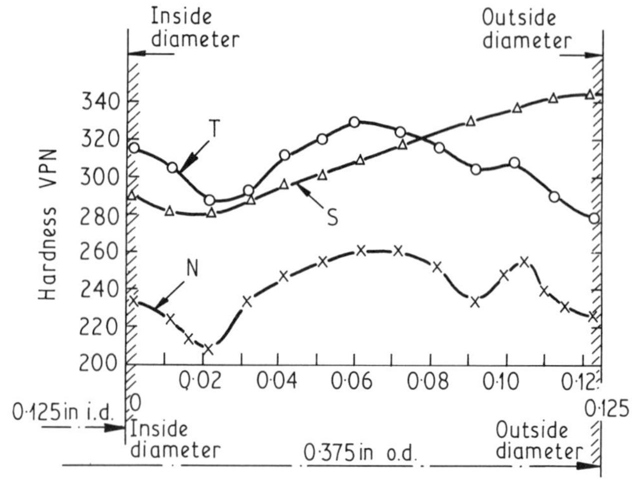

Fig. 6: Microhardness traverses (200gm load) from bore to outside diameter for three types of 304 tubing

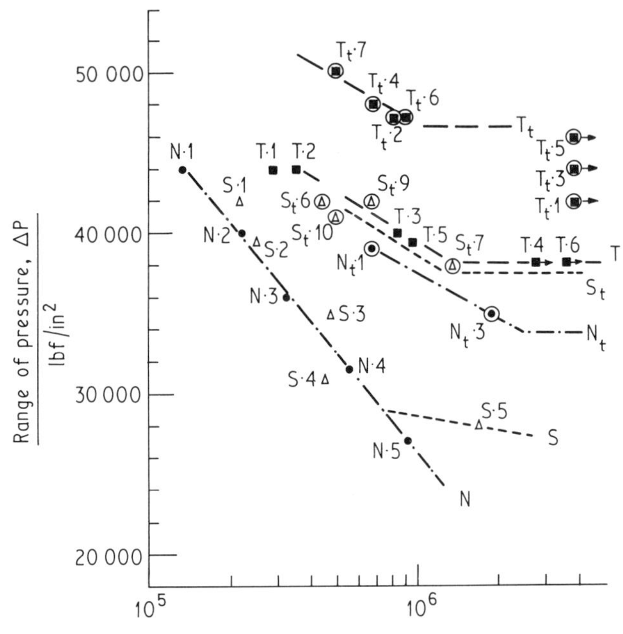

Fig. 7: Fatigue strength of three types of 304 tubing in the as-received (no suffix) and pressure-tested condition (60,000 Lbf/in²) (with suffix t)

FATIGUE STRENGTH AND MODE OF FRACTURE OF HIGH PRESSURE TUBING MADE FROM LOW-ALLOY HIGH-STRENGTH STEELS

J. ROGAN, BTech, DIC, PhD
Department of Mechanical Engineering, Imperial College of Science and Technology, London

SYNOPSIS This paper summarises the results of a very large number of cyclic high pressure fatigue tests on a wide range of commercially available thick-walled high pressure tubing. The tube materials are medium Carbon-Nickel-Chromium-Molybdenum steels heat treated to ultimate tensile strength levels in the range 140,000 to 195,000 Lbf/in^2 (950 to 1350 MN/m^2). The sizes of tubing are 1 to 2 inch (25 to 50 mm) inside diameters with overall diameter ratios ranging from 2.25 to 2.9.

It is shown that the mode of fatigue failure of certain types of tubing can lead to rapid extension of the original fatigue crack along the tube which could produce dangerous end-to-end failures in service conditions.

INTRODUCTION

1. Previous work upon factors affecting the fatigue strength of thick walled cylinders has mainly involved tests on carefully machined specimens, usually cut from bar stock.

2. This paper deals with factors affecting the fatigue behaviour of commercially available high strength steel tubing of the type used by the petrochemical industry in plants producing low density polyethylene. The polymerisation process requires high static pressures often with a superimposed cyclic pressure. The magnitude and frequency of this cyclic pressure or bump cycle varies from one process to another but typically the pressure may be cycled from 45,000 to 30,000 Lbf/in^2 several times a minute. Thus in one year's operation the reactors may be subjected to several million pressure cycles, and this means that the fatigue properties of the components of the reactors are very important. The resistance to fatigue failure should be as high as possible and if fatigue failures do occur the mode of failure must not lead to massive leakage of the contents of the reactor since this could produce an explosion.

3. In experiments designed to study the fatigue behaviour of actual high pressure tubing it is not permissible to machine the inside diameter nor to turn down the outside diameter to some convenient size. Such machining would interfere with the fatigue strength and mode of failure of the tubing.

4. The author designed and built large capacity high pressure fatigue machines capable of testing 2 inch bore (50 mm) tubing at pressures cycling from any minimum pressure up to 70,000 Lbf/in^2 maximum pressure at speeds of about 360 cpm (6 HZ). These are much modified versions of the original "Bristol" machines designed by Morrisson, Crossland and Parry. Still larger machines are presently being built.

5. This paper gives some of the results which have been obtained on these machines over the past eight years and to the writer's knowledge these are the first published results of a systematic study of the fatigue properties of actual high strength high pressure tubing.

6. Many factors which can only be mentioned briefly in this report will be discussed in detail in future papers.

FATIGUE STRENGTH

7. The fatigue strength of the whole range of tubing has been measured in repeated pressure and at high mean pressures.

8. The method of manufacturing the tubing does not have a significant effect upon fatigue strength; it can however have a profound effect upon the mode of fatigue failure as will be seen later.

Non-autofrettaged Tubing

9. The fatigue strength of non-autofrettaged tubing is influenced by the strength level to which the tubing has been heat treated and by conditions at the inside diameter such as surface finish, surface defects and residual stresses imparted during the tube finishing process.

10. In general terms increasing the static tensile strength level from say 140,000 to 190,000 Lbf/in^2 does not produce any increase in the non-autofrettaged fatigue strength because of the difficulty in achieving a good surface finish at the bore of the higher strength tube. Indeed the higher strength tube can have a markedly lower non-autofrettaged fatigue strength if the tubing contains bore defects produced either during tube manufacture or by corrosion or oxidation during subsequent fabrication processes.

11. Residual stresses at the bore have a pronounced effect upon non-autofrettaged fatigue

strength. The author has never found detrimental residual stresses in any length of unused high pressure tubing. The stress relief treatments which form part of all the tube manufacturing processes reduce residual stresses to a negligibly low value. Thus the fatigue strength of non-autofrettaged tubing in the as-received condition is not improved by further stress relief treatments in vacuum furnaces in the laboratory even if we slightly re-temper the tubing which guarantees the removal of all residual stresses.

12. One major tube manufacturer uses a severe grit blasting technique for bore finishing. This produces a marked increase in the non-autofrettaged fatigue strength but in practice the benefits of bore blasting can be lost if the coverage is not 100% of the bore surface (99% is not good enough) or if the tubing is stress relieved after a subsequent fabrication process such as bending.

13. The fatigue results for a selection of non-autofrettaged high pressure tubing are shown in figure 1 for repeated pressure and figure 2 for cycling about a mean pressure of 30,000 lbf/in^2.

14. The highest fatigue strength curve A is obtained with medium tensile strength tubing which has a good bore surface finish or with bore-blasted tubing. The results for curve B represent average bore finish with medium strength or good bore finish with higher strength tubing while curve C represents mediocre bore finish for all strength levels. Curve D shows the low fatigue strength of higher strength tubing with poor bore surface finish and is also typical of bore-blasted tubing in which oxidation pits have formed during stress relief and during storage.

15. It can also be seen that the range of pressure at the fatigue limit for tests about a high mean pressure is only slightly lower than for repeated pressure. This is especially true for the medium strength tubing where slight localised yielding has occurred during the high mean pressure tests.

16. Thus the reduction in fatigue strength due to an increase in mean stress is much less for medium strength tubing than for higher strength tubing.

17. From these results it is clear that the benefits of heat treating tubing to higher strength levels can only be realised if the tubing is autofrettaged.

Autofrettaged Tubing

18. Autofrettage produces a marked increase in fatigue strength particularly for the higher strength tubing. This can be seen in figures 3 and 4 for repeated pressure tests and high mean pressure tests respectively. Figure 5 compares the non-autofrettaged and autofrettaged fatigue strength for a wide range of tubing.

19. In these fatigue curves the effect of the degree of autofrettage is shown by comparing for example the fatigue strength of tube Y, non-autofrettaged, YA modest autofrettage and YAA optimum autofrettage. The relevant pressure vs. expansion curves for the two autofrettage conditions is shown in figure 6

20. Increasing the degree of autofrettage produces progressively smaller increases in the fatigue strength of the thick-walled tubing as tested on the fatigue machines at 50°C which is their normal operating temperature.

21. One often reads in the literature about "reversed yielding" which is interpreted by some to mean that autofrettage beyond a certain degree would lead to a reduction in the magnitude of the compressive residual stresses at the tube bore, with a consequent reduction in fatigue strength.

22. After a very searching study of the effects of the degree of autofrettage upon fatigue strength, the fact is that it is impossible to over-autofrettage plain bore thick-walled cylinders from the viewpoint of reducing the room temperature fatigue strength. Tests on a piece of 1 in. bore 2.6 in. O.D. tubing cut into 5 ft. lengths which were autofrettaged to different degrees showed a progressive increase in fatigue strength for autofrettage pressures up to 80% of burst pressure. For higher autofrettage pressures the fatigue strength was a constant high value and even specimens cut from a length of the tubing which had been pressurised to burst showed the same value of fatigue strength.

23. Rather than worry about "reversed yielding", the main concern should be about the stability of the residual stresses induced by autofrettage in thick-walled tubing which is subsequently subjected to cyclic high pressures at operating temperatures of 300 to 350°C. The author has not done high pressure fatigue tests at these temperatures (this is the next area of investigation) but by measurements of loss of residual stresses and reduction in fatigue strength in high pressure tubing removed from service from several different polyethylene plants around the world, it is clear that there is an OPTIMUM autofrettage for tubing of any given strength level and diameter ratio. Autofrettaging beyond this optimum level can lead to loss of residual stresses in subsequent service conditions with a consequential serious reduction in fatigue strength.

24. This is an important aspect of high pressure engineering about which more will be written elsewhere. It is surprising that so little has been published on residual stresses due to autofrettage since the work by Franklin and Morrison in 1960.

25. The fatigue strength of properly autofrettaged high-strength low-alloy steel tubing is extremely high and the results shown in figures 3 to 5 enable reactors to be designed with a considerable degree of confidence.

MODE OF FAILURE

26. In recent years the tendency in designing new tubular process plants has been to use larger and larger bore tubing. At first sight this simply entails scaling up the sizes of tubing which has performed satisfactorily in existing plants. Thus a tube with 1 in. bore, diameter ratio 2.8, wall thickness 0.9 in. becomes for example a 1.5 in. bore, diameter ratio 2.8, wall thickness 1.35 in. tube.

27. When testing some "scaled-up" tubing it was discovered that if one continued to cycle the internal pressure inside a specimen which had already failed by a fatigue crack growing from the bore to the outside diameter, it was sometimes possible to produce a rapid extension of the crack along the whole length of the tubular specimen.

28. This axial extension of the original fatigue crack can occur after only a few hundred or a few thousand of these post-failure pressure cycles. The technique of post-failure pressure cycling gives a qualitative measure of those factors which govern the resistance of the tubing to fast axial crack extension. These factors are directly associated with the tube manufacturing process and with the wall thickness of the tubing.

29. Figure 7 shows typical examples of the thousands of fatigue failures in tubing tested at Imperial College. The fatigue cracks are perfectly stable and no amount of post-failure pressure cycling will cause the fatigue cracks to extend along the tube. All that happens is that the shear lip where the crack has broken through to the outside diameter becomes slightly larger. Such fatigue cracks obtained on the laboratory fatigue machines correspond to stable fatigue cracks that occur in plant service. The small amount of ethylene leaking from the tube into the surrounding cooling water jacket is easily detected and the plant is safely shut down.

30. Figure 8 shows examples of fatigue cracks in tubing made by three different manufacturers. In each case the initial fatigue crack has suddenly extended in a fast brittle manner along the whole length of the specimen.

31. Thus in tube N, the initial crack a-a grew suddenly to b-b and then split from end-to-end C-C after a few hundred post-failure pressure cycles. Tubes R and XA failed by a similar rapid axial extension of the initial fatigue crack from a-a to b-b.

32. In plant service a rapid axial extension of a fatigue crack could split the 30 ft. (10 metre) tubes from end-to-end leading to a massive escape of the reactor contents which could produce a disastrous explosion.

33. Thus larger bore tubes have to be designed around the capabilities of the tube manufacturers. This usually will involve using lower diameter ratios for the large bore tubing so that the tube wall thickness is not too great for the achievement of good fracture toughness properties throughout the whole wall of the tube.

34. The technique of post-failure pressure cycling is still the only way of assessing the susceptability to fast axial crack extension. Work is in hand to try to develop a cheaper and more fundamental way of measuring these properties.

CONCLUSIONS

35. Fatigue results for a wide range of commercially available high strength low alloy steel tubing have been presented for as-received and for autofrettaged conditions.

36. The concept of an optimum autofrettage treatment for stability of residual stresses in actual service conditions has been introduced and the probability of loss of residual stress and reduction in fatigue strength of tubing autofrettaged beyond this optimum degree has been emphasised.

37. The dangers involved in scaling up tube dimensions without taking account of the limitations of each tube manufacturing process have been discussed and more will be said in future reports.

38. The damage that can be done to the fatigue properties of tubing by introducing bore defects either during tube manufacture or, much more commonly, during subsequent fabrication processes and inadequate storage have been pointed out.

39. The need to design high pressure tubing which has a high fatigue strength and a safe mode of fatigue failure has been repeatedly emphasised. A growing tendency in some quarters to go back twenty years and base designs upon some static burst pressure divided by some arbitrary factor of ignorance is to be deplored.

ACKNOWLEDGEMENTS

The whole of this work has been carried out in the Mechanical Engineering Department of Imperial College, and the author is grateful to the Head of Department, Professor Sir Hugh Ford, F.R.S., for his encouragement.

All of this work has been sponsored by research grants and contracts from industrial concerns in Britain, United States, France and Germany. In particular the generous financial support of Dart Industries Inc., Chemical Group, and the interest of their President, Mr. Ralph Knight, is gratefully acknowledged.

The technical excellence of the staff and students associated with this work is warmly appreciated, especially that of Mr. P. S. J. Crofton and Mr. G. E. Eles.

Finally, throughout these years the author has had the advice and encouragement of his friend and colleague, Dr. K. E. Bett of the Department of Chemical Engineering and Technology, for which he is ever grateful.

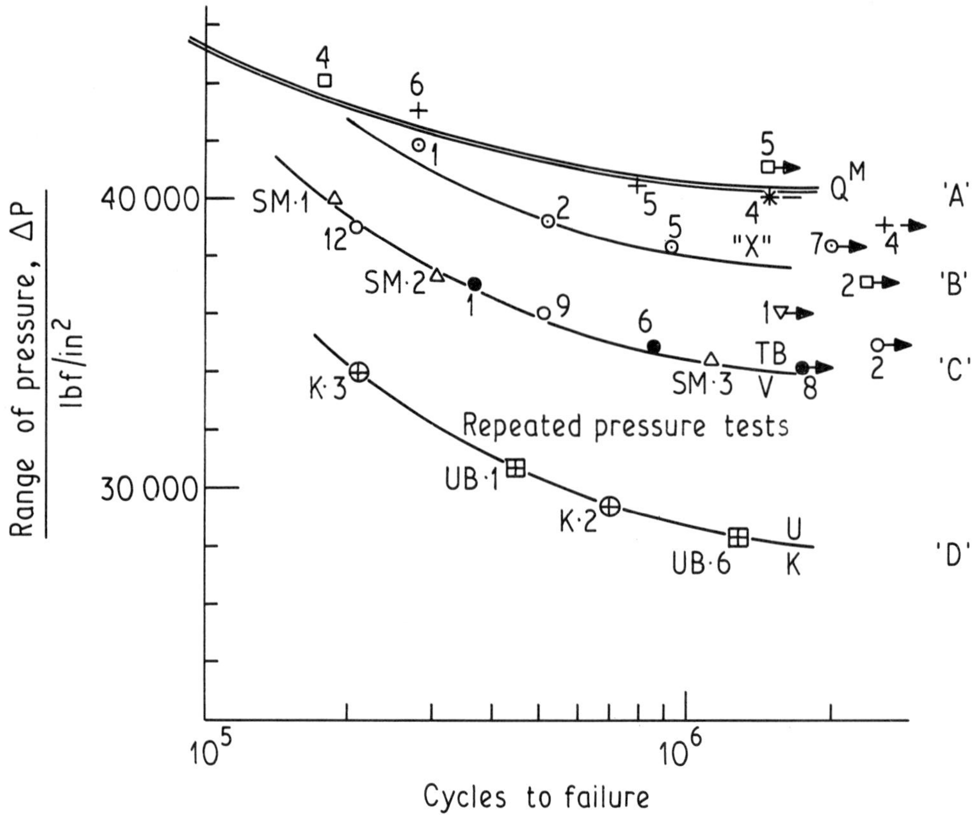

Fig. 1: Fatigue results for as-received tubing tested in repeated pressure

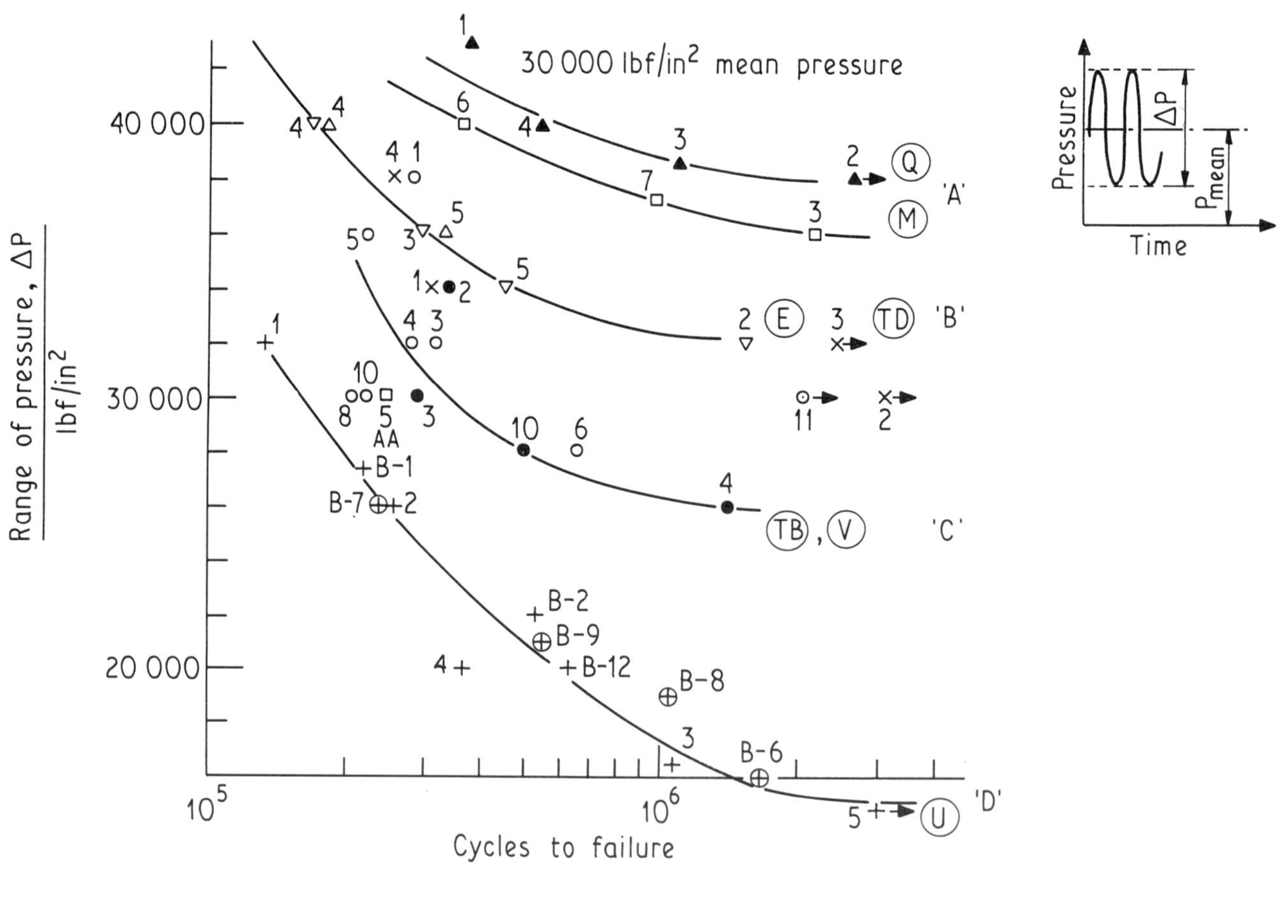

Fig. 2: Fatigue results for as-received tubing tested about a mean pressure of 30,000 Lbf/in²

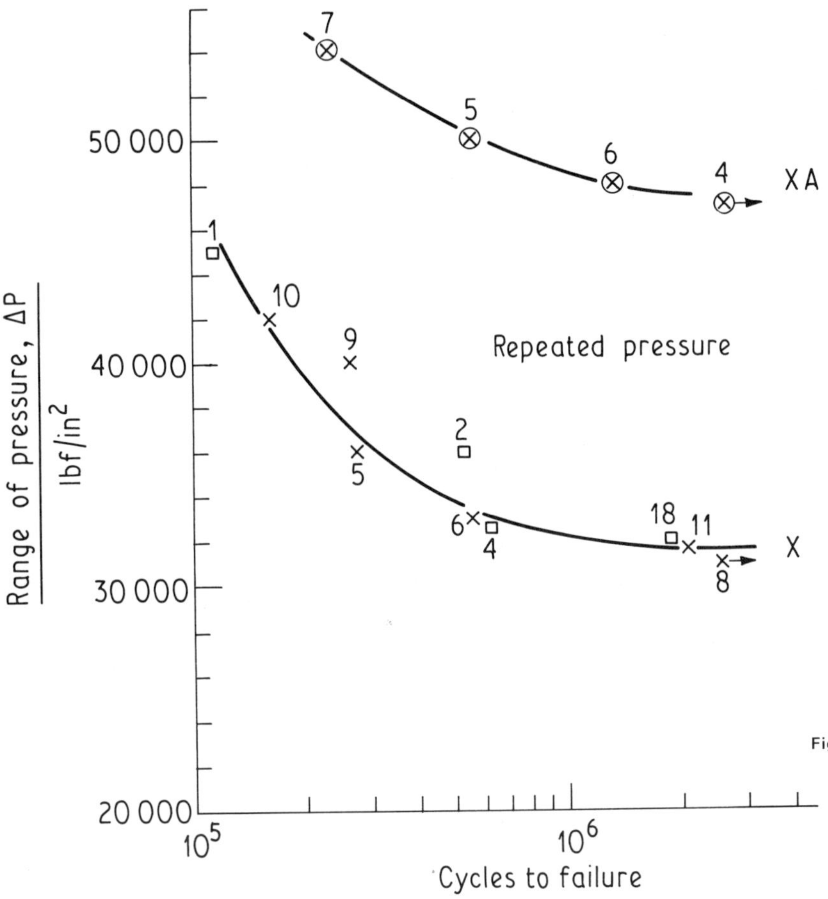

Fig. 3: Effect of autofrettage upon fatigue strength in repeated pressure tests
Tube X, medium strength, poor bore finish
K = 2.6

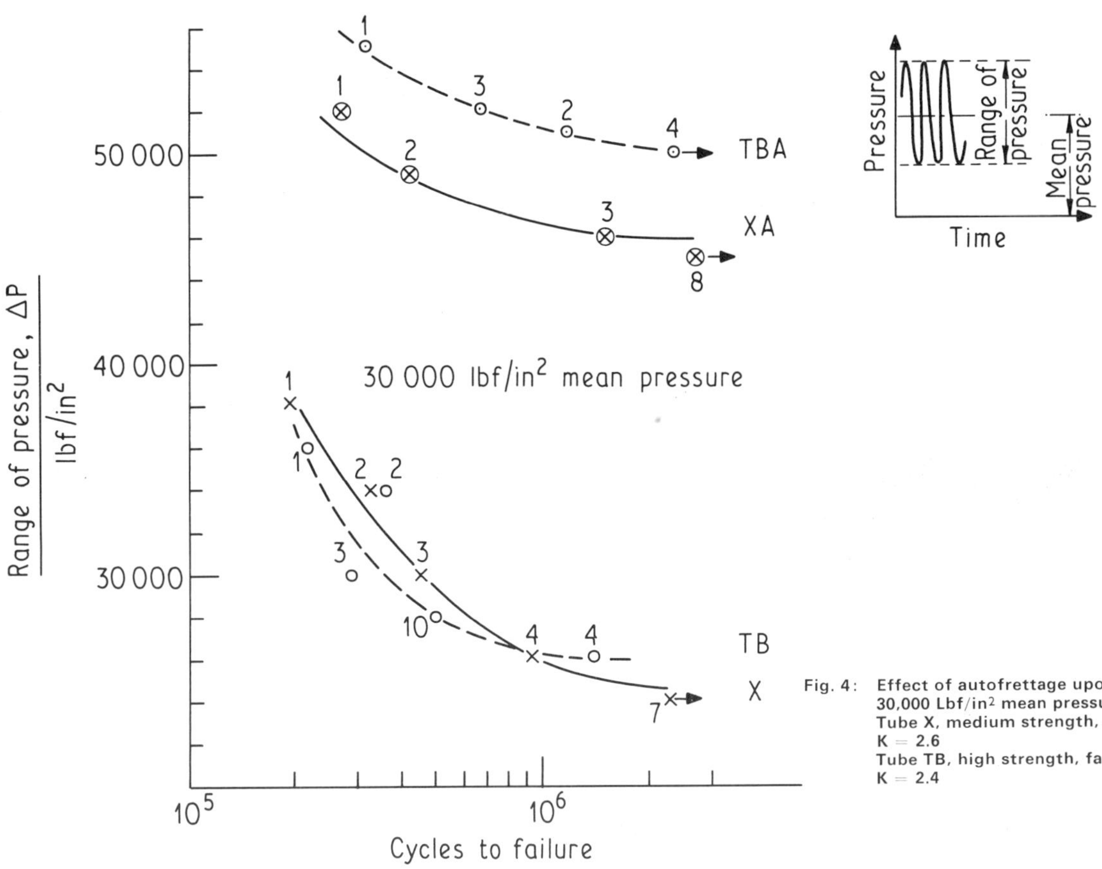

Fig. 4: Effect of autofrettage upon fatigue strength at 30,000 Lbf/in² mean pressure
Tube X, medium strength, poor bore finish.
K = 2.6
Tube TB, high strength, fairly good bore finish.
K = 2.4

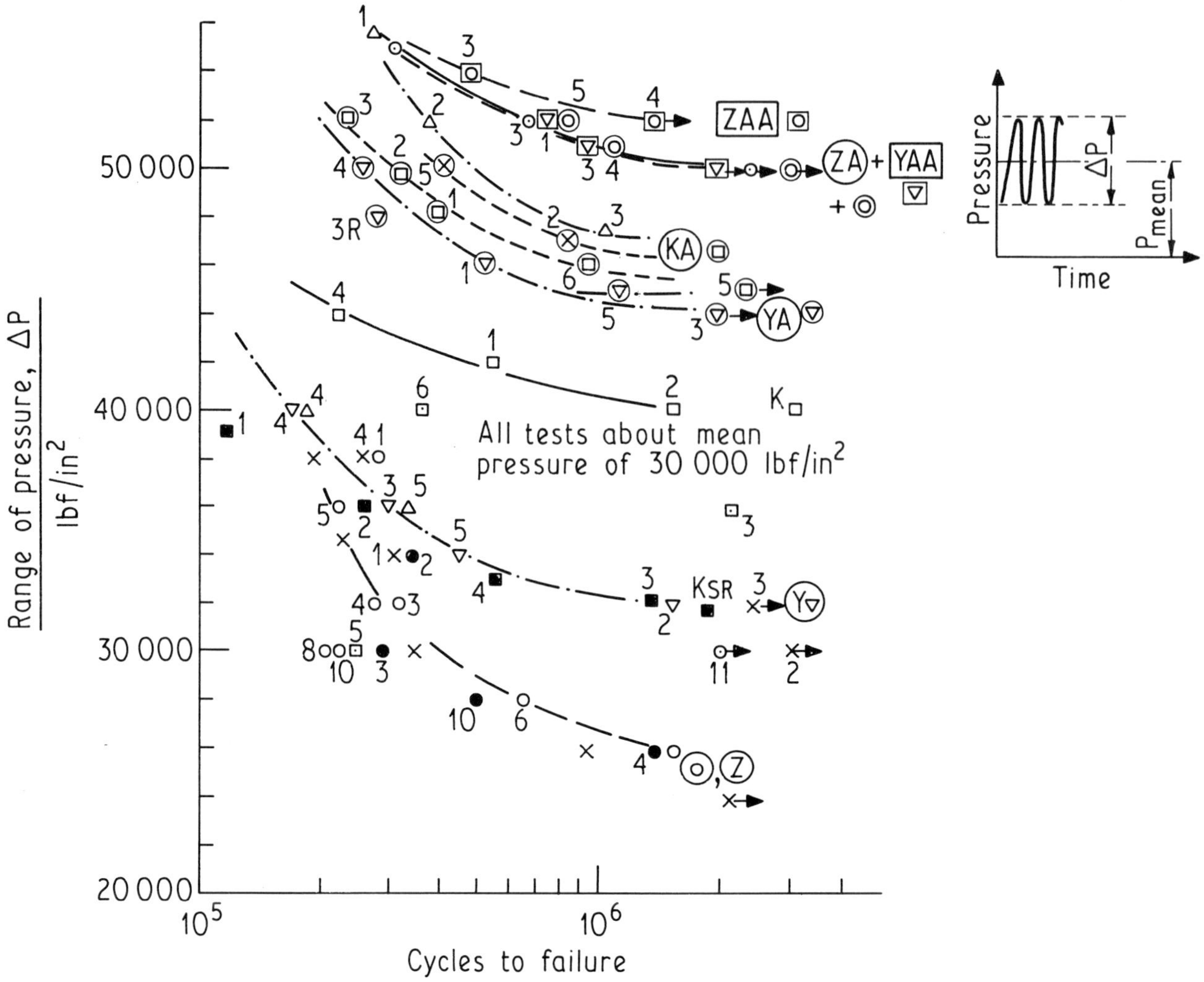

Fig. 5: As-received and autofrettaged fatigue strength at 30,000 Lbf/in² mean pressure for a selection of good quality tubing

A is autofrettaged, AA is "optimum" autofrettaged, Tube K is bore blasted tubing and the results K_{SR} are for as-received tubing subjected to a low temperature 500°C. stress relief in vacuum.

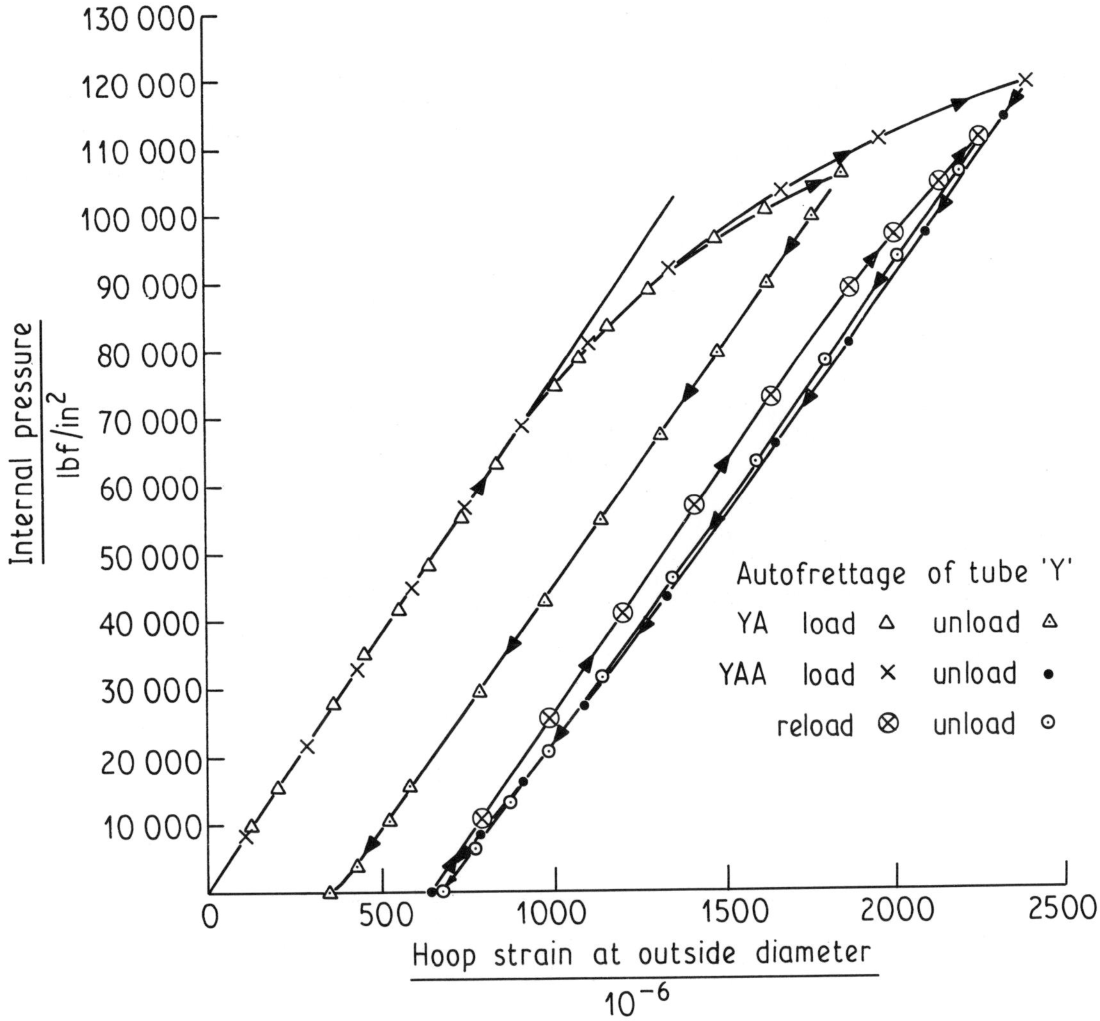

Fig. 6: Pressure-expansion curve during autofrettage of pieces of tube Y to different degrees of autofrettage

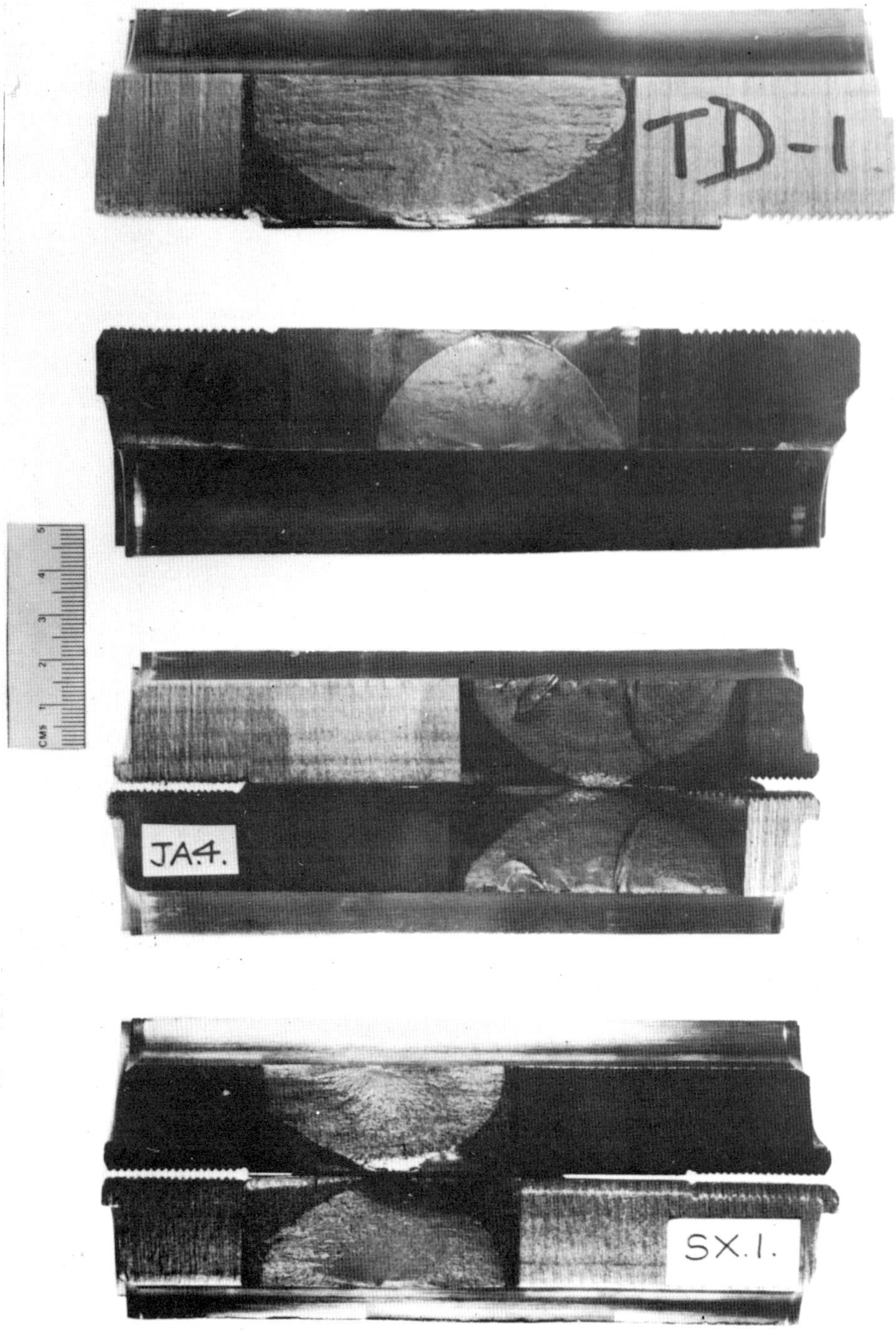

Fig. 7: Typical stable fatigue cracks in thick-walled tubing

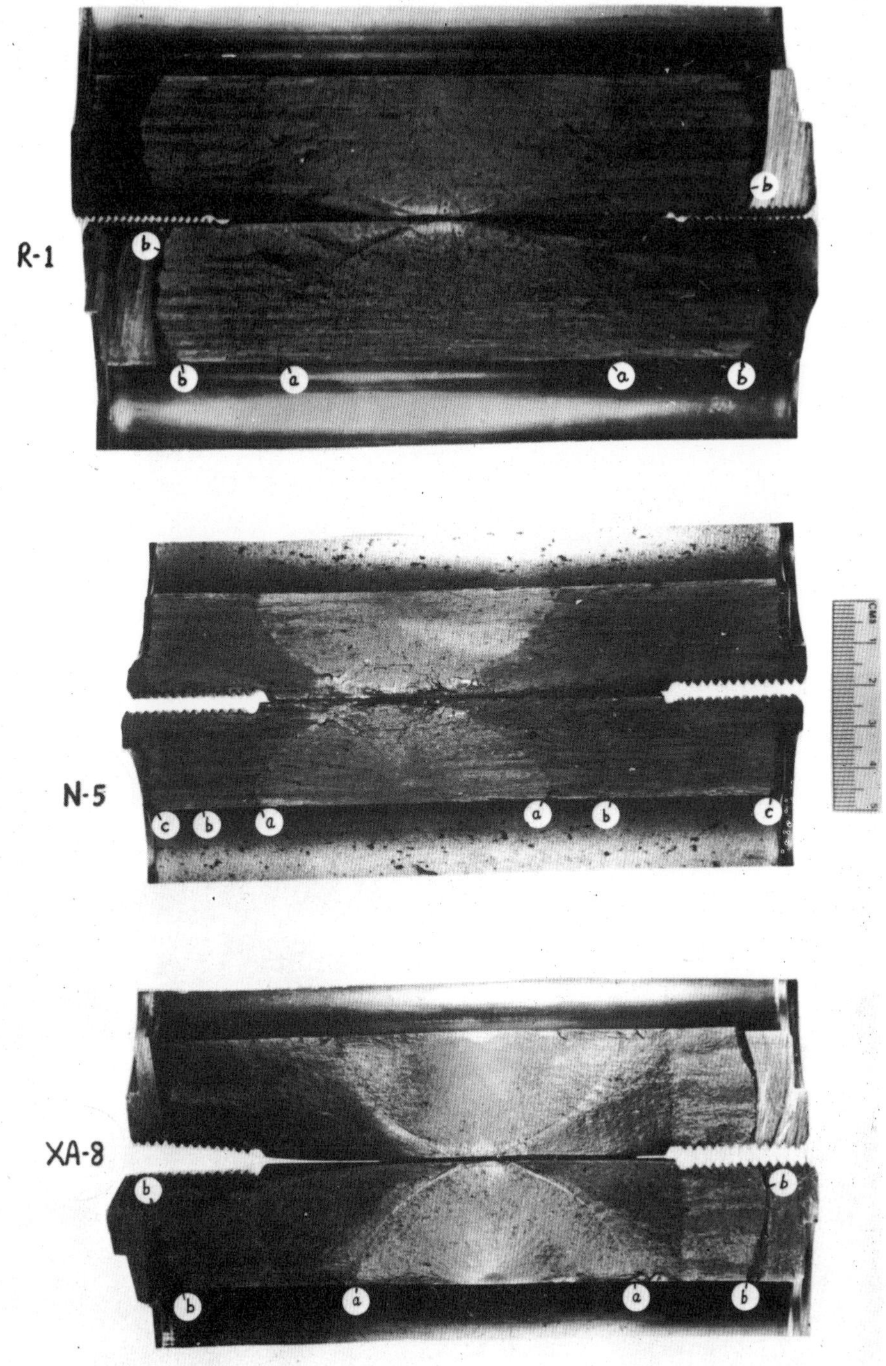

Fig. 8: Fast axial crack extension during post-failure pressure cycling. All test specimens from each length of tube, made by three different manufacturers, showed this unstable mode of failure

FATIGUE DESIGN OF CONTAINER FOR VERY HIGH PRESSURE

M. NISHIHARA, Y. YAMAGUCHI and S. HATTORI
Kobe Steel Ltd., Kobe, Japan

SYNOPSIS The criterion governing the fatigue strength of finite life thick walled cylinders subjected to very high internal pressures, was studied. It was found that the fatigue strength of simple monobloc cylinders was well correlated with the repeated tension fatigue strength of transverse specimens in terms of the maximum difference of principal stresses at bore. When the mean hoop stress was decreased by prestressing, however, the fatigue strength of a cylinder was markedly increased and became coincident with that of uniaxial specimens in terms of the maximum principal stress. A fatigue design concept of very high pressure containers for production use was also proposed.

1. INTRODUCTION

Repeated very high pressures have been used recently in certain metal working processes, for example, hydrostatic extrusion. The fatigue design of a very high pressure container is one of the most important aspects in the application of these processes to industry. Much information on the fatigue behaviour of thick walled cylinders has already been obtained through the extensive works of Morrison, Crossland, Parry, Davidson, Haslam (1)-(5) and others. But, for designing a high pressure container made from very high strength steel, it is still difficult to evaluate the effects of diameter ratio and mean hoop stress especialy in the range of comprehensive stress

The main objects of the present work are to study the effects of cylinder material, diameter ratio and mean stress on the fatigue strength of elastic open ended thick walled cylinder subjected to very high internal pressures and to obtain a simplified criterion for the fatigue design of containers for high pressure.

2. TEST PROCEDURE

Thick walled cylinders made from a high strength steel SNCM8, a very high strength maraging steel KMS18-17 and a hard tool steel SKD11 were tested under repeated internal pressures up to 1.47GPa. The chemical composition and mechanical properties of the cylinder materials are listed in Tables 1 and 2, respectively. The bore of the test cylinders was 20mm and the diameter ratio varied from 1.4 to 3.2. In order to clarify the effects of mean hoop stress, some tests were carried out by using cylinders which were statically fluid-supported in order to give a certain value of mean hoop stress. A diagram of the fatigue test rig is shown in Fig. 1. Stress conditions were limited to the elastic range in the fatigue tests. Castor oil was used as a pressure transmitting medium. Uniaxial test specimen samples taken from the cylinder materials were also tested at various mean stresses.

3. UNIAXIAL FATIGUE STRENGTH OF LONGITUDINAL SPECIMEN

Fig. 2 shows the effects of mean stress on the uniaxial fatigue strength of specimens cut longitudunally from the cylinder materials. The fatigue strength increased with comprehensive mean stress, the effect being larger in the case of the harder cylinder materials. The fatigue data of transverse specimens of KMS 18-17 would be identical to those in Fig. 2, because the material does not show anisotropy in fatigue strength as shown in Table 2.

4. FATIGUE STRENGTH OF CYLINDER

The results of fatigue tests on simple monobloc and statically fluid-supported cylinders made from SNCM8, KMS18-17 and SKD11 are shown in Fig. 3. The fatigue strength of cylinders in terms of hoop stress at bore increases markedly with decrease in maximum hoop stress. It is also seen that the fatigue strength of simple monobloc cylinders increases slightly with decrease in diameter ratio.

4.1. EFFECTS OF DIAMETER RATIO

Figure 4 shows the dependence of the fatigue strength of simple monobloc cylinders made from SNCM8 and MKS18-17 on the diameter ratio.

It is generally assumed that the fatigue crack nucleation under multi-axial stress conditions is mainly governed by the range of maximum shear stress $\Delta\tau_{max}$, and the crack propagation rate da/dN is described by the well known equation $da/dN = A\cdot(\Delta K_I)^m$, where ΔK_I is the range of stress intensity factor, and A and m are materials constants. The stress intensity factor K_I for a semi-elliptical crack at the inner bore of a thick walled cylinder subjected to internal pressure is approximately given by Underwood (6) as

$$K_I = \frac{1.1}{\Phi}(\sigma_\theta + p_i)(\pi a)^{1/2} = \frac{1.1}{\Phi}(2\cdot\tau_{max})(\pi a)^{1/2} \qquad (1)$$

where Φ is the elliptical integral function, σ_θ is the hoop stress at the bore, p_i is the internal pressure and a is the crack depth.

If Φ remains constant during crack propagation, the range of the maximum difference of principal stresses or the range of maximum shear stress at the crack propagation life of N is given as

$$\sigma_{\theta \cdot max} + p_{i \cdot max} = 2\Delta\tau_{max} = \frac{\Phi}{1.1\pi^{1/2}}\left\{\frac{1}{AN}\int_{a_i}^{a_f} a^{-\frac{m}{2}}\,da\right\}^{\frac{1}{m}} = \text{const.} \quad (2)$$

where $\sigma_{\theta \cdot max}$ is the maximum hoop stress at bore corresponding to the maximum internal pressure $p_{i \cdot max}$, a_i is the initial crack depth and a_f is the final crack depth.

The fatigue strength of a cylinder with a diameter ratio of 1.0 may be regarded as the repeated tension fatigue strength of a transverse specimen of cylinder material, provided that the effects of curvature and the chemical effects of pressure medium are neglected. Consequently, the fatigue strength of a simple monobloc cylinder subjected to internal pressures may be estimated by the following equation, which is in a simple form compared with that derived by Haslam (4)

$$\sigma_{\theta \cdot max} + p_{i \cdot max} = \frac{2k^2}{k^2-1}\,p_{i \cdot max} = \frac{2k^2}{k^2+1}\,\sigma_{\theta \cdot max} = \sigma_{T.O} \quad (3)$$

where $\sigma_{T.O}$ is the repeated tension fatigue strength of a transverse specimen cut from the cylinder material and k is diameter ratio of the cylinder.

The curves in Fig. 4 are calculated by equation (3) where $\sigma_{T.O}$ of KMS18-17 is determined from Fig. 2 and that of SNCM8 is estimated to be 82% of the repeated tension fatigue strength of the longitudinal specimen owing to its anisotropy in fatigue strength. The curves are well fitted by the experimental data.
Equation (3) is apparently an expression of the "maximum shear stress theory". However the uniaxial fatigue data of longitudinal specimens and torsional fatigue data do not correlate with those of cylinders subjected to repeated internal pressures.
To investigate the physical meaning of equation (3), a SNCM8 cylinder in which a thin copper alloy liner was fitted, was tested. Even when a liner failed, the fatigue test was continued by replacing it. Thus, the bore of the cylinder was perfectly protected against pressure medium.
The experimental results are plotted in Fig.4 and a remarkable increase of fatigue strength in terms of maximum hoop stress is observed. In spite of the existence of a compressive radial stress at the bore, the fatigue strength of the cylinder seems to be coincident with $\sigma_{T.O}$.
Judging from the experimental results mentioned above, the pressure medium has detrimental effects on the fatigue strength of the cylinder which is attributed to penetration of cracks at the bore. Equation (3) might be called an "extended principal stress theory", in which the mechanical effects of pressure medium are taken into account.

4.2. EFFECTS OF MEAN HOOP STRESS

Figure 5 obtained from Fig. 3 shows the relationship between mean hoop stress $\sigma_{\theta \cdot m}$ and hoop stress range $\Delta\sigma_\theta$ for cylinders made from KMS18-17. When the mean hoop stress decreases, the fatigue strength of a cylinder markedly increases and asymptotically approaches to that of an uniaxial test specimen in terms of the maximum principal stress. This suggests that the effects of the penetration of the pressure medium can be practically eliminated with a high compressive mean hoop stress. Using this result, a fracture mechanical study was reported by the present authors (7). Similar results to Fig. 5 were obtained from the fatigue tests on cylinders made from SNCM8 and SKD11.

5. FRACTURE OF CYLINDER BY FATIGUE

Figure 6 shows the appearances of cylinders failed by fatigue. The fatigue crack of a simple monobloc cylinder propagates in radial direction until it reaches the critical crack depth derived from the fracture toughness K_{IC} of the cylinder materials. When the critical crack depth is larger than the wall thickness, the cylinder will fail by moderate cracking. When it is smaller than wall thickness, however, the cylinder will burst sometimes catastrophically. Even a cylinder with a thin wall (inner diameter: 20mm, diameter ratio: 1.4) made from KMS18-17 maraging steel failed by bursting. When the maximum hoop stress was sufficiently decreased with static supporting pressure, the fatigue crack of a test cylinder with the diameter ratio of 3.2 and the same bore (20mm) subjected to the repeated pressures of 1.47GPa propagated through the whole wall thickness without bursting (Fig. 6-a). The life of a cylinder made from SKD11 tool steel always ended in bursting at any high compressive mean hoop stress because of its very low fracture toughness (Fig. 6-6). Further, when the cylinder was strengthened in hoop direction by static supporting pressure, circumferential cracking of cylinders made from SKD11 tool steel as sometimes observed at their sealing positions (Fig. 6-c). A stress analysis shows that this cracking will be caused by axialtensile stresses at the sealing position of the inner bore (8). This cracking was effectively suppressed by applying a relatively low compressive axial stress.

6. FATIGUE DESIGN OF VERY HIGH PRESSURE CONTAINER

6.1. A DESIGN CRITERION

From the above information, a simple criterion for fatigue design of a very high pressure container with a finite life was derived. Figure 7 is the schematic diagram of the fatigue design criterion. The line AB shows the fatigue strength of an uniaxial specimen cut transversely from the cylinder material. The line OA shows the stress condition for an uniaxial specimen under repeated tension, or for a monobloc cylinder under repeated internal pressure. Under the stress condition on the line OA, the fatigue strength of a cylinder can be evaluated using equation (3). The fatigue strength of a cylinder under the stress condition of $\sigma_{\theta \cdot max}=0$, is assumed to be the same as the repeated compression fatigue strength of an uniaxial specimen at the point B. In the range of $\sigma_{\theta \cdot m} < \frac{1}{2}\Delta\sigma_\theta$, the straight line AB will give a conservative fatigue strength of a cylinder with a diameter ratio k_n under various mean hoop stress. This relationship is formulated as follows.

$$\left.\begin{aligned} &\Delta\sigma_\theta = \frac{2(\alpha-\beta)}{\alpha+\beta}\sigma_{\theta \cdot m} + \frac{2\alpha\beta}{\alpha+\beta}\sigma_{T.O}, \\ &|\sigma_{\theta \cdot m}| \le \tfrac{1}{2}\Delta\sigma_\theta, \\ &\text{where} \quad \alpha = \frac{k^2+1}{2k^2}, \quad \beta = \frac{1+M}{1-M} \\ &\text{-M: gradient of the line AB in Fig. 7} \end{aligned}\right\} \quad (4)$$

The stress condition for a cylinder to be elastic during pressure cycling is a prerequisite for this design criterion.

6.2. PRESTRESSING

As already discussed, compressive mean hoop stresses induced by prestressing would be essential for designing a very high pressure container with a long fatigue life. Practical ways of the prestressing are autofrettage, shrink- or press-fitting of two or more cylinders with interference, wire or strip winding, fluid support, and combinations of these methods. A high residual compressive hoop stress is easily induced by autofrettage, which is a common method of prestressing a container. It is noticed, however, that the residual stress initially induced by autofrettage can not necessarily be maintained during service. Figure 8 shows the changes in the residual hoop stress at the bore of autofrettaged cylinders made from SNCM8 and KMS18-17 during pressure cycling. The relaxation of the residual compressive hoop stress depends on the operating pressures, the number of pressure cycles and the cylinder materials. The residual stress is rapidly relaxed at the early stage of pressure cycling and flattens out to a lower level after pressure cycles of about 10 per cent of the total life. The relaxation rate increases at higher operating pressures. This tendency is more noticeable on cylinders made from KMS18-17 maraging steel compared with SNCM8 steel. In spite of the high residual stress initially induced by autofrettage, only low settled residual stress is effective in improving the fatigue life of high pressure container. On the other hand, the residual stress initially induced by fluid support is easily maintained.

6.3. A VERY HIGH PRESSURE CONTAINER WITH LOOSE LINER

A container for a 500ton (4.9MN) hydrostatic extrusion press (maximum working pressure: 1.47GPa), which was designed and manufactured at KOBE STEEL, LTD. in 1973, has an inner cylinder (inner diameter: 48mm, diameter ratio: 3.2, length: 1.800mm) supported by a static fluid pressure of 490MPa. But, for such a very high pressure, the inner cylinder was designed to have a finite fatigue life on the order of 10 cycles at most. To reduce the maintenance cost of a very high pressure container with a finite fatigue life, a new design of container was developed by the present authors. This container is composed of an outer mantle and a fluid supported inner cylinder with a thin loose liner (Fig.9). The thin liner made from very high strength steel is fitted in the inner cylinder with a limited clearance prior to the application of the fluid support. Subsequently, the inner cylinder is fluid-supported to the extent that the clearance is eliminated and the thin liner is effectively prestressed. Even when the liner fails at the end of fatigue life, the safety of the container is well assured. By removing the supporting pressure, the thin liner can be easily replaced. The outer mantle and the inner cylinder could be easily designed to have infinite fatigue life. This new design of container leads to the reduction of maintenance costs of the container.

7. CONCLUSION

A simple criterion was proposed for fatigue design of thick walled cylinders with finite life which is subjected to repeated internal pressures. This may be applied to the practical design of container which remains elastic under service conditions.

8. ACKNOWLEDGEMENT

The authors wish to express their appreciation to Dr. S. Denoh and Mr. H. Yanai for their efforts in experimentation.

REFERENCES

1. Morrison, J.L.M., Crossland, B. and Parry, J.S.C., "Strength of Thick Cylinders Subjected to Repeated Internal Pressure." Proc. Inst. of Mech. Eng., 1960, Vol.174, p.95.

2. Parry, J.S.C., "Fatigue of Thick Cylinders: Further Practical Information." Proc. Inst. of Mech. Eng., 1966, Vol. 180, p.387.

3. Davidson, T.E., Eisenstadt, R. and Reiner, A.N., "Fatigue Characteristics of Open-End Thick-Walled Cylinders Under Cyclic Internal Pressure." Trans. ASME, 1963, 85 (Ser.D), p.555.

4. Haslam, G.H., "The Fatigue Limit of Cylinders Subjected to Repeated Internal Pressures." High Temp. High Pres., 1969, Vol. 1, No.6, p.705.

5. Haslam, G.H., "Estimation of The Fatigue Life of A Thick-Walled Cylinder Subjected to Repeated Internal Pressure." Jnl. Mech. Eng. Sci., 1971, Vol. 13, No. 3, p.130.

6. Underwood, J.H., "Stress Intensity Factors for Internally Pressurized Thick Wall Cylinders." ASTM STP513, 1972, p.59.

7. Yamaguchi, Y., Hattori, S. and Yanai, H., "Effects of Diameter Ratio and Mean Stress on the Fatigue Strength of Thick Walled Cylinders under Very High Internal Pressures." Trans. Japan Society of Mechanical Engineers, to be published.

8. Yamaguchi, Y., Hattori, S. and Yanai, H., "Fatigue Strength of Thick Wall Cylinders Subjected to Very High Internal Pressures." Jnl. Materials Science Japan, Vol.23, No.252, 1974, p.789.

Table 1. Chemical composition and heat treatment of steels tested

Steels	Chemical composition (%)												Heat treatment
	C	Si	Mn	P	S	Ni	Cr	Mo	V	Co	Ti	Al	
SNCM8 ★)	.40	.20	.75	.015	.015	1.80	.80	.20	–	–	–	.030	Oil-hardened from 850°C Tempered at 625°C
KMS18-17 +)	.01	.06	.04	.007	.005	17.80	–	4.86	–	7.50	.78	.075	Solution annealed at 815°C Age-hardened at 500°C
SKD11 ★)	1.55	.36	.38	.023	.008	.11	11.54	.83	.26	–	–	–	Air-cooled from 1020°C Tempered at 575°C

★) Japanese Industrial Standard, air melt.

+) Kobe Steel Maraging Steel, remelt in vacuum.

Table 2. Mechanical properties of steels tested

Steels	Static properties				Uniaxial fatigue strength ($\sigma_m = 0$), MPa			
	VHN	Ultimate tensile strength, GPa	0.2% Proof stress, GPa	Reduction of area, %	3×10^4 cycles		1×10^6 cycles	
					Longitudinal	Transverse	Longitudinal	Transverse
SNCM8	360	1.186	1.108	59.4	569	510	471	422
KMS18-17	550	1.726	1.687	59.8	804	834	520	530
SKD11	750	–	2.403	–	804	608	657	490

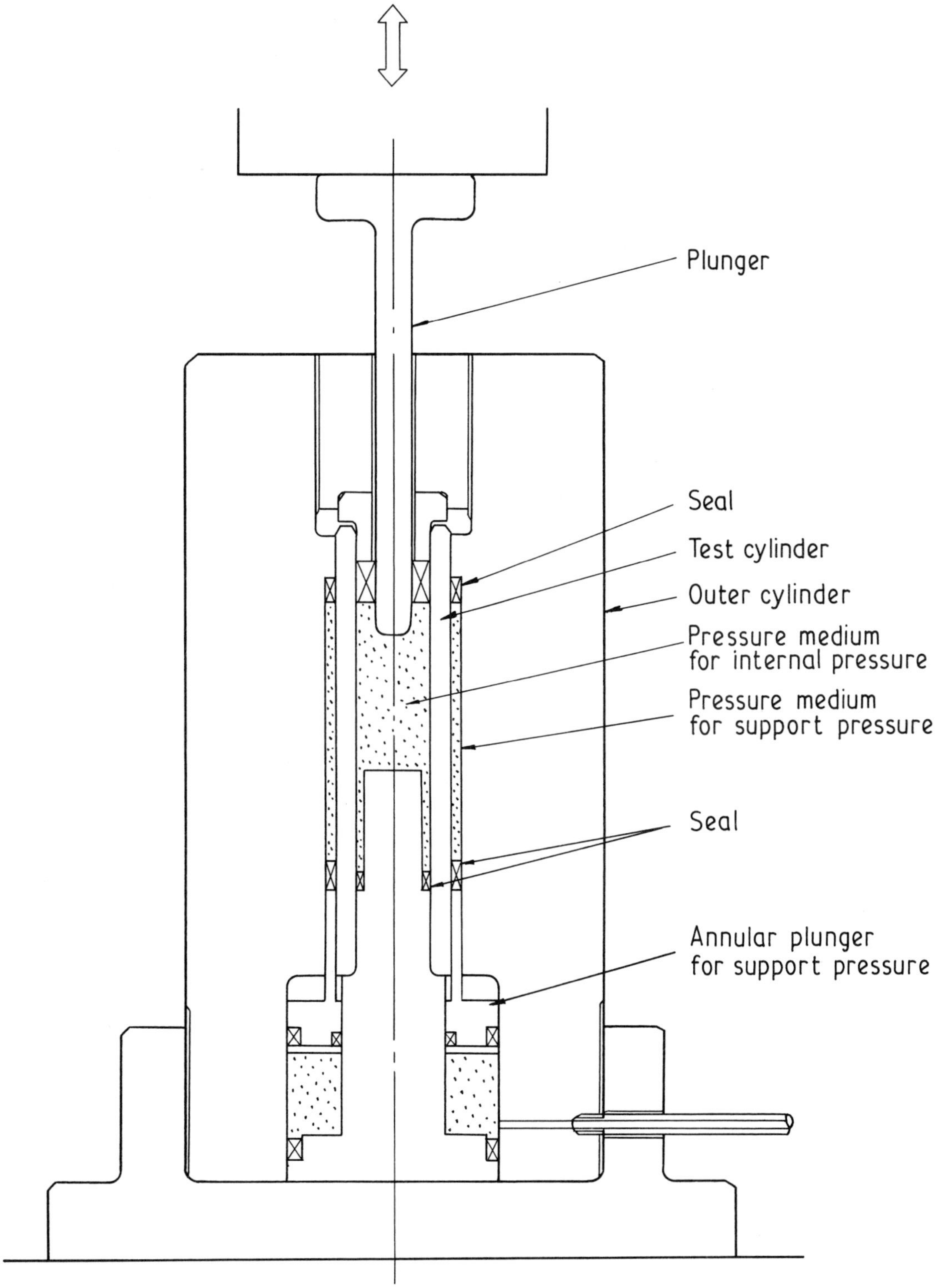

Fig. 1: Fatigue test rig for fluid-supported cylinders

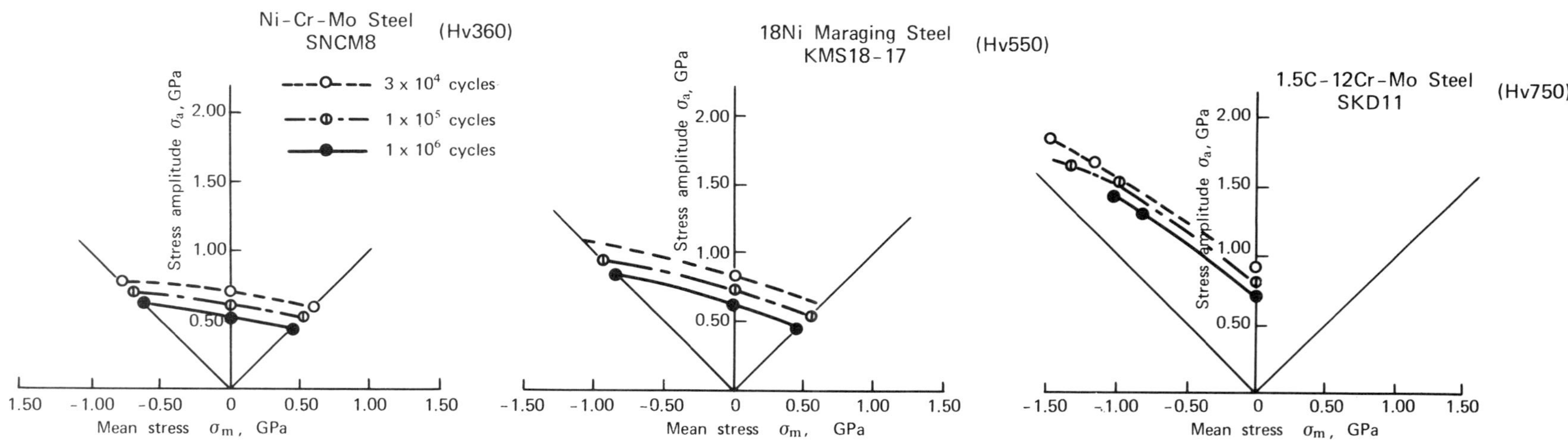

Fig. 2: Uniaxial fatigue strength of high strength steels

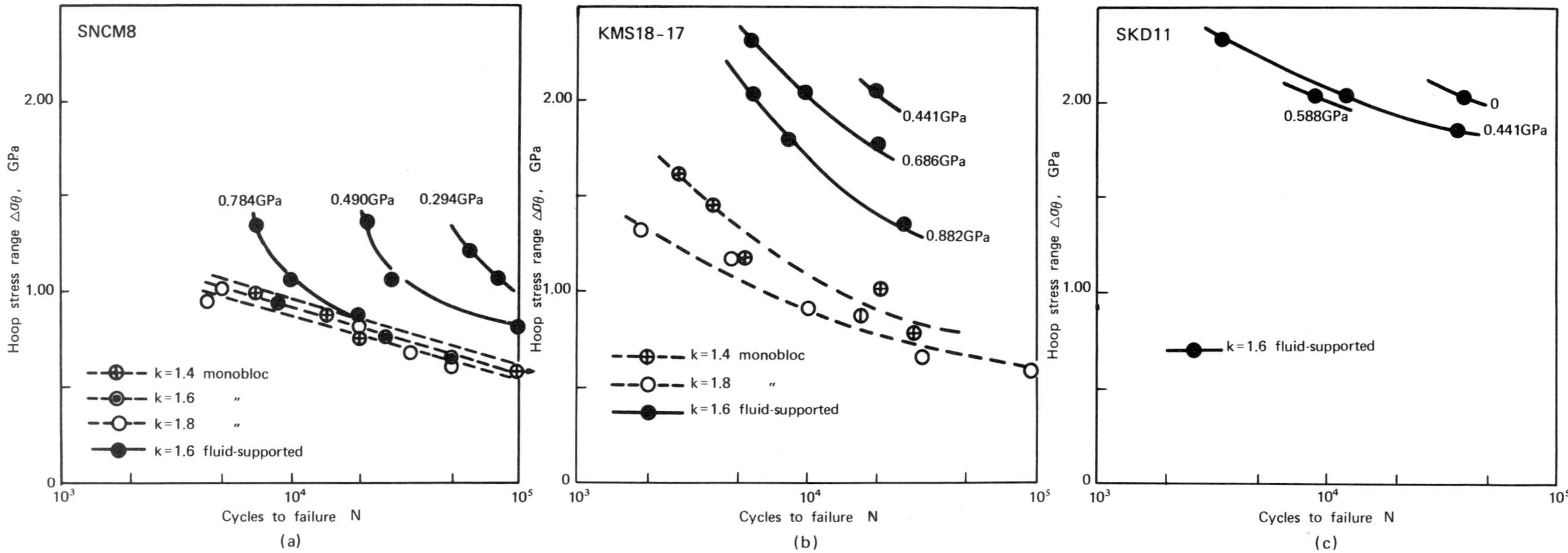

Suffixes are values of maximum hoop stress at bore of fluid-supported cylinders.
k is diameter ratio.

Fig. 3: Fatigue strength of cylinder subjected to internal pressures

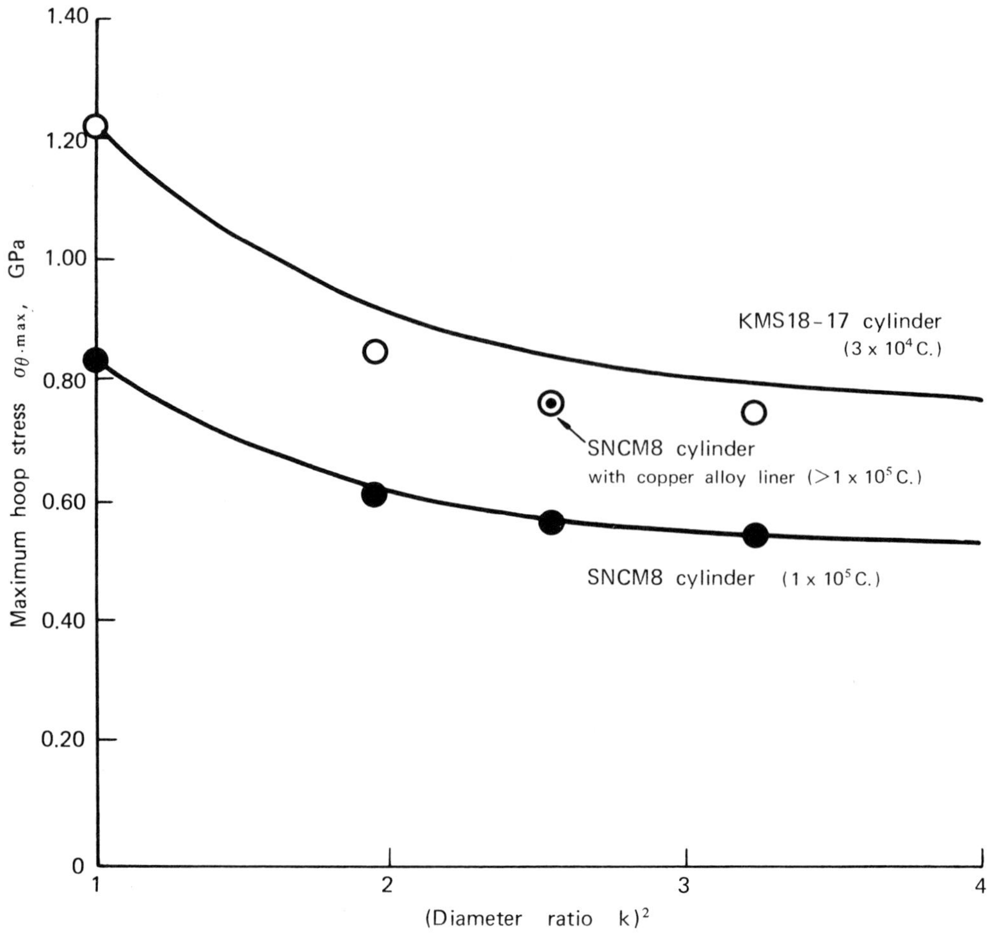

Fig. 4: **Effects of diameter ratio on fatigue strength of cylinders**

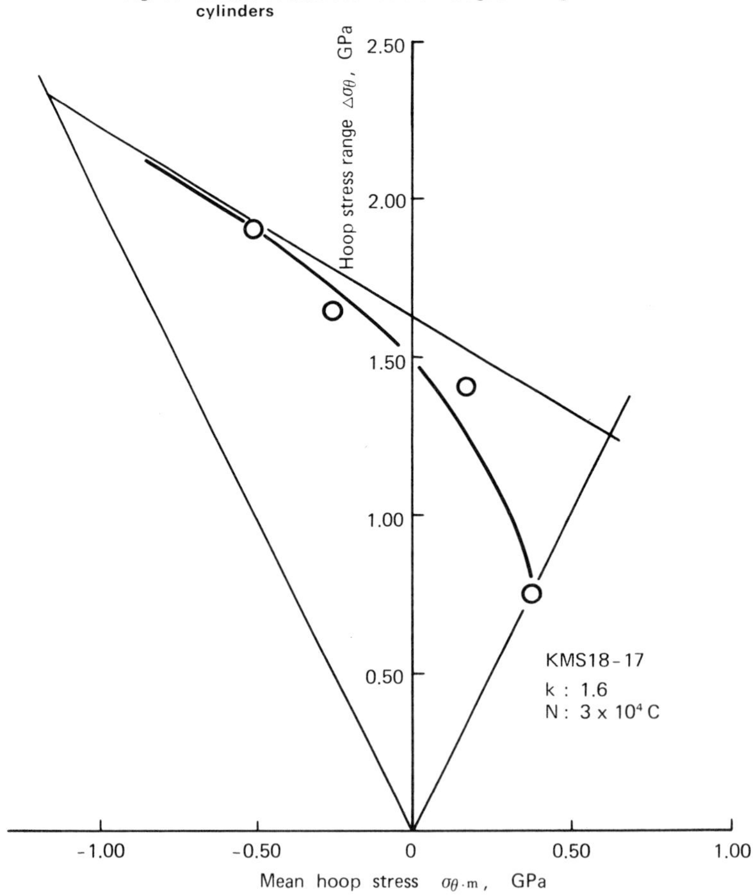

Fig. 5: **Effects of mean hoop stress on fatigue strength of cylinders**

(a)

Maraging Steel (KMS18-17), k=3.2
Cycles to failure : 12000 cycles
Repeated pressure : 1.470GPa
Supporting pressure : 0.490GPa

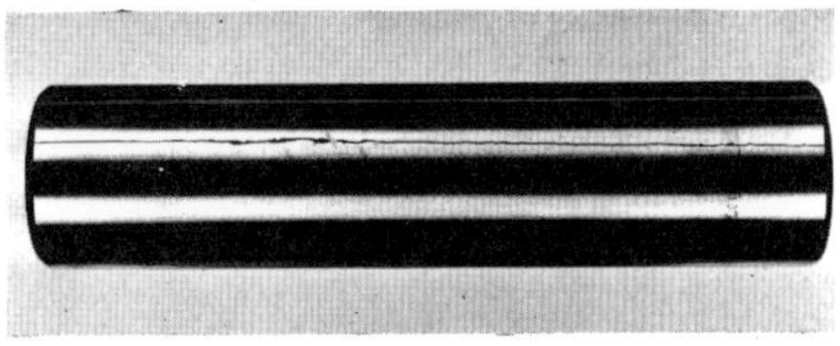

(b)

1.5C-12Cr-1Mo Steel (SKD11), k=1.6
Cycles to failure : 3300 cycles
Repeated pressure : 1.014GPa
Supporting pressure : 0.570GPa

(c)

1.5C-12Cr-1Mo Steel (SKD11), k=1.6
Cycles to failure : 41000 cycles
Repeated pressure : 0.895GPa
Supporting pressure : 0.617GPa

Fig. 6: Cylinders failed by fatigue

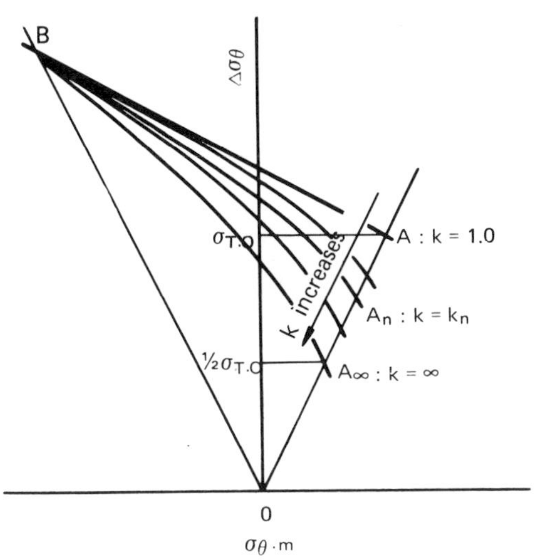

(a)

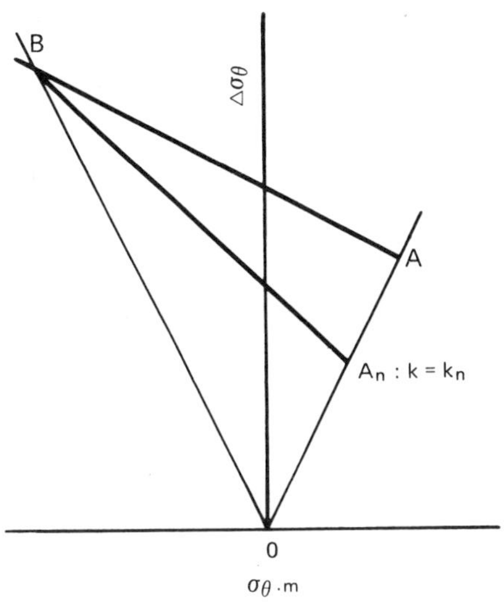

(b)

Fig. 7: Schematic diagram of fatigue design of cylinders

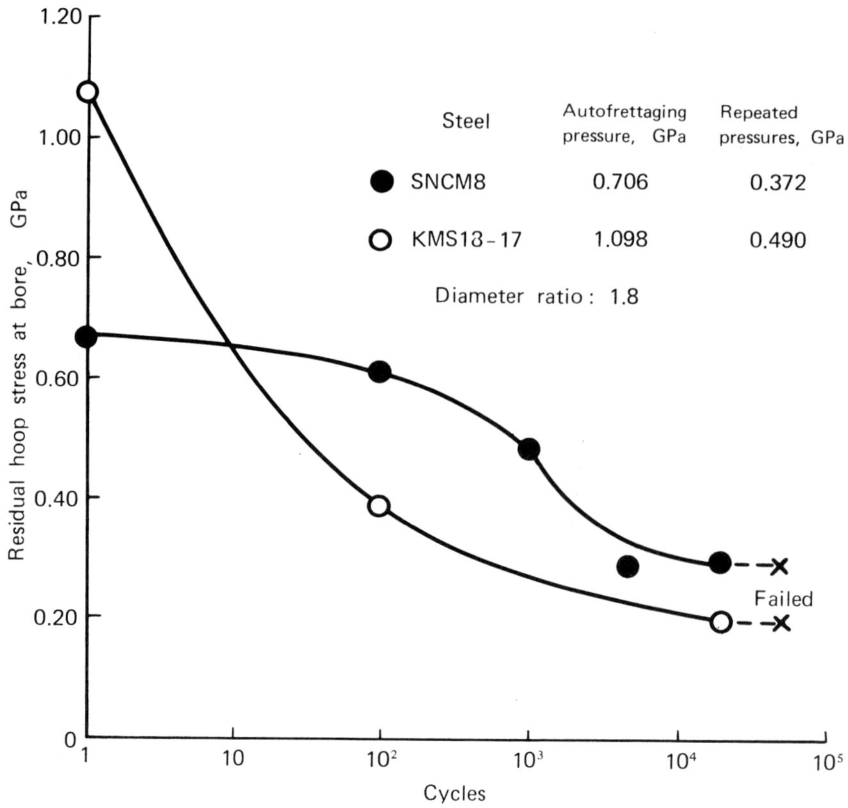

Fig. 8: Reduction of residual stress of autofrettaged cylinders during pressure cycling

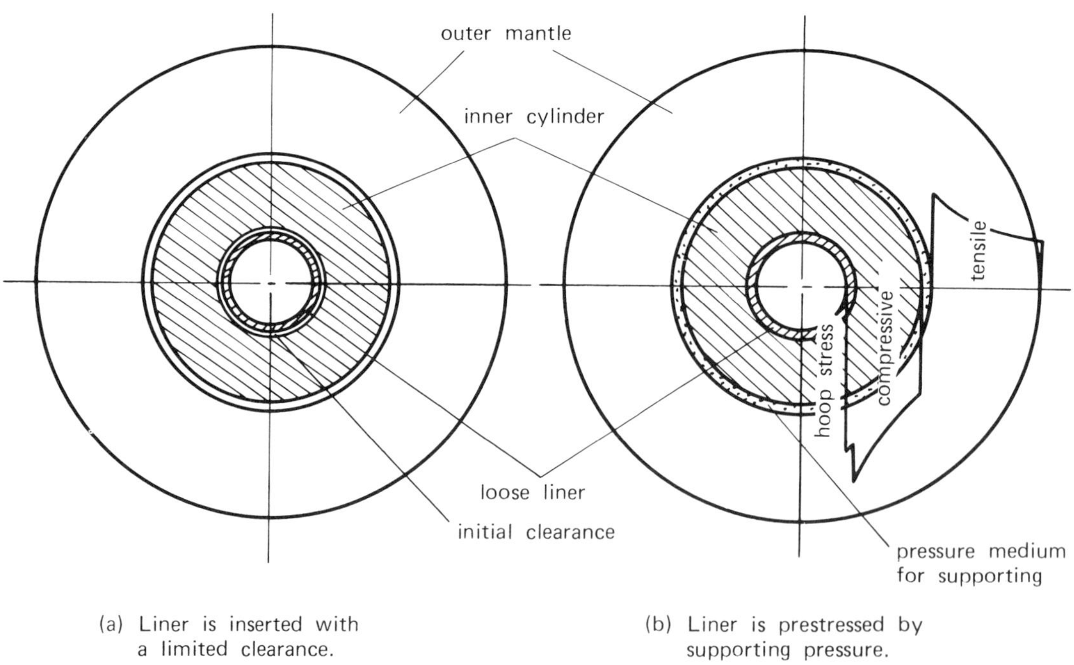

Fig. 9: Fluid-supported cylinder with thin loose liner

FATIGUE BEHAVIOUR OF CONTAINER COMPONENTS SUBMITTED TO INTERNAL PULSATING PRESSURE

P. HESTERMANS
Institut Belge des Hautes Pressions, Sterrebeek-Belgium

SYNOPSIS—The fatigue behaviour of the inner components of high pressure containers has been studied. A test rig composed of a fatigue press and a compound cylinder has been constructed. The compound cylinder is made of two concentric cylinders separated by a pressurized fluid ring. The specimen is the inner cylinder and is submitted to a pulsating internal pressure, while it is supported by a static pressure. The first results obtained did show the important effect of a tensile axial stress: most of the specimens did fail through circumferential cracking. A device was constructed to apply an axial force to the specimen. Some results obtained are described.

1. INTRODUCTION

To develop hydrostatic extrusion into an industrial process, containers able to sustain repeated pressure of 2–3 G Pa (20–30 kbars) are needed. Fatigue data for materials used as components for the inner parts of these containers are scarce. Results on the fatigue behaviour of thick cylinders were first reported by Morrison *et al.*[1] who worked on monobloc cylinders made of medium strength steels at relatively low pressure. Since then, other studies on the subject appeared in the literature: Burns *et al.*,[2] Jones *et al.*,[3] Haslam,[4] Harvey *et al.*,[5] Tomkins,[6] Nishihara *et al.*[7] The principle of a supporting fluid pressure was used systematically by some authors.[5,7]

Several criteria were proposed to describe the fatigue behaviour of thick-walled cylinders. Morrison *et al.*[1] have shown experimentally that the range of bore shear stress was the factor controlling the fatigue life of thick-walled cylinders. Other authors[8] have shown that the choice of a criteria was dependent on the region of the mortality curve concerned. Nishihara[7] correlated his results with the maximum principal stress (the hoop stress at the bore). Recently Turner[9] reviewed the possibility of applying linear elastic fracture mechanics to the case of thick cylinders. The difficulty encountered is to make a proper choice of the governing factor. Jones *et al.*[3] and Tomkins[6] proposed to consider as factor the maximum plastic shear strain at the bore, while Haslam[4] proposed as driving stress a modified hoop stress taking into account the fluid penetration in the cracks. Recently, Tomkins[6b] suggested that the life be related to the maximal shear stress following the relation:

$$N \propto \frac{1}{(\tau)^3}$$

2. EXPERIMENTAL APPARATUS

The test rig was composed of the specimen placed in an outer cylinder which contains the supporting pressure and means to supply this pressure (up to 1·5 G Pa). The cyclic internal pressure was generated by the movement of a plunger inside the specimen. The motion of the plunger was activated by a jack fed by an hydraulic pulsator (Losenhausen Werke, Germany). The general arrangement of the apparatus is given in Fig. 1. Details on the high pressure container are given in Fig. 2.

The outer cylinder was autofrettaged, and was made of tempered high strength Ni Cr Mo steel. The heavy end plates were made in the same steel. The bore was 66·51 mm and could accommodate specimens up to 60 mm. The standard bore of the specimens was 20 mm. The plunger was 12 mm diameter and moved at about 6 to 10 cycles per minute. The fluid used both inside and outside the specimen was diethylhexyl sebacate. There was a replenishing hole with a check valve in the bottom plug of the specimen bore.

Pressure readings were made with a manganin gauge having an electrical connection through the bottom plug. Seals on the supporting pressure ring were made of graphitized nylon U-rings machined in our workshop. Seals on the bottom plug of the specimen bore and along the moving piston were of the same type, their extrusion through the clearances being prevented by conventional beryllium copper mitre rings. Graphitized nylon U-rings were found superior to the conventional 'O ring'.

Several specimens did fail circumferentially and to prevent this type of fracture, a compressive force was subsequently applied. Therefore, a jack was inserted between the lower cover and the cylinder. The jack plunger could develop an axial stress of about 1 G Pa. This arrangement is shown in Fig. 3.

3. MATERIALS AND SPECIMENS

Four materials have been tested. They are:

—A Cr-Mo-Va steel, tempered to 55 R_c.
—A molybdenum tool steel tempered to 63–65 R_c (BS 4659: BM1).
—A 18 Ni(350) Maraging steel tempered to 59–61 R_c.
—Tungsten carbide with various percentages of cobalt binder.

Standard specimens were 120 mm long and 20 mm ID The OD was 40 mm (k = 2) or 60 mm (k = 3). Some k = 2 specimens were tested when placed with a small clearance in a sleeve, about 40 mm ID and 60 mm OD, the pressure then being applied to the OD of the outer cylinder. All specimens were ground at the bore, the outer surface and the flat faces.

4. FATIGUE TESTS RESULTS

4.1. *Tests on* Cr Mo Va *steel*

Five specimens were tested. The results can be found in Table 1. In all cases, the external pressure was such that the Lamé hoop stress at the bore was compressive (−1·8 G Pa). The two cylinders with a diameter ratio of 3 lasted about 1200 cycles, with an internal pulsating pressure of 2 G Pa. The fracture modes of the two cylinders were, however, different. Although specimen no. 9 ruptured longitudinally from the middle section to the end, the other specimen ruptured catastrophically with cracks extending in all directions. This may be due to different heat treatments. The first specimen with a diameter ratio of two was deformed plastically, the internal pressure being obviously too high. Specimen no. 10 fractured longitudinally from end to end. The last specimen was tested at a lower internal pressure (0·9 G Pa), but after 4500 cycles, testing was discontinued.

TABLE 1

Cr Mo Va steel specimens

Spec. no.	k	P_{ext} G Pa	P_{int} G Pa	N Cycles
8	3	0·8	2·0	1180
9	3	0·8	2·0	1188
3	2	0·675	1·8	100
10	2	0·675	1·2	2467
11	2	0·675	0·9	(4464)

4.2. *Tests on tool steels*

Several kinds of tool steels were tested, but it was only possible to run fatigue tests on specimens of a molybdenum tool steel tempered to 63–65 R_c. With the other kinds, we observed failures on the first application of the internal pressure. In some cases, failure was even observed with external pressure only. Generally, fracture was circumferential at about 20 mm from both extremities. To avoid such fracture, it was decided to apply an axial force. As the axial load required to compress a specimen having a diameter ratio of 3 was excessive, a specimen with a diameter ratio of 2, was inserted in a Maraging sleeve, to make a compound cylinder with an overall diameter ratio of 3. Two small discs were placed at each end of the specimen and an axial compressive stress was generated when the cover bolts were tightened. This specimen (no. 23) was submitted to an external pressure of 1·17 G Pa and to an internal pressure of 1·92 G Pa, and failed after 146 pressure cycles with a longitudinal fracture from end to end. However, another specimen (no. 24) placed under similar conditions, ruptured after 7 pressure cycles. In this case, some faint longitudinal cracks, extending less than 15 mm from one end were observed.

To improve the control of the axial stress, the hydraulic jack already described was constructed and used with specimen no. 25, which lasted 275 cycles, but failed circumferentially at about ⅓ of its height.

The specimen no. 24 described above was ground to a length of 90 mm, to remove the end cracks. The specimen was machined to the shape given in Fig. 4. This arrangement has been designed after Semerchan *et al.*[10] Under the pressure conditions given in Table 2, the specimen lasted 222 cycles but ruptured at the conical recess, when the conical alumina insulator around the manganin lead was extruded.

TABLE 2

Molybdenum tool steel specimens

Spec. no.	k*	P_i G Pa	P_e** G Pa	σ_z G Pa	N Cycles
23	2	1·92	1·17	(?)	146
24	2	1·92	1·25	(?)	7
25	2	1·92	1·25	−0·53	275
38	2	2·1	1·25	−0·53	222

* Specimens placed in a sleeve.
** Pressure acting on the sleeve.

4.3. *Tests on* 18 Ni (350) *Maraging steel*

More success was achieved with specimens made in this alloy. The results are given in Table 3. Two specimens of diameter ratio of 3 were tested. With an external supporting pressure of 1·156 G Pa, they failed after 1172 and 952 pressure cycles to 2·07 G Pa. The first specimen ruptured circumferentially at about mid length. Although a small axial stress was imposed on the second specimen, by means of the tightening of the cover bolts, the fracture is circumferential and occurred at about 20 mm from one end. Three specimens with a diameter ratio of 2 were tested at various internal pressures. They were inserted in a sleeve of 40 mm ID and 60 mm OD, and compressed axially prior to the application of the external pressure. Fatigue tests were conducted with internal pressures of 2·1 (no. 26), 2·3 (no. 28) and 2·0 G Pa (no. 31), the external pressure being 1·22 G Pa except for no. 26 where the external pressure was 1·19 G Pa. The type of fracture obtained revealed a symmetric thumbnail crack at the outside surface. At the bore, a number of straight axial cracks could be detected, with some cross-links between cracks. The fatigue lives were 1563 cycles (no. 26), 1219 cycles (no. 28) and 1697 cycles (no. 31). The base pressure

TABLE 3

Maraging 18 Ni(350) specimens

Spec. no.	k	P_i Pa	P_e G Pa	σ_z G Pa	N Cycles
10	3	2·07	1·15	0	1178
19	3	2·07	1·15	(?)	952
26	2*	2·1	1·19**	−0·53	1563
28	2*	2·3	1·22**	−0·53	1219
31	2*	2·0	1·22**	−0·53	1607
34	2*	max 2·5 min 0·5	1·22**	−0·53	1903

* Specimen placed in a sleeve.
** Pressure acting on the sleeve.

for specimen no. 34 was increased to 0·56 G Pa instead of the base pressure of 0·02 G Pa of the replenishing device. The specimen was submitted to cycles with an upper pressure of 2·5 G Pa. Failure was observed after 1903 cycles. The last specimen (no. 33) was of the shorter type of Fig. 4. It was submitted to 2682 cycles. The failure mode was the same as for specimen nos. 28 and 31.

4.4. *Tests on tungsten carbide specimens*

The results can be found in Table 4. Three specimens with a diameter ratio of 1·732 were tested. The fluid

TABLE 4

Tungsten carbide specimens

Spec. no.	k	P_e G Pa	P_i G Pa	σ_z G Pa	N Cycles
15	1·732	0·7	1·05	0	—
16	1·732	0·7	0·8	0	22
			0·7		110
17	1·732	1·0	—	0	—
27	2*	1·25**	2·5	−0·53	1
29	2*	1·25**	0·75	−0·53	—
30	2*	1·25**	1·5	−1·0	—
32	2*	1·25**	2·0	−1·0	32
37	2*	1·25**	2·0	−1·0	36

* Specimen placed in a sleeve.
** Pressure acting on the sleeve.

pressure was applied directly to the outer surface, with annular plugs between the outer diameter and the bore of the external container. One specimen (no. 17) failed circumferentially when only the external pressure was applied, another (no. 15) failed the first time the internal pressure was applied. A third specimen (no. 16) could be submitted to 110 cycles at 7 kb, followed by 22 cycles at 8 kb. Both cylinders failed circumferentially at about 20 mm from their flat faces.

A specimen with a diameter ratio of 2 (no. 22) was also tested with the supporting pressure applied directly to the outer surface, with annular plugs between the outer diameter and the bore of the external container. When the external supporting pressure was generated for the first time, the specimen was broken into 3 pieces, with a circumferential crack at about $\frac{1}{3}$ of its height, and the smaller part broken longitudinally in two.

It was thus decided to insert all further specimens tested in a Maraging sleeve, 40 mm ID and 60 mm OD, and to apply an axial force on the 40 mm OD specimen. Two cylinders were tested with an external pressure of 1·25 G Pa applied to the sleeve, and an axial compressive stress of 0·53 G Pa. The first cylinder (no. 27) was pressurized internally to 2·5 G Pa, but failed on the second pressure cycle at 0·75 G Pa, separating off a disc about 18 mm thick. The second cylinder failed when the internal pressure was raised for the first time at 0·75 G Pa. Two discs, each about 10 mm thick were separated from both ends. Even with an axial compressive stress of 1 G Pa, another specimen (no. 30) failed when the internal pressure reached 1·5 G Pa for the first time, separating off a disc about 20 mm thick from one end.

Two of these specimens (nos. 27 and 30) were then re-used. The first one (no. 27) was ground on its broken face and inserted in the Maraging sleeve, with two Maraging pieces, about 10 mm long at each end. An axial compressive stress of 1 G Pa and an external pressure of 1·25 G Pa were applied. This specimen (no. 32) was submitted to a cycling internal pressure of 2 G Pa, and failed after 32 pressure applications. The fracture is in two planes at 90°, one circumferential at mid length, the other following a longitudinal plane.

The last specimen (no. 37) was ground with the ends as in Fig. 4. With an external supporting pressure of 1·25 G Pa applied on the sleeve, and an axial compressive stress of 1 G Pa, it failed after 36 cycles at 2 G Pa, rupturing rather catastrophically, with cracks in all directions, but with a main longitudinal crack.

5. DISCUSSION

The present results indicate that two types of behaviour can be found in test materials for inner components. It was found that one of the ductile materials was not too sensitive to induced tensile axial stresses, and specimens failed longitudinally, perpendicular to the main stress. When a higher pressure was applied, which was the case with the 18 Ni (350) Maraging specimens, circumferential cracking was also found, but after cycling. The application of an axial load however is able to circumvent this type of failure. The same observation was also made by Nishihara[7] on 18 Ni (250) Maraging specimens, and on tool steels. He has shown, following the work of Shibahara[12] that there are several tensile stresses close to the seals and also suggested that an axial compression be applied to the specimens.

Circumferential fracture observed on tool steels and tungsten carbide specimens, even when only the supporting external pressure is applied, suggest that tensile axial stresses are induced, which can lead to rupture. A rough estimate of the tensile axial stress that could be induced by the seals gives values that are not too high, but the presence of internal plugs can also hinder the bore contraction under external pressure and induce bending stresses. Some specimens tested with plugs having a generous clearance, failed when submitted to their first pressure cycle, so that one has to concede that the external seals play the dominating role.

Another possible explanation would be that the extremities of the specimens, being not submitted either to the internal or the external pressure, induce a complex state of stress in their neighbourhood. One has observed, on the one hand, that with the tool steel specimens, discs were separated from the central body at both ends, and showed some faint longitudinal cracks. On the other hand, shortened tool steel and tungsten carbide specimens, could be submitted to a small number of cycles and did fracture differently. Thus it seems that for the most brittle materials a suitable specimen shape or a suitable arrangement of seals must be found. This could be determined by an analysis of the state of stress, i.e. by the finite element or some other method of analysis.

Very short lives were found with the materials studied. For the ductile materials and in particular with Maraging, the compressive hoop stress was put to the highest possible value. When the internal pressure was applied, the hoop stress was however slightly tensile, but even with a comprehensive hoop stress, the cyclic life was only about 2000 cycles. The life duration with the brittle materials was even shorter, for instance with tungsten carbide, specimen no. 16 operated with the hoop stress varying between $\sigma_t^+ = -0{\cdot}56$ G Pa and $\sigma_t^- = -2{\cdot}1$ G Pa. The specimens no. 32 and 37 operated between $-0{\cdot}62$ G Pa and $-3{\cdot}11$ G Pa.

As Tomkins suggested[6b] the driving force for fatigue could well be the amplitude of the shear stress, reached when the internal pressure is applied.

It seems that for ductile material, there are no mechanical problems to be solved, and eventually we can use improved materials such as 18 Ni (350) Maraging subjected to some thermomechanical treatment, or 13 Ni (400) Maraging. For the brittle materials, some mechanical problems remain, and the question of the support of the extremities, or the design of the seal region should find a solution.

6. ACKNOWLEDGEMENT

We are grateful to the members of the 'Committee for the study of Hydrostatic Extrusion'. (Arbed Felten and Guillaume, Bekaert NV, Comité des Matériaux Nouveaux, CRIF, LVD Co, PRB Metallurgia, Presses Raskin, Usines E. Henricot) for having supported this study. This work was performed with the help of IRSIA.

References

1. Morrison, G. L. M., Crossland, B. and Parry, J. S. C. 'Strength of thick cylinders subjected to repeated internal pressure.' Proc. Instn. Mech. Engrs., 1970, 1974, 95.
2. Burns, D. J. and Parry, J. S. C. 'Effect of mean shear stress on the fatigue behaviour of thick walled cylinders.' Proc. Instn. Mech. Engrs., 1967, 183 (Part 3 C), 72.
3. Jones, P. M. and Tomkins, B. 'Low endurance fatigue of very thick walled cylinders.' Proc. Instn. Mech. Engrs., 1968, 183, (Part 3 C), 81.
4. Haslam, G. H. (a) 'The fatigue limit of cylinders subjected to repeated internal pressures.' High Temp.-High Press., 1969, 1, (6), 705. (b) 'Fatigue of cylinders subjected to pulsating internal oil pressure.' J. Engg. Ind., 1972, 94, (2), 289.
5. Harvey, D. C. and Lengyel, B. 'Fatigue of open ended thick cylinders supported by static external fluid pressure.' High Temp.-High Press., 1973, 5, (5), 481.
6. Tomkins, B. (a) 'Fatigue failure criteria for thick walled cylindrical pressure vessels.' Press. Ves. and Piping, 1973, 1, 37–39. (b) 'Design of very high pressure vessels for repeated pressure applications.' 2nd International Conference on Press. Vessels, San Antonio, October 1973.
7. Nishihara, M. and Yamaguchi, Y. 'Fatigue strength of thick walled cylinders under very high internal pressures.' 11th Annual Meeting of EHPRG—London 1973.
8. Davidson, T. E., Eisenstadt, R. and Reiner, A. N. 'Fatigue characteristics of open-end thick walled cylinders under cyclic internal pressure.' J. Basic Engg., 1963, 85, 555.
9. Turner, C. E. 'Fracture Mechanics and its application to high pressure vessels.' High Temp.-High Press., 1974, 6, 1.
10. Semerchan, A. A., Antanovitch, A. A., Popov, V. V., Sadkov, Yu. A., Plotnikov, M. A. and Kurshin, A. P. 'Sealing the bottoms in ultra-high pressure vessels.' Prib. i Tekhn. Eksper., 1972, (5), 214.
11. Shibahara, M. et al. 'Problems of the Finite Hollow Cylinders under Axially Symmetrical Deformation.' Metal. Trans. J.S.M.E., 1968, 34, (259), 388.

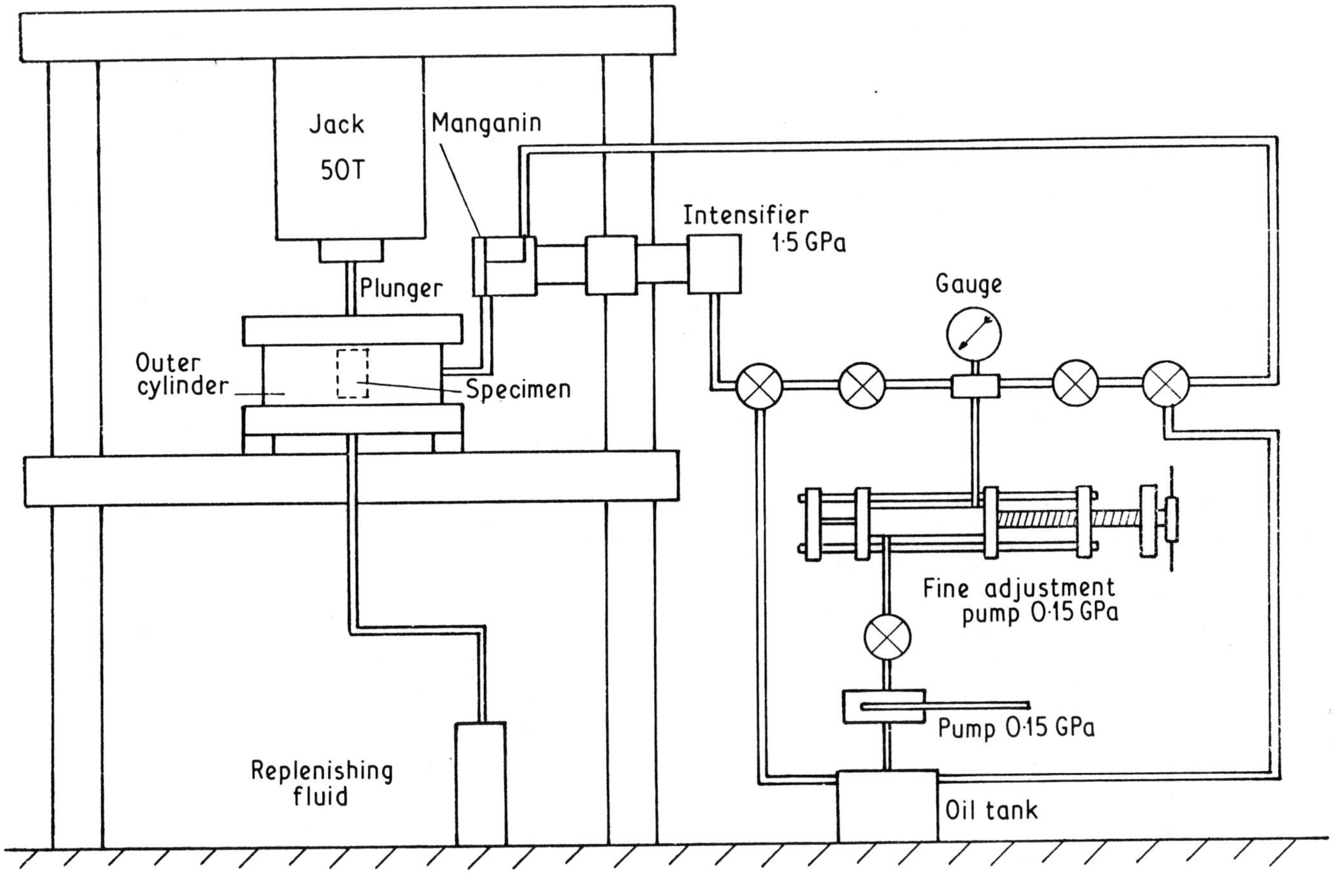

Jack
50T
Manganin
Intensifier
1·5 GPa
Plunger
Gauge
Outer
cylinder
Specimen
Fine adjustment
pump 0·15 GPa
Pump 0·15 GPa
Replenishing
fluid
Oil tank

Fig. 1:

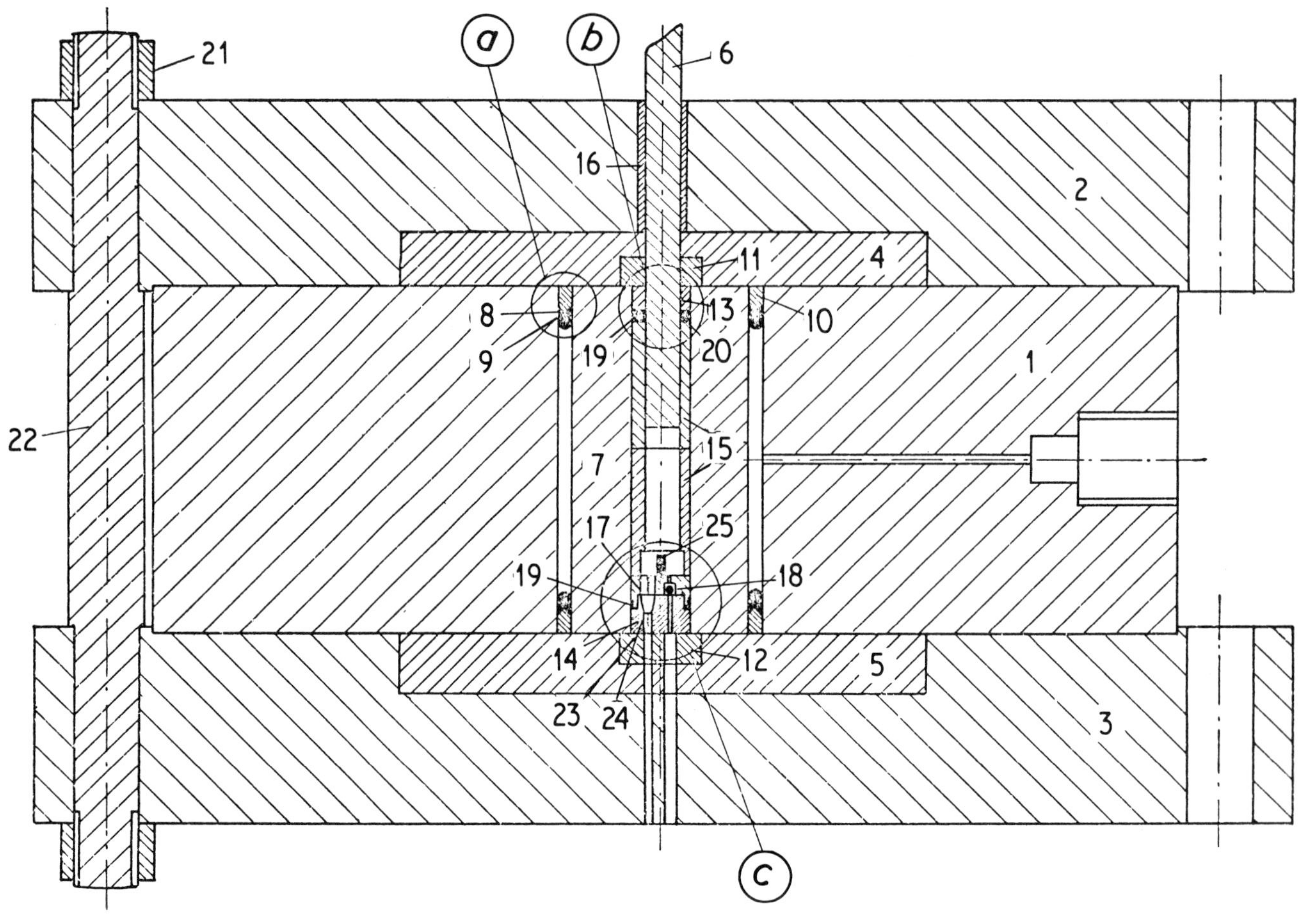

1. Autofrettaged container
2,3. Covers
4,5. Thrust receiving plates
6. Plunger
7. Specimen tested
8,9. Mitre rings
10. Packing retaining ring
11,12. Thrust receiving ring
13. Packing retaining ring
14. Tool steel plug
15. Bronze sleeve (with radial holes)
16. Plunger guide
17. Ball valve box
18. Ball valve
19,20. Mitre rings
21. Cover tightening nuts
22. Tie rods
23. Manganin feed through
24. Insulating alumina cone
25. Manganin gauge

Fig. 2.

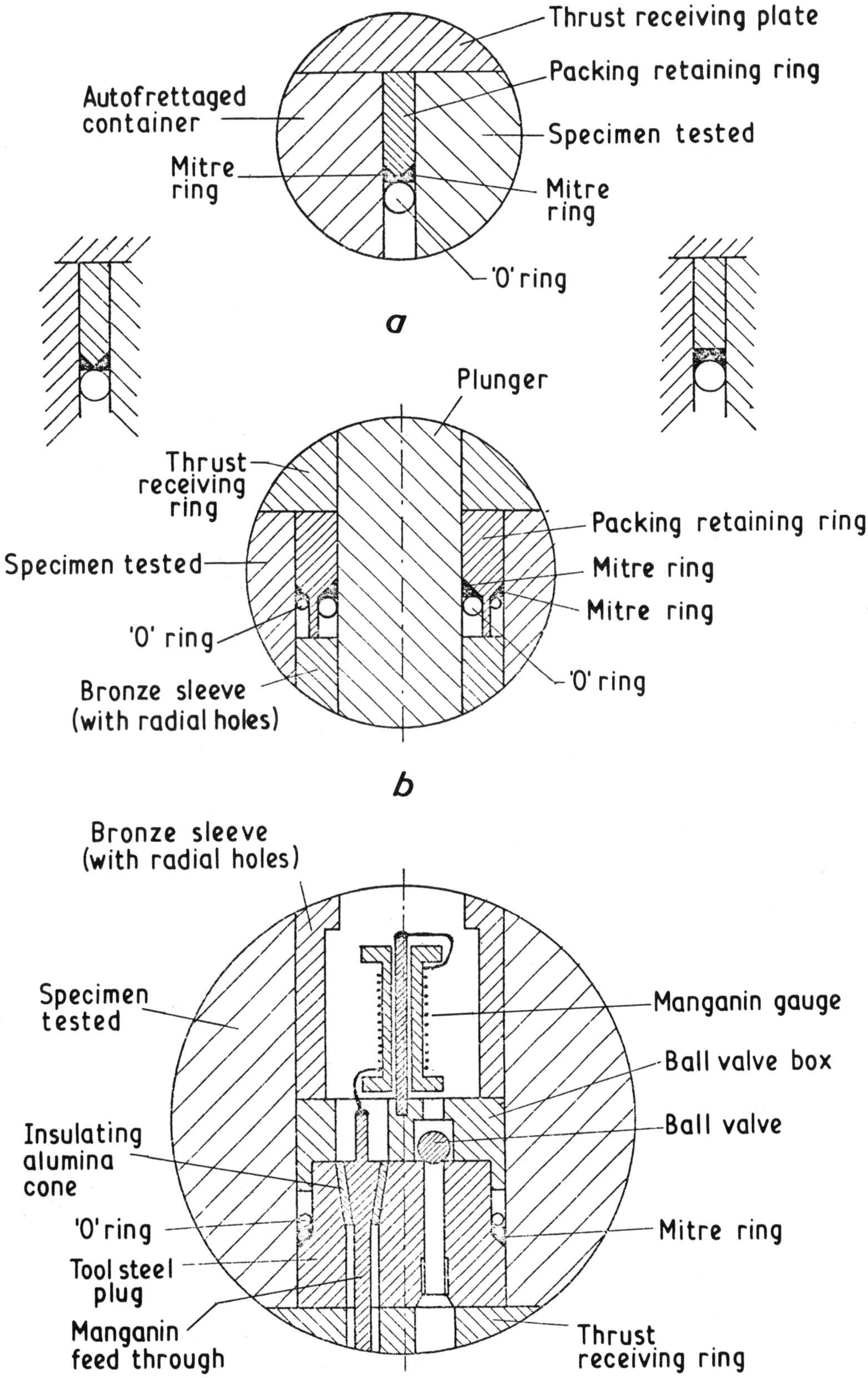
Thrust receiving plate
Packing retaining ring
Autofrettaged container
Specimen tested
Mitre ring
Mitre ring
'O' ring
a
Plunger
Thrust receiving ring
Packing retaining ring
Specimen tested
Mitre ring
Mitre ring
'O' ring
'O' ring
Bronze sleeve (with radial holes)
b
Bronze sleeve (with radial holes)
Specimen tested
Manganin gauge
Ball valve box
Ball valve
Insulating alumina cone
'O' ring
Mitre ring
Tool steel plug
Manganin feed through
Thrust receiving ring
c

Fig. 2a, 2b, 2c

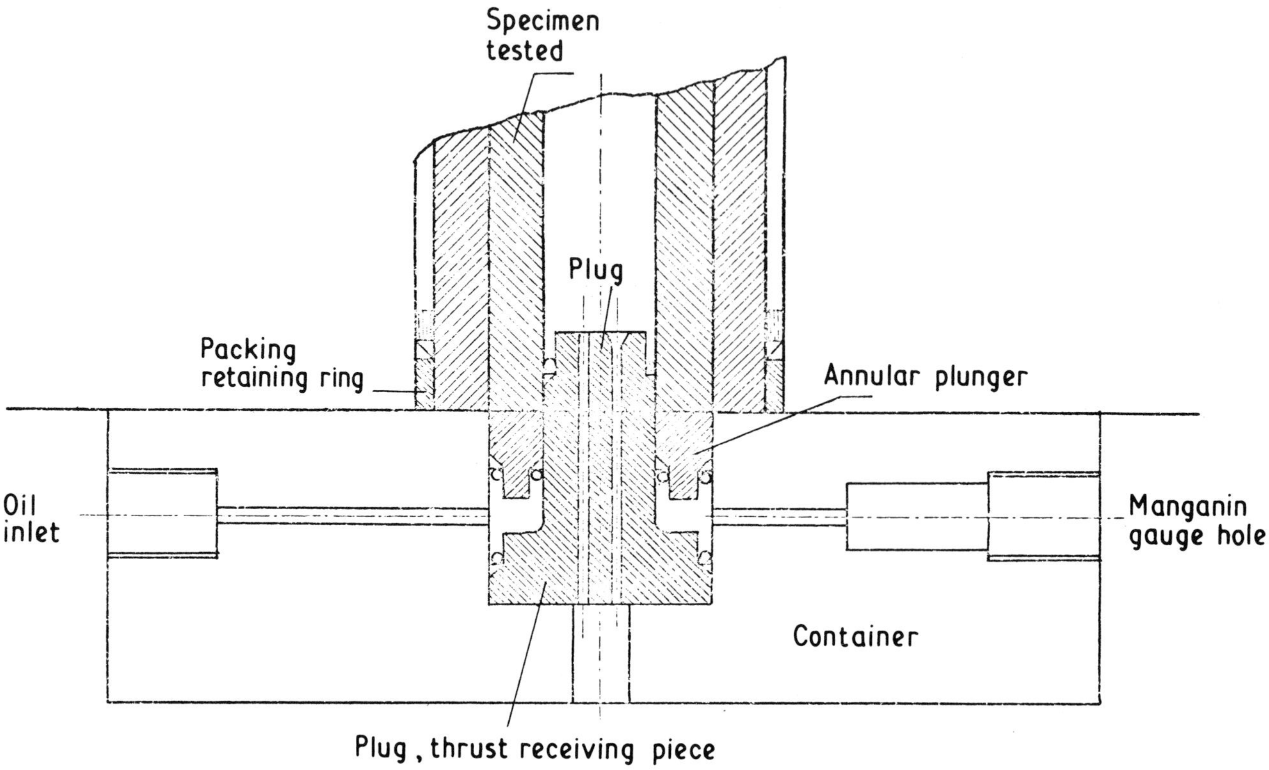

Specimen
tested
Plug
Packing
retaining ring
Annular plunger
Oil
inlet
Manganin
gauge hole
Container
Plug , thrust receiving piece

Fig. 3

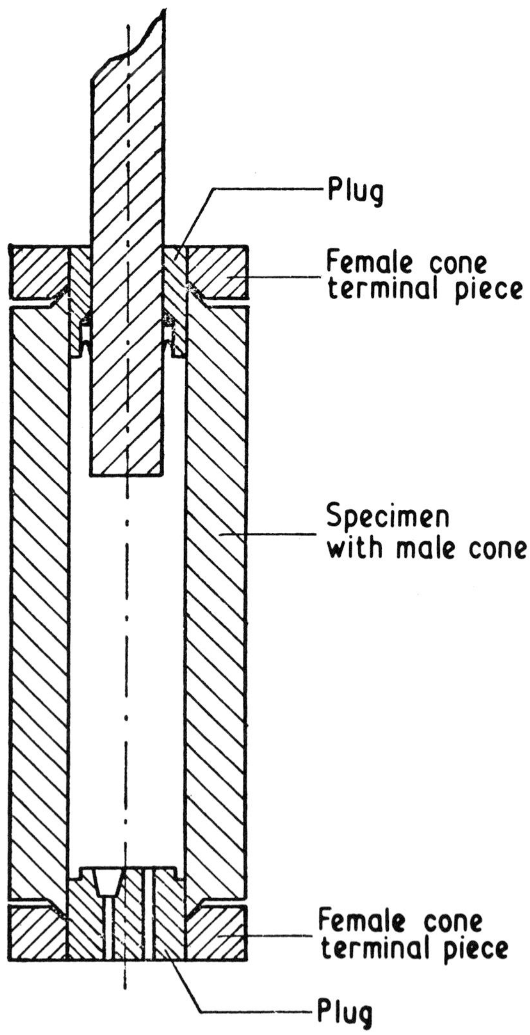

Plug
Female cone
terminal piece
Specimen
with male cone
Female cone
terminal piece
Plug

Fig. 4

A NOVEL METHOD FOR DETERMINING THE FATIGUE LIFE OF THICK WALLED CYLINDERS. DESIGN OF THE MACHINE AND INITIAL RESULTS

A. J. W. BUCKLAND, BSc, MSc and BELA LENGYEL, PhD, DIC

SYNOPSIS Existing methods of fatigue testing thick walled cylinders are examined and the design and development of a machine is described, which simulates tests with both internal and external cyclic pressures, with high or zero base pressures, using significantly smaller loads and power than existing machines. The method also simulates the effect of the moving seal commonly used in practice because it subjects the test cylinder to cyclic seal stresses and it allows the convenient measurement of seal forces for use in the development of alternative seal designs.

INTRODUCTION

1. In recent years the industrial application of hydrostatic extrusion, isostatic compaction, the "conventional" cold extrusion of steel and other high pressure processes has received increasing attention. The containers used in these processes are subjected to high cyclic pressures and often compound vessels are used which must be designed with fatigue consideration in mind. As a consequence, great efforts have been made to find fatigue criteria on which the design of thick walled cylinders, monobloc and compound, could be based. Experiments have been carried out on monobloc cylinders under various loadings, to simulate the stress conditions in compound containers.

2. Normally the arrangement illustrated in Fig. 1 is used in fatigue experiments on thick-walled open-ended cylinders (ref. 1). The test cylinder (1) is sealed at both ends (3,4) and the reciprocating plunger (2) develops a pulsating pressure in fluid (A). This arrangement becomes expensive and troublesome at very high pressures, because the pressure acting on the end face of the plunger results in large loads, high energy consumption and expensive equipment. Furthermore, several parts of the testing machine are subjected to the same number of fatigue cycles as the specimen itself, which could and often do lead to machine failures. The loads can be reduced to only a limited extent by decreasing the plunger diameter, owing to a practical limit on the permissible length/diameter ratio of the plunger. Similar considerations apply also to existing arrangements used for testing cylinders with "closed ends" (ref. 2).

3. In practical applications the pressure in the cylinder reduces to atmospheric along the seals when fluid is the pressure transmitting medium, or it steeply drops at the billet/punch interface in conventional extrusion, with a consequent change in the Lamé stresses to a quite different stress distribution (ref. 3). In conventional extrusion open ended compound containers are used with high strength steel cylinders shrunk onto a hard liner often made of tool steel or tungsten carbide. Since these liner materials have very low strengths both under static and cyclic loading when tensile stresses are present, such as the axial tensile stresses at the pressure discontinuities, it is important to develop criteria for designing against static and fatigue failure under such conditions. Furthermore, the pressure discontinuity at the billet/punch interface travels axially along the container bore as the material is extruded, rather than being stationary as at the seals in the arrangement illustrated in Fig. 1. In the former case axial tensile stresses could lead to fractures in the diametral plane anywhere in the container bore along the punch travel. Clearly, a conventional experimental rig is not suitable to test the fatigue resistance of containers under this type of loading condition.

4. At high pressures seal friction is of great interest, not only because if often amounts to a significant proportion of the total load (ref. 4), but it could also be indicative of the service performance of a moving seal. In the conventional arrangement illustrated in Fig. 1, the seal forces due to seal friction can not be easily separated from the end load on the plunger, consequently seal parameters could not be conveniently examined.

5. It would be clearly advantageous to have a testing machine which would satisfy three design criteria. The first, would eliminate the need for the application of large reciprocating pressures and loads. The second, would reproduce the effect of the "travelling pressure discontinuity" on the fatigue life of containers. The third, would allow the recording of frictional seal forces during the test, thus monitoring seal performance.

6. This report describes the design of a machine which attempts to satisfy these design criteria and outlines the first experiences in its operation.

DESIGN CONSIDERATIONS

7. The basic arrangement of the new method is illustrated in Fig. 2. Specimen (1) slides along

stationary plunger (2). Constant pressure is generated by an outside source and introduced to high pressure zone (A) between seals (3,4), through central drillings in the mandrel. The seals are located on the central mandrel and remain static with respect to it. The two extreme positions of test cylinder (1) are shown in full and in dotted lines respectively. As the cylinder moves from one extreme position to the other, each wall section is subjected in turn to the effect of pressure and then relieved of it again, i.e. each element of material experiences cyclic stress conditions. In this design only the relative displacement of test cylinder (1) and plunger (2) is important, whether the cylinder moves and the plunger is stationary, or vice versa, the results are identical.

8. In the arrangement illustrated in Fig. 2 the pressure is static, consequently work is expanded to overcome seal friction only. The frictional forces are only a small proportion of the force which would be required to generate the same pressure by reciprocating plunger (2) (ref. 4), thus the equipment becomes lighter and cheaper. In addition to seal friction, power is required only to replenish the pressure losses caused by leakage past the seals, thus good seal design is of paramount importance in reducing the total power requirement.

9. The bore stresses along the length of a cylinder of 25 mm bore diameter and 1.75 diameter ratio are illustrated in Fig. 3 for the arrangement shown in Fig. 1 (ref. 3). There are two points of interest in this diagram. The hoop stress line has a discontinuity in the seal region before it rises to the theoretical value and the axial stress has finite values around the seals,it falling to the theoretical zero towards the middle of the specimen only. It has been shown that, under usual conditions of testing, such testpieces always fail near the mid-section in fatigue, indicating that seal stresses have no effect on fatigue strength, consequently only the Lamé stresses are of interest. These stresses rise and fall smoothly with pressure as a function of time, the rate of increase and decrease depends on the speed of plunger (2) and on the compressibility/ pressure relationship of pressurised substance (A).

10. In the new arrangement, as each element of the specimen material passes over the seal, it experiences these variable hoop and axial stresses during part of the cycle. The axial tensile stress could become the decisive parameter determining fatigue strength, particularly in cases when an external pressure is also superimposed on the specimen, as will be described in connection with Fig. 4 and 5.

11. This moving seal effect is a significant feature of the present design since, in compound containers most frequently used in conventional direct extrusion, there is no feasible alternative to a pressure discontinuity travelling along the container wall, a condition reproduced in this arrangement.

12. Fatigue tests under combined internal and external pressures are particularly useful when fatigue data for the design of compound containers are required. One such conventional case is when both internal and external pressures act on the test cylinder and pulsate in phase. Usually either two separate pressure sources are needed or, in the so-called floating liner design, the specimen is allowed to move axially under the effect of the internal pressure to generate the external pressure. Both alternatives are expensive and cumbersome.

13. Fig. 4 illustrates the equivalent arrangement for the new method. Both pressures (A,B) are static and, as test cylinder (1) is moved from one extreme position to the other, it is subjected section by section to the internal and external pressures simultaneously, between seals (2,3) and (4,5). As a result each element of the material experiences cyclic stresses equivalent to the conventional case when pulsating internal and external pressures are applied to the cylinder wall.

14. Another important experimental situation is when the outer pressure is static and the bore pressure is pulsating. The equivalent new arrangement is shown in Fig. 5. Between seals (4) and (5), specimen (1) is subjected to static outer pressure (B) alone. As the cylinder moves downwards each part would be subjected to bore pressure (A) as it enters the region between seals 2 and 3 and when it passes beyond that region the pressure at its bore will be relieved. Consequently, this portion experiences stress conditions equivalent to the conventional case when outside static and cyclic bore pressures are applied to the test cylinder. Similar arrangement can be devised to stimulate the effect of a static bore pressure combined with a cyclic outside pressure.

15. A further arrangement is when, in the conventional case, the bore pressure cycles between non-zero minimum and maximum values. This could be made even more complex by adding a static outside pressure. The equivalent new arrangement is illustrated in Fig. 6. Once more all pressures are static and pressure (C) the base pressure, is the smaller of the two bore pressures. Pressure (B) is the static outside pressure. Between seals (2) and (3) the test cylinder is subjected to pressures (B) and (C) only and, as the cylinder moves downwards, it is subjected section by section also to bore pressure (A) between seals (1) and (2).A similar arrangement could be devised for the equivalent to the conventional case when a static bore pressure is combined with an outside pressure which pulsates between two non-zero values.

DESIGN DETAILS AND DEVELOPMENT OF THE EXPERIMENTAL EQUIPMENT

16. An assembly drawing showing the constructional details of the new equipment is shown in Fig. 7, equivalent to the principle shown diagramatically in Fig. 4. Here the test cylinder is subjected to high bore pressure with an external support pressure proportional to the bore pressure. It was designed for 15 kb specimen bore pressure and a simultaneously applied 7.5 kb support pressure as a maximum.

17. The bore pressure is introduced into specimen (1) from an external supply through an axial hole in central mandrel (2), which has a central pressure drilling and a cross bore in the pressure region. This pressure is contained between

internal seals (3) which are held at a fixed distance apart by seal spacer (4). The spacer is a steel tube with cross drillings to allow the pressurising fluid to reach the bore of the test cylinder. The seals are supported by two steel tubes (5) locating on the central mandrel. The latter is a composite high pressure tube capable of withstanding the design pressure of 15 kb. The end loads produced by pressure on the seals is taken by end plates (6) which are held together by four tie rods.

18. The external pressure is introduced to vessel (7) from an outside supply through a cross drilling and contained between seals (8) which are situated on the same diametral plane as the internal seals. The end loads produced by the seals are taken by seal supports (9). Vessel (7) is held in position by mild steel tubes (10). These serve a triple function as supports, safety guards and cooling jackets. The specimen is moved by reciprocating ram (11) which is connected to three equi-spaced push rods (12). The other end of the rods are attached to the specimen through a threaded collar.

19. It was felt at an early stage of the design that it would be useful to provide facilities for preloading the seals. For this purpose a dual jacking system is located in the base plate and load is applied to the seals by an internal jack). (13). The pressure is supplied by a small hand pump connected to the cross drilling in cylinder (14), and also used for applying end load to the external seals with supply pressure introduced through a drilling in the base plate. These jacks were designed to overcome the end loads on the seals at the highest design pressure.

20. Fig. 8 is an overall view of the apparatus, with the pressure take-off at the bottom and a twin intensifier system partly visible on the left. The reciprocating ram is at the top while end plates (6) of Fig. 7 can be seen at the centre, attached to the welded frame of the machine. Between the end plates are the tie rods, and between these is the specimen fully shrouded in a cooling jacket.

21. Fig. 9 illustrates this centre portion of the equipment with the cooling jacket removed, exposing the push rods at the top, which are attached to the specimen through an internally threadded collar. A thermocouple is attached to the outside surface of the specimen, to monitor its temperature.

22. Fig. 10 shows the same centre portion of the equipment with the specimen removed so that the central mandrel, seals and seal supports become visible.

23. It has been clear from the outset that, while the proposed design offers a reduction in force and power, a price must be paid for this gain because significantly longer specimens and machine stroke are needed than in the conventional method. While an increase in the specimen length would be acceptable, the long operating strokes represent the generation of more frictional heat at the seals which could lead to damage to the seals and specimen unless seal friction is reduced considerably.

24. In the initial tests the life of a Hallprene seal (ref. 1) was 20,000 cycles at 2.5 kb bore pressure. Seal wear is very rapid and usually preferential on one side, so that the mitre ring becomes misaligned and causes a side thrust as the seal extrudes past it. This is followed by cold welding between the test cylinder and the hardened steel seal support with seizure and tearing of the cylinder bore, which then destroys the seal in one or two strokes. The resultant sudden loss of pressure often causes severe damage to other parts of the equipment also.

25. Conditions are improved by the use of aluminium bronze mitre rings and by the introduction of a lead sandwich into the seal to act as a high pressure lubricant, which has a tendency to extrude past the anti-extrusion rings and have a reduced angle to decrease side thrust and friction on the cylinder bore. This compound seal is illustrated in Fig. 7 Detail A.

26. Misalignment problems are cured and the possibility of seizure reduced by the introduction of copper bearings over the seal supports and aluminium bronze bearings with locating shoulders and 0.6 mm clearance on the cylinder wall also help to reduce the side thrust. The seal material is now bronze filled PTFE, which has good frictional and heat transfer properties and it also works satisfactorily under the more severe conditions of high sliding speeds and temperatures. It is also a much harder material so less susceptible to extrusion but, to effect initial sealing, it must be pre-loaded by an outside source.

27. Assessments of seal friction are of great assistance in the development of the compound seal. The force required to reciprocate the specimen is measured in terms of the angle of the anti-extrusion ring at various pressures as tests progress, seal forces are recorded in terms of leakage rate, the fluid leakage being measured in terms of the number of strokes completed by the press before the automatic pressure control system cuts in, i.e. the number of strokes for a set pressure-fall is recorded.

28. The present findings are that the seal forces are lower at the beginning of the test and the rate of oil leakage higher but, after a few hundred cycles, the leakage rate is gradually reduced to almost half the initial value and the forces increase by some 30%. Over the next about 2,000 cycles the leakage rate increases to a value somewhat lower than that measured initially and the forces decrease by a similar amount. A possible explanation is that, after some initial wear of the anti-extrusion ring, the area of contact between it and the specimen increases, reducing the leakage rate and increasing seal friction. As the test progresses and the anti-extrusion ring wears away radially, its area of contact with the cylinder wall is reduced with a resultant increase in leakage and reduction in frictional force. In this tertiary period the rate of reduction in the frictional force is more rapid for anti-extrusion rings with a 45° angle than for those with a smaller angle.

CONCLUSIONS

29. Experiments have already proved that this apparatus is a useful tool for monitoring the service performance of seals, which could be of great help in seal design and development. However, it is essential to carry out a great deal more work, particularly on the development of a low friction seal, before this machine could also become a fully useful experimental tool for long life fatigue work on thick walled cylinders. In addition to the compound seal described in the preceding section, a controlled clearance metal seal is being developed on the principle of the Morrison seal to ensure low heat generation, which in turn would allow fast cycling speeds by using a mechanical drive in place of the present reciprocating ram.

30. This experiment has a number of practical advantages. First, the machine is small and light, consequently initial and running cost are smaller than for a conventional machine. Second, the number of stress cycles on each material element could be increased to two for each stroke of the machine so that, assuming the same rate of operation, the number of stress cycles becomes twice than that for the conventional machine. Third, the effect of stress distributions associated with a moving pressure discontinuity can be studied on the fatigue life of thick walled cylinders, which should be invaluable in the design of compound extrusion containers. Fourth, the fluid pressure is static, eliminating fatigue problems in the components of the apparatus. The pressure can be conveniently measured anywhere in the generating system. The use of a small pressure zone is feasible, reducing forces and equipment size. Pressure control is simple.

ACKNOWLEDGEMENTS

31. The financial assistance of the Science Research Council in this work is acknowledged. The authors thank Professor J.M. Alexander for use of experimental facilities in the Metalworking Laboratory of the Department of Mechanical Engineering, Imperial College.

REFERENCES

1. HARVEY, D.C. and LENGYEL, B. High Temperatures-High Pressures, 1973, 5, 515.

2. MORRISON, J.L.M., CROSSLAND, B., and PARRY, J.S.C. Proc. Inst. Mech. Engrs. 1960, 174, 95.

3. LENGYEL, B. and HARVEY, D.C. Proc. 12th Int. Machine Tool Design and Research Conf., 1972.

4. HARVEY, D.C. and LENGYEL, B. High Temperatures-High Pressures, 1970, 2, 631.

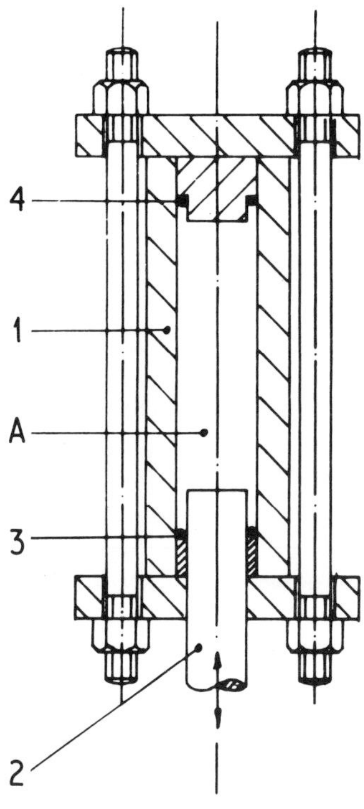

Fig. 1: Conventional arrangement for testing monobloc cylinders

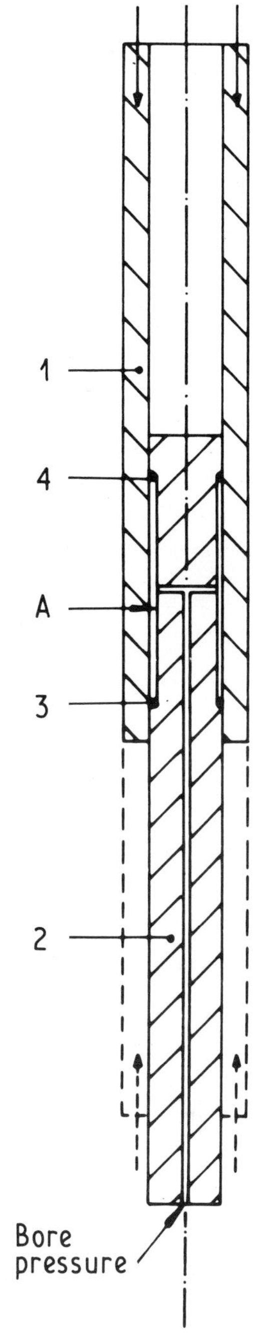

Fig. 2: New arrangement for testing monobloc cylinders

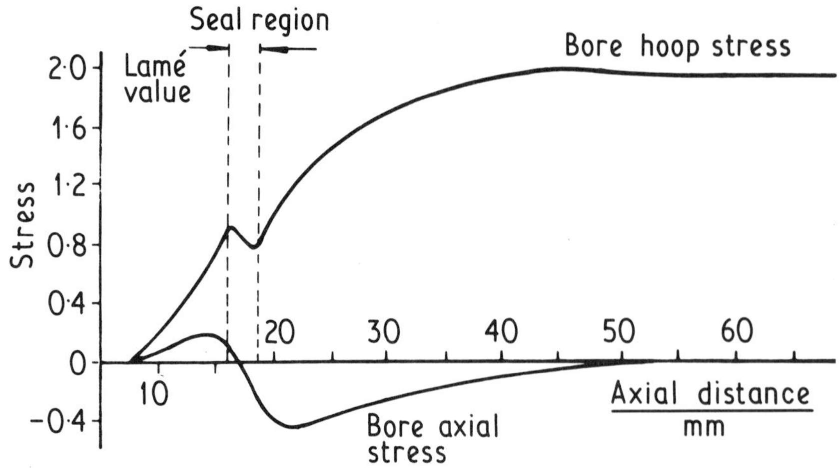

Fig. 3: Stresses in test cylinder with unit bore pressure

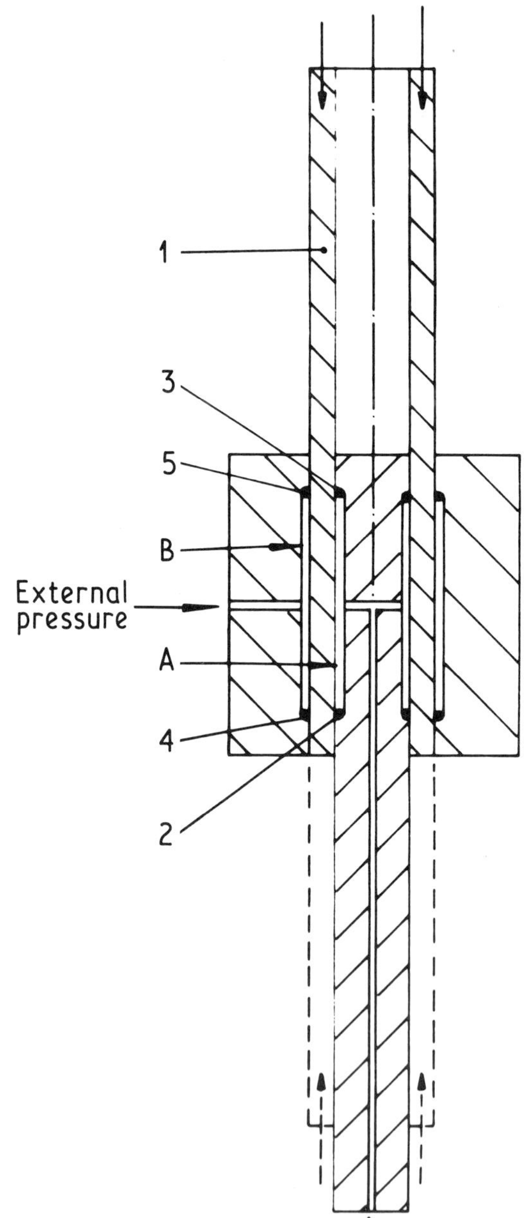

Fig. 4: Arrangement for bore pressure with proportional external support pressure

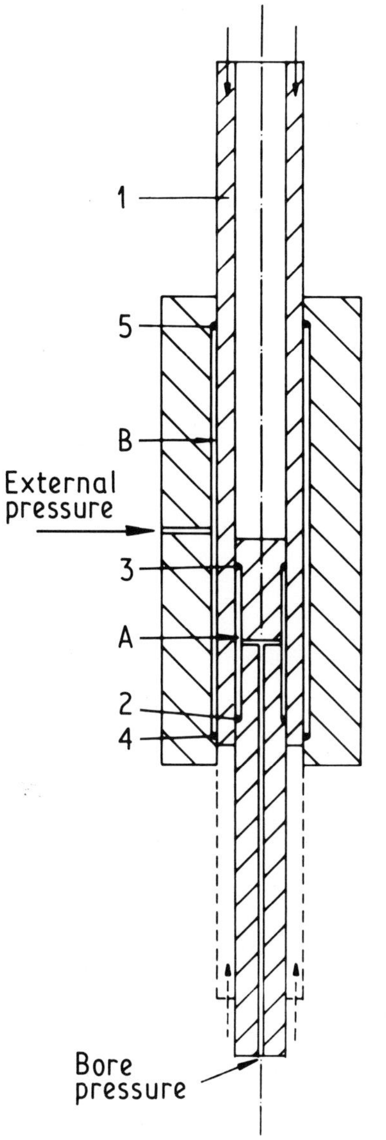

Fig. 5: Arrangement for bore pressure with static external support pressure

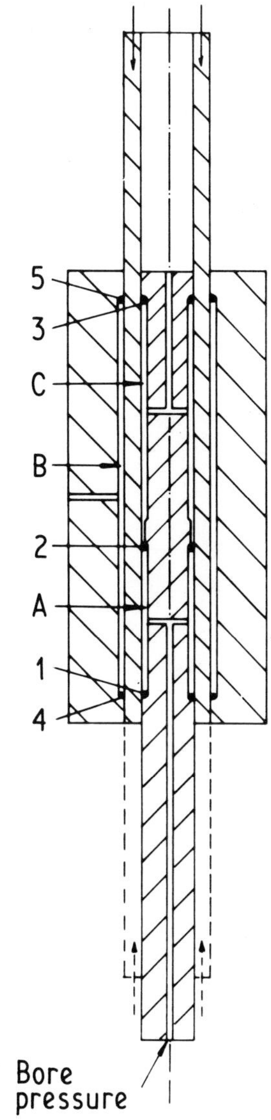

Fig. 6: Arrangement for bore pressure with static support pressure and high bore base pressure

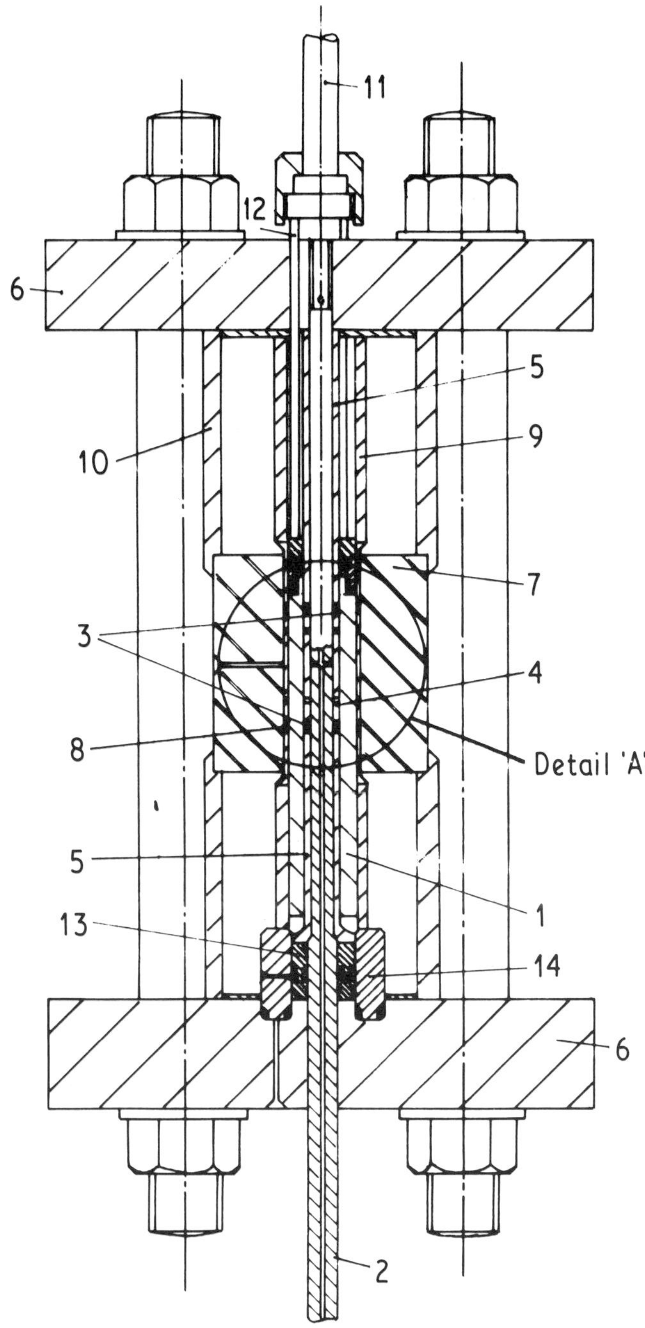

Fig. 7: Assembly drawing of new machine

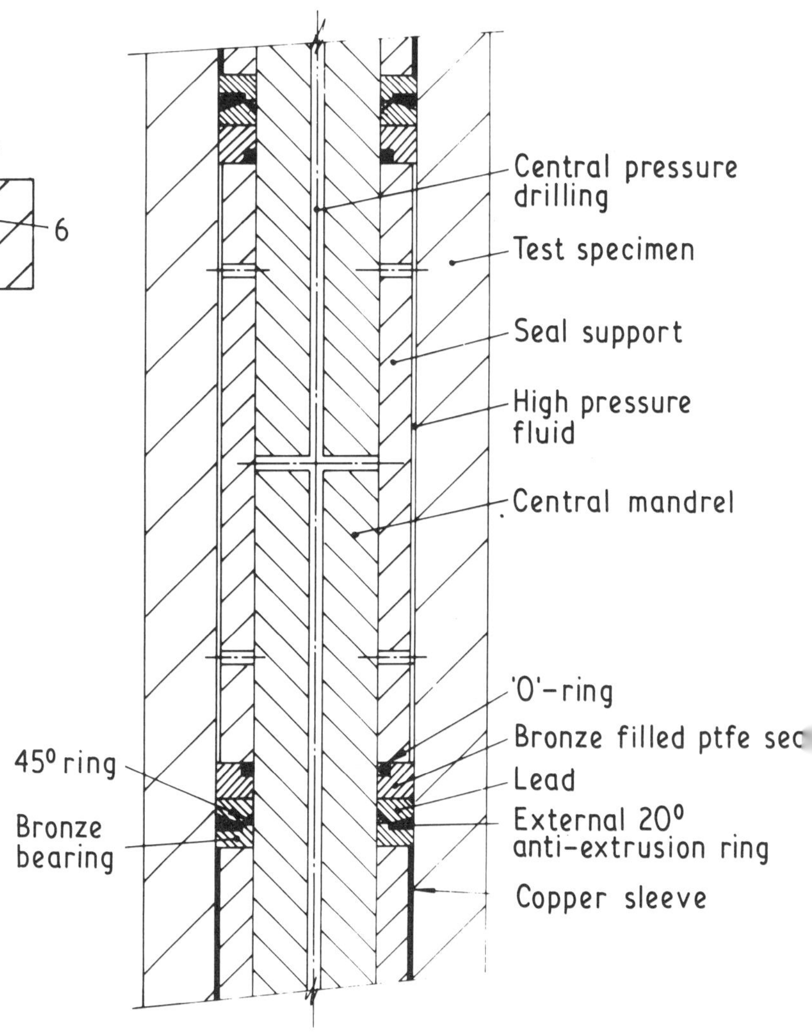

Detail 'A'

Fig. 8: General view of the machine

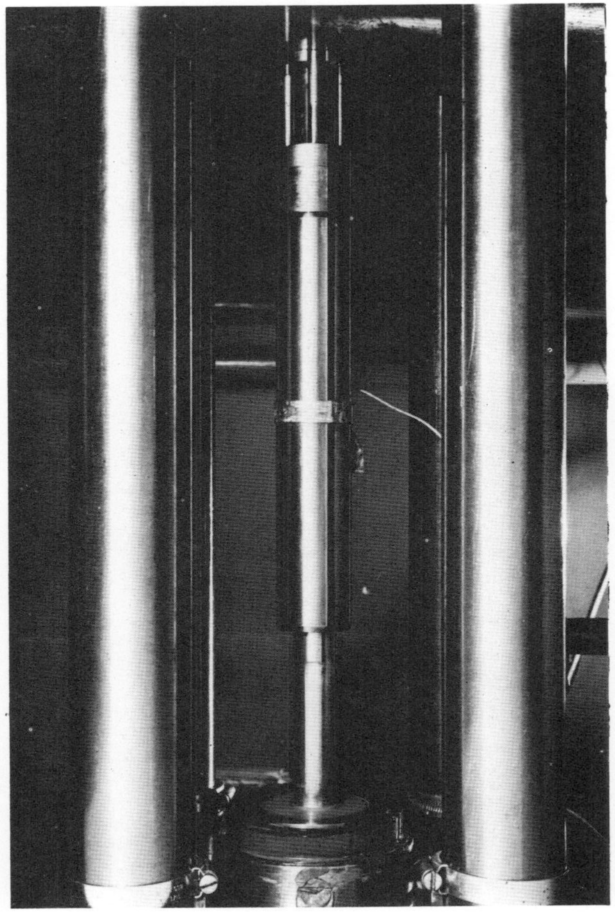

Fig. 9: Specimen with drive arrangement

Fig. 10: Central mandrel, seals and seal supports

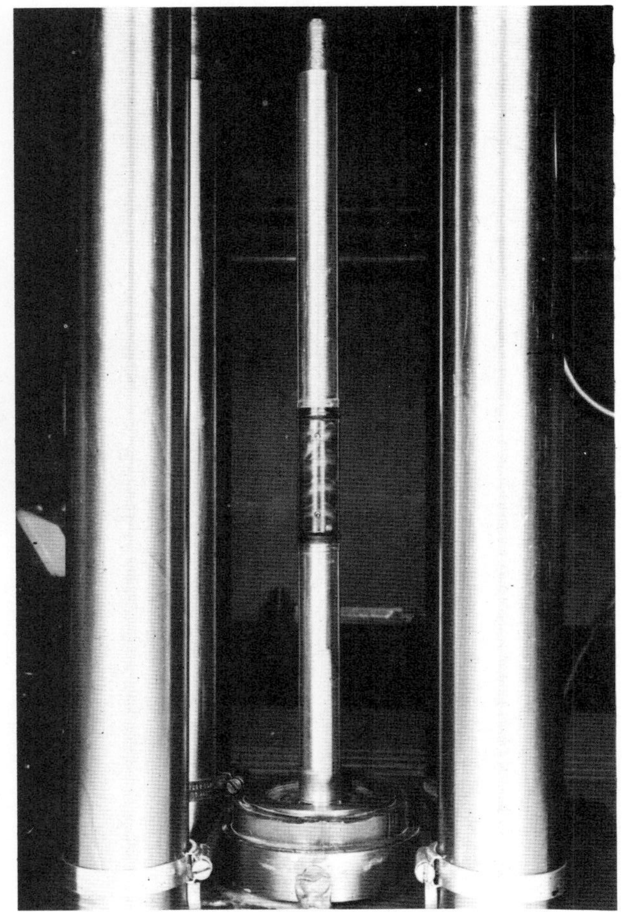

EFFECT OF INTERNAL SURFACE FLAW ON THE FATIGUE STRENGTH OF A THICK-WALLED CYLINDER UNDER CYCLIC INTERNAL PRESSURE

K. NISHIOKA and K. HIRAWAKA
Central Research Laboratories, Sumito Metal Ind., Ltd., Japan

SYNOPSIS. An empirical expression for the stress intensity factor applicable to the semi-elliptical surface crack originating along the axial direction of the bore surface of internally pressurized cylinders has been experimentally determined. The empirical expressions are applied to determine a critical pressure corresponding to the fatigue limit of a thick-walled cylinder with an internal surface flaw from the linear fracture mechanism approach. The effect of autofrettage on the flawed cylinder is also obtained.

1. INTRODUCTION

Fracture of internally pressurized thick-walled cylinders usually initiates at inner surface flaws, such as tool marks, large inclusions or corrosion pits, which propagates to final failure by fatigue during service.

In such cases the majority of the fatigue process consists of crack propagation. The objective of the present work is to find the means by which the effect of the internal flaws on the fatigue strength of a thick-walled cylinder subjected to cyclic internal pressure can be evaluated theoretically. It is thought that a fatigue crack propagation analysis based on linear elastic fracture mechanics offers a means of doing so. A complete fracture mechanics solution of these cracks is not, however, available due to the three dimensional complexity of curved front surface flaws. Recently, a fracture mechanics solution has been proposed by Underwood (1). Little experimental data, however, exists to evaluate the limited analytical results which have been proposed. In addition, the question of whether or not pressure acts on the crack surface under cyclic pressurization should be clarified by experiments.

Firstly, in the present paper, the applicability of fracture mechanics to surface (part through) crack propagation due to fatigue is examined. Next, an experiment made to determine the stress intensity factor for a pressurized semi-elliptical surface crack from a fracture mechanics approach is described.

2 EXPERIMENTAL PROCEDURE

2.1 Experimental method

The method is based on the fact that the range in stress intensity factor, ΔK, controls the mean fatigue crack growth rate, da/dN. The relationship is given by

$$da/dN = C\ (\Delta K)^m \qquad (1)$$

where C and m are empirical constants.

First, C and m for the material used are determined from a fatigue test of crack geometries with known stress intensity calibrations. Next, a thick-walled cylinder of the same material is tested in fatigue, measuring the crack growth rate, crack length and internal pressure.

It is assumed here that the constants C and m will also hold true for the internally pressurized cylinders, since the stress intensity factor accounts for all the geometrical dependence. Thus, substitution of the observed crack growth rates of internally pressurized cylinders into Eq. (1) gives experimental values for the range in stress intensity factor which may be used to establish an empirical relationship between K, crack geometry, and internal pressures.

To confirm the above assumption that the stress intensity factor controls the fatigue crack growth rate in the case of a part through crack as well as a through crack, subsurface crack growth rates in a plate subjected to cyclic bending or tensile stress were analysed by the linear fracture mechanics.

2.2 Test Apparatus

The pressure system is of 5000 kg/cm^2 capacity with a cyclic rate of up to 30 cycles/minute and consists of an intensifier type pumping system which feeds pressure to the specimens. A closed loop servo control system was used to control the wave form of the pressure variation to be sinusoidal from atmospheric level to a pre-determined maximum pressure. The high pressure fluid was a turbine oil.

2.3 Specimens and procedure

The material used for the test was quenched and tempered Ni-Cr-Mo steel (AISI 4340). The chemical composition and mechanical properties are listed in Table 1.

The cylinder for testing under cyclic internal pressure is shown in Fig. 1. The internal diameter is 27mm, and the diameter ratio, R, is 2.3. After heat treatment, the bores were honed to a surface finish of 0.1μ and a notch was made in the bore surface in the longitudinal direction by an electric discharge cutting method.

The baseline fatigue test specimen is shown in Fig. 2. It consists of approximately one half of a disc cut from a thick walled cylinder. Kendall and Hussian (2) successfully employed this method. This specimen has the advantage of enabling the crack growth rate to be determined in the same direction as the crack observed in a cylinder. The compliance of the specimen was measured by means of a displacement gauge fitted at points on a line between the centres of the loading points.

Fatigue cracks in an internally pressurized cylinder can well be represented by a semi-ellipse with major axis, 2c, along the inner surface and a semi-minor axis, a, in the plane of the crack in the direction of expected crack growth. Since it is difficult to determine sub-surface crack extension, a heat tinting method was employed. At predetermined intervals, the specimen was removed from the fatigue machine and heated, for example, at 400°C for 4 hours to make a colour marking on the crack surface which shows the crack front at that number of cycles. With successive heat treatments of 400°C x 1h and 250°C x 1h, crack growth rates and crack shapes were determined. The cylinder specimen with an internal flaw was autofrettaged to a pressure of 8,000 kgf/cm^2 which is sufficient to strain the material at the bore of cylinder beyond the yield point. Sachs method showed that a compressive residual stress of about 60 kgf/cm^2 was left at the bore of cylinder remote from a flaw and that the compressive residual stress was extended for one-third of the wall thickness. The autofrettaged cylinders were then subjected to ring compression load as well as cyclic pressure.

3. EXPERIMENTAL RESULTS AND DISCUSSION

3.1 Baseline fatigue test

The da/dN - ΔK relationship is found to be

$$da/dN = 3 \times 10^{-10} (\Delta K)^{3.0} \qquad (2)$$

where ΔK has the unit of $kgf/mm^{-3/2}$ and da/dN of mm/cycle.

3.2 Fatigue crack shape

A typical crack shape is shown in Fig 3. A semi-elliptical crack originating along the inner surface is extending to the outer surface. Fig 3. shows the aspect ratios, a/2c, of the crack extended from an initial surface flaw of length 10mm, (2c = 10mm). The aspect ratio is small in the initial stage of crack growth and becomes nearly constant at 0.4 after the crack extends to 5mm.

3.3 Crack growth rate of pressurized cylinder

The fatigue crack growth rate obtained is shown in Fig. 4 (a). Substitution of the observed crack growth rate into equation (2) yields an empirical expression for K of the internally pressurized thick walled cylinder. The results are presented in Fig. 4 (b) as a plot of $K_1/P\sqrt{\pi a}$ as a function of a/t, where t is the wall thickness and P the pressure.

3.4. Analytical approach

The stress intensity factor for the pressurized cylinder with semi-elliptical surface flaw may be obtained approximately by adding the familiar K_1 for semi-elliptical surface flaw in a finite plate in tension, to K_1 for pressurized semi-elliptical surface flaw in finite plate (3-5). The result becomes

$$K_1 = Me(P + \sigma_\theta)\sqrt{\pi a/Q} \qquad (3)$$

where

$$Me = 1 + 0.12(1 - a/2c), \quad Q = \phi^2 - 0.212(\sigma_\theta/\sigma_y) \qquad (4)$$

$$\& \quad \phi = \int_0^{\pi/2} \{1 - \frac{c^2 - a^2}{c^2} \sin^2\theta\}^{1/2} d\theta \qquad (5)$$

σ_θ is tentatively taken as the Lame tangential stress in the wall. The relation is plotted in Fig. 4(b). As shown in the figure, the agreement between experiment and the equation (3) is relatively good for shallow cracks (a/t < 0.4).

This agreement indicates that Eq. (3) can be used to estimate the effect of variations in crack geometry and pressure. In addition, it is noted that the full effect of pressure in the crack is assured because the effect is taken into account in Eq. (3).

3.5 Effect of flaw on the fatigue limit

The combination of equation (3) and the threshold stress intensity factor, K_{th}, for fatigue crack growth of this material is expected to give the fatigue limit of a cylinder with an internal flaw. K_{th} was determined by a cyclic ring compression fatigue test to be 17 $kgf/mm^{3/2}$ and the critical pressure calculated from Eq.(3) is shown in Fig. 5. The fatigue test results of the thick-walled cylinder with internal flaw are compared in Fig 5, which shows good agreement between the calculation and the experiment.

3.6 Effect of autofrettage

The fatigue test results on autofrettaged cylinders are shown in Table 2. The results indicate that autofrettaging cylinders with an internal flaw greatly increases their fatigue strength and make it possible to compensate for the detrimental effect of the internal flaws.

3.7 Applicability of fracture mechanics to surface cracks

Fig. 6(a) shows the shapes of surface cracks, which are obtained under cyclic bending and tension stresses, with respect to the ratio of the K value at the point of maximum crack depth to that when the crack intersects the free surface. Fig. 6(b) shows the variation of semi-elliptical crack shape (Aspect ratio a/2c), and the crack growth rate at the point of maximum crack depth which is compared with the growth rate of through cracks.

These figures indicate that the surface crack growth in fatigue is controlled by the stress intensity factor at the tip of the crack front.

5. CONCLUSIONS

The following are the main findings of the above experiment.

(1) An empirical expression for stress intensity factor applicable to the semi-elliptical surface crack originating along the axial direction of bore surface of internally pressurized cylinders has been experimentally determined.

(2) The empirical relation can be well expressed by adding the familiar K_1 for a semi elliptical surface flaw in a finite plate in tension, to K_1 for a pressurized semi-elliptical surface flaw in a finite plate.

(3) It is assured that the pressure is fully acting on the crack surface during cyclic pressurization of thick-walled cylinders.

(4) The critical pressure which corresponds to the fatigue limit of an internally pressurized thick-walled cylinder with internal flaw can be determined from the above K_1 expression using the material constant, K_{th} the threshold stress intensity factor for fatigue crack growth.

(5) Autofrettage can make it possible to compensate for the detrimental effect of internal flaws. However its effect on plain cylinders without flaw is very limited.

6. References

(1) Underwood, J.H., 'Stress Intensity Factors for internally Pressurized Thick-Wall Cylinders', ASTM STP 513, 1972, 59-70.

(2) Kendall, D.P., and Hussain, A.M., 'A New Fracture Toughness Test Method for Thick Walled Cylinder Material', Exp. Mechs. Vol.12, 1972, 184-194.

(3) Irwin, G.R., 'Crack-Extension Force for a Part Through Crack in a Plate', Trans. of ASME, Vol.29, Ser.E, 1962, 651-654.

(4) Paris, P.C., and Sih, G.C., 'Stress Analysis of Cracks', ASTM STP 381, 1964, 30-81.

(5) Kobayashi, A.S. and Moss, W.L. 'Stress Intensity Magnification Factors for Surface-flawed Tension Plate and Notched Round Tension Bar', Fracture, 31-45, Chapman and Hall Ltd.

7. Acknowledgements

The authors express their gratitude for the support and encouragement given by Dr. N. Oda, Manager of Central Research Laboratories. Sumitomo Metal Ind., Ltd.

The authors also express their appreciation for the effort made in this experiment by Mr I. Kitaura of our laboratories.

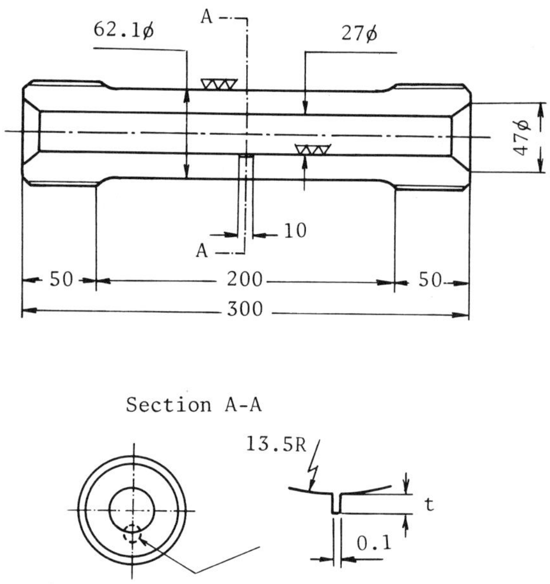

Fig. 1: Test cylinder for cyclic internal pressure

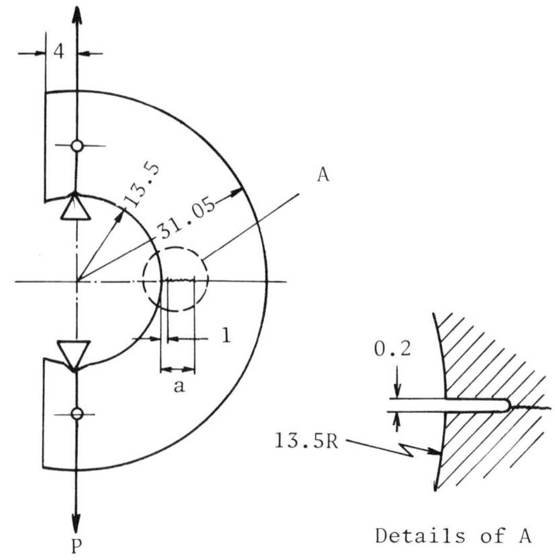

Fig. 2: Baseline fatigue test specimen

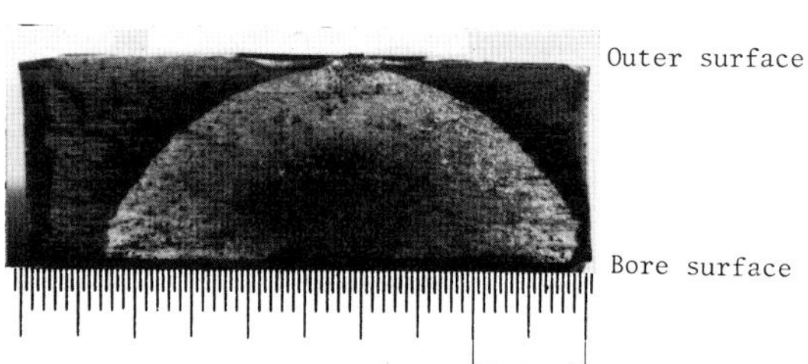

(a)

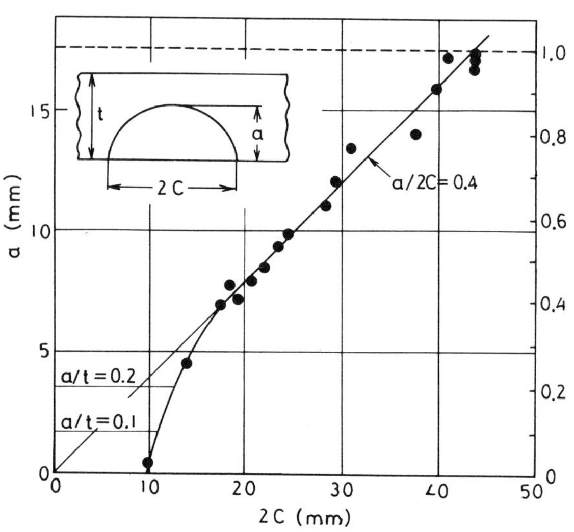

(b)

Fig. 3: Fatigue crack shape in pressurized cylinder

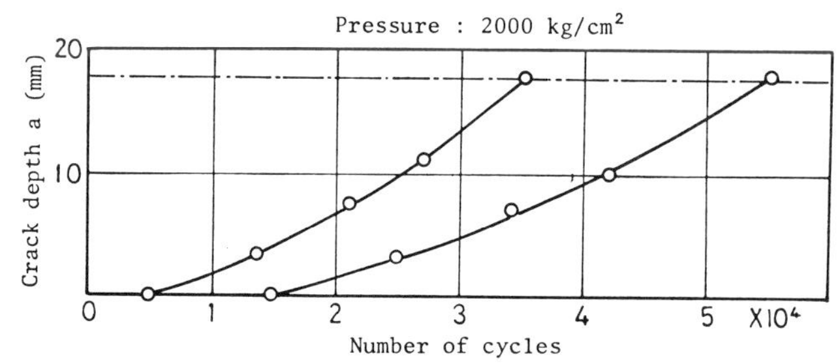

(a) Fatigue crack growth rate

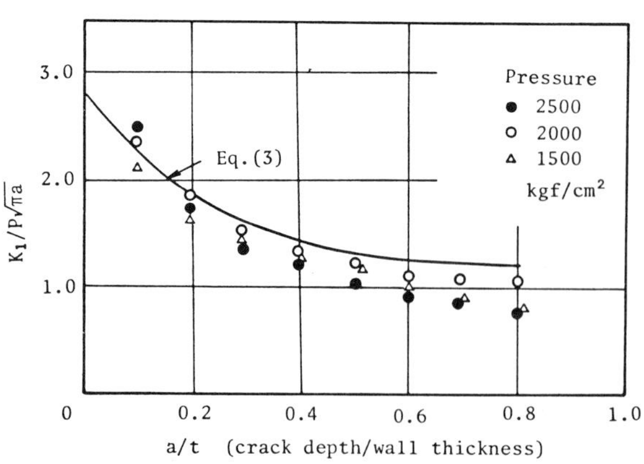

(b) Empirical K_1 for pressurized cylinder

Fig. 4: Fatigue crack growth rate and K_1 for pressurized cylinders

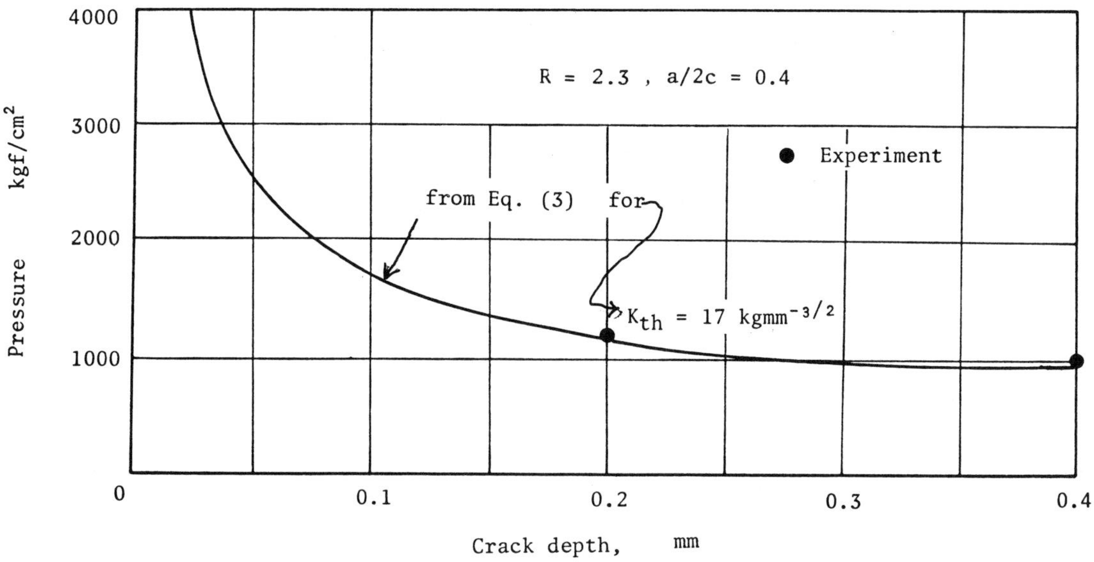

Fig. 5: Critical pressure corresponding to the fatigue limit of a thick walled cylinder

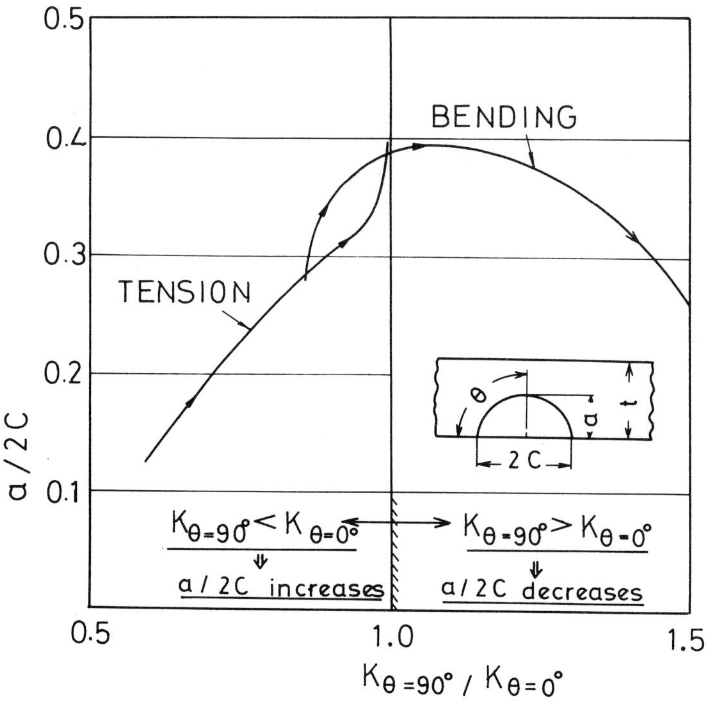

(a) Variation of semielliptical crack shape (Aspect ratio a/2c)

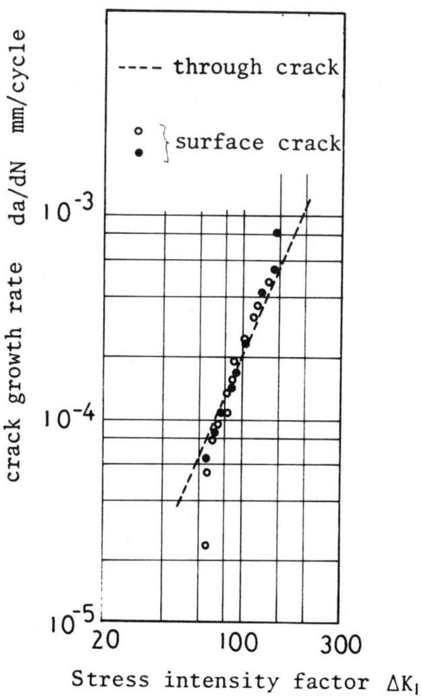

(b) Crack growth rate of surface crack and through crack

Fig. 6: Applicability of fracture mechanics to surface crack growth

Table 1 Chemical composition and mechanical properties

C	Si	Mn	P	S	Ni	Cr	Mo
0.41	0.29	0.87	0.009	0.009	1.85	0.87	0.29

0.2% proof stress kgf/mm^2	Tensile strength kgf/mm^2	Elongation %	Reduction of area %
94.2	107.7	19.0	59.3

Table 2 (a) Effect of autofrettaging(Cyclic internal pressure)

Treatment	Pressure kgf/mm^2	Flaw depth mm	No. of cycles
As quench & tempered	4000	0.4	14892
	"	0.2	18310
Autofrettage	"	0.4	50682
"	"	0.2	70080

(b) Effect of autofrettaging(Ring compression fatigue)

Treatment	Flaw depth mm	Figue limit kg/mm^2	Increased rate %
As quench & tempered	0	72	-
	0.2	27	-
	0.4	18	-
Autofrettage	0	72	0
	0.2	62	130
	0.4	60	250

FAILURE PROBLEMS IN HIGH PRESSURE EQUIPMENT

B. TOMKINS, MA, PhD
United Kingdom Atomic Energy Authority, The Reactor Group, Reactor Fuel Element Laboratories,
Springfields, Preston, Lancs.

SYNOPSIS Three types of failure problem are encountered in high pressure equipment: fatigue, stress corrosion and fast fracture. It is clear that conventional material mechanical properties are of limited value in the assessment of these problems and more relevant material parameters are discussed.

INTRODUCTION

Well designed high pressure equipment can safely contain the high pressure medium on first pressure application. Most equipment failures occur only after a period of service operation so that a time dependent element in the relevant fracture process is significant. For situations involving repeated pressure applications, the most common such element is fatigue. A less common factor is stress corrosion cracking in the presence of a hostile environment. In situations where overloads can occur in service, fast fracture is an ever present hazard in components made from high strength, low toughness materials.

Time dependent fracture implies the nucleation and growth of cracks, so the parameters controlling the rate of these two processes dominate failure. These parameters are of two types: the applied stress-strain field, often the local one, and component materials properties. In the present paper, emphasis is placed on the latter. Now the conventional materials properties of static strength (yield and ultimate) and ducility (elongation and reduction in area) do not relate directly to time dependent fracture. The strength parameters and uniform elongation partially describe the static material constitutive relations and the reduction in area describes the final state of a fast fracture process which has usually involved both the nucleation and propagation of internal cracks. When fast fracture resistance is considered, a property often specified is the Charpy or Izod energy value. This is a fracture parameter but again involves both the nucleation and rapid propagation of a crack from a notch root. It also involves a more dynamic material rupture initially than is encountered in most fast fractures in service.

In the paper, the processes of fatigue, stress corrosion and fast fracture will be examined briefly as they apply in high pressure situations. The materials parameters involved will be discussed in some detail and their relationship, if any, to the more conventional properties shown. The failure experience drawn upon is that encountered during the development of high pressure metalworking processes at RFL (Reactor Fuel Element Laboratories), Springfields and in this respect it gives an indicative rather than exhaustive view of the topics.

FATIGUE FAILURE

The essence of fatigue failure at ambient temperature is the initiation at a free surface, and subsequent propagation into the bulk of the material, of a dominant crack. The distinction between these two processes is difficult to make because, in many materials, cracks propagate steadily from a size of a few micrometers (Ref. 1). At such a size, the discrete slip processes which nucleate cracks are still operative. For practical purposes, it is most useful to confine initiation to the necessary material and stress conditions for nucleating a crack which will subsequently grow. In the limited life region of the fatigue curve (see Fig. 1) initiation is extremely rapid so that this part of the S-N (or ϵ-N) relationship is determined by crack propagation (Ref. 2).

Because cracks initiate naturally by slip processes, i.e. plastic deformation, it is reasonable to assume that a plasticity criterion will define the minimum applied cyclic stress levels below which initiation is impossible. Now it is well known that individual grains or microstructural elements of material yield at applied stresses which can be well below the macroscopic yield stress (Y). It is also known that crack initiation occurs within such a micro-unit and Esin and Jones (Ref.3) were able to correlate the microyield stress (Y_m) with the endurance limit stress for several steels tested under fully reversed (R = -1) loading. Previously, Feltner and Morrow (Ref. 4) had shown a correlation of the endurance limit with a cyclic yield stress (Y_c) for 4340 steel. This cyclic yield is identical to the microyield stress as it represents the lower stress limit for non-recoverable inelastic deformation to occur in the material, i.e. $Y_m = Y_c$. Thus, a necessary condition for fatigue failure to occur is that Y_c is exceeded in either a positive or negative direction. Tomkins (Ref. 5) has shown that for a multiaxial situation in a pressure vessel steel (En25), the Von Mises criterion, which applies to general yield, also applies to microyield. In such cases, a generalised necessary condition for crack initiation would be,

$$\pm \bar{\sigma} \geq Y_c \quad \ldots (1)$$

where $\bar{\sigma}$ is the applied effective stress.

This would be expected to define the endurance limit stress ($\bar{\sigma}_e$) for R = -1 cycling. Table 1 shows comparative values of Y, Y_c and $\bar{\sigma}_e$ for several steels. As expected a good correlation exists between Y_c and $\bar{\sigma}_e$. The data on H-11 and maraging steel illustrates the lack of correlation between static and cyclic yielding properties and shows incidentally why maraging steel has such a low fatigue strength. The UTS gives a better order for the steels but still shows no clear correlation with $\bar{\sigma}_e$.

In the limited life region, the fatigue behaviour is dominated by crack propagation. Once a crack is present, it has a specific orientation to the applied stress-strain field so that certain components of this field control the local crack stress field and hence its growth rate. Cracks tend to initiate in the maximum shear stress direction and hence early growth (usually designated stage I (Ref. 6)) is controlled by the range of applied maximum shear stress. Later growth is normal to the maximum tensile stress (Stage II) and it is this stress which then controls growth. These stress components define the extent of localised plastic deformation at the crack tip which results in crack advance (Ref. 7). The materials parameter involved is the mean flow stress in this zone. This stress is the flow stress (T) of cycled material which is approximately equal to the static UTS for many steels. Thus, above the endurance limit, materials with a higher UTS have a longer endurance for cycling at a given applied stress level. This statement is true however over only a limited region as at very high applied stress levels, in the low cycle fatigue regime, unstable crack advance leads to very high growth rates and low endurances in high strength steels.

FATIGUE CRACK INITIATION IN HIGH PRESSURE VESSELS

In repeatedly pressurised high pressure equipment, which has been well designed to eliminate cracking outside the high pressure areas, fatigue cracks often initiate in vessel bores. In these areas, the stress system is triaxial although the axial components of stress and strain are often small when compared with the radial and hoop components. For a closed-ended thick walled cylinder of diameter ratio (W = o.d./i.d.). Crossland and Bones (Ref. 8) have shown that the elastic bore stress situation is equivalent to a shear stress,

$$\tau = \frac{PW^2}{(W^2 - 1)} \qquad \ldots (2)$$

with a superimposed hydrostatic tension,

$$\sigma_h = \frac{P}{(W^2 - 1)} \qquad \ldots (3)$$

Consideration of simple internally pressurised thick walled cylinders in open, closed and fixed ended situations shows that the effective bore stress ($\bar{\sigma}$) under pressure (P) is approximately given by,

$$\bar{\sigma} = \tau\sqrt{3} \qquad \ldots (4)$$

Equation (4) is strictly true for the closed ended condition.

In considering the fatigue behaviour of thick walled cylinders with different diameter ratios in a range of materials Morrison, Crossland and Parry (Ref. 9) and Parry (Ref. 10) have shown that the fatigue limit obeys a maximum bore shear stress criterion, which is consistent with equation (4). In Table 2, the limiting value of $\tau\sqrt{3}$ is compared with the value of Y_c for the steels listed in Table 1. A good correlation exists between Y_c and $\tau\sqrt{3}$ in agreement with the crack initiation condition of equation (1). However as Morrison et al (Ref. 9) point out, there is a discrepancy between pressurised cylinder data, which involves a repeated stress condition (R = 0), and fully reversed data used to establish equation (1). In terms of stress range, cylinder data implies that only ½ the fully reversed stress range is needed to initiate cracks which will subsequently grow. For repeated tension tests (R = 0) it is known that the fatigue limit is much greater than Y_c. It is also known that for pressurised cylinders cycled with a high mean pressure, the range of bore shear stress ($\Delta\tau$) can be much less than $\bar{\sigma}_e/\sqrt{3}$ (Ref. 10). Thus, although equation (1) represents a necessary condition for fatigue failure, it is not the only condition.

The peculiar feature about repeated pressure tests is the presence of a fluid medium at the bore surface where cracks initiate. Thephysical effect of fluid pressure able to enter the slip steps which occur locally once Y_c is exceeded is to produce a much higher local stress field for the enbryo crack. This local stress field seems able to propagate the crack when the general stress field is still too low. Support for a physical fluid effect as the additional stress field comes from the fact that by protecting the bore from fluid penetration the fatigue strength of cylinders is increased (Ref. 9).

Despite the 'peculiarity' of environment assisted growth in thick cylinder fatigue, it is clear that as in simple fatigue, the material property which dominates crack initiation is the cyclic yield stress Y_c.

FATIGUE CRACK PROPAGATION IN HIGH PRESSURE VESSELS

The propagation rate of a crack initiated in a vessel bore will depend on its orientation to the applied stress field and the magnitude of that stress field. For an unprotected situation, where pressurising fluid is allowed to enter the crack, a stage II crack growing normal to the hoop stress (σ_θ) will have a crack growth equation given by (5),

$$\frac{da}{dN} = A\,\Delta\gamma\,\frac{\Delta K^2}{T^2} \qquad ---(5)$$

where ΔK is the mode I range of stress intensity factor for a driving stress (σ_θ + p). $(\Delta K/T)^2$ gives an estimate of the crack tip plastic zone size and $\Delta\gamma$ is the applied shear strain range in the vessel wall at the relevant crack depth. Equation (5) indicates a crack growth rate proportional to current crack size (a). This means that most of the endurance of a vessel in this region is occupied by crack growth in the bore region where σ_θ and $\Delta\gamma$ are the bore values of these parameters. For elastic conditions Equation (5) can be given in terms of bore shear stress range components as,

$$\frac{da}{dN} = A_1\,\frac{\Delta\tau^3 a}{GT^2} \qquad \ldots (6)$$

where G is the shear modulus.

An integration of Equation (6) with bore values inserted should give a good estimate of the endurance of a vessel with naturally initiated bore cracks as,

$$\Delta\tau^3 N_f = \frac{GT^2}{A_1} \ln (a_f/a_o) \qquad \ldots (7)$$

and N_f the endurance where a_f and a_o are final and initial crack sizes. From this equation it can be seen that one would expect ($\Delta\tau$ - N_f) endurance curves for vessels to have a slope of $\frac{1}{3}$. Curves for monobloc vessels from three of the steels whose properties are given in Table 1 are shown in Fig. 2, and it can be seen that the third power law for $\Delta\tau$ applies. Taking comparitive stress levels for a given endurance (say 2 x 10^5 cycles) shows that the shear stress range is proportional to $(T)^{\frac{2}{3}}$ in accordance with equation (7). At higher stress levels than those shown in Fig. 2, plastic deformation will occur and equation (5) would be dominated by the shear strain range term which gives a much higher power in stress range than 3.

In some high pressure components cracks may be initiated by some other means (e.g. thermal shock in gun tubes (Ref. 11)) or there may be significant pre-existing defects (Ref. 12). In this case, for elastic conditions specific material fatigue crack growth data can be used in the form

$$\frac{da}{dN} = A_2 \, \Delta K^m \qquad \ldots (8)$$

along with stress intensity solutions for a cracked ring such as that of Bowie and Freese (Ref. 13). These solutions take account of the cylinder wall stress field and pressure in the crack, and also the length of crack (a) relative to vessel wall thickness (t). Such an approach has been used successfully by both Crymble, Goldthorpe and Austin (Ref. 11) and Nishioka and Hirakawa (Ref. 12). As m is usually of the order 2-4, the similarity between equations (5) and (8) can be seen. For many materials, the constant A_2 is known to have the dependence on elastic modulus expected from equation (5) (Ref. 14), but the dependence on T is less well established. It is to be expected, however, that at lower stress levels than those involved in the naturally initiated case, the relevant flow stress parameter is less than the UTS, which confuses the material dependence (Ref. 15).

In summary, it can be said that the material UTS is a good guide to fatigue crack growth resistance and hence endurance in the limited life region for undefected vessels. For defected vessels, where propagation may occur at lower stress (pressure) levels, the useful materials strength parameter is less well defined.

STRESS CORROSION FAILURE

The susceptibility of steels to certain forms of environmental attack is well understood, particularly when an aqueous environment is involved. In the presence of a tensile stress field, an acceleration of corrosion processes can occur due to the action of plastic deformation in the material which can break protective films and expose new material surface. Therefore, in low strength metals excessive pitting by stress corrosion tends to be associated with stresses approaching yield in accordance with the need for some plastic deformation. High strength steels and other alloys are often immune to environmental cracking by reasons of their high yield (Ref. 16). However if a defect or crack is present, or forms due to the operation of another fracture process, crack extension under a static load can occur at very low stress levels in even mild aqueous environments. The presence of a crack means that even under low applied stresses yielding occurs at the crack tip and a high stress condition for stress corrosion is satisfied. In high strength steel-water situations the mechanism of crack advance involves hydrogen embrittlement (Ref. 17). Considerable work on cracked specimens of high strength steel has shown that for a given material and environment condition the critical condition for crack advance is defined by a stress intensity factor (K_{ISCC}). No stress corrosion crack growth will occur if the applied tensile stress and current crack length do not satisfy the conditions, $K \geq K_{ISCC}$. K_{ISCC} is not a simple material property as it is environment dependent. For a high strength steel therefore static load, slow crack extension can occur for $K_{ISCC} \leq K \leq K_c$ (where K_c is the fracture toughness, usually the plane strain toughness K_{IC} in the case of components). K_{IC} unlike K_{ISCC} is a simple material parameter independent of environment (Ref. 18).

Therefore, in the case of stress corrosion failure, as fatigue, one must think of the initiation and growth of cracks. The conditions for each process are quite different. Certainly for lower strength metals, initiation involves the need for high stresses, of the order of the yield. In higher strength metals, initiation is more difficult at operating stresses but growth of a crack or defect can occur at alarmingly low stresses. In the next section the failure of a high pressure container under static loading will be described in detail as an illustration of stress corrosion failure in high strength steel, high pressure components.

THE STRESS CORROSION FAILURE OF A HIGH PRESSURE VESSEL

The vessel which failed was a simple duobloc with taper fit to give bore precompression. It was an experimental assembly with both inner and outer rings made from H-11 steel. The vessel dimensions are shown in Fig. 3. The failure involved disintegration of the outer ring some six months after assembly and whilst the vessel was still in its delivery crate. The vessel had never been internally pressurised. The outer ring material specification and stresses are given in Table 3. Both ultrasonic and dye penetrant tests had been carried out on the outer ring after assembly. For the time between these tests and failure the vessel had been covered with preservative grease and paper and crated in a wooden packing case.

Inspection of the outer ring fragments showed that the fracture origin was at a point on the ring outer surface (Fig. 4). At the origin a distinct pit could be seen surrounded by a darkened area (Fig. 5). It was soon established that pitting had occurred on the outer surface in the vicinity of the fracture origin (Fig. 5b). Such pits existed beneath hardened grease patches which were extremely adherent. Sectioning revealed that the fracture origin was definitely a corrosion pit some 0.25 mm deep. The further darkened area seen in Fig. 5a and indicating slow crack extension was to a depth of some 1.50 mm.

Knowing the outer fibre ring hoop stress it is possible to examine these features and see their consistency with a stress corrosion failure. The formation of the initial corrosion pit which acted as a crack initiator is puzzling and indicates the presence under the protective grease of an active corrosive agent. As mentioned earlier water itself would not be expected to cause pitting in this steel at such low stress levels. The origin of this corrosive agent is not known but the fact that pitting was confined to a small area could perhaps indicate that the vessel was splashed with something sometime between assembly and packing. If the pit at the fracture origin is considered as a crack, its maximum stress intensity is 9.1 $MNm^{-3/2}$. Such a value compares with a minimum K_{ISCC} of 17 $MNm^{-3/2}$ found by Johnson and Willner (Ref. 19) for a 1590 MN/m^2 yield H-11 steel. However the fracture toughness for this steel was 43 $MNm^{-3/2}$ compared with 25 $MNm^{-3/2}$ for that of the failed ring, so that K_{ISCC} for the ring material could be less than 17 $MNm^{-3/2}$. The presence of enough water beneath the protective coatings to give slow crack growth is surprising but Johnson and Willner (Ref. 19) have shown that for H-11 steel in relative humidity as low as 1% the K_{ISCC} is only 25% above the minimum value. The final critical crack depth indication at 1.50 mm would give a stress intensity of 22.2 $MNm^{-3/2}$ which compares very well with the measured K_{IC} for the ring material.

Thus the ring failure does fit the pattern of a stress corrosion failure once an initial crack (pit) had been formed. The importance of the material/environment parameter K_{ISCC} is underlined by this failure and great care should be taken in the use of high strength steels in pressure vessels which have contact with an aqueous medium. This type of material/environment information is very difficult to produce for a given situation and too little is known of the physical processes involved in this type of stress corrosion failure. However, some lessons can be drawn from this failure experience.

(1) Care must be taken at all times following component manufacture so that initiating defects are not produced. Such defects could be scratches, impact marks or corrosion pits.

(2) A careful check should be made on the environment which the component is likely to see when stressed (either during service or as a result of prestress).

(3) Outer rings of multi-ring vessels should always be of a high toughness material. In the present vessel the outer ring had such a low toughness because it had been previously manufactured, but not used, as a component for another vessel. The stored energy in large prestressed components is so large that items from even medium toughness materials can fragment on failure.

It must be noted that in terms of material properties the standard mechanical properties of strength and ductility give no clue as to a material's resistance to stress corrosion failure. Even the plane strain fracture toughness is no guide as to the possibility of sub-critical crack growth. The amount of data available on the relevant material/environment parameter K_{ISCC} is very limited.

FAST FRACTURE

The previous example of a stress corrosion failure indicated how well a measure of fracture toughness could define the <u>final</u> fracture condition. However, this applied to the outer ring of a compound vessel where the stress field was effectively simple tension. In the case of a bore crack or defect in a pressurised vessel, some difficulty has been encountered in equating the estimated critical stress intensity to the plane strain fracture toughness K_{IC} (Ref. 20). For a thick walled component the plane strain condition is usually satisfied and good analytical (Ref. 13) and finite element (Ref. 20) solutions now exist which take account of the cylindrical geometry of a vessel and the effect of pressure in the crack, but use of the material K_{IC} tends to underestimate the critical condition. In their paper to this Conference, Crymble, Goldthorpe and Austin (Ref. 11) show that use of pressurised rings leads to more consistent results although they still appreciably underestimate the critical crack size for the smaller crack lengths. In our own work on fatigue failure in very high pressure vessels, up to 15 kb, fast fracture terminated most of the tests. An examination of the critical crack sizes in these tests showed a similar trend to that observed in the fracture experiments mentioned earlier i.e. the critical crack size was longer than expected. The most striking example involved a monobloc vessel of H-11 steel with W ratio of 7.2 repeatedly pressurised to 12.5 kb. At this pressure, some bore autofrettage occurred and the maximum hoop stress (σ_θ) was estimated at 238 MN/m^2. The critical crack was 7.5 mm deep and 15 mm long and the material fracture toughness was 27.5 $MNm^{-3/2}$. Using an approximate method for such a short crack, the stress intensity is:

$$K = 1.12\ (\sigma_\theta + p)\ (\pi\ a/Q)^{\frac{1}{2}} \qquad \ldots (9)$$

where Q is a crack shape parameter estimated at 2.5. The estimated K at failure is 162 $MNm^{-3/2}$ compared with the K_{IC} value of 27.5 $MNm^{-3/2}$.

It is thus pertinent to ask whether, in strongly biaxial stress fields, K_{IC} is a unique material parameter which alone can define the mode I fracture condition. Recently Rice (Ref. 21) has shown that the plastic zone size (D) can be significantly affected by a biaxial stress field applied to a cracked body. Also, Chipperfield and Knott (Ref. 22), from the metallurgical side, have shown how fast fracture in high strength steels proceeds by shear fracture to second phase particles along the flow lines which radiate from the crack tip. Conditions for such fracture might be expected to depend on the normal stress across these flow lines and hence vary with biaxiality. At first glance, the Rice analysis would indicate an increase in both D and CTOD with the biaxial state obtaining at the vessel bore over that, for example, estimated using the simple condition represented by equation (9) or the more detailed elastic analysis of Bowie and Freese (Ref. 13). In conjunction with a K_{IC} criterion, this would give the opposite trend to that observed. A simple examination of the crack tip flow line situation is shown in Fig. 6. For small CTOD values of an edge crack, the <u>local</u> crack tip flow lines radiated at $\pm 45^\circ$ from the tip. A crude estimate of the normal stress across the flow lines is

$$\sigma_n = \frac{\sigma + \sigma_r}{2}$$

where σ is the local applied stress normal to the crack and σ_r is the vessel radial stress at the crack tip location. To satisfy equilibrium in this region,

$$\sigma \simeq 2\,k + \sigma_r \qquad \ldots (10)$$

where k is the material shear flow stress. Therefore,

$$\sigma_n \simeq k + \sigma_r \qquad \ldots (11)$$

For short cracks compared with the vessel wall thickness, $\sigma_r \simeq -p$ and so σ_n is a function of applied pressure. σ_n is positive when $p < k$ and negative when $p > k$. For the monobloc case cited earlier, $\sigma_n \simeq -k$ which could be consistent with a strongly delayed fast fracture condition. For longer cracks in elastically stressed vessels, $\sigma_n \rightarrow k$ and the normal fracture condition might be expected to hold. This is consistent with the recent observations of Crymble, Goldthorpe and Austin (Ref. 11) in their ring tests.

Although there are these indications that K_{IC} is not a sufficient parameter to define the unstable fracture condition at the bore of a pressurised component, it does give a conservative estimate. Further work is needed in this area but it is certain that K_{IC} as a materials parameter is a better guide than the ill defined notch impact parameters such as Charpy or Izod.

CONCLUDING REMARKS

This brief review of potential fracture modes in high pressure equipment has shown how conventional engineering mechanical properties fail in the main to indicate the probability of failure. The initiation of fatigue cracks is dependent on the micro- or cyclic yield (Y_c) which can be much lower than expected for strongly cyclically softening high strength steels. The propagation of naturally initiated fatigue cracks, in the absence of a hostile environment, is however roughly dependent on the material UTS, the cyclic flow stress (T) being well estimated by the static UTS. For low stress fatigue crack growth in defected components, there seems to be less dependence on material strength. Ductility is not a guide to fatigue resistance.

The complexities of the material-environment interaction dominate stress corrosion failure. Although the parameter K_{ISCC} defines the critical stress intensity for stress corrosion crack growth, it is not a simple material constant but depends on the particular material-environmental combination. The susceptibility of high strength steels to corrosion pitting and hence crack initiation is an important area for study.

The final remarks on fast fracture indicate the possibility for high strength materials of a lack of uniqueness of the plane strain fracture toughness K_{IC} in multiaxial situations. Indications are that use of K_{IC} will give a conservative estimate of the fast fracture condition for high pressure vessels. In both fast fracture and fatigue, the mechanical effect of pressure in the crack must be taken into account for bore cracks or defects.

The practical illustrations used in the paper are only indicative of the wide range of high pressure components which can fail. However, they do illustrate the peculiarities which high pressure can introduce into failure processes. Great care must be taken in the use of the best designed equipment and adequate guarding provided at all times.

ACKNOWLEDGMENTS

The author acknowledges the work of Mr G R Kirland and Mr T Ingham in the analysis of the stress corrosion failure. This paper is published by permission of Mr R V Moore, Member for Reactors, U.K.A.E.A.

REFERENCES

1. GROSSKREUTZ J C. Fatigue crack propagation in aluminium single crystals. J.Appl.Phys. 1962, 33, 1787-1792.

2. TOMKINS B. Fatigue failure. Proc. Conf. on the Mechanics and Physics of Fracture (Cambridge) 1975 Paper 23.

3. ESIN A and JONES W J D. A theory of fatigue based on the microstructural accumulation of strain energy. Nucl.Eng. and Design 1966, 4, 292-298.

4. FELTNER C E and MORROW J D. Microplastic strain hysteresis energy as a criterion for fatigue fracture. ASME J. Basic Eng. 1961, 15-22.

5. TOMKINS B. Fatigue failure criteria for thick-walled cylindrical pressure vessels. Pres.Ves. and Piping 1973, 1, 37-59.

6. FORSYTH P J E. A two-stage process of fatigue crack growth. Proc. Crack Propagation Symp. (Cranfield) 1961, 76-94.

7. TOMKINS B. Fatigue crack propagation - an analysis. Phil. Mag. 1968, 18, 1041-1066.

8. CROSSLAND B and BONES J A. Behaviour of thick-walled steel cylinders subjected to internal pressure. Proc. Instn. Mech. Engrs. 1958, 172, 777.

9. MORRISON J L M, CROSSLAND B and PARRY J S C. Strength of thick cylinders subjected to repeated internal pressure. Proc. Instn. Mech. Engrs. 1960, 174, 95.

10. BURNS D J and PARRY J S C. Effect of mean shear stress on the fatigue behaviour of thick-walled cylinders. Proc. Instn. Mech. Engrs. 1968, 182, 72.

11. CRYMBLE T G, GOLDTHORPE B D and AUSTIN B A. The application of linear elastic fracture mechanics to fatigue and fracture of thick wall cylinders. Second Internat. Conf. on High Press. Engg. pub. Inst. Mech. Eng 1976. p

12. NISHIOKA K and HIRAKAWA K. Effect of internal surface flow on the fatigue strength of thick-walled cylinder under cyclic internal pressure. Second Internat. Conf. on High Press. Engg. pub. Inst. Mech. Eng 1976. p

13. BOWIE O L and FREESE C E. Elastic analysis for a radial crack in a circular ring. Eng. Fract. Mech. 1972, 4, 315.

14. BARSOM J M. Fatigue crack propagation in steels of various yield strengths. ASME Paper 71-PVP-12 1971.

15. TOMKINS B. The development of fatigue crack propagation models for engineering applications at elevated temperatures. ASME J. Eng. Mtls. and Tech. 1975 to be published.

16. BROWN B F. A new stress-corrosion cracking test for high-strength alloys. ASTM Mater.Res. Stand. 1966, 66, 129.

17. JOHNSON H H and PARIS P C. Sub-critical flow growth. Eng. Fract. Mech. 1968, 1, 3-45.

18. PETERSON M H, BROWN B F, NEWBEGIN R L and GROOVER R E. Stress corrosion cracking of high strength steels and titanium alloys in chloride solutions at ambient temperature. Corrosion 1967, 23, 142.

19. JOHNSON H H and WILLNER A M. Moisture and stable crack growth in a high strength steel. Appl. Mater. Res. 1965, 4, 34.

20. SHANNON R W E and AUSTIN B A. The application of fracture mechanics to failure of thick-walled cylinders. Instn. Mech. Engrs. Proc. Conf. 1971, Paper C32/71.

21. RICE J R. Limitations to the small scale yielding approximation for crack tip plasticity. J. Mech. Phys. Solids 1974, 22, 17-26.

22. CHIPPERFIELD C G and KNOTT J F. Microstructure and toughness of structural steels. Metals Tech. 1975, 2, 45-51.

23. PARRY J S C. Fatigue of thick cylinders: further practical information. Proc. Instn. Mech. Engrs. 1965-66, 180, 387-404.

24. BLASS J J and FINDLAY W N. Discussion to Ref. 23.

25. AUSTIN B A and CROSSLAND B. Low-endurance fatigue strength of thick-walled cylinders: development of a testing machine and preliminary results. Proc. Instn. Mech. Engrs. 1965-66, 180, 118.

26. VAN SWAM L F P. Fatigue behaviour of maraging steel 300. D.Sc. Thesis MIT. 1973.

TABLE 1

Static and cyclic strength parameters for four steels

Material	Static yield or 0.2% proof stress, Y (MN/m^2)	Cyclic yield, Y_c (MN/m^2)	Effective endurance limit stress amplitude, σ_e (MN/m^2)	Ultimate tensile strength (MN/m^2)
Mild steel	225	207	207	518
En25 (660)*	841	465	448	1103
Maraging 300	1868 (Ref. 26)	600†	676 (Ref. 26)	1917 (Ref. 26)
H-11 (520)*	1310	862	724	2130

*Refers to tempering temperature (°C)

†Estimated value

TABLE 2

Effective cylinder endurance limited stress compared with material cyclic yield stress for four steels

	Cyclic yield, Y_c (MN/m^2)	Effective cylinder bore stress limit, $\tau\sqrt{3}$ (MN/m^2)	
Mild steel	207	233	(Ref. 9)
En25 (660)*	465	498	(Ref. 9)
Maraging 300	600†	602	(Ref. 23)
H-11 (520)*	862	-	

*Refers to tempering temperature (°C)

†Estimated value.

TABLE 3

Material and stress specifications for failed outer ring

Material	H-11 steel
Compoosition	0.4% C, 5% Cr, 1.3% Mo, 0.55% V
0.2% proof stress	1514 MN/m^2
Uniform elongation	4%
Reduction of area	8%
Plane strain fracture toughness	25 $MNm^{-3/2}$
W ratio	1.66
Outer fibre hoop stress	440 MN/m^2
Stored energy	550,000 J

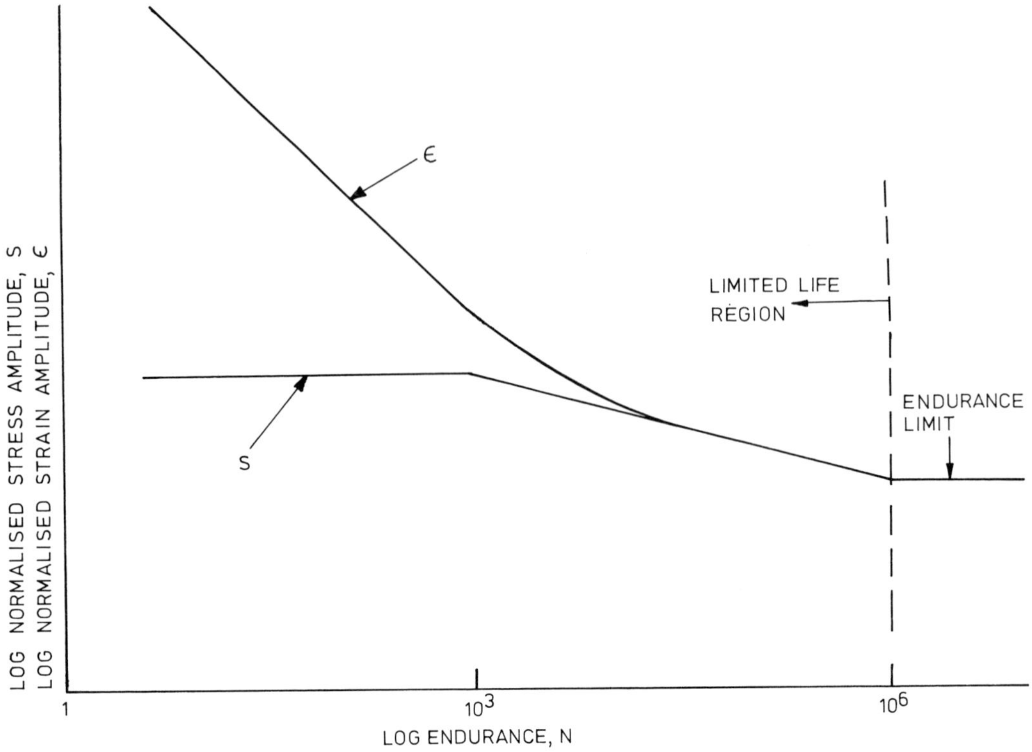

Fig. 1: Schematic S-N and E-N relationships

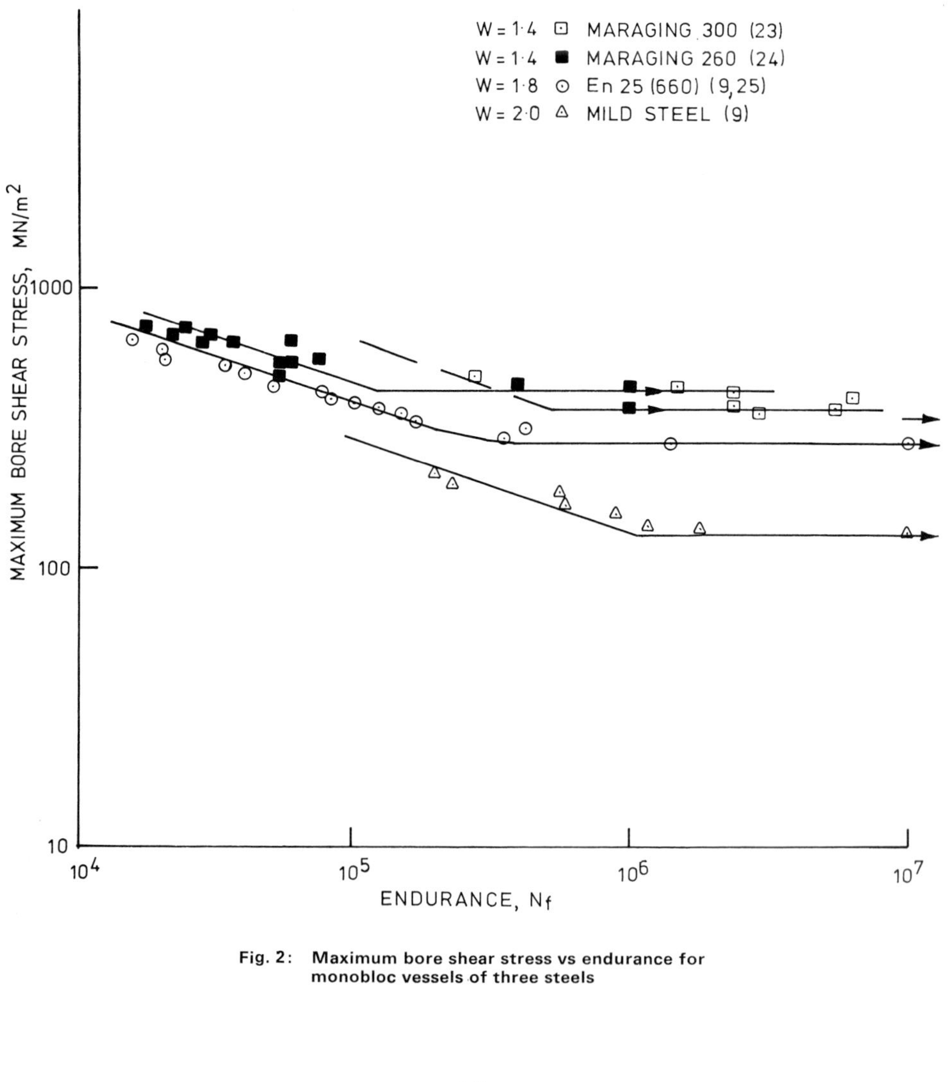

Fig. 2: Maximum bore shear stress vs endurance for monobloc vessels of three steels

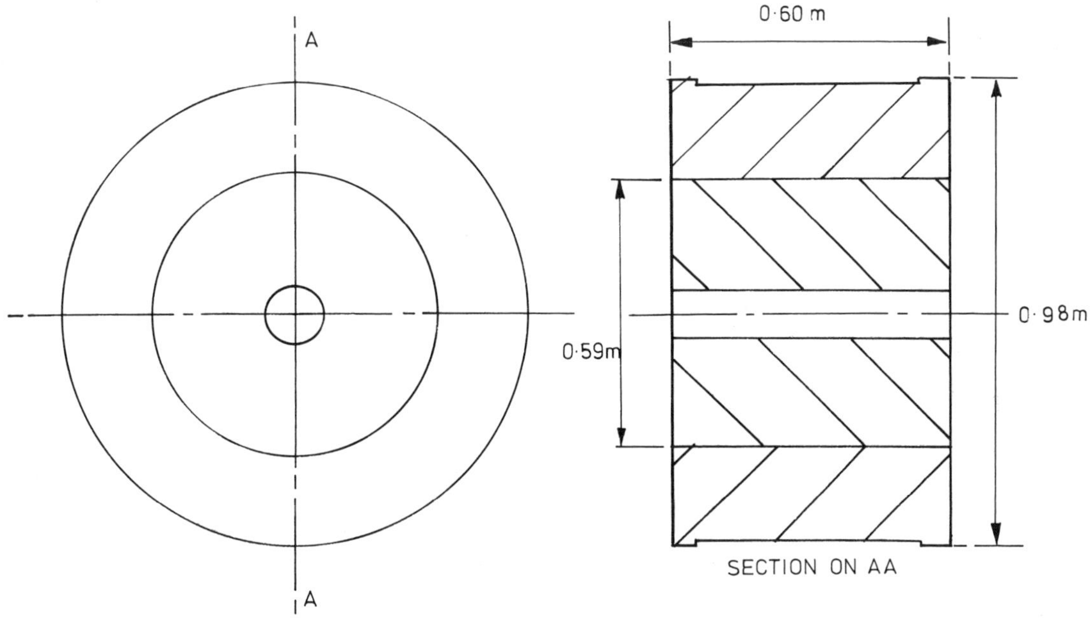

Fig. 3: Details of failed duobloc container

Fig. 4: Piece of container outer ring showing failure site

Fig. 5(a): Pit at fracture origin on surface of outer ring

(×20)

Fig. 5(b): Surface pitting in vicinity of fracture origin

(×20)

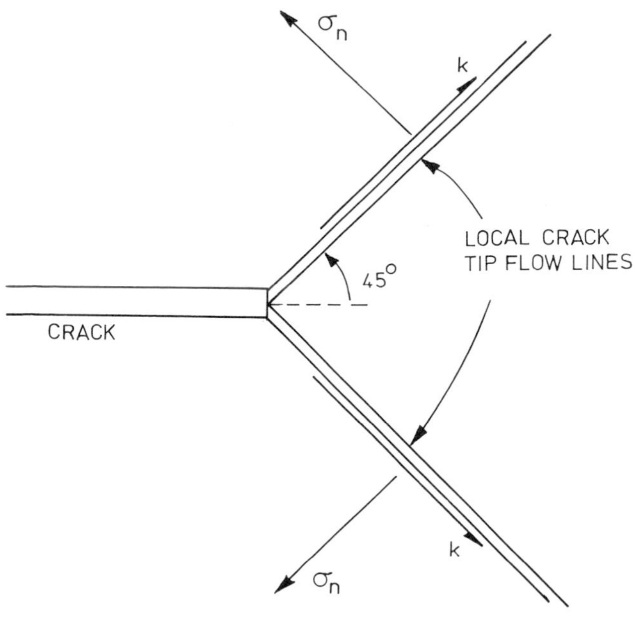

Fig. 6: Simple crack tip flow field relevant to pressure vessel fracture

THE APPLICATION OF LINEAR ELASTIC FRACTURE MECHANICS TO FATIGUE AND FRACTURE OF THICK WALL CYLINDERS

T. G. CRYMBLE, BSc, PhD*, B. D. GOLDTHORPE, BSc, MSc** and B. A. AUSTIN, BSc, PhD*
* Mechanical Engineering Department, The Queen's University of Belfast
** Royal Armament Research and Development Establishment, Sevenoaks, Kent

INTRODUCTION

1. The practical importance of having an understanding of the fatigue and fracture behaviour of thick-wall cylinders is evident both in the commercial and military fields. Thus, for example, the economic viability of hydrostatic extrusion and compaction processes utilizing extremely high pressures is, to a large extent, dependent on designing a vessel with a predictably long life. Similarly, because of the requirements for reduced weight (to facilitate air-transportability) and increased ballistic performance, the margins available to achieve safety and longevity of modern gun tubes have been severely reduced. Because of the higher working stresses, the designers of such vessels are using more-or-less conventional steels, heat-treated to greater yield strength levels than hitherto.

2. By and large, these harder steels do not possess significantly greater resistance to cyclic crack growth and, other things being equal, they exhibit a lowering of fracture toughness. The decrease in critical crack size arising from higher working stresses and reduced toughness is bad enough in so far as it shortens fatigue life; more worrying perhaps, is the possibility of a transition to a dangerous mode of failure. Without a doubt, many of the catastrophic failures of high pressure vessels which have occured in the past, could have been avoided by the application of a reliable procedure for predicting fatigue and fracture behaviour. Such a procedure depends primarily on accurate stress-intensity relationships for pressurized thick-wall cylinders. The determination of these stress-intensity factors and their application in estimating fatigue life constitute the basis of the present paper.

Crack Characteristics

3. The problem of crack initiation cannot of course be dealt with using fracture mechanics. However, for situations where working stresses are extremely high, the crack initiation phase may be relatively short. Indeed, in many cases it may be advisable to assume the existence of a crack like flaw of a magnitude corresponding to that which would just be detected reliably by non-destructive examination. In the case of gun tubes, the initiation phase is virtually non-existent because of the peculiar thermal conditions which give rise to craze cracking at the bore surface within a very few number of rounds.

4. Crack shape is obviously an important parameter. The most common type of vessel fatigue crack is that which originates at a point defect and propagates radially to form a semi-ellipse whose major axis is parallel to the cylinder bore. Because of the complex three dimensional nature of such a crack in a pressurized thick-wall cylinder, a fundamental analysis of this case has not been feasible. Several empirical adaptations e.g. Underwood (1) of the basic equation

$$K_1 = \alpha\, 1.12\sigma\, (\pi a)^{\frac{1}{2}} \qquad (1)$$

relating to an edge cracked plate, have been presented; α is a function of the ratio of minor to major crack axes $\frac{a}{b}$, and also of $\frac{a}{T}$ where T is the wall thickness. However, the authors intend to confine their attention in this paper to the straight front (i.e. constant depth) crack. Apart from the fact that the straight crack represents the limiting case of weakness, it is more amenable to analysis. Hence, a number of experimental and analytical stress-intensity solutions [Bowie and Freese (2), Underwood et al (3), Shannon and Austin (4), Boyle (5)] are already available for the case of a cylinder containing a single crack. Furthermore, the straight-front crack is typical of those which originate at the rifled bore of a gun tube.

5. Perhaps even more important (certainly so for gun tubes) than crack shape is the number of cracks occurring in a vessel. For the long straight-front crack the effect of multiplicity is evident even from the earlier results for multiple cracks radiating uniformly from a hole in an infinite plate [Bowie (6), Westmann (7), Kutter (8)]. In particular, Kutter's results indicate that the worst case is represented by two diametrally-opposed equal-depth cracks. An approximate procedure due to Goldthorpe (9), based on the above infinite plate solutions, has provided stress-intensity calibrations for 10, 20 and 40 equi-distant cracks in a cylinder with a diameter ratio of 2.0. Goldthorpe's 40 crack solution predicts stress intensities very much less than those for the single crack case. Finite element [Shannon and Austin (4)], and boundary collocation [Boyle (5)] computations for two diametrally opposed equal - depth cracks in a cylinder whose diameter ratio W, is 2.0 confirms a substantial increase in stress intensity relative to that for the single crack case. Needless to say, the single and multiple crack solutions tend to converge at small crack depths.

Present Work

6. The research which is now being reported may

conveniently be divided into two sections:

(a) The experimental determination of stress-intensity relationships
(b) The application of stress-intensity factors in estimating fatigue life.

7. Previous experimental work [Shannon and Austin (4)] has involved the performance of fracture tests on long model vessels containing straight-front cracks generated by pressure cycling from shallow starter notches extending nearly the full length of the bore. The direct application of this idea to multiple crack cases was not feasible since it would be virtually impossible to achieve propagation, to a uniform depth of an array of cracks. The method finally arrived at, involved the use of ring specimens, 48 mm bore and 32 mm axial thickness with a diameter ratio of 2.0. The crack from which fracture is intended to occur exists initially as a narrow slot carefully hacksawn to within 3 - 4 mm of the final desired crack length. A fatigue crack is thengenerated from this notch by applying repeated tensile loading along a diameter of the ring normal to slot as shown in Fig. 1. After pre-cracking, further narrow slots are sawn to the depth of the fatigue crack (Fig. 2). Since the contribution of the fully sawn notches is solely to create the required stress distribution in the region of the fatigue crack it is felt that the system of slots behaves effectively as an array of cracks.

8. A special rig for testing these ring specimens has been developed (Fig. 3). Fluid sealing is accomplished by rubber O rings housed in end closures and bearing on the flat end faces of the specimen around a circumference beyond the extent of the slots. Even with the provision of conventional anti-extrusion rings, it is clear that the reliable operation of the sealing system depends on maintaining contact between the specimen and closure faces. Obviously, if no provision is made to prevent the separation which would normally occur due to elastic displacement of the reaction frame, seal failure could occur. In the present design this separation is prevented by incorporating a displacement compensating device. This depends on the creation of a force slightly in excess of that tending to separate the specimen and closure faces. Thus, the elastic displacement of the reaction frame due to the pressure end load is taken up by movement of the compensating piston. Even with this provision slight extrusion of the chamfer ring occurs; this may be due to localized elastic distortion. Overall the ring test equipment has performed very well. Most importantly, pre-cracking from being difficult and tedious is now straightforward and expeditious. Furthermore, the fatigue cracks obtained by direct load cycling (Fig. 4) are straighter than those obtained by pressure-cycling long, shallow-notched cylinders. Specimen manufacture is considerably simplified and substantial economy of material is effected. There is, however, one feature of the ring test which causes some concern. Due to the design of the sealing system, the internal pressure acts on the substantial area of the specimen end faces within the O ring. This axial compression may be significant in encouraging undesirable plasticity effects.

9. Results have been obtained for the cases of one, two, ten and twenty cracks respectively. The ring material was gun-steel hardened by oil quenching from 830°C and tempering at 350°C to give a yield strength of 1400 MN m^{-2} and a fracture toughness of 71.7 MN $m^{-\frac{3}{2}}$. Analysis of this material gave the following typical composition.

C	Si	Mn	Ni	Cr	Mo	S	P	V
.39	.35	.52	3.24	.98	.70	.011	.009	.20

The results for the one and two crack cases are compared in Fig. 5 with the well verified boundary collocation results of Bowie and Freese (2) and Boyle (5). Semi-empirical solutions obtained by Goldthorpe (9) for 10, 20 and 40 cracks are given in Fig. 6 together with the experimental results for 10 and 20 cracks. The data have been nomalized with respect to equation 1, applied to a very long shallow crack where α = 1 and the stress σ is replaced by a term which represents the sum of the internal pressure and the corresponding bore hoop stress. Thus, if equation 1 is correct, for shallow cracks at any rate, all results should tend to a value of 1.0 at very small crack depths.

10. The good agreement between the boundary collocation solutions and the ring fracture results for the single crack case indicates that the ring test is significantly better than that due to Shannon and Austin (4) using long cylinders. Considering the single crack results, the discrepancy between the Bowie-Freese solution and the ring test results is greatest for short cracks. Bearing in mind the A.S.T.M. crack length recommendation for valid K_{Ic} testing, the increase in strength beyond analytically predicted values at low a/T values is hardly surprising. An improvement in the test results for short cracks could only be effected by testing very much larger specimens or using a material with a lower ratio of K_{Ic} to yield strength.

11. One of the most important applications of fracture mechanics is to provide a means of predicting the fatigue and final failure of structures from small scale laboratory tests with the stress intensity calibration providing the essential connecting link. To make these predictions it is necessary to know the fracture toughness of the material and the fatigue crack growth characteristics. Fatigue data is often expressed in the form

$$\frac{da}{dn} = C\,(\Delta K)^m \qquad (2)$$

where ΔK is the range of applied stress-intensity. A relationship of this form gives a good fit to experimental data over the middle range of ΔK values but is not normally very good at high or low values. For a number of gun-steels tested an expression which has been found to be applicable over a wider range of ΔK values has the form

$$\frac{da}{dn} = Ae^{B\Delta K} - C \qquad (3)$$

where $\frac{da}{dn}$ is in mm/cycle, ΔK is in $MNm^{-\frac{3}{2}}$

and $A = 40.6 \times 10^6$ $B = 0.064$ $C = 61.0 \times 10^6$

Expressing the K calibration in the form

$$K_I = P\,\sqrt{a}\,f\left(\frac{a}{T}\right) \qquad (4)$$

and substituting K_I, for ΔK in eqtn (3) (for the case where the stress-intensity ranges between 0 and K_I) and integrating given the fatigue life

$$n = T\int_{a_o}^{a_c} \frac{d\,\frac{a}{T}}{A\exp\left[BP\sqrt{a}\,f\left(\frac{a}{T}\right)\right] - C}$$

where a_o is the initial flaw depth and a_c is the critical crack depth obtained by substitution of the K_{Ic} value in equation (4).

12. These relationships have been used to demonstrate the effect of crack configuration and initial crack depth on fatigue life shown in Fig. 7. These results are based on a pressure of 334 MNm^{-2} and a K_{Ic} of 110 $MNm^{-\frac{3}{2}}$, the cylinder diameter ratio and internal diameter were 2.0 and 101.6 mm respectively. It can be seen that for any initial crack depth the fatigue life for a single crack is much shorter than the fatigue life for forty cracks, the factor varying between 4 and 25 depending on the initial crack depth. The effect of initial crack depth is also shown to be important; for a single crack, the fatigue life is extremely sensitive and for 40 cracks much less sensitive to initial crack depth. These results are particularly critical in an application such as a gun tube where craze cracking occurs to a depth of 0.5 to 1.5 mm and where the number of cracks may vary, depending on the operating conditions. Thus, the fatigue life may change dramatically simply from the effects of crack geometry without consideration of material properties.

13. The effect of K_{Ic} on fatigue life is shown in Fig. 8. This illustrates that K_{Ic} has very little effect on fatigue life for the practical range of K_{Ic} values 80-140 $MNm^{-\frac{3}{2}}$ with either of the crack configurations considered. It is only when the K_{Ic} falls to extremely low levels that a serious reduction in fatigue life is observed. There are two important implications of these results. In the first instance, with the normal range of toughness values, the most important effect of high toughness is to prevent catastrophic fracture; and ideally ensure a leak-before-break failure. Secondly, under conditions where serious embrittlement of the steel has occured, the resulting large reduction in toughness will in addition, cause a substantial deterioration in fatigue life.

14. It is evident from the stress intensity calibrations presented in this paper that the mode of failure is controlled not only by material properties but also by geometric considerations. This is demonstrated in Fig. 9 where the variation in critical crack depth with K_{Ic} value is shown for two crack models. The internal pressure is again taken as 334 MNm^{-2}, the internal diameter as 101.6 mm and the diameter ratio as 2.0. The extremely large effect of crack configuration on critical crack depth is apparent; it is important to note that for this example there is no possibility of obtaining a leak-before-break failure with a single crack. With multiple cracking, however, it is certainly possible to approach leak-before-break within a practical range of toughness values.

15. Conclusions

Analytical and semi-empirical K-calibrations for thick wall cylinders have been verified by a new experimental technique. This technique has proved to be relatively simple to operate and to have many advantages and applications as a testing technique.

16. The K calibrations available for thick wall cylinders have been shown to be extremely sensitive to the number of cracks present. The fatigue life and failure mode are therefore often critically dependent on the actual crack configuration and this must be a major consideration in assessing the possible fatigue and failure of thick wall pressure vessels.

References

1. Underwood, J. M. Stress-intensity factors for internally pressurized thick-wall cylinders. ASTM Spec. Tech. Publ. 513, 1972, p.59.

2. Bowie, O. L. and Freese, C. E. Elastic analysis for a radial crack in a circular ring, Engineering Fracture Mechanics, 1972, Vol. 4, p.315.

3. Underwood, J. M., Lasselle, R. R., Scanlon, R.D. and Hussain, M. A. A compliance K calibration for a pressurized thick-wall cylinder with a radial crack, Tech. Report WVT-7026, 1970, Watervliet Arsenal, Watervliet, N.Y.

4. Shannon, R. W. E. and Austin, B. A. Paper C32/71, in Practical applications of fracture mechanics to pressure vessel technology, Inst. of Mech. Engs., London, 1971.

5. Boyle, E. F. The calculation of elastic and elastic-plastic crack extension forces, Ph.D. Thesis, The Queen's University of Belfast, 1972.

6. Bowie, O. L. Analysis of an infinite plate containing radial cracks originating at the boundary of an internal circular hole, Jn. of Maths and physics, April 1956, Vol. 25, p.60.

7. Westmann, R. A. Pressurized star crack, Jn. of Maths and Physics, July 1964, Vol. 43, p.191.

8. Kutter, M. K. Stress analysis of a pressurized circular hole with radial cracks in an infinite elastic plate, Int. Jn. of Fracture Mech. 1970, Vol. 6, No. 3, p.233.

9. Goldthorpe, B. D. Paper 17, British Steel Corporation Conference, Mechanics and Mechanisms of Crack Growth, Cambridge, 1973.

Fig. 1: Ring fatigue pre-cracking arrangement

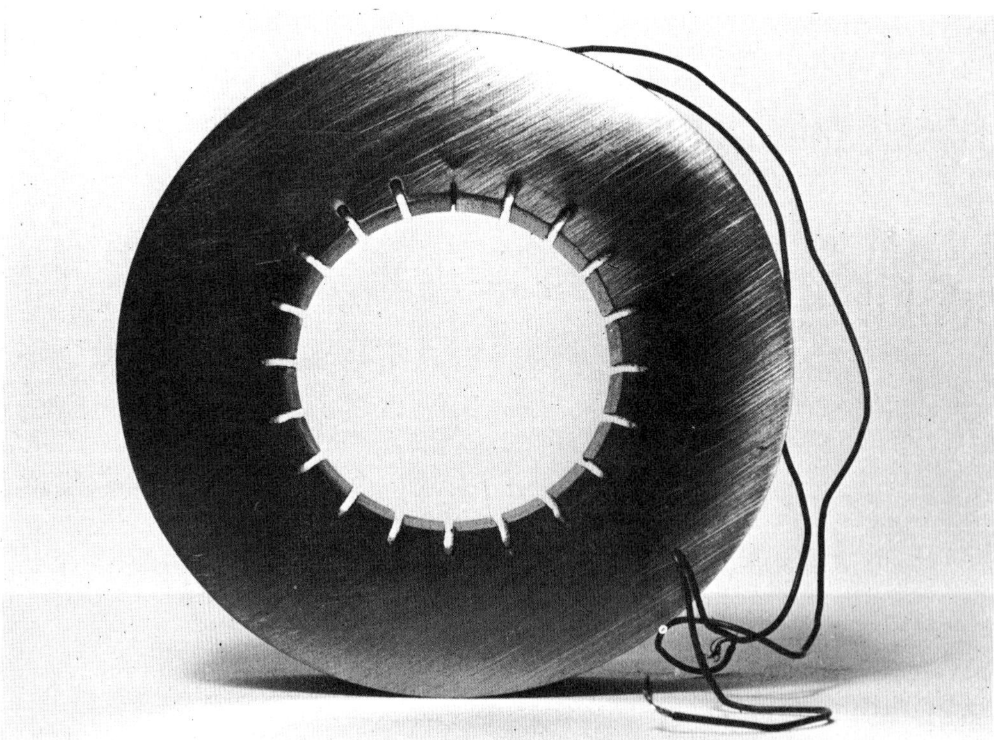

Fig. 2: An example of a multiply cracked ring

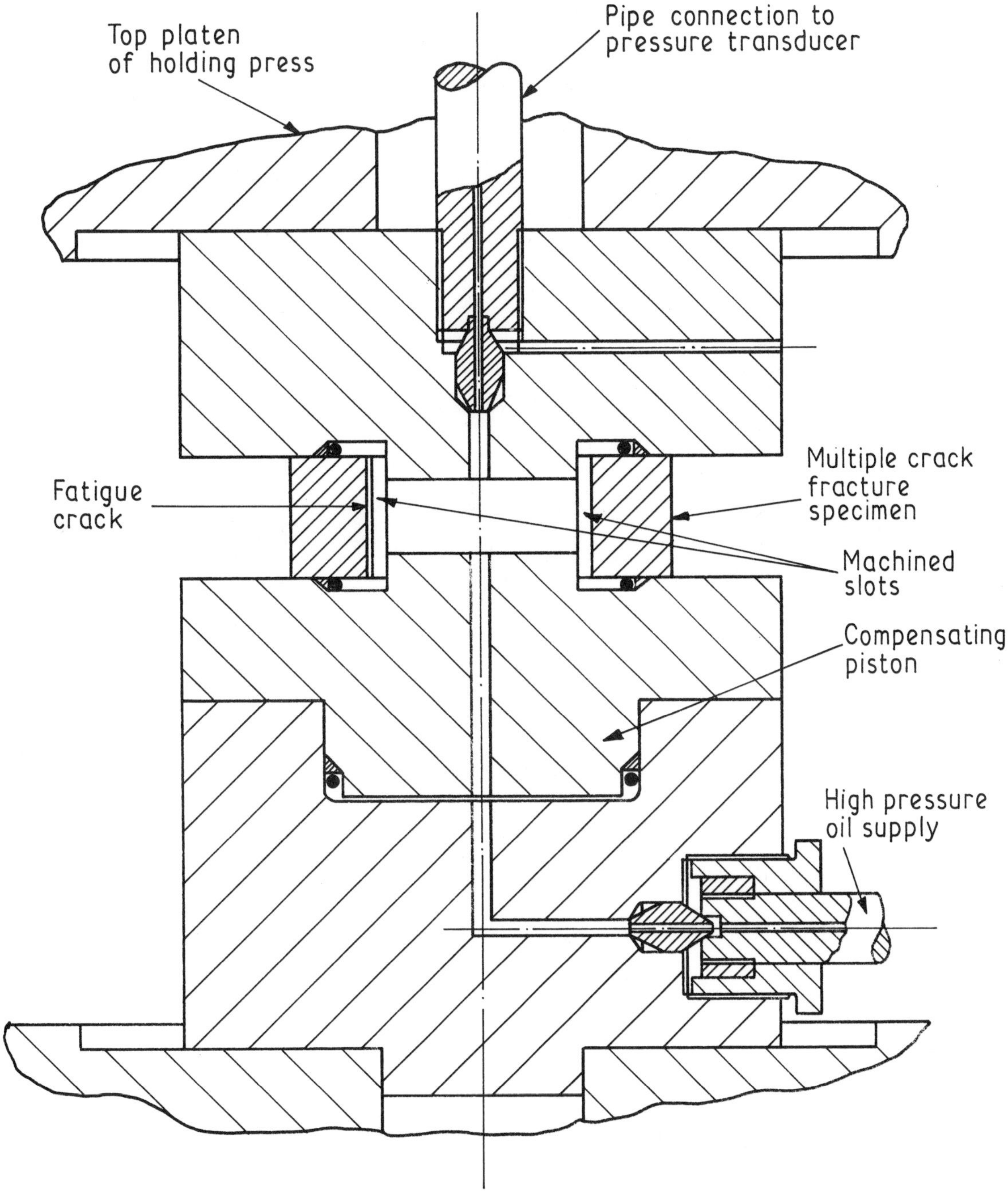

Fig. 3: General arrangement of the ring fracture test rig

Fig. 4: Examples of the fatigue cracks generated in ring specimens

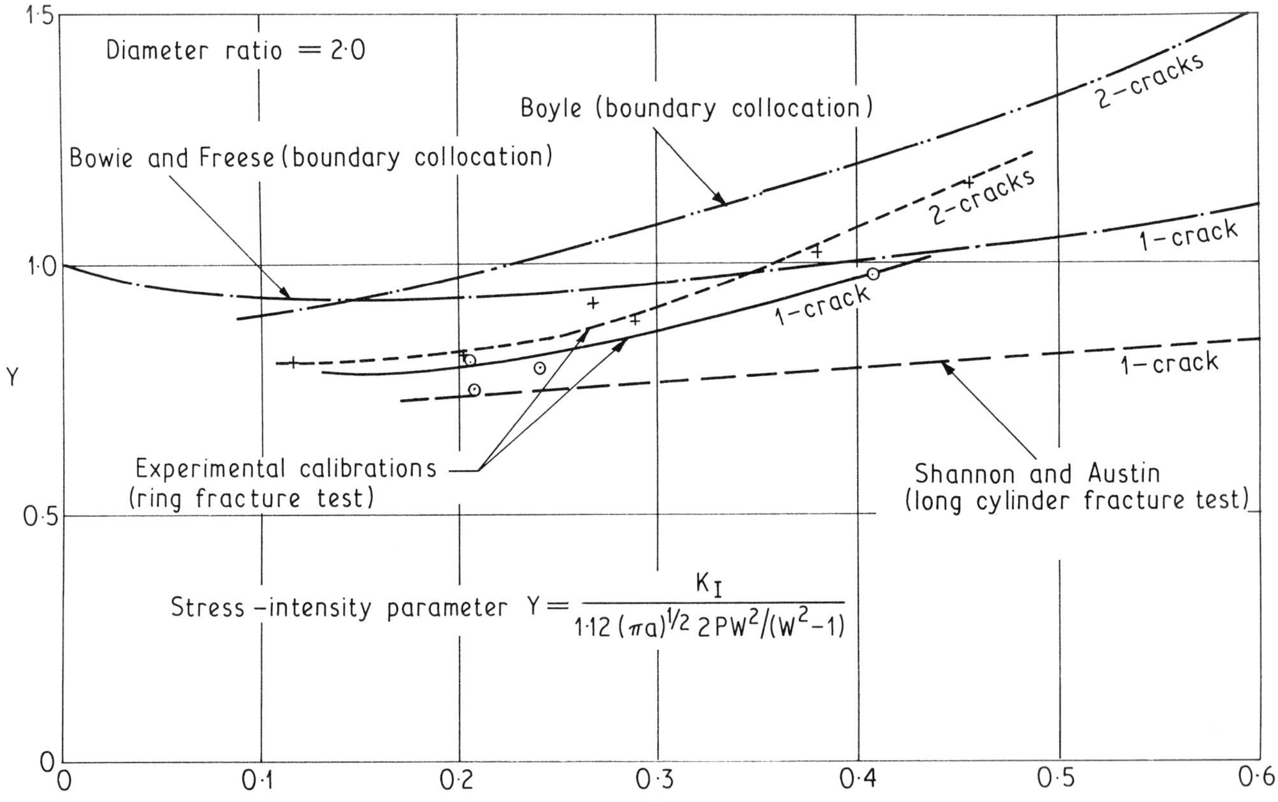

Fig. 5: Comparison of stress-intensity calibrations for cylinders containing one or two straight-front bore cracks

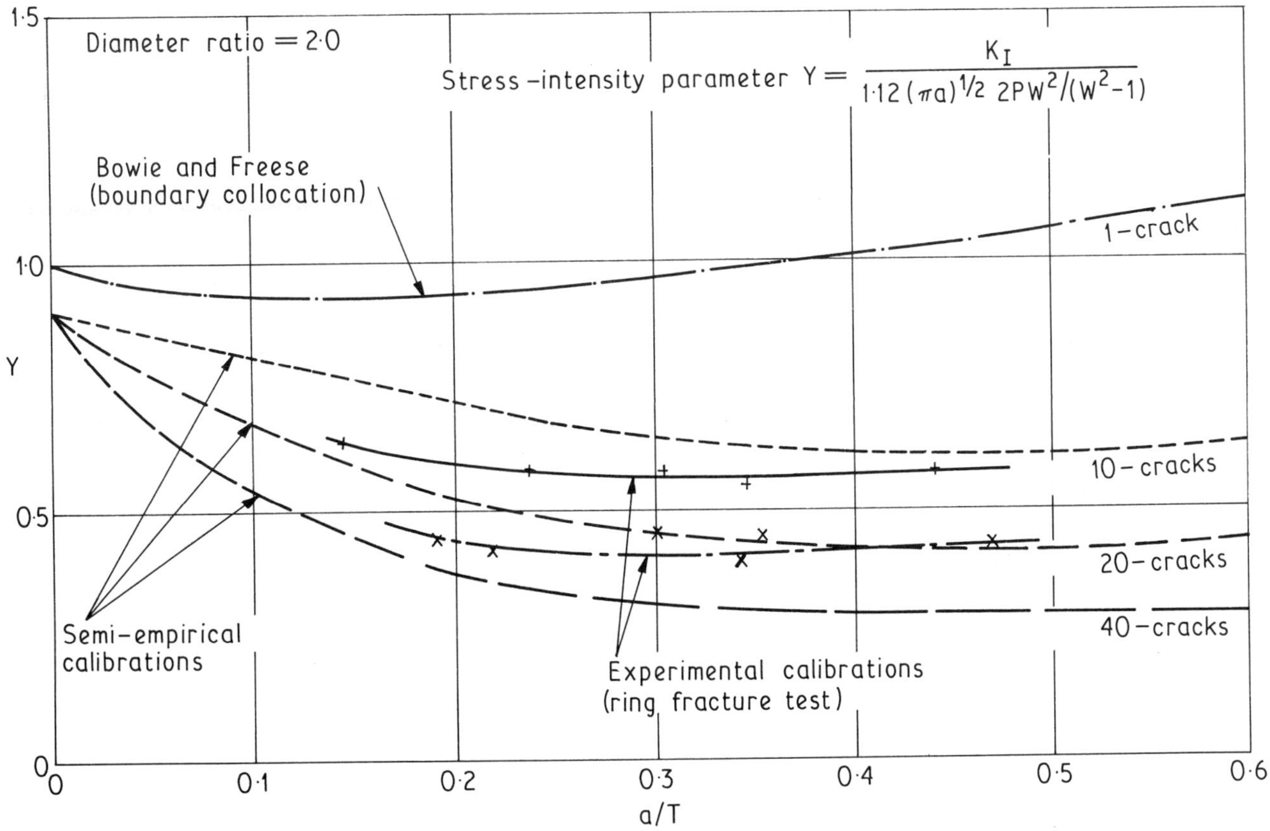

Fig. 6: Comparison of stress-intensity calibrations for cylinder containing multiple bore cracks

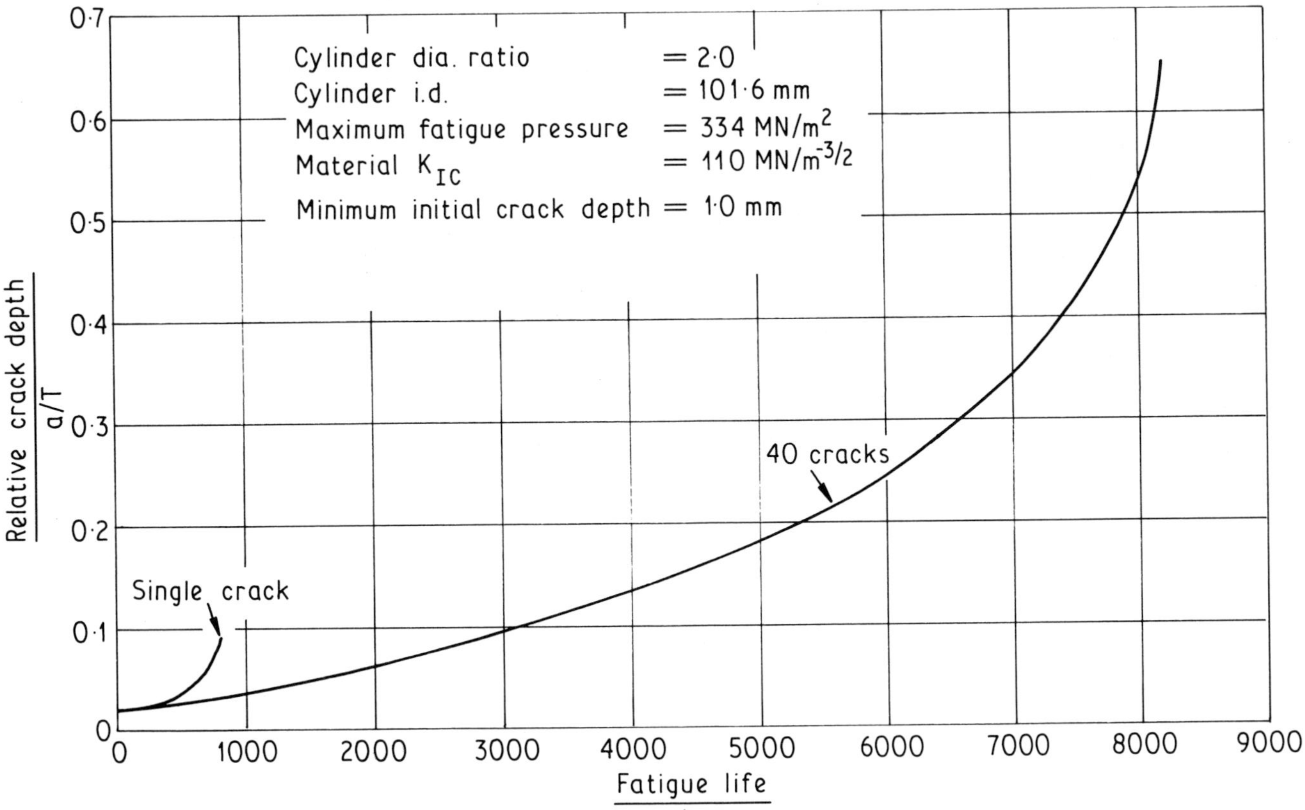

Fig. 7: Fatigue crack growth

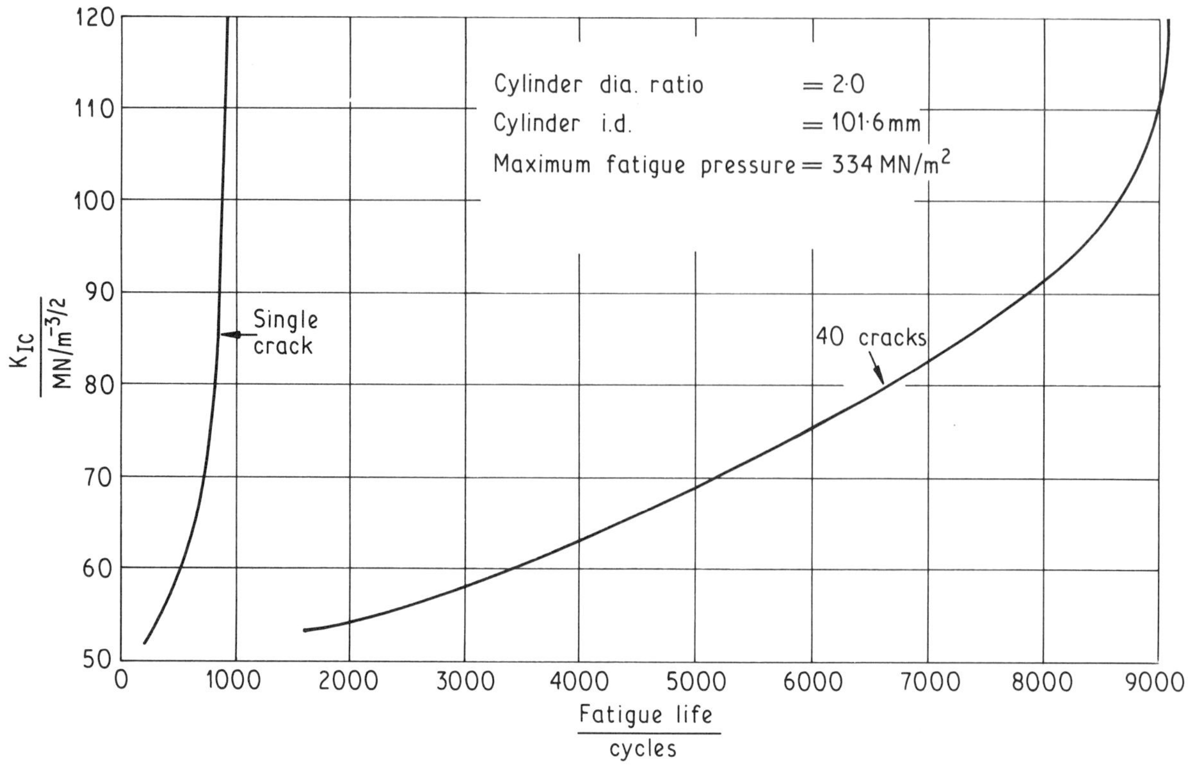

Fig. 8: Effect of K_{IC} on fatigue life

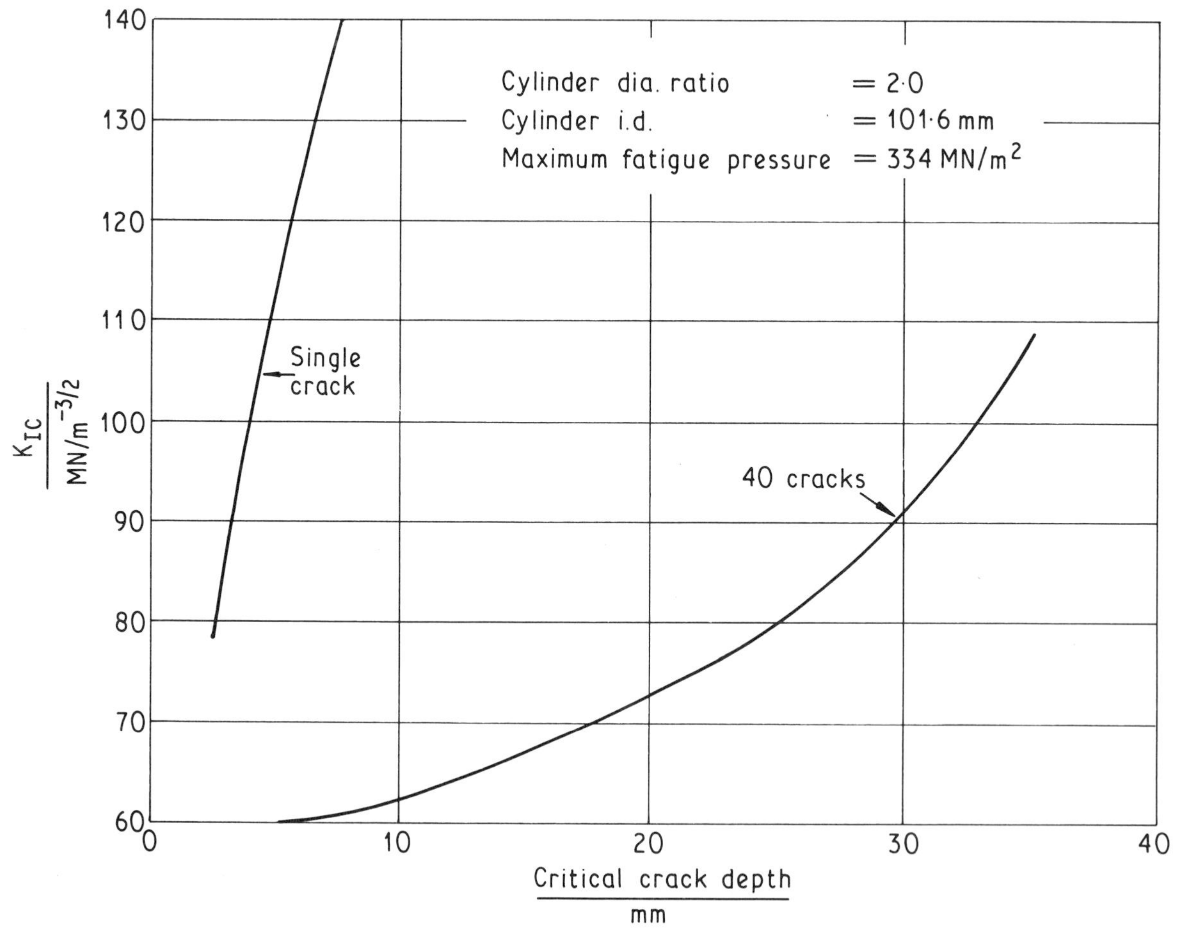

Fig. 9: Effect of K_{IC} on critical crack depth

A COMMENT RELATED TO THE CRYMBLE, GOLDTHORPE AND AUSTIN RING FRACTURE TESTS

J. H. UNDERWOOD, MSc
US Army Benet Weapons Laboratory, Watervliet, NY, USA

1. The ring fracture test is a useful test for determining the fracture pressure of cylinders with straight cracks, and it is certainly easier to perform than conventional tests. The only problem with the test is that mentioned by the authors, that is the effect on the fracture process of the internal pressure which acts on the end faces of the specimen. During the discussion sessions at the Second International Conference on High Pressure Engineering, Dr Austin, Dr Tomkins and Prof Crossland all expressed concern related to the general problem of the effects of pressure on crack-related fracture. The objective here is to consider one aspect of the pressure effect in the ring fracture tests reported by Crymble, Goldthorpe and Austin and to indicate how this consideration can apply to other pressure vessel fracture problems.

2. The effects of pressure on crack-related fracture can be divided into at least three categories. First, the direct mechanical effect of the pressure acting on the crack faces is quite well understood. For geometries with pressure loading and for which a reliable elastic K_I solution is available, such as the cylinder problem solved by Bowie and Freese (ref. 2), the mechanical effect of pressure in the crack is included in the solution. Second, pressure can no doubt change the micro-mechanisms of deformation and fracture at the crack tip and thus change the observed macroscopic fracture behaviour of materials. A third category of pressure effects on fracture and the area of interest here is the different macro-yielding behaviour which can occur in pressure loaded specimens compared with uniaxially loaded specimens. The effect of the different yielding behaviour on the crack-tip plastic zone size can cause significant changes in fracture behaviour.

3. The axial compression present in the ring tests had a significant effect on the yielding behaviour of the ring specimen and thus on its fracture behaviour. This can be demonstrated as follows. Plane strain K_{Ic}-type fracture occurs only when the crack-tip plastic zone is small relative to crack size and other specimen dimensions. The expression for plastic zone size, r_y for plane strain conditions is

$$r_y = \frac{1}{6\pi}\left\{\frac{K_I}{(\sigma_\theta)_y}\right\}^2 \qquad (1)$$

where K_I is the opening mode stress intensity and $(\sigma_\theta)_y$ is the tangential stress which causes yielding in the ring specimen. The uniaxial yield strength for the material can not be used directly in equation (1) because of the axial compression present in the specimen. Using the maximum shear stress yield criterion, for simplicity, gives

$$\frac{(\sigma_\theta)_y - \sigma_z}{2} = \frac{\sigma_y}{2}$$

and this leads to

$$(\sigma_\theta)_y = \sigma_y - P \qquad (2)$$

where σ_y is the uniaxial yield stress and $P = -\sigma_z$ is the axial compressive stress. Equations (1) and (2) were used to calculate r_y using the values of K_{Ic}, σ_y and P from the Crymble, Goldthorpe and Austin paper. The values of r_y normalized by crack length are compared in the Table below with the ratio of Y measured in the ring fracture tests to Y taken directly from the K_I solutions. The dimensionless parameter Y is the same as used by Crymble *et al.*, that is

$$Y = \frac{K_I}{1{\cdot}12(\pi a)^{\frac{1}{2}} 2PW^2/(W^2-1)}.$$

	One crack		Two cracks	
$\frac{a}{T}$	$\frac{Y_{meas.}}{Y_{ref.\,2}}$	$\frac{r_y}{a}$	$\frac{Y_{meas.}}{Y_{ref.\,5}}$	$\frac{r_y}{a}$
0·1	0·88	0·102	0·91	0·103
0·2	0·83	0·043	0·84	0·041
0·3	0·85	0·026	0·84	0·025
0·4	0·96	0·018	0·88	0·017
0·5	1·07	0·013	0·96	0·013

The significant feature of the comparison is that as the size of the plastic zone decreases relative to crack length the measured fracture behaviour approaches that expected from the elastic K_I solutions. A minimum crack length requirement for a K_{Ic} test is included in the British and the American K_{Ic} test method and is

$$a > 2{\cdot}5\left[\frac{K_{Ic}}{\sigma_y}\right]^2.$$

Combining this requirement with equation (1) results in

$$\frac{r_y}{a} < 0{\cdot}022.$$

Note that the four test results with $r_y/a < 0{\cdot}022$ are in good agreement with the expected result.

4. The pressure effect in the ring tests can be summarized as follows. The axial pressure on the specimen favours yielding in the tangential direction and produces a larger than expected plastic zone. This then leads to the higher than expected fracture pressure, particularly for tests with short cracks. This same general phenomenon can also occur due to the radial compressive stress present in a pressurized cylinder. The tangential stress required for yielding of a pressurized cylinder can be written, using the shear stress criterion, as

$$(\sigma_\theta)_y = \sigma_y - P\left\{\frac{(r_2/r)^2 - 1}{(r_2/r_1)^2 - 1}\right\} \qquad (3)$$

where the second term is the radial stress and r_2, r_1 and r are outer radius, inner radius and radial position in the wall, respectively. When $(\sigma_\theta)_y$ from equation (3) is used to determine r_y from equation (1), the value of r_y can be much larger than that for the uniaxial stress situation, and the larger plastic zone will cause a higher than expected fracture pressure in a cylinder. A rough guide for estimating the higher than expected fracture pressure of cylinders is the following, based on the discussion here and on K_{Ic} test results from similar materials. For $r_y/a \cong 0{\cdot}02$ fracture pressures of up to 10% above those calculated from known K_{Ic} values can be expected. For $r_y/a > 0{\cdot}02$ larger increases in fracture pressure can be expected.

CLOSURES

FATIGUE OF THREADED END CLOSURES OF HIGH PRESSURE VESSELS

M. G. GRAY, Research Assistant
R. J. PICK, Associate Professor
Department of Mechanical Engineering, University of Waterloo, Waterloo, Ontario, Canada

ABSTRACT

The results of fatigue tests on a series of AISI 4340 high pressure cylinders with threaded end closures are presented and compared to the stress levels predicted by a theoretical analysis. Buttress and V threads with an undercut ahead of the first thread were studied during the crack initiation and propagation stages. The load distribution in the threads was predicted using the analysis technique of D.G. Sopwith, the stress levels were subsequently predicted using a finite element technique.

1.0 INTRODUCTION

The advantages of threaded end closures of high pressure vessels are numerous; low cost, simplicity of design and manufacture. However, detailed stress calculation and prediction of fatigue life is not possible using simplified techniques more applicable to nut-bolt connections. Early work by Sopwith (1) supported by the experimental data of Gray (2) allows the most detailed calculation of thread loading by considering the thread as a tapered cantilever. Although this technique appears attractive in predicting the thread load distributions, extension to predict stress levels is very approximate and will not allow consideration of details used to reduce stress levels in critical regions, e.g. an undercut ahead of the first thread.

A finite element analysis can be applied to screw threads to provide a more accurate analysis of stress levels and to consider cylinder modifications.

This paper will compare the load and stress levels predicted by Sopwith's technique and finite element analysis with the fatigue lives of a series of cylinder-threaded plug fatigue specimens. It is intended to be a preliminary report of a study in progress to investigate a technique for the prediction of the fatigue life of high pressure screw threads.

2.0 EXPERIMENTAL INVESTIGATION

To determine the fatigue life of a screw thread representative of those used in high pressure applications the specimen of figure 1 has been designed. The thread is loaded by the internal pressure generated within the thick walled cylinder through the stroking of a ram driven by a hydraulic actuator (figure 2). The ram is cycled between pressure limits, the pressure being sensed by strain gauges on the cylinder. The maximum design pressure is 50,000 psi, the lower minimum 2,000 psi, a slight pressure always being retained to prevent seal leakage. Seal leakage is minimal, however and recharging of the system is easily accomplished through a check valve mounted in the plug (figure 3).

The fatigue specimens were rough machined from a 3 inch diameter rod of AISI 4340 steel. The specimens were heat-treated to a Brinell hardness of 331 and a reported yield strength of 160,000 psi. After heat treatment the threads were precision ground to ASME standards. Care was taken to maximize the thread root radius. A small undercut was machined ahead of the first thread, the detailed dimensions are shown on figure 4. Both Buttress threads and 60° V-threads were tested, between pressures of 15,000 psi, 20,000 psi and 30,000 psi maximum and 2,000 psi minimum. Four strain gauges were mounted 90° apart on the outer diameter of the cylinder over the first thread. These were monitored continuously during testing and gave an indication of the initiation of a fatigue crack. Visual inspection for the crack during interruptions in the test were unsuccessful. However striations on the fracture surface, indicating crack growth, always correlated with strain gauge indications. It is not known how sensitive the strain gauges are to the crack initiation or the minimum detectable crack size. However plotting of the strain levels allows extrapolation to predict the point of crack propagation. Figure 5 shows a typical plot.

Replicas of the threads were made from a mixture of carbon black and sulphur before the specimens were tested. These will be used in future studies to determine the influence of machining tolerances and will be compared with the threads after testing to determine if gross plasticity has occured. For pressures of 20,000 psi and 30,000 psi the plug was difficult to remove indicating plastic distortion of the threads; tests at 15,000 psi gave little if any plasticity.

3.0 FINITE ELEMENT ANALYSIS

A finite element technique developed by Pick and Burns for the analysis of screw threads has been described previously (3). Through the use of a special interaction analysis at the thread interface the thread load distributions as well as stress levels can be predicted from an exact model of the thread-plug geometry. Although accurate this approach had several shortcomings; it requires considerable preparation of input data and long computing times. Of more importance, because of the numerous elements used in the model and the large computer memory required, only five or six threads can be represented. This appeared sufficient for Buttress and Acme threads as the majority of the load is transferred across the first few threads. However, it appears that when the thread has a large flank angle such as a V-thread or reverse buttress thread, the increased sliding between the cylinder and the plug distributes the load more evenly along the thread. Representing this with only five or six threads in contact was not adequate.

This finite element approach has been modified to make use of a matrix reduction scheme devised by one of the authors (4). Use of the reduced matrices in a partitioning program (5) allows the consideration of more threads while reducing the computing costs and data preparation.

However from the point of view of a designer such an approach is too complex at a preliminary design stage. A simpler approach as described in this paper predicts the load distribution in the threads using the analysis of D.G. Sopwith (1) and the stress levels using an ordinary finite element approach.

Sopwith approximated the thread by a tapered cantilever beam of width equal to the helical length of the thread. Although such an analysis does not allow for bending in the cylinder or nut or local geometric details it is easily used and must be considered a useful design tool. Load distributions predicted by the complete finite element model of Pick and Burns (3) and experimentally by Gray (2) and Warnke (6) show reasonable agreement. This would indicate that the detailed geometry of the plug-cylinder has only a minor inlfuence on the thread load distribution. Unfortunately prediction of stress levels by Sopwith's analysis requires the use of a stress intensification factor not readily available. As an alternative the authors have used a simple finite element analysis of the cylinder applying the loads predicted by Sopwith's analysis to predict the stress levels. To properly model the detailed geometry of the threads numerous elements are required and therefore a simple constant strain axisymetric triangular element has been used. The threads are modelled as axisymetric rings. Although approximate such programs are common place and are easily used by the designer. In this study the total analysis contained 2083 elements, 1170 nodes with elements as small as .007 inches concentrated in each thread. The following approximations have been used.

a) As the elements represent constant stress/strain this was assumed to act at the element centroid. Stress levels at the surface of the material can be obtained by extrapolating the centroid stress values to the surface.

b) The load predicted by Sopwith's analysis is continuous along the thread helix. It was averaged into a constant axisymetric point load for each axisymetric ring representing a thread.

c) The load on the thread was assumed to be a point load acting at the mid-depth of the thread.

The authors are currently studying the sensitivity of the solution with respect to these assumptions and the cylinder geometry. It is hoped that a much less refined finite element model will still give useable results.

4.0 RESULTS

Table 1 gives the results of the fatigue tests showing the number of cycles to crack initiation and final fracture. Figures 6 and 7 show a typical fracture which was consistent in size, shape and point of initiation in all specimens except specimen number 7.

The thread begins at the undercut and is of reduced cross-section up to one thread pitch from the undercut; 360° on the helix from the undercut. The fatigue crack initiated 180° beyond this point which can be considered as being ahead of the first full profile thread, half way along its helix. Specimen number 7, the exception, had the fatigue crack initiate 720° from the start of the thread helix. This point can be considered as being ahead of the second full thread profile at the beginning of its helix, as the first thread has a full profile at this position.

Using Sopwith's analysis, loads of 1.95 and 2.75 times the average thread load were calculated to be acting on the first thread of the V and Buttress thread respectively. Considering these loads the finite element analysis predicted the maximum stresses at the point where the fatigue crack appeared to initiate. The stress levels for the element at this point are shown in figure 8 for an internal pressure of 15,000 psi.

For pressures of 20,000 psi and 30,000 psi the finite element analysis will increase the stress levels proportionally to pressure as the analysis is linearly elastic. Stresses above yield would be apparent locally at the root of the first thread. An elastic-plastic analysis is possible, but not justified. It is doubtful if cyclic plasticity could be accurately modelled due to the excessive computing costs and a lack of knowledge of the Bauschinger Effect in this material. If elastic shakedown occurs then the fatigue behaviour will be primarily influenced by the stress/strain ranges which can be considered elastic after the first cycle. An elastic plastic finite element analysis of the first cycle of loading could be used to predict mean stresses assuming elastic unloading. However as the effect of mean stress is secondary this is not justified and the mean stress can be estimated simply by assuming an elastic-perfectly plastic material.

The stresses shown on fig 8 must be considered as those influencing crack initiation. The crack propagation stage will be influenced by the general stress field into which the crack is growing. In this example the Buttress thread had, in general, higher stress levels suggesting faster crack propagation. However, further studies are required on the effect of the crack on the general stress distribution before this can be stated conclusively.

5.0 DISCUSSION

Studying figure 8 one must conclude that the V-thread has lower stress levels. This appears to be primarily due to the reduced load level on the first V-thread as predicted by Sopwith's analysis. The Buttress thread has a load on the first thread approximately 40% higher than the V thread. The stress levels are approximately 45% higher suggesting that the primary effect of thread geometry is on load distribution as opposed to stress concentration at the root of the thread. This conclusion was also reached by Warnke (6) considering the load and stress distribution predicted by Sopwith's technique.

The lower load and stress levels in the V thread can be considered as a summation of three effects; a larger root radius giving lower stress concentrations; the radial component of the tooth force creating compressive stresses; the greater flank angle and larger tooth flexibility allowing more load sharing amongst the teeth through axial movement of the plug with respect to the cylinder. The latter appears to be the most important aspect and the most difficult to treat analytically.

As the load predicted by Sopwith's method is critical to the analysis, the effects not modelled by this technique must be discussed. The undercut ahead of the first thread is not represented by Sopwith. However, in this experimental study the undercut was small and not considered important with respect to load distribution. Pick and Burns (3) indicate that only a large undercut has a significant effect on the thread load distribution in a Buttress thread.

Bending in the cylinder due to the thread loads and internal pressure is also not represented by Sopwith's analysis. Those threads with large flank angles such as the V thread will have increased bending in the cylinder due to the larger forces created. This bending will produce some axial bending in the threads but primarily cause radial displacements, allowing axial motion of the plug and redistribution of the load in the thread. Although these effects are complex, one would believe that increased movement between the plug and the cylinder would allow an improvement in the load distribution. The displacements predicted by the finite element analysis indicate larger radial displacements for the V thread which should be beneficial in reducing the loads in the first threads.

The prediction of loads by Sopwith's analysis considers the thread a perfect helix. However Pick and Burns (3) indicate that errors in the thread helix or pitch have a significant effect on the load distribution. In the region of the first thread there exists the possibility of higher machining tolerances as the thread and the undercut are produced by different machining operations. The critical region in which crack initiation occurs is usually in this transition region between the undercut and thread profile. Although the authors have taken great care to eliminate machining tolerances they undoubtedly are responsible for some experimental scatter. Work is currently underway to investigate this effect.

In general, from the analysis and considering the assumptions made, the V thread should have the best fatigue life due to the lower thread loading. Experimental investigations (7) of the fatigue of thread like projections under similar loads reinforce this statement, concluding that for similar loads, the V thread has a longer fatigue life than the Buttress.

However, even allowing for the expected experimental scatter, the fatigue results of Table 1 do not indicate this advantage of the V-thread. At high pressure levels (30,000 psi) there appears to be little difference in the fatigue life of V threads or buttress threads. Considering the stress levels at this pressure, cyclic plasticity probably occurs eliminating any effect of thread geometry and stress concentrations. However at lower pressures, elastic shakedown probably occurs and the fatigue life is primarily influenced by the cyclic stress range after the first cycle. At 20,000 psi and 15,000 psi one would expect some indication of longer fatigue lives in the V threads.

As experimental scatter in the data can be due to the nature of low cycle fatigue and the effect of machining tolerances, one cannot conclude that the analysis technique is invalid. However it is surprising that the Buttress thread always had a longer fatigue life. This is even more prominent if one considers that the failure of Specimen 7 was not representative possibly due to machining tolerances.

It is hoped that further experimental data will clarify this discrepancy and studies are continuing particularly with respect to machining tolerances.

CONCLUSION

An experimental investigation of the fatigue lives of both Buttress and V threads indicates that unless gross plasticity occurs the buttress has a similar or longer fatigue life than the V thread. Traditionally the Buttress thread has been assumed to be the best thread for high axial load applications. However, the prediction of the thread loads using Sopwith's technique and the stress levels by finite element techniques indicate that the buttress thread is more highly stressed than the V thread. This contrary prediction must be investigated and resolved before one can hope to predict the fatigue life of a high pressure screw thread by this method.

REFERENCES

1) Sopwith D.G.: The Distribution of Load in Screw Threads, Proc. Instn. Mech. Engrs., Applied Mech. Proc. Vol. 159, 1948, p. 373.

2) Gray, R.M.: Preliminary Investigation of Load Distribution in Threaded Connections, Report No. WAL 730/562-47, 1954, Purdue Research Foundation, Contract No. DA 11-022-ORD-17.

3) Pick, R.J. and Burns D.J.: "Finite Element Analysis of Threaded end Closures of Thick-Walled Vessels", Proc 3rd International Conf. on High Pressure, Aviemore, Scotland, 1970.

4) Pick, R.J.: Computer-aided Design of Thick-Walled High Pressure Vessels, Ph.D. Thesis, University of Waterloo, 1970.

5) Zienkiewicz, O.C.: The Finite Element Method, McGraw Hill, 1971.

6) Warnke, E.G.: Design and Fatigue Life of Openable High Pressure Vessels, Proc. Instn. of Mech. Engrs., Vol. 182, Part 3C.

7) Weigle, R.E., Lasselle, R.R., Purtell, J.P., Experimental Investigation of the Fatigue Behaviour of Thread-Type Projections, Experimental Mechanics, May 1963.

TABLE 1

TEST	THREAD TYPE	PRESSURE RANGE (PSI)	NUMBER OF CYCLES TO CRACK INITIATION	NUMBER OF CYCLES TO FAILURE
1	V-thread	2,000 - 30,000	7,500	23,377
2	V-thread	2,000 - 30,000	7,000	22,335
3	Buttress	2,000 - 30,000	8,000	23,972
4	Buttress	2,000 - 20,000	66,000	208,192
5	V-thread	2,000 - 20,000	26,000	74,197
6	V-thread	2,000 - 20,000	25,000	79,885
7*	Buttress	2,000 - 20,000	35,000	76,130
8	Buttress	2,000 - 15,000	No Crack Initiation at 2.5×10^6 cycles	
9	V-thread	2,000 - 15,000	76,000	367,985

*Failure ahead of second thread

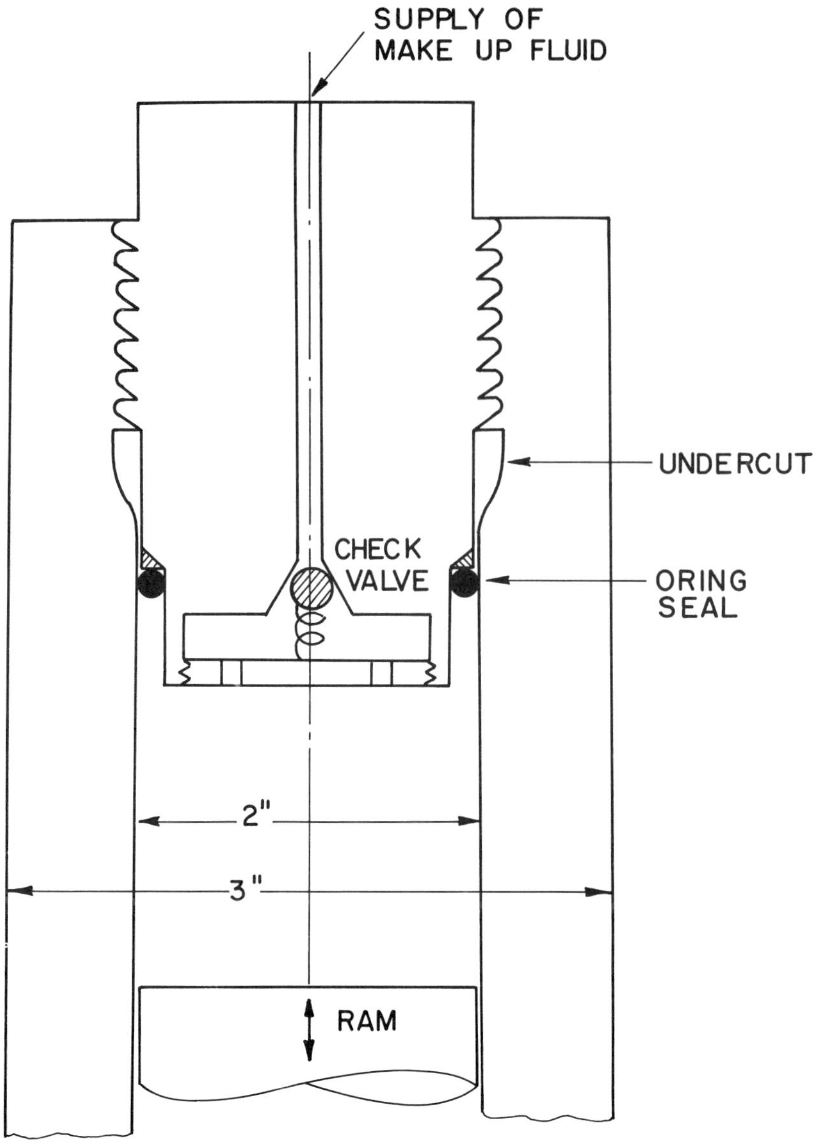

Fig. 1: Threaded end closure of thick walled cylinder

Fig. 2:

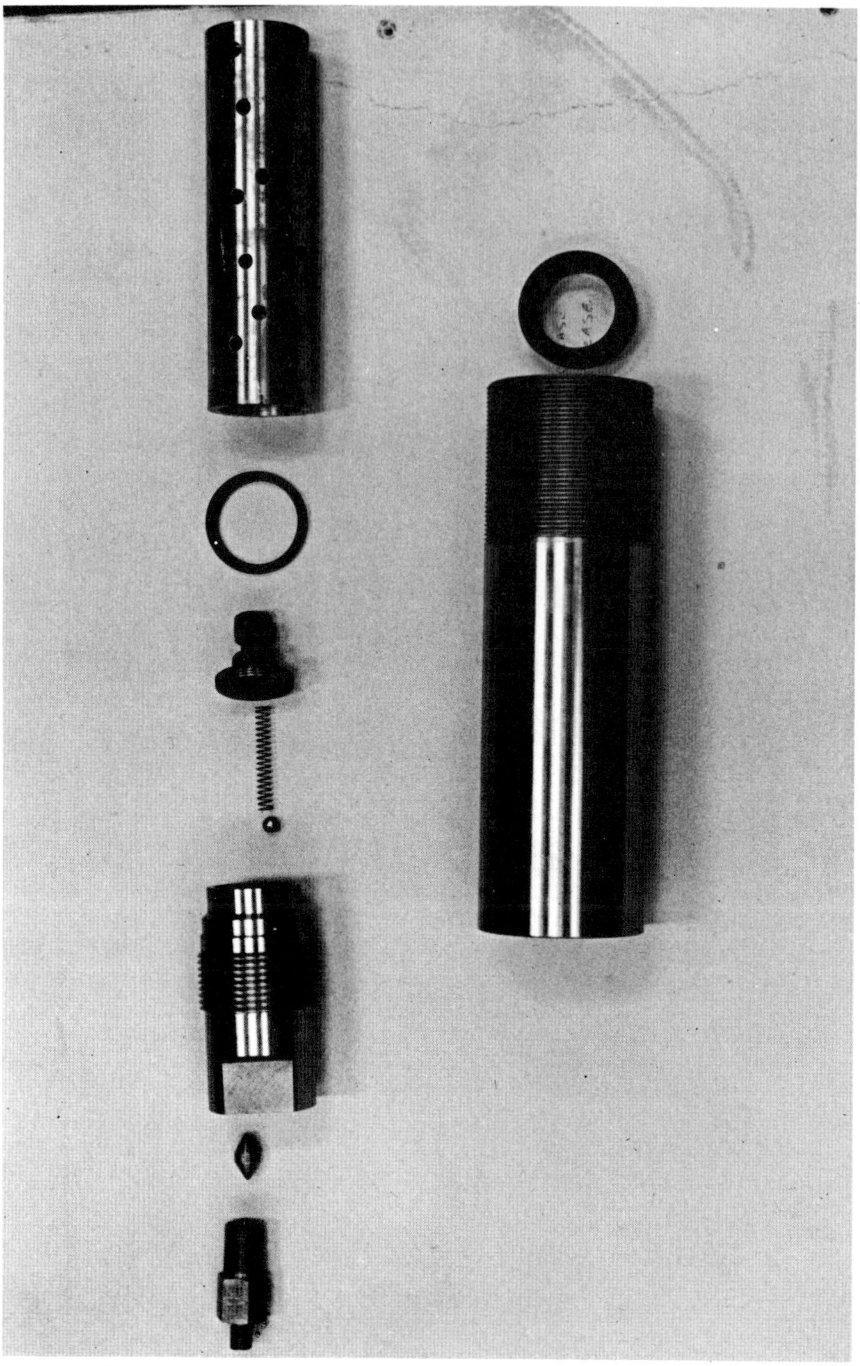

Fig. 3:

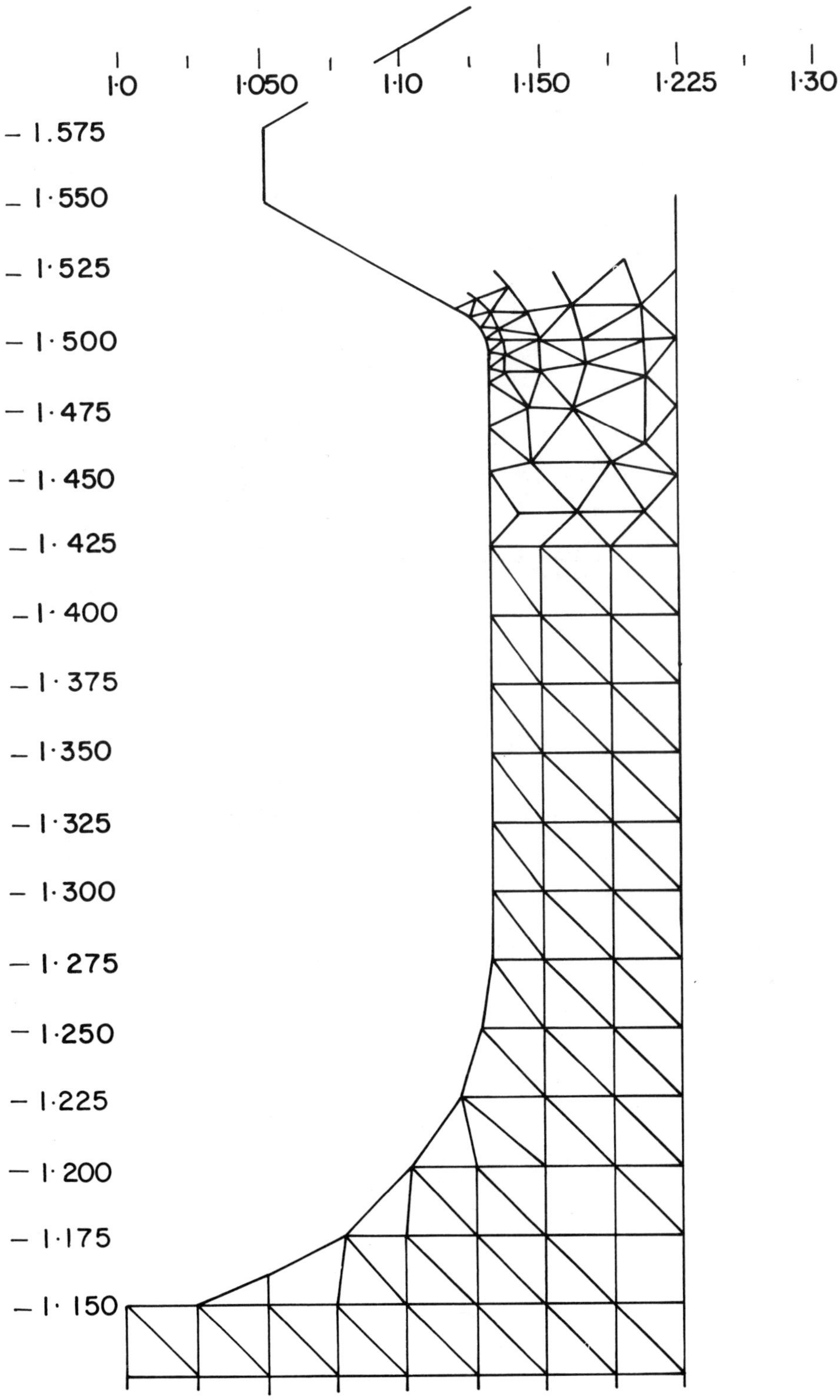

Fig. 4: V-Thread detail

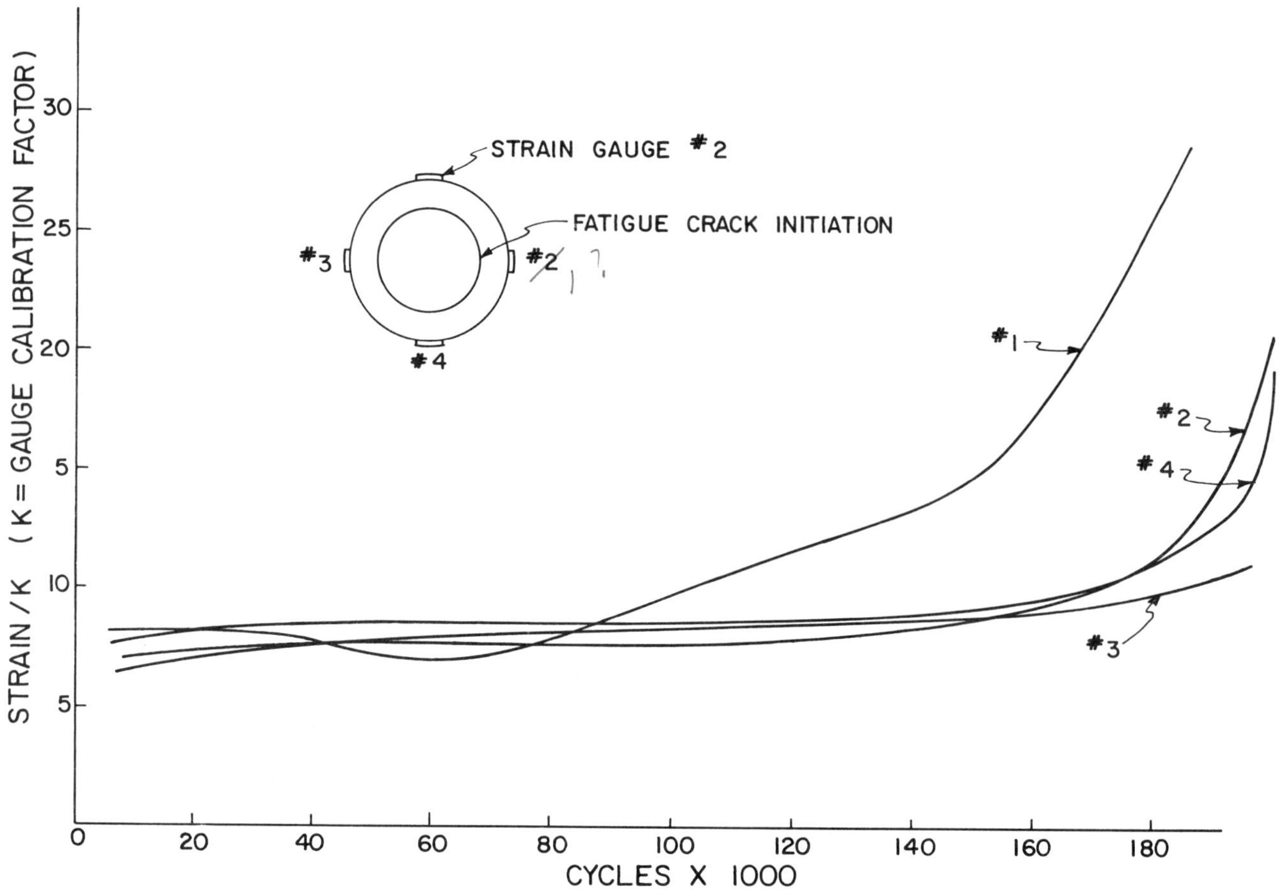

Fig. 5: Strain gauge results test No. 4

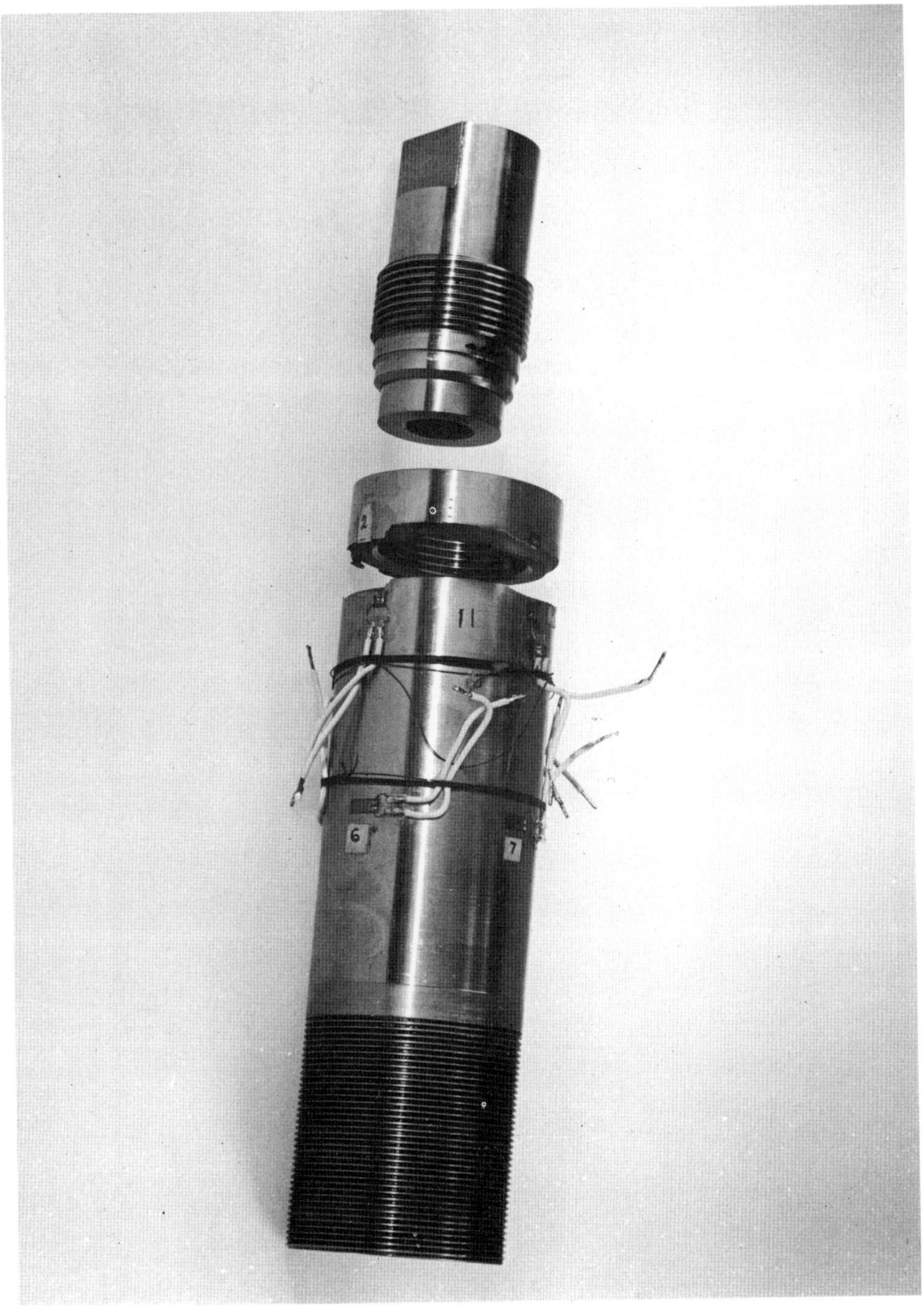

Fig. 6:

. Fig. 7:

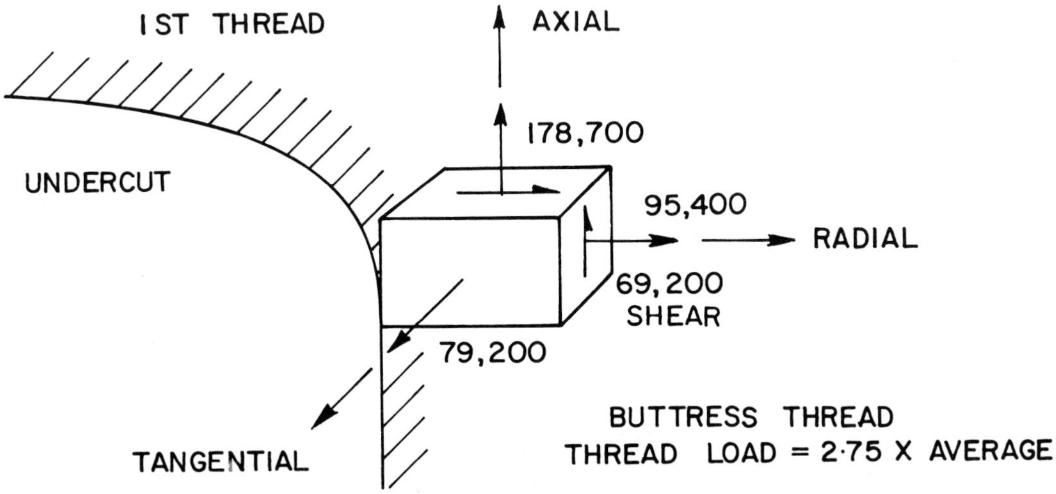

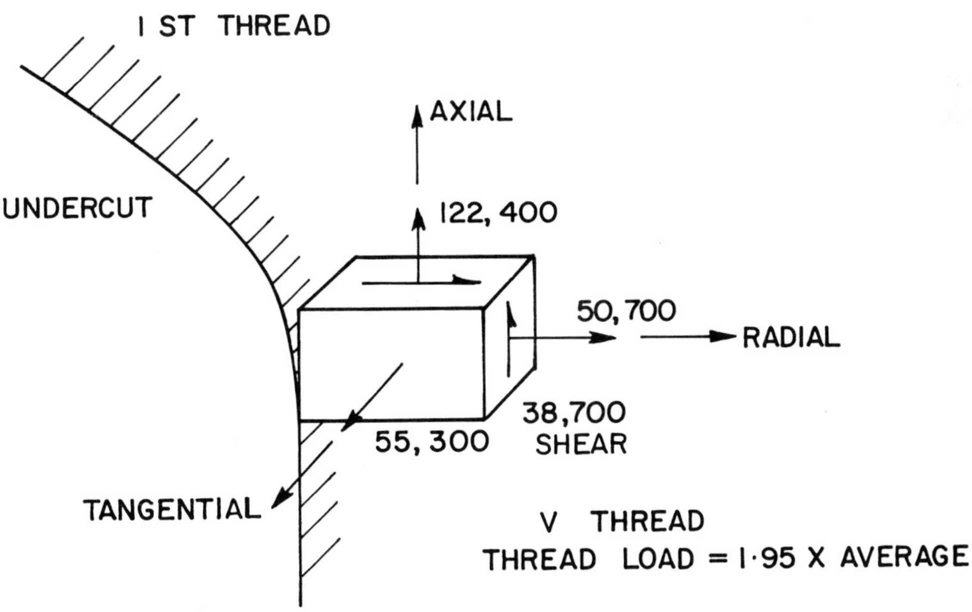

Fig. 8: Critical stress levels (PSI) at internal pressure of 15,000 PSI

THREADED CLOSURES FOR LARGE HIGH PRESSURE VESSELS

DONALD M. FRYER and CHARLES W. SMITH, Jr.

SYNOPSIS—This paper describes convenient, safe, and economical criteria for designing pressure vessel closures greater than 300 mm (approximately 12″) i.d. and 830 bars (12 000 p.s.i.) pressure.

INTRODUCTION

1. The objective of this article is to describe convenient, safe, and economical criteria for designing large pressure vessel closures.
2. For discussion purposes, we will define large as greater than 300 mm (approximately 12″) and 830 bars (12 000 p.s.i.) pressure.

THE PROBLEM

3. Most large vessel applications include generally a cycle in which the vessel must be charged, closed, heated or cooled, pressurized, returned to ambient temperature and atmospheric pressure, and opened. However, superimposed on these, the designer must consider and incorporate such parameters as:

(a) What is the required operating temperature?
(b) Is the temperature changed before, after, or during pressurizing?
(c) Is the pressurized part of the cycle essentially static, or is there dynamic loading such as from a system compressor?
(d) How long does one cycle last?
(e) How frequently must the cycle be repeated?
(f) What total cycle life is required?
(g) Is there an operating space restriction?
(h) What construction materials are acceptable to the process?

EVALUATION

4. In selecting a closure configuration, one can quickly rule out some of the types traditionally successful in smaller and/or lower pressure vessels.
5. As an example, consider the simple flanged design with a compressed seal illustrated in Fig. 1. This closure is easy to design, produce, and maintain; and lends itself to a wide selection of seal materials including 'O' rings. However, with a simple calculation of end load, and a selection of screw size and quantity with acceptable tensile stresses; the torque required to preload properly each screw to seal a 300 mm i.d., 830 bars vessel would approach 275 kg m (2000 ft/lb); and even if this were a reasonable torque level, the number of screws would make assembly a lengthy procedure.
6. Likewise, the popular set screw closure shown in Fig. 2, will also be time consuming and require very high torques. Even though it has the desirable advantages over the flanged design of being more compact and having its fasteners loaded in shear, the set screw closure also is unacceptable for large high pressure vessels.

POSSIBLE SOLUTIONS

7. It quickly becomes obvious that to be most effective and practical, large closures must be ones in which we utilize the pressure rather than trying to overwhelm it. To accomplish this, some form of self sealing must be used.
8. At this point, the closure designer has two areas to consider and resolve. One is the joint seal itself, and the other is resisting the end load.
9. Over the years, a wide variety of seal types have been used successfully. Some of these, Delta-ring, Double-cone ring, Bridgman, Modified Bridgman, Wave-ring, and single and double elastomeric 'O' rings are illustrated in Figs. 3 through 9.
10. None of these is universally best. Selection must be based on diameter, temperature, acceptable materials, and manufacturing and maintenance capability. Also, many geometric variations are possible depending on the application and the designer's ingenuity.

RESISTING THE END LOAD

11. The method selected to contain the force from the pressure on the closure components will generally be some version of one of the following. Again, individual designer ingenuity can generate many variations.

(a) A flanged arrangement, as discussed earlier, since with a self energizing seal the torque problem disappears. Also flanges are not limited to the conventional flat form of Fig. 1. Figure 10 shows a variation used in conjunction with a Bridgman seal.

Flanges, however, are still undesirable for large vessels because they are bulky and time consuming.

(b) An internally threaded closure. This was illustrated in Figs. 5 and 6 showing Bridgman seals, and is seen also in Fig. 11 showing a dual 'O' ring seal held in place by a cover and nut. The threaded closure principle will be discussed in more detail later.

(c) External split-clamps as illustrated in Fig. 12. This utilizes two clamp sections held together by horizontal fasteners. It is very nice in that it has smooth lines, generous radii, and lends itself to automatic opening; but like the standard flange configuration, it takes up a lot of space; and requires large expensive forgings for the clamps.

(d) Yoke or Press Frame. Figure 13 shows one typical arrangement.

The frames are made in a variety of ways including large forged platens and tie rods, picture frames cut from steel plates, and hoops. While this offers utmost simplicity in producing the vessel and its covers, and lends itself readily to automatic manipulation; space and added cost may be drawbacks.

(e) Shear Pin. This, like the split-clamp and press frame is also a threadless principle as is shown in Fig. 14. It is used extensively for Isostatic Presses; and because of its simplicity, is quite easy to automate.

The end load is carried by the pin in double shear, and with good dimensional control is transmitted into the shell with minimal bending stress added to the longitudinal stress. The concept, however, is size limited in that the pin must be designed based on maximum bending strain rather than strength. Since increasing vessel diameter has an exponential affect on the pin, its size quickly becomes awkward to manage.

Two examples illustrating the range of successful large pin closure vessels are:

1. 610 mm (24″) inside diameter at 1380 bars (20 000 p.s.i.) with a 610 mm diameter pin.
2. 305 mm (12″) inside diameter at 4140 bars (60 000 p.s.i.) with a 305 mm diameter pin.

VESSELS WITH INTERNALLY THREADED CLOSURES

12. After considering all the pros and cons, many designers will inevitably choose some form of internally threaded vessel closure since this will satisfy an optimum combination of minimum shell size, minimum space and reasonable ease of opening. And, the best designs will usually evolve to something either extremely compact, such as shown back in Figs. 5, 6, and 11, or having a minimum number of parts such as in Fig. 15.

13. Notice in Fig. 15, there are two designs on opposite sides of the centreline; however, they are both single piece closures except for the seals. One is a solid plug with a single 'O' ring seal which stays on the plug; and the other is a minimum weight cover which seals against a dual 'O' ring carrier.

THREADED CLOSURE SAFETY

14. The only one of the original objectives not yet accounted for is safety, and this requires the designer to consider carefully the various stress situations in the vessel closure area.

15. One frequently under-considered aspect is bending, or more clearly, the vessel's tendency to bell-mouth under load, which if extreme, could allow the closure parts to be expelled. Irrespective of thread form used, this stress will be a maximum in the undercut area below the bottom thread. This, for fatigue resistance, is why well designed vessels will have a generously sized undercut with a good surface finish.

16. It can be shown by classical analyses, and has been verified by finite element analysis, that the magnitude of the bending stress can approach three times the average longitudinal tensile stress in the undercut area. Thus the combined bending and longitudinal stress at the undercut surface can be estimated easily by the simple equation:

$$\sigma_{cbl} = \frac{4P(d_i)^2}{(d_o)^2 - (d_{uc})^2}$$

where P = Pressure
d_i = Inside Diameter
d_o = Outside Diameter
d_{uc} = Undercut Diameter

17. The other major safety consideration is to evaluate sufficiently the thread stresses. The traditional method involves merely calculating the average shear stress using the equation:

$$\tau = \frac{2(\pi/4)(d_i)^2 P}{\pi d_\tau l_\tau}$$

where d_τ and l_τ are the shear diameter and thread length respectively.

18. As an example, consider a 457 mm (18″ i.d.), 3100 bars (45 000 p.s.i.) vessel with a 610 mm (24″) thread shear diameter and a 318 mm (12.5″) thread length.

19. The average shear stress is only 17.1 kg/mm^2 (24 350 p.s.i.) and the average load per thread is 472 160 kg (1 040 000 lb).

20. A more sophisticated analysis, however, will show the true load distribution as seen in Fig. 16. The first thread actually carries almost four (4) times the theoretical average load, and failure to recognize this high first thread condition can lead to fatigue failure.

WHAT THREAD TO USE

21. Frequently designers will select either the Acme or buttress thread form for large vessel closures. These, in their classical form, are illustrated superimposed on each other in Fig. 17.

22. The age old debate of which form is better will not be resolved here. However, there are a number of major differences some of which favour Acme while others favour buttress.

23. The load side flank angle on the buttress thread is usually only 5° to 7°. Thus radial force on the shell is kept to a minimum, whereas, the standard Acme thread has a $14\frac{1}{2}$° angle. Some designers feel, however, that starting friction (even with well lubricated threads) will be sufficiently high so as to neutralize the theoretical difference in radial force transmitted to the shell.

24. The thread thickness through the pitch diameter is essentially the same, so the average shear stress also is the same for both forms. The flank angle on the reverse side of the Acme thread is the same as the load side, so more opening force capability is available in the event of a stuck closure.

25. The thread thickness is greater at the root of the buttress form, therefore, this thread's bending stress is

less. Remember, from Fig. 16, the first thread is actually loaded almost four (4) times as much as a theoretically average thread so an improved bending strength would seem important. However, as shown in Fig. 18, a modified Acme thread lends itself to a greater reduction in stress concentration than does a modified buttress thread because a more generous root radius can be obtained.
26. Many feel this modified Acme thread stress concentration advantage more than offsets the relative bending strength advantage of the buttress thread.

AN IDEAL SOLUTION

27. Since the choice between Acme and buttress will probably never be resolved, let us consider an alternative. There is a patented thread design—the Gasche Resilient Thread—which reduces the high first thread loading.
28. This design incorporates a semi-circular helical groove machined on the nut and a matching semi-circular helical groove in the vessel shell. Wound in the nut groove is a specially designed high alloy spring; thus the nut can be engaged with the body as with any thread. The resiliency of the spring allows significant equalizing of thread loading, and using the same vessel example as Fig. 15 will change the thread distribution as shown in Fig. 19. Examples of this thread are shown in Figs. 20 and 21.
29. By further refinement of this concept, as shown in Fig. 22, it is possible to obtain an even better distribution of the load. This is done by tapering the nut a small amount. The direction of the taper creates a slightly larger gap between the nut and the body at the bottom than at the top. This causes the upper threads to carry a greater share of the load because of the varying load area on the spring.
30. This, plus the resilient effect of the spring, produces the desirable result of almost equal load distribution along the axis of the thread, as illustrated in Fig. 23. By way of comparison, all the load distribution curves are combined in Fig. 24. The improvements obtainable with the resilient thread are clearly visible. Photo-elastic studies, and independent stress analyses, have been made to verify this improved load distribution.
31. Two examples illustrating successful large Resilient Thread vessels are:

1. 914 mm (36″ i.d.) at 1035 bars (15 000 p.s.i.):
   This, incidentally, is the vessel shown in Fig. 21.
2. 457 mm (18″ i.d.) at 3450 bars (50 000 p.s.i.).

32. These vessels both are used for hot isostatic pressing.

AUTOMATIC VESSEL OPENING

33. Many devices have been devised and built to rotate, support and lift the closure parts of an internally threaded large vessel. One such structure, recently completed is shown in Fig. 25. This particular vessel has a 1524 mm (60″ i.d.) and is rated at 1035 bars (15 000 p.s.i.).
34. One obvious objective of automating a closure is to physically assist operating personnel. Another, however, when production rates are important, is to decrease turn-around time between operating cycles.
35. A method, well proven in improving turn-around time, is the interrupted thread technique as shown in Fig. 26.
36. This technique is normally employed with the buttress thread and even though there does not have to be any sacrifice in safety, the vessel and closure components must have increased lengths to compensate for the lost thread section. Of course, this will require greater cost and more operating head room.
37. Another method for speeding up the closure process utilizes the Resilient Thread. The nut and thread section of the vessel are each machined in a conical configuration as seen in Fig. 27. Only several turns instead of the same number as there are threads, are needed to completely free the cover from the vessel. Figure 28 shows a comparison between straight and conical Resilient covers; and Fig. 29 shows a conical Resilient cover poised above a large vessel. While this design may not improve cycle time quite as much as the interrupted thread, it does not adversely affect operating space requirements. Also, this closure will not be as self damaging if the cover rotating and lifting mechanism should malfunction.

CONCLUSION

38. Using the principle of resiliency is a theoretically sound as well as proven means of reducing peak vessel thread stresses and improving overall closure versatility.
39. Thus, the feasibility is improved for designing and manufacturing large vessels to operate productively and safely at high pressure.

References

1. Frederick, D. D. 'Pressure Vessel Closure.' Machine Design Magazine.
2. Gasche, Fred. 'Design of Reactors and Closures.' Industrial and Engineering Chemistry, May 1956.
3. Strelzoff, S., Pan, L. D. and Miller, E. J. 'Trends in Pressure Vessels and Closures.' Chemical Engineering, October 1968.
4. Warnke, E. G. 'Design and Fatigue Life of Openable High Pressure Vessels.' Atomic Weapons Research Establishment, June 1967.
5. Wieger, J. W. 'Stress Analysis of Pressure Chamber for Isostatic Press.' Midvale-Heppenstall Company, October 1970.
6. Lessells & Associates, Inc. 'Tapered Resilient Thread Analysis.' Technical Report No. 885, January 1965.
7. Frederick, D. D. 'Resilient Vessel Closures and Connections for High Pressure—Temperature Service.' Chemical Engineering Progress, February 1968.
8. Comings, E. W. 'High Pressure Technology.' McGraw-Hill, 1956.

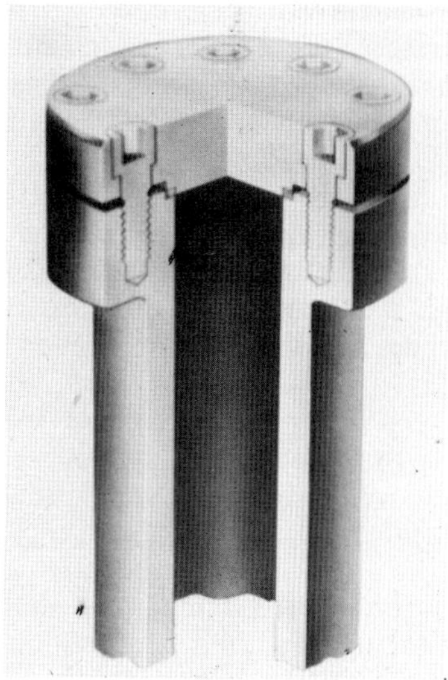

Fig. 1: Bolted closure pressure vessels

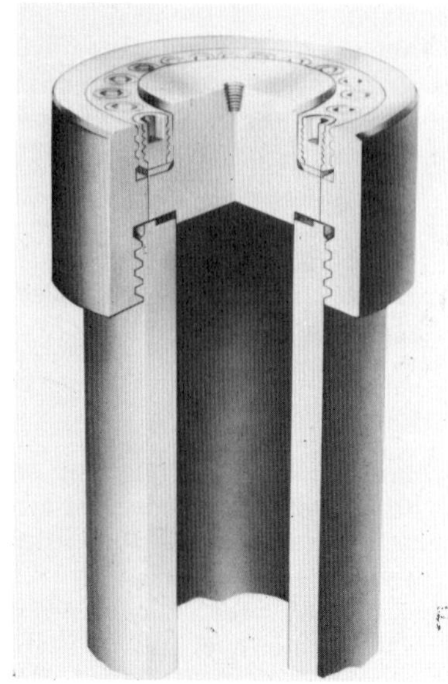

Fig. 2: Set screw pressure vessels

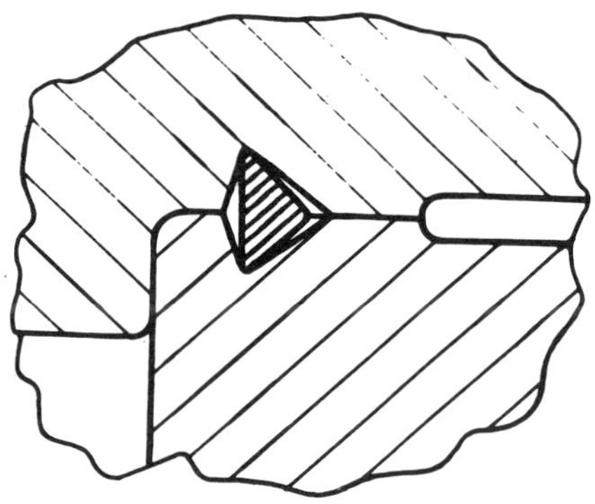

Fig. 3: Delta ring seal

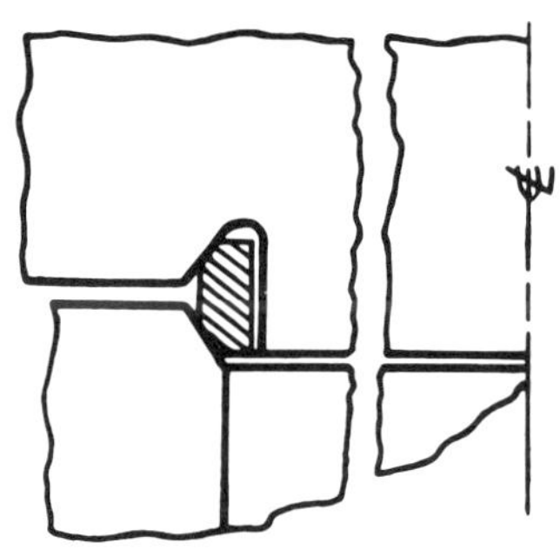

Fig. 4: Double-cone ring seal

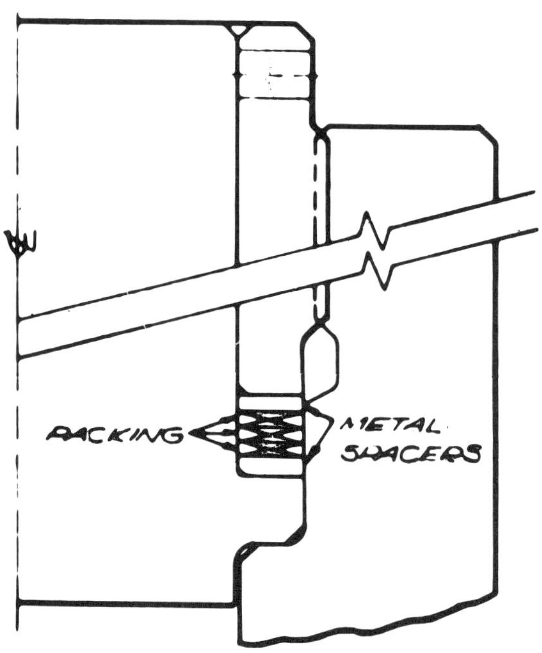

Fig. 5: Bridgman seal

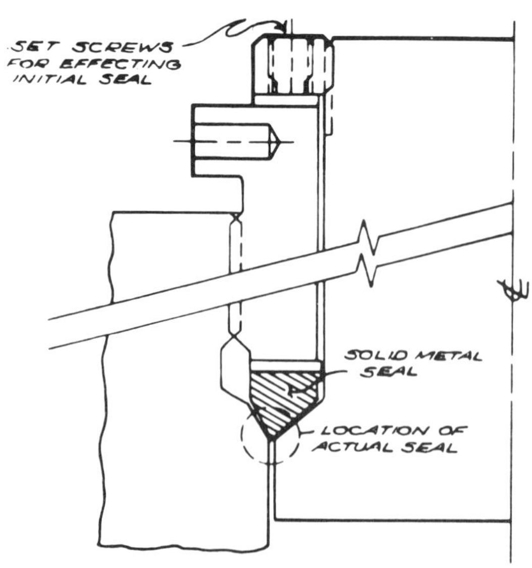

Fig. 6: Modified Bridgman seal

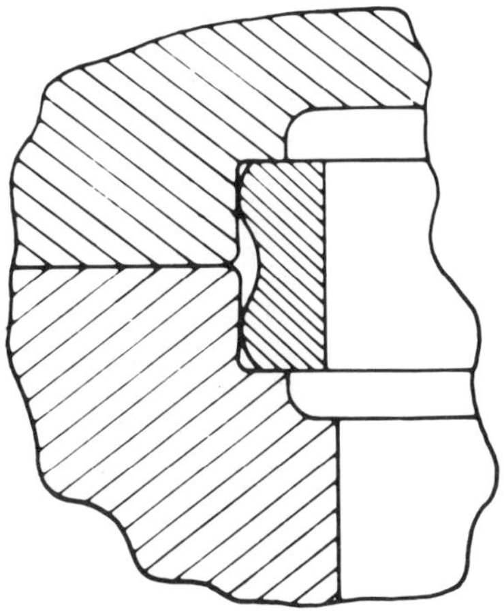

Fig. 7: Wave ring seal

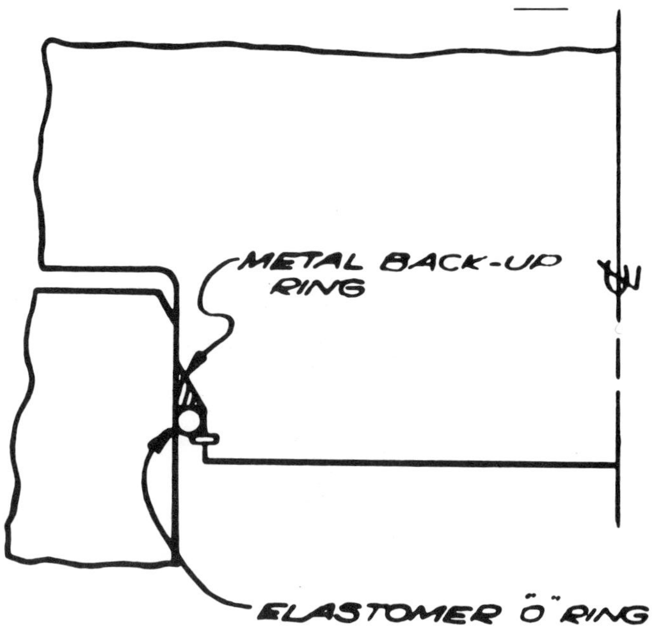

Fig. 8: Single 'O' ring seal

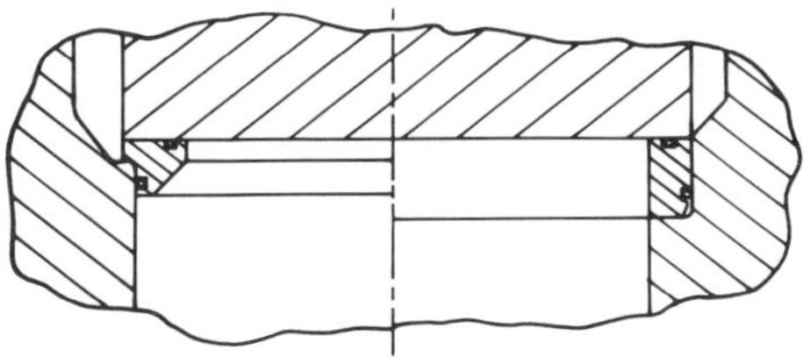

Fig. 9: Double 'O' ring seal

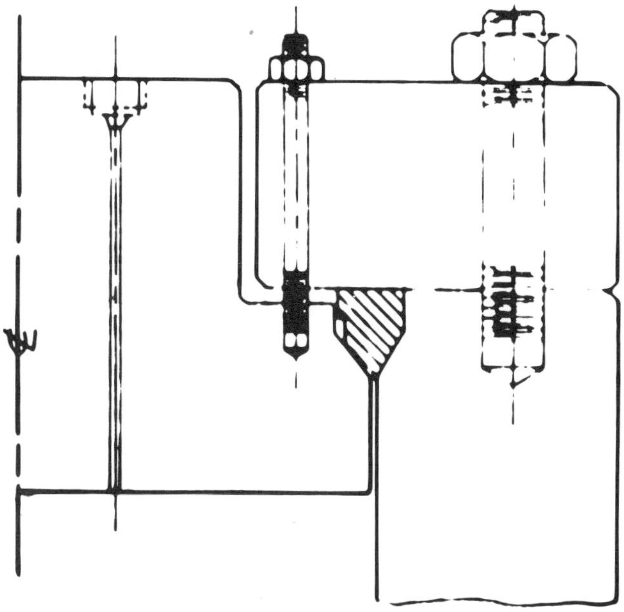

Fig. 10: Flanged closure

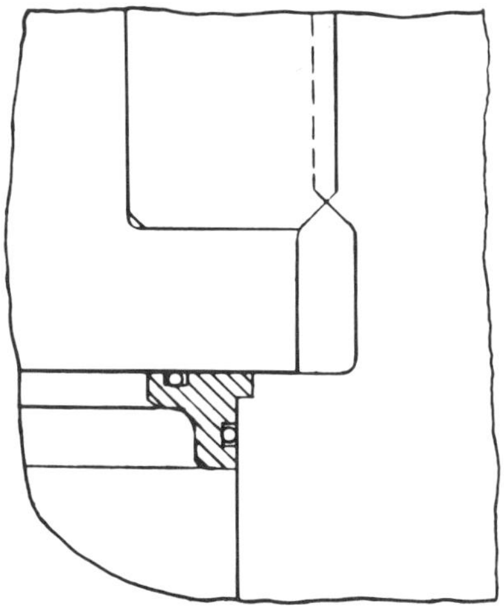

Fig. 11: Internally threaded closure

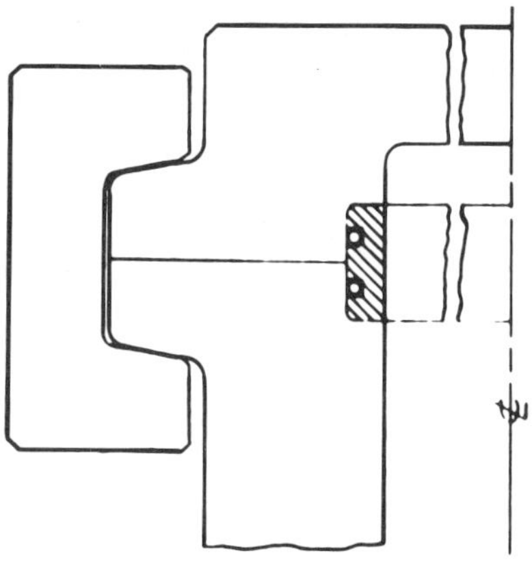

Fig. 12: External split-clamps closure

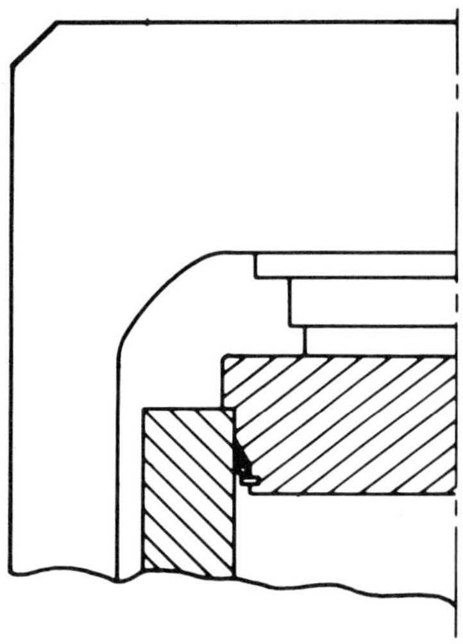

Fig. 13: Yoke or press frame closure

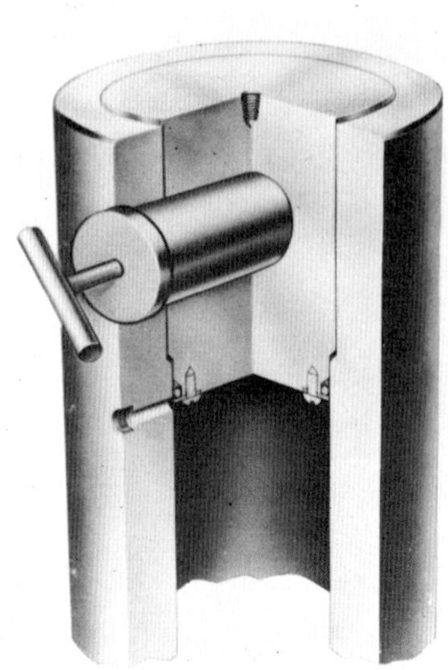

Fig. 14: Threadless pin pressure vessel closure

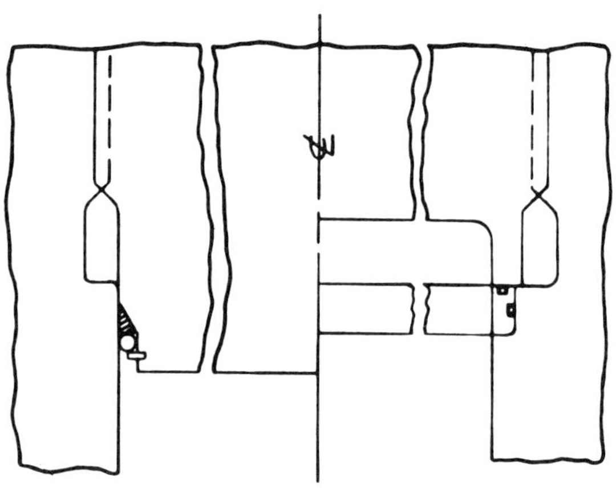

Fig. 15: Additional internally threaded closures

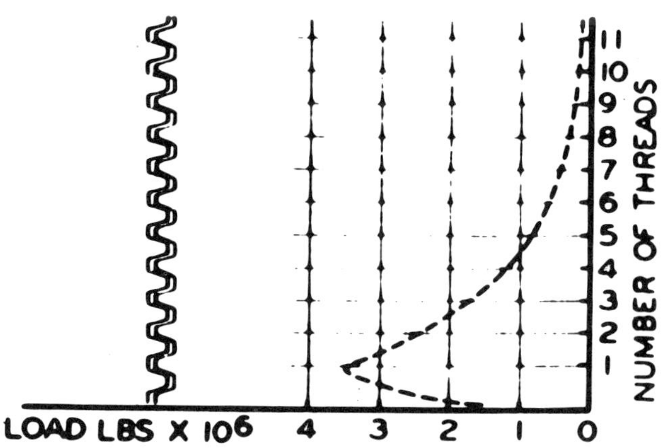

Fig. 16: Thread load distribution curve. With conventional thread, load is concentrated on bottom threads

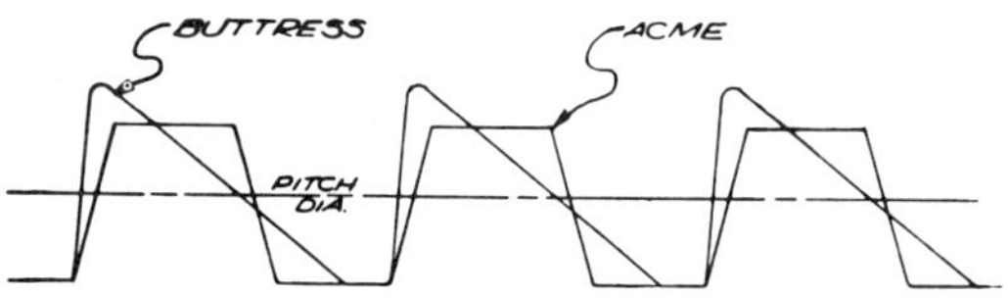

Fig. 17: Acme or buttress thread form

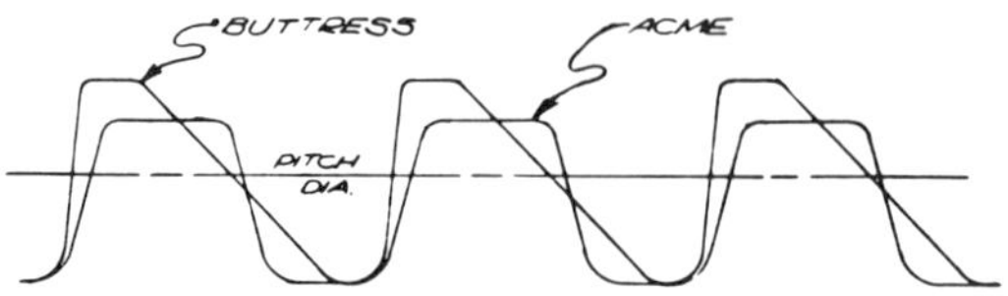

Fig. 18: Acme or buttress thread form

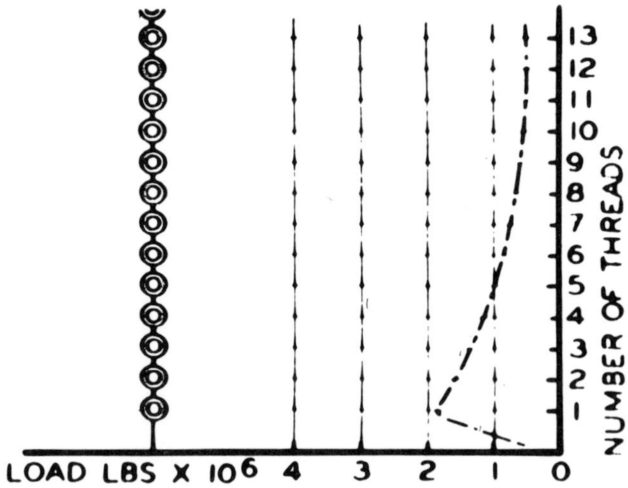

Fig. 19: Thread load distribution curve. Use of resilient thread greatly reduces the load concentration

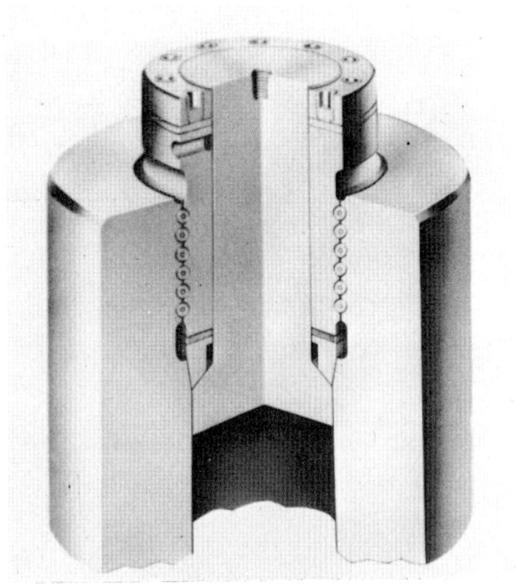

Fig. 20: AE self-sealing pressure vessels with Gasche resilient thread

Fig. 21: AE self-sealing pressure vessel with Gasche resilient thread

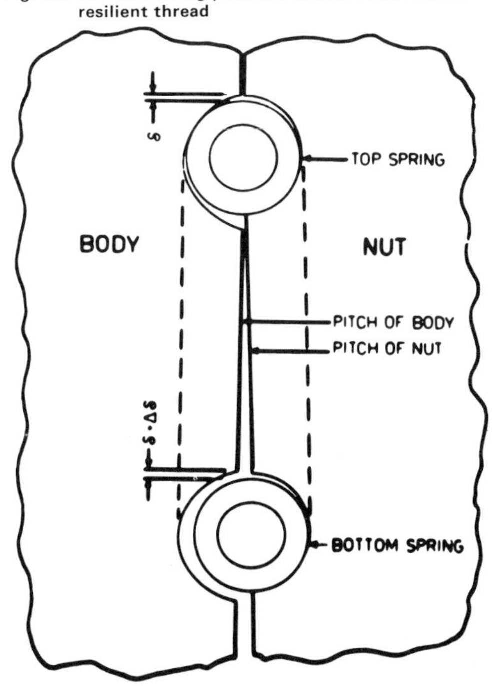

Fig. 22: Tapered resilient thread principle

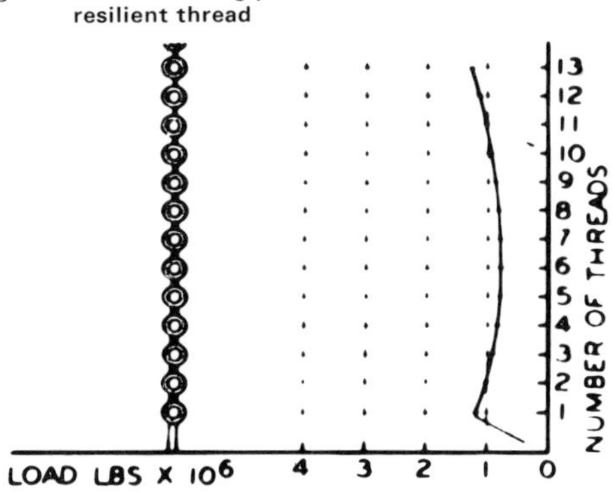

Fig. 23: Thread load distribution curve. Use of tapered resilient thread ensures nearly uniform distribution of load

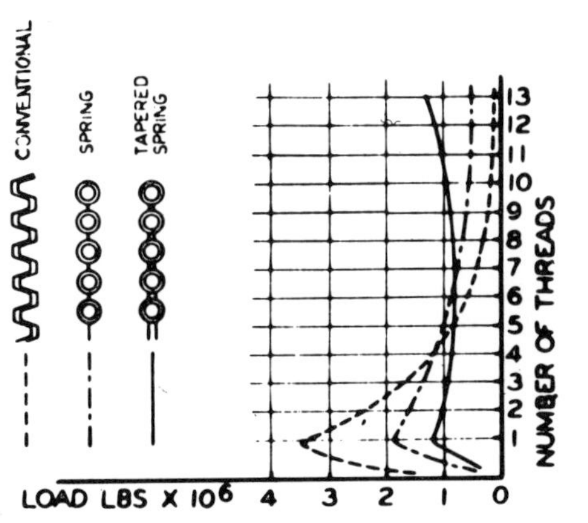

Fig. 24: Thread load distribution curves superimposed

Fig. 25: Automatic vessel opening equipment

Fig. 26: The interrupted thread technique

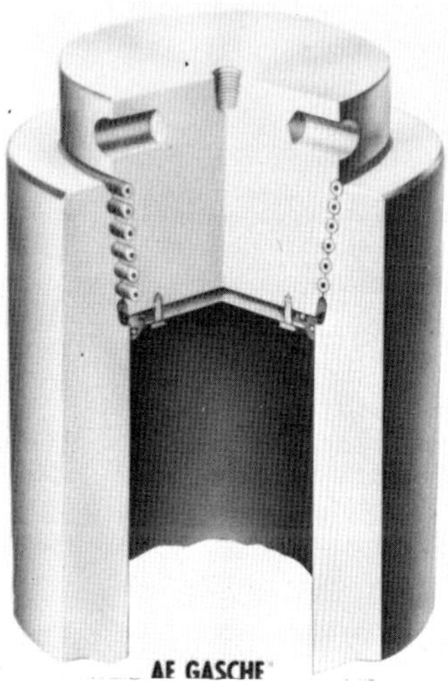

Fig. 27: AE Gasche resilient thread pressure vessel (conical configuration)

Fig. 28: Comparison between straight and conical resilient covers

Fig. 29: Conical resilient cover poised above a large vessel

SAFETY

INSPECTION AND SAFETY ASPECTS OF DESIGN

D. E. WITKIN
National Forge Company

INTRODUCTION

Safety, in the context of High Pressure Design, is a many faceted concept. Since safety must consider the behavior of the actual high pressure plant, the designer is faced with the nebulous "what if" consideration in creating a design which will safely withstand the normal risks and rigors of operation and maintenance. Assuming that the competent design engineer will properly perform an adequate stress analysis, and select material where strength gives reasonable factors of safety, safety remains dependent upon operation, installation, equipment, and material considerations and a comprehensive discussion of all of these is manifestly impossible in the available time. It is the purpose of this paper to point out several areas where the designer can make a pressure vessel more safe by careful consideration of what might occur after the drawings and stress analyses leave the design office and the vessel is built and placed into operation.

High Pressure Polyethylene Reactor

One such example is the design of a closure for a high pressure polyethylene reactor.

End closures on many such high pressure polyethylene reactors consist of a self-energizing seal ring, a cover, and a split clamp ring which co-operates with either an external thread of a stubby flange on both reactor body and cover. It is my purpose to show that closures which are designed and manufactured to standards suitable for this class of pressure vessel can, under certain circumstances, lead to unsafe conditions which could have disastrous consequences.

It is further my purpose to indicate some areas where the designer, if he is aware and appreciative of what might be, can, by suitable design, make the integrity of the reactor less sensitive to mishandling and erroneous assembly. This can make the difference between safe (Code level factors of safety) and potential failure conditions.

To provide a concrete example rather than these kind of generalities, consider a closure system as depicted in Figure 1.

Deriving the force (F') on the clamp bolts (Appendix I), it is seen that

$$F' = \frac{F_e \operatorname{Tan}(\alpha - \rho)}{\pi}$$

where F_e = the hydrostatic end load
α = flange angle
ρ = friction angle (arctan coef of friction)

Considering a flange angle (α) of 10° and the coefficient of friction of dry clean steel as 0.39 (Marks Handbook pg. 3-36 and 3-37) making the friction angle $\rho = 21.3^\circ$, it is obvious that $\alpha - \rho$ is equal to -11.3°, but since frictional effects cannot be exerted in a reverse direction, the effect is that of a self-locking taper and thus the end load transmitted to the bolts is zero. This fulfills all reasonable design condition requirements but:

Suppose an assembly mechanic decides to put a Molykote lubricant on the flange to facilitate assembly, then the coefficient of friction is reduced to 0.033 with a friction angle of 1.9°. Under these circumstances there will be a load imposed upon the bolts. In a typical case this might be of the order of magnitude of 420,000 pounds or in terms of stress on two - 2" bolts it is 75,000 psi. This is well above allowable stress levels and when added to the normal service stress such as those imposed by radial growth of the reactor, stress due to thermal differences, preload, etc., can well exceed the ultimate strength of the bolting. The designer should, therefore, be aware that such anomalies can occur and design so that if they do, their effects are minimized. In this case, the solution is obvious. Select a flange angle such that if inadvertent lubrication takes place, the loads imposed upon the selected bolting are sufficiently low so as to still have a structure which can safely withstand the pressure loads. From the brief analytical presentation above, a suitable angle is obviously 2°. This makes the angle practically self-locking even with perfect molybdenum disulfide lubricant.

Another area of consideration is the bolting itself. In the properly designed clamp the only function of the bolting is to force the clamp sections together into a proper assembly. The forces required to do this increase, if as outlined above, the flange surfaces are kept unlubricated. A significant bolt preload can be developed due to such assembly techniques. With this type of closure, however, bolt preload is additive to any other imposed loads contrary to the case of a conventional flanged connection.

It is obvious that such a preload is not desired and should not be imposed upon the bolting. To the logical designer looking for the most efficient way to contain the required pressure within a structure bolting should only carry those loads which are created by the internal pressure.

Thus procedures should be imposed wherein the bolts are backed off after the clamps are assembled and then retightened uniformly with a low nominal torque just sufficient to insure that they are snug. This then allows the bolts to withstand the imposed service stresses such as that caused by radial growth of the reactor body. This stress can be expressed as

$$s_2 = \frac{2 P R_i^2}{\frac{3L (R_o^2 - R_i^2)}{2\pi R_o} + \frac{A}{b R_o}\left[(R_o^2 + R_i^2) - \nu(R_o^2 - R_i^2)\right]}$$

and for a typical high pressure reactor be of a magnitude of 40,000 psi.

To design considering the safety aspects, however, it is naive to expect that such instructions will always be followed. Thus the design of bolting should be such that they can withstand this order of stress in addition to inadvertent excessive preload. DO NOT RELY ON THE ASSEMBLY TO BE DONE AS SPECIFIED OR CONVERSELY DESIGN SO THAT MIS-ASSEMBLY CANNOT CAUSE A CATASTROPHE.

Finally consider the obvious. Use open rather than cap nuts to insure that proper thread engagement is made. Inserting a depth gage into a gage hole to check thread engagement while a good drafting room practice, will not be done in the field, but direct visual evidence of partial nut engagement will usually be noted and corrected.

In short, when a closure is designed, every effort to make it insensitive to improper assembly should be made, because this generally will be the true margin of safety in operation. I do not wish to imply that any polyethylene producer uses incompetent operational and maintenance personnel, on the contrary, they are generally a highly competent crew. The point is that try as we might, we can never overcome Murphy's famous laws and thus for safety's sake we should design so that we can circumvent them.

These few examples typify the sort of things which can and inevitably will occur to cause a failure of a pressure vessel which was otherwise adequately designed with factors of safety following accepted engineering practice. There are, of course, many other examples which suggest themselves to the high pressure designer.

High Pressure Seal

An example to illustrate another point is typified in the design of a high pressure seal for a pressure vessel which is used as an isostatic press, as illustrated in Fig. 2.

The safety lesson here, is to design such that if a failure should occur, its consequences are minimized. One such design feature is the location and design of the high pressure seal. The seal engagement should always be kept as short as possible (consistent with reliable sealing) so that if separation of the closure should occur, the amount of energy transferred to it will be minimized An excessively long seal engagement provides additional time for the internal pressure to act upon the projectile, increasing the impulse and, hence, increasing the momentum imparted to it, or it can be considered in terms of energy and the greater distance over which force can be exerted increases the imparted energy.

In this example, the pressure vessel is 24" in diameter and 96" long, giving a void volume of 24.1 ft^3. When containing water at 80,000 psi, the stored energy is 27.3 x 10^6 ft. lbs. Taking a simplistic approach, for illustrative purposes, consider that energy is transferred to the closure as long as there is seal engagement (actually energy transfer takes place for some additional distance at a decreasing rate), then with a 1" seal engagement only approximately 11% of the total stored energy can be transferred to a projectile while if the engagement were increased to 3", as used in some designs, then there is a potential for approximately 33% of the total energy to be imparted to the projectile. This represents an obvious increase to the hazard in the surrounding area.

The designer should approach this in two ways. First, design a closure which has a minimum amoun of elastic deflection during pressurization. For example, a yoke type structure can have substantially greater eleastic deflection in the direction of the longitudinal axis of the pressure vessel than would be experienced with a more conventional threaded closure, solely due to the differences in the length of the elastically deflected members. In the first case, the length is that of the entire pressure vessel, while in the latter case it is only the undercut region between the threads and seal which need be considered. In use and function, one is as good as the other as long as the seal movement is known and allowed for in design. However, it is obvious that if there is greater elastic deformation, thus greater seal movement, then the seal must be placed deeper into engagemen in the pressure vessel. From the standpoint of safety, such deeper engagement should be minimized. Thus in design the pressure vessel can be made more safe by creating a structure which will minimize the seal engagement. Several ways of doing this are:

1. Use of threaded closures over yoke type closures to reduce elastic deflections.

2. If yoke structures are required, then the design should incorporate a means of controlling the relative movements of yoke and vessel body to avoid the necessity of allowing for random motion. One such way used by National Forge and patented in the U.S., the U.K., and many European countries, consists of designing the bore such that the upper and lower seals are of different diameters and using the internal pressure to keep the pressure vessel lifted tightly against the upper closure, thus eliminating the need for allowance for seal movement due to elastic deflections.

3. Design the supporting structures in any case to have minimal elastic deflection in the axial direction consistent with other flexibility requirements.

Again, the point is, careful attention to the details of design can significantly enhance the safety of the high pressure vessel in ways which are not considered by existing codes and standards.

Material

Since another session of this conference addresses itself to material, there is no attempt here to discuss this matter except to the extent that no discussion of safety aspects of design can afford to ignore it.

The key operative words are ductility and fracture toughness. Since the design is based upon the assumption of ductile behavior and the factors of safety are appropriate to such an assumption, it is obvious that means to insure such ductility are essential. Specification of chemistry and strength levels alone are not sufficient. Elongation, reduction of area, impact tests reporting energy levels, fracture appearance, and lateral expansion are minimum requirements along with adequate NDT examinations (both ultrasonic and magnetic particle). K_{Ic} data, if available, is useful, but in many cases will not be available. In any case, a key point is that the test specimens be taken in regions which are representative of the highly stressed parts of the pressure vessel (e.g. the bore) and that they be from prolongations of the actual pressure vessel forging. The mass effects in heavy section metallurgy preclude any consideration of separately treated test metal.

In this area there can be no short cuts, but reliance only upon adequate testing of adequate samples is the only safe way to design. Further, one should avoid pushing any particular alloy to the limit of its range. There is a potential loss of ductility and toughness margins when an alloy is pushed to its limit. From the standpoint of safety, use a richer alloy or a thinner section, or a lower strength level as the basis of design. Some materials which we have found suitable over the past 15 years are summarized in Appendix II. A complete specification for these materials, including sampling, NDT, etc., is presently under consideration by the ASTM and should be issued within the next year.

Inspection

Safety is enhanced when the selected factor of safety becomes truly a factor of safety rather than a factor of ignorance. The more items which can be removed from the area of ignorance, the more safe any design becomes. In the area of design, modern stress analyzing methods provide extremely accurate mathematical models and predict stress levels with a remarkably high degree of accuracy. When such techniques are used (e.g. Finite Element Methods using computer codes such as ANSYS) the design goes a long way toward removing half the ignorance element from the factor of safety. Inspection is the basic tool for removing the other half. Here too the designer plays an essential role. Consideration must be given to the inspectability of the design. For example, an integrally forged pressure vessel body can be examined by ultrasonic means, by magnetic particle means, by acoustical emission means, radiographic means, by direct measurement. If the mechanical properties of the steel were adequately determined (see Material above) then one can confirm and reconfirm the integrity of the structure at any time. This can insure that the structure is, in fact, as the designer assumed it to be - the essential point of inspection.

With composite structures, the process becomes more complex. Many factors can be inspected only once during an in-process inspection procedure during manufacture such as the degree of prestress obtained from a shrink fit or from a wire wrapping. Changes in these conditions cannot be readily detected, nor can re-inspection be made to verify that the initial conditions were met. Thus there is greater uncertainty which for the same factor of safety must translate into less safety.

The best assurance of safety still rests with the designer who should through his design efforts include the details large and small which will enhance the safety of the pressure vessel by permitting inspection procedures to the greatest extent possible and by as many different techniques, should they be required.

From a manufacturing view point, there are two points to be made which are perhaps beyond normal procedures, but which should always be applied to the construction of high pressure vessels. The first is that all applications of pressure to the vessel during its construction (autofrettage, hydrostatic tests, etc.) should be done under well instrumented conditions, with strain gages applied to critical as well as referenced areas. This serves to validate the mathematical model used for the stress analysis and gives assurance that the results of the stress analysis can be relied upon. The second, is to maintain detailed records of the dimensional inspection so that changes which occur in service can be readily detected. Also germane is the monitoring of pressurization procedures already mentioned. Key dimensions should be measured before and after to assure that the plastic deformation is, in fact, as predicted. If not, adequate explanations should be documented by additional investigations.

APPENDIX I

DERIVATION OF BOLT LOAD

Appendix I takes the case illustrated in Fig. 1 and utilizing the same nomenclature therein, derives the loading which would be felt on the bolts which restrain the split clamp ring which holds the end closure against the hydrostatic end load. It is merely an analytical consideration of the theoretical force components in various directions as they are transferred across a slanted pressure interface. The net result is an expression for the force on the bolts as a function of the end load on the closure, the angle of the flange, and the coefficient of friction between the flange surfaces.

$$dF_c = dF_e \tan(\alpha - \rho)$$

where

$$\tan \rho = f_o = \text{coef. of friction (static)}$$

$$dF_s = dF_c \cos \theta$$

$$dF_e = \frac{F_e}{2\pi} d\theta$$

$$dF_c = \frac{F_e}{2\pi} \tan(\alpha - \rho)\, d\theta$$

$$F_s = 2 \int_{\theta=0}^{\theta=60} \frac{F_e}{2\pi} \tan(\alpha - \rho) \cos\theta \, d\theta \; (2\text{–}60^\circ \text{ segm.})$$

$$= 2 \frac{F_e}{2\pi} \tan(\alpha - \rho) \Big[\sin\theta \Big]_{\theta=0}^{\theta=60}$$

$$= \frac{F_e}{\pi} = \tan(\alpha - \rho) \sin 60^\circ$$

Force on (2) bolts (one connection) from two sloping edges (on head and body)

$$F' = \frac{\not{2} F_s}{\not{2} \cos 30} = \frac{F_e}{\pi} \tan(\alpha - \rho) \frac{\not{\sin 60^\circ}}{\not{\cos 30^\circ}}$$

$$= \frac{F_e}{\pi} \tan(\alpha - \rho)$$

<u>List of Symbols</u>

F'	-	force on clamping bolts
F_e	-	end load on closure due to hydrostatic forces
α	-	flange angle
ρ	-	friction angle (arc tangent coefficient of friction)
s	-	stress
P	-	internal pressure
R_i	-	inside radius of pressure vessel
R_o	-	outside radius of pressure vessel
L	-	length of bolt
ν	-	Poisson's ratio
A	-	area of the bolt
b	-	width of one-half of the clamp ring
R_p	-	bolt circle radius

Appendix II - Summary of Material Properties

TABLE I – CHEMICAL REQUIREMENTS

	Type I %	Type 2 %	Type 3 %
Carbon	0.35 max.	0.40 max.	0.40 max.
Manganese	0.90 max.	0.90 max.	0.90 max.
Phosphorus	0.015 max.	0.015 max.	0.015 max.
Sulfur	0.015 max.	0.015 max.	0.015 max.
Silicon	0.35 max.	0.35 max.	0.35 max.
Nickel	1.5 to 2.25	2.25 to 3.00	3.00 to 4.00
Chromium	0.80 to 1.20	0.80 to 1.10	0.80 to 1.40
Molybdenum	0.20 to .30	0.30 to 0.50	0.40 to 0.80
Vanadium	0.10 max.	0.20 max.	0.20 max.

TABLE 2 – PERMISSIBLE VARIATIONS IN CHECK ANALYSIS [a]

Element	Limit or Maximum of Specified Range per cent	Permissible Variation over the Maximum Limit or Under the Minimum Limit, per cent, by Cross-Sectional Areas			
		200 sq. in. and under	Over 200 to 400 sq. in., incl.	Over 400 to 800 sq. in., incl.	Over 800 to 1600 sq. in., incl.
Carbon	0.40 and under	0.03	0.04	0.05	0.06
Manganese	0.90 and under	0.04	0.05	0.06	0.07
Silicon	0.35 and under	0.02	0.03	0.04	0.04
Nickel	1.5 to 2.25, incl.	0.05	0.05	0.05	0.05
	2.25 to 4.00, incl.	0.07	0.07	0.07	0.07
Chromium	0.80 to 1.40	0.06	0.06	0.07	0.07
Molybdenum	0.80 and under	0.04	0.05	0.06	0.07
Vanadium	0.20 and under	0.02	0.02	0.02	0.03

[a] *Check analysis for phosphorus and sulfur shall conform to the requirements of Table 1.*

TABLE 3 – TENSILE REQUIREMENTS

	Class I	Class 2	Class 3	Class 4	Class 5
Tensile Strength Min. PSI	115,000	135,000	155,000	175,000	190,000
Yield Strength Min. (.2% Offset)PSI	100,000	120,000	140,000	160,000	180,000
Elongation 2 In., Min., Per Cent	18	16	16	15	12
Reduction of Area, Min., Per Cent	50	45	40	35	30

TABLE 4 – CHARPY V-NOTCH IMPACT REQUIREMENTS AT 40F [a]

	Classes 1 & 2	Class 3	Classes 4 & 5
Minimum average value of set of three specimens, ft-lb [b]	35	30	25
Minimum value of one specimen ft-lb	30	25	20

[a] *These Charpy values are for tests made on standard 10-mm square specimens. If sub-size impact specimens are used, the required minimum ft-lb values shall be determined by multiplying the ft-lb values in Table 4 by 5/6 for 7.5 by 10-mm specimens, and by 2/3 for 5 by 10-mm specimens.*

[b] *Not more than one specimen from a set may be below this value.*

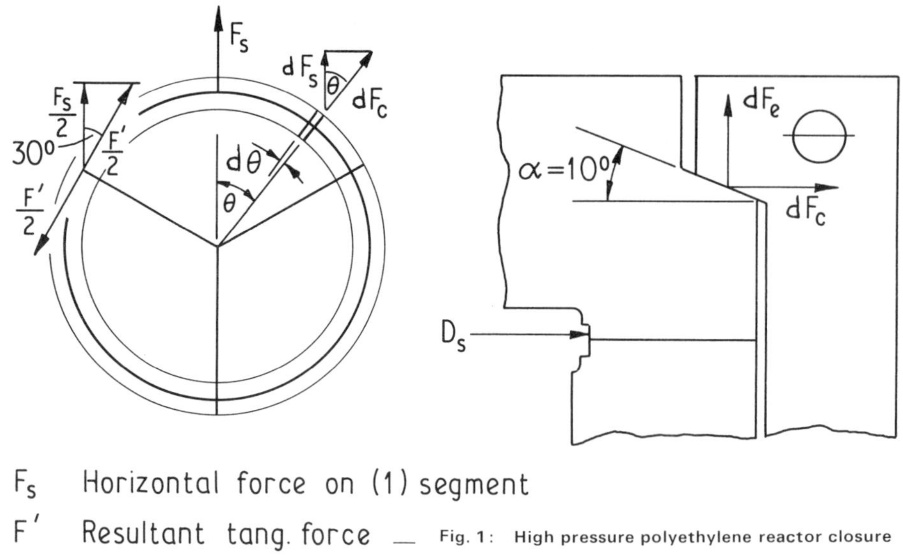

Fig. 1: High pressure polyethylene reactor closure

The Back-Up Ring, O-Ring and Retaining Ring should remain in position in the cover when closure is raised. Operator should make routine check prior to closing vessel.

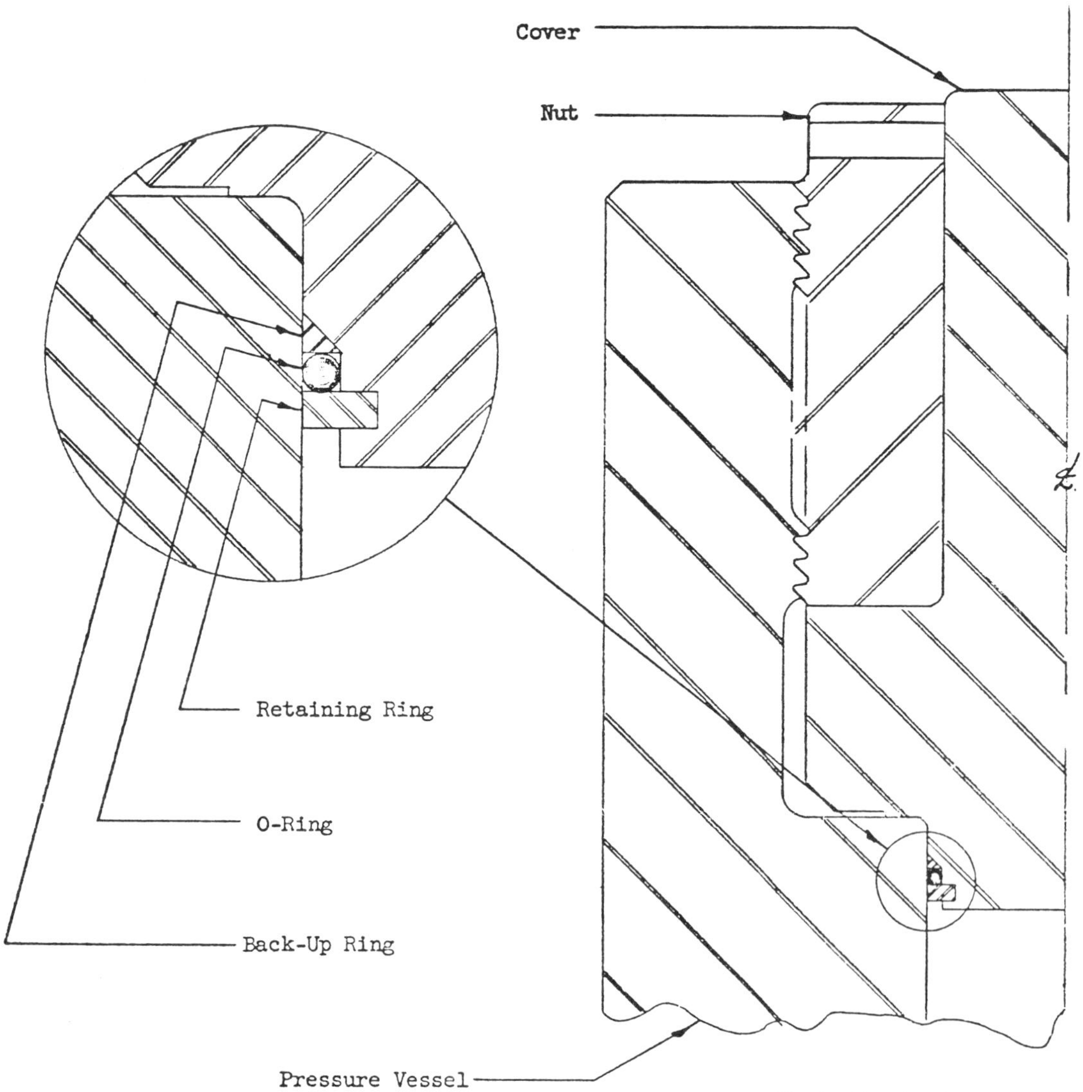

Fig. 2: Vessel seal arrangement

ADDITIONAL SAFETY DEVICES FOR PRESSURE VESSELS

H. D. B. RAES
National Forge Europe—Pressure Systems Division
Sint-Niklaas, Belgium

SYNOPSIS.

A protective shield for a pressure vessel which includes a ductile member for absorbing kinetic energy released if the pressure vessel fails by a portion of the pressure vessel breaking-off and tending to fly away from the remainder of the pressure vessel. The shield will undergo a substantial deformation, without breaking and will absorb the energy of the broken parts, which are constrained to move over a limited distance and are prevented from endangering life or limb, or causing damage to property. All kinetic energy becomes absorbed at stresses far below the rupture strength of the shield construction.

Energy Release.

A pressure vessel pressurized by gas or fluid contains energy that can be expressed by

$$E = \int p\,d\,v$$

Assuming an ideal gas and an adiabatic expansion, we have :

$$PV^{\gamma} = \text{constant}$$

For real gasses we can still use the relation :

$$PV^{n} = \text{constant}$$

by dividing into a number of smaller intervals of pressure (ref. 1).

For a fluid we can express the energy by :

$$E = \frac{1}{2}\beta . V . p$$

where β = compressibility of the fluid at the concerned pressure.
V = inside volume of container.
p = pressure.

β is a function of pressure and can be expressed as

$$\beta = \frac{1}{B}\,p$$

and $$B = \frac{(p - po)}{V - V_o}\,V_o$$

is the bulk modulus of the liquid.

The bulk modulus B is not completely a constant but the use of an average value for the pressure range is acceptable.

This leads to

$$E = \frac{1}{2} \cdot \frac{1}{B} \cdot V \cdot p^2$$

Figure 1 illustrates the energy contained for a vessel pressurized by water and by argon as a function of volume and pressure.

We assume that with a sudden rupture of the pressure container the total stored energy becomes free as expansion energy.
This assumption is very conservative as it requires a complete brittle fracture where no energy is lost in deformation of parts or by friction.
The elastic strain energy released by the pressure container has been omitted since it can be shown to be negligible in comparison with the fluid energy.
In some cases, the release of chemical energy has to be taken in consideration. Especially at the higher pressures, the total released energy can be extremely high, demanding not only extremely careful design, but also making additional safety precautions desirable.

Safety Aspects.

To obtain maximum safety on pressure vessels following points are necessary :

1) Accurate and computer aided vessel design.
   It is absolutely necessary that the stress pattern in the critical areas of the vessel are known including all possible effects of stress raisers.

2) Sound material technology and manufacturing control.
   The material specification is a very important point in the vessel design.

Strength, elasticity, ductility and toughness have to be combined in the most ideal way.
Reliable control during manufacturing has to assure that the material meets the required specifications.

3) In service or periodic inspection of the pressure vessel.
In many cases, the origin of eventual problems can be detected by regular inspection.
The vessel construction should be such that it allows for an easy inspection.

Vessels designed, manufactured and inspected according to well-known and proven techniques have been demonstrated to be very safe. Unfortunately, also a few vessels have been installed, which fell short of one or more of the requirements stated above, and these have resulted in disastrous failures.
But even vessels which can be considered as completely safe in all aspects may fail by unintentional misuse, damage or accident. Although failures of pressure vessels have been rare, the high internal pressures currently utilized in such vessels have increased greatly both the probability of failure of such vessels and the chances of serious damage to life and property in the event of the vessel's failure.
When high pressure vessels have failed, broken parts of the vessels or of the devices used in conjunction with such vessels have tended to fly-away at high velocities and with tremendous kinetic energy from the remainder of the pressure vessels. Because of their velocity and energy, such broken parts have had the obvious potential for causing serious injuries to persons and property in the vicinity of the pressure vessels.
The serious consequences of failure require that besides a safe vessel construction and reliable inspection, additional safety devices should be available to act in the event that such a failure would happen.
Based on many years of experience and on the latest modern techniques, N.F. is building vessels where utmost care is given to all of the above mentioned points. The many hundreds of vessels installed by NATIONAL FORGE have also proven to be very safe. However, for the above mentioned reasons and also for psychological reasons, N.F. has designed for its vessel a special safety device, which, in case of hypothetical failure, will minimize the effects of a catastrophic failure.

Conditions for additional safety device.

To be acceptable as a good additional safety device, following points have to be fulfilled :

1) The safety device should be additional and be no part of the normal stress carrying construction of the vessel.
2) In case of hypothetical failure, all energy of moving parts has to be absorbed in a calculated and controlled way.
3) The safety device should remain unstressed during normal operation of the vessel, and not be subjected to fatigue.
4) The safety device should not be subjected to corrosion or deterioration.

N.F. energy absorption protection shield.

The shield is designed to take up and control the movement of the broken parts whilst at the same time absorbing the kinetic energy of these parts. This protective shield is basically a cage structure, mounted around the pressure vessel in such a way that during normal operation of the pressure vessel, the shield does not become stressed and is not subject to fatigue.
Upon a fracture of the pressure vessel, we can assume that the total kinetic energy of the fragments is roughly equal to the total expansion energy released until the seal is broken, after which the fluid will have a free expansion. The broken parts are constrained by the shield which will allow the parts to move upwards over a limited and calculated distance, while the kinetic energy will be completely absorbed by the deformation of the shield. At the start, the deformation forces are very low, but they gradually increase, eliminating any impulsive loading effects.
The protection shield basically consists of 2 annular rings placed in annular recesses at both ends of the pressure vessel (fig. 2). Both rings are connected by means of a number of metal straps provided around the periphery of the vessel. Each of the metal straps is formed with convex segments along its length. At both ends, the metal straps are connected to the rings by welding. The straps are made out of a ductile material, such as mild steel. Enough clearance is provided between the rings and the end faces of the annular recess, so that, due to elastic elongation of the vessel under normal operation, the shield does not become stressed. As soon as the shield experiences stresses as a result of a fracture, the absorption of energy from the separated parts starts slowly by elastic deformation of the convex segments of the straps, followed by plastic deformation of the same segments, and further followed by an elastic and a plastic deformation of the straightened straps. Fig. 3 shows the absorption of energy in function of the displacement for a typical application.
Number and thickness of straps and number and size of segments are calculated according to the maximum energy to be absorbed. The shield is constructed such that the metal straps will not rupture in the course of absorbing the maximum possible energy from a fracture of the pressure vessel and annular rings and welds are designed to be stronger than the straps.

From the above description it is obvious that the protection shield will essentially work in the case of broken parts moving in the axial direction, such as for a fracture at the vessel head, which would be the most likely to happen and would be also the most catastrophic failure.
A crack in the cylindrical part would under influence of the higher tangential stresses in this region appear as an opening through the bore, without having parts fragmenting. A thin cylindrical plate, placed between vessel wall and metal straps, will protect in this case against unlikely small parts and high pressure fluid jets.
Tests on the absorption of energy based on this principle have been done at the Laboratory of Prof. Soete at the University of Ghent, and have proven satisfactory performance.
An alternative construction of the described protection shield is shown on Fig. 4. Here the formed metal straps are replaced by a number of rods which contain a bushing spaced between one or two annular rings and the end of the rods. The energy absorption is accomplished here by crumpling the bushings. A practical application of the described protection shield is shown on fig. 5. It concerns here a vessel for isostatic pressing, having an internal diameter of 900 mm and a length of 2 500 mm, with an operative pressure of 200×10^6 N/m^2. The vessel makes a complete cycle every 15 minutes.

Conclusion :

The large amounts of energy stored in high pressure vessels can cause disastrous effects in case of failure.
Such vessels therefore require an extremely safe conception on design, stress analysis, material technology, workmanship and inspection.
Besides this, additional safety devices in case of hypothetical failure are desirable. The N.F. protection shield absorbs the kinetic energy of the broken parts torn off in the event of a fracture, and allows these parts only to move over a short distance, preventing them from causing further damage.
This shield is not a stress carrying device and is not taken into account in the calculation of the pressure vessel.
It is thus a fully extra feature, mounted around the vessel for being an additional safety in case of hypothetical failure.

Acknowledgment

We would like to thank Pr. W. Soete and his co-operators of the "Laboratory of Strength of Materials and Welding Technology" of the University of Ghent, Belgium, for the excellent help and additional advices which we have received during our tests.

Patent Application No. 59985/73 in Great-Britain.

References.

1. Manning W.R.D. and Labrow S. "High Pressure Engineering" (Leonard Hill Books).

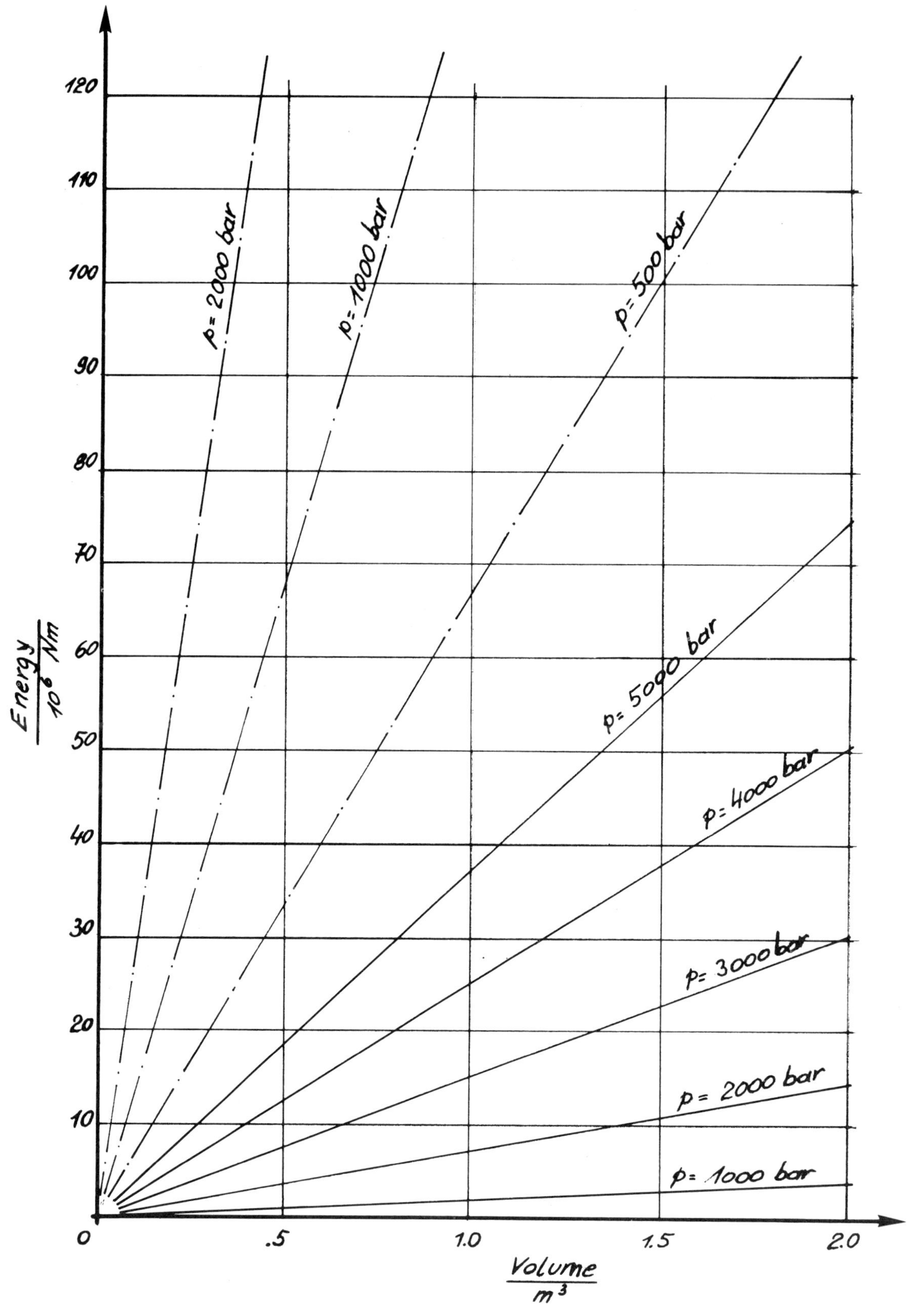

Fig. 1: Expansion energy stored in a pressure vessel for argon — — and for water——

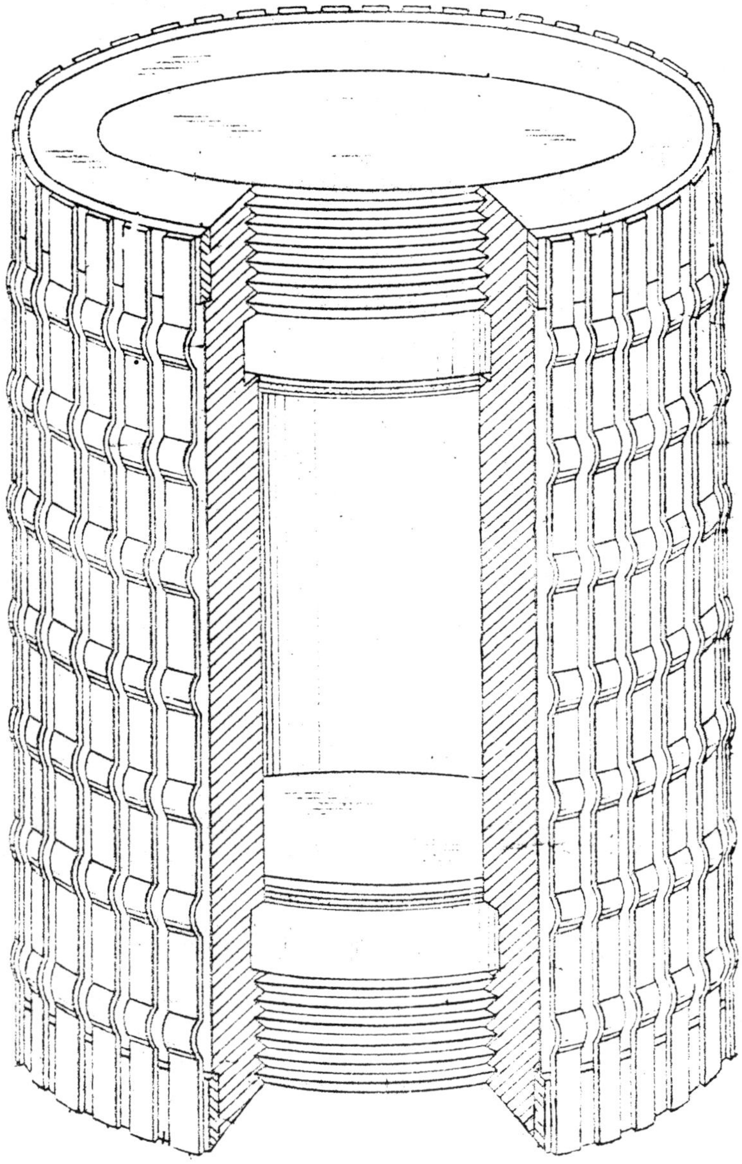

Fig. 2: National Forge protection shield, execution with straps and convex segments

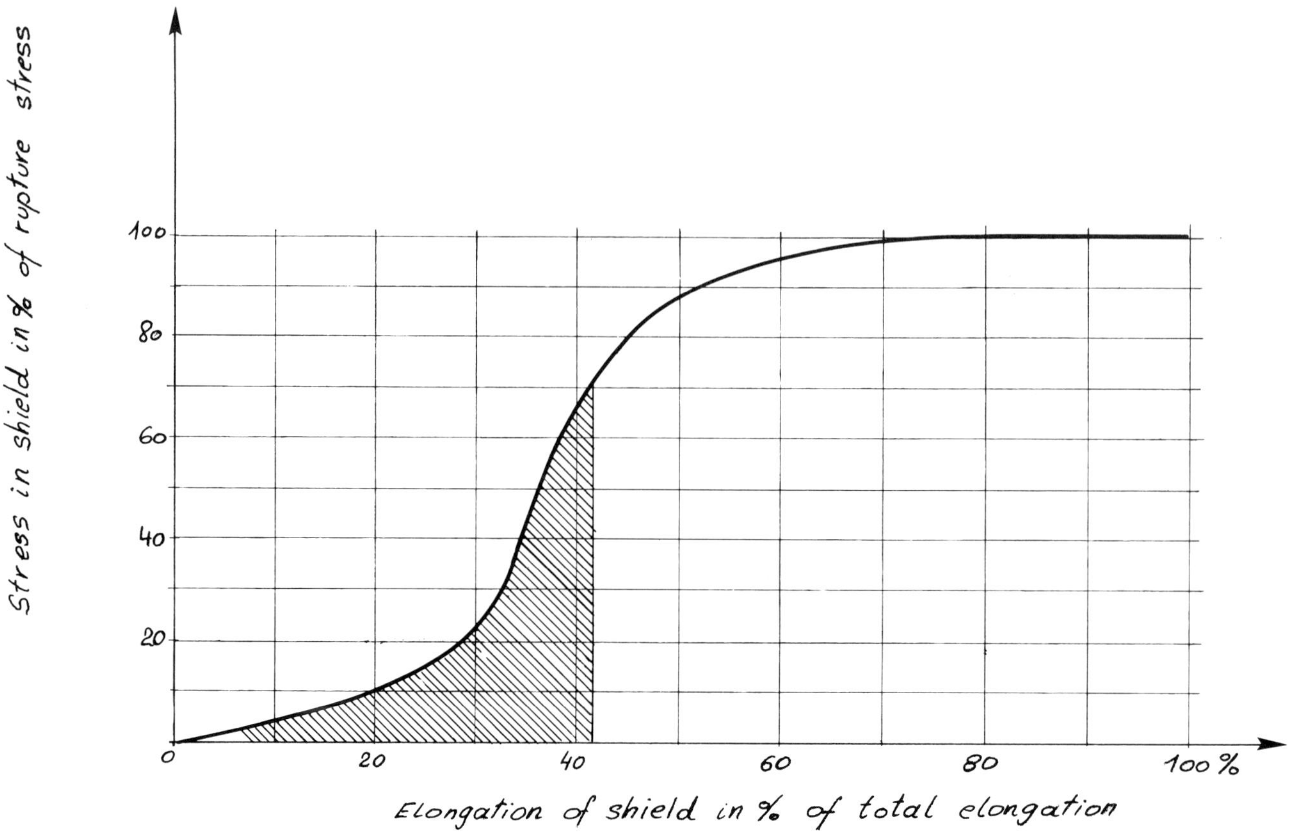

Fig. 3: Absorption of kinetic energy in function of total elongation of the shield

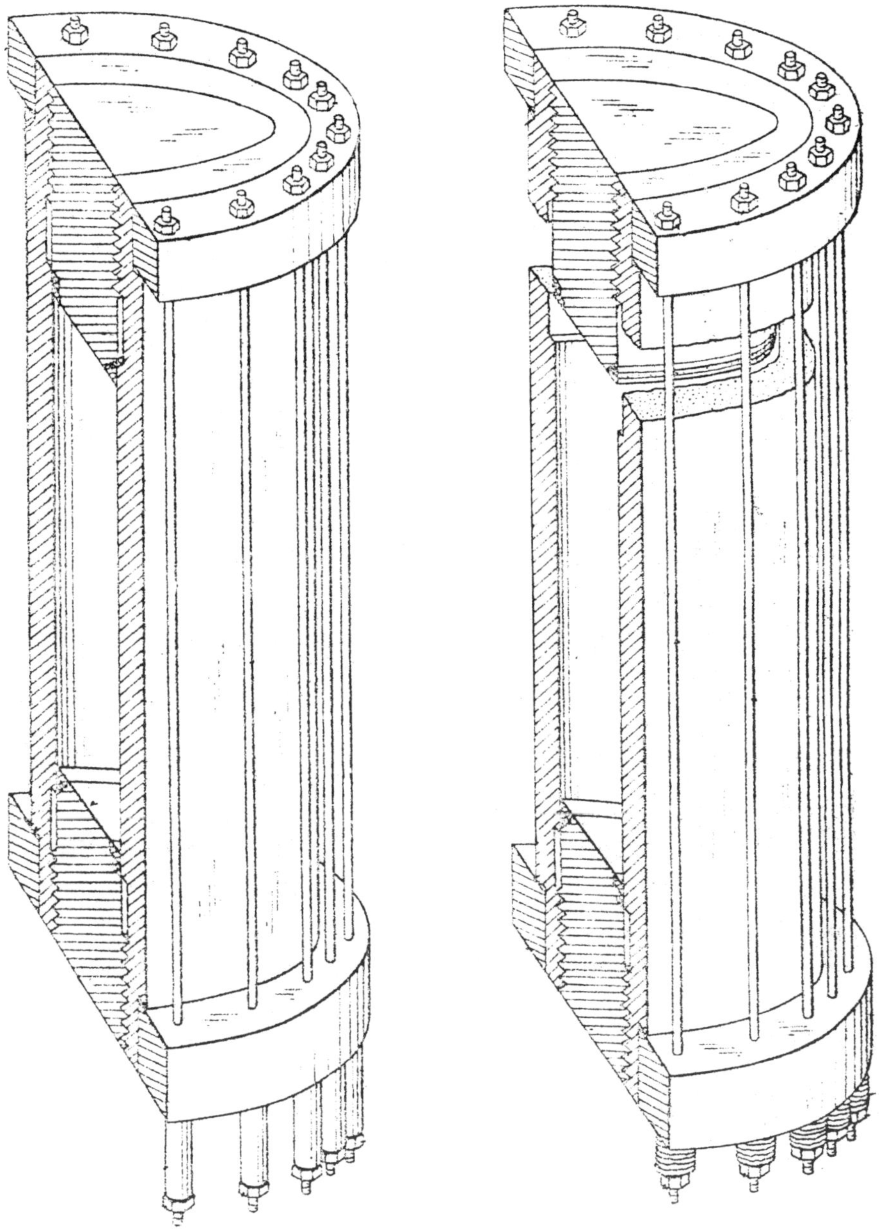

Fig. 4: Protection shield, execution with crumpling bushings

Fig. 5: Practical application of National Forge protection shield. Vessel for 2000 bars, 900 mm I.D. by 2,500 mm inside length

REVIEW OF DESIGN FOR ADEQUATE GAS RELIEF FROM VESSELS AT HIGH PRESSURE

R. STEPHENSON, MSc, CEng, MIMechE
Imperial Chemical Industries (Plastics Division) Limited, Welwyn Garden City, Hertfordshire, England

SYNOPSIS. This paper gives a brief review of the various aspects that have to be considered when designing for safe explosion-relief of high pressure containers. The main theme that is discussed is the sizing of the relief vent-port and the vent-pipe for the situation where pressure relief must be fast and disposal of the explosion products must be completed without hazard.

INTRODUCTION

1. The safety of gas containers under pressure relies upon the integrity of the material of construction and the adequacy of the design. However, in order to ensure that the pressure, for which the material and the design has been specified, is not exceeded then pressure vessels must be fitted with adequate means of explosion relief and fast evacuation or dispersion of the products of an explosion to a safe area. Also, there must be due regard for the design of the area where high pressure gas vessels are working so that proper protection is afforded to personnel and equipment.

Neglect of these considerations promotes the conditions for a potential tragedy. They must therefore be taken fully into account at the outset of the design and development stage of high pressure equipment, no matter how small or temporary the proposed installation.

The procedure for safe design of equipment associated with hazard in the Petrochemicals Industry is to consider that the worst will happen. In the context of this paper this would imply that an explosion in a high pressure container will occur; and the vessel is likely to burst under the maximum release of energy from the explosive material inside that vessel.

In this paper the stage is considered where the explosion has happened and the problem is to preserve the integrity of the pressure vessel. Safe design would demand that the vessel should not fail under prescribed working conditions throughout a given working life. But, in the event of an inadvertent rise in pressure, adequate means of pressure relief and a method for fast evacuation or dispersion of the products of an explosion to a safe area must be fitted.

RELIEF-VENT AREA

2. The fast relief of over-pressure in high pressure systems is the most important aspect of safety-vent design. The cause of over-pressure that is of main interest here is the explosion of gas, either the self-heating and self-accelerating decomposition or the progressive reaction of explosive burning in a gas, explained by Groothuizen et al (1); the detonation and its effects are not considered here but reference can be made to the paper by Munday (2).

Clearly the vessel relief vent that is designed for relief of a fast explosion can be assumed to be equally efficient for the relief of a slow rise of over-pressure, which may occur as a result of an operational fault in a vessel pressurising system.

The rate of pressure rise of an explosion inside a closed container depends primarily upon the properties of the contained gas. Figure 1 shows two simple examples of low pressure explosions of methane and hydrogen in air and how their pressure versus time curves and rates of pressure-rise differ.

However the volume of the container, the induced gas turbulence, and the intensity of ignition can also have considerable influence on the violence of the explosion. Figure 2 shows the sort of effect that the turbulence of exploding gas may have in a vented vessel. The three sketches represent three stages in the development of an explosion in a vessel with the relief vent failing in the second sketch. The rush of gas from the vessel gives rise to more violence in the explosion with the onset of an extended combustion front due to turbulence and this is shown in the lower sketch.

Comparisons of methods of venting various low pressure gases from enclosures have been conducted by Heinrich (3), Donat (4) and Bartknect (5), but the results only seem to emphasise the scant knowledge that is available in the high pressure field. There appears to be no accurate method for predicting theoretically the violence of an explosion in a vented vessel at any pressure. The only sure way of assessing the violence to be expected in the high pressure system explosion is to carry out experimental controlled tests on the small-scale vessels and to relate these laboratory results to the larger vessels.

One approach to the safe conduct of small-scale tests is to provide the experimental vessel with an over-sized relief vent in the first instance. Such proportions as to be considered adequate, and even conservative for controlled experiments in the laboratory, can be found by reference to the American Petroleum Institute Guide for pressure relief and depressurising systems (6), where the ratio of vent area to vessel vapour volume is suggested as being no less than 33 mm^2/litre.

After the preliminary venting experiments have established a reference pressure versus time curve then the vent area can be progressively reduced over a series of tests, A smaller vent area is selected that will give adequate fast pressure relief without causing large pressure overshoot beyond the maximum setting of the relief device. The final pressure versus time curve for the small scale tests will provide a record which serves as a characteristic curve for the effect that the ignition-energy, pressure, and turbulence, have on the gas discharge for that geometry. Some provision however, to reduce the influence of heat transfer on the explosion phenomena, must be made in the small scale tests, as there would obviously be a significant difference in the heat losses between small and large high pressure vessels per unit mass of gas contained.

Figure 3 shows the expected characteristics of pressure versus time curves for vented explosions. Clearly there are limits to how slow a release should be because relief that is too slow leads to excessive pressure over-shoot. Relief that is too fast leads to problems related to high gas velocities and turbulence, either in the vessel or in the discharging gas, and excessive forces on the relief vent assembly and on the vent pipe assembly.

When considering the pressure relief of low pressure containers the scale up of the relief area from the small scale vessel to the full size vessel has been suggested by Donat (7) to follow the cubic law of explosions:

$$\frac{A_2}{V_2} = f_1 \left[\frac{V_1}{V_2}\right]^{1/3} \qquad 1$$

where A_2 = vent area of full-sized vessel
V_1 = volume of small-scale vessel
V_2 = volume of full-sized vessel
f_1 = specific opening vent area to vessel volume = A_1/V_1
A_1 = vent area of small-scale vessel

An alternative, and much more conservative approach to scale-up would be to use a linear relationship:

$$\frac{A_2}{V_2} = f_1 \qquad 2$$

Many writers assert that the vent area determined in the small scale experimental vessel must be converted linearly, ie in direct proportion to the full scale vessel. However experimental work on moderate pressure relief from explosions refute this argument in favour of the cubic law.

It is reasonable to suppose that venting at the high pressures considered here may also follow the cubic law for scaling up but more work needs to be done at high pressures to verify whether this is so. It is often advantageous to restrict the maximum vent-area of a large high pressure container to within well defined limits because of problems associated with excessive discharge rates, as has already been suggested. There may be a need to avoid discharge rates that could give rise to large pressure differentials inside the vessel causing damage to internal structures, such as a thin walled inner shell which could collapse or break loose causing a relief vent blockage for example; and side thrusts on the vessel and reactions on pipework supports must also be kept to reasonable levels.

To summarise, it would appear that the designer has apparently no clear theoretical guide from the literature for sizing explosion relief-vents for very high pressure containers, but a way may be found with empirical data. Now to look at the problem of providing a means of closing the relief vent for vessels on work.

RELIEF DEVICE

3. A relief valve, a dump valve or a bursting device is used to close the relief vent of a pressure vessel under normal working conditions. The problems associated with maintaining a reliable and leak-proof relief valve at high pressure, added to the cumbersome bulk of the preloading mechanism required usually forbid its use; while the dump valve is normally to be found only under well specified conditions for an over-pressure with moderate pressure rise which justify the expense of a temperature rise signal to be made available to operate the valve movement via a servo-control.

The most simple and reliable method for relieving high pressure is the bursting device of which there are three main types: The bursting disc, the bursting cap, and the bursting tube; Figure 4 shows the three types diagramatically. First the bursting tube suggested by Manning (8) which can be made reliable enough to fail within 2½% of the chosen bursting pressure. It can be more expensive than the other two arrangements and may require more care in assembly but can clearly be used for large vent sizes at higher pressures. If pairs of tubes are machined together then one can be used as a means of accurately calibrating the other by bursting it under pressure. This technique of calibrating which can be used with confidence on double ended tubes can hardly be used on single ended tubes and these must be calibrated against precise measurements taken after machining.

Bursting caps, described by Donat (4), provide an extremely simple and effective arrangement for large bore high-pressure relief-vent closure. The bursting cap is a short blind-ended tube fitted over the vent port. Excessive pressure causes the end cap to fail in tension, and the point of failure can be predetermined by a machined annular groove cut into the outer side-wall of the cap. The cap is propelled forward on rupture (and at the high pressures and small diameters considered in this context of 100 MPa +, the inertia effects have negligible effection speed of response). However provision must be made to catch end caps to prevent them passing out through the vent pipe and becoming dangerous missiles.

Bursting discs, which are widely used for low and medium pressures have an operating limitation (imposed by the need to keep the disc membrane stresses low) and this is defined approximately by Marston Excelsior (9) as the product of vent diameter and pressure rating which rarely exceeds 1000 mm MPa (or more simply: 10^6 m Pa). Nevertheless where they can be used then they provide reliable protection at comparatively low cost, and maintenance problems are minimal. Reliability is normally accepted as ± 5% if the "operating" temperature is accurately known.

Often discs are attached to a sleeve or liner assembly in a vessel wall, as described by Ziefle (10), and this is shown in Figure 5. This protects the vessel body or disc holder from corrosive high temperature and high velocity gas. Needless to say the velocity of gas in the narrow relief port is high (local sonic conditions are reached) and the corollary from this is that heat and mass transfer is also high (especially in the recirculating flow regions at the bore inlet); eg this condition would promote hydrogen diffusion into the steel of the vessel wall, in the absence of a sleeve, if free hydrogen were present in the gas. The sleeve or liner can be renewed as a routine with disc replacement; the position of the disc at the vessel bore end of the sleeve should be noted where the disc is operating near to temperature of the vessel contents so that it acts as a temperature sensitive pressure-relief device.

A similar device to the bursting-disc is the shear-disc, described by Lake and Inglis (11), in which the central portion of the disc is considerably thicker than the perimeter clamped in the disc-holder. Over-pressure does not cause bending of the central portion to failure, as in the bursting-disc, but causes shear failure at the change of section near the rim. The failure then leaves a full bore opening with the central section projected out by the escaping gas. Accurate machining and measurement are needed to maintain predicted failure pressures however, due to the fact that a complicated stress pattern and yielding/failure mechanism can be readily influenced by slight changes in geometry at the change of section.

VENT PIPE

4. The products of a vessel explosion are normally discharged into a vent pipe. The vent pipe must serve to transfer discharge products from the pressure vessel to a place of safety. At the same time its design must not impede the flow from the pressure vessel and so cause build up of back pressure on the relief vent; and it must withstand shock load. The first step in vent pipe design is to calculate the maximum mass flow rate that the pipe must contain. As this maximum flow will almost always be the initial discharge of the vessel contents immediately after the relief vent ruptures there will normally be no need to consider the complication of calculating the overall discharge with time for vessel pressure relief.

The discharge for a high-pressure vessel will be choked flow so that the standard equation for choked flow through a sharp-edge orifice can be used with an appropriate discharge coefficient, ie assuming adiabatic let-down of gas through an orifice with a discharge coefficient of 0.6

$$w = D_c A \left[\left(\frac{2}{m+1} \right)^{\frac{m+1}{m-1}} m P_o \rho_o \right]^{\frac{1}{2}} \qquad 3$$

Brodkey (12)

where w = maximum discharge rate from vessel
m = mean adiabatic index of expansion
P_o = vessel gas pressure
ρ_o = vessel gas density
D_c = discharge coefficient

It must be noted here that the index of adiabatic expansion, m, for a gas at high pressure, is not the index associated with the ideal gas (ie the ratio of the specific heats) but one appropriate to that particular gas at that vessel pressure and temperature as the relief vent blows.

The gas discharging into the blast-pipes can set up considerable shock loads on the pipe-work which will be seen as impact reaction forces, Boyle (13). There can be no doubt that the best design of vent pipe is one that is straight and does not require joints or bellows, but the opportunity for providing such a design is rare.

There is, at present, nothing published that can be taken as a definitive method for analysing the pressures and loads to be encountered in a contorted pipe with a sudden gas discharge. The best that can be achieved in analysing the situation is to follow the method for calculating steady state pipe flow. [Brodkey (12) and Streeter and Wylie (14).] Pipe support reactions can then be deduced from the derived values of momentum changes and pressure-drops, at changes of pipe section and bends, with a 100% increase allowed due to shock loading.

Using this method the pipe flow is assumed to be adiabatic friction flow (no heat transfer through the pipe walls). The end-to-end pressure ratio, P_E/P_I, can be related to the pipe friction losses, $\sum K$, and inlet Mach number, M_I, by the equations:

$$\frac{P_E}{P_I} = M_I^2 \left[\frac{\gamma - 1}{\gamma + 1} \left(\frac{2}{M_I^2 (\gamma - 1)} + 1 \right) \right]^{\frac{1}{2}} \qquad 4a$$

$$\text{and} \sum_I^E K = \frac{1}{\gamma} \left[\frac{1}{M_I^2} - 1 \right] + \frac{\gamma + 1}{2\gamma} \ln \left[\frac{(\gamma + 1) M_I^2}{2 + (\gamma - 1) M_I^2} \right]$$

$$M_E = 1 \qquad 4b$$

The gas is assumed to be accelerated by isentropic expansion from the total gas pressure, P_T, to a stream pressure, P_I, that can be related to the local gas velocity at that point (from equations 3) by the equation:

$$\frac{P_T}{P_I} = \left[\left(\frac{\gamma - 1}{2}\right) M_I^2 + 1\right]^{\frac{\gamma}{\gamma - 1}} \qquad 5$$

where $\gamma = c_p/c_v$, the average index of adiabatic expansion

The total pressure-drop and mass flow situation can now be described for flow into and along a pipe using these two sets of equations 4 and 5; the two graphs in Figure 6 show how these equations are plotted for the value of $\gamma = 1.4$.

The value of the velocity head loss, $\sum K$, can represent a total of any number of velocity head losses representing, pipe-bends, pipe-junctions, or changes in pipe diameter, as well as straight pipe frictional losses, Miller (15).

GAS COOLING

5. In most instances the discharged products resulting from an explosion in a vessel will be at very high temperature. It is convenient to consider the gas in the vent pipe as being the same temperature as the gas up-stream in the vessel when designing the vent pipe. However to discharge hot and inflammable gases (such as a hydrocarbon with possibly hot entrained carbon) into the atmosphere without applying some means for cooling the gas is clearly dangerous, Coevert et al (16).

Figure 7 shows a method of cooling the gas which is to arrange for automatic drench of the vented products with water in the vent pipe, Durand and Marzais (17). The amount of water required for a given situation can be simply calculated by proportioning the maximum energy of the explosion, of all the contents of the high pressure system protected by the relief mechanism, to the energy required to raise the water to steam at vent pipe pressure.

The method for automatically releasing the water has not been detailed in Figure 7 but a proprietory quick-acting valve such as the type developed by the Summerfield Research Station (18), which operates with a small over-pressure on a trigger device, can be used for this purpose.

Cooling water flowing into the discharging gas stream is almost certain to be broken up into drops by the impinging gas stream. Using the two-phase flow analysis of Golan and Stenning (19) and Collier et al (20) the hold up of water in the gas, H, suggests mist flow of small droplets entrained in the gas. This is important to note because the surface area available for heat transfer is most likely to be a controlling factor. The size of these droplets can be only estimated because the accuracy of the method remains in doubt. The estimated final drop size d, is given by a method due to Hughmark (21) which provides in turn, an estimated interfacial area, a, using a relation from Calderbank (22)

$$a = \frac{6H}{d} \qquad 6$$

The heat transfer may now be calculated to make dry steam at vent-pipe mean pressure.

Having calculated the heat transfer required, there remains the question of heat transfer rate, this must be such as to reduce high gas temperatures (1000°C+) to a safe value by the time the gas discharges from the vent-pipe. Eckert and Drake (23) give a heat transfer coefficient for condensing steam as between 5KW/m²°C and 30 KW/m²°C, while the two-phase flow calculation, for the air/water mixtures to be considered in vent-pipe drench systems, gives a heat transfer coefficient of between 2.5 KW/m²°C and 25 KW/m²°C using the method of Collier et al (24). Using the most conservative arrangement of heat transfer rate and interfacial area then total heat transfer rates may be calculated.

The use of water brings its own penalty in the form of higher pipe-flow pressure drop for two phase flow which can be in the region of 40% higher for mist flow, Collier et al (20).

DISPOSAL

6. The situation arising from the discharge into the atmosphere of products which are denser than air has been studied by Bodurtha et al (Fig.8). This condition is critical to safety if large masses of inflammable gas are allowed to accumulate at ground level or the condition causes an unacceptable concentration of pollution on the ground (from wet or dry carbon for example). The main conclusion from this work is that high vent pipe velocities, that arise from a minimum practical pipe diameter, are required to reduce the concentration of inflammable or pollution product. However the equations for calculating for hazards in the vent pipe area are not at all satisfactory for general application and there is much to be said for directing the discharge horizontally into an area which is remote and isolated from personnel and equipment, or into a catch pot or quench, as described by Boyle (13) and in API521 (6).

CONCLUSIONS

7. The information now available for designing relief arrangements for high pressure vessels, working in excess of 100 MPa, allows only just adequate relief systems. Designers could benefit from more active Research into the safe venting from violent explosions and decomposition at high pressures along the lines of work now currently in progress for explosions from low pressure containers.

REFERENCES

(1) Groothuizen, Th M et al, "Phenomenology, Test Methods and Case Histories of Explosions in Liquids and Solids", 1st Int Loss Prevention Symp, The Hague/Delft, 28-30 May 1974, pp 239-251.

(2) Munday, G "Detonations in Vessels and Pipelines". The Chem Engr April 1971, pp 135-144/152.

(3) Heinrich, HJ "The Design of Pressure-Release Vents for the Protection of Plants from Explosion Hazard in the Chemical Industry" (in German), Chemie, Eng Techn 38 11 1966 pp 1125-1133.

(4) Donat, C, "Use of Bursting Safety Devices for Slow and Rapid Increases in Pressure" (in German), Chemie Ing Techn 45 11 1973 pp 790-796.

(5) Bartknecht W "The Course of Gas and Dust Explosions and their Control" 1st Int Loss Prevention Symp, the Hague/Delft 28-30 May 1974, pp 159-174.

(6) API RP521, "Guide for Pressure Relief and Depressuring Systems" American Petroleum Institute Publication.

(7) Donat, C, "Explosion Pressure Relief with Bursting Discs and Explosion Hatches", 2nd Int Symp, on Prevention of Occupational Risks in the Chemical Industry, Frankfurt 21-23 June 1973, pp 289-308.

(8) Manning, W R D, "Tubular Bursting Elements" Trans Inst Chem Eng 31 1953, pp 138-142.

(9) Marston Excelsior, "Marston Bursting Discs" Marston Excelsior Limited catalogue 1974, p 12.

(10) Ziefle, R G, "Design Approaches to Safe High Pressure Polyethylene" Safety in High Pressure Plants Symposium, A Inst Chem Eng 21-24, May 1972 paper 43a.

(11) Lake G G and Inglis, N P "The Design and Manufacture of Bursting Discs" Proc Inst Mech, Eng 142, Feb 1940 (4) pp 365-378.

(12) Brodkey R S "The Phenomena of Fluid Motions" Addison-Wesley 1967, pp 181-199.

(13) Boyle, W J, "Sizing Relief Area for Polymerisation Reactors" Chem Eng, Prog 63, 8, 1967 pp 61-66.

(14) Streeter, V L and Wylie, E B "Fluid Mechanics" 6th Edit, McGraw-Hill, 1975, pp 364-365.

(15) Miller, D S Internal Flow; A Guide to Losses in Pipe and Duct Systems" British Hydromechanics Research Association 1973.

(16) Coevert, K, Groothuizen, Th M, Pasman, H J and Trense, R W, "Explosions of unconfined Vapour Clouds" 1st Int Loss Prevention Symp The Hague/Delft, 28-30 May 1974 pp 145-157.

(17) Durand, P and Marzais, JJ, "Explosion Preventive Device" 11th Annual Meeting of EHPRG April 1973 London, p 17.

(18) Summerfield Research Station, "Pressure Sensitive Valve SPV 100, Imperial Metal Industries Catalogue.

(19) Golan, LP and Stenning, A H "Two-Phase Vertical Flow Maps" Proc Inst Mech Eng 184 (3C) 1970, pp 108-114.

(20) Collier, J G, Ward, J A, Claxton, K T and Chisholm, D "Two-Phase Pressure Drop and Void Fractions in Tubes" HTFS, DR 15, AERE Harwell.

(21) Hughmark, G A "Drop Breaking in Turbulent Pipe Flow" AICh E Jnl, 17 (4) July 1971, p 1000.

(22) Uhl, U W and Gray J B (Ed) "Mixing, Vol 2, Chap 1 "Mass Transfer"by P H Calderbank, Academic Press, 1967, p 3.

(23) Eckert, E R G and Drake, R M "Heat and Mass Transfer 2nd Edit, "McGraw-Hill, 1959, p 12.

(24) Collier, J G, Lacey, P M C and Polling D J "Designing for two-phase Flow ; Part V" NEL Report No 389, January 1968, Figure 10.

(25) Bodurtha, F T, Palmer, P A, and Walsh, W H, "Discharge of Heavy Gases from Relief Valves" Chem Eng Prog 69 4 1973 pp 37-41.

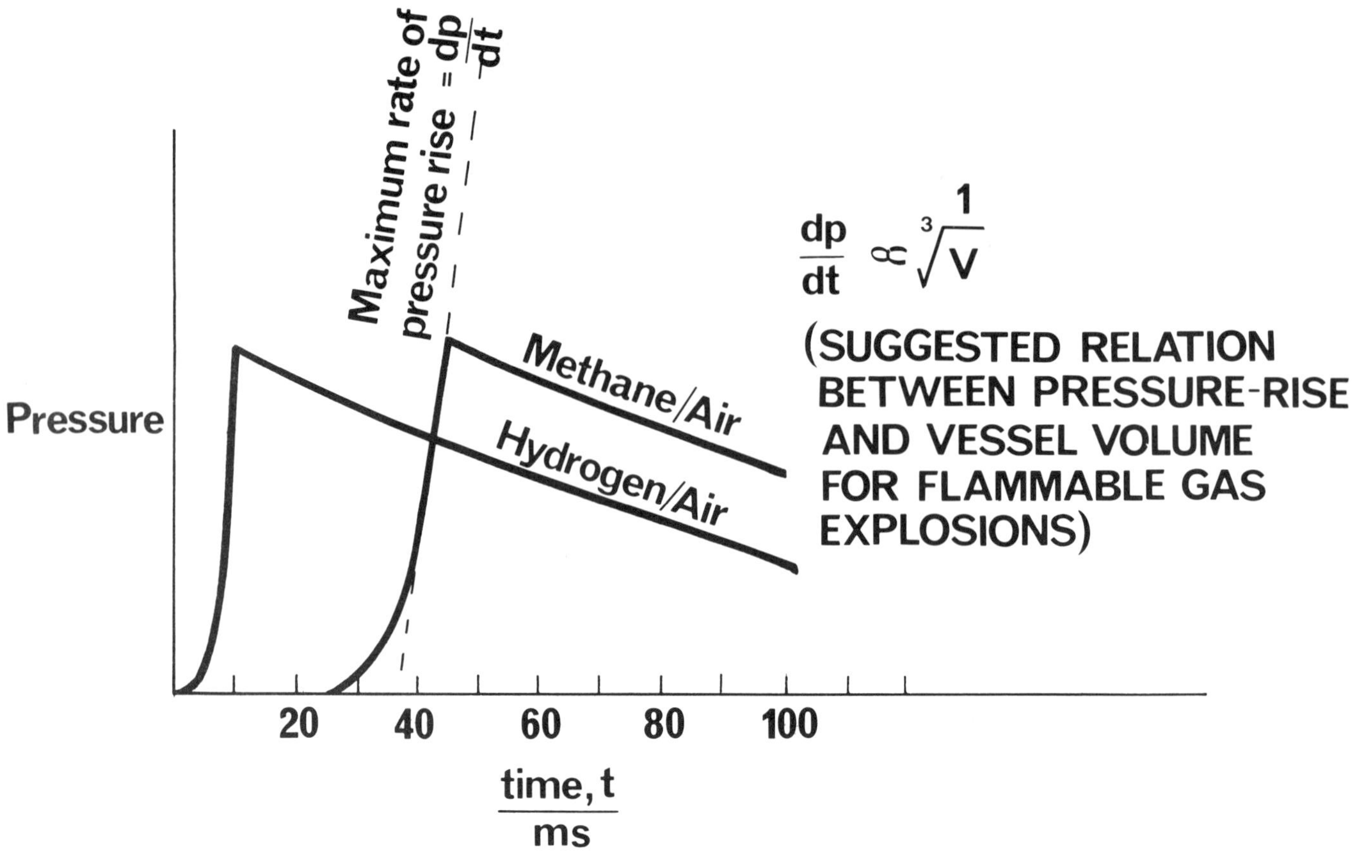

Fig. 1: Course of the explosion of methane and hydrogen (Bartknecht/Donat)

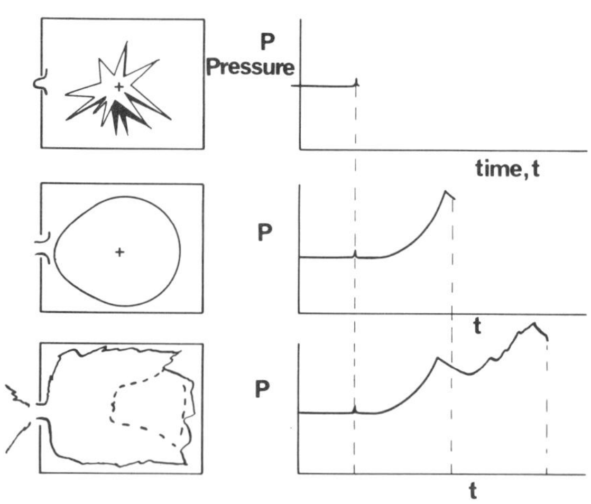

Fig. 2: Diagrammatic effect of turbulence on pressure relief (Heinrich, 1966)

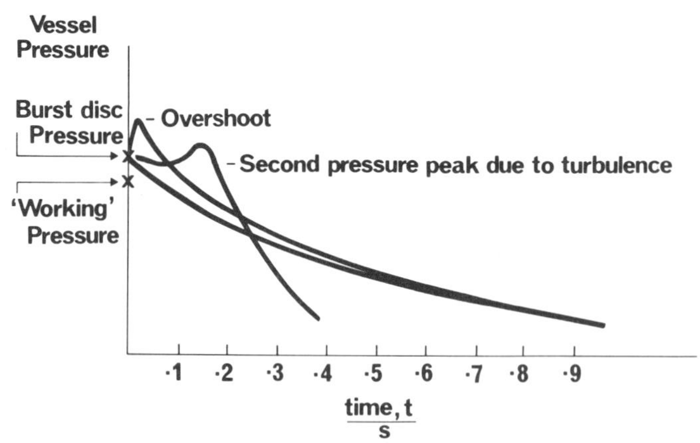

Fig. 3: Pressure decay curves (Heinrich, 1966)

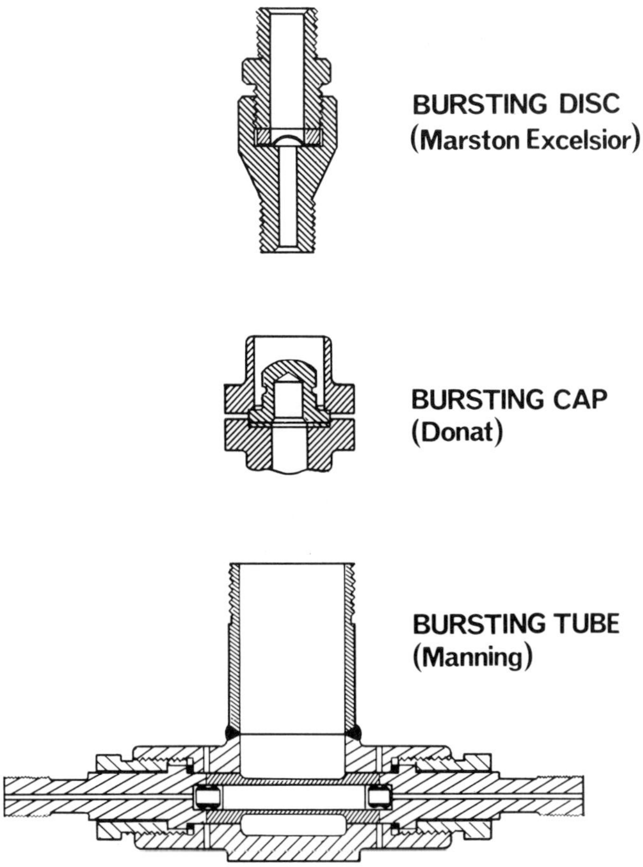

Fig. 4: Typical bursting devices

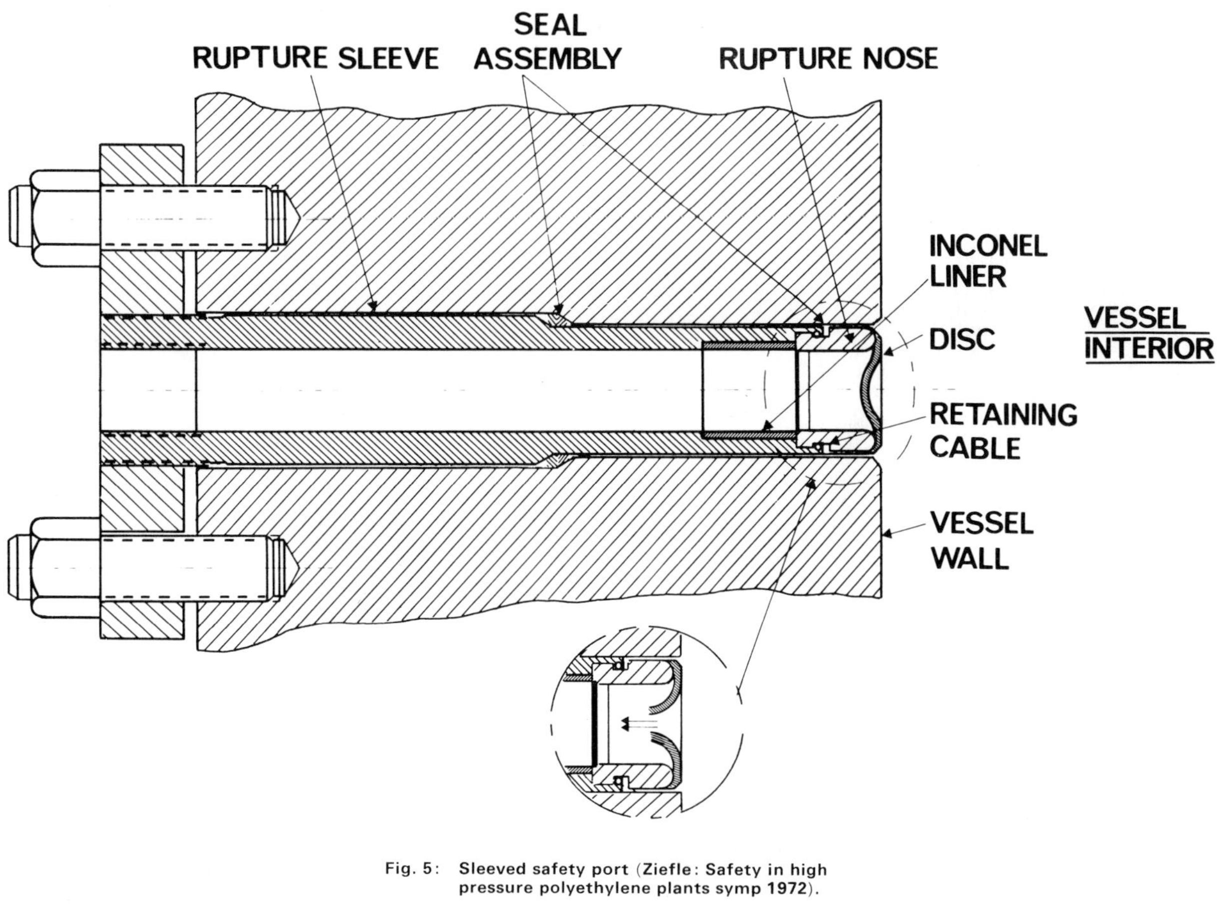

Fig. 5: Sleeved safety port (Ziefle: Safety in high pressure polyethylene plants symp 1972).

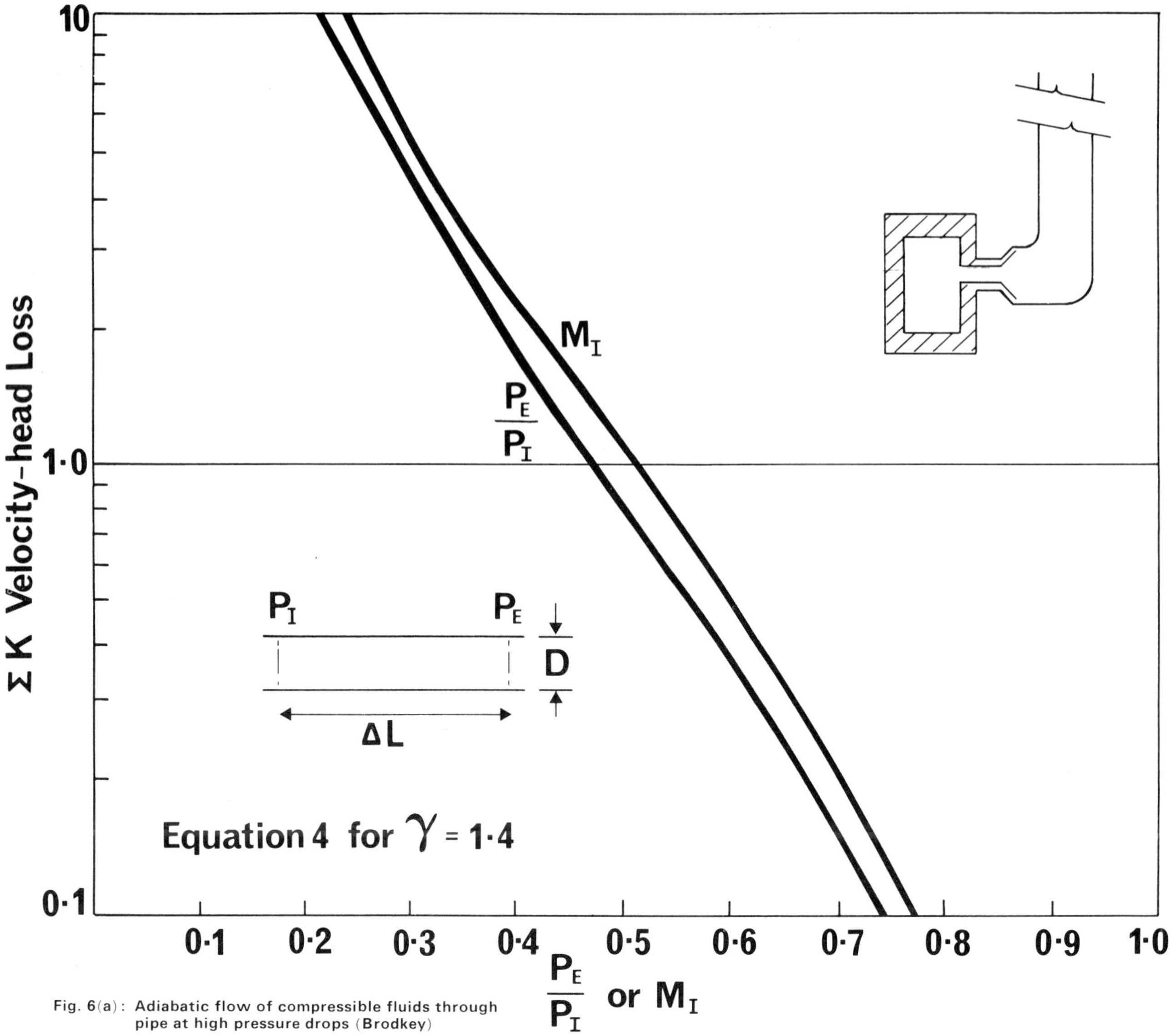

Fig. 6(a): Adiabatic flow of compressible fluids through pipe at high pressure drops (Brodkey)

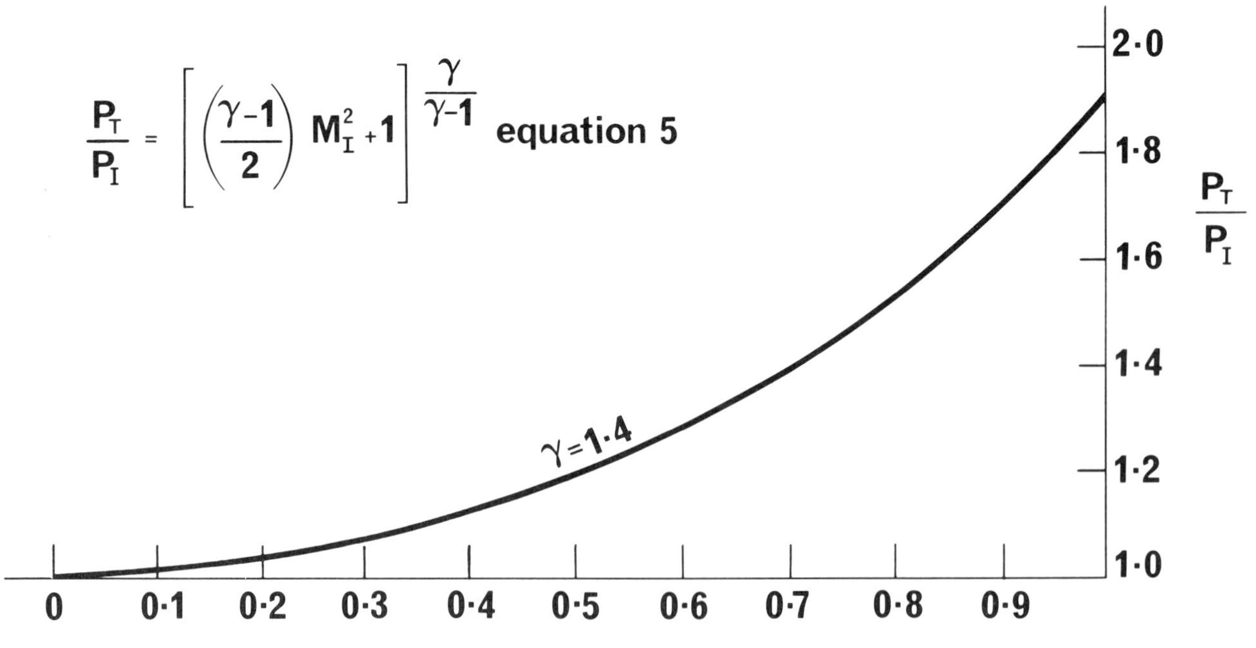

Fig. 6(b): Pressure drop and final mach number for isentropic expansion

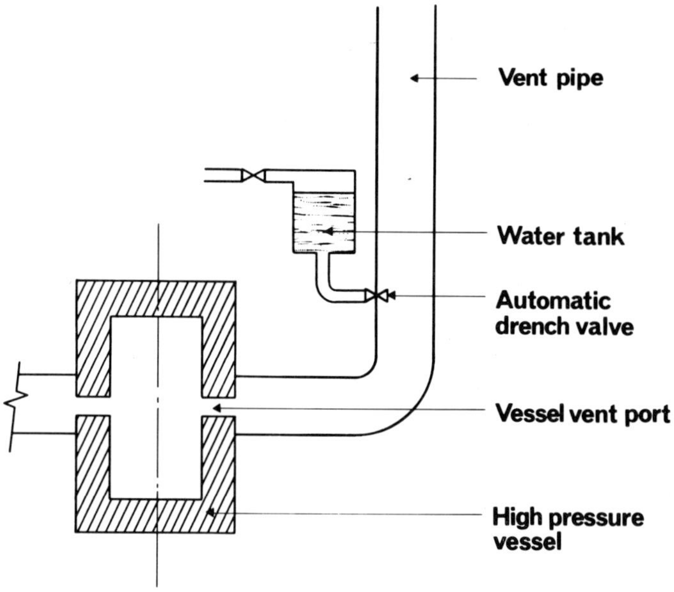

Fig. 7: Water drenching

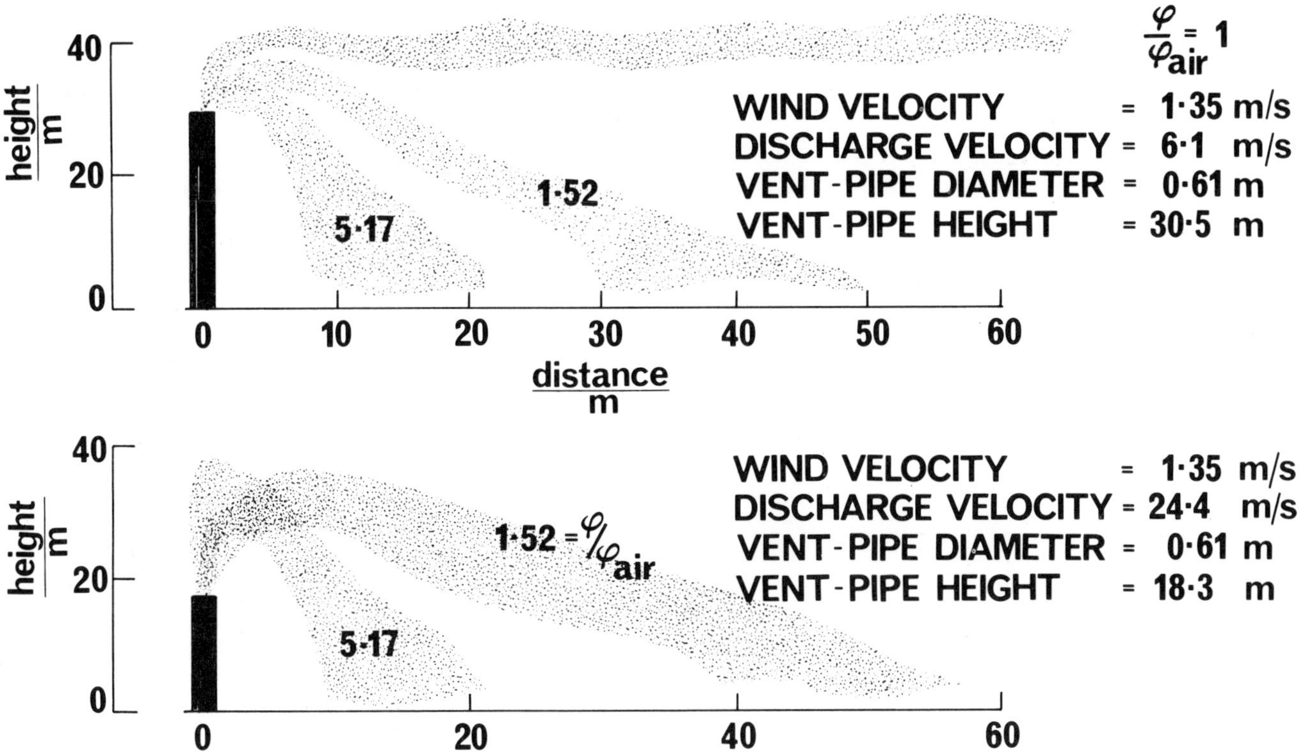

Fig. 8: Study of plume behaviour from a blast pipe (Bordurtha, Palmer and Walsh)

SOME SAFETY ASPECTS ON HYDROSTATIC EXTRUSION IN PRODUCTION PLANTS

KURT ZANDER,
Process Technology, Sales Promotion, Documentation Industrial Division, QUINTUS Department, ASEA, Västeras

SYNOPSIS

During recent years the development of the hydrostatic extrusion technique has, to a great extent, been aimed at various applications in industrial production. With regard to the use of oil as the pressure medium and its setting under high pressure in the extrusion plant, the required means of safety for the operating personnel will be located to the extrusion press and to the run-out arrangements for the extruded product.

INTRODUCTION

1. During recent years the development of the hydrostatic extrusion technique has, to a great extent, been aimed at various applications in industrial production. A few production plants have been put into operation, and in the next years it can be expected that many new plants will be built, especially for the purpose of producing profiles and tubes in aluminium, aluminium alloys, copper, copper alloys, and also composites of non-ferrous metals. As the hydrostatic extrusion technique generally includes the use of a very high oil pressure in the order of 1000-1500 MPa, particular attention has to be paid to the creation and handling of this pressure, and to the possible influence the pressure may have on the extruded product. In this lecture such arrangements of safety will be discussed which are related to the use of the high oil pressure for the extrusion process. Other safety arrangements based on more traditional manufacturing proceedings may also be necessary.

HYDROSTATIC EXTRUSION PLANTS

2. A hydrostatic extrusion plant comprises both conventional and high pressure equipment. Generally, the equipment is arranged in a line, the lay-out of which is made to suit the major portion of the processing of the manufactured product. Sometimes it is desired to make a mix of products in the same high pressure apparatus (extrusion press) with only minor changes in the auxiliary equipment. The extrusion press may be arranged for vertical or horizontal extrusion, the latter being the most common for production purposes, as it generally provides a more convenient factory floor plan and flow of processed material.

3. Figure 1 illustrates in principle the lay-out of a hydrostatic extrusion plant for the production of a composite product. In this case the extrusion is performed with the billet at room temperature. The billet preparation equipment comprises machines for tapering the billet front ends, cleaning and degreasing, assembly of the composite billet, and transportation to the extrusion press. The product reception equipment comprises guiding and cooling section, a coiler for those products which have smaller cross sections and a line with a cutter for the reception of straights.

4. Hydrostatic extrusion can also be performed with billets of elevated temperature, in which case a furnace for heating the billets - and possibly a second furnace for heating the dies - are placed in the line, immediately prior to the press. Non-ferrous billets heated to 550°C are today pressed in commercial production; the extrusion of harder materials, such as steel, will require still higher pre-heating temperatures.

5. With regard to the use of oil and its position under high pressure in the extrusion plant, the required means of safety for the operating personnel will be located on the extrusion press and on the run-out arrangements for the extruded product.

ENERGY IN THE CONTAINER

6. Hydrostatic extrusion presses for production purposes are today operating at pressures up to approximately 1500 MPa. The total internal volume of the high pressure containers reach the order of 0.05 m^3. This internal volume should be used as efficiently as possible. Then when under pressure a container is mainly filled with the billet and the die, and a minor portion of the total volume is filled with the compressed oil. An estimate of the contained energy in a typical extrusion container, assuming that half of the volume is filled with compressed oil, gives a value of the order of 5.5MN mm. of stored energy than in a cold isostatic production press, of normal size. The extrusion container contains less. However, 5MN m Nm is high enough to constitute a serious danger to the operating personnel, should it be released suddenly, for instance due to failure or faulty operation of the pressure containing system.

A momentarily release of this amount of energy has an effect corresponding to approximately that of an explosion of 1 kg of TNT. Therefore, satisfactory protection for the personnel working around hydrostatic extrusion presses should be incorporated connection with the installation of the press.

FAILURES IN THE CONTAINER SYSTEM

7. As mentioned above , hydrostatic extrusion presses for production purposes are mostly of horizontal design, and the following comments are made with horizontal presses in mind.

8. Extrusion presses basically a pressure intensifier, in which the main ram forces a plunger into the high pressure container in order to generate the pressure.

9. The high pressure container is generally made to a multi-layer design, with one or more conical liners pressed into the central bore of a cylindrical vessel. The vessel may be made from a solid steel forging. The liners are generally prestressed to a degree such that during the extrusion process they are still under compression in order to obtain a beneficial influence on their fatigue resistance. Quite apart from the damage to the cylindrical vessel or liner, the failure of a container can have serious consequences.

10. In case of the cylindrical vessels made from solid forgings the effect of brittle failure, occuring as a secondary effect from a repeated number of liner failures, must be considered. Broken parts from the vessel may then form missiles. It is therefore recommended that such vessels are carefully shielded, for instance by an appropriately designed cover, or by a well anchored protection wall.

11. The vessel may also be built in such a way that a sudden release of energy is made impossible. In the QUINTUS wire-wound high pressure container this design philosophy has been adopted. The vessel is made of a steel forging around which is wrapped by numerous layers of pre-stressed wire. In order to achieve the best possible fatigue resistance, the steel forging is tangentially prestressed by the wire mantle to such an extent that it is in compression even at the maximum extrusion pressure. At both ends of the forging the wire mantle is retained within flanges. The entire vessel is held between heavy steel flanges which are located within the press frame. Should a crack occur in the forging, to propagation cannot continue through the wire layers. Fragments from the forging will not form missiles, but will be safely contained by the winding. Escaping pressure medium expands slowly through the wire mantle, thereby loosing its energy. Other safety characteristics of the wire-wound design were described at the 2nd International Conference on Pressure Vessel Technology, San Antonio, Texas, 1973.

12. Failure of the inner container liner has to be considered as an occasion within manufacturing maintenance routines. It is not possible, with the use of present days materials, to reach an infinite life of the liner, and it is therefore regarded as a wear part. The most violent mode of liner failure is when a crack occurs across the section of the liner. The pressure medium will then enter into the crack and tear the liner into two tube-shaped parts with a force which is greater than the press force. One of these parts will rest against the die support, while the other is accelerated backwards over the high pressure plunger. The plunger itself is accelerated forwards, i.e. in the opposite direction, due to the pressure drop in the container, being pushed by the energy in the oil behind the main ram of the press.

13. The taper of the liner in the high pressure vessel is normally directed such that in case of a liner failure, the container will remain against the die support. The press frame has for a moment to carry the normal load from the plunger (main ram), as well as the extra load from the extrusion pressure acting over the broken liner section. In practice, this means more than double the rated load of the frame, if the liner fails at the maximum extrusion pressure.

14. Obviously, a reliable protection has to be arranged for the press operators against the above type of failure. One of the possible protection arrangements is built into the QUINTUS hydrostatic extrusion presses. The protection is combined with the back-up system for the seal for the high pressure plunger Fig.2. The seal is supported by a rigid back-up plate, and this plate is in turn supported by four back-up rams which extend backwards into the main ram of the press. Normally, these rams are connected to the main ram pressure and give automatically the necessary back-up force for the seal. When a liner is broken, the back-up rams act as shock absorbers which effectively brake the movements of the broken liner and the main ram and hence the high pressure plunger. In addition, the transmission of large reaction forces to the press frame are suppressed.

15. The area between the main ram and the high pressure container is shielded by a protection cover, which prevents pieces from a broken liner being projected outside the press. A mild steel plate lined with boards is used for this purpose. The plate thickness is chosen with the effect of a failure of the high pressure plunger also in mind. The board lining prevents ricochetting of splinters. Fig.3 illustrates the plate arrangement.

16. The cover plate will also give protection against high pressure oil issuing from a leaking seal. The protection cover is arranged for easy removal to facilitate inspection and service.

COMPLETE EXTRUSION

17. Complete extrusion referes to the ease with which the billet is totally extruded, sometimes at speed, through the oil leaving no discard. In extrusion through a die with conical entry the extrus ion ratio gradually decreases as the back end of the billet passes through the cone. This results in to an acceleration of the rear portion of the extruding billet. Should the product be long, for instance a flexible wire, the accelerating force cannot be transmitted to the entire mass of the product. So the rear end of the product concertinas. Extruded products which have the shape of bars or tubes, on the

other hand, are more likely to be given an enhanced exit speed. In such a case they must therefore be retarded at the end of the run out line of the press. The volume of the pressure medium and the compressibility determined set the amount of energy which can be put into the product. Should the compression volume be larger than the volume of the conical part of the die, the total amount of the available energy cannot be transmitted to the product, as the pressure medium can expand freely through the die as soon as the product has left the die opening.

18. Thus, the maximum amount of energy which can be put into an escaping product occurs when a bar seals off the die opening until the pressure has decreased to zero. This is the case when a stationary mandrel of a tube extrusion tool breaks close to the mandrel head. If the breakage occurs in the beginning of the process, before any extrusion has taken place there will be negligable losses due to friction at the die aperture. Should the mandrel have a volume larger than the compression volume of the container oil, all the available energy will be transmitted to it. As the mandrel is a piece of steel with a concentrated mass, it can be compared with a projectile. From a safety point of view, a mandrel breakage is therefore the most serious case of complete extrusion.

19. Suppose that all the available energy is transmitted to the escaping product, in this case a steel mandrel. Further, suppose that the compressibility of the pressure medium is constant. The momentary pressure P in the extrusion container can then be expressed as

$$P = Pe - \frac{Am \cdot X}{Ve} \beta$$

where P_e is the extrusion pressure at the moment of breakage
A_m is the area of the mandrel
X is the travelling distance of the mandrel
V_e is the volume of the pressure medium
β is the compressibility of the pressure medium

When P = 0 all the energy in the extrusion container is consumed. The mandrel has then travelled the distance X

$$X = \frac{Pe \cdot Ve}{Am \cdot \beta}$$

If the mandrel is longer than X. then as the compressibility is assumed set to be constant, the energy E can be estimated.

$$E = \tfrac{1}{2}Pe \cdot Am \cdot X$$

The exit speed is $v = \sqrt{\frac{2E}{m_m}}$

where m_m is the mass of the mandrel.

20. In order to determine the extent of the required axial protection, an estimate of the breakthrough force of a projectile which hits a steel plate as to be made the thickness of a steel plate, to withstand perforation by a certain projectile with 50% probability, can be estimated. From "Containing Ballistic Fragments by Rodney F. Recht, Denver Research Institute, University of Denver" The estimate is valid for a plate with a hardness of about 300 Brinell, and for a projectile which has equal length and diameter ratio of unity.

$$t = \left[\frac{v \cdot m^{1/4}}{50}\right]^{4/3}$$

where v is the speed in m/s
t is the thickness in mm
m is the mass in kg

For a given P_e, V_e, & B The largest thickness required occurs when a mandrel has a minimum mass, but still a volume big enough to absorb all the available energy.

21. Using the above estimates for an extrusion pressure of 1500 MPa, a volume of pressure medium of 0.02 m^3, and the compressibility β

$$= 8 \cdot 10^9 \ N/m^2$$

the minimum mass for the mandrel will be 29.5 kg, (assuming the mandrel has a density of $8{,}10^3$ Kg/m^3,) the exit speed is 435 m/s, and the required thickness of the plate 51 mm. Considering that a plate does not normally have a hardness of 300 Brinell and that the mandrel is several times longer than its diameter, the thickness should be increased. In order to avoid ricochets from projectiles with less energy it is desirable to have a barrier of sand in front of the plate. The thickness of the plate can then be decreased.

22. Radial or lateral protection is required to guard against prevent completely extruded bars being projected towards the side and also to cover ricocheting projectiles. Figure 4 illustrates the run-out line for an extrusion press for bars. It is designed for 20 m long bars and comprises concrete walls and roof. The bars are taken out near the bottom of one of the long sides. Outside the outlet opening there is a protection wall which can asorb the energy of fast moving parts. This plant can also be adapted for tube extrusion, and the rear end of the run-out line has a wall containing a steel plate and a sand barrier.

23. If only tubes are required, these are normally extruded into a guide tube. The guide tube can then be bent upwards immediately after leaving the press and the protection walls can be located adjacent to the press. Figure 5 illustrates this arrangement.

FIRE

24. If the hydrostatic extrusion is performed with billets at elevated temperature, the pressure medium may start to burn when it comes into contact with the hot billet. In the QUINTUS hydrostatic extrusion presses a separate plunger is used to clamp the billet during the retraction of the billet loading manipulator. The container is then open and some pressure medium flushes over the hot billet. The medium used is castor oil which has a fairly high flash point. However, during the flushing operation it starts to burn with a flame directed axially from the container towards the die support and is then bent upwards. Figure 6 illustrates the maximum extent of the flame.

The burning continues during the closing operation of the container and is automatically extinguished when the container is closed. Billets preheated to 600°C have been run in production without experiencing of excessive fire hazard. However, normal preventive precautions should certainly be taken with regard to the fire hazard.

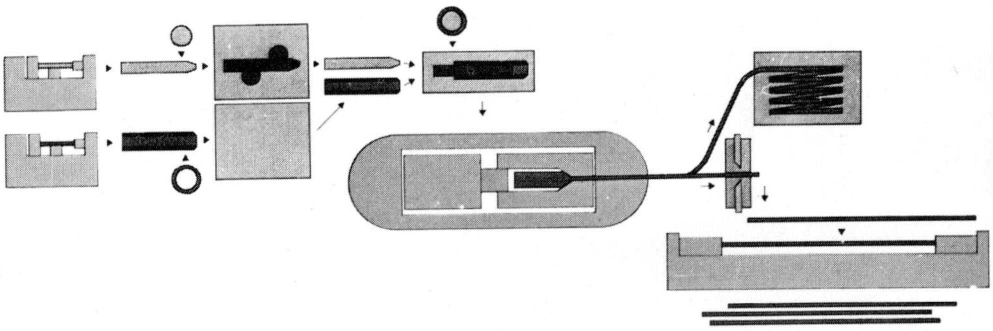

Fig. 1: Schematical lay-out of a hydrostatic extrusion plant for coiled and straight products

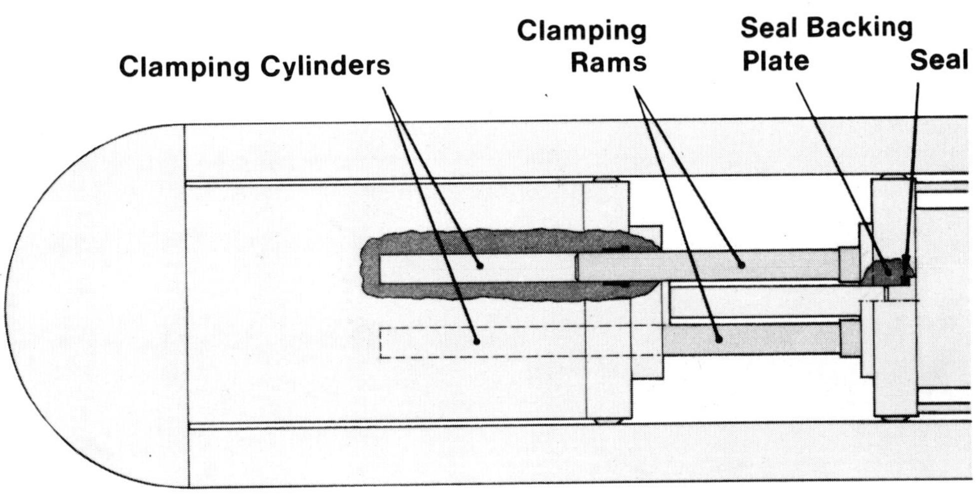

Fig. 2: Principle of a clamping ram system which acts as shock absorber in case of liner failure on complete extrusion

Fig. 3: The press is equipped with a protection plate which during the extrusion covers the area between the main ram and the high pressure container

Fig. 4: Run-out line for 20m straight bars

Fig. 5: The guide tube at the exit of a tube extrusion press is bent upwards in an s-curve, the first bend of the s being inside protection walls. The walls are located adjacent to the press

Fig. 6: Castor oil used as a pressure medium burns when it comes into contact with heated billets. The fire is extinguished as the high pressure container is closed

THE HPTA HIGH PRESSURE SAFETY CODE

G. SAVILLE, Department of Chemical Engineering, Imperial College, London
B. G. COX, Metals Research Ltd., Royston, Herts.

This code, prepared by members of the High Pressure Technology Association, and published by the Association, aims at giving advice to the user of high-pressure equipment as to safe methods of operation. It is not intended as a replacement of the various standards or codes of practice which have a statutory, or near statutory, status in certain more restricted pressure vessel fields such as steam boiler manufacture and operation.

The code has been written with laboratory, pilot plant and small industrial users primarily in mind. The form of the recommendations given is therefore dictated by the fact that these users will not normally have available the extensive test facilities or the expertise of the large industrial high pressure organisation.

Safety at high pressure is, in part, the result of good design and any safety code must consider certain aspects of pressure vessel design. However, the user of high pressure equipment, particularly if he is relatively inexperienced, is much more concerned with assessing the suitability of a vessel which he already has on hand and for this he requires something simpler than a full design code. It is our view that assessment of this sort is possible in many straightforward situations provided that the criteria used are sufficiently conservative.

In view of the difficulty in assessing a vessel made from one of the very high strength steels, where the most likely mode of failure is brittle fracture, we decided that it would be unwise to allow the inexperienced user to make his own assessment but to direct him towards seeking advice from an expert in the field.

On the other hand, vessels made from ductile, medium strength steels, provided they are not misused, are unlikely to fail in a brittle manner and we believe that it is permissible for the user to make his own assessment of this class of vessel. For the purposes of the Code, a ductile steel is defined as one which simultaneously has an ultimate tensile strength of less than 10 kbar, a tensile yield strength of not more than 0.85 of the U.T.S., a reduction in area of more than 30% and a Charpy V-notch impact value of more than 27J, although there are, in addition, certain size restrictions specified.

We base the criteria for safe operation of a pressure vessel on the yield and bursting pressures, both of which are quoted in terms of tensile rather than shear properties since values of these are more readily obtained by the user. We use the maximum shear stress criterion for the yield pressure and the empirical mean diameter formulae for the ultimate bursting pressure, both chosen because they provide a conservative result. Subsections of the code also cover the various complications which can arise, such as fatigue, creep, effect of cross-bores, etc. and give some idea of their quantitive effect.

The Safety Code, in addition to dealing with the assessment of pressure vessels for high pressure duty, has sections on 'Safe operating procedures', 'High-pressure systems', 'Fittings and ancillary devices' and 'Design of Barricades'. The first three of these consider the general arrangement and safe operation of a high pressure plant or installation. The last deals with the question of how to provide suitable protection for plant operators should an accident occur at high pressure. Barricade design is very badly covered in the scientific and engineering literature and much that is found there is often difficult to reconcile with experience. We have therefore reassessed the information available; and for assistance in this we are particularly indebted to Mr W.G. High; and present the formulae needed to determine the size and wall thickness needed of a barricade if it is to protect the useragainst both fragments and shock waves from an exploding pressure vessel.